[美] Gilbert Strang（吉尔伯特·斯特朗） 著

Introduction to Linear Algebra

(Sixth Edition)

线性代数（第6版）

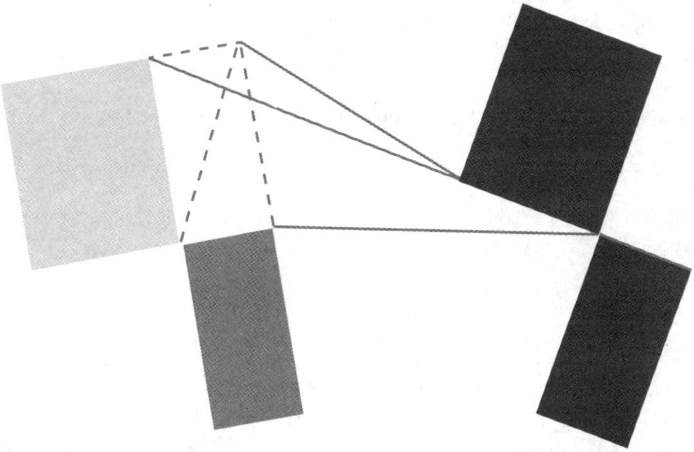

U0387622

清華大學出版社
北 京

北京市版权局著作权合同登记号　图字：01-2023-6162

图书在版编目（CIP）数据

线性代数：第6版 = Introduction to Linear Algebra, Sixth Edition : 英文 / （美）吉尔伯特·斯特朗（Gilbert Strang）著. -- 北京 : 清华大学出版社, 2024. 7. -- ISBN 978-7-302-66807-7

Ⅰ. O151.2

中国国家版本馆 CIP 数据核字第 2024FD6609 号

责任编辑： 佟丽霞
封面设计： 何凤霞
责任印制： 刘海龙

出版发行： 清华大学出版社
网　　址： https://www.tup.com.cn, https://www.wqxuetang.com
地　　址： 北京清华大学学研大厦 A 座　　**邮　　编：** 100084
社 总 机： 010-83470000　　**邮　　购：** 010-62786544
投稿与读者服务： 010-62776969, c-service@tup.tsinghua.edu.cn
质量反馈： 010-62772015, zhiliang@tup.tsinghua.edu.cn
印 装 者： 小森印刷霸州有限公司
经　　销： 全国新华书店
开　　本： 170mm×230mm　　**印　　张：** 27.5
版　　次： 2024 年 7 月第 1 版　　**印　　次：** 2024 年 7 月第 1 次印刷
定　　价： 108.00 元

产品编号：105188-01

Preface

One goal of this Preface can be achieved right away. You need to know about the video lectures for MIT's Linear Algebra course **Math 18.06**. Those videos go with this book, and they are part of MIT's OpenCourseWare. The direct links to linear algebra are

https://ocw.mit.edu/courses/18-06-linear-algebra-spring-2010/

https://ocw.mit.edu/courses/18-06sc-linear-algebra-fall-2011/

On YouTube those lectures are at **https://ocw.mit.edu/1806videos** and **/1806scvideos**

The first link brings the original lectures from the dawn of OpenCourseWare. Problem solutions by graduate students (really good) and also a short introduction to linear algebra were added to the new 2011 lectures. And the course today has a new start—the crucial ideas of linear independence and the column space of a matrix have moved near the front.

I would like to tell you about those ideas in this Preface.

Start with two column vectors $\boldsymbol{a}_1$ and $\boldsymbol{a}_2$. They can have three components each, so they correspond to points in 3-dimensional space. The picture needs a center point which locates the zero vector:

$$\boldsymbol{a}_1 = \begin{bmatrix} 2 \\ 3 \\ 1 \end{bmatrix} \qquad \boldsymbol{a}_2 = \begin{bmatrix} 1 \\ 4 \\ 2 \end{bmatrix} \qquad \textbf{zero vector} = \begin{bmatrix} 0 \\ 0 \\ 0 \end{bmatrix}.$$

The vectors are drawn on this 2-dimensional page. But we all have practice in visualizing three-dimensional pictures. Here are $\boldsymbol{a}_1, \boldsymbol{a}_2, 2\boldsymbol{a}_1$, and the vector sum $\boldsymbol{a}_1 + \boldsymbol{a}_2$.

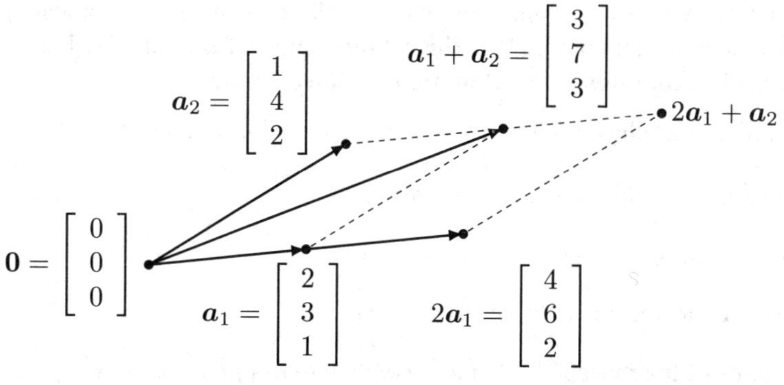

That picture illustrated two basic operations—adding vectors $\boldsymbol{a}_1 + \boldsymbol{a}_2$ and multiplying a vector by 2. Combining those operations produced a **"linear combination"** $2\boldsymbol{a}_1 + \boldsymbol{a}_2$:

$$\textbf{Linear combination} = c\boldsymbol{a}_1 + d\boldsymbol{a}_2 \ \textbf{ for any numbers } c \textbf{ and } d$$

Those numbers c and d can be negative. In that case $c\boldsymbol{a}_1$ and $d\boldsymbol{a}_2$ will reverse their directions: they go right to left. Also very important, c and d can involve fractions. Here is a picture with a lot more linear combinations. **Eventually we want all vectors $c\boldsymbol{a}_1 + d\boldsymbol{a}_2$.**

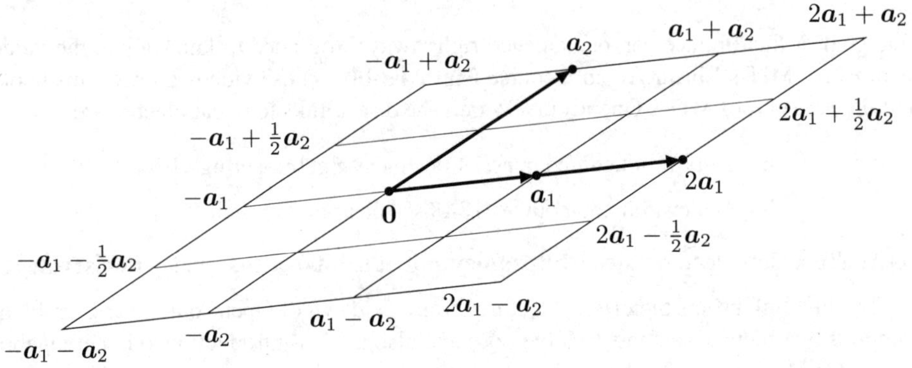

Here is the key! **The combinations $c\boldsymbol{a}_1 + d\boldsymbol{a}_2$ fill a whole plane.** It is an infinite plane in 3-dimensional space. By using more and more fractions and decimals $\boldsymbol{c}$ and $\boldsymbol{d}$, we fill in a complete plane. *Every point on the plane is a combination of $\boldsymbol{a}_1$ and $\boldsymbol{a}_2$.*

Now comes a fundamental idea in linear algebra: **a matrix**. The matrix A holds **n column vectors $\boldsymbol{a}_1, \boldsymbol{a}_2, \ldots, \boldsymbol{a}_n$**. At this point our matrix has two columns $\boldsymbol{a}_1$ and $\boldsymbol{a}_2$, and those are vectors in 3-dimensional space. So the matrix has **three rows and two columns.**

3 by 2 matrix
m **= 3 rows**
n **= 2 columns**

$$A = \begin{bmatrix} & & \\ \boldsymbol{a}_1 & & \boldsymbol{a}_2 \\ & & \end{bmatrix} = \begin{bmatrix} 2 & 1 \\ 3 & 4 \\ 1 & 2 \end{bmatrix}$$

The combinations of those two columns produced a plane in three-dimensional space. There is a natural name for that plane. **It is the column space of the matrix**. For any A, **the column space of A contains all combinations of the columns.**

Here are the four ideas introduced so far. You will see them all in Chapter 1.

1. **Column vectors** $\boldsymbol{a}_1$ and $\boldsymbol{a}_2$ in three dimensions
2. **Linear combinations** $c\boldsymbol{a}_1 + d\boldsymbol{a}_2$ of those vectors
3. **The matrix A** contains the columns $\boldsymbol{a}_1$ and $\boldsymbol{a}_2$
4. **Column space of the matrix** = *all linear combinations of the columns* = *plane*

Now we include 2 more columns in A

The 4 columns are in 3-dimensional space

$$A = \begin{bmatrix} 2 & 1 & 3 & 0 \\ 3 & 4 & 7 & 0 \\ 1 & 2 & 3 & -1 \end{bmatrix}$$

Linear algebra aims for an understanding of every column space. Let me try this one.

Columns 1 and 2 produce the same plane as before (same $\boldsymbol{a}_1$ and $\boldsymbol{a}_2$)

Column 3 contributes nothing new **because $\boldsymbol{a}_3$ is on that plane :** $\boldsymbol{a}_3 = \boldsymbol{a}_1 + \boldsymbol{a}_2$

Column 4 is **not on the plane** : Adding in $c_4\boldsymbol{a}_4$ raises or lowers the plane

The column space of this matrix A is the **whole 3-dimensional space** : all points !

You see how we go a column at a time, left to right. Each column can be **independent** of the previous columns or it can be a **combination** of those columns. To produce every point in 3-dimensional space, you need three independent columns.

Matrix Multiplication $A = CR$

Using the words "linear combination" and "independent columns" gives a good picture of that 3 by 4 matrix A. Column **3** is a linear combination : **column 1 + column 2. Columns 1, 2, 4 are independent.** The only way to produce the zero vector as a combination of the independent columns $1, 2, 4$ is to multiply all those columns by **zero.**

We are so close to a key idea of Chapter 1 that I have to go on. Matrix multiplication is the perfect way to write down what we know. From the 4 columns of A we pick out the independent columns $\boldsymbol{a}_1, \boldsymbol{a}_2, \boldsymbol{a}_4$ in the column matrix C. **Every column of R tells us the combination of $\boldsymbol{a}_1, \boldsymbol{a}_2, \boldsymbol{a}_4$ in C that produces a column of A. A equals C times R** :

$$A = \begin{bmatrix} 2 & 1 & \mathbf{3} & 0 \\ 3 & 4 & \mathbf{7} & 0 \\ 1 & 2 & \mathbf{3} & -1 \end{bmatrix} = \begin{bmatrix} \mathbf{2} & \mathbf{1} & 0 \\ \mathbf{3} & \mathbf{4} & 0 \\ \mathbf{1} & \mathbf{2} & -1 \end{bmatrix} \begin{bmatrix} 1 & 0 & \mathbf{1} & 0 \\ 0 & 1 & \mathbf{1} & 0 \\ 0 & 0 & \mathbf{0} & 1 \end{bmatrix} = CR$$

Column 3 of A is dependent on columns 1 and 2 of A, and column 3 of R shows how. Add the independent columns 1 and 2 of C to get column $\boldsymbol{a}_3 = \boldsymbol{a}_1 + \boldsymbol{a}_2 = (\mathbf{3}, \mathbf{7}, \mathbf{3})$ of A.

Matrix multiplication : Each column j of CR is C times column j of R

Section 1.3 of the book will *multiply a matrix times a vector* (two ways). Then Section 1.4 will *multiply a matrix times a matrix.* This is the key operation of linear algebra. It is important that there is more than one good way to do this multiplication.

I am going to stop here. The normal purpose of the Preface is to tell you about the big picture. The next pages will give you two ways to organize this subject—especially the first seven chapters that more than fill up most linear algebra courses. Then come optional chapters, leading to the most active topic in applications today : **deep learning.**

The Four Fundamental Subspaces

You have just seen how the course begins—with the columns of a matrix A. There were two key steps. One step was to take all combinations $c\boldsymbol{a}_1 + d\boldsymbol{a}_2 + e\boldsymbol{a}_3 + f\boldsymbol{a}_4$ of the columns. This led to the **column space of A**. The other step was to **factor the matrix into C times R**. That matrix C holds a full set of **independent columns**.

I fully recognize that this is only the Preface to the book. You have had zero practice with the column space of a matrix (and even less practice with C and R). But the good thing is : Those are the right directions to start. Eventually, *every matrix will lead to four fundamental spaces*. Together with the column space of A comes the **row space—all combinations of the rows**. When we take all combinations of the n columns and all combinations of the m rows—those combinations fill up "spaces" of vectors.

The other two subspaces complete the picture. Suppose the row space is a plane in three dimensions. Then there is one special direction in the 3D picture—**that direction is perpendicular to the row space**. That perpendicular line is the **nullspace** of the matrix. We will see that the vectors in the nullspace (perpendicular to all the rows) solve $A\boldsymbol{x} = \mathbf{0}$: the most basic of linear equations.

And if vectors perpendicular to all the rows are important, so are the vectors perpendicular to all the columns. Here is the picture of the **Four Fundamental Subspaces**.

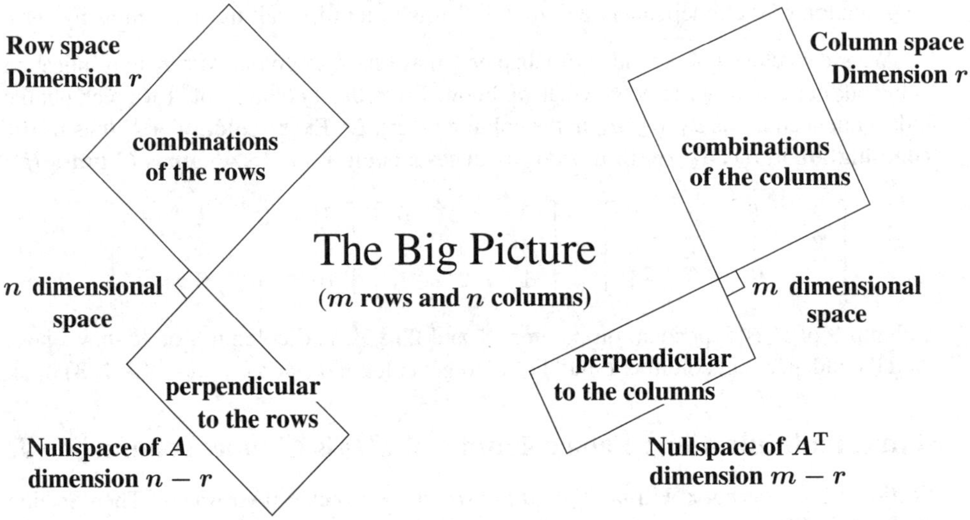

The Four Fundamental Subspaces : An m by n matrix with r independent columns.

This picture of four subspaces comes in Chapter 3. The idea of perpendicular spaces is developed in Chapter 4. And special "basis vectors" for all four subspaces are discovered in Chapter 7. That step is the final piece in the **Fundamental Theorem of Linear Algebra**. The theorem includes an amazing fact about any matrix, square or rectangular:

The number of independent columns equals the number of independent rows.

Five Factorizations of a Matrix

Here are the organizing principles of linear algebra. When our matrix has a special property, these factorizations will show it. Chapter after chapter, they express the key idea in a direct and useful way.

The usefulness increases as you go down the list. **Orthogonal matrices** are the winners in the end, because their columns are perpendicular unit vectors. That is perfection.

$$\text{2 by 2 Orthogonal Matrix} = \begin{bmatrix} \cos\theta & -\sin\theta \\ \sin\theta & \cos\theta \end{bmatrix} = \text{Rotation by Angle } \theta$$

Here are the five factorizations from Chapters 1, 2, 4, 6, 7 :

1	$\boldsymbol{A = CR}$	$= \boldsymbol{R}$ combines independent columns in $\boldsymbol{C}$ to give all columns of $\boldsymbol{A}$
2	$\boldsymbol{A = LU}$	$=$ Lower triangular $\boldsymbol{L}$ times Upper triangular $\boldsymbol{U}$
4	$\boldsymbol{A = QR}$	$=$ Orthogonal matrix $\boldsymbol{Q}$ times Upper triangular $\boldsymbol{R}$
6	$\boldsymbol{S = Q\Lambda Q^{\mathrm{T}}}$	$=$ (Orthogonal $\boldsymbol{Q}$) (Eigenvalues in $\boldsymbol{\Lambda}$) (Orthogonal $\boldsymbol{Q^{\mathrm{T}}}$)
7	$\boldsymbol{A = U\Sigma V^{\mathrm{T}}}$	$=$ (Orthogonal $\boldsymbol{U}$) (Singular values in $\boldsymbol{\Sigma}$) (Orthogonal $\boldsymbol{V^{\mathrm{T}}}$)

May I call your attention to the last one ? It is the **Singular Value Decomposition (SVD)**. It applies to every matrix A. Those factors U and V have perpendicular columns—all of length one. Multiplying any vector by U or V leaves a vector of the same length—so computations don't blow up or down. And Σ is a positive diagonal matrix of "singular values". If you learn about eigenvalues and eigenvectors in Chapter 6, *please* continue a few pages to singular values in Section 7.1.

Deep Learning

For a true picture of linear algebra, applications have to be included. Completeness is totally impossible. At this moment, the dominating direction of applied mathematics has one special requirement: *It cannot be entirely linear* !

One name for that direction is "deep learning". It is an extremely successful approach to a fundamental scientific problem : **Learning from data**. In many cases the data comes in a matrix. Our goal is to look inside the matrix for the connections between variables. Instead of solving matrix equations or differential equations that express known input-output rules, *we have to find those rules.* The success of deep learning is to build a function $\boldsymbol{F}(\boldsymbol{x}, \boldsymbol{v})$ with inputs $\boldsymbol{x}$ and $\boldsymbol{v}$ of two kinds :

The vectors $\boldsymbol{v}$ describes the features of the training data.

The matrices $\boldsymbol{x}$ assign weights to those features.

The function $\boldsymbol{F}(\boldsymbol{x}, \boldsymbol{v})$ is close to the correct output for that training data $\boldsymbol{v}$.

When $\boldsymbol{v}$ changes to unseen test data, $\boldsymbol{F}(\boldsymbol{x}, \boldsymbol{v})$ stays close to correct.

This success comes partly from the form of the learning function $\boldsymbol{F}$, which allows it to include vast amounts of data. In the end, a linear function $\boldsymbol{F}$ would be totally inadequate. The favorite choice for $\boldsymbol{F}$ is **piecewise linear**. This combines simplicity with generality.

Applications in the Book and on the Website

I hope this book will be useful to you long after the linear algebra course is complete. It is all the applications of linear algebra that make this possible. Matrices carry data, and other matrices *operate on that data*. The goal is to "see into a matrix" by understanding its eigenvalues and eigenvectors and singular values and singular vectors. And each application has special matrices—here are four examples:

Markov matrices M	Each column is a set of probabilities adding to 1.
Incidence matrices A	Graphs and networks start with a set of nodes. The matrix A tells the *connections* (edges) between those nodes.
Transform matrices F	The Fourier matrix uncovers the *frequencies* in the data.
Covariance matrices C	The **variance** is key information about a random variable. *The covariance explains dependence between variables.*

We included those applications and more in this Sixth Edition. For the crucial computation of matrix weights in deep learning, Chapter 9 presents the ideas of **optimization**. This is where linear algebra meets calculus: **derivative = zero** becomes a matrix equation at the minimum point because $\boldsymbol{F}(\boldsymbol{x})$ has many variables.

Several topics from the Fifth Edition gave up their places but not their importance. Those sections simply moved onto the Web. The website for this new Sixth Edition is

math.mit.edu/linearalgebra

That website includes sample sections from this new edition and solutions to all Problem Sets. These sections (and more) are saved online from the Fifth Edition:

Fourier Series **Norms and Condition Numbers**

Iterative Methods and Preconditioners **Linear Algebra for Cryptography**

Here is a small touch of linear algebra—three questions before this course gets serious:

1. Suppose you draw three straight line segments of lengths $\boldsymbol{r}$ and $\boldsymbol{s}$ and $\boldsymbol{t}$ on this page. What are the conditions on those three lengths to allow you to make the segments into a triangle? In this question you can choose the directions of the three lines.

2. Now suppose the directions of three straight lines $\boldsymbol{u}, \boldsymbol{v}, \boldsymbol{w}$ are fixed and different. But you could stretch those lines to $\boldsymbol{au}, \boldsymbol{bv}, \boldsymbol{cw}$ with any numbers $\boldsymbol{a}, \boldsymbol{b}, \boldsymbol{c}$. Can you always make a closed triangle out of the three vectors $\boldsymbol{au}$, $\boldsymbol{bv}$, $\boldsymbol{cw}$?

3. Linear algebra doesn't stay in a plane! Suppose you have **four lines** $\boldsymbol{u}, \boldsymbol{v}, \boldsymbol{w}, \boldsymbol{z}$ in different directions in 3-dimensional space. Can you always choose the numbers $\boldsymbol{a}, \boldsymbol{b}, \boldsymbol{c}, \boldsymbol{d}$ (zeros not allowed) so that $\boldsymbol{au} + \boldsymbol{bv} + \boldsymbol{cw} + \boldsymbol{dz} = \boldsymbol{0}$?

For typesetting this book, maintaining its website, offering quality textbooks to Indian fans, I am grateful to Ashley C. Fernandes of Wellesley Publishers (www.wellesleypublishers.com)

gilstrang@gmail.com Gilbert Strang

Table of Contents

1 Vectors and Matrices

Linear algebra is about vectors $\boldsymbol{v}$ and matrices A. Those are the basic objects that we can add and subtract and multiply (when their shapes match correctly). The first vector $\boldsymbol{v}$ has two components $v_1 = 2$ and $v_2 = 4$. The vector $\boldsymbol{w}$ is also 2-dimensional.

$$\boldsymbol{v} = \begin{bmatrix} v_1 \\ v_2 \end{bmatrix} = \begin{bmatrix} 2 \\ 4 \end{bmatrix} \qquad \boldsymbol{w} = \begin{bmatrix} w_1 \\ w_2 \end{bmatrix} = \begin{bmatrix} 1 \\ 3 \end{bmatrix} \qquad \boldsymbol{v} + \boldsymbol{w} = \begin{bmatrix} 3 \\ 7 \end{bmatrix}$$

The **linear combinations of $\boldsymbol{v}$ and $\boldsymbol{w}$ are the vectors** $c\boldsymbol{v} + d\boldsymbol{w}$ for all numbers c and d:

$$\textbf{The linear combinations} \quad c\begin{bmatrix} 2 \\ 4 \end{bmatrix} + d\begin{bmatrix} 1 \\ 3 \end{bmatrix} = \begin{bmatrix} 2c + 1d \\ 4c + 3d \end{bmatrix} \quad \textbf{fill the } \boldsymbol{xy} \textbf{ plane}$$

The **length** of that vector $\boldsymbol{w}$ is $||\boldsymbol{w}|| = \sqrt{10}$, the square root of $w_1^2 + w_2^2 = 1 + 9$. The **dot product** of $\boldsymbol{v}$ and $\boldsymbol{w}$ is $\boldsymbol{v} \cdot \boldsymbol{w} = v_1 w_1 + v_2 w_2 = (2)(1) + (4)(3) = 14$. In Section 1.2, $\boldsymbol{v} \cdot \boldsymbol{w}$ will reveal the angle between those vectors.

The big step in Section 1.3 is to introduce a **matrix**. This matrix $\boldsymbol{A}$ contains our two column vectors. The vectors have two components, so the matrix is 2 by 2:

$$A = \begin{bmatrix} \boldsymbol{v} & \boldsymbol{w} \end{bmatrix} = \begin{bmatrix} 2 & 1 \\ 4 & 3 \end{bmatrix}.$$

When a matrix multiplies a vector, we get a combination $c\boldsymbol{v} + d\boldsymbol{w}$ of its columns:

$$\textbf{A times} \begin{bmatrix} c \\ d \end{bmatrix} = \begin{bmatrix} 2 & 1 \\ 4 & 3 \end{bmatrix} \begin{bmatrix} c \\ d \end{bmatrix} = \begin{bmatrix} 2c + 1d \\ 4c + 3d \end{bmatrix} = c\boldsymbol{v} + d\boldsymbol{w}.$$

And when we look at **all combinations $A\boldsymbol{x}$** (with every c and d), those vectors produce the **column space of the matrix A**. Here that column space is a plane.

With three vectors, the new matrix $\boldsymbol{B}$ has columns $\boldsymbol{v}, \boldsymbol{w}, \boldsymbol{z}$. In this example $\boldsymbol{z}$ is a combination of $\boldsymbol{v}$ and $\boldsymbol{w}$. So the column space of $\boldsymbol{B}$ is still the xy plane. The vectors $\boldsymbol{v}$ and $\boldsymbol{w}$ are **independent** but $\boldsymbol{v}, \boldsymbol{w}, \boldsymbol{z}$ are **dependent**. A combination produces zero:

$$\boldsymbol{B} = \begin{bmatrix} \boldsymbol{v} & \boldsymbol{w} & \boldsymbol{z} \end{bmatrix} = \begin{bmatrix} 2 & 1 & 3 \\ 4 & 3 & 7 \end{bmatrix} \text{ has } \boldsymbol{B} \begin{bmatrix} 1 \\ 1 \\ -1 \end{bmatrix} = \boldsymbol{v} + \boldsymbol{w} - \boldsymbol{z} = \begin{bmatrix} 2 \\ 4 \end{bmatrix} + \begin{bmatrix} 1 \\ 3 \end{bmatrix} - \begin{bmatrix} 3 \\ 7 \end{bmatrix} = \begin{bmatrix} 0 \\ 0 \end{bmatrix}$$

The final goal is to understand **matrix multiplication AB = A times each column of B.**

1.1 Vectors and Linear Combinations

1 $2\boldsymbol{v} - 3\boldsymbol{w}$ is a **linear combination** $c\boldsymbol{v} + d\boldsymbol{w}$ of the vectors $\boldsymbol{v}$ and $\boldsymbol{w}$.

2 For $\boldsymbol{v} = \begin{bmatrix} 4 \\ 1 \end{bmatrix}$ and $\boldsymbol{w} = \begin{bmatrix} 2 \\ -1 \end{bmatrix}$ that combination is $2\begin{bmatrix} 4 \\ 1 \end{bmatrix} - 3\begin{bmatrix} 2 \\ -1 \end{bmatrix} = \begin{bmatrix} 2 \\ 5 \end{bmatrix}$.

3 All combinations $c\begin{bmatrix} 4 \\ 1 \end{bmatrix} + d\begin{bmatrix} 2 \\ -1 \end{bmatrix}$ fill the xy plane. They produce every $\begin{bmatrix} x \\ y \end{bmatrix}$.

4 The vectors $c\begin{bmatrix} 1 \\ 1 \\ 2 \end{bmatrix} + d\begin{bmatrix} 0 \\ 0 \\ 4 \end{bmatrix}$ fill a **plane** in xyz space. $\begin{bmatrix} 1 \\ 2 \\ 3 \end{bmatrix}$ **is not on that plane.**

Calculus begins with numbers x and functions $f(x)$. Linear algebra begins with vectors $\boldsymbol{v}, \boldsymbol{w}$ and their linear combinations $c\boldsymbol{v} + d\boldsymbol{w}$. Immediately this takes you into two or more (possibly many more) dimensions. But linear combinations of vectors $\boldsymbol{v}$ and $\boldsymbol{w}$ are built from just two basic operations:

Multiply a vector $\boldsymbol{v}$ by a number $\quad 3\boldsymbol{v} = 3\begin{bmatrix} 2 \\ 1 \end{bmatrix} = \begin{bmatrix} 6 \\ 3 \end{bmatrix}$

Add vectors $\boldsymbol{v}$ and $\boldsymbol{w}$ of the same dimension: $\boldsymbol{v} + \boldsymbol{w} = \begin{bmatrix} 2 \\ 1 \end{bmatrix} + \begin{bmatrix} 4 \\ 3 \end{bmatrix} = \begin{bmatrix} 6 \\ 4 \end{bmatrix}$

Those operations come together in a **linear combination** $c\boldsymbol{v} + d\boldsymbol{w}$ of $\boldsymbol{v}$ and $\boldsymbol{w}$:

Linear combination
$c = 5$ and $d = -2$ $\quad 5\boldsymbol{v} - 2\boldsymbol{w} = 5\begin{bmatrix} 2 \\ 1 \end{bmatrix} - 2\begin{bmatrix} 4 \\ 3 \end{bmatrix} = \begin{bmatrix} 10 \\ 5 \end{bmatrix} - \begin{bmatrix} 8 \\ 6 \end{bmatrix} = \begin{bmatrix} 2 \\ -1 \end{bmatrix}$

That idea of a linear combination opens up two key questions:

1 Describe all the combinations $c\boldsymbol{v} + d\boldsymbol{w}$. Do they fill a plane or a line?

2 Find the numbers c and d that produce a specific combination $c\boldsymbol{v} + d\boldsymbol{w} = \begin{bmatrix} 2 \\ 0 \end{bmatrix}$

We can answer those questions here in Section 1.1. But linear algebra is not limited to 2 vectors in 2-dimensional space. As long as we stay linear, the problems can get bigger (more dimensions) and harder (more vectors). The vectors will have m components instead of 2 components. We will have n vectors $\boldsymbol{v}_1, \boldsymbol{v}_2, \ldots, \boldsymbol{v}_n$ instead of 2 vectors. **Those n vectors in m-dimensional space will go into the columns of an m by n matrix A:**

m rows
n columns
m by n matrix

$$A = \begin{bmatrix} \boldsymbol{v}_1 & \boldsymbol{v}_2 & \cdots & \boldsymbol{v}_n \end{bmatrix}$$

Let me repeat the two key questions using A, and then retreat back to $m = 2$ and $n = 2$:

1	**Describe all the combinations** $A\boldsymbol{x} = x_1\boldsymbol{v}_1 + x_2\boldsymbol{v}_2 + \cdots + x_n\boldsymbol{v}_n$ **of the columns**
2	**Find the numbers x_1 to x_n** that produce a desired output vector $A\boldsymbol{x} = \boldsymbol{b}$

Linear Combinations $cv + dw$

Start from the beginning. A vector $\boldsymbol{v}$ in 2-dimensional space has *two components*. To draw $\boldsymbol{v}$ and $-\boldsymbol{v}$, use an arrow that begins at the zero vector:

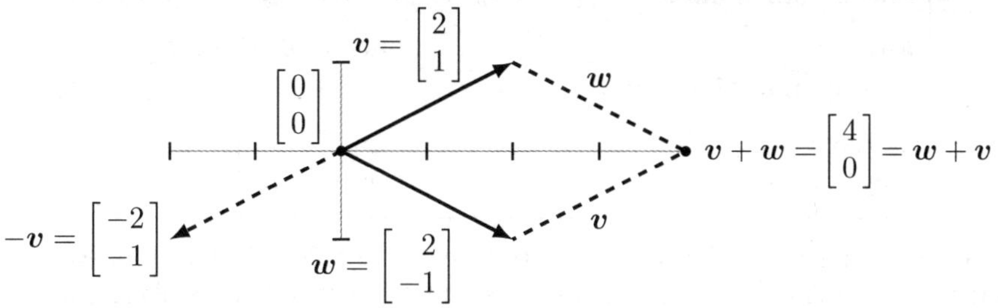

The vectors $c\boldsymbol{v}$ (for all numbers c) fill an infinitely long line in the xy plane. If $\boldsymbol{w}$ is not on that line, then the vectors $d\boldsymbol{w}$ fill a second line. We aim to see that **the linear combinations $c\boldsymbol{v} + d\boldsymbol{w}$ fill the plane**. Combining points on the $\boldsymbol{v}$ line and the $\boldsymbol{w}$ line gives all points.

Here are four different linear combinations—*we can choose any numbers c and d*:

$$\begin{aligned} 1\boldsymbol{v} + 1\boldsymbol{w} &= \text{sum of vectors} \\ 1\boldsymbol{v} - 1\boldsymbol{w} &= \text{difference of vectors} \\ 0\boldsymbol{v} + 0\boldsymbol{w} &= \textit{zero vector} \\ c\boldsymbol{v} + 0\boldsymbol{w} &= \text{vector } c\boldsymbol{v} \text{ in the direction of } \boldsymbol{v} \end{aligned}$$

Solving Two Equations

Very often linear algebra offers the choice of a picture or a computation. The picture can start with $\boldsymbol{v} + \boldsymbol{w}$ and $\boldsymbol{w} + \boldsymbol{v}$ (those are equal). Another important vector is $\boldsymbol{w} - \boldsymbol{v}$, going backwards on $\boldsymbol{v}$. The Preface has a picture with many more combinations—starting to fill a plane. But if we aim for a particular vector like $c\boldsymbol{v} + d\boldsymbol{w} = \begin{bmatrix} 8 \\ 2 \end{bmatrix}$, it will be better to compute the exact numbers c and d. Here are two ways to write down this problem.

$$\boxed{\text{Solve} \quad c\begin{bmatrix} 2 \\ 1 \end{bmatrix} + d\begin{bmatrix} 2 \\ -1 \end{bmatrix} = \begin{bmatrix} 8 \\ 2 \end{bmatrix}. \qquad \text{This means} \quad \begin{aligned} \mathbf{2c + 2d} &= \mathbf{8} \\ \mathbf{c - d} &= \mathbf{2} \end{aligned}.}$$

The rules for solution are simple but strict. We can multiply equations by numbers (*not zero!*) and we can subtract one equation from another equation. Experience teaches that the key is to produce zeros on the left side of the equations. One zero will be enough!

$\frac{1}{2}$(equation 1) is $\boldsymbol{c + d = 4}$

Subtract this from equation 2

Then c is eliminated →

$$\begin{aligned} \mathbf{2c + 2d} &= \mathbf{8} \\ \mathbf{0c - 2d} &= \mathbf{-2} \end{aligned}$$

The second equation gives $\boldsymbol{d = 1}$. Going upwards, the first equation becomes $\boldsymbol{2c + 2 = 8}$. Its solution is $\boldsymbol{c = 3}$. This combination is correct:

$$\mathbf{3}\begin{bmatrix} 2 \\ 1 \end{bmatrix} + \mathbf{1}\begin{bmatrix} 2 \\ -1 \end{bmatrix} = \begin{bmatrix} 8 \\ 2 \end{bmatrix}. \qquad \text{In matrix form} \begin{bmatrix} \mathbf{2} & \mathbf{2} \\ \mathbf{1} & \mathbf{-1} \end{bmatrix}\begin{bmatrix} \mathbf{3} \\ \mathbf{1} \end{bmatrix} = \begin{bmatrix} \mathbf{8} \\ \mathbf{2} \end{bmatrix}$$

Column Way, Row Way, Matrix Way

Column way **Linear combination**	$c\begin{bmatrix} v_1 \\ v_2 \end{bmatrix} + d\begin{bmatrix} w_1 \\ w_2 \end{bmatrix} = \begin{bmatrix} b_1 \\ b_2 \end{bmatrix}$
Row way **Two equations for** c **and** d	$\begin{aligned} v_1c + w_1d &= b_1 \\ v_2c + w_2d &= b_2 \end{aligned}$
Matrix way **2 by 2 matrix**	$\begin{bmatrix} v_1 & w_1 \\ v_2 & w_2 \end{bmatrix}\begin{bmatrix} c \\ d \end{bmatrix} = \begin{bmatrix} b_1 \\ b_2 \end{bmatrix}$

If the points $\boldsymbol{v}$ and $\boldsymbol{w}$ and the zero vector $\mathbf{0}$ are not on the same line, there is exactly one solution $\boldsymbol{c}, \boldsymbol{d}$. Then the linear combinations of $\boldsymbol{v}$ and $\boldsymbol{w}$ exactly fill the xy plane. The vectors $\boldsymbol{v}$ and $\boldsymbol{w}$ are **"linearly independent"**. The 2 by 2 matrix $A = \begin{bmatrix} \boldsymbol{v} & \boldsymbol{w} \end{bmatrix}$ is "*invertible*".

Can Elimination Fail ?

Elimination fails to produce a solution only when the equations don't have a solution in the first place. This can happen when the vectors $\boldsymbol{v}$ and $\boldsymbol{w}$ *lie on the same line through the center point* $(0, 0)$. Those vectors are not independent.

The reason is clear: All combinations of $\boldsymbol{v}$ and $\boldsymbol{w}$ will then *lie on that same line*. If the desired vector $\boldsymbol{b}$ is *off the line*, then the equations $c\boldsymbol{v} + d\boldsymbol{w} = \boldsymbol{b}$ have no solution:

Example $\quad \boldsymbol{v} = \begin{bmatrix} 1 \\ 2 \end{bmatrix} \quad \boldsymbol{w} = \begin{bmatrix} 3 \\ 6 \end{bmatrix} \quad \boldsymbol{b} = \begin{bmatrix} 1 \\ 0 \end{bmatrix} \qquad \begin{aligned} \mathbf{1c + 3d} &= \mathbf{1} \\ \mathbf{2c + 6d} &= \mathbf{0} \end{aligned}$

Those two equations can't be solved. To eliminate the 2 in the second equation, we multiply equation 1 by 2. Then elimination subtracts $2c + 6d = \mathbf{2}$ from the second equation $2c + 6d = \mathbf{0}$. **The result is** $\mathbf{0} = \mathbf{-2}$: *impossible*. We not only eliminated $\boldsymbol{c}$, we also eliminated $\boldsymbol{d}$.

With $\boldsymbol{v}$ and $\boldsymbol{w}$ on the same line, combinations of $\boldsymbol{v}$ and $\boldsymbol{w}$ fill that line but not a plane. When $\boldsymbol{b}$ is not on that line, no combination of $\boldsymbol{v}$ and $\boldsymbol{w}$ equals $\boldsymbol{b}$. The original vectors $\boldsymbol{v}$ and $\boldsymbol{w}$ are **"linearly dependent"** because $\boldsymbol{w} = 3\boldsymbol{v}$.

Linear combinations of two *independent vectors* $\boldsymbol{v}$ and $\boldsymbol{w}$ in two-dimensional space can produce any vector $\boldsymbol{b}$ in that plane. Then these equations have a solution:

2 equations
2 unknowns $\qquad c\boldsymbol{v} + d\boldsymbol{w} = \boldsymbol{b} \qquad \begin{aligned} cv_1 + dw_1 &= b_1 \\ cv_2 + dw_2 &= b_2 \end{aligned}$

Summary The combinations $c\boldsymbol{v} + d\boldsymbol{w}$ fill the x-y plane unless $\boldsymbol{v}$ is in line with $\boldsymbol{w}$.

Important Here is a different example where elimination seems to be in trouble. But we can easily fix the problem and go on.

$$c\begin{bmatrix} \mathbf{0} \\ \mathbf{1} \end{bmatrix} + d\begin{bmatrix} \mathbf{2} \\ \mathbf{3} \end{bmatrix} = \begin{bmatrix} \mathbf{2} \\ \mathbf{7} \end{bmatrix} \quad \text{or} \quad \begin{aligned} 0 + 2d &= 2 \\ c + 3d &= 7 \end{aligned}$$

That zero looks dangerous. But we only have to **exchange equations to find** d **and** c:

$$\begin{aligned} c + 3d &= 7 \\ 0 + 2d &= 2 \end{aligned} \quad \text{leads to} \quad \boldsymbol{d} = 1 \text{ and } \boldsymbol{c} = 4$$

Vectors in Three Dimensions

Suppose $\boldsymbol{v}$ and $\boldsymbol{w}$ have three components instead of two. Now they are vectors in three-dimensional space (which we will soon call $\mathbf{R}^3$). We still think of points in the space and arrows out from the zero vector. And we still have linear combinations of $\boldsymbol{v}$ and $\boldsymbol{w}$:

$$\boldsymbol{v}=\begin{bmatrix}2\\3\\1\end{bmatrix}\qquad \boldsymbol{w}=\begin{bmatrix}1\\1\\0\end{bmatrix}\qquad \boldsymbol{v}+\boldsymbol{w}=\begin{bmatrix}3\\4\\1\end{bmatrix}\qquad c\boldsymbol{v}+d\boldsymbol{w}=\begin{bmatrix}2c+d\\3c+d\\c+0\end{bmatrix}$$

But there is a difference! The combinations of $\boldsymbol{v}$ and $\boldsymbol{w}$ **do not fill** the whole 3-dimensional space. If we only have 2 vectors, their combinations can at most fill a 2-dimensional plane. It is not a case of linear dependence, it is just a case of not enough vectors.

We need *three* independent vectors if we want their combinations to fill 3-dimensional space. Here is the simplest choice $\boldsymbol{i}, \boldsymbol{j}, \boldsymbol{k}$ for those three independent vectors:

$$\boldsymbol{i}=\begin{bmatrix}1\\0\\0\end{bmatrix}\qquad \boldsymbol{j}=\begin{bmatrix}0\\1\\0\end{bmatrix}\qquad \boldsymbol{k}=\begin{bmatrix}0\\0\\1\end{bmatrix}\qquad c\boldsymbol{i}+d\boldsymbol{j}+e\boldsymbol{k}=\begin{bmatrix}c\\d\\e\end{bmatrix}\qquad (1)$$

Now $\boldsymbol{i}, \boldsymbol{j}, \boldsymbol{k}$ go along the x, y, z axes in three-dimensional space $\mathbf{R}^3$. We can easily write any vector $\boldsymbol{v}$ in $\mathbf{R}^3$ as a combination of $\boldsymbol{i}, \boldsymbol{j}, \boldsymbol{k}$:

Vector form

Matrix form

$$\boldsymbol{v}=\begin{bmatrix}v_1\\v_2\\v_3\end{bmatrix}=v_1\boldsymbol{i}+v_2\boldsymbol{j}+v_3\boldsymbol{k}\qquad \begin{bmatrix}\mathbf{1}&0&0\\0&\mathbf{1}&0\\0&0&\mathbf{1}\end{bmatrix}\begin{bmatrix}v_1\\v_2\\v_3\end{bmatrix}=\begin{bmatrix}v_1\\v_2\\v_3\end{bmatrix}$$

That is the 3 by 3 **identity matrix** $\boldsymbol{I}$. Multiplying by I leaves every vector unchanged. I is the matrix analog of the number 1, because $I\boldsymbol{v}=\boldsymbol{v}$ for every $\boldsymbol{v}$.

Notice that we could not take a linear combination of a 2-dimensional vector $\boldsymbol{v}$ and a 3-dimensional vector $\boldsymbol{w}$. They are not in the same space.

How Do We Know It is a Plane?

Suppose $\boldsymbol{v}$ and $\boldsymbol{w}$ are nonzero vectors with three components each. Assume they are *independent*, so the vectors point in different directions: $\boldsymbol{w}$ is not a multiple $c\boldsymbol{v}$. *Then their linear combinations fill a plane inside* 3*-dimensional space.* The surface is flat. Here is one way to see that this is true:

> Look at any two combinations $c\boldsymbol{v}+d\boldsymbol{w}$ and $C\boldsymbol{v}+D\boldsymbol{w}$. Halfway between those points is $\boldsymbol{h}=\frac{1}{2}(c+C)\boldsymbol{v}+\frac{1}{2}(d+D)\boldsymbol{w}$. This is another combination of $\boldsymbol{v}$ and $\boldsymbol{w}$. So our surface has the basic property of a plane: Halfway between any two points on the surface is **another point $\boldsymbol{h}$ on the surface**. The surface must be flat!
>
> Maybe that reasoning is not complete, even with the exclamation point. We depended on intuition for the properties of a plane. Another proof is coming in Section 1.2.

Problem Set 1.1

1 Under what conditions on a, b, c, d is $\begin{bmatrix} c \\ d \end{bmatrix}$ a multiple m of $\begin{bmatrix} a \\ b \end{bmatrix}$? Start with the two equations $c = ma$ and $d = mb$. By eliminating m, find one equation connecting a, b, c, d. You can assume no zeros in these numbers.

2 Going around a triangle from $(0,0)$ to $(5,0)$ to $(0,12)$ to $(0,0)$, what are those three vectors $\boldsymbol{u}, \boldsymbol{v}, \boldsymbol{w}$? What is $\boldsymbol{u}+\boldsymbol{v}+\boldsymbol{w}$? What are their lengths $||\boldsymbol{u}||$ and $||\boldsymbol{v}||$ and $||\boldsymbol{w}||$? The length squared of a vector $\boldsymbol{u} = (u_1, u_2)$ is $||\boldsymbol{u}||^2 = u_1^2 + u_2^2$.

Problems 3–10 are about addition of vectors and linear combinations.

3 Describe geometrically (line, plane, or all of $\mathbf{R}^3$) all linear combinations of

$$\text{(a)} \begin{bmatrix} 1 \\ 2 \\ 3 \end{bmatrix} \text{ and } \begin{bmatrix} 3 \\ 6 \\ 9 \end{bmatrix} \qquad \text{(b)} \begin{bmatrix} 1 \\ 0 \\ 0 \end{bmatrix} \text{ and } \begin{bmatrix} 0 \\ 2 \\ 3 \end{bmatrix} \qquad \text{(c)} \begin{bmatrix} 2 \\ 0 \\ 0 \end{bmatrix} \text{ and } \begin{bmatrix} 0 \\ 2 \\ 2 \end{bmatrix} \text{ and } \begin{bmatrix} 2 \\ 2 \\ 3 \end{bmatrix}$$

4 Draw $\boldsymbol{v} = \begin{bmatrix} 4 \\ 1 \end{bmatrix}$ and $\boldsymbol{w} = \begin{bmatrix} -2 \\ 2 \end{bmatrix}$ and $\boldsymbol{v}+\boldsymbol{w}$ and $\boldsymbol{v}-\boldsymbol{w}$ in a single xy plane.

5 If $\boldsymbol{v}+\boldsymbol{w} = \begin{bmatrix} 5 \\ 1 \end{bmatrix}$ and $\boldsymbol{v}-\boldsymbol{w} = \begin{bmatrix} 1 \\ 5 \end{bmatrix}$, compute and draw the vectors $\boldsymbol{v}$ and $\boldsymbol{w}$.

6 From $\boldsymbol{v} = \begin{bmatrix} 2 \\ 1 \end{bmatrix}$ and $\boldsymbol{w} = \begin{bmatrix} 1 \\ 2 \end{bmatrix}$, find the components of $3\boldsymbol{v}+\boldsymbol{w}$ and $c\boldsymbol{v}+d\boldsymbol{w}$.

7 Compute $\boldsymbol{u}+\boldsymbol{v}+\boldsymbol{w}$ and $2\boldsymbol{u}+2\boldsymbol{v}+\boldsymbol{w}$. How do you know $\boldsymbol{u}, \boldsymbol{v}, \boldsymbol{w}$ lie in a plane?

These lie in a plane because $\boldsymbol{w} = c\boldsymbol{u} + d\boldsymbol{v}$. **Find** c **and** d

$$\boldsymbol{u} = \begin{bmatrix} 1 \\ 2 \\ 3 \end{bmatrix} \quad \boldsymbol{v} = \begin{bmatrix} -3 \\ 1 \\ -2 \end{bmatrix} \quad \boldsymbol{w} = \begin{bmatrix} 2 \\ -3 \\ -1 \end{bmatrix}.$$

8 Every combination of $\boldsymbol{v} = (1,-2,1)$ and $\boldsymbol{w} = (0,1,-1)$ has components that add to _____. Find c and d so that $c\boldsymbol{v}+d\boldsymbol{w} = (3,3,-6)$. Why is $(3,3,6)$ impossible?

9 In the xy plane mark all nine of these linear combinations:

$$c\begin{bmatrix} 2 \\ 1 \end{bmatrix} + d\begin{bmatrix} 0 \\ 1 \end{bmatrix} \quad \text{with} \quad c = 0,1,2 \quad \text{and} \quad d = 0,1,2.$$

10 (Not easy) How could you decide if the vectors $\boldsymbol{u} = (1,1,0)$ and $\boldsymbol{v} = (0,1,1)$ and $\boldsymbol{w} = (a,b,c)$ are linearly independent or dependent?

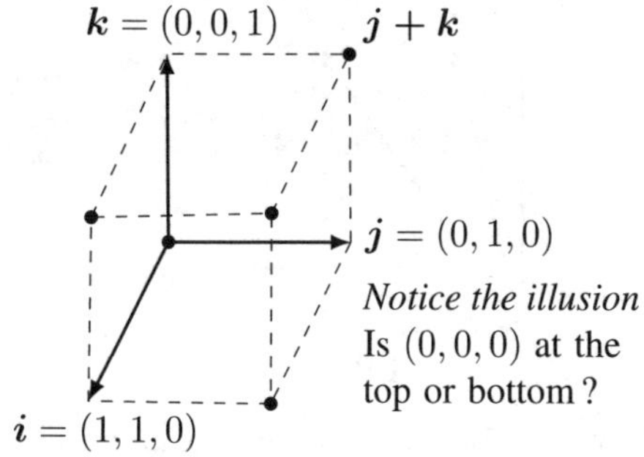

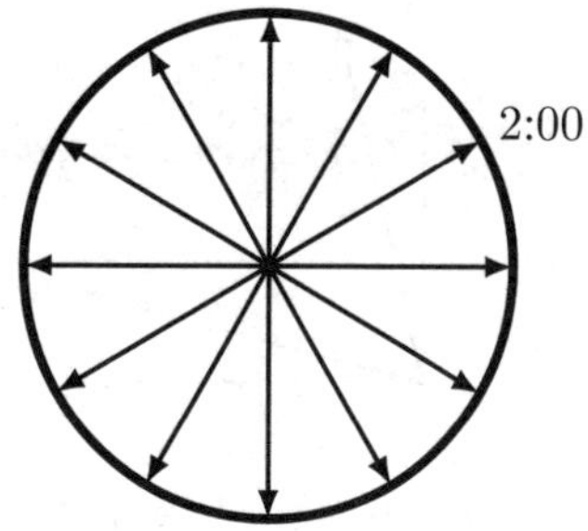

Figure 1.1: Unit cube from $\boldsymbol{i}, \boldsymbol{j}, \boldsymbol{k}$ and twelve clock vectors: all lengths = 1.

11 If three corners of a parallelogram are $(1,1)$, $(4,2)$, and $(1,3)$, what are all three of the possible fourth corners? Draw those three parallelograms.

Problems 12–15 are about special vectors on cubes and clocks in Figure 1.1.

12 Four corners of this unit cube are $(0,0,0)$, $(1,0,0)$, $(0,1,0)$, $(0,0,1)$. What are the other four corners? Find the coordinates of the center point of the cube. The center points of the six faces have coordinates ____. The cube has how many edges ?

13 *Review Question.* In xyz space, where is the plane of all linear combinations of $\boldsymbol{i} = (1,0,0)$ and $\boldsymbol{i} + \boldsymbol{j} = (1,1,0)$?

14 (a) What is the sum $\boldsymbol{V}$ of the twelve vectors that go from the center of a clock to the hours 1:00, 2:00, ..., 12:00?

(b) If the 2:00 vector is removed, why do the 11 remaining vectors add to 8:00?

(c) The components of that 2:00 vector are $\boldsymbol{v} = (\cos\theta, \sin\theta)$? What is θ ?

15 Suppose the twelve vectors start from 6:00 at the bottom instead of $(0,0)$ at the center. The vector to 12:00 is doubled to $(0,2)$. The new twelve vectors add to ____.

16 Draw vectors $\boldsymbol{u}$, $\boldsymbol{v}$, $\boldsymbol{w}$ so that their combinations $c\boldsymbol{u} + d\boldsymbol{v} + e\boldsymbol{w}$ fill only a line. Find vectors $\boldsymbol{u}$, $\boldsymbol{v}$, $\boldsymbol{w}$ in 3D so that their combinations $c\boldsymbol{u} + d\boldsymbol{v} + e\boldsymbol{w}$ fill only a plane.

17 What combination $c\begin{bmatrix}1\\2\end{bmatrix} + d\begin{bmatrix}3\\1\end{bmatrix}$ produces $\begin{bmatrix}14\\8\end{bmatrix}$? Express this question as two equations for the coefficients c and d in the linear combination.

Problems 18–19 go further with linear combinations of $\boldsymbol{v}$ and $\boldsymbol{w}$ (see Figure 1.2a).

18 Figure 1.2a shows $\frac{1}{2}\boldsymbol{v} + \frac{1}{2}\boldsymbol{w}$. Mark the points $\frac{3}{4}\boldsymbol{v} + \frac{1}{4}\boldsymbol{w}$ and $\frac{1}{4}\boldsymbol{v} + \frac{1}{4}\boldsymbol{w}$ and $\boldsymbol{v} + \boldsymbol{w}$. Draw the line of all combinations $c\boldsymbol{v} + d\boldsymbol{w}$ that have $c + d = 1$.

19 Restricted by $0 \le c \le 1$ and $0 \le d \le 1$, shade in all the combinations $c\boldsymbol{v} + d\boldsymbol{w}$. Restricted only by $c \ge 0$ and $d \ge 0$ draw the "cone" of all combinations $c\boldsymbol{v} + d\boldsymbol{w}$.

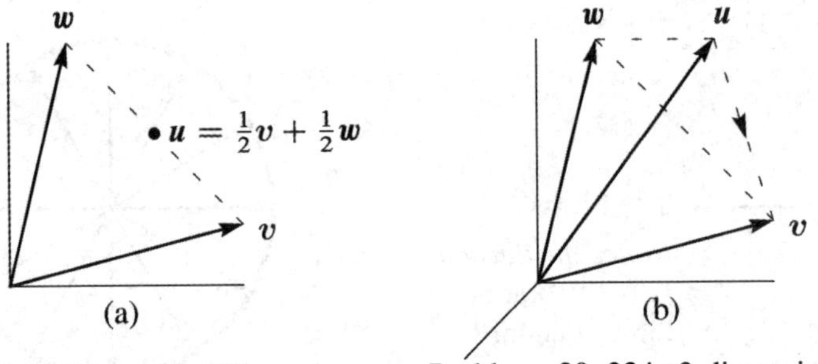

Figure 1.2: Problems **18–19** in a plane Problems **20–23** in 3-dimensional space

Problems 20–23 deal with $\boldsymbol{u}, \boldsymbol{v}, \boldsymbol{w}$ in three-dimensional space (see Figure 1.2b).

20 Locate $\frac{1}{3}\boldsymbol{u}+\frac{1}{3}\boldsymbol{v}+\frac{1}{3}\boldsymbol{w}$ and $\frac{1}{2}\boldsymbol{u}+\frac{1}{2}\boldsymbol{w}$ in Figure 1.2 b. Challenge problem: Under what restrictions on c, d, e, will the combinations $c\boldsymbol{u}+d\boldsymbol{v}+e\boldsymbol{w}$ fill in the dashed triangle? To stay in the triangle, one requirement is $c \geq 0, d \geq 0, e \geq 0$.

21 The three dashed lines in the triangle are $\boldsymbol{v}-\boldsymbol{u}$ and $\boldsymbol{w}-\boldsymbol{v}$ and $\boldsymbol{u}-\boldsymbol{w}$. Their sum is _____. Draw the head-to-tail addition around a plane triangle of $(3,1)$ plus $(-1,1)$ plus $(-2,-2)$.

22 Shade in the pyramid of combinations $c\boldsymbol{u}+d\boldsymbol{v}+e\boldsymbol{w}$ with $c \geq 0, d \geq 0, e \geq 0$ and $c+d+e \leq 1$. Mark the vector $\frac{1}{2}(\boldsymbol{u}+\boldsymbol{v}+\boldsymbol{w})$ as inside or outside this pyramid.

23 If you look at *all* combinations of those $\boldsymbol{u}, \boldsymbol{v}$, and $\boldsymbol{w}$, is there any vector that can't be produced from $c\boldsymbol{u}+d\boldsymbol{v}+e\boldsymbol{w}$? Different answer if $\boldsymbol{u}, \boldsymbol{v}, \boldsymbol{w}$ are all in _____.

Challenge Problems

24 How many corners $(\pm 1, \pm 1, \pm 1, \pm 1)$ does a cube of side 2 have in 4 dimensions? What is its volume? How many 3D faces? How many edges? Find one edge.

25 Find *two different combinations* of the three vectors $\boldsymbol{u}=(1,3)$ and $\boldsymbol{v}=(2,7)$ and $\boldsymbol{w}=(1,5)$ that produce $\boldsymbol{b}=(0,1)$. Slightly delicate question: If I take any three vectors $\boldsymbol{u}$, $\boldsymbol{v}$, $\boldsymbol{w}$ in the plane, will there always be two different combinations that produce $\boldsymbol{b}=(0,1)$?

26 The linear combinations of $\boldsymbol{v}=(a,b)$ and $\boldsymbol{w}=(c,d)$ fill the plane unless _____. Find four vectors $\boldsymbol{u}$, $\boldsymbol{v}$, $\boldsymbol{w}$, $\boldsymbol{z}$ with four nonzero components each so that their combinations $c\boldsymbol{u}+d\boldsymbol{v}+e\boldsymbol{w}+f\boldsymbol{z}$ produce all vectors in four-dimensional space.

27 Write down three equations for c, d, e so that $c\boldsymbol{u}+d\boldsymbol{v}+e\boldsymbol{w}=\boldsymbol{b}$. Write this also as a matrix equation $A\boldsymbol{x}=\boldsymbol{b}$. Can you somehow find c, d, e for this $\boldsymbol{b}$?

$$\boldsymbol{u}=\begin{bmatrix}2\\-1\\0\end{bmatrix} \quad \boldsymbol{v}=\begin{bmatrix}-1\\2\\-1\end{bmatrix} \quad \boldsymbol{w}=\begin{bmatrix}0\\-1\\2\end{bmatrix} \quad \boldsymbol{b}=\begin{bmatrix}1\\0\\0\end{bmatrix}.$$

1.2 Lengths and Angles from Dot Products

1 The "dot product" of $\boldsymbol{v} = \begin{bmatrix} 1 \\ 2 \end{bmatrix}$ and $\boldsymbol{w} = \begin{bmatrix} 4 \\ 6 \end{bmatrix}$ is $\boldsymbol{v} \cdot \boldsymbol{w} = (1)(4) + (2)(6) = 4 + 12 = \mathbf{16}$.

2 The length squared of $\boldsymbol{v} = (1, 3, 2)$ is $\boldsymbol{v} \cdot \boldsymbol{v} = 1 + 9 + 4 = 14$. **The length is** $||\boldsymbol{v}|| = \sqrt{\mathbf{14}}$.

3 $\boldsymbol{v} = (1, 3, 2)$ is **perpendicular** to $\boldsymbol{w} = (4, -4, 4)$ because $\boldsymbol{v} \cdot \boldsymbol{w} = \mathbf{0}$.

4 The angle $\theta = 45^\circ$ between $\boldsymbol{v} = \begin{bmatrix} 1 \\ 0 \end{bmatrix}$ and $\boldsymbol{w} = \begin{bmatrix} 1 \\ 1 \end{bmatrix}$ has $\cos\theta = \dfrac{\boldsymbol{v} \cdot \boldsymbol{w}}{||\boldsymbol{v}|| \; ||\boldsymbol{w}||} = \dfrac{1}{(1)(\sqrt{2})}$.

5 All angles have $|\cos\theta| \le 1$. All vectors have **Schwarz inequality** $|\boldsymbol{v} \cdot \boldsymbol{w}| \le ||\boldsymbol{v}|| \; ||\boldsymbol{w}||$ | **Triangle inequality** $||\boldsymbol{v} + \boldsymbol{w}|| \le ||\boldsymbol{v}|| + ||\boldsymbol{w}||$.

The most useful multiplication of vectors $\boldsymbol{v}$ and $\boldsymbol{w}$ is their **dot product** $\boldsymbol{v} \cdot \boldsymbol{w}$. We multiply the first components $v_1 w_1$ and the second components $v_2 w_2$ and so on. Then we **add those results** to get a single number $\boldsymbol{v} \cdot \boldsymbol{w}$:

$$\boxed{\text{The dot product of } \boldsymbol{v} = \begin{bmatrix} v_1 \\ v_2 \end{bmatrix} \text{ and } \boldsymbol{w} = \begin{bmatrix} w_1 \\ w_2 \end{bmatrix} \text{ is } \boldsymbol{v} \cdot \boldsymbol{w} = v_1 w_1 + v_2 w_2.} \qquad (1)$$

If the vectors are in n-dimensional space with n components each, then

$$\textbf{Dot product} \qquad \boxed{\boldsymbol{v} \cdot \boldsymbol{w} = v_1 w_1 + v_2 w_2 + \cdots + v_n w_n = \boldsymbol{w} \cdot \boldsymbol{v}} \qquad (2)$$

The dot product $\boldsymbol{v} \cdot \boldsymbol{v}$ **tells us the squared length** $||\boldsymbol{v}||^2 = v_1^2 + \cdots + v_n^2$ **of a vector.** In two dimensions, this is the Pythagoras formula $a^2 + b^2 = c^2$ for a right triangle. The sides have $a^2 = v_1^2$ and $b^2 = v_2^2$. The hypotenuse has $||\boldsymbol{v}||^2 = v_1^2 + v_2^2 = a^2 + b^2$.

To reach n dimensions, we can add one dimension at a time. Figure 1.2 shows $\boldsymbol{v} = (1, 2)$ in two dimensions and $\boldsymbol{w} = (1, 2, 3)$ in three dimensions. Now the right triangle has sides $(1, 2, 0)$ and $(0, 0, 3)$. Those vectors add to $\boldsymbol{w}$. The first side is in the xy plane, the second side goes up the perpendicular z axis . For this triangle in 3D with hypotenuse $\boldsymbol{w} = (1, 2, 3)$, the law $a^2 + b^2 = c^2$ becomes $(1^2 + 2^2) + (3^2) = 14 = ||\boldsymbol{w}||^2$.

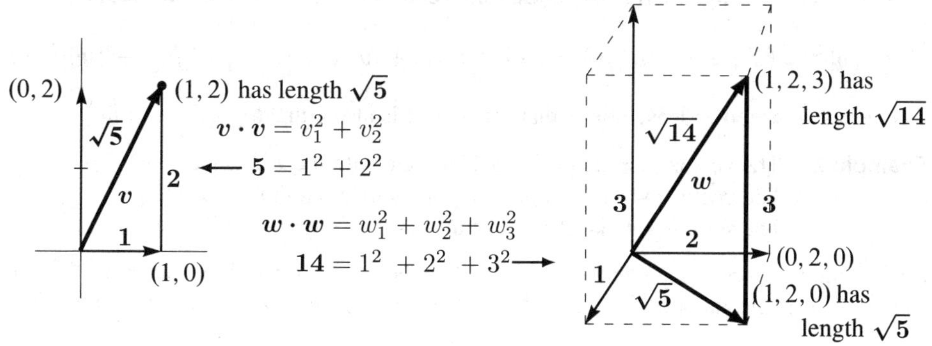

Figure 1.3: The length $\sqrt{\boldsymbol{v} \cdot \boldsymbol{v}} = \sqrt{5}$ in a plane and $\sqrt{\boldsymbol{w} \cdot \boldsymbol{w}} = \sqrt{14}$ in three dimensions.

The length of a four-dimensional vector would be $\sqrt{v_1^2+v_2^2+v_3^2+v_4^2}$. Thus the vector $(1,1,1,1)$ has length $\sqrt{1^2+1^2+1^2+1^2} = 2$. This is the diagonal through a unit cube in four-dimensional space. That diagonal in n dimensions has length $\sqrt{n}$.

We use the words **unit vector** when the length of the vector is 1. Divide $\boldsymbol{v}$ by $||\boldsymbol{v}||$.

> **A unit vector $\boldsymbol{u}$ has length $||\boldsymbol{u}|| = 1$. If $\boldsymbol{v} \neq \boldsymbol{0}$ then $\boldsymbol{u} = \dfrac{\boldsymbol{v}}{||\boldsymbol{v}||}$ is a unit vector.**

Example 1 The standard unit vector along the x axis is written $\boldsymbol{i}$. In the xy plane, the unit vector that makes an angle "theta" with the x axis is $\boldsymbol{u} = (\cos\theta, \sin\theta)$:

$$\textbf{Unit vectors} \quad \boldsymbol{i} = \begin{bmatrix} 1 \\ 0 \end{bmatrix} \quad \text{and} \quad \boldsymbol{u} = \begin{bmatrix} \cos\theta \\ \sin\theta \end{bmatrix}. \text{ Notice } \boldsymbol{i}\cdot\boldsymbol{u} = \cos\theta.$$

$\boldsymbol{u} = (\cos\theta, \sin\theta)$ is a unit vector because $\boldsymbol{u}\cdot\boldsymbol{u} = \cos^2\theta + \sin^2\theta = 1$.

In four dimensions, one example of a unit vector is $\boldsymbol{u} = \left(\frac{1}{2}, \frac{1}{2}, \frac{1}{2}, \frac{1}{2}\right)$. Or you could start with the vector $\boldsymbol{v} = (1, 5, 5, 7)$. Then $||\boldsymbol{v}||^2 = 1 + 25 + 25 + 49 = 100$. So $\boldsymbol{v}$ has length 10 and $\boldsymbol{u} = \boldsymbol{v}/10$ is a unit vector.

The word "**unit**" is always indicating that some measurement equals "one". The unit price is the price for one item. A unit cube has sides of length one. A unit circle is a circle with radius one. Now we see that a "unit vector" has length = 1.

Perpendicular Vectors

Suppose the angle between $\boldsymbol{v}$ and $\boldsymbol{w}$ is 90°. Its cosine is zero. That produces a valuable test $\boldsymbol{v}\cdot\boldsymbol{w} = 0$ for perpendicular vectors.

> **Perpendicular vectors have $\boldsymbol{v}\cdot\boldsymbol{w} = 0$. Then $||\boldsymbol{v}+\boldsymbol{w}||^2 = ||\boldsymbol{v}||^2 + ||\boldsymbol{w}||^2$.** (3)

This is the most important special case. It has brought us back to 90° angles and lengths $a^2 + b^2 = c^2$. The algebra for **perpendicular vectors** ($\boldsymbol{v}\cdot\boldsymbol{w} = 0 = \boldsymbol{w}\cdot\boldsymbol{v}$) is easy :

$$||\boldsymbol{v}+\boldsymbol{w}||^2 = (\boldsymbol{v}+\boldsymbol{w})\cdot(\boldsymbol{v}+\boldsymbol{w}) = \boldsymbol{v}\cdot\boldsymbol{v} + \boldsymbol{v}\cdot\boldsymbol{w} + \boldsymbol{w}\cdot\boldsymbol{v} + \boldsymbol{w}\cdot\boldsymbol{w} = ||\boldsymbol{v}||^2 + ||\boldsymbol{w}||^2. \quad (4)$$

Two terms were zero. Please notice that $||\boldsymbol{v}-\boldsymbol{w}||^2$ is also equal to $||\boldsymbol{v}||^2 + ||\boldsymbol{w}||^2$.

Example 2 The vector $\boldsymbol{v} = (1,1)$ is at a 45° angle with the x axis
The vector $\boldsymbol{w} = (1,-1)$ is at a $-45°$ angle with the x axis
The sum $\boldsymbol{v}+\boldsymbol{w}$ is $(2,0)$. The difference $\boldsymbol{v}-\boldsymbol{w}$ is $(0,2)$.

So the angle between $(1,1)$ and $(1,-1)$ is 90°. Their dot product is $\boldsymbol{v}\cdot\boldsymbol{w} = 1 - 1 = 0$. This right triangle has $||\boldsymbol{v}||^2 = 2$ and $||\boldsymbol{w}||^2 = 2$ and $||\boldsymbol{v}-\boldsymbol{w}||^2 = ||\boldsymbol{v}+\boldsymbol{w}||^2 = 4$.

Example 3 The vectors $\boldsymbol{v} = (4,2)$ and $\boldsymbol{w} = (-1,2)$ have a *zero* dot product:

Dot product is zero
Vectors are perpendicular
$$\begin{bmatrix}4\\2\end{bmatrix} \cdot \begin{bmatrix}-1\\2\end{bmatrix} = -4+4=0.$$

Put a weight of 4 at the point $x = -1$ (left of zero) and a weight of 2 at the point $x = 2$ (right of zero). The x axis will balance on the center point like a see-saw. The weights balance because the dot product is $(4)(-1)+(2)(2)=0$.

This example is typical of engineering and science. The vector of weights is $(w_1, w_2) = (4,2)$. The vector of distances from the center is $(v_1, v_2) = (-1,2)$. The weights times the distances, w_1v_1 and w_2v_2, give the "moments". The equation for the see-saw to balance is $\boldsymbol{w} \cdot \boldsymbol{v} = w_1v_1 + w_2v_2 = 0 =$ zero dot product.

Example 4 The unit vectors $\boldsymbol{v} = (1,0)$ and $\boldsymbol{u} = (\cos\theta, \sin\theta)$ have $\boldsymbol{v} \cdot \boldsymbol{u} = \cos\theta$. Now we are connecting the dot product to the angle between vectors.

Cosine of the angle $\boldsymbol{\theta}$ The cosine formula is easy to remember for unit vectors:

If $\lvert\lvert\boldsymbol{v}\rvert\rvert = 1$ and $\lvert\lvert\boldsymbol{u}\rvert\rvert = 1$, the angle θ between $\boldsymbol{v}$ and $\boldsymbol{u}$ has $\mathbf{cos}\,\boldsymbol{\theta} = \boldsymbol{v} \cdot \boldsymbol{u}$.

In mathematics, zero is always a special number. For dot products, it means that *these two vectors are perpendicular.* The angle between them is 90°. The clearest example of perpendicular vectors is $\boldsymbol{i} = (1,0)$ along the x axis and $\boldsymbol{j} = (0,1)$ up the y axis. Again the dot product is $\boldsymbol{i} \cdot \boldsymbol{j} = 0+0=0$. The cosine of 90° is zero.

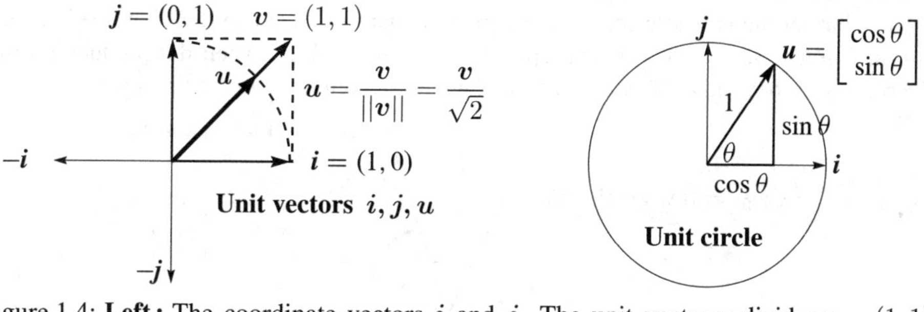

Figure 1.4: **Left:** The coordinate vectors $\boldsymbol{i}$ and $\boldsymbol{j}$. The unit vector $\boldsymbol{u}$ divides $\boldsymbol{v} = (1,1)$ by its length $\lvert\lvert\boldsymbol{v}\rvert\rvert = \sqrt{2}$. **Right:** The unit vector $\boldsymbol{u} = (\cos\theta, \sin\theta)$ is at angle θ with $\boldsymbol{i}$.

Example 5 Dot products enter in economics and business. We have three goods to buy. Their prices for each unit are (p_1, p_2, p_3)—this is the price vector $\boldsymbol{p}$. The quantities we buy are (q_1, q_2, q_3). *Buying q_1 units at the price p_1 brings in q_1p_1*. The total cost adds up quantities q times prices p. ***This is the dot product q · p in three dimensions***:

$$\textbf{\textit{Cost}} = (q_1, q_2, q_3) \cdot (p_1, p_2, p_3) = q_1p_1 + q_2p_2 + q_3p_3 = \textbf{\textit{dot product}}.$$

A zero dot product means that "the books balance". Total sales equal total purchases if $\boldsymbol{q} \cdot \boldsymbol{p} = 0$. Then $\boldsymbol{p}$ is perpendicular to $\boldsymbol{q}$ (in three-dimensional space). A supermarket with thousands of goods goes quickly into high dimensions.

Spreadsheets have become essential in management. They compute linear combinations and dot products. What you see on the screen is a matrix.

The Angle Between Two Vectors

We know that perpendicular vectors have $\boldsymbol{v} \cdot \boldsymbol{w} = 0$. The dot product is zero when the angle is 90°. Our next step is to connect all dot products to angles. The dot product $\boldsymbol{v} \cdot \boldsymbol{w}$ finds the angle between any two nonzero vectors $\boldsymbol{v}$ and $\boldsymbol{w}$.

Example 6 The unit vectors $\boldsymbol{v} = (\cos\alpha, \sin\alpha)$ and $\boldsymbol{w} = (\cos\beta, \sin\beta)$ have $\boldsymbol{v} \cdot \boldsymbol{w} = \cos\alpha\cos\beta + \sin\alpha\sin\beta$. In trigonometry this is the formula for $\cos(\beta - \alpha)$. Figure 1.5 shows that the angle between the unit vectors $\boldsymbol{v}$ and $\boldsymbol{w}$ is $\beta - \alpha$.

The dot product $\boldsymbol{w} \cdot \boldsymbol{v}$ ***equals*** $\boldsymbol{v} \cdot \boldsymbol{w}$. The order of $\boldsymbol{v}$ and $\boldsymbol{w}$ makes no difference.

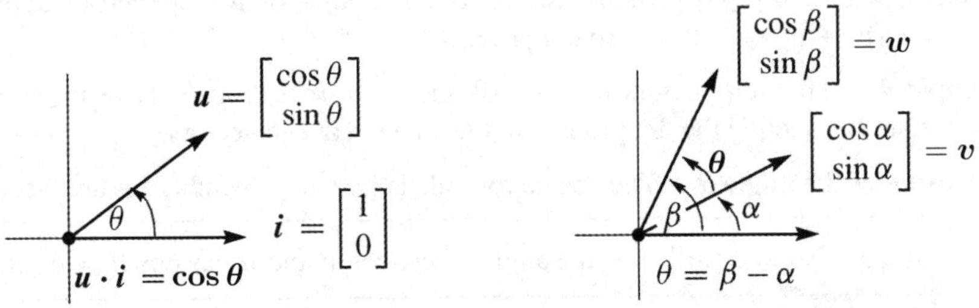

Figure 1.5: Unit vectors: $\boldsymbol{u} \cdot \boldsymbol{i} = \cos\theta$. The angle between the vectors is θ.

Suppose $\boldsymbol{v} \cdot \boldsymbol{w}$ is **not zero**. It may be positive, it may be negative. The sign of $\boldsymbol{v} \cdot \boldsymbol{w}$ immediately tells whether we are below or above a right angle. The angle is less than 90° when $\boldsymbol{v} \cdot \boldsymbol{w}$ is positive. The angle is above 90° when $\boldsymbol{v} \cdot \boldsymbol{w}$ is negative.

The borderline is where vectors are perpendicular to $\boldsymbol{v}$. On that dividing line between plus and minus, $\boldsymbol{w}_2 = (1, -3)$ is perpendicular to $\boldsymbol{v} = (3, 1)$. Their dot product is zero. Then $\boldsymbol{w}_3$ goes beyond a 90° angle with $\boldsymbol{v}$. The test becomes $\boldsymbol{v} \cdot \boldsymbol{w}_3 < 0$: *negative*.

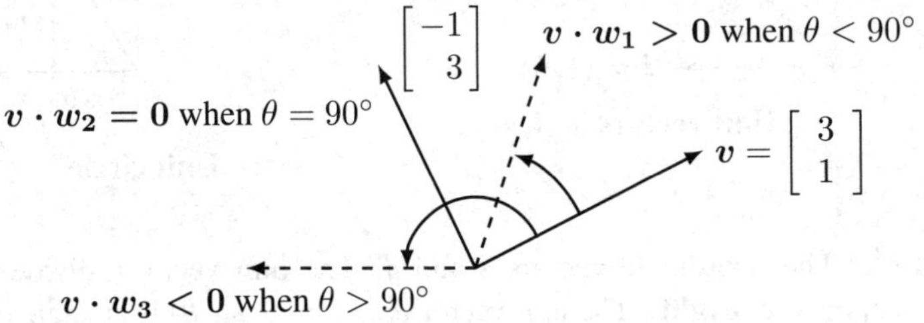

Figure 1.6: Small angle $\boldsymbol{v} \cdot \boldsymbol{w_1} > \mathbf{0}$. Right angle $\boldsymbol{v} \cdot \boldsymbol{w_2} = \mathbf{0}$. Large angle $\boldsymbol{v} \cdot \boldsymbol{w_3} < \mathbf{0}$.

The dot product reveals the exact angle $\boldsymbol{\theta}$. To repeat: For unit vectors $\boldsymbol{u}$ and $\boldsymbol{U}$, ***the dot product*** $\boldsymbol{u} \cdot \boldsymbol{U}$ ***is the cosine of*** θ. This remains true in n dimensions.

Remember that $\cos\theta$ is never greater than 1. It is never less than -1. *The dot product of unit vectors is between* -1 *and* 1. **The cosine of** θ **is revealed by** $\boldsymbol{u} \cdot \boldsymbol{U}$.

What if $\boldsymbol{v}$ *and* $\boldsymbol{w}$ *are not unit vectors*? Divide by their lengths to get $\boldsymbol{u} = \boldsymbol{v}/\|\boldsymbol{v}\|$ and $\boldsymbol{U} = \boldsymbol{w}/\|\boldsymbol{w}\|$. Then the dot product of those unit vectors $\boldsymbol{u}$ and $\boldsymbol{U}$ gives $\cos\theta$.

COSINE FORMULA If $\boldsymbol{v}$ and $\boldsymbol{w}$ are nonzero vectors then $\dfrac{\boldsymbol{v} \cdot \boldsymbol{w}}{\|\boldsymbol{v}\|\,\|\boldsymbol{w}\|} = \cos\theta$. (5)

Whatever the angle, this dot product of $\boldsymbol{v}/||\boldsymbol{v}||$ with $\boldsymbol{w}/||\boldsymbol{w}||$ never exceeds one. That is the ***"Schwarz inequality"*** $|\boldsymbol{v} \cdot \boldsymbol{w}| \leq ||\boldsymbol{v}|| \, ||\boldsymbol{w}||$ for dot products—or more correctly the Cauchy-Schwarz-Bunyakowsky inequality. It was found in France and Germany and Russia (and maybe elsewhere—it is the most important inequality in mathematics).

Since $|\cos\theta|$ never exceeds 1, the cosine formula in (5) gives two great inequalities.

SCHWARZ INEQUALITY	$\lvert\boldsymbol{v} \cdot \boldsymbol{w}\rvert \leq \lVert\boldsymbol{v}\rVert \, \lVert\boldsymbol{w}\rVert$
TRIANGLE INEQUALITY	$\lVert\boldsymbol{v} + \boldsymbol{w}\rVert \leq \lVert\boldsymbol{v}\rVert + \lVert\boldsymbol{w}\rVert$

The triangle inequality comes directly from the Schwarz inequality !

$$||\boldsymbol{v}+\boldsymbol{w}||^2 = \boldsymbol{v}\cdot\boldsymbol{v}+\boldsymbol{v}\cdot\boldsymbol{w}+\boldsymbol{w}\cdot\boldsymbol{v}+\boldsymbol{w}\cdot\boldsymbol{w} \leq ||\boldsymbol{v}||^2 + 2\,||\boldsymbol{v}||\,||\boldsymbol{w}|| + ||\boldsymbol{w}||^2. \quad (6)$$

The square root is $||\boldsymbol{v}+\boldsymbol{w}|| \leq ||\boldsymbol{v}|| + ||\boldsymbol{w}||$. **Side 3 cannot exceed Side 1 + Side 2.**

Example 7 Find $\cos\theta$ for $\boldsymbol{v} = \begin{bmatrix} 2 \\ 1 \end{bmatrix}$ and $\boldsymbol{w} = \begin{bmatrix} 1 \\ 2 \end{bmatrix}$ and check both inequalities.

The dot product is $\boldsymbol{v} \cdot \boldsymbol{w} = \mathbf{4}$. Both $\boldsymbol{v}$ and $\boldsymbol{w}$ have length $\sqrt{\mathbf{5}}$. So $||\boldsymbol{v}|| \, ||\boldsymbol{w}|| = \mathbf{5}$.

$$\cos\theta = \frac{\boldsymbol{v}\cdot\boldsymbol{w}}{||\boldsymbol{v}|| \, ||\boldsymbol{w}||} = \frac{4}{\sqrt{5}\sqrt{5}} = \frac{4}{5}.$$

The Schwarz inequality is $4 < 5$. By the triangle inequality, side 3 $= ||\boldsymbol{v}+\boldsymbol{w}||$ is less than side 1 + side 2. With $\boldsymbol{v}+\boldsymbol{w} = (3,3)$ the three side lengths are $\sqrt{\mathbf{18}} < \sqrt{\mathbf{5}} + \sqrt{\mathbf{5}}$. Square this inequality to get $18 < 20$. This confirms the triangle inequality.

Example 8 The dot product of $\boldsymbol{v} = (a,b)$ and $\boldsymbol{w} = (b,a)$ is $2ab$. Their lengths are $||\boldsymbol{v}|| = ||\boldsymbol{w}|| = \sqrt{a^2+b^2}$. The Schwarz inequality $\boldsymbol{v}\cdot\boldsymbol{w} \leq ||\boldsymbol{v}|| \, ||\boldsymbol{w}||$ is $\mathbf{2ab \leq a^2 + b^2}$.

Any numbers a^2 and b^2 have *geometric mean* $|\boldsymbol{ab}| \leq$ *arithmetic mean* $\frac{1}{2}\left(\boldsymbol{a}^2+\boldsymbol{b}^2\right)$. The proof is that $\frac{1}{2}\left(\boldsymbol{a}^2+\boldsymbol{b}^2\right) - ab = \frac{1}{2}(a-b)^2$ is a perfect square : never negative !

A Plane in 3 Dimensions

Suppose $\boldsymbol{n}$ is a unit vector with three components n_1, n_2, n_3. Look at all vectors $\boldsymbol{w}$ in $\mathbf{R}^3$ that are perpendicular to $\boldsymbol{n}$ (so $\boldsymbol{w}\cdot\boldsymbol{n} = 0$) :

The vectors $\boldsymbol{w}$ with $\boldsymbol{w}\cdot\boldsymbol{n} = 0$ fill a 2-dimensional plane in $\mathbf{R}^3$

The whole plane is perpendicular to its "normal vector $\boldsymbol{n}$". The equation of a 2-dimensional plane in 3-dimensional space is $n_1w_1 + n_2w_2 + n_3w_3 = 0$. For the "$xy$ plane" the normal vector $\boldsymbol{n}$ going straight upward has components $0, 0, 1$. So the equation of the xy plane is just $w_3 = 0$ or $z = 0$, which we already knew.

■ WORKED EXAMPLES ■

1.2 A For the vectors $\boldsymbol{v} = (3,4)$ and $\boldsymbol{w} = (4,3)$ test the Schwarz inequality on $\boldsymbol{v} \cdot \boldsymbol{w}$ and the triangle inequality on $\|\boldsymbol{v} + \boldsymbol{w}\|$. Find $\cos\theta$ for the angle between $\boldsymbol{v}$ and $\boldsymbol{w}$.

Solution The dot product is $\boldsymbol{v} \cdot \boldsymbol{w} = (3)(4) + (4)(3) = 24$. The length of $\boldsymbol{v}$ is $\|\boldsymbol{v}\| = \sqrt{9+16} = 5$ and also $\|\boldsymbol{w}\| = 5$. The sum $\boldsymbol{v} + \boldsymbol{w} = (7,7)$ has length $7\sqrt{2} < 10$.

Schwarz inequality $|\boldsymbol{v} \cdot \boldsymbol{w}| \le \|\boldsymbol{v}\|\,\|\boldsymbol{w}\|$ is $24 < 25$.

Triangle inequality $\|\boldsymbol{v} + \boldsymbol{w}\| \le \|\boldsymbol{v}\| + \|\boldsymbol{w}\|$ is $7\sqrt{2} < 5 + 5$.

Cosine of angle $\cos\theta = \dfrac{24}{25}$ Thin angle from $\boldsymbol{v} = (3,4)$ to $\boldsymbol{w} = (4,3)$

1.2 B Which $\boldsymbol{v}$ and $\boldsymbol{w}$ give *equality* $|\boldsymbol{v} \cdot \boldsymbol{w}| = \|\boldsymbol{v}\|\,\|\boldsymbol{w}\|$ and $\|\boldsymbol{v} + \boldsymbol{w}\| = \|\boldsymbol{v}\| + \|\boldsymbol{w}\|$?

Equality: One vector is a multiple of the other as in $\boldsymbol{w} = c\boldsymbol{v}$. Then the angle is $0°$ or $180°$. In this case $|\cos\theta| = 1$ and $|\boldsymbol{v} \cdot \boldsymbol{w}|$ *equals* $\|\boldsymbol{v}\|\,\|\boldsymbol{w}\|$. If the angle is $0°$, as in $\boldsymbol{w} = 2\boldsymbol{v}$, then $\|\boldsymbol{v} + \boldsymbol{w}\| = \|\boldsymbol{v}\| + \|\boldsymbol{w}\|$ (both sides give $3\|\boldsymbol{v}\|$). This $\boldsymbol{v}, 2\boldsymbol{v}, 3\boldsymbol{v}$ triangle is flat.

1.2 C Find a unit vector $\boldsymbol{u}$ in the direction of $\boldsymbol{v} = (3,4)$. Find a unit vector $\boldsymbol{U}$ that is perpendicular to $\boldsymbol{u}$. There are two possibilities for $\boldsymbol{U}$.

Solution For a unit vector $\boldsymbol{u}$, divide $\boldsymbol{v}$ by its length $\|\boldsymbol{v}\| = 5$. For a perpendicular vector $\boldsymbol{V}$ we can choose $(-4,3)$ or $(4,-3)$. For a *unit* vector $\boldsymbol{U}$, divide $\boldsymbol{V}$ by its length $\|\boldsymbol{V}\| = 5$.

1.2 D We want to explain the angle formula $\boldsymbol{v} \cdot \boldsymbol{w} = \|\boldsymbol{v}\|\,\|\boldsymbol{w}\| \cos\theta$. This is $ab\cos\theta$ in the Law of Cosines $c^2 = a^2 + b^2 - 2ab\cos\theta$.

Use Pythagoras in the left triangle !

$$
\begin{aligned}
c^2 &= (a\sin\theta)^2 + (b - a\cos\theta)^2 \\
&= a^2\sin^2\theta + b^2 - 2ab\cos\theta + a^2\cos^2\theta \\
&= \boldsymbol{a^2 + b^2 - 2ab\cos\theta}
\end{aligned}
$$

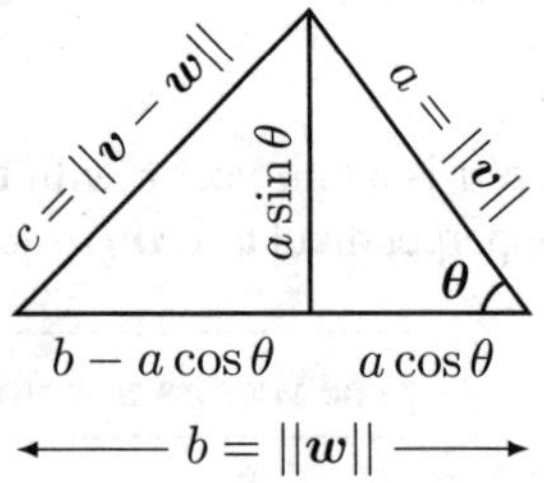

The picture shows the sides $a = \|\boldsymbol{v}\|$, $b = \|\boldsymbol{w}\|$, $c = \|\boldsymbol{v} - \boldsymbol{w}\|$ and two right triangles.

Compare with $\|\boldsymbol{v} - \boldsymbol{w}\|^2 = \|\boldsymbol{v}\|^2 + \|\boldsymbol{w}\|^2 - 2\boldsymbol{v} \cdot \boldsymbol{w}$ to match $\boldsymbol{v} \cdot \boldsymbol{w}$ with $\|\boldsymbol{v}\|\,\|\boldsymbol{w}\| \cos\theta$.

Problem Set 1.2

1 Calculate the dot products $\boldsymbol{u} \cdot \boldsymbol{v}$ and $\boldsymbol{u} \cdot \boldsymbol{w}$ and $\boldsymbol{u} \cdot (\boldsymbol{v}+\boldsymbol{w})$ and $\boldsymbol{w} \cdot \boldsymbol{v}$:

$$\boldsymbol{u}=\begin{bmatrix}-.6\\ .8\end{bmatrix} \qquad \boldsymbol{v}=\begin{bmatrix}4\\3\end{bmatrix} \qquad \boldsymbol{w}=\begin{bmatrix}1\\2\end{bmatrix}.$$

2 Compute the lengths $\|\boldsymbol{u}\|$ and $\|\boldsymbol{v}\|$ and $\|\boldsymbol{w}\|$ of those vectors. Check the Schwarz inequalities $|\boldsymbol{u} \cdot \boldsymbol{v}| \leq \|\boldsymbol{u}\|\,\|\boldsymbol{v}\|$ and $|\boldsymbol{v} \cdot \boldsymbol{w}| \leq \|\boldsymbol{v}\|\,\|\boldsymbol{w}\|$.

3 Find unit vectors in the directions of $\boldsymbol{v}$ and $\boldsymbol{w}$ in Problem 1, and find the cosine of their angle θ. Choose vectors $\boldsymbol{a}, \boldsymbol{b}, \boldsymbol{c}$ that make 0°, 90°, and 180° angles with $\boldsymbol{w}$.

4 For any *unit* vectors $\boldsymbol{v}$ and $\boldsymbol{w}$, find the dot products (actual numbers) of

(a) $\boldsymbol{v}$ and $-\boldsymbol{v}$ (b) $\boldsymbol{v}+\boldsymbol{w}$ and $\boldsymbol{v}-\boldsymbol{w}$ (c) $\boldsymbol{v}-2\boldsymbol{w}$ and $\boldsymbol{v}+2\boldsymbol{w}$

5 Find unit vectors $\boldsymbol{u}_1$ and $\boldsymbol{u}_2$ in the directions of $\boldsymbol{v}=(1,3)$ and $\boldsymbol{w}=(2,1,2)$. Find unit vectors $\boldsymbol{U}_1$ and $\boldsymbol{U}_2$ that are perpendicular to $\boldsymbol{u}_1$ and $\boldsymbol{u}_2$.

6 (a) Describe every vector $\boldsymbol{w}=(w_1,w_2)$ that is perpendicular to $\boldsymbol{v}=(2,-1)$.

(b) All vectors perpendicular to $\boldsymbol{V}=(1,1,1)$ lie on a _____ in 3 dimensions.

(c) The vectors perpendicular to both $(1,1,1)$ and $(1,2,3)$ lie on a _____.

7 Find the angle θ (from its cosine) between these pairs of vectors:

(a) $\boldsymbol{v}=\begin{bmatrix}1\\ \sqrt{3}\end{bmatrix}$ and $\boldsymbol{w}=\begin{bmatrix}1\\0\end{bmatrix}$ (b) $\boldsymbol{v}=\begin{bmatrix}2\\2\\-1\end{bmatrix}$ and $\boldsymbol{w}=\begin{bmatrix}2\\-1\\2\end{bmatrix}$

(c) $\boldsymbol{v}=\begin{bmatrix}1\\ \sqrt{3}\end{bmatrix}$ and $\boldsymbol{w}=\begin{bmatrix}-1\\ \sqrt{3}\end{bmatrix}$ (d) $\boldsymbol{v}=\begin{bmatrix}3\\1\end{bmatrix}$ and $\boldsymbol{w}=\begin{bmatrix}-1\\-2\end{bmatrix}$.

8 True or false (give a reason if true or find a counterexample if false):

(a) If $\boldsymbol{u}=(1,1,1)$ is perpendicular to $\boldsymbol{v}$ and $\boldsymbol{w}$, then $\boldsymbol{v}$ is parallel to $\boldsymbol{w}$.

(b) If $\boldsymbol{u}$ is perpendicular to $\boldsymbol{v}$ and $\boldsymbol{w}$, then $\boldsymbol{u}$ is perpendicular to $\boldsymbol{v}+2\boldsymbol{w}$.

(c) If $\boldsymbol{u}$ and $\boldsymbol{v}$ are perpendicular unit vectors then $\|\boldsymbol{u}-\boldsymbol{v}\|^2=2$. *Yes*!

9 The slopes of the arrows from $(0,0)$ to (v_1,v_2) and (w_1,w_2) are v_2/v_1 and w_2/w_1. **Suppose the product** v_2w_2/v_1w_1 **of those slopes is** -1. Show that $\boldsymbol{v} \cdot \boldsymbol{w}=0$ and the vectors are perpendicular. (The line $y=4x$ is perpendicular to $y=-\frac{1}{4}x$.)

10 Draw arrows from $(0,0)$ to the points $\boldsymbol{v}=(1,2)$ and $\boldsymbol{w}=(-2,1)$. Multiply their slopes. That answer is a signal that $\boldsymbol{v} \cdot \boldsymbol{w}=0$ and the arrows are _____.

11 If $\boldsymbol{v} \cdot \boldsymbol{w}$ is negative, what does this say about the angle between $\boldsymbol{v}$ and $\boldsymbol{w}$? Draw a 3-dimensional vector $\boldsymbol{v}$ (an arrow), and show where to find all $\boldsymbol{w}$'s with $\boldsymbol{v} \cdot \boldsymbol{w} < 0$.

12 With $\boldsymbol{v} = (1, 1)$ and $\boldsymbol{w} = (1, 5)$ choose a number c so that $\boldsymbol{w} - c\boldsymbol{v}$ is perpendicular to $\boldsymbol{v}$. Then find the formula for c starting from *any* nonzero $\boldsymbol{v}$ and $\boldsymbol{w}$.

13 Find nonzero vectors $\boldsymbol{u}, \boldsymbol{v}, \boldsymbol{w}$ that are perpendicular to $(1, 1, 1, 1)$ and to each other.

14 The geometric mean of $x = 2$ and $y = 8$ is $\sqrt{xy} = 4$. The arithmetic mean is larger: $\frac{1}{2}(x+y) =$ ______. This would come from the Schwarz inequality for $\boldsymbol{v} = (\sqrt{2}, \sqrt{8})$ and $\boldsymbol{w} = (\sqrt{8}, \sqrt{2})$. Find $\cos\theta$ for this $\boldsymbol{v}$ and $\boldsymbol{w}$.

15 **How long is the vector** $\boldsymbol{v} = (1, 1, \ldots, 1)$ **in 9 dimensions**? Find a unit vector $\boldsymbol{u}$ in the same direction as $\boldsymbol{v}$ and a unit vector $\boldsymbol{w}$ that is perpendicular to $\boldsymbol{v}$.

16 What are the cosines of the angles α, β, θ between the vector $(1, 0, -1)$ and the unit vectors $\boldsymbol{i}, \boldsymbol{j}, \boldsymbol{k}$ along the axes? Check the formula $\cos^2\alpha + \cos^2\beta + \cos^2\theta = 1$.

Problems 17–20 lead to the main facts about lengths and angles in triangles.

17 The vectors $\boldsymbol{v} = (4, 2)$ and $\boldsymbol{w} = (-1, 2)$ are two sides of a right triangle. Check the Pythagoras formula $a^2 + b^2 = c^2$ which is for ***right triangles only***:

$$(\text{length of } \boldsymbol{v})^2 + (\text{length of } \boldsymbol{w})^2 = (\text{length of } \boldsymbol{v} + \boldsymbol{w})^2.$$

18 (Rules for dot products) These equations are simple but useful:

(1) $\boldsymbol{v} \cdot \boldsymbol{w} = \boldsymbol{w} \cdot \boldsymbol{v}$ **(2)** $\boldsymbol{u} \cdot (\boldsymbol{v} + \boldsymbol{w}) = \boldsymbol{u} \cdot \boldsymbol{v} + \boldsymbol{u} \cdot \boldsymbol{w}$ **(3)** $(c\boldsymbol{v}) \cdot \boldsymbol{w} = c(\boldsymbol{v} \cdot \boldsymbol{w})$

Use **(1)** and **(2)** with $\boldsymbol{u} = \boldsymbol{v} + \boldsymbol{w}$ to prove $\|\boldsymbol{v} + \boldsymbol{w}\|^2 = \boldsymbol{v} \cdot \boldsymbol{v} + 2\boldsymbol{v} \cdot \boldsymbol{w} + \boldsymbol{w} \cdot \boldsymbol{w}$.

19 The "Law of Cosines" comes from $(\boldsymbol{v} - \boldsymbol{w}) \cdot (\boldsymbol{v} - \boldsymbol{w}) = \boldsymbol{v} \cdot \boldsymbol{v} - 2\boldsymbol{v} \cdot \boldsymbol{w} + \boldsymbol{w} \cdot \boldsymbol{w}$:

Cosine Law $$\|\boldsymbol{v} - \boldsymbol{w}\|^2 = \|\boldsymbol{v}\|^2 - 2\|\boldsymbol{v}\|\,\|\boldsymbol{w}\|\cos\theta + \|\boldsymbol{w}\|^2.$$

Draw a triangle with sides $\boldsymbol{v}$ and $\boldsymbol{w}$ and $\boldsymbol{v} - \boldsymbol{w}$. Which of the angles is θ?

20 The ***triangle inequality*** says: (length of $\boldsymbol{v} + \boldsymbol{w}$) $\leq$ (length of $\boldsymbol{v}$) + (length of $\boldsymbol{w}$). Problem 18 found $\|\boldsymbol{v} + \boldsymbol{w}\|^2 = \|\boldsymbol{v}\|^2 + 2\boldsymbol{v} \cdot \boldsymbol{w} + \|\boldsymbol{w}\|^2$. Increase that $\boldsymbol{v} \cdot \boldsymbol{w}$ to $\|\boldsymbol{v}\|\,\|\boldsymbol{w}\|$ to show that $\|$**side 3**$\|$ cannot exceed $\|$**side 1**$\| + \|$**side 2**$\|$:

Triangle inequality $$\|\boldsymbol{v} + \boldsymbol{w}\|^2 \leq (\|\boldsymbol{v}\| + \|\boldsymbol{w}\|)^2 \quad \textbf{or} \quad \boxed{\|\boldsymbol{v} + \boldsymbol{w}\| \leq \|\boldsymbol{v}\| + \|\boldsymbol{w}\|}$$

21 The Schwarz inequality $|\boldsymbol{v} \cdot \boldsymbol{w}| \leq \|\boldsymbol{v}\|\,\|\boldsymbol{w}\|$ by algebra instead of trigonometry:

(a) Multiply out both sides of $(v_1w_1 + v_2w_2)^2 \leq \left(v_1^2 + v_2^2\right)\left(w_1^2 + w_2^2\right)$.

(b) Show that the difference between those two sides equals $(v_1w_2 - v_2w_1)^2$. This cannot be negative since it is a square—so the inequality is true.

22 One-line proof of the inequality $|\boldsymbol{u} \cdot \boldsymbol{U}| \leq 1$ for unit vectors (u_1, u_2) and (U_1, U_2):

$$|\boldsymbol{u} \cdot \boldsymbol{U}| \leq |u_1|\,|U_1| + |u_2|\,|U_2| \leq \frac{u_1^2 + U_1^2}{2} + \frac{u_2^2 + U_2^2}{2} = 1.$$

Put $(u_1, u_2) = (.6, .8)$ and $(U_1, U_2) = (.8, .6)$ in that whole line and find $\boldsymbol{u} \cdot \boldsymbol{U}$.

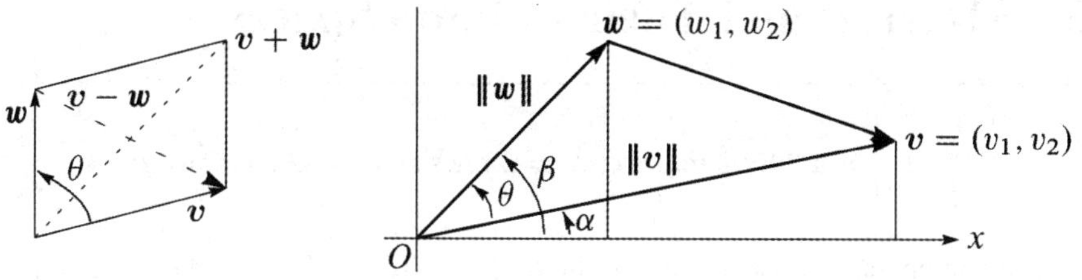

23 (a) Why is $|\cos\theta|$ never greater than 1 ? (b) Find $\cos\theta$ in an equilateral triangle.

24 Show that the squared diagonal lengths $||\boldsymbol{v}+\boldsymbol{w}||^2 + ||\boldsymbol{v}-\boldsymbol{w}||^2$ in a parallelogram add to the sum of the four squared side lengths $2||\boldsymbol{v}||^2 + 2||\boldsymbol{w}||^2$.

25 (*Recommended*) If $||\boldsymbol{v}|| = 5$ and $||\boldsymbol{w}|| = 3$, what are the smallest and largest possible values of $||\boldsymbol{v}-\boldsymbol{w}||$? What are the smallest and largest possible values of $\boldsymbol{v}\cdot\boldsymbol{w}$?

Challenge Problems

26 Can three vectors in the xy plane have $\boldsymbol{u}\cdot\boldsymbol{v} < 0$ and $\boldsymbol{v}\cdot\boldsymbol{w} < 0$ and $\boldsymbol{u}\cdot\boldsymbol{w} < 0$? I don't know how many vectors in xyz space can have all negative dot products. (Four of those vectors in the plane would certainly be impossible . . .).

27 Find 4 perpendicular unit vectors of the form $(\pm\frac{1}{2}, \pm\frac{1}{2}, \pm\frac{1}{2}, \pm\frac{1}{2})$: Choose + or −.

28 Using $\boldsymbol{v} = \mathsf{randn}(3,1)$ in MATLAB, create a random unit vector $\boldsymbol{u} = \boldsymbol{v}/\|\boldsymbol{v}\|$. Using $V = \mathsf{randn}(3,30)$ create 30 more random unit vectors $\boldsymbol{U}_j$. What is the average size of the dot products $|\boldsymbol{u}\cdot\boldsymbol{U}_j|$? In calculus, the average is $\int_0^\pi |\cos\theta| d\theta/\pi = 2/\pi$.

29 In the xy plane, when could four vectors $\boldsymbol{v}_1, \boldsymbol{v}_2, \boldsymbol{v}_3, \boldsymbol{v}_4$ not be the four sides of a quadrilateral ? We could not allow them to have any 4 lengths.

1.3 Matrices and Their Column Spaces

1 $A = \begin{bmatrix} 1 & 2 \\ 3 & 4 \\ 5 & 6 \end{bmatrix}$ is a **3 by 2 matrix**: $m = 3$ rows and $n = 2$ columns. Rank 2.

2 The 3 components of $A\boldsymbol{x}$ are dot products of the 3 rows of A with the vector $\boldsymbol{x}$:

Row at a time
A times $\boldsymbol{x}$

$$\begin{bmatrix} 1 & 2 \\ 3 & 4 \\ 5 & 6 \end{bmatrix} \begin{bmatrix} 7 \\ 8 \end{bmatrix} = \begin{bmatrix} 1\cdot 7 + 2\cdot 8 \\ 3\cdot 7 + 4\cdot 8 \\ 5\cdot 7 + 6\cdot 8 \end{bmatrix} = \begin{bmatrix} 23 \\ 53 \\ 83 \end{bmatrix}.$$

3 $A\boldsymbol{x}$ is also a **combination of the columns** of A $\quad \begin{bmatrix} 1 & 2 \\ 3 & 4 \\ 5 & 6 \end{bmatrix} \begin{bmatrix} \mathbf{7} \\ \mathbf{8} \end{bmatrix} = \mathbf{7} \begin{bmatrix} 1 \\ 3 \\ 5 \end{bmatrix} + \mathbf{8} \begin{bmatrix} 2 \\ 4 \\ 6 \end{bmatrix}.$

4 The **column space of A** contains **all combinations** $A\boldsymbol{x} = x_1\boldsymbol{a}_1 + x_2\boldsymbol{a}_2$ of the columns.

5 **Rank one matrices**: All columns of A (and all combinations $A\boldsymbol{x}$) are on **one line**.

Sections 1.1 and 1.2 explained the mechanics of vectors—linear combinations, dot products, lengths, and angles. We have vectors in $\mathbf{R}^2$ and $\mathbf{R}^3$ and every $\mathbf{R}^n$.

Section 1.3 begins the algebra of m by n matrices: our true goal. A typical matrix A is a rectangle of m times n numbers—**m rows and n columns**. If m equals n then A is a "square matrix". The examples below are 3 by 3 matrices.

$$\begin{bmatrix} \mathbf{1} & 0 & 0 \\ 0 & \mathbf{1} & 0 \\ 0 & 0 & \mathbf{1} \end{bmatrix} \qquad \begin{bmatrix} \mathbf{2} & 0 & 0 \\ 0 & \mathbf{4} & 0 \\ 0 & 0 & \mathbf{5} \end{bmatrix} \qquad \begin{bmatrix} 2 & 1 & -3 \\ \mathbf{0} & 4 & 7 \\ \mathbf{0} & \mathbf{0} & 5 \end{bmatrix} \qquad \begin{bmatrix} 2 & \mathbf{1} & \mathbf{-3} \\ \mathbf{1} & 4 & \mathbf{7} \\ \mathbf{-3} & \mathbf{7} & 5 \end{bmatrix}$$

Identity matrix	**Diagonal** matrix	**Triangular** matrix	**Symmetric** matrix

We often think of the columns of A as vectors $\boldsymbol{a}_1, \boldsymbol{a}_2, \ldots, \boldsymbol{a}_n$. Each of those n vectors is in m-dimensional space. In this example the $\boldsymbol{a}$'s have $m = 3$ components each:

$m = 3$ rows
$n = 4$ columns
3 by 4 matrix A

$$\boldsymbol{A} = \begin{bmatrix} \boldsymbol{a}_1 & \boldsymbol{a}_2 & \boldsymbol{a}_3 & \boldsymbol{a}_4 \end{bmatrix} = \begin{bmatrix} \mathbf{-1} & \mathbf{1} & 0 & 0 \\ 0 & \mathbf{-1} & \mathbf{1} & 0 \\ 0 & 0 & \mathbf{-1} & \mathbf{1} \end{bmatrix}$$

This is a "difference matrix" because A times $\boldsymbol{x}$ produces a vector $A\boldsymbol{x}$ of differences like $x_2 - x_1$. *How does an m by n matrix A multiply an n by 1 vector $\boldsymbol{x}$?* There are two ways to the same answer—we work with the rows of A or we work with the columns.

> The **row picture** of $A\boldsymbol{x}$ will come from **dot products** of $\boldsymbol{x}$ with the **rows of A**.
>
> The **column picture** will come from **linear combinations** of the **columns of A**.

Row picture of Ax Each row of A multiplies the column vector $\boldsymbol{x}$. Those multiplications *row times column* are dot products ! The first dot product comes from row 1 of A:

$$(\textbf{row 1}) \cdot \boldsymbol{x} = (-1,1,0,0) \cdot (x_1,x_2,x_3,x_4) = \boldsymbol{x_2} - \boldsymbol{x_1}.$$

It takes m times n small multiplications to find the $m = 3$ dot products that go into $A\boldsymbol{x}$.

Three rows Three dot products

$$A\boldsymbol{x} = \begin{bmatrix} -1 & 1 & 0 & 0 \\ 0 & -1 & 1 & 0 \\ 0 & 0 & -1 & 1 \end{bmatrix} \begin{bmatrix} x_1 \\ x_2 \\ x_3 \\ x_4 \end{bmatrix} = \begin{bmatrix} \text{row } 1 \cdot \boldsymbol{x} \\ \text{row } 2 \cdot \boldsymbol{x} \\ \text{row } 3 \cdot \boldsymbol{x} \end{bmatrix} = \begin{bmatrix} \boldsymbol{x_2 - x_1} \\ \boldsymbol{x_3 - x_2} \\ \boldsymbol{x_4 - x_3} \end{bmatrix} \quad (1)$$

Notice well that each row of A has the same number of components as the vector $\boldsymbol{x}$. Four columns of A multiply x_1 to x_4. Otherwise multiplying $A\boldsymbol{x}$ would be impossible.

Column picture of $A\boldsymbol{x}$ The matrix A times the vector $\boldsymbol{x}$ is a **combination of the columns of $\boldsymbol{A}$**. The n columns are multiplied by the n numbers in $\boldsymbol{x}$. Then add those column vectors $x_1\boldsymbol{a}_1, \ldots, x_n\boldsymbol{a}_n$ to find the same vector $A\boldsymbol{x}$ as in equation (1):

$$\boxed{A\boldsymbol{x} = x_1(\text{column } \boldsymbol{a}_1) + x_2(\text{column } \boldsymbol{a}_2) + x_3(\text{column } \boldsymbol{a}_3) + x_4(\text{column } \boldsymbol{a}_4)} \quad (2)$$

This combination of n columns involves exactly the same multiplications as dot products of $\boldsymbol{x}$ with the m rows. But it is higher level ! We have a vector equation instead of three dot products. You see the same result $A\boldsymbol{x}$ in equation (1) above and equation (3) below.

Combination of columns

$$A\boldsymbol{x} = x_1 \begin{bmatrix} -1 \\ 0 \\ 0 \end{bmatrix} + x_2 \begin{bmatrix} 1 \\ -1 \\ 0 \end{bmatrix} + x_3 \begin{bmatrix} 0 \\ 1 \\ -1 \end{bmatrix} + x_4 \begin{bmatrix} 0 \\ 0 \\ 1 \end{bmatrix} = \begin{bmatrix} \boldsymbol{x_2 - x_1} \\ \boldsymbol{x_3 - x_2} \\ \boldsymbol{x_4 - x_3} \end{bmatrix} \quad (3)$$

Let me admit something right away. If I have numbers in A and $\boldsymbol{x}$, and I want to compute $A\boldsymbol{x}$, then I tend to *use dot products*: the row picture. But if I want to *understand* $A\boldsymbol{x}$, the column picture is better. "The column vector $A\boldsymbol{x}$ is a combination of the columns of A."

We are aiming for a picture of not just one combination $A\boldsymbol{x}$ of the columns (from a particular $\boldsymbol{x}$). What we really want is a picture of **all combinations of the columns** (from multiplying A by all vectors $\boldsymbol{x}$). This figure shows one combination $2\boldsymbol{a}_1 + \boldsymbol{a}_2$ and then it tries to show the plane of all combinations $x_1\boldsymbol{a}_1 + x_2\boldsymbol{a}_2$ (for every x_1 and x_2).

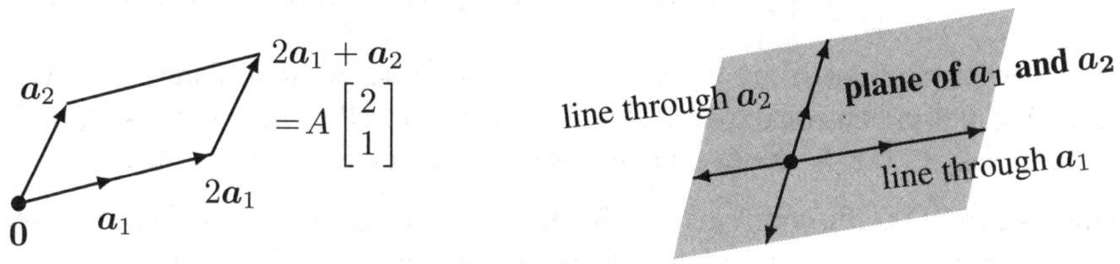

Figure 1.7: A linear combination of $\boldsymbol{a}_1$ and $\boldsymbol{a}_2$. All linear combinations fill a plane.

The next important words are **independence**, **dependence**, and **column space**.

Here is a key point ! Columns of A might or might not contribute anything new. They might be combinations of earlier columns (which we already included). Example 1 shows 3 columns that give 3 new directions. Those columns are "independent":

Example 1 Independent columns $A_1 = \begin{bmatrix} 1 & 0 & 0 \\ 2 & 4 & 0 \\ 3 & 5 & 6 \end{bmatrix}$ *Each column gives a new direction.* Their combinations fill 3D space $\mathbf{R}^3$.

If we look at all combinations of the columns, we see all vectors (b_1, b_2, b_3) in 3D space. The first column $x_1(1,2,3)$ allows us to match any number b_1. Then $x_2(0,4,5)$ leaves b_1 alone and we can match any number b_2. Finally $x_3(0,0,6)$ doesn't touch b_1 and b_2 and allows us to match any b_3. We have found x_1, x_2, x_3 so that $A_1\boldsymbol{x} = \boldsymbol{b}$.

Independence means: The only combination of columns that produces $A\boldsymbol{x} = (0,0,0)$ is $\boldsymbol{x} = (0,0,0)$. The columns are **independent** when each new column is a vector that we don't already have as a combination of previous columns. Independence will be important.

Example 2 Dependent columns $A_2 = \begin{bmatrix} 1 & 2 & 3 \\ 1 & 4 & 5 \\ 6 & 0 & 6 \end{bmatrix}$ **Column 1 + column 2 = column 3** **Their combinations don't fill 3D space** $1+2=3$ $1+4=5$ $6+0=6$

The opposite of independent is "*dependent*". These three columns of A_2 are **dependent**. Column 3 is in the plane of columns 1 and 2. Nothing new from column 3.

I usually test independence going from left to right. The column $(1,1,6)$ is no problem. Column 2 is *not* a multiple of column 1 and $(2,4,0)$ gives a new direction. But column 3 is the sum of columns 1 and 2. The third column vector $(3,5,6)$ is not independent of $(1,1,6)$ and $(2,4,0)$. That matrix A_2 only has two independent columns.

If I went from right to left, I would start with independent columns 3 and 2. Then column 1 is a combination (column 3 minus column 2). Either way we find that the three columns are in the **same plane**: two independent columns produce a plane in 3D.

That plane is the column space of this matrix: Plane = all combinations of the columns.

Dependent columns in Example 2 column 1 + column 2 − column 3 is $(\mathbf{0}, \mathbf{0}, \mathbf{0})$.

Example 3 $A_3 = \begin{bmatrix} 1 & 3 & 4 \\ 2 & 6 & 8 \\ 5 & 15 & 20 \end{bmatrix}$ Now $\boldsymbol{a}_2$ is 3 times $\boldsymbol{a}_1$. And $\boldsymbol{a}_3$ is 4 times $\boldsymbol{a}_1$. **Every pair of columns is dependent.**

This example is important. You could call it an extreme case. All three columns of A_3 lie on the **same line** in 3-dimensional space. That line consists of all column vectors $(c, 2c, 5c)$—all the multiples of $(1,2,5)$. Notice that $c = 0$ gives the center point $(0,0,0)$.

That line in 3D is the column space for this matrix A_3. The line contains all vectors $A_3\boldsymbol{x}$. By allowing every vector $\boldsymbol{x}$, we fill in the column space of A_3—and here we only filled one line. That is almost the smallest possible column space.

The **column space** $\mathbf{C}(A)$ contains all vectors $A\boldsymbol{x}$: **All combinations of the columns.**

Thinking About the Column Space of A

"Vector spaces" are a central topic. Examples are coming unusually early. They give you a chance to see what linear algebra is about. The combinations of all columns produce the column space, but you only need independent columns. So we start with column 1, and go from left to right in identifying independent columns. Here are two examples $\boldsymbol{A_4}$ and $\boldsymbol{A_5}$.

$$\boldsymbol{A_4} = \begin{bmatrix} \mathbf{1} & \mathbf{1} & \mathbf{1} & \mathbf{1} \\ 0 & \mathbf{1} & \mathbf{1} & \mathbf{1} \\ 0 & 0 & \mathbf{1} & \mathbf{1} \\ 0 & 0 & 0 & \mathbf{1} \end{bmatrix} \qquad \boldsymbol{A_5} = \begin{bmatrix} \mathbf{1} & \mathbf{1} & 0 & 0 \\ 0 & \mathbf{1} & \mathbf{1} & 0 \\ 0 & 0 & \mathbf{1} & \mathbf{1} \\ \mathbf{1} & 0 & 0 & \mathbf{1} \end{bmatrix}$$

$\boldsymbol{A_4}$ has four independent columns. For example, column 4 is not a combination of columns $1, 2, 3$. There are no dependent columns in A_4. Triangular matrices like A_4 are easy *provided the main diagonal has no zeros*. Here the diagonal is $1, 1, 1, 1$.

$\boldsymbol{A_5}$ is not so easy. Columns 1 and 2 and 3 are independent. The big question is whether column 4 is a combination of columns $1, 2, 3$. To match the final 1 in column 4, that combination will have to start with column 1.

To cancel the 1 in the top left corner of A_5, we need *minus* the second column. Then we need *plus* column 3 so that -1 and $+1$ in row 2 will also cancel. Now we see what is true about this matrix A_5 : Only 3 independent columns because

$$\textbf{Column 4 of } \boldsymbol{A_5} = \textbf{Column 1} - \textbf{Column 2} + \textbf{Column 3.} \tag{4}$$

The next step is to "visualize" the column space—all combinations of the four columns. That word is in quotes because the task may be impossible. I don't think that drawing a 4-dimensional figure would help (possibly this is wrong). A_4 is a good place to start.

Do you see why $\mathbf{C}(A_4)$ *is the whole space* $\mathbf{R}^4$? If we look to algebra, we see that every vector $\boldsymbol{v}$ in $\mathbf{R}^4$ is a combination of the columns. Here is the combination :

$$\boldsymbol{v} = \begin{bmatrix} v_1 \\ v_2 \\ v_3 \\ v_4 \end{bmatrix} = (v_1 - v_2)\begin{bmatrix} 1 \\ 0 \\ 0 \\ 0 \end{bmatrix} + (v_2 - v_3)\begin{bmatrix} 1 \\ 1 \\ 0 \\ 0 \end{bmatrix} + (v_3 - v_4)\begin{bmatrix} 1 \\ 1 \\ 1 \\ 0 \end{bmatrix} + (v_4)\begin{bmatrix} 1 \\ 1 \\ 1 \\ 1 \end{bmatrix} \tag{5}$$

This says : Every $\boldsymbol{v}$ is in the column space. **We solved the four equations $\boldsymbol{A_4 x} = \boldsymbol{v}$!**

> The short useful word **"span"** describes all the linear combinations of a set of vectors.
> So **the span of the columns of A** (independent or not) **is the column space**.

Geometrically, here is one way to look at A_4. The first column $(1, 0, 0, 0)$ is responsible for a line in 4-dimensional space. That line contains every vector $(c_1, 0, 0, 0)$. The second column is responsible for another line, containing every vector $(c_2, c_2, 0, 0)$. *If you add every vector* $(c_1, 0, 0, 0)$ *to every vector* $(c_2, c_2, 0, 0)$, you get a 2-dimensional plane inside 4-dimensional space. That plane is the *span* of columns 1 and 2.

That was the first two columns. The main rule of linear algebra is *keep going*. The last two columns give two more directions in $\mathbf{R}^4$. *The four columns are independent.* At the end, equation (5) shows how **every 4-dimensional vector is a combination of the four columns of A_4**. The column space of A_4 is all of $\mathbf{R}^4$.

If we attempt the same plan for the matrix A_5, the first 3 columns cooperate. But column 4 of A_5 is a combination of columns $1, 2, 3$. Those three columns combine to give a *three-dimensional subspace* inside $\mathbf{R}^4$. Column 4 happens to be *in that subspace.*

That three-dimensional subspace is the whole column space $\mathbf{C}(A_5)$. We can only solve $A_5\boldsymbol{x} = \boldsymbol{v}$ when $\boldsymbol{v}$ is in $\mathbf{C}(A_5)$. The matrix A_5 only has three independent columns.

I always write $\mathbf{C}(A)$ for the column space of A. When A has m rows, the columns are vectors in m-dimensional space $\mathbf{R}^m$. The column space might fill all of $\mathbf{R}^m$ or it might not. For $m = 3$, here are all four possibilities for column spaces in 3-dimensional space :

1.	The whole space $\mathbf{R}^3$	A has 3 independent columns
2.	A plane in $\mathbf{R}^3$ going through $(0, 0, 0)$	2 independent columns
3.	A line in $\mathbf{R}^3$ going through $(0, 0, 0)$	1 independent column
4.	The single point $(0, 0, 0)$ in $\mathbf{R}^3$	A is a matrix of zeros !

Here are simple matrices to show those four possibilities for the column space $\mathbf{C}(A)$:

$$\begin{bmatrix} \mathbf{1} & 0 & 0 \\ 0 & \mathbf{1} & 0 \\ 0 & 0 & \mathbf{1} \end{bmatrix} \qquad \begin{bmatrix} \mathbf{1} & 0 & 0 \\ 0 & \mathbf{1} & 0 \\ 0 & 0 & 0 \end{bmatrix} \qquad \begin{bmatrix} \mathbf{1} & 0 & 0 \\ 0 & 0 & 0 \\ 0 & 0 & 0 \end{bmatrix} \qquad \begin{bmatrix} 0 & 0 & 0 \\ 0 & 0 & 0 \\ 0 & 0 & 0 \end{bmatrix}$$

$\mathbf{C}(A) = \mathbf{R}^3 = xyz$ space $\quad \mathbf{C}(A) = xy$ plane $\quad \mathbf{C}(A) = x$ axis $\quad \mathbf{C}(A) =$ one point $(0, 0, 0)$

Author's note The words "column space" did not appear in Chapter 1 of the 5th edition. I thought the idea of a *space* was too important to come so soon. Now I think that the best way to understand such an important idea is to see it early and often. It is examples more than definitions that make ideas clear—in mathematics as in life.

Here is a succession of questions. With practice in the next section 1.4, you will find the keys to the answers. They give a real understanding of any matrix A.

1. How many columns of A are independent ? That number r is the **"rank"** of A.

2. Which are the first r independent columns ? They are a **"basis"** for the column space.

3. What combinations of those r basic columns produce the remaining $n - r$ columns ?

4. Write any A as an m by r column matrix C times an r by n matrix R : $\boldsymbol{A = CR}$.

5. (Amazing) The r rows of R are a basis for the **row space** of A : combinations of rows.

Section 1.4 will explain how to multiply those matrices C and R. The result is $A = CR$. C contains columns from A. Please notice that the rows of R *do not* come directly from A.

Matrices of Rank One

Now we come to the building blocks for all matrices. *Every matrix of rank r is the sum of r matrices of rank one.* For a rank one matrix, all column vectors lie along the same line. That one line through $(0,0,0)$ is the whole column space of the rank one matrix.

Example $A_6 = \begin{bmatrix} 1 & 3 & -2 \\ 4 & 12 & -8 \\ 2 & 6 & -4 \end{bmatrix}$ has rank $\boldsymbol{r}$ **= 1**. All columns: same direction!

Columns 2 and 3 are multiples $3\boldsymbol{a}_1$ and $-2\boldsymbol{a}_1$ of the first column $\boldsymbol{a}_1 = (1,4,2)$. The column space $\mathbf{C}(A_6)$ is only a line containing all vectors $c\boldsymbol{a}_1 = (c, 4c, 2c)$.

Here is a wonderful fact about any rank one matrix. You may have noticed the rows of A_6. **All the rows are multiples of one row. When the column space is a single line in m-dimensional space, the row space is a single line in n-dimensional space.** All rows of this matrix A_6 are multiples of ____.

An example like A_6 raises a basic question. **If all columns are in the same direction, why does it happen that all rows are in the same direction?** To find an answer, look first at this 2 by 2 matrix. Column 2 is m times column 1:

$A = \begin{bmatrix} a & ma \\ b & mb \end{bmatrix}$	Is row 2 a multiple of row 1?

Yes! The second row (b, mb) is $\frac{b}{a}$ times the first row (a, ma). If the column rank is 1, then the row rank is 1. To cover every possibility we have to check the case when $a = 0$. Then row 1 $= \begin{bmatrix} 0 & 0 \end{bmatrix} = 0$ times row 2.

Our 2 by 2 proof is complete. Let me look at this 3 by 3 matrix. Rank $= 1$ if $a \neq 0$.

$A = \begin{bmatrix} a & ma & pa \\ b & mb & pb \\ c & mc & pc \end{bmatrix}$	Column 2 is m times column 1 Column 3 is p times column 1 Rows 2 and 3 are b/a and c/a times row 1

This matrix does not have two independent columns. Is it the same for the rows of A? Is row 2 in the same direction as row 1? *Yes.* Is row 3 in the same direction as row 1? *Yes. The rule still holds.* The row rank of this A is also 1 (equal to the column rank).

Now jump from rank one matrices to all matrices. At this point we could make a guess: It looks possible that **row rank equals column rank for every matrix**. If A has r independent columns, then A has r independent rows. A wonderful fact!

I believe that this is the first great theorem in linear algebra. So far we have only seen the case of rank one matrices. The next section 1.4 will explain matrix multiplication AB and lead us toward an understanding of "row rank = column rank" for all matrices.

Problem Set 1.3

This section introduced column spaces. But we don't yet have a computational system to decide independence or dependence of column vectors. So these problems stay with whole numbers and small matrices.

1 Describe the column space of these matrices : a point, a line, a plane, all of 3D.

$$A_1 = \begin{bmatrix} 2 & 2 \\ 1 & 1 \\ 5 & 6 \end{bmatrix} \quad A_2 = \begin{bmatrix} 1 & 0 & 0 \\ 1 & 1 & 0 \\ 1 & 1 & 1 \end{bmatrix} \quad A_3 = \begin{bmatrix} 1 & 5 \\ 2 & 10 \\ 1 & 5 \end{bmatrix} \quad A_4 = \begin{bmatrix} 0 & 0 \\ 0 & 0 \\ 0 & 0 \end{bmatrix}$$

2 Find a combination of the columns that produces $(0, 0, 0)$: column space $=$ *plane*. The trivial combination (zero times every column) is not allowed.

Which columns are dependent on earlier columns ?

$$A_1 = \begin{bmatrix} 1 & 2 & 3 \\ 4 & 5 & 6 \\ 7 & 8 & 9 \end{bmatrix} \quad A_2 = \begin{bmatrix} 1 & 4 & 7 \\ 2 & 5 & 8 \\ 3 & 6 & 9 \end{bmatrix}$$

3 Describe the column spaces in $\mathbf{R}^3$ of B and C :

$$B = \begin{bmatrix} 1 & 2 \\ 2 & 1 \\ 3 & 3 \end{bmatrix} \quad C = \begin{bmatrix} B & -B \end{bmatrix} \quad \text{(3 rows, 4 columns)}$$

4 Multiply $A\boldsymbol{x}$ and $B\boldsymbol{y}$ and $I\boldsymbol{z}$ using dot products as in $(\textbf{rows of } \boldsymbol{A}) \cdot \boldsymbol{x}$:

$$A\boldsymbol{x} = \begin{bmatrix} 2 & 1 & 2 \\ 4 & 2 & 4 \\ 0 & 1 & 0 \end{bmatrix} \begin{bmatrix} 1 \\ 2 \\ 5 \end{bmatrix} \quad B\boldsymbol{y} = \begin{bmatrix} 1 & 0 & 0 \\ 1 & 1 & 0 \\ 1 & 1 & 1 \end{bmatrix} \begin{bmatrix} 4 \\ 4 \\ 10 \end{bmatrix} \quad I\boldsymbol{z} = \begin{bmatrix} 1 & 0 & 0 \\ 0 & 1 & 0 \\ 0 & 0 & 1 \end{bmatrix} \begin{bmatrix} z_1 \\ z_2 \\ z_3 \end{bmatrix}$$

5 Multiply the same A times $\boldsymbol{x}$ and B times $\boldsymbol{y}$ and I times $\boldsymbol{z}$ using combinations of the **columns** of A and B and I, as in $A\boldsymbol{x} = 1(\text{column } 1) + 2(\text{column } 2) + 5(\text{column } 3)$.

6 In Problem 5, how many independent columns does A have ? How many independent columns in B ? How many independent columns in $A + B$?

7 Can you find A and B (both with two independent columns) so that $A + B$ has
(a) 1 independent column (b) No independent columns (c) 4 independent columns

8 The "column space" of a matrix contains all combinations of the columns. Describe the column spaces in $\mathbf{R}^3$ of A and B and C :

$$A = \begin{bmatrix} 1 & 0 & 0 \\ 0 & 1 & 0 \\ 0 & 0 & 1 \end{bmatrix} \quad B = \begin{bmatrix} 2 & 4 \\ 1 & 2 \\ 2 & 4 \end{bmatrix} \quad C = \begin{bmatrix} 1 & 0 & 1 & 2 \\ 0 & 2 & 2 & 4 \\ 0 & 2 & 2 & 4 \end{bmatrix}$$

9 Find a 3 by 3 matrix A with 3 independent columns and all nine entries $= 1$ or 2. What is the maximum possible number of 1's with independent columns ?

10 Complete A and B so that they are rank one matrices. What are the column spaces of A and B? What are the row spaces of A and B?

$$A = \begin{bmatrix} 3 & \\ 5 & 15 \end{bmatrix} \qquad B = \begin{bmatrix} 1 & 2 & -5 \\ 4 & & \end{bmatrix}$$

11 Suppose A is a 5 by 2 matrix with columns $\boldsymbol{a}_1$ and $\boldsymbol{a}_2$. We include one more column to produce B (5 by 3). Do A and B have the same column space if

(a) the new column is the zero vector? (b) the new column is $(1,1,1,1,1)$?

(c) the new column is the difference $\boldsymbol{a}_2 - \boldsymbol{a}_1$?

12 Explain this important sentence. It connects column spaces to linear equations.

> $A\boldsymbol{x} = \boldsymbol{b}$ has a solution vector $\boldsymbol{x}$ if the vector $\boldsymbol{b}$ is in the column space of A.

The equation $A\boldsymbol{x} = \boldsymbol{b}$ looks for a combination of columns of A that produces $\boldsymbol{b}$. What vector will solve $A\boldsymbol{x} = \boldsymbol{b}$ for these right hand sides $\boldsymbol{b}$?

$$\begin{bmatrix} 1 & 3 \\ 2 & 4 \end{bmatrix} \begin{bmatrix} x_1 \\ x_2 \end{bmatrix} = \begin{bmatrix} 4 \\ 6 \end{bmatrix} \quad \text{or} \quad \begin{bmatrix} -2 \\ -2 \end{bmatrix} \quad \text{or} \quad \begin{bmatrix} 1 \\ 1 \end{bmatrix}$$

13 Find two 3 by 3 matrices A and B with the same column space = the plane of all vectors perpendicular to $(1,1,1)$. What is *usually* the column space of $A + B$?

14 Which numbers q would leave A with two independent columns?

$$A = \begin{bmatrix} 1 & 0 & 2 \\ 3 & 1 & 9 \\ 5 & 0 & q \end{bmatrix} \qquad A = \begin{bmatrix} 1 & 4 & 7 \\ 2 & 5 & 8 \\ 3 & 6 & q \end{bmatrix} \qquad A = \begin{bmatrix} 1 & 1 & 2 \\ 2 & 2 & 4 \\ 0 & 0 & q \end{bmatrix}$$

15 Suppose A times $\boldsymbol{x}$ equals $\boldsymbol{b}$. If you add $\boldsymbol{b}$ as an extra column of A, explain why the rank r (number of independent columns) stays the same.

16 *True or false*

(a) If the 5 by 2 matrices A and B have independent columns, so does $A + B$.

(b) If the m by n matrix A has n independent columns, then $m \geq n$.

(c) A random 3 by 3 matrix almost surely has 3 independent columns.

17 If A and B have rank 1, what are the possible ranks of $A + B$? Give an example of each possibility. Why is rank 4 impossible?

18 Find the linear combination $3\boldsymbol{s}_1 + 4\boldsymbol{s}_2 + 5\boldsymbol{s}_3 = \boldsymbol{b}$. Then write $\boldsymbol{b}$ as a matrix-vector multiplication $S\boldsymbol{x}$, with 3, 4, 5 in $\boldsymbol{x}$. Compute the three dot products (row of S) $\cdot\, \boldsymbol{x}$:

$$\boldsymbol{s}_1 = \begin{bmatrix} 1 \\ 1 \\ 1 \end{bmatrix} \quad \boldsymbol{s}_2 = \begin{bmatrix} 0 \\ 1 \\ 1 \end{bmatrix} \quad \boldsymbol{s}_3 = \begin{bmatrix} 0 \\ 0 \\ 1 \end{bmatrix} \quad \text{go into the columns of } S.$$

19 If (a, b) is a multiple of (c, d) with $abcd \neq 0$, *show that* (a, c) *is a multiple of* (b, d). This is surprisingly important; two columns are falling on one line. You could use numbers first to see how a, b, c, d are related. The question will lead to:

$$\text{If } \begin{bmatrix} a & b \\ c & d \end{bmatrix} \text{ has dependent rows, then it also has dependent columns.}$$

20 Solve these equations $S\boldsymbol{y} = \boldsymbol{b}$ with $\boldsymbol{s}_1, \boldsymbol{s}_2, \boldsymbol{s}_3$ in the columns of the sum matrix S:

$$\begin{bmatrix} 1 & 0 & 0 \\ 1 & 1 & 0 \\ 1 & 1 & 1 \end{bmatrix} \begin{bmatrix} y_1 \\ y_2 \\ y_3 \end{bmatrix} = \begin{bmatrix} 1 \\ 1 \\ 1 \end{bmatrix} \quad \text{and} \quad \begin{bmatrix} 1 & 0 & 0 \\ 1 & 1 & 0 \\ 1 & 1 & 1 \end{bmatrix} \begin{bmatrix} y_1 \\ y_2 \\ y_3 \end{bmatrix} = \begin{bmatrix} 1 \\ 4 \\ 9 \end{bmatrix}.$$

The sum of the first 3 odd numbers is _____. The sum of the first 10 is _____.

21 Solve these three equations for y_1, y_2, y_3 in terms of c_1, c_2, c_3:

$$S\boldsymbol{y} = \mathbf{c} \qquad \begin{bmatrix} 1 & 0 & 0 \\ 1 & 1 & 0 \\ 1 & 1 & 1 \end{bmatrix} \begin{bmatrix} y_1 \\ y_2 \\ y_3 \end{bmatrix} = \begin{bmatrix} c_1 \\ c_2 \\ c_3 \end{bmatrix}.$$

Write the solution $\boldsymbol{y}$ as a matrix A times the vector $\mathbf{c}$. A is the "*inverse matrix*" S^{-1}. Are the columns of S independent or dependent?

22 The three rows of this square matrix A are dependent. Then linear algebra says that the three columns must also be dependent. Find $\boldsymbol{x} \neq \mathbf{0}$ that solves $A\boldsymbol{x} = \mathbf{0}$:

$$A = \begin{bmatrix} 1 & 2 & 3 \\ 3 & 5 & 6 \\ 4 & 7 & 9 \end{bmatrix} \qquad \begin{array}{l} \text{Row } 1 + \text{ row } 2 = \text{ row } 3 \\ A \text{ has only two independent rows} \\ \text{Then only two independent columns} \end{array}$$

23 Which numbers c give dependent columns? Then a combination of columns is zero.

$$\begin{bmatrix} 1 & 1 & 0 \\ 3 & 2 & 1 \\ 7 & 4 & c \end{bmatrix} \qquad \begin{bmatrix} 1 & 0 & c \\ 1 & 1 & 0 \\ 0 & 1 & 1 \end{bmatrix} \qquad \begin{bmatrix} c & c & c \\ 2 & 1 & 5 \\ 3 & 3 & 6 \end{bmatrix} \qquad \begin{bmatrix} c & 1 \\ 4 & c \end{bmatrix}$$

24 If the columns combine into $A\boldsymbol{x} = \mathbf{0}$ then each row of A has $\mathbf{row} \cdot \boldsymbol{x} = 0$:

$$\text{If} \quad \begin{bmatrix} \\ \boldsymbol{a}_1 & \boldsymbol{a}_2 & \boldsymbol{a}_3 \\ \\ \end{bmatrix} \begin{bmatrix} x_1 \\ x_2 \\ x_3 \end{bmatrix} = \begin{bmatrix} 0 \\ 0 \\ 0 \end{bmatrix} \qquad \textbf{then by rows} \begin{bmatrix} \boldsymbol{r}_1 \cdot \boldsymbol{x} \\ \boldsymbol{r}_2 \cdot \boldsymbol{x} \\ \boldsymbol{r}_3 \cdot \boldsymbol{x} \end{bmatrix} = \begin{bmatrix} 0 \\ 0 \\ 0 \end{bmatrix}.$$

The three rows also lie in a plane. Why is that plane perpendicular to $\boldsymbol{x}$?

1.4 Matrix Multiplication AB and CR

1 To multiply AB we need *row length for* A = *column length for* B.

2 The number in row i, column j of AB is **(row i of A)** $\cdot$ **(column j of B)**.

3 By columns: **A times column j of B produces column j of AB.**

4 Usually AB is different from BA. But always $(\boldsymbol{AB})\,\boldsymbol{C} = \boldsymbol{A}\,(\boldsymbol{BC})$.

5 If A has r independent columns in C, then $\boldsymbol{A} = \boldsymbol{CR} = (\boldsymbol{m} \times \boldsymbol{r})\,(\boldsymbol{r} \times \boldsymbol{n})$.

We know how to multiply a matrix A times a column vector $\boldsymbol{x}$ or $\boldsymbol{b}$. This section moves to matrix-matrix multiplication: **a matrix A times a matrix B**. The new rule builds on the old one, when the matrix B has columns $\boldsymbol{b}_1, \boldsymbol{b}_2, \ldots, \boldsymbol{b}_p$. We just multiply A times each of those p columns of B to find the p columns of AB.

Column j of AB equals A times column j of B

$$\textbf{If } B = \begin{bmatrix} \boldsymbol{b}_1 & \cdots & \boldsymbol{b}_p \end{bmatrix} \textbf{ then } AB = \begin{bmatrix} A\boldsymbol{b}_1 & \cdots & A\boldsymbol{b}_p \end{bmatrix} \tag{1}$$

To see that clearly, start with a 2 by 2 "exchange matrix" for B. So B has two columns $\boldsymbol{b}_1$ and $\boldsymbol{b}_2$. We multiply A times each column to produce a column of AB:

$$A\boldsymbol{b}_1 = \begin{bmatrix} 1 & 2 \\ 3 & 4 \end{bmatrix}\begin{bmatrix} \mathbf{0} \\ \mathbf{1} \end{bmatrix} = \begin{bmatrix} \mathbf{2} \\ \mathbf{4} \end{bmatrix} \quad A\boldsymbol{b}_2 = \begin{bmatrix} 1 & 2 \\ 3 & 4 \end{bmatrix}\begin{bmatrix} \mathbf{1} \\ \mathbf{0} \end{bmatrix} = \begin{bmatrix} \mathbf{1} \\ \mathbf{3} \end{bmatrix} \quad \boldsymbol{AB} = \begin{bmatrix} \mathbf{1} & \mathbf{2} \\ \mathbf{3} & \mathbf{4} \end{bmatrix}\begin{bmatrix} \mathbf{0} & \mathbf{1} \\ \mathbf{1} & \mathbf{0} \end{bmatrix} = \begin{bmatrix} \mathbf{2} & \mathbf{1} \\ \mathbf{4} & \mathbf{3} \end{bmatrix}$$

For this matrix B, the result of multiplying AB is to *exchange the columns of* A.

There is more to see when we multiply the same A by a full 2 by 2 matrix B:

$$\begin{bmatrix} 1 & 2 \\ 3 & 4 \end{bmatrix}\begin{bmatrix} \mathbf{5} & \mathbf{6} \\ \mathbf{7} & \mathbf{8} \end{bmatrix} \text{ has } A\boldsymbol{b}_1 = \begin{bmatrix} 1 & 2 \\ 3 & 4 \end{bmatrix}\begin{bmatrix} \mathbf{5} \\ \mathbf{7} \end{bmatrix} \text{ and } A\boldsymbol{b}_2 = \begin{bmatrix} 1 & 2 \\ 3 & 4 \end{bmatrix}\begin{bmatrix} \mathbf{6} \\ \mathbf{8} \end{bmatrix}$$

Here is the point. We can multiply $A\boldsymbol{b}_1$ (matrix times vector) the *row way* or the *column way*. The row way uses dot products of $\boldsymbol{b}_1$ with *every row of* A:

$$\begin{matrix} \textbf{Row way} \\ \textbf{Dot products} \end{matrix} \qquad A\boldsymbol{b}_1 = \begin{bmatrix} 1 & 2 \\ 3 & 4 \end{bmatrix}\begin{bmatrix} 5 \\ 7 \end{bmatrix} = \begin{bmatrix} \text{row } 1 \cdot \boldsymbol{b}_1 \\ \text{row } 2 \cdot \boldsymbol{b}_1 \end{bmatrix} = \begin{bmatrix} 1 \cdot 5 + 2 \cdot 7 \\ 3 \cdot 5 + 4 \cdot 7 \end{bmatrix} = \begin{bmatrix} \mathbf{19} \\ \mathbf{43} \end{bmatrix} \tag{2}$$

The column way uses a combination of the *columns* of A to find $A\boldsymbol{b}_1$. Same result:

$$\begin{matrix} \textbf{Column way} \\ \textbf{Combine columns} \end{matrix} \quad A\boldsymbol{b}_1 = \begin{bmatrix} 1 & 2 \\ 3 & 4 \end{bmatrix}\begin{bmatrix} 5 \\ 7 \end{bmatrix} = \mathbf{5}\begin{bmatrix} 1 \\ 3 \end{bmatrix} + \mathbf{7}\begin{bmatrix} 2 \\ 4 \end{bmatrix} = \begin{bmatrix} 5 \\ 15 \end{bmatrix} + \begin{bmatrix} 14 \\ 28 \end{bmatrix} = \begin{bmatrix} \mathbf{19} \\ \mathbf{43} \end{bmatrix} \tag{3}$$

Both ways use the same 4 multiplications. With numbers like these, I think most people choose the row way. **To multiply AB, take the dot product of each row of A with each column of B.** When A has 2 rows and B has 2 columns, that means 4 dot products.

When A is m by n and B is n by p, then $\boldsymbol{AB}$ **is** $\boldsymbol{m}$ **by** $\boldsymbol{p}$. So we need mp dot products.

Row way
Rows of A

$$\boldsymbol{AB} = \begin{bmatrix} \text{row } 1 \cdot \text{col } 1 & \text{row } 1 \cdot \text{col } 2 \\ \text{row } 2 \cdot \text{col } 1 & \text{row } 2 \cdot \text{col } 2 \end{bmatrix} = \begin{bmatrix} \mathbf{19} & \mathbf{22} \\ \mathbf{43} & \mathbf{50} \end{bmatrix} \qquad (4)$$

Now compute AB the **column way**: combinations of columns of A. This is a vector operation and it produces whole columns of AB. Equation (3) found the first column. Now we find 22 and 50 in the second column of AB from A times $\boldsymbol{b}_2$:

Column way
for $\boldsymbol{Ab_2}$

$$A\boldsymbol{b}_2 = \begin{bmatrix} 1 & 2 \\ 3 & 4 \end{bmatrix} \begin{bmatrix} 6 \\ 8 \end{bmatrix} = \mathbf{6} \begin{bmatrix} 1 \\ 3 \end{bmatrix} + \mathbf{8} \begin{bmatrix} 2 \\ 4 \end{bmatrix} = \begin{bmatrix} 6 \\ 18 \end{bmatrix} + \begin{bmatrix} 16 \\ 32 \end{bmatrix} = \begin{bmatrix} \mathbf{22} \\ \mathbf{50} \end{bmatrix} \qquad (5)$$

Equations (3) and (5) gave the same two columns of AB as equation (4). Both ways use *the same* 8 *multiplications*; only the order is different. To multiply an $\boldsymbol{m}$ **by** $\boldsymbol{n}$ matrix A times an $\boldsymbol{n}$ **by** $\boldsymbol{p}$ matrix B, we can count the small multiplications: AB is $\boldsymbol{m}$ **by** $\boldsymbol{p}$.

Row way $\boldsymbol{mp}$ dot products in AB, $\boldsymbol{n}$ multiplications each: $\boldsymbol{mnp}$ small multiplications

Column way $\boldsymbol{p}$ columns in AB, $\boldsymbol{mn}$ multiplications each: $\boldsymbol{mnp}$ small multiplications

The actual speed will depend on how the matrices are stored. I think column storage is usual. Please note that it is faster to move large pieces of a matrix from storage rather than individual numbers. In a big multiplication, matrix-matrix operations using BLAS 3 (Level 3 Basic Linear Algebra Subprograms) are the best. The comparison with Level 1 (*vector-vector*) and Level 2 (*matrix-vector*) is online at **netlib.org/blas/**.

So far we have used (row) · (column) dot products and (matrix) (column) $A\boldsymbol{b}_j$ in multiplying AB. The other two ways are (row) (matrix) and (column) (row), coming soon. All four ways use the same mnp multiplications in varying orders to find AB.

If A and B are 2 by 2, that means $n^3 = 8$ small multiplications for AB.† See below.

$\boldsymbol{AB}$ is usually different from $\boldsymbol{BA}$

For $B = \begin{bmatrix} 0 & 1 \\ 1 & 0 \end{bmatrix}$, AB exchanged the columns of A. **But $\boldsymbol{BA}$ exchanges the rows of $\boldsymbol{A}$!**

$$\boldsymbol{AB} = \begin{bmatrix} \mathbf{2} & \mathbf{1} \\ \mathbf{4} & \mathbf{3} \end{bmatrix} \qquad \boldsymbol{BA} = \begin{bmatrix} 0 & 1 \\ 1 & 0 \end{bmatrix} \begin{bmatrix} 1 & 2 \\ 3 & 4 \end{bmatrix} = \begin{bmatrix} \mathbf{3} & \mathbf{4} \\ \mathbf{1} & \mathbf{2} \end{bmatrix} \qquad (6)$$

Matrix multiplication is not commutative. In general $\boldsymbol{BA} \neq \boldsymbol{AB}$. Multiply A on the left for row operations on A, and multiply on the right by B for column operations on A.

Question Why does squaring the exchange matrix give $B^2 = \begin{bmatrix} 0 & 1 \\ 1 & 0 \end{bmatrix} \begin{bmatrix} 0 & 1 \\ 1 & 0 \end{bmatrix} = I$?

† Strassen noticed that **7 multiplications** are enough for 2 by 2 matrices, at the cost of extra additions. For n by n matrices this reduces the multiplication count to $\boldsymbol{n^c}$, where $c = \log_2 7$ instead of the usual $c = \log_2 8 = 3$. Hard work has now reduced c even more. Certainly c cannot go below 2, because all of the n^2 entries in A and B must be used. Finding the smallest exponent c is an extremely tough unsolved problem.

AB times $C = A$ times BC

For matrix multiplication, **this associative law is true**. We are not willing to give up this extremely useful law. We can multiply AB first or we can multiply BC first.
The matrices stay in the order A, B, C and their sizes must be right for multiplication:

$\boldsymbol{A}$ **is** $\boldsymbol{m \times n}$ $\quad \boldsymbol{B}$ **is** $\boldsymbol{n \times p}$ $\quad \boldsymbol{C}$ **is** $\boldsymbol{p \times q}$. Then $\boldsymbol{AB}$ **is** $\boldsymbol{m \times p}$ and $\boldsymbol{(AB)C}$ **is** $\boldsymbol{m \times q}$.

We can test the law using the exchange matrix B on the rows and the columns of A:

$$(\boldsymbol{BA})\boldsymbol{B} = \begin{bmatrix} 0 & 1 \\ 1 & 0 \end{bmatrix} \begin{bmatrix} 1 & 2 \\ 3 & 4 \end{bmatrix} \begin{bmatrix} 0 & 1 \\ 1 & 0 \end{bmatrix} = \begin{bmatrix} \mathbf{3} & \mathbf{4} \\ \mathbf{1} & \mathbf{2} \end{bmatrix} \begin{bmatrix} 0 & 1 \\ 1 & 0 \end{bmatrix} = \begin{bmatrix} \mathbf{4} & \mathbf{3} \\ \mathbf{2} & \mathbf{1} \end{bmatrix}$$

$$\boldsymbol{B}(\boldsymbol{AB}) = \begin{bmatrix} 0 & 1 \\ 1 & 0 \end{bmatrix} \begin{bmatrix} 1 & 2 \\ 3 & 4 \end{bmatrix} \begin{bmatrix} 0 & 1 \\ 1 & 0 \end{bmatrix} = \begin{bmatrix} 0 & 1 \\ 1 & 0 \end{bmatrix} \begin{bmatrix} \mathbf{2} & \mathbf{1} \\ \mathbf{4} & \mathbf{3} \end{bmatrix} = \begin{bmatrix} \mathbf{4} & \mathbf{3} \\ \mathbf{2} & \mathbf{1} \end{bmatrix}$$

So row operations on A can come *before or after* column operations on A.

Notice the meaning of $(AB)C = A(BC)$ when C is just a column vector $\boldsymbol{x}$. If that vector $\boldsymbol{x}$ has a single 1 in component j, then the associative law is $(AB)\boldsymbol{x} = A(B\boldsymbol{x})$. This tells us how to multiply matrices! The left side is **column** $\boldsymbol{j}$ **of** $\boldsymbol{AB}$. The right side is $\boldsymbol{A}$ **times column** $\boldsymbol{j}$ **of** $\boldsymbol{B}$. So their equality is exactly the rule for matrix multiplication that we saw in equation (1). It is simply the right rule.

Let me bring together the important facts about ABC and also A times $B + C$:

$$\boxed{\textbf{Associative}\ \ (\boldsymbol{AB})\boldsymbol{C} = \boldsymbol{A}(\boldsymbol{BC})\ \ \text{and}\ \ \textbf{Distributive}\ \ \boldsymbol{A}(\boldsymbol{B}+\boldsymbol{C}) = \boldsymbol{AB}+\boldsymbol{AC}} \quad (7)$$

Review of AB

Dot products $\quad$ (Row $\boldsymbol{i}$ of $\boldsymbol{A}$)$\cdot$(Col $\boldsymbol{j}$ of $\boldsymbol{B}$)$=(\boldsymbol{AB})_{\boldsymbol{ij}}$ = number in row i, col j of AB

Combine columns $\quad$ (Matrix $\boldsymbol{A}$) (Column $\boldsymbol{b}_j$ of $\boldsymbol{B}$) = vector in column j of AB

With numbers (the usual way), mp dot products produce the m by p matrix AB.
With vectors (the big picture), p combinations $A\boldsymbol{b}_j$ produce the p columns of AB.

For computing by hand, I would use the row way to find each number in AB. I visualize multiplication by columns: **The columns** $\boldsymbol{Ab_j}$ **in** $\boldsymbol{AB}$ **are combinations of columns of** $\boldsymbol{A}$.

Rank One Matrices and $A = CR$

All columns of a rank one matrix lie on the same line. That line is the column space of A. Examples in Section 1.3 pointed to a remarkable fact: *The rows also lie on a line.* When all the columns of A are in the same column direction, then all the rows of A are in the same row direction. Here is a new example of this extreme case: **rank** $\boldsymbol{r} = \mathbf{1}$.

Example 1 $\quad A = \begin{bmatrix} 1 & 2 & 10 & 100 \\ 3 & 6 & 30 & 300 \\ 2 & 4 & 20 & 200 \end{bmatrix} = \begin{matrix} \text{rank one matrix} \\ \text{one independent column} \\ \text{one independent row!} \end{matrix}$

All columns are multiples of $(1,3,2)$. All rows are multiples of $\begin{bmatrix} 1 & 2 & 10 & 100 \end{bmatrix}$. **Only one independent row when there is only one independent column.** *Why is this true*? Another example: **Matrix of all 1's = (Column of 1's)** times **(Row of 1's).**

Our approach is through matrix multiplication. We factor A into C times R. For this very special matrix, C has one column and R has one row. $\boldsymbol{CR}$ **is** $(3 \times 1)\,(1 \times 4)$.

$$\textbf{Rank} = \textbf{1} \qquad \boldsymbol{A} = \begin{bmatrix} 1 & 2 & 10 & 100 \\ 3 & 6 & 30 & 300 \\ 2 & 4 & 20 & 200 \end{bmatrix} = \begin{bmatrix} \mathbf{1} \\ \mathbf{3} \\ \mathbf{2} \end{bmatrix} \begin{bmatrix} \mathbf{1} & \mathbf{2} & \mathbf{10} & \mathbf{100} \end{bmatrix} = \boldsymbol{CR} \tag{8}$$

The dot products (row of C)$\cdot$(column of R) are small multiplications like 1 times 1. The last dot product is 2 times 100. We are following the dot product rule! This is multiplication of thin matrices CR. 12 small multiplications produce the 12 numbers in A.

The rows of A are numbers $1, 3, 2$ times the (only) row $\begin{bmatrix} 1 & 2 & 10 & 100 \end{bmatrix}$ of R. By factoring this special A into **one column times one row**, the conclusion jumps out:

If the column space of A is a line, the row space of A is also a line.

One column in C, one row in R. Our next goal is to allow $\boldsymbol{r}$ **columns in** $\boldsymbol{C}$ and to find $\boldsymbol{r}$ **rows in** $\boldsymbol{R}$. And to see $\boldsymbol{A} = \boldsymbol{CR}$. **That number $\boldsymbol{r}$ is the "rank" of $\boldsymbol{A}$.**

C Contains the First r Independent Columns of A

Suppose we go from left to right, looking for independent columns in any matrix A:

> If column 1 of A is not all zero, put it into the matrix C
>
> If column 2 of A is not a multiple of column 1, put it into C
>
> If column 3 of A is not a combination of columns 1 and 2, put it into C. *Continue.*

At the end C will have r columns taken from A. That number r is the **rank of $\boldsymbol{A}$ and $\boldsymbol{C}$.** The n columns of A might be dependent. The r columns of C will surely be **independent**.

Independent	*No column of C is a combination of previous columns*
columns	*No combination of columns gives* $C\boldsymbol{x} = \mathbf{0}$ *except* $\boldsymbol{x}$ = *all zeros*

Those r independent columns in C combine to give all n columns in A.

$C\boldsymbol{x} = \mathbf{0}$ means that x_1(column 1 of C) + x_2(column 2 of C) + $\cdots$ = *zero vector.* With independent columns, $C\boldsymbol{x} = \mathbf{0}$ only happens if *all x's are zero.* Otherwise we can divide by the last nonzero coefficient x and that column would be a combination of the earlier columns—which our construction forbids. C always has independent columns.

$$\textbf{Example 2} \qquad A = \begin{bmatrix} \mathbf{2} & 6 & \mathbf{4} \\ \mathbf{4} & 12 & \mathbf{8} \\ \mathbf{1} & 3 & \mathbf{5} \end{bmatrix} \quad \text{leads to} \quad C = \begin{bmatrix} \mathbf{2} & \mathbf{4} \\ \mathbf{4} & \mathbf{8} \\ \mathbf{1} & \mathbf{5} \end{bmatrix} \qquad \textbf{Rank } \boldsymbol{r} = \mathbf{2}$$

Columns 1 and 3 go into C. Column 2 is 3 times column 1: *not independent, not in* C.

Matrix Multiplication C times R

R tells how to produce all columns of A from the columns of C. Then $A = CR$. The first column of A is actually in C, so the first column of R just has 1 and 0. The third column of A comes second in C, so the third column of R just has 0 and 1.

$$\begin{array}{c}\textbf{Notice } \boldsymbol{I}\\ \textbf{inside } \boldsymbol{R}\\ \textbf{Rank } \boldsymbol{r=2}\end{array} \qquad \boldsymbol{A=CR} \text{ is } \begin{bmatrix} 2 & 6 & 4 \\ 4 & 12 & 8 \\ 1 & 3 & 5 \end{bmatrix} = \begin{bmatrix} \mathbf{2} & \mathbf{4} \\ \mathbf{4} & \mathbf{8} \\ \mathbf{1} & \mathbf{5} \end{bmatrix} \begin{bmatrix} \mathbf{1} & ? & \mathbf{0} \\ \mathbf{0} & ? & \mathbf{1} \end{bmatrix}. \tag{9}$$

Two columns of A went straight into C, so *part of R is the identity matrix.* The question marks are in column 2 because column 2 of A is *not in C*. It is a dependent column. Column 2 of A is 3 times column 1, so *that number* 3 *goes into R.*

$$\begin{array}{l} A \text{ is } m \times n \\ C \text{ is } m \times r \\ R \text{ is } r \times n \end{array} \qquad \boxed{\boldsymbol{A=CR} \text{ is } \begin{bmatrix} 2 & 6 & 4 \\ 4 & 12 & 8 \\ 1 & 3 & 5 \end{bmatrix} = \begin{bmatrix} \mathbf{2} & \mathbf{4} \\ \mathbf{4} & \mathbf{8} \\ \mathbf{1} & \mathbf{5} \end{bmatrix} \begin{bmatrix} \mathbf{1} & \mathbf{3} & \mathbf{0} \\ \mathbf{0} & \mathbf{0} & \mathbf{1} \end{bmatrix}} \tag{10}$$

That example is typical of $A = CR$. We review the descriptions of C and R.

1. $\boldsymbol{C}$ contains a full set of r **independent columns** (chosen left to right) in A
2. $\boldsymbol{R} = \begin{bmatrix} \boldsymbol{I} & \boldsymbol{F} \end{bmatrix}$ contains the **identity matrix $\boldsymbol{I}$** in the same r columns that held C.
3. **The dependent columns of $\boldsymbol{A}$ are combinations $\boldsymbol{CF}$** of the independent columns in C.

That matrix F goes into the other $n - r$ columns of $\boldsymbol{R} = \begin{bmatrix} \boldsymbol{I} & \boldsymbol{F} \end{bmatrix}$. $A = CR$ becomes

$$\boldsymbol{A = C\begin{bmatrix} I & F \end{bmatrix} = \begin{bmatrix} C & CF \end{bmatrix}} = \begin{bmatrix} \textbf{indep cols of } \boldsymbol{A} & \textbf{dep cols of } \boldsymbol{A} \end{bmatrix} \textbf{ (in correct order)}$$

$\boldsymbol{C}$ has the same column space as $\boldsymbol{A}$. $\boldsymbol{R}$ has the same row space as $\boldsymbol{A}$. Here $\boldsymbol{F} = \begin{bmatrix} \mathbf{-1} \\ \mathbf{2} \end{bmatrix}$

$$\begin{array}{c}\textbf{Example 3}\\ \textbf{of } \boldsymbol{A=CR}\\ \textbf{Rank 2}\end{array} \qquad \begin{bmatrix} 1 & 2 & 3 \\ 4 & 5 & 6 \\ 7 & 8 & 9 \end{bmatrix} = \begin{bmatrix} 1 & 2 \\ 4 & 5 \\ 7 & 8 \end{bmatrix} \begin{bmatrix} \mathbf{1} & \mathbf{0} & \mathbf{-1} \\ \mathbf{0} & \mathbf{1} & \mathbf{2} \end{bmatrix} \tag{11}$$

When a column of $\boldsymbol{A}$ goes into $\boldsymbol{C}$, a column of $\boldsymbol{I}$ goes into $\boldsymbol{R}$.

$$\boxed{\textbf{Column } \boldsymbol{j} \textbf{ of } \boldsymbol{A} = \boldsymbol{C} \text{ times } \textbf{column } \boldsymbol{j} \textbf{ of } \boldsymbol{R}. \qquad \textbf{Row } \boldsymbol{i} \textbf{ of } \boldsymbol{A} = \textbf{row } \boldsymbol{i} \textbf{ of } \boldsymbol{C} \text{ times } \boldsymbol{R}.}$$

If all columns of A are independent, then $C = A$. What matrix is R? **Answer $\boldsymbol{R = I}$.**

Chapter 1 finds $\boldsymbol{C}$ (independent columns of A) before R. Chapter 3 will find R first.
Here column 3 of A is the 2nd independent column in C. Then column 3 of R is $\begin{matrix}\mathbf{0}\\ \mathbf{1}\end{matrix}$
$\boldsymbol{A}=\begin{bmatrix}1 & 2 & \mathbf{3} & 4\\ 1 & 2 & \mathbf{4} & 5\end{bmatrix}=\begin{bmatrix}\mathbf{1} & \mathbf{3}\\ \mathbf{1} & \mathbf{4}\end{bmatrix}\begin{bmatrix}\mathbf{1} & 2 & \mathbf{0} & 1\\ \mathbf{0} & 0 & \mathbf{1} & 1\end{bmatrix}=\boldsymbol{CR}$ All three ranks $=2$
$\boldsymbol{R}$ **tells how to recover all columns of** $\boldsymbol{A}$ **from the independent columns in** $\boldsymbol{C}$.

Here is an informal proof that ***the row rank of A equals the column rank of A***

1. The r columns of C are independent (they are chosen that way from A)
2. Every column of A is a combination of those r columns of C (this is $A = CR$)
3. The r rows of R are independent (they contain the r by r matrix I)
4. Every row of A is a combination of the r rows of R (this is $A = CR$ by rows !)

How to Find the Matrix $\boldsymbol{R}$

Up to now you have had very little help in discovering the matrix R in $A = CR$. If you could tell that column 3 of this matrix A is a combination of columns 1 and 2, then the numbers x and y in that combination will go into column 3 of R:

Example 4 $$\boldsymbol{A}=\begin{bmatrix}1 & 3 & 4\\ 2 & 4 & 2\\ 3 & 7 & 6\end{bmatrix} \qquad x\begin{bmatrix}1\\2\\3\end{bmatrix}+y\begin{bmatrix}3\\4\\7\end{bmatrix}\overset{?}{=}\begin{bmatrix}4\\2\\6\end{bmatrix}. \qquad (12)$$

But even for this small matrix, we can't immediately see x and y. So we don't know the rank of A (2 or 3 **?**). There has to be a good way to discover x and y.

That good way is elimination. It will be the key algorithm in Chapter 2 for square matrices and again in Chapter 3 for all matrices. We want to introduce it now for this matrix.

The idea is to simplify A by "*row operations*". That will simplify the equations for x and y. We will **eliminate the 2 and 3** in column 1 of A. To do that, **subtract 2 times row 1 from row 2 of** $\boldsymbol{A}$ and also **subtract 3 times row 1 from row 3.** The matrix A changes to B.

$$\boldsymbol{B}=\begin{bmatrix}1 & 3 & 4\\ 0 & -2 & -6\\ 0 & -2 & -6\end{bmatrix} \qquad x\begin{bmatrix}1\\0\\0\end{bmatrix}+y\begin{bmatrix}3\\-2\\-2\end{bmatrix}\overset{?}{=}\begin{bmatrix}4\\-6\\-6\end{bmatrix} \qquad (13)$$

We only did what is legal. *Subtracting an equation from an equation leaves a new equation.* The new equation is $-2y = -6$, so we know $\boldsymbol{y = 3}$. Then if $\boldsymbol{x = -5}$ the top equation becomes $-5 + 9 = 4$, which is correct. The original equations (12) are solved by $-5, 3$:

$$\begin{matrix}x=-5\\ y=+3\end{matrix} \qquad -\mathbf{5}\begin{bmatrix}1\\2\\3\end{bmatrix}+\mathbf{3}\begin{bmatrix}3\\4\\7\end{bmatrix}=\begin{bmatrix}4\\2\\6\end{bmatrix} \qquad \textbf{Column 3 of } \boldsymbol{A} \textbf{ is dependent}$$

So -5 and 3 are the numbers we needed in column 3 of R. All the ranks are $r = 2$:

$$\boldsymbol{A} = \begin{bmatrix} 1 & 3 & 4 \\ 2 & 4 & 2 \\ 3 & 7 & 6 \end{bmatrix} = \begin{bmatrix} \mathbf{1} & \mathbf{3} \\ \mathbf{2} & \mathbf{4} \\ \mathbf{3} & \mathbf{7} \end{bmatrix} \begin{bmatrix} \mathbf{1} & \mathbf{0} & \mathbf{-5} \\ \mathbf{0} & \mathbf{1} & \mathbf{3} \end{bmatrix} = \boldsymbol{CR}. \tag{14}$$

There is more to see in this example. The elimination process that reduced A to B is called *row reduction*. I will complete it from B to U, to make the matrix even simpler. Just subtract row 2 of B from row 3 of B to see a **row of zeros in $\boldsymbol{U}$:**

$$\boldsymbol{A} \to \boldsymbol{B} = \begin{bmatrix} 1 & 3 & 4 \\ 0 & -2 & -6 \\ 0 & -2 & -6 \end{bmatrix} \to \boldsymbol{U} = \begin{bmatrix} \mathbf{1} & 3 & 4 \\ 0 & \mathbf{-2} & -6 \\ \mathbf{0} & \mathbf{0} & \mathbf{0} \end{bmatrix} = \begin{matrix} \textbf{upper triangular} \\ \textbf{matrix } \boldsymbol{U} \end{matrix}. \tag{15}$$

That zero row is a clear signal: the row rank is also 2. Chapter 2 will stop with U. Chapter 3 will **eliminate upward** to produce more zeros. **We end up with $\boldsymbol{R_0}$ and $\boldsymbol{R}$**:

$$\boldsymbol{U} = \begin{bmatrix} 1 & 3 & 4 \\ 0 & -2 & -6 \\ 0 & 0 & 0 \end{bmatrix} \to \begin{bmatrix} 1 & 3 & 4 \\ 0 & 1 & 3 \\ 0 & 0 & 0 \end{bmatrix} \to \begin{bmatrix} \mathbf{1} & 0 & \mathbf{-5} \\ 0 & \mathbf{1} & \mathbf{3} \\ 0 & 0 & 0 \end{bmatrix} = \boldsymbol{R_0}$$

All rows of R_0 are combinations of the original rows of A
That zero row of R_0 shows that A has rank $r = 2$
The 2 by 2 identity matrix shows that columns $1, 2$ of A are independent (in C)
Removing the zero row of $\boldsymbol{R_0}$ leaves the desired matrix $\boldsymbol{R}$ in $\boldsymbol{A} = \boldsymbol{CR}$
Elimination in Chapter 3 will be a systematic way to find $\boldsymbol{R}$

Key facts	The r columns of C are a **basis** for the column space of A: **dimension $\boldsymbol{r}$**
$\boldsymbol{A} = \boldsymbol{CR}$	The r rows of R are a **basis** for the row space of A: **dimension $\boldsymbol{r}$**

Those words **"basis"** and **"dimension"** will be properly defined later in Section 3.4.

Chapter 1 *starts with independent columns of* A, placed in C.
Chapter 3 *starts with the rows of* A, *and combines them into* R.

We are emphasizing CR because both matrices are so important. $\boldsymbol{C}$ contains r independent columns of A. $\boldsymbol{R}$ tells how to combine those columns to give all columns of A. (R contains I, because r columns of A are already in C.) Chapter 3 will produce R *directly from* A *by elimination*, the most used algorithm in computational mathematics.

$A = CR$ will be the key to a fundamental problem: *Solving linear equations* $A\boldsymbol{x} = \boldsymbol{b}$.

Columns of A times Rows of B . . . Columns of C times Rows of R

Before this chapter ends, I want to add this message. There is another way to multiply matrices (producing the same matrix AB or CR as always). This way is not so well known, but it is powerful. **The new way multiplies columns of A times rows of B.**

$$\boldsymbol{AB} = \begin{bmatrix} | & & | \\ \boldsymbol{a}_1 & \cdots & \boldsymbol{a}_n \\ | & & | \end{bmatrix} \begin{bmatrix} - & \boldsymbol{b}_1^* & - \\ & \vdots & \\ - & \boldsymbol{b}_n^* & - \end{bmatrix} = \boldsymbol{a}_1\boldsymbol{b}_1^* + \boldsymbol{a}_2\boldsymbol{b}_2^* + \cdots + \boldsymbol{a}_n\boldsymbol{b}_n^*. \tag{16}$$

columns $\boldsymbol{a}_k$ rows $\boldsymbol{b}_k^*$ Add columns $\boldsymbol{a}_k$ times rows $\boldsymbol{b}_k^*$

Those matrices $\boldsymbol{a}_k\boldsymbol{b}_k^*$ are called *outer products*. We recognize that they have *rank one*: **column times row**. They are entirely different from dot products (**rows times columns**). If A is an m by n matrix and B is an n by p matrix, then columns of A times rows of B adds up to the *same answer* AB as dot products of rows of A and columns of B.

AB involves the same mnp small multiplications but in a new order !

(Row) · (Column) mp dot products, n multiplications each **total mnp**

(Column) (Row) n rank one matrices, mp multiplications each **total mnp**

Columns × Rows $$\begin{bmatrix} 1 & 4 \\ 2 & 5 \\ 3 & 6 \end{bmatrix} \begin{bmatrix} 7 & 8 & 9 \\ 10 & 11 & 12 \end{bmatrix} = \begin{bmatrix} 1 \\ 2 \\ 3 \end{bmatrix} \begin{bmatrix} 7 & 8 & 9 \end{bmatrix} + \begin{bmatrix} 4 \\ 5 \\ 6 \end{bmatrix} \begin{bmatrix} 10 & 11 & 12 \end{bmatrix}$$

Rank 1 + Rank 1 $$= \begin{bmatrix} 7 & 8 & 9 \\ 14 & 16 & 18 \\ 21 & 24 & 27 \end{bmatrix} + \begin{bmatrix} 40 & 44 & 48 \\ 50 & 55 & 60 \\ 60 & 66 & 72 \end{bmatrix} = \begin{bmatrix} \mathbf{47} & \mathbf{52} & \mathbf{57} \\ \mathbf{64} & \mathbf{71} & \mathbf{78} \\ \mathbf{81} & \mathbf{90} & \mathbf{99} \end{bmatrix} = \boldsymbol{AB}$$

This example has $mnp = (3)\,(2)\,(3) = 18$. At the start of the second line you see the 18 multiplications (in two 3 by 3 matrices). Then 9 additions give the correct answer AB.

As we learned in this section, the rank of AB is 2. *Two independent columns, not three. Two independent rows, not three.* The next chapter uses different words. *AB has no inverse matrix: it is not invertible.* And in Chapter 5: *The determinant of AB is zero.*

Note about the matrix R

We were amazed to learn that the row matrix R in $A = CR$ is already a famous matrix in linear algebra! It is essentially the **"reduced row echelon form"** of the original A. MATLAB calls it **rref** (A) and includes $m - r$ zero rows. With the zero rows, we call it $\boldsymbol{R}_0$.

The factorization $\boldsymbol{A} = \boldsymbol{CR}$ is a big step in linear algebra. The Problem Set will look closely at the matrix R, its form is remarkable. R has the identity matrix in r columns. Then C multiplies each column of R to produce a column of A. **$\boldsymbol{R}_0$ comes in Chapter 3.**

Example 5 $$A = \begin{bmatrix} \boldsymbol{a}_1 & \boldsymbol{a}_2 & 3\boldsymbol{a}_1 + 4\boldsymbol{a}_2 \end{bmatrix} = \begin{bmatrix} \boldsymbol{a}_1 & \boldsymbol{a}_2 \end{bmatrix} \begin{bmatrix} \mathbf{1} & 0 & \mathbf{3} \\ 0 & \mathbf{1} & \mathbf{4} \end{bmatrix} = \boldsymbol{CR}.$$

Here $\boldsymbol{a}_1$ and $\boldsymbol{a}_2$ are the independent columns of A. The third column is dependent—a combination of $\boldsymbol{a}_1$ and $\boldsymbol{a}_2$. Therefore it is in the plane produced by columns 1 and 2. All three matrices A, C, R have rank $r = 2$.

We can try that new way (**columns** $\times$ **rows**) to quickly multiply CR in Example 5:

Columns of C times rows of R
$$CR = \boldsymbol{a}_1 \begin{bmatrix} 1 & 0 & 3 \end{bmatrix} + \boldsymbol{a}_2 \begin{bmatrix} 0 & 1 & 4 \end{bmatrix} = \begin{bmatrix} \boldsymbol{a}_1 & \boldsymbol{a}_2 & 3\boldsymbol{a}_1 + 4\boldsymbol{a}_2 \end{bmatrix} = A$$

(3 by 2) (2 by 4) = (3 by 4) **Four Ways to Multiply $AB = C$**

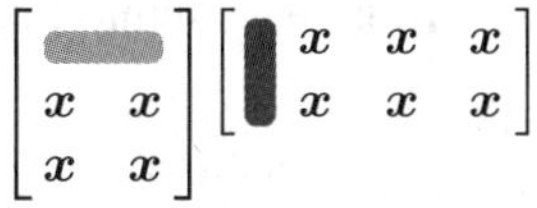

(Row i of A) $\cdot$ **(Column k of B)** = **Number C_{ik}**
$i = 1$ to 3 $k = 1$ to 4 **12 numbers**

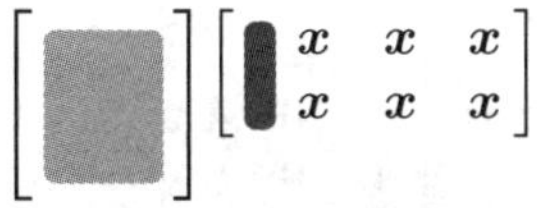

$\boldsymbol{A}$ times (**Column k** of B) = **Column k of C**
$k = 1$ to 4 **4 columns**

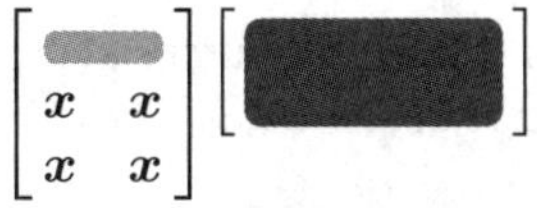

(**Row i** of A) times $\boldsymbol{B}$ = **Row i of C**
$i = 1$ to 3 **3 rows**

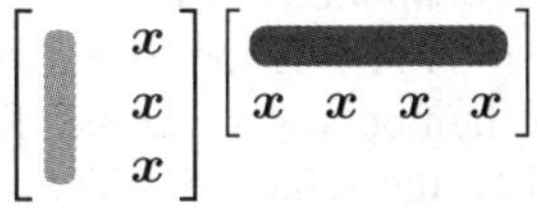

(**Column j** of A) (**Row j** of B) = **Rank 1 Matrix**
$j = 1$ to 2 **2 matrices**

Dot product way, Column way, Row way, Columns times rows

Problem Set 1.4

1 Rewrite this four-way table for $AB = C$ when A is m by n and B is n by p. How many dot products and columns and rows and rank one matrices go into AB? In all four cases the total count of small multiplications is mnp.

2 If all columns of $A = \begin{bmatrix} \boldsymbol{a} & \boldsymbol{a} & \boldsymbol{a} \end{bmatrix}$ contain the same $\boldsymbol{a} \neq \mathbf{0}$, what are C and R?

3 Multiply A times B (3 examples) using *dot products*: (each row) $\cdot$ (each column).

$$\begin{bmatrix} 1 & 0 & 0 \\ 1 & 1 & 0 \\ 1 & 1 & 1 \end{bmatrix} \begin{bmatrix} 1 & 0 & 0 \\ -1 & 1 & 0 \\ 1 & -1 & 1 \end{bmatrix} \qquad \begin{bmatrix} 1 & 2 & 3 \end{bmatrix} \begin{bmatrix} 4 \\ 5 \\ 6 \end{bmatrix} \qquad \begin{bmatrix} 4 \\ 5 \\ 6 \end{bmatrix} \begin{bmatrix} 1 & 2 & 3 \end{bmatrix}$$

4 Test the truth of the associative law $(AB)C = A(BC)$.

(a) $\begin{bmatrix} 1 & 1 \end{bmatrix} \begin{bmatrix} 1 \\ 1 \end{bmatrix} \begin{bmatrix} 1 & 1 & 1 \end{bmatrix}$ (b) $\begin{bmatrix} 1 & 2 \\ 0 & 1 \end{bmatrix} \begin{bmatrix} 1 & 3 \\ 0 & 1 \end{bmatrix} \begin{bmatrix} 1 & 4 \\ 0 & 1 \end{bmatrix}$

5 Why is it impossible for a matrix A with 7 columns and 4 rows to have 5 independent columns? This is not a trivial or useless question.

6 Going from left to right, put each column of A into the matrix C if that column is not a combination of earlier columns:

$$A = \begin{bmatrix} 2 & -2 & 1 & 6 & 0 \\ 1 & -1 & 0 & 2 & 0 \\ 3 & -3 & 0 & 6 & 1 \end{bmatrix} \qquad C = \begin{bmatrix} 2 & & & \\ 1 & & & \\ 3 & & & \end{bmatrix}$$

7 Find R in Problem 6 so that $A = CR$. If your C has r columns, then R has r rows. The 5 columns of R tell how to produce the 5 columns of A from the columns in C.

8 This matrix A has 3 independent columns. So C has the same 3 columns as A. What is the 3 by 3 matrix R so that $A = CR$? What is different about $B = CR$?

$$\text{Upper triangular} \quad A = \begin{bmatrix} 2 & 2 & 2 \\ 0 & 4 & 4 \\ 0 & 0 & 6 \end{bmatrix} \qquad B = \begin{bmatrix} 2 & 2 & 2 \\ 0 & 0 & 4 \\ 0 & 0 & 6 \end{bmatrix}$$

9 Suppose A is a random 4 by 4 matrix. The probability is 1 that the columns of A are "independent". In that case, what are the matrices C and R in $A = CR$?

Note Random matrix theory has become an important part of applied linear algebra—especially for very large matrices when even multiplication AB is too expensive. An example of "*probability* 1" is choosing two whole numbers at random. The probability is 1 that they are different. But they could be the same! Problem 10 is another example of this type.

10 Suppose A is a random 4 by 5 matrix. With probability 1, what can you say about C and R in $A = CR$? In particular, which columns of A (going into C) are probably independent of previous columns, when you go from left to right?

11 Create your own example of a 4 by 4 matrix A of rank $r = 2$. Then factor A into $CR = (4 \text{ by } 2)\,(2 \text{ by } 4)$.

12 Factor these matrices into $A = CR = (m \text{ by } r)\,(r \text{ by } n)$: all ranks equal to r.

$$A_1 = \begin{bmatrix} 1 & 2 & 3 \\ 1 & 3 & 4 \end{bmatrix} \quad A_2 = \begin{bmatrix} 0 & 1 & 2 & 3 \\ 0 & 1 & 3 & 5 \end{bmatrix} \quad A_3 = \begin{bmatrix} 2 & 1 & 3 \\ 6 & 3 & 9 \end{bmatrix} \quad A_4 = \begin{bmatrix} 1 & 0 & 0 & 4 \\ 0 & 2 & 2 & 0 \end{bmatrix}$$

13 Starting from $C = \begin{bmatrix} 1 \\ 3 \end{bmatrix}$ and $R = \begin{bmatrix} 2 & 4 \end{bmatrix}$ compute CR and RC and CRC and RCR.

14 Complete these 2 by 2 matrices to meet the requirements printed underneath:

$$\begin{bmatrix} 3 & 6 \\ 5 & \end{bmatrix} \qquad \begin{bmatrix} 6 & 7 \\ 7 & \end{bmatrix} \qquad \begin{bmatrix} 2 & \\ 3 & 6 \end{bmatrix} \qquad \begin{bmatrix} 3 & 4 \\ & -3 \end{bmatrix}$$

rank one — orthogonal columns — rank 2 — $A^2 = I$

15 Suppose $A = CR$ with independent columns in C and independent rows in R. Explain how each of these logical steps follows from $A = CR = (m \text{ by } r)\,(r \text{ by } n)$.

1. Every column of A is a combination of columns of C.
2. Every row of A is a combination of rows of R. What combination is row 1 ?
3. The number of columns of C = the number of rows of R (needed for CR).
4. *Column rank equals row rank.* The number of independent columns of A equals the number of independent rows in A.

16 (a) The vectors $AB\boldsymbol{x}$ produce the column space of AB. Show why this vector $AB\boldsymbol{x}$ is also in the column space of A. (Is $AB\boldsymbol{x} = A\boldsymbol{y}$ for some vector $\boldsymbol{y}$?) Conclusion : The column space of A *contains* the column space of AB.

(b) Choose nonzero matrices A and B so the column space of AB contains only the zero vector. This is the smallest possible column space.

17 True or false, with a reason (not easy) :

(a) If 3 by 3 matrices A and B have rank 1, then AB will always have rank 1.

(b) If 3 by 3 matrices A and B have rank 3, then AB will always have rank 3.

(c) Suppose $AB = BA$ for every 2 by 2 matrix B. Then $A = \begin{bmatrix} c & 0 \\ 0 & c \end{bmatrix} = cI$ for some number c. Only those matrices $A = cI$ commute with every B.

18 This section mentioned a special case of the law $(\boldsymbol{AB})\boldsymbol{C} = \boldsymbol{A}(\boldsymbol{BC})$.

$$A = C = \text{ exchange matrix } \begin{bmatrix} 0 & 1 \\ 1 & 0 \end{bmatrix} \qquad B = \begin{bmatrix} 1 & 2 \\ 3 & 4 \end{bmatrix}.$$

(a) First compute AB (row exchange) and also BC (column exchange).

(b) Now compute the double exchanges : $(AB)C$ with rows first and $A(BC)$ with columns first. Verify that those double exchanges produce the same ABC.

19 Test the column-row matrix multiplication in equation (16) to find AB and BA :

$$AB = \begin{bmatrix} 1 & 0 & 0 \\ 1 & 1 & 0 \\ 1 & 1 & 1 \end{bmatrix} \begin{bmatrix} 1 & 1 & 1 \\ 0 & 1 & 1 \\ 0 & 0 & 1 \end{bmatrix} \qquad BA = \begin{bmatrix} 1 & 1 & 1 \\ 0 & 1 & 1 \\ 0 & 0 & 1 \end{bmatrix} \begin{bmatrix} 1 & 0 & 0 \\ 1 & 1 & 0 \\ 1 & 1 & 1 \end{bmatrix}$$

20 How many small multiplications for $(AB)C$ and $A(BC)$ if those matrices have sizes $ABC = (4 \times 3)\,(3 \times 2)\,(2 \times 1)$? The two counts are different.

Thoughts on Chapter 1

Most textbooks don't have a place for the author's thoughts. But a lot of decisions go into starting a new textbook. This chapter has intentionally jumped right into the subject, with discussion of independence and rank. There are so many good ideas ahead, and they take time to absorb, so why not get started ? Here are two questions that influenced the writing.

What makes this subject easy ? All the equations are linear.

What makes this subject hard ? So many equations and unknowns and ideas.

Book examples are small size. But if we want the temperature at many points of an engine, there is an equation at every point: easily $n = 1000$ unknowns.

I believe the key is to work right away with matrices. $A\boldsymbol{x} = \boldsymbol{b}$ is a perfect format to accept problems of all sizes. The linearity is built into the symbols $A\boldsymbol{x}$ and the rule is $A(\boldsymbol{x} + \boldsymbol{y}) = A\boldsymbol{x} + A\boldsymbol{y}$. Each of the m equations in $A\boldsymbol{x} = \boldsymbol{b}$ represents a flat surface:

$2x + 5y - 4z = 6$ is a plane in three-dimensional space

$2x + 5y - 4z + 7w = 9$ is a 3D plane (*hyperplane* ?) in four-dimensional space

Linearity is on our side, but there is a serious problem in visualizing 10 planes meeting in 11-dimensional space. Hopefully they meet along a line: dimension $11 - 10 = 1$. An 11th plane should cut through that line at one point (which solves all 11 equations). What the textbook and the notation must do is to keep the counting simple

Here is what we expect for a random m by n matrix A:

$\boldsymbol{m < n}$ Probably many solutions to the m equations $A\boldsymbol{x} = \boldsymbol{b}$

$\boldsymbol{m = n}$ Probably one solution to the n equations $A\boldsymbol{x} = \boldsymbol{b}$

$\boldsymbol{m > n}$ Probably no solution: too many equations with only n unknowns in $\boldsymbol{x}$

But this count is not necessarily what we get! Columns of A can be combinations of previous columns: nothing new. An equation can be a combination of previous equations. **The rank $\boldsymbol{r}$ tells us the real size of our problem**, from independent columns and rows. The beautiful formula is $\boldsymbol{A = CR} = (m \times r)\,(r \times n)$: three matrices of rank r.

Notice: The columns of A that go into C must multiply the matrix I inside R.

We end with the great **associative law** $\boldsymbol{(AB)\,C = A\,(BC)}$. Suppose C has 1 column:

AB has columns $A\boldsymbol{b}_1, \ldots, A\boldsymbol{b}_n$ and then $(AB)\boldsymbol{c}$ equals $c_1 A\boldsymbol{b}_1 + \cdots + c_n A\boldsymbol{b}_n$.

$B\boldsymbol{c}$ has one column $c_1\boldsymbol{b}_1 + \cdots + c_n\boldsymbol{b}_n$ and then $A(B\boldsymbol{c}) = A\,(c_1\boldsymbol{b}_1 + \cdots + c_n\boldsymbol{b}_n)$.

Linearity gives equality of those two sums. This proves $(AB)\,\boldsymbol{c} = A\,(B\boldsymbol{c})$.

The same is true for every column of C. Therefore $\boldsymbol{(AB)\,C = A\,(BC)}$.

Notice that over and over—for $A\boldsymbol{x}$ and AB and CR—we write about linear combinations of columns of A or C.

Not about dot products with the rows!

2 Solving Linear Equations $Ax = b$

The matrices in this chapter are square: n by n. $\boldsymbol{Ax = b}$ **gives** $\boldsymbol{n}$ **equations** (one from each row of A). Those equations have $\boldsymbol{n}$ **unknowns** in the vector $\boldsymbol{x}$. Often but not always there is one solution $\boldsymbol{x}$ for each $\boldsymbol{b}$. In this case A has an **inverse** $\boldsymbol{A^{-1}}$ with $A^{-1}A = I$ and $AA^{-1} = I$. Multiplying $Ax = b$ by A^{-1} produces the symbolic solution $x = A^{-1}b$.

This chapter aims to find that solution x. **But we don't compute** $\boldsymbol{A^{-1}}$. (That would solve $Ax = b$ for every possible b.) We go forward column by column, assuming that A has independent columns. We only stop if this proves wrong. At the end $Ax = b$ has changed to a **triangular system** $\boldsymbol{Ux = c}$, and now the solution x is easy to find.

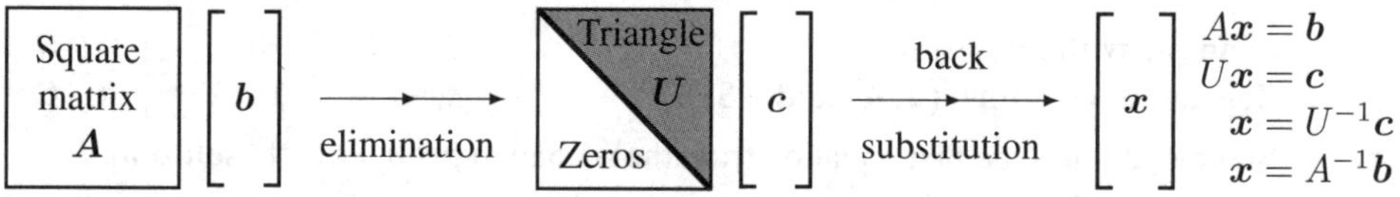

Here is an idea that goes back thousands of years (to China). Each step of "elimination" produces a zero in the matrix. The original A changes slowly into an **upper triangular** $\boldsymbol{U}$. We may need row exchanges. This is not exciting, it is just the natural way to simplify A.

To describe all the steps we need matrices. This is the point of linear algebra! A simple elimination matrix E_{ij} produces a zero where row i meets column j $(i > j)$. Overall, an elimination matrix E multiplies A to give $EA = U$. And we multiply U by an **inverse matrix** $L = E^{-1}$ to come back to A. Here are key matrices in this chapter:

Coefficient matrix $\boldsymbol{A}$	Upper triangular $\boldsymbol{U}$	Lower triangular $\boldsymbol{L}$
Elimination matrix $\boldsymbol{E_{ij}}$	Overall elimination $\boldsymbol{E}$	Inverse matrix $\boldsymbol{A^{-1}}$
Permutation matrix $\boldsymbol{P}$	Transpose matrix $\boldsymbol{A^{\mathrm{T}}}$	Symmetric matrix $\boldsymbol{S = S^{\mathrm{T}}}$

Our goal is to explain all the steps from $\boldsymbol{A}$ **to** $\boldsymbol{EA = U}$ **to** $\boldsymbol{A = E^{-1}U = LU}$ **to** x. (If the steps fail, this signals that $Ax = b$ has no solution for most b.) Every computer system has a code to find the triangular U and then the solution x. Those codes are used so often that elimination adds up to the greatest cost in all of scientific computing.

But the codes are highly engineered and we don't know a better way to solve $Ax = b$. Section 2.5 introduces difference matrices from Computational Science and Engineering.

2.1 Elimination and Back Substitution

1 Elimination subtracts ℓ_{ij} times row j from row i, leave a zero in row i.

2 $\boldsymbol{A x}=\boldsymbol{b}$ becomes $U\boldsymbol{x}=\boldsymbol{c}$ (or else $A\boldsymbol{x}=\boldsymbol{b}$ is proved to have no solution).

3 Then $U\boldsymbol{x}=\boldsymbol{c}$ is solved by **back substitution** because U is upper triangular.

This chapter explains a systematic way to solve $A\boldsymbol{x}=\boldsymbol{b}$: *$n$ equations for n unknowns.* The n by n matrix A is given and the n by 1 column vector $\boldsymbol{b}$ is given. There may be *no vector* $\boldsymbol{x}=(x_1,x_2,\ldots,x_n)$ that solves $A\boldsymbol{x}=\boldsymbol{b}$, or there may be exactly *one solution*, or there may be *infinitely many* solution vectors $\boldsymbol{x}$. Our job is to decide among these three possibilities and to find all solutions. Here are the possibilities with $n=2$.

1 **Exactly one solution to $A\boldsymbol{x}=\boldsymbol{b}$.** In this case A has independent columns. The rank of A is 2. The only solution to $A\boldsymbol{x}=\mathbf{0}$ is $\boldsymbol{x}=0$. ***A* has an *inverse matrix* $\boldsymbol{A^{-1}}$.**

Example with one solution $(\boldsymbol{x},\boldsymbol{y})=(\mathbf{1},\mathbf{1})$ $\quad 2x+3y=5 \quad \begin{bmatrix} \mathbf{2} & \mathbf{3} \\ \mathbf{4} & \mathbf{2} \end{bmatrix}$
Independent columns $(\mathbf{2},\mathbf{4})$ and $(\mathbf{3},\mathbf{2})$ $\quad 4x+2y=6$

2 **No solution to $A\boldsymbol{x}=\boldsymbol{b}$.** In this case $\boldsymbol{b}$ is not a combination of the columns of A. In other words $\boldsymbol{b}$ is not in the column space of A. The rank of A is 1.

Example with no solution $\quad 2x+3y=\ \ 6 \quad \begin{bmatrix} \mathbf{2} & \mathbf{3} \\ \mathbf{4} & \mathbf{6} \end{bmatrix}$
Dependent columns $(\mathbf{2},\mathbf{4})$ and $(\mathbf{3},\mathbf{6})$ $\quad 4x+6y=15$

Subtract 2 times the first equation from the second to get $0=3$. **No solution.**

3 There will be **infinitely many solutions** to $A\boldsymbol{X}=\mathbf{0}$ when the columns of A are not independent. This is the meaning of **dependent columns**—many ways to produce the zero vector $\boldsymbol{b}=\mathbf{0}$. We can multiply $\boldsymbol{X}$ by any number α.

If there is one solution to $A\boldsymbol{x}=\boldsymbol{b}$ then we can add any solution to $A\boldsymbol{X}=\mathbf{0}$. All the vectors $\boldsymbol{x}+\alpha\boldsymbol{X}$ solve the same equations, so we have many solutions.

$$\text{For any number } \alpha \qquad A(\boldsymbol{x}+\alpha\boldsymbol{X})=A\boldsymbol{x}+\alpha A\boldsymbol{X}=\boldsymbol{b}+\mathbf{0}=\boldsymbol{b}. \tag{1}$$

Example with infinitely many solutions $\quad 2x+3y=\ \ 6 \quad \begin{bmatrix} \mathbf{2} & \mathbf{3} \\ \mathbf{4} & \mathbf{6} \end{bmatrix}$
***A* has dependent columns: $\boldsymbol{b}$ is in $\mathbf{C}(\boldsymbol{A})$** $\quad 4x+6y=12$

Those equations $A\boldsymbol{x}=\boldsymbol{b}$ are solved by $x=0, y=2$. But there are more solutions because $\boldsymbol{X}=(\mathbf{3},-\mathbf{2})$ solves $A\boldsymbol{X}=\mathbf{0}$. Then $2\boldsymbol{X}=(6,-4)$ also solves $A(2\boldsymbol{X})=\mathbf{0}$. All vectors $\alpha\boldsymbol{X}$ can be added to the particular solution $\boldsymbol{x}=(0,2)$ to produce more solutions:

$$\boldsymbol{x}+\alpha\boldsymbol{X}=(\mathbf{0}+\mathbf{3}\alpha,\mathbf{2}-\mathbf{2}\alpha) \textbf{ is a line of solutions to our two equations } A\boldsymbol{x}=\boldsymbol{b}.$$

This chapter will start with $U\boldsymbol{x}=\boldsymbol{c}$: one solution, easy to find. Then we explain how $A\boldsymbol{x}=\boldsymbol{b}$ leads to $U\boldsymbol{x}=\boldsymbol{c}$. When this fails, a row exchange may save it. When row exchanges also fail, A has no inverse matrix A^{-1}. Its columns are dependent (cases $\mathbf{2}-\mathbf{3}$).

Back Substitution to Solve $Ux = c$

This chapter will give a systematic way to decide between those possibilities $\mathbf{1, 2, 3}$: One solution, no solution, infinitely many solutions. This system is called **elimination**. It simplifies the matrix A without changing any solution $\boldsymbol{x}$ to the equation $A\boldsymbol{x} = \boldsymbol{b}$. We do the same operations to both sides of the equation, and those operations are reversible. Elimination keeps all solutions $\boldsymbol{x}$ and creates no new ones.

Let me show you the ideal result. Elimination produces an upper triangular matrix. That matrix is called U. **Then $A\boldsymbol{x} = \boldsymbol{b}$ leads to $U\boldsymbol{x} = \boldsymbol{c}$**, which we easily solve:

Apply elimination to $A\boldsymbol{x}=\boldsymbol{b}$ (next page)
The result is $U\boldsymbol{x} = \boldsymbol{c}$ (here)
Back substitution now finds $\boldsymbol{x}$

$$U\boldsymbol{x} = \boldsymbol{c} \text{ is } \begin{bmatrix} \mathbf{2} & 3 & 4 \\ 0 & \mathbf{5} & 6 \\ 0 & 0 & \mathbf{7} \end{bmatrix} \begin{bmatrix} x_1 \\ x_2 \\ x_3 \end{bmatrix} = \begin{bmatrix} \mathbf{19} \\ \mathbf{17} \\ \mathbf{14} \end{bmatrix}$$

That letter $\boldsymbol{U}$ stands for **upper triangular**. The matrix has all zeros below its diagonal. Highly important: **The "pivots" $\mathbf{2, 5, 7}$ on that main diagonal of $\boldsymbol{U}$ are not zero.** Then we can solve the equations by going from bottom to top: **find $\boldsymbol{x_3}$ then $\boldsymbol{x_2}$ then $\boldsymbol{x_1}$**.

Back substitution	The last equation $\mathbf{7}x_3 = \mathbf{14}$ gives $\boldsymbol{x_3} = \mathbf{2}$
Work upwards	The next equation $\mathbf{5}x_2 + \mathbf{6(2)} = \mathbf{17}$ gives $\boldsymbol{x_2} = \mathbf{1}$
Upwards again	The first equation $\mathbf{2}x_1 + \mathbf{3(1)} + \mathbf{4(2)} = \mathbf{19}$ gives $\boldsymbol{x_1} = \mathbf{4}$
Conclusion	The only solution to this example $U\boldsymbol{x} = \boldsymbol{c}$ is $\boxed{\boldsymbol{x} = (\mathbf{4, 1, 2}).}$
Special note	In solving for x_1, x_2, x_3 we needed to divide by the pivots $\mathbf{2, 5, 7}$.

These pivots were probably not on the diagonal of the original matrix A (which we haven't seen). The pivots $2, 5, 7$ were discovered when "elimination" produced the lower triangular zeros in U. This crucial step from A to U is still to be explained! We have displayed the final back substitution step, next we explain elimination.

Equations $A\boldsymbol{x}=\boldsymbol{b}$ Elimination to $U\boldsymbol{x}=\boldsymbol{c}$ Back substitution to $\boldsymbol{x}=U^{-1}\boldsymbol{c}=A^{-1}\boldsymbol{b}$

Note We would not allow the number zero to be a pivot. That would destroy our plan because an equation like $0x_1 = 2$ or $0x_2 = 5$ or $0x_3 = 8$ has no solution. Back substitution will break down with a zero in any pivot position (on the diagonal of U). **The test for independent columns in $\boldsymbol{A}$ is $\boldsymbol{n}$ nonzero pivots in $\boldsymbol{U}$ (after possible row exchanges).**

Every square matrix A with independent columns (full rank) can be reduced to a triangular matrix U with nonzero pivots. *This is our job.* It is possible that we may need to put the equations $A\boldsymbol{x} = \boldsymbol{b}$ in a different order. We start with the usual case when elimination goes from A to U. Then back substitution as above finds the solution vector $\boldsymbol{x}$ to $A\boldsymbol{x} = \boldsymbol{b}$.

From A to U and b to c: Elimination in Each Column

First comes a matrix A (independent columns) that will require no row exchanges. We will apply elimination matrices E_{21} then E_{31} then E_{32}. A and $\boldsymbol{b}$ will change to U and $\boldsymbol{c}$.

The starting matrix is A
The first pivot is 2
The right side is $\boldsymbol{b}$

$$A = \begin{bmatrix} \mathbf{2} & 3 & 4 \\ 4 & 11 & 14 \\ 2 & 8 & 17 \end{bmatrix} \qquad \boldsymbol{b} = \begin{bmatrix} 19 \\ 55 \\ 50 \end{bmatrix} \tag{2}$$

E_{21} multiplies equation 1 by 2 and subtracts from equation 2. You see the new zero.

$$E_{21} = \begin{bmatrix} 1 & 0 & 0 \\ -2 & 1 & 0 \\ 0 & 0 & 1 \end{bmatrix} \qquad \boldsymbol{E_{21}A} = \begin{bmatrix} \mathbf{2} & 3 & 4 \\ \mathbf{0} & 5 & 6 \\ 2 & 8 & 17 \end{bmatrix} \qquad \boldsymbol{E_{21}b} = \begin{bmatrix} 19 \\ \mathbf{17} \\ 50 \end{bmatrix} \tag{3}$$

This produced the desired zero in column 1. It changed equation 2. To produce another zero, *we subtract row* 1 *from row* 3 *using* E_{31}. This completes elimination in column 1:

$$E_{31} = \begin{bmatrix} 1 & 0 & 0 \\ 0 & 1 & 0 \\ \mathbf{-1} & 0 & 1 \end{bmatrix} \qquad \boldsymbol{E_{31}E_{21}A} = \begin{bmatrix} \mathbf{2} & 3 & 4 \\ \mathbf{0} & \mathbf{5} & 6 \\ \mathbf{0} & 5 & 13 \end{bmatrix} \qquad \begin{bmatrix} 19 \\ 17 \\ \mathbf{31} \end{bmatrix} \tag{4}$$

Move now to column 2 and row 2 (the second pivot row). The pivot is 5, on the diagonal. To eliminate the 5 below it, multiply row 2 by the number 1 and subtract from row 3.

$$E_{32} = \begin{bmatrix} 1 & 0 & 0 \\ 0 & 1 & 0 \\ 0 & \mathbf{-1} & 1 \end{bmatrix} \qquad U = \begin{bmatrix} \mathbf{2} & 3 & 4 \\ \mathbf{0} & \mathbf{5} & 6 \\ \mathbf{0} & \mathbf{0} & \mathbf{7} \end{bmatrix} \qquad \boldsymbol{c} = \begin{bmatrix} 19 \\ 17 \\ 14 \end{bmatrix} \tag{5}$$

$E_{32}E_{31}E_{21}A = U$ is triangular. $\boldsymbol{x} = (\mathbf{4, 1, 2})$ solved $U\boldsymbol{x} = \boldsymbol{c}$ on page 41 and $\boldsymbol{x} = (\mathbf{4, 1, 2})$ solves $A\boldsymbol{x} = \boldsymbol{b}$ here. Since U has $2, 5, 7$ on its diagonal we know that back substitution will succeed. The columns of U are independent (and therefore the columns of the original A were independent, as we will see). The matrices A and U have full rank.

We can summarize the elimination steps when no row exchanges are involved.

> Use the first equation to produce zeros in column 1 below the first pivot.
>
> Use the new second equation to clear out column 2 below pivot 2 in row 2.
>
> *Continue to column* 3. The expected result is an upper triangular matrix U.

Elimination on A produces U. **The same steps were applied to the right hand side $\boldsymbol{b}$.** Those steps produce a new right hand side $\boldsymbol{c}$. The new equations $U\boldsymbol{x} = \boldsymbol{c}$ (equivalent to the old equations $A\boldsymbol{x} = \boldsymbol{b}$) are solved by back substitution (previous page): $\boldsymbol{x} = (\mathbf{4, 1, 2})$.

Possible Breakdown of Elimination

Elimination might fail. *Zero can appear in a pivot position.* Subtracting that zero from lower rows will not clear out the column below the unwanted zero. Here is an example:

Zero in pivot 2 from elimination in column 1
$$A = \begin{bmatrix} 2 & 3 & 4 \\ 4 & \mathbf{6} & \mathbf{14} \\ 2 & \mathbf{8} & \mathbf{17} \end{bmatrix} \rightarrow \begin{bmatrix} 2 & 3 & 4 \\ 0 & \mathbf{0} & \mathbf{6} \\ 0 & \mathbf{5} & \mathbf{13} \end{bmatrix} = B$$

The cure is simple if it works. **Exchange row 2 with the zero for row 3 with the 5.** Then the second pivot is **5** and we can clear out the second column below that pivot. Elimination continues to U as normal after the row exchange by the matrix P.

Row exchange Successful
$$PB = \begin{bmatrix} 1 & 0 & 0 \\ 0 & 0 & 1 \\ 0 & 1 & 0 \end{bmatrix} \begin{bmatrix} 2 & 3 & 4 \\ 0 & \mathbf{0} & \mathbf{6} \\ 0 & \mathbf{5} & \mathbf{13} \end{bmatrix} = \begin{bmatrix} 2 & 3 & 4 \\ 0 & \mathbf{5} & \mathbf{13} \\ 0 & \mathbf{0} & \mathbf{6} \end{bmatrix}$$

For this small example, the row exchange is all we need. It produced U with nonzero pivots $2, 5, 6$. Normally there are more columns and rows to work on, before we reach U.

Caution! That row exchange was a success. This is what we hope for, to reach U with no zeros on its main diagonal. (The pivots $2, 5, 6$ are on the diagonal.) But a slightly different matrix A^* would lead to a bad situation: **no pivot is available in column 2.**

Dependent columns
U^* is not invertible
A^* is not invertible
$$\rightarrow \quad A^* = \begin{bmatrix} 2 & 3 & 4 \\ 4 & \mathbf{6} & 14 \\ 2 & \mathbf{3} & 17 \end{bmatrix} \rightarrow \begin{bmatrix} 2 & 3 & 4 \\ 0 & \mathbf{0} & 6 \\ 0 & \mathbf{0} & 13 \end{bmatrix} = \boldsymbol{U^*} \quad (6)$$

At this point elimination is helpless in column 2. *No second pivot.* This misfortune tells us that **the matrix A^* did not have full rank**. Column 2 of U^* is in the same direction as column 1 of U^*. Column 2 of A^* is in the same direction as column 1 of A^*.

You see how dependent columns are systematically identified by elimination. They can't escape a zero in the pivot. Then there will be nonzero solutions $\boldsymbol{X}$ to $A^*\boldsymbol{X} = \mathbf{0}$. The columns of U^* (and A^*) are not independent.

This example has column 2 $= \frac{3}{2}$ column 1. The solution vector $\boldsymbol{X}$ is $(3, -2, 0)$. *The equation $A^*\boldsymbol{x} = \boldsymbol{b}$ may or may not be solvable, depending on $\boldsymbol{b}$*: probably not.

Dependent or Independent Columns

This A^* looks like a failure of elimination: No second pivot. But it was a success because the problem was identified: dependent columns. The beauty of aiming for a triangular matrix U or U^* is that the diagonal entries tell us everything.

A triangular matrix U has full rank exactly when its main diagonal has no zeros.

In that case (square matrix with nonzero pivots) the columns of U are independent. Also the rows are independent. We can see this directly because elimination has simplified the original A to the triangular U.

How do we know that a zero on the diagonal of U^* leads to dependent columns?

$$\boldsymbol{U}^* = \begin{bmatrix} * & * & * & * \\ 0 & * & * & * \\ 0 & 0 & \mathbf{0} & * \\ 0 & 0 & 0 & * \end{bmatrix} =$$

Upper triangular with an extra zero on its diagonal
This matrix is **singular** (not full rank 4) (no inverse)
The first three columns are dependent
The last two rows are dependent

The Row Picture and the Column Picture

The next pictures will show the three possibilities for $A\boldsymbol{x} = \boldsymbol{b}$: **No solution or a line of solutions or one solution**. There are two ways to see this. We start with the *rows of* A and we graph the two equations: **the row picture**. We have trouble if the lines don't meet.

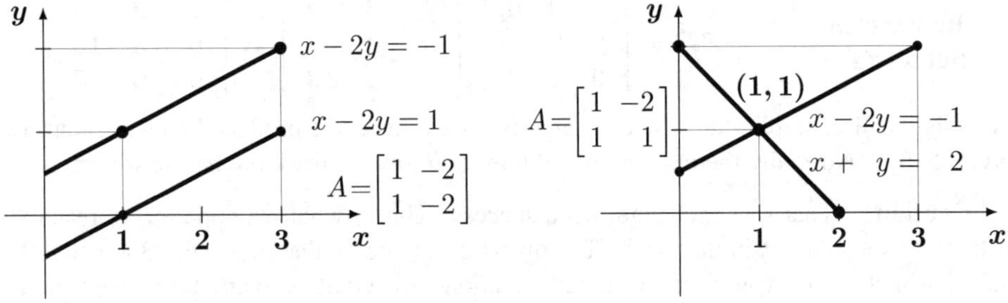

Figure 2.1: Parallel lines mean **no solution**. One line twice means **a line of solutions**. Intersecting lines give **one solution**. The solution is where the lines meet.

If we had three equations for x, y, and z, those two lines would change to three planes. The three planes meet at a single point in 3-dimensional space. This row picture becomes hard to draw. The column picture is much easier in three or more dimensions.

The **column picture** just shows column vectors: columns of A and also the vector $\boldsymbol{b}$. We are not looking for points where these vectors meet. The goal of $A\boldsymbol{x} = \boldsymbol{b}$ is to **combine the columns of $\boldsymbol{A}$** so as to produce the vector $\boldsymbol{b}$.

This is always possible when the columns of A (n vectors in n-dimensional space) are *independent*. Then the column space of A contains all vectors $\boldsymbol{b}$ in $\mathbf{R}^n$. There is exactly one combination $A\boldsymbol{x}$ of the columns that equals $\boldsymbol{b}$. Elimination finds that solution $\boldsymbol{x}$.

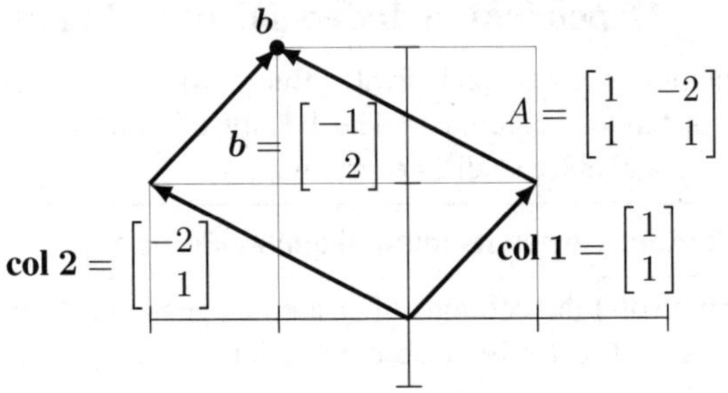

The columns of A are independent
Column 1 + Column 2 = $\boldsymbol{b}$
Then the solution is $x_1 = \mathbf{1}, x_2 = \mathbf{1}$
Construct $\boldsymbol{b}$ from the columns!
The bottom point is $(0, 0)$

Figure 2.2: **Column picture**. The vector $\boldsymbol{b}$ is a combination $A\boldsymbol{x}$ of the columns of A.

Examples of Elimination and Permutation

This chapter will go on to express the whole process using matrices. An elimination matrix E will act on $A\boldsymbol{x} = \boldsymbol{b}$. In case zero appears in a pivot position, use a permutation matrix P. The final result is an upper triangular U and a new right hand side $\boldsymbol{c}$. Then $U\boldsymbol{x} = \boldsymbol{c}$ is solved by back substitution.

In reality a computer takes those steps ($\boldsymbol{x} = A\backslash\boldsymbol{b}$ in MATLAB). But it is good to solve a few examples—*not too many*—by hand. You see the steps to $U\boldsymbol{x} = \boldsymbol{c}$ and then to the solution $\boldsymbol{x}$. This page contains a variety of examples, hopefully to show the way.

$$\boldsymbol{A} = \begin{bmatrix} 2 & 4 & -2 \\ 4 & 9 & -3 \\ -2 & -3 & 7 \end{bmatrix} \xrightarrow{E_{21}} \begin{bmatrix} 2 & 4 & -2 \\ \mathbf{0} & 1 & 1 \\ -2 & -3 & 7 \end{bmatrix} \xrightarrow{E_{31}} \begin{bmatrix} 2 & 4 & -2 \\ \mathbf{0} & 1 & 1 \\ \mathbf{0} & 1 & 5 \end{bmatrix} \xrightarrow{E_{32}} \begin{bmatrix} 2 & 4 & -2 \\ \mathbf{0} & 1 & 1 \\ \mathbf{0} & \mathbf{0} & 4 \end{bmatrix} = \boldsymbol{U}$$

Those elimination steps E_{21} and E_{31} and E_{32} produced zeros in positions $(2,1)$ and $(3,1)$ and $(3,2)$. The matrices E have -2 and $+1$ and -1 in those positions. The same steps E_{21}, E_{31}, E_{32} must be applied to the right hand side $\boldsymbol{b}$, to keep the equations correct.

$$\boldsymbol{b} = \begin{bmatrix} 2 \\ 8 \\ 10 \end{bmatrix} \to E_{21}\boldsymbol{b} = \begin{bmatrix} 2 \\ 4 \\ 10 \end{bmatrix} \to E_{31}E_{21}\boldsymbol{b} = \begin{bmatrix} 2 \\ 4 \\ 12 \end{bmatrix} \to E_{32}E_{31}E_{21}\boldsymbol{b} = E\boldsymbol{b} = \boldsymbol{c} = \begin{bmatrix} 2 \\ 4 \\ 8 \end{bmatrix}.$$

There is a simple way to make sure that operations on the matrix A (left side of equations) are also executed on $\boldsymbol{b}$ (right side of equations). The good way is to **include $\boldsymbol{b}$ as an extra column with $\boldsymbol{A}$**. The combination $\begin{bmatrix} A & \boldsymbol{b} \end{bmatrix}$ is called an **augmented matrix**.

$$\begin{bmatrix} A & \boldsymbol{b} \end{bmatrix} = \begin{bmatrix} 2 & 4 & -2 & \mathbf{2} \\ 4 & 9 & -3 & \mathbf{8} \\ -2 & -3 & 7 & \mathbf{10} \end{bmatrix} \xrightarrow{E} \begin{bmatrix} 2 & 4 & -2 & \mathbf{2} \\ \mathbf{0} & 1 & 1 & \mathbf{4} \\ \mathbf{0} & \mathbf{0} & 4 & \mathbf{8} \end{bmatrix} = \begin{bmatrix} U & \boldsymbol{c} \end{bmatrix}. \quad (7)$$

Now we include an example that requires a permutation matrix P. It will exchange equations and avoid zero in the pivot. The new matrix A needs P to improve column 2.

Exchange rows 2 and 3
$$A = \begin{bmatrix} 1 & 1 & 1 \\ 2 & 2 & 3 \\ 0 & 4 & 5 \end{bmatrix} \xrightarrow{E} \begin{bmatrix} 1 & 1 & 1 \\ 0 & \mathbf{0} & 1 \\ 0 & \mathbf{4} & 5 \end{bmatrix} \xrightarrow{\boldsymbol{P}} \begin{bmatrix} 1 & 1 & 1 \\ 0 & \mathbf{4} & 5 \\ 0 & \mathbf{0} & 1 \end{bmatrix} = U$$

That permutation P_{23} exchanged rows 2 and 3 when it was needed to avoid a zero pivot. But we could have exchanged rows 2 and 3 at the start. Then E_{21} and E_{31} change places.

In the final description $PA = LU$ of elimination on A, all the E's will be moved to the right side. **Each matrix in $\boldsymbol{E_{32}E_{31}E_{21}}$ is inverted**. Those inverses come in **reverse order $\boldsymbol{L} = E_{21}^{-1}E_{31}^{-1}E_{32}^{-1}$**. The overall equation is $\boldsymbol{PA = LU}$. Often no permutations are needed and elimination produces $\boldsymbol{A = LU}$: the best equation of all, in Section 2.2.

Section 2.4 will return to understand all the possible permutations of n rows. *There are $n\,!$ permutation matrices P, including $P = I$ for no row exchanges.*

Problem Set 2.1

Problems 1–10 are about elimination on 2 by 2 systems.

1 What multiple ℓ_{21} of equation 1 should be subtracted from equation 2 ?

$$\begin{aligned} 2x + 3y &= 1 \\ 10x + 9y &= 11. \end{aligned}$$

After elimination, write down the upper triangular system and circle the two pivots. Use back substitution to find x and y (and check that solution).

2 If equation 1 is added to equation 2, which of these are changed: the lines in the row picture, the vectors in the column picture, the coefficient matrix, the solution ?

3 What multiple of equation 1 should be *subtracted* from equation 2 ?

$$\begin{aligned} 2x - 4y &= 6 \\ -x + 5y &= 0. \end{aligned}$$

After this elimination step, solve the triangular system. If the right side changes to $(-6, 0)$, what is the new solution?

4 What multiple ℓ of equation 1 should be subtracted from equation 2 to remove c ?

$$\begin{aligned} ax + by &= f \\ cx + dy &= g. \end{aligned}$$

The first pivot is a (assumed nonzero). Elimination produces what formula for the second pivot ? What is y ? The second pivot is missing when $ad = bc$: singular.

5 Choose a right side which gives no solution and another right side which gives infinitely many solutions. What are two of those solutions ?

Singular system $\qquad 3x + 2y = 10 \quad \text{and} \quad 6x + 4y =$ ____

6 Choose a coefficient b that makes this system singular. Then choose a right side g that makes it solvable. Find two solutions in that singular but solvable case.

$$\begin{aligned} 2x + by &= 16 \\ 4x + 8y &= g. \end{aligned}$$

7 For which numbers a does elimination break down (1) permanently (2) temporarily ? Solve for x and y after fixing the temporary breakdown by a row exchange.

$$\begin{aligned} ax + 3y &= -3 \\ 4x + 6y &= 6. \end{aligned}$$

8 For which three numbers k does elimination break down ? Which is fixed by a row exchange ? Is the number of solutions 0 or 1 or ∞ ? **Draw the 3 row pictures.**

3 pictures from
3 particular k's
$$\begin{aligned} kx + 3y &= 6 \\ 3x + ky &= -6. \end{aligned}$$

9 What test on b_1 and b_2 decides whether these two equations allow a solution? How many solutions will they have? Draw the column pictures for $\boldsymbol{b} = (1, 2)$ and $(1, 0)$.

$$\begin{aligned} 3x - 2y &= b_1 \\ 6x - 4y &= b_2. \end{aligned}$$

10 Draw the lines $x+y = 5$ and $x+2y = 6$ and the equation $y=$ _____ that comes from elimination. Which line $5x - 4y = c$ goes through the solution of these equations?

Problems 11–20 study elimination on 3 by 3 systems (and possible failure).

11 (Recommended) A system of linear equations *can't have* exactly two solutions.

(a) If (x, y, z) and (X, Y, Z) are two solutions, what is another solution?

(b) If 25 planes meet at two points, where else do they meet?

12 Reduce to upper triangular form by row operations. Then find x, y, z.

$$\begin{aligned} 2x + 3y + z &= 8 \\ 4x + 7y + 5z &= 20 \\ -2y + 2z &= 0 \end{aligned} \qquad\qquad \begin{aligned} 2x - 3y \quad &= 3 \\ 4x - 5y + z &= 7 \\ 2x - y - 3z &= 5 \end{aligned}$$

13 Which number d forces a row exchange, and what is the triangular system (not singular) for that d? Which d makes this system singular (no third pivot)?

$$\begin{aligned} 2x + 5y + z &= 0 \\ 4x + dy + z &= 2 \\ y - z &= 3. \end{aligned}$$

14 Which number b leads later to a row exchange? Which b leads to a missing pivot? In that singular case find a nonzero solution x, y, z.

$$\begin{aligned} x + by \quad &= 0 \\ x - 2y - z &= 0 \\ y + z &= 0. \end{aligned}$$

15 (a) Construct a 3 by 3 system that needs 2 row exchanges to become triangular.
(b) Construct a system that needs a row exchange and breaks down later.

16 If rows 1 and 2 are the same, how far can you get with elimination (allowing row exchange)? If columns 1 and 2 are the same, which pivot is missing?

$$\textbf{Equal rows}\quad \begin{aligned} 2x - y + z &= 0 \\ 2x - y + z &= 0 \\ 4x + y + z &= 2 \end{aligned} \qquad\qquad \begin{aligned} 2x + 2y + z &= 0 \\ 4x + 4y + z &= 0 \\ 6x + 6y + z &= 2. \end{aligned}\quad \textbf{Equal columns}$$

17 Construct a 3 by 3 example that has 9 different coefficients on the left side, but rows 2 and 3 become zero in elimination. How many solutions to your system with $\boldsymbol{b} = (1, 10, 100)$ and how many with $\boldsymbol{b} = (0, 0, 0)$?

18 Which number q makes this system singular and which right side t gives it infinitely many solutions? Find the solution that has $z = 1$.

$$\begin{aligned} x + 4y - 2z &= 1 \\ x + 7y - 6z &= 6 \\ 3y + qz &= t. \end{aligned}$$

19 For which two numbers a will elimination fail on $A = \begin{bmatrix} a & 2 \\ a & a \end{bmatrix}$?

20 For which three numbers a will elimination fail to give three pivots?

$$A = \begin{bmatrix} a & 2 & 3 \\ a & a & 4 \\ a & a & a \end{bmatrix} \text{ is singular for three values of } a.$$

21 Look for a matrix that has row sums 4 and 8, and column sums 2 and s: The four equations below are solvable only if $s =$ _____. Find two matrices with the correct row and column sums. Write down the 4 by 4 system $\boldsymbol{Ax} = \boldsymbol{b}$ with $\boldsymbol{x} = (a, b, c, d)$ and make A triangular by elimination.

$$\text{Matrix} \begin{bmatrix} a & b \\ c & d \end{bmatrix} \qquad \begin{matrix} a + b = 4 & a + c = 2 \\ c + d = 8 & b + d = s \end{matrix} \qquad \begin{matrix} \text{4 equations} \\ \text{4 unknowns} \end{matrix}$$

22 Create a MATLAB command `A(2, : ) = ...` for the new row 2, to subtract 3 times row 1 from the existing row 2 if the 3 by 3 matrix A is already known.

23 Find experimentally the average 1st and 2nd and 3rd pivot sizes from MATLAB's $[L, U] = \mathbf{lu}\,(\mathbf{rand}\,(3))$ with random entries between 0 and 1. The average of $U(1, 1)$ is above $\frac{1}{2}$ because **lu** picks the largest pivot in column 1.

24 If the last corner entry is $A(5, 5) = 11$ and the last pivot of A is $U(5, 5) = 4$, what different entry $A(5, 5)$ would have made A singular?

25 Suppose elimination takes A to U without row exchanges. Then row j of U is a combination of which rows of A? If $\boldsymbol{Ax} = \mathbf{0}$, is $\boldsymbol{Ux} = \mathbf{0}$? If $\boldsymbol{Ax} = \boldsymbol{b}$, is $\boldsymbol{Ux} = \boldsymbol{b}$? If A starts out lower triangular, what is the upper triangular U?

26 Start with 100 equations $\boldsymbol{Ax} = \mathbf{0}$ for 100 unknowns $\boldsymbol{x} = (x_1, \ldots, x_{100})$. Suppose elimination reduces the 100th equation to $0 = 0$, so the system is "singular".

(a) Singular systems $\boldsymbol{Ax} = \mathbf{0}$ have infinitely many solutions. This means that some linear combination of the 100 ***columns*** of A is _____.

(b) Invent a 100 by 100 singular matrix with no zero entries.

(c) Describe in words the row picture and column picture of your $\boldsymbol{Ax} = \mathbf{0}$.

2.2 Elimination Matrices and Inverse Matrices

1 Elimination multiplies A by $E_{21}, \ldots, E_{n1}$ then $E_{32}, \ldots, E_{n2}$ as A becomes $\boldsymbol{EA = U}$.

2 In reverse order, the inverses of the E's multiply U to recover $A = E^{-1}U$. **This is $\boldsymbol{A = LU}$.**

3 $\boldsymbol{A^{-1}A = I}$ and $\boldsymbol{(LU)^{-1} = U^{-1}L^{-1}}$. Then $A\boldsymbol{x} = \boldsymbol{b}$ becomes $\boldsymbol{x} = A^{-1}\boldsymbol{b} = \boldsymbol{U^{-1}L^{-1}b}$.

All the steps of elimination can be done with matrices. Those steps can also be *undone* (inverted) with matrices. For a 3 by 3 matrix we can write out each step in detail—almost word for word. But for real applications, matrices are a much better way.

The basic elimination step subtracts a multiple $\boldsymbol{\ell_{ij}}$ of equation $\boldsymbol{j}$ from equation $\boldsymbol{i}$. We always speak about *subtractions* as elimination proceeds. If the first pivot is $a_{11} = 3$ and below it is $a_{21} = -3$, we could just add equation 1 to equation 2. That produces zero. But we stay with subtraction: *subtract* $\ell_{21} = -1$ *times equation* 1 *from equation* 2. Same result. The inverse step is addition. Equation (10) to (11) at the end shows it all.

Here is the matrix that subtracts 2 times row 1 from row 3: Rows 1 and 2 stay the same.

Elimination matrix $\boldsymbol{E_{ij} = E_{31}}$
Row 3, column 1, multiplier 2

$$E_{31} = \begin{bmatrix} \mathbf{1} & 0 & 0 \\ 0 & \mathbf{1} & 0 \\ \mathbf{-2} & 0 & \mathbf{1} \end{bmatrix}$$

If no row exchanges are needed, then three elimination matrices E_{21} and E_{31} and E_{32} will produce three zeros below the diagonal. This changes A to the triangular U:

$$\boxed{\boldsymbol{E = E_{32}E_{31}E_{21}} \qquad \boldsymbol{EA = U} \textbf{ is upper triangular}} \tag{1}$$

The number ℓ_{32} is affected by the ℓ_{21} and ℓ_{31} that came first. We subtract ℓ_{32} times *row* 2 *of* U (the final second row, not the original second row of A). This is the E_{32} step that produces zero in row 3, column 2 of U. E_{32} gives the last step of 3 by 3 elimination.

Example 1 E_{21} and then E_{31} subtract multiples of row 1 from rows 2 and 3 of A:

$$E_{31}E_{21}A = \begin{bmatrix} 1 & 0 & 0 \\ 0 & 1 & 0 \\ \mathbf{-2} & 0 & 1 \end{bmatrix} \begin{bmatrix} 1 & 0 & 0 \\ \mathbf{1} & 1 & 0 \\ 0 & 0 & 1 \end{bmatrix} \begin{bmatrix} \mathbf{3} & 1 & 0 \\ \mathbf{-3} & 1 & 1 \\ \mathbf{6} & 8 & 4 \end{bmatrix} = \begin{bmatrix} \mathbf{3} & 1 & 0 \\ \mathbf{0} & 2 & 1 \\ \mathbf{0} & 6 & 4 \end{bmatrix} \quad \begin{matrix} \textbf{two new} \\ \textbf{zeros in} \\ \textbf{column 1} \end{matrix} \tag{2}$$

To produce a zero in column 2, E_{32} subtracts $\ell_{32} = 3$ times the **new row 2** from row 3:

$$(E_{32})\,(E_{31}E_{21}A) = \begin{bmatrix} 1 & 0 & 0 \\ 0 & 1 & 0 \\ 0 & \mathbf{-3} & 1 \end{bmatrix} \begin{bmatrix} 3 & 1 & 0 \\ 0 & \mathbf{2} & 1 \\ 0 & \mathbf{6} & 4 \end{bmatrix} = \begin{bmatrix} 3 & 1 & 0 \\ \mathbf{0} & \mathbf{2} & 1 \\ \mathbf{0} & \mathbf{0} & \mathbf{1} \end{bmatrix} = \boldsymbol{U} \quad \begin{matrix} \boldsymbol{U} \textbf{ has zeros} \\ \textbf{below the} \\ \textbf{main diagonal} \end{matrix} \tag{3}$$

Notice again: E_{32} is subtracting 3 times the row $0, 2, 1$ and not the original row of A. At the end, the pivots $3, 2, 1$ are on the main diagonal of U: zeros below that diagonal.

The **inverse** of each matrix E_{ij} **adds back $\boldsymbol{\ell_{ij}}$(row $\boldsymbol{j}$)** to row i. This leads to the inverse of their product $E = E_{32}E_{31}E_{21}$. That inverse of E is special. *We call it* L.

The Facts About Inverse Matrices

Suppose A is a square matrix. We look for an "***inverse matrix***" A^{-1} of the same size, so that $\boldsymbol{A^{-1}}$ **times** $\boldsymbol{A}$ **equals** $\boldsymbol{I}$. Whatever A does, A^{-1} undoes. Their product is the identity matrix—which does nothing to a vector, so $A^{-1}Ax = x$. *But A^{-1} might not exist.*

The n by n matrix A needs n independent columns to be invertible. Then $A^{-1}A = I$.

What a matrix mostly does is to multiply a vector. Multiplying $Ax = b$ by A^{-1} gives $A^{-1}Ax = A^{-1}b$. ***This is*** $\boldsymbol{x = A^{-1}b}$. The product $A^{-1}A$ is like multiplying by a number and then dividing by that number. Numbers have inverses if they are not zero. Matrices are more complicated and interesting. The matrix A^{-1} is called **"*A* inverse"**.

> **DEFINITION** The matrix A is ***invertible*** if there exists a matrix A^{-1} that "inverts" A:
>
> **Two-sided inverse** $\quad \boldsymbol{A^{-1}A = I}$ and $\boldsymbol{AA^{-1} = I}$. $\quad$ (4)

Not all matrices have inverses. This is the first question we ask about a square matrix: Is A invertible? Its columns must be independent. We don't mean that we actually calculate A^{-1}. In most problems we never compute it! Here are seven "notes" about A^{-1}.

Note 1 ***The inverse exists if and only if elimination produces n pivots*** (row exchanges are allowed). Elimination solves $Ax = b$ without explicitly using the matrix A^{-1}.

Note 2 The matrix A cannot have two different inverses. Suppose $BA = I$ and also $AC = I$. Then $B = C$, according to this "proof by parentheses" = associative law.

$$B(AC) = (BA)C \quad \text{gives} \quad BI = IC \quad \text{or} \quad \boldsymbol{B = C}. \tag{5}$$

This shows that a *left inverse* B (multiplying A from the left) and a *right inverse* C (multiplying A from the right to give $AC = I$) must be the *same matrix*.

Note 3 If A is invertible, the one and only solution to $Ax = b$ is $x = A^{-1}b$:

$$\boxed{\textbf{\textit{Multiply}} \quad Ax = b \quad \textbf{\textit{by}} \quad \boldsymbol{A^{-1}}. \quad \textbf{\textit{Then}} \quad x = A^{-1}Ax = \boldsymbol{A^{-1}b}.}$$

Note 4 (Important) ***Suppose there is a nonzero vector x such that*** $Ax = \mathbf{0}$. *Then A has dependent columns. It cannot have an inverse.* No matrix can bring $\mathbf{0}$ back to x.

If A is invertible, then $Ax = \mathbf{0}$ only has the zero solution $x = A^{-1}\mathbf{0} = \mathbf{0}$.

Note 5 A square matrix is invertible if and only if its columns are independent.

Note 6 A 2 by 2 matrix is invertible if and only if the number $ad - bc$ is not zero:

$$\textbf{2 by 2 Inverse} \qquad \begin{bmatrix} a & b \\ c & d \end{bmatrix}^{-1} = \frac{1}{ad-bc}\begin{bmatrix} d & -b \\ -c & a \end{bmatrix}. \tag{6}$$

This number $ad - bc$ is the **determinant** of A. A matrix is invertible if its determinant is not zero (Chapter 5). The test for n pivots is usually decided before the determinant appears.

Note 7 A triangular matrix has an inverse provided no diagonal entries d_i are zero:

$$\text{If} \quad A = \begin{bmatrix} \boldsymbol{d_1} & \times & \times & \times \\ 0 & \bullet & \times & \times \\ 0 & 0 & \bullet & \times \\ 0 & 0 & 0 & \boldsymbol{d_n} \end{bmatrix} \quad \text{then} \quad A^{-1} = \begin{bmatrix} \boldsymbol{1/d_1} & \times & \times & \times \\ 0 & \bullet & \times & \times \\ 0 & 0 & \bullet & \times \\ 0 & 0 & 0 & \boldsymbol{1/d_n} \end{bmatrix}$$

Example 2 The 2 by 2 matrix $A = \left[\begin{smallmatrix} 1 & 2 \\ 1 & 2 \end{smallmatrix}\right]$ is not invertible. It fails the test in Note 6, because $ad = bc$. It also fails the test in Note 4, because $A\boldsymbol{x} = \mathbf{0}$ when $\boldsymbol{x} = (2, -1)$. It fails to have two pivots as required by Note 1. Its columns are clearly dependent.

Elimination turns the second row of this matrix A into a zero row. No pivot.

Example 3 Three of these matrices are invertible, and three are singular. Find the inverse when it exists. Give reasons for noninvertibility (zero determinant, too few pivots, nonzero solution to $A\boldsymbol{x} = \mathbf{0}$) for the other three. The matrices are in the order $\boldsymbol{A}, \boldsymbol{B}, \boldsymbol{C}, \boldsymbol{D}, \boldsymbol{S}, \boldsymbol{T}$:

$$\begin{bmatrix} 4 & 3 \\ 8 & 6 \end{bmatrix} \quad \begin{bmatrix} 4 & 3 \\ 8 & 7 \end{bmatrix} \quad \begin{bmatrix} 6 & 6 \\ 6 & 0 \end{bmatrix} \quad \begin{bmatrix} 6 & 6 \\ 6 & 6 \end{bmatrix} \quad \begin{bmatrix} 1 & 0 & 0 \\ 1 & 1 & 0 \\ 1 & 1 & 1 \end{bmatrix} \quad \begin{bmatrix} 1 & 1 & 1 \\ 1 & 1 & 0 \\ 1 & 1 & 1 \end{bmatrix}$$

Solution The three matrices with inverses are $\boldsymbol{B}, \boldsymbol{C}, \boldsymbol{S}$:

$$\boldsymbol{B}^{-1} = \frac{1}{4}\begin{bmatrix} 7 & -3 \\ -8 & 4 \end{bmatrix} \quad \boldsymbol{C}^{-1} = \frac{1}{36}\begin{bmatrix} 0 & 6 \\ 6 & -6 \end{bmatrix} \quad \boldsymbol{S}^{-1} = \begin{bmatrix} 1 & 0 & 0 \\ -1 & 1 & 0 \\ 0 & -1 & 1 \end{bmatrix}$$

A is not invertible because its determinant is $4 \cdot 6 - 3 \cdot 8 = 24 - 24 = 0$. D is not invertible because it has only one pivot; row 2 becomes zero when row 1 is subtracted. T has two equal rows (and the second column minus the first column is zero). In other words $T\boldsymbol{x} = \mathbf{0}$ has the nonzero solution $\boldsymbol{x} = (-1, 1, 0)$. *Not invertible.*

The Inverse of a Product AB

For two nonzero numbers a and b, the sum $a + b$ might or might not be invertible. The numbers $a = 3$ and $b = -3$ have inverses $\frac{1}{3}$ and $-\frac{1}{3}$. Their sum $a + b = 0$ has no inverse. But the product $ab = -9$ does have an inverse, which is $\frac{1}{3}$ times $-\frac{1}{3}$.

For matrices A and B, the situation is similar. Their *product* AB has an inverse if and only if A and B are separately invertible (and the same size). The important point is that $\boldsymbol{A^{-1}}$ **and** $\boldsymbol{B^{-1}}$ **come in reverse order**:

> **If A and B are invertible (same size) then the inverse of AB is $B^{-1}A^{-1}$.**
>
> $$(\boldsymbol{AB})^{-1} = \boldsymbol{B}^{-1}\boldsymbol{A}^{-1} \qquad (\boldsymbol{AB})(\boldsymbol{B}^{-1}\boldsymbol{A}^{-1}) = \boldsymbol{AIA}^{-1} = \boldsymbol{AA}^{-1} = \boldsymbol{I} \quad (7)$$

We moved parentheses to multiply BB^{-1} first. Similarly $B^{-1}A^{-1}$ times AB equals I.

$B^{-1}A^{-1}$ illustrates a basic rule of mathematics: Inverses come in reverse order. It is also common sense: If you put on socks and then shoes, the first to be taken off are the _____. The same reverse order applies to three or more matrices:

$$\textbf{Reverse order} \qquad (ABC)^{-1} = C^{-1}B^{-1}A^{-1} \tag{8}$$

Example 4 ***Inverse of an elimination matrix***. If E subtracts 5 times row 1 from row 2, then E^{-1} *adds* 5 times row 1 to row 2:

$$\begin{array}{c}\textbf{\textit{E} subtracts}\\ \boldsymbol{E^{-1}}\textbf{ adds}\end{array} \qquad E = \begin{bmatrix} 1 & 0 & 0 \\ \mathbf{-5} & 1 & 0 \\ 0 & 0 & 1 \end{bmatrix} \quad \text{and} \quad E^{-1} = \begin{bmatrix} 1 & 0 & 0 \\ \mathbf{5} & 1 & 0 \\ 0 & 0 & 1 \end{bmatrix}$$

Multiply EE^{-1} to get the identity matrix I. Also multiply $E^{-1}E$ to get I. We are adding and subtracting the same 5 times row 1. If $AC = I$ then for square matrices $CA = I$.

For square matrices, an inverse on one side is automatically an inverse on the other side.

Example 5 Suppose F subtracts 4 times row 2 from row 3, and F^{-1} adds it back:

$$F = \begin{bmatrix} 1 & 0 & 0 \\ 0 & 1 & 0 \\ 0 & -4 & 1 \end{bmatrix} \quad \text{and} \quad F^{-1} = \begin{bmatrix} 1 & 0 & 0 \\ 0 & 1 & 0 \\ 0 & 4 & 1 \end{bmatrix}.$$

Now multiply F by the matrix E in Example 4 to find FE. Also multiply E^{-1} times F^{-1} to find $(FE)^{-1}$. Notice the orders FE and $E^{-1}F^{-1}$!

$$FE = \begin{bmatrix} 1 & 0 & 0 \\ \mathbf{-5} & 1 & 0 \\ \mathbf{20} & \mathbf{-4} & 1 \end{bmatrix} \quad \textbf{is inverted by} \quad E^{-1}F^{-1} = \begin{bmatrix} 1 & 0 & 0 \\ \mathbf{5} & 1 & 0 \\ \mathbf{0} & \mathbf{4} & 1 \end{bmatrix}. \tag{9}$$

The result is beautiful and correct. The product FE contains "20" but its inverse doesn't. E subtracts 5 times row 1 from row 2. Then F subtracts 4 times the *new row* 2 (changed by row 1) from row 3. ***In this order FE, row 3 feels an effect of size 20 from row 1.***

In the order $E^{-1}F^{-1}$, that effect does not happen. First F^{-1} adds 4 times row 2 to row 3. After that, E^{-1} adds 5 times row 1 to row 2. There is no 20, because row 3 doesn't change again. ***In this order $E^{-1}F^{-1}$, row 3 feels no effect from row 1.***

This is why we choose $A = LU$, to go back from the triangular U to the original A. The multipliers fall into place perfectly in the lower triangular L: Equation (11) below.

The elimination order is FE. The inverse order is $\boldsymbol{L = E^{-1}F^{-1}}$.
The multipliers 5 and 4 fall into place below the diagonal of 1's in L.

L is the Inverse of E

E is the product of all the elimination matrices E_{ij}, taking A into its upper triangular form $EA = U$. We are assuming for now that no row exchanges are involved ($P = I$). The difficulty with E is that multiplying all the separate elimination steps E_{ij} does not produce a good formula. But the inverse matrix E^{-1} becomes beautiful when we multiply the inverse steps E_{ij}^{-1}. Remember that those steps come in the *opposite order*.

With $n = 3$, the complication for $E = E_{32}E_{31}E_{21}$ is in the bottom left corner:

$$\boldsymbol{E} = \begin{bmatrix} 1 & & \\ 0 & 1 & \\ 0 & -\ell_{32} & 1 \end{bmatrix} \begin{bmatrix} 1 & & \\ 0 & 1 & \\ -\ell_{31} & 0 & 1 \end{bmatrix} \begin{bmatrix} 1 & & \\ -\ell_{21} & 1 & \\ 0 & 0 & 1 \end{bmatrix} = \begin{bmatrix} 1 & & \\ -\ell_{21} & 1 & \\ (\boldsymbol{\ell_{32}\ell_{21}} - \boldsymbol{\ell_{31}}) & -\ell_{32} & 1 \end{bmatrix} \quad (10)$$

Watch how that confusion disappears for $\boldsymbol{E^{-1} = L}$. Reverse order is the good way:

$$\boldsymbol{E^{-1}} = \begin{bmatrix} 1 & & \\ \ell_{21} & 1 & \\ 0 & 0 & 1 \end{bmatrix} \begin{bmatrix} 1 & & \\ 0 & 1 & \\ \ell_{31} & 0 & 1 \end{bmatrix} \begin{bmatrix} 1 & & \\ 0 & 1 & \\ 0 & \ell_{32} & 1 \end{bmatrix} = \begin{bmatrix} 1 & & \\ \boldsymbol{\ell_{21}} & 1 & \\ \boldsymbol{\ell_{31}} & \boldsymbol{\ell_{32}} & 1 \end{bmatrix} = \boldsymbol{L} \quad (11)$$

All the multipliers $\boldsymbol{\ell_{ij}}$ appear in their correct positions in $\boldsymbol{L}$. The next section will show that this remains true for all matrix sizes. **Then $\boldsymbol{EA = U}$ becomes $\boldsymbol{A = LU}$.**

Equation (11) is the key to this chapter: Each $\boldsymbol{\ell_{ij}}$ is in its place for $\boldsymbol{E^{-1} = L}$.

Problem Set 2.2 (more questions than needed)

0 If you exchange columns 1 and 2 of an invertible matrix A, what is the effect on A^{-1}?

Problems 1–11 are about elimination matrices.

1 Write down the 3 by 3 matrices that produce these elimination steps:

(a) E_{21} subtracts 5 times row 1 from row 2.

(b) E_{32} subtracts -7 times row 2 from row 3.

(c) P exchanges rows 1 and 2, then rows 2 and 3.

2 In Problem 1, applying E_{21} and then E_{32} to $\boldsymbol{b} = (1, 0, 0)$ gives $E_{32}E_{21}\boldsymbol{b} =$ _____. Applying E_{32} before E_{21} gives $E_{21}E_{32}\boldsymbol{b} =$ _____. When E_{32} comes first, row _____ feels no effect from row _____.

3 Which three matrices E_{21}, E_{31}, E_{32} put A into triangular form U?

$$A = \begin{bmatrix} 1 & 1 & 0 \\ 4 & 6 & 1 \\ -2 & 2 & 0 \end{bmatrix} \quad \text{and} \quad E_{32}E_{31}E_{21}A = EA = U.$$

Multiply those E's to get one elimination matrix E. What is $E^{-1} = L$?

4 Include $\boldsymbol{b} = (1, 0, 0)$ as a fourth column in Problem 3 to produce $[\,A \;\; \boldsymbol{b}\,]$. Carry out the elimination steps on this augmented matrix to solve $A\boldsymbol{x} = \boldsymbol{b}$.

5 Suppose $a_{33} = 7$ and the third pivot is 5. If you change a_{33} to 11, the third pivot is ____. If you change a_{33} to ____, there is no third pivot.

6 If every column of A is a multiple of $(1, 1, 1)$, then $A\boldsymbol{x}$ is always a multiple of $(1, 1, 1)$. Do a 3 by 3 example. How many pivots are produced by elimination?

7 Suppose E subtracts 7 times row 1 from row 3.

(a) To *invert* that step you should ____ 7 times row ____ to row ____.

(b) What "inverse matrix" E^{-1} takes that reverse step (so $E^{-1}E = I$)?

(c) If the reverse step is applied first (and then E) show that $EE^{-1} = I$.

8 The ***determinant*** of $M = \left[\begin{smallmatrix} a & b \\ c & d \end{smallmatrix}\right]$ is det $M = ad - bc$. Subtract ℓ times row 1 from row 2 to produce a new M^*. Show that det $M^* =$ det M for every ℓ. When $\ell = c/a$, *the product of pivots equals the determinant*: $(a)(d - \ell b)$ equals $ad - bc$.

9 (a) E_{21} subtracts row 1 from row 2 and then P_{23} exchanges rows 2 and 3. What matrix $M = P_{23}E_{21}$ does both steps at once?

(b) P_{23} exchanges rows 2 and 3 and then E_{31} subtracts row 1 from row 3. What matrix $M = E_{31}P_{23}$ does both steps at once? Explain why the M's are the same but the E's are different.

10 (a) What matrix adds row 1 to row 3 and *at the same time* row 3 to row 1 ?

(b) What matrix adds row 1 to row 3 and *then* adds row 3 to row 1 ?

11 Create a matrix that has $a_{11} = a_{22} = a_{33} = 1$ but elimination produces two negative pivots without row exchanges. (The first pivot is 1.)

12 For these "permutation matrices" find P^{-1} by trial and error (with 1's and 0's) :

$$P = \begin{bmatrix} 0 & 0 & 1 \\ 0 & 1 & 0 \\ 1 & 0 & 0 \end{bmatrix} \quad \text{and} \quad P = \begin{bmatrix} 0 & 1 & 0 \\ 0 & 0 & 1 \\ 1 & 0 & 0 \end{bmatrix}.$$

13 Solve for the first column (x, y) and second column (t, z) of A^{-1}. Check AA^{-1}.

$$A = \begin{bmatrix} 10 & 20 \\ 20 & 50 \end{bmatrix} \qquad \begin{bmatrix} 10 & 20 \\ 20 & 50 \end{bmatrix}\begin{bmatrix} x \\ y \end{bmatrix} = \begin{bmatrix} 1 \\ 0 \end{bmatrix} \quad \text{and} \quad \begin{bmatrix} 10 & 20 \\ 20 & 50 \end{bmatrix}\begin{bmatrix} t \\ z \end{bmatrix} = \begin{bmatrix} 0 \\ 1 \end{bmatrix}.$$

14 Find an upper triangular U (not diagonal) with $U^2 = I$. Then $U^{-1} = U$.

15 (a) If A is invertible and $AB = AC$, prove quickly that $B = C$.

(b) If $A = \left[\begin{smallmatrix} 1 & 1 \\ 1 & 1 \end{smallmatrix}\right]$, find two different matrices such that $AB = AC$.

16 (Important) If A has row 1 + row 2 = row 3, show that A is not invertible:

(a) Explain why $A\boldsymbol{x} = (0, 0, 1)$ cannot have a solution. Add eqn 1 + eqn 2.

(b) Which right sides (b_1, b_2, b_3) might allow a solution to $A\boldsymbol{x} = \boldsymbol{b}$?

(c) In the elimination process, what happens to equation 3?

17 If A has column 1 + column 2 = column 3, show that A is not invertible:

(a) Find a nonzero solution $\boldsymbol{x}$ to $A\boldsymbol{x} = \boldsymbol{0}$. The matrix is 3 by 3.

(b) Elimination keeps columns $1 + 2 = 3$. Explain why there is no third pivot.

18 Suppose A is invertible and you exchange its first two rows to reach B. Is the new matrix B invertible? How would you find B^{-1} from A^{-1}?

19 (a) Find invertible matrices A and B such that $A + B$ is not invertible.

(b) Find singular matrices A and B such that $A + B$ is invertible.

20 If the product $C = AB$ is invertible (A and B are square), then A itself is invertible. Find a formula for A^{-1} that involves C^{-1} and B.

21 If the product $M = ABC$ of three square matrices is invertible, then B is invertible. (So are A and C.) Find a formula for B^{-1} that involves M^{-1} and A and C.

22 If you add row 1 of A to row 2 to get B, how do you find B^{-1} from A^{-1}?

23 Prove that a matrix with a column of zeros cannot have an inverse.

24 Multiply $\begin{bmatrix} \mathrm{a} & \mathrm{b} \\ \mathrm{c} & \mathrm{d} \end{bmatrix}$ times $\begin{bmatrix} \mathrm{d} & -\mathrm{b} \\ -\mathrm{c} & \mathrm{a} \end{bmatrix}$. What is the inverse of each matrix if $ad \neq bc$?

25 (a) What 3 by 3 matrix E has the same effect as these three steps? Subtract row 1 from row 2, subtract row 1 from row 3, then subtract row 2 from row 3.

(b) What single matrix L has the same effect as these three reverse steps? Add row 2 to row 3, add row 1 to row 3, then add row 1 to row 2.

26 If B is the inverse of A^2, show that AB is the inverse of A.

27 Show that $\mathsf{A} = 4 * \mathsf{eye}(4) - \mathsf{ones}(4,4)$ is *not* invertible: Multiply $\mathsf{A} * \mathsf{ones}(4,1)$.

28 There are sixteen 2 by 2 matrices whose entries are 1's and 0's. How many of them are invertible?

29 Change I into A^{-1} as elimination reduces A to I (the Gauss-Jordan idea).

$$\begin{bmatrix} A & I \end{bmatrix} = \begin{bmatrix} 1 & 3 & 1 & 0 \\ 2 & 7 & 0 & 1 \end{bmatrix} \quad \text{and} \quad \begin{bmatrix} A & I \end{bmatrix} = \begin{bmatrix} 1 & 4 & 1 & 0 \\ 3 & 9 & 0 & 1 \end{bmatrix}$$

30 Could a 4 by 4 matrix A be invertible if every row contains the numbers $0, 1, 2, 3$ in some order? What if every row of B contains $0, 1, 2, -3$ in some order?

31 Find A^{-1} and B^{-1} (*if they exist*) by elimination on $[\,A \;\; I\,]$ and $[\,B \;\; I\,]$:

$$A = \begin{bmatrix} 2 & 1 & 1 \\ 1 & 2 & 1 \\ 1 & 1 & 2 \end{bmatrix} \quad \text{and} \quad B = \begin{bmatrix} 2 & -1 & -1 \\ -1 & 2 & -1 \\ -1 & -1 & 2 \end{bmatrix}.$$

32 **Gauss-Jordan elimination** acts on $[\,U \;\; I\,]$ to find the matrix $[\,I \;\; U^{-1}\,]$:

$$\text{If } U = \begin{bmatrix} 1 & a & b \\ 0 & 1 & c \\ 0 & 0 & 1 \end{bmatrix} \quad \text{then} \quad U^{-1} = \begin{bmatrix} & & \\ & & \\ & & \end{bmatrix}.$$

33 True or false (with a counterexample if false and a reason if true): A is square.

(a) A 4 by 4 matrix with a row of zeros is not invertible.

(b) Every matrix with 1's down the main diagonal is invertible.

(c) If A is invertible then A^{-1} and A^2 are invertible.

34 (Recommended) Prove that A is invertible if $a \neq 0$ and $a \neq b$ (find the pivots or A^{-1}). Then find three numbers c so that C is not invertible:

$$A = \begin{bmatrix} a & b & b \\ a & a & b \\ a & a & a \end{bmatrix} \qquad C = \begin{bmatrix} 2 & c & c \\ c & c & c \\ 8 & 7 & c \end{bmatrix}.$$

35 This matrix has a remarkable inverse. Find A^{-1} by elimination on $[\,A \;\; I\,]$. Extend to a 5 by 5 "alternating matrix" and guess its inverse; then multiply to confirm.

$$\text{Invert } A = \begin{bmatrix} 1 & -1 & 1 & -1 \\ 0 & 1 & -1 & 1 \\ 0 & 0 & 1 & -1 \\ 0 & 0 & 0 & 1 \end{bmatrix} \text{ and solve } Ax = \begin{bmatrix} 1 \\ 1 \\ 1 \\ 1 \end{bmatrix}.$$

36 Suppose the matrices P and Q have the same rows as I but in any order. They are "permutation matrices". Show that $P - Q$ is singular by solving $(P - Q)\,x = \mathbf{0}$.

37 Find and check the inverses (assuming they exist) of these **block matrices**:

$$\begin{bmatrix} I & 0 \\ C & I \end{bmatrix} \quad \begin{bmatrix} A & 0 \\ C & D \end{bmatrix} \quad \begin{bmatrix} 0 & I \\ I & D \end{bmatrix}.$$

38 How does elimination from A to U on a 3 by 3 matrix tell you if A is invertible?

39 If $A = I - uv^{\mathrm{T}}$ then $A^{-1} = I + uv^{\mathrm{T}}(1 - v^{\mathrm{T}}u)^{-1}$. Show that $AA^{-1} = I$ except $Au = \mathbf{0}$ when $v^{\mathrm{T}}u = 1$.

2.3 Matrix Computations and $A = LU$

> **1** The elimination steps from A to U cost $\frac{1}{3}n^3$ multiplications and subtractions.
>
> **2** Each right side $\boldsymbol{b}$ costs only n^2: forward to $U\boldsymbol{x} = \boldsymbol{c}$, then back-substitution for $\boldsymbol{x}$.
>
> **3** Elimination without row exchanges factors A into LU (**two proofs of $A = LU$**).

How would you compute the inverse of an n by n matrix A? Before answering that question I have to ask: Do you really want to know A^{-1}? It is true that the solution to $A\boldsymbol{x} = \boldsymbol{b}$ (which we do want) is given by $\boldsymbol{x} = A^{-1}\boldsymbol{b}$. Computing A^{-1} and multiplying $A^{-1}\boldsymbol{b}$ is a very slow way to find $\boldsymbol{x}$. We should understand A^{-1} even if we don't use it.

Here is a simple idea for A^{-1}. That matrix is the solution to $AA^{-1} = I$. The identity matrix has n columns $\boldsymbol{e}_1, \boldsymbol{e}_2, \ldots, \boldsymbol{e}_n$. Then $AA^{-1} = I$ is really n equations $A\boldsymbol{x}_1 = \boldsymbol{e}_1$ to $A\boldsymbol{x}_n = \boldsymbol{e}_n$ for the n columns $\boldsymbol{x}_k$ of A^{-1}. Three equations if the matrices are 3 by 3:

$$\boldsymbol{AA^{-1}=I} \quad A\begin{bmatrix} \boldsymbol{x}_1 & \cdots & \boldsymbol{x}_n \end{bmatrix} = \begin{bmatrix} \boldsymbol{e}_1 & \cdots & \boldsymbol{e}_n \end{bmatrix} \qquad A\begin{bmatrix} \boldsymbol{x}_1 & \boldsymbol{x}_2 & \boldsymbol{x}_3 \end{bmatrix} = \begin{bmatrix} 1 & 0 & 0 \\ 0 & 1 & 0 \\ 0 & 0 & 1 \end{bmatrix}. \quad (1)$$

We are solving n equations and they have the same coefficient matrix A. So we can solve them together. This is called "Gauss-Jordan elimination". Instead of a matrix $\begin{bmatrix} A & \boldsymbol{b} \end{bmatrix}$ augmented by one right hand side $\boldsymbol{b}$, we have a matrix $\begin{bmatrix} A & I \end{bmatrix}$ augmented by n right hand sides (the columns of I). Then as A changes to I, elimination produces $\begin{bmatrix} \boldsymbol{I} & \boldsymbol{A^{-1}} \end{bmatrix}$.

$$\begin{bmatrix} \boldsymbol{A} & \boldsymbol{I} \end{bmatrix} = \left[\begin{array}{rrr|rrr} \mathbf{1} & 0 & 0 & 1 & 0 & 0 \\ \mathbf{-1} & \mathbf{1} & 0 & 0 & 1 & 0 \\ \mathbf{0} & \mathbf{-1} & \mathbf{1} & 0 & 0 & 1 \end{array}\right] \rightarrow \left[\begin{array}{rrr|rrr} 1 & 0 & 0 & 1 & 0 & 0 \\ \mathbf{0} & 1 & 0 & 1 & 1 & 0 \\ \mathbf{0} & -1 & 1 & 0 & 0 & 1 \end{array}\right]$$

Gauss-Jordan elimination
Solve $AX = I \Rightarrow X = A^{-1}$
Slower than solving $A\boldsymbol{x} = \boldsymbol{b}$

$$\rightarrow \left[\begin{array}{rrr|rrr} 1 & 0 & 0 & \mathbf{1} & \mathbf{0} & \mathbf{0} \\ \mathbf{0} & 1 & 0 & \mathbf{1} & \mathbf{1} & \mathbf{0} \\ \mathbf{0} & \mathbf{0} & 1 & \mathbf{1} & \mathbf{1} & \mathbf{1} \end{array}\right] = \begin{bmatrix} \boldsymbol{I} & \boldsymbol{A^{-1}} \end{bmatrix}$$

The key point is that the elimination steps on A only have to be done once. The same steps are applied to the right hand side—but now $AA^{-1} = I$ has n right hand sides. The n solutions $\boldsymbol{x}_i$ to $A\boldsymbol{x}_i = \boldsymbol{e}_i$ go into the n columns of A^{-1}. Then Gauss-Jordan takes $\begin{bmatrix} A & I \end{bmatrix}$ into $\begin{bmatrix} I & A^{-1} \end{bmatrix}$. Here elimination is equivalent to multiplication by A^{-1}.

In this example A subtracts rows and A^{-1} adds. $AA^{-1} = I$ is linear algebra's version of the Fundamental Theorem of Calculus: *Derivative of integral of $f(x)$ equals $f(x)$.*

The Cost of Elimination

A very practical question is cost—or computing time. We can solve 1000 equations on a PC. What if $n = 100{,}000$? (*Is A dense or sparse*?) Large systems come up all the time in scientific computing, where a three-dimensional problem can easily lead to a million unknowns. We can let the calculation run overnight, but we can't leave it for 100 years.

Reducing A to U The first stage of elimination produces zeros below the first pivot in column 1. To find each of the new entries below row 1 requires one multiplication and one subtraction. *We will count this first stage as n^2 multiplications and n^2 subtractions.* The count is actually $n^2 - n$, because row 1 does not change.

The next stage clears out the second column below the second pivot. The working matrix is now of size $n-1$. Estimate this stage by $(n-1)^2$ multiplications and subtractions. The matrices are getting smaller as elimination goes forward. The rough count to reach U is the sum of squares $n^2 + (n-1)^2 + \cdots + 2^2 + 1^2$.

There is an exact formula $\frac{1}{3}n\left(n+\frac{1}{2}\right)(n+1)$ for this sum of squares. When n is large, the $\frac{1}{2}$ and the 1 are not important. *The number that matters is $\frac{1}{3}n^3$.* The sum of squares is like the integral of x^2 ! The integral from 0 to n is $\frac{1}{3}n^3$:

Reducing A to U requires about $\frac{1}{3}n^3$ multiplications and $\frac{1}{3}n^3$ subtractions.

What about the right side b? Going forward, we subtract multiples of b_1 from the lower components $b_2, \ldots, b_n$. This is $n-1$ steps. The second stage takes only $n-2$ steps, because b_1 is not involved. The last stage of forward elimination (**b to c**) takes one step.

Now start back substitution. Computing x_n uses one step (divide by the last pivot). The next unknown uses two steps. When we reach x_1 it will require n steps ($n-1$ substitutions of the other unknowns, then division by the first pivot). The total count on the right side, from b to c to x —*forward to the bottom and back to the top*— is exactly n^2:

$$[(n-1)+(n-2)+\cdots+1] \;+\; [1+2+\cdots+(n-1)+n] = \boldsymbol{n^2}. \qquad (2)$$

To see that sum, pair off $(n-1)$ with 1 and $(n-2)$ with 2. The pairings leave n terms, each equal to n. That makes n^2. Going from b to c to x costs a lot less than the left side !

Back substitution ***Each right side needs n^2 multiplications and n^2 subtractions.***

How long does it take to solve $Ax = b$? For a random matrix of order $n = 1000$, a typical time on a PC is 1 second. The time is multiplied by about 8 when n is multiplied by 2. For professional codes go to **netlib.org**.

According to this n^3 rule, matrices that are 10 times as large (order 10,000) will take a thousand seconds. Matrices of order 100,000 will take a million seconds. This is too expensive without a supercomputer, but remember that these matrices are full. Most large matrices in practice are sparse (many zero entries). In that case $A = LU$ is much faster. **Please see "Computing Time" on the website math.mit.edu/linearalgebra**.

Proving $A = LU$

Elimination is expressed by $EA=U$ and inverted by $LU=A$. Equation (11) in Section 2.2 showed how the multipliers ℓ_{ij} fall exactly into the right positions in E^{-1} which is L.

Why should we want to find a proof for every size n? The reason is: A proof means that we have not just seen that pattern and believed it and liked it, but understood it.

The Great Factorization $A = LU$

Let me review the forward steps of elimination. They start with a matrix A and they end with an upper triangular matrix U. Every elimination step E_{ij} produces a lower triangular zero. Those steps E_{ij} subtract ℓ_{ij} times equation j from equation i below it. Row exchanges (permutations) are coming soon but not yet.

To invert one elimination step E_{ij}, we add instead of subtracting:

$$E_{31} = \begin{bmatrix} 1 & & \\ 0 & 1 & \\ -\ell_{31} & 0 & 1 \end{bmatrix} \quad \text{and} \quad L_{31} = \text{inverse of } E_{31} = \begin{bmatrix} 1 & & \\ 0 & 1 & \\ \ell_{31} & 0 & 1 \end{bmatrix}.$$

Equation (10) in Section 2.2 multiplied $E_{32}E_{31}E_{21}$ with a messy result:

$$E = \begin{bmatrix} 1 & & \\ 0 & 1 & \\ 0 & -\ell_{32} & 1 \end{bmatrix} \begin{bmatrix} 1 & & \\ 0 & 1 & \\ -\ell_{31} & 0 & 1 \end{bmatrix} \begin{bmatrix} 1 & & \\ -\ell_{21} & 1 & \\ 0 & 0 & 1 \end{bmatrix} = \begin{bmatrix} 1 & & \\ -\ell_{21} & 1 & \\ (\ell_{32}\ell_{21} - \ell_{31}) & -\ell_{32} & 1 \end{bmatrix}.$$

Equation (11) showed how inverses in reverse order $E_{21}^{-1}E_{31}^{-1}E_{32}^{-1}$ produced perfection:

$$\boldsymbol{E^{-1}} = \begin{bmatrix} 1 & & \\ \ell_{21} & 1 & \\ 0 & 0 & 1 \end{bmatrix} \begin{bmatrix} 1 & & \\ 0 & 1 & \\ \ell_{31} & 0 & 1 \end{bmatrix} \begin{bmatrix} 1 & & \\ 0 & 1 & \\ 0 & \ell_{32} & 1 \end{bmatrix} = \begin{bmatrix} \mathbf{1} & & \\ \boldsymbol{\ell_{21}} & \mathbf{1} & \\ \boldsymbol{\ell_{31}} & \boldsymbol{\ell_{32}} & \mathbf{1} \end{bmatrix} = \boldsymbol{L}$$

Then elimination $EA = U$ becomes $A = E^{-1}U = LU$ if we run it backward from U to A. These pages aim to show the same result for any matrix size n. The formula $A = LU$ is one of the great matrix factorizations in linear algebra.

Here is one way to understand why L has all the ℓ_{ij} in position, with no mix-up.

The key reason why A equals LU: Ask yourself about the pivot rows that are subtracted from lower rows. Are they the original rows of A? *No*, elimination probably changed them. Are they rows of U? *Yes*, the pivot rows never change again. When computing the third row of U, we subtract multiples of earlier rows of U (*not rows of the original* A):

$$\text{Row 3 of } U = (\text{Row 3 of } A) - \ell_{31}(\text{Row 1 of } U) - \ell_{32}(\text{Row 2 of } U). \tag{3}$$

Rewrite this equation to see that the row $[\,\ell_{31} \quad \ell_{32} \quad 1\,]$ is multiplying the matrix U:

$$\textbf{Row 3 of } \boldsymbol{A} = \ell_{31}(\text{Row 1 of } U) + \ell_{32}(\text{Row 2 of } U) + 1(\text{Row 3 of } U). \tag{4}$$

This is exactly row 3 ***of*** $A = LU$. That row of L holds ℓ_{31}, ℓ_{32}, 1. All rows look like this, whatever the size of A. With no row exchanges, we have $\boldsymbol{A = LU}$.

Second Proof of $A = LU$: Multiply Columns Times Rows

I would like to present another proof of $A = LU$. The idea is to see elimination as removing *one column of L times one row of U from A*. The problem becomes one size smaller.

Elimination begins with pivot row = row 1 of A. We multiply that pivot row by the numbers ℓ_{21} and ℓ_{31} and eventually ℓ_{n1}. Then we subtract from row 2 and row 3 and eventually row n of A. By choosing $\ell_{21} = a_{21}/a_{11}$ and $\ell_{31} = a_{31}/a_{11}$ and eventually $\ell_{n1} = a_{n1}/a_{11}$, the subtraction leaves zeros in column 1.

$$\text{Step 1 removes } \begin{bmatrix} 1\ (\text{row } 1) \\ \ell_{21}(\text{row } 1) \\ \ell_{31}(\text{row } 1) \\ \ell_{41}(\text{row } 1) \end{bmatrix} \text{ from } A \text{ to leave } \boldsymbol{A_2} = \begin{bmatrix} 0 & 0 & 0 & 0 \\ 0 & \times & \times & \times \\ 0 & \times & \times & \times \\ 0 & \times & \times & \times \end{bmatrix}.$$

Key point: **We removed a rank one matrix: column times row**. It was the column $\boldsymbol{\ell_1} = (1, \ell_{21}, \ell_{31}, \ell_{41})$ times row 1 of A—the first pivot row $\boldsymbol{u}_1$.

Now we face a similar problem for A_2. And we take a similar step to reach A_3 :

$$\text{Step 2 removes } \begin{bmatrix} 0\ (\text{row 2 of } A_2) \\ 1\ (\text{row 2 of } A_2) \\ \ell_{32}(\text{row 2 of } A_2) \\ \ell_{42}(\text{row 2 of } A_2) \end{bmatrix} \text{ from } A_2 \text{ to leave } \boldsymbol{A_3} = \begin{bmatrix} 0 & 0 & 0 & 0 \\ 0 & 0 & 0 & 0 \\ 0 & 0 & \times & \times \\ 0 & 0 & \times & \times \end{bmatrix}.$$

Row 2 of A_2 was the second pivot row = second row $\boldsymbol{u}_2$ of U. We removed a column $\boldsymbol{\ell}_2 = (0, 1, \ell_{32}, \ell_{42})$ times that second pivot row. Continuing in the same way, every step removes a column $\boldsymbol{\ell}_j$ of L times a pivot row $\boldsymbol{u}_j$ of U. Now add those pieces back together:

$$\boldsymbol{A} = \boldsymbol{\ell}_1\boldsymbol{u}_1 + \boldsymbol{\ell}_2\boldsymbol{u}_2 + \cdots + \boldsymbol{\ell}_n\boldsymbol{u}_n = \begin{bmatrix} \boldsymbol{\ell}_1 & \cdots & \boldsymbol{\ell}_n \end{bmatrix} \begin{bmatrix} \boldsymbol{u}_1 \\ \vdots \\ \boldsymbol{u}_n \end{bmatrix} = \boldsymbol{LU} \tag{5}$$

columns times rows

That last crucial step was column-row multiplication of L times U. This was introduced at the very end of Chapter 1. The Problem Set for this section will review this important way to multiply LU—by adding up rank-one matrices (columns of L times rows of U).

Notice that U is upper triangular. The pivot row $\boldsymbol{u}_k$ begins with $k - 1$ zeros. And L is lower triangular with 1's on its main diagonal. Column $\boldsymbol{\ell}_k$ also begins with $k - 1$ zeros.

Elimination Without Row Exchanges

The next section is going to allow row exchanges P_{ij}. They are necessary to move zeros out of the pivot positions. Before we go there, we can answer this basic question: *When is $A = LU$ possible with no row exchanges and no zeros in the pivots?*

Answer **All upper left k by k submatrices of A must be invertible** (sizes $k = 1$ to n).

The reason is that elimination is also factoring every one of those k by k corners of A. Those corner matrices A_k agree with $L_k U_k$ (k by k corners of L and U are invertible).

$$\begin{bmatrix} \boldsymbol{A_k} & * \\ * & * \end{bmatrix} = \begin{bmatrix} \boldsymbol{L_k} & 0 \\ * & * \end{bmatrix} \begin{bmatrix} \boldsymbol{U_k} & * \\ \boldsymbol{0} & * \end{bmatrix} \text{ tells us that } \boldsymbol{A_k = L_k U_k}. \quad (6)$$

So each corner matrix A_k must be invertible to reach the final form $A = LU$.

Problem Set 2.3

Problems 1–8 compute the factorization $A = LU$ (and also $A = LDU$).

1 (Important) Forward elimination changes $\left[\begin{smallmatrix} 1 & 1 \\ 1 & 2 \end{smallmatrix}\right]\boldsymbol{x} = \boldsymbol{b}$ to a triangular $\left[\begin{smallmatrix} 1 & 1 \\ 0 & 1 \end{smallmatrix}\right]\boldsymbol{x} = \boldsymbol{c}$:

$$\begin{aligned} x + y &= 5 \\ x + 2y &= 7 \end{aligned} \quad \longrightarrow \quad \begin{aligned} x + y &= 5 \\ y &= 2 \end{aligned} \qquad \begin{bmatrix} 1 & 1 & 5 \\ 1 & 2 & 7 \end{bmatrix} \longrightarrow \begin{bmatrix} 1 & 1 & 5 \\ 0 & 1 & 2 \end{bmatrix}$$

That step subtracted $\ell_{21} =$ _____ times row 1 from row 2. The reverse step *adds* ℓ_{21} times row 1 to row 2. The matrix for that reverse step is $L =$ _____. Multiply this L times the triangular system $\left[\begin{smallmatrix} 1 & 1 \\ 0 & 1 \end{smallmatrix}\right]\boldsymbol{x}_1 = \left[\begin{smallmatrix} 5 \\ 2 \end{smallmatrix}\right]$ to get _____ $=$ _____. In letters, L multiplies $U\boldsymbol{x} = \boldsymbol{c}$ to give _____.

2 Write down the 2 by 2 triangular systems $L\boldsymbol{c} = \boldsymbol{b}$ and $U\boldsymbol{x} = \boldsymbol{c}$ from Problem 1. Check that $\boldsymbol{c} = (5, 2)$ solves the first one. Find $\boldsymbol{x}$ that solves the second one.

3 What matrix E puts A into triangular form $EA = U$? Multiply by $E^{-1} = L$ to factor A into LU:

$$A = \begin{bmatrix} 2 & 1 & 0 \\ 0 & 4 & 2 \\ 6 & 3 & 5 \end{bmatrix}.$$

4 What two elimination matrices E_{21} and E_{32} put A into upper triangular form $E_{32}E_{21}A = U$? Multiply by E_{32}^{-1} and E_{21}^{-1} to factor A into $LU = E_{21}^{-1}E_{32}^{-1}U$:

$$A = \begin{bmatrix} 1 & 1 & 1 \\ 2 & 4 & 5 \\ 0 & 4 & 0 \end{bmatrix}.$$

5 What three elimination matrices E_{21}, E_{31}, E_{32} put A into its upper triangular form $E_{32}E_{31}E_{21}A = U$? Multiply by E_{32}^{-1}, E_{31}^{-1} and E_{21}^{-1} to factor A into L times U:

$$A = \begin{bmatrix} 1 & 0 & 1 \\ 2 & 2 & 2 \\ 3 & 4 & 5 \end{bmatrix} \quad \text{and} \quad L = E_{21}^{-1}E_{31}^{-1}E_{32}^{-1}.$$

6 A and B are symmetric across the diagonal (because $4 = 4$). Find their triple factorizations LDU and say how U is related to L for these symmetric matrices:

$$\textbf{Symmetric} \qquad A = \begin{bmatrix} 2 & 4 \\ 4 & 11 \end{bmatrix} \quad \text{and} \quad B = \begin{bmatrix} 1 & 4 & 0 \\ 4 & 12 & 4 \\ 0 & 4 & 0 \end{bmatrix}.$$

7 (*Recommended*) Compute L and U for the symmetric matrix A:

$$A = \begin{bmatrix} a & a & a & a \\ a & b & b & b \\ a & b & c & c \\ a & b & c & d \end{bmatrix}.$$

Find four conditions on a, b, c, d to get $A = LU$ with four pivots.

8 This nonsymmetric matrix will have the same L as in Problem **7**:

$$\textbf{Find } L \textbf{ and } U \textbf{ for} \qquad A = \begin{bmatrix} a & r & r & r \\ a & b & s & s \\ a & b & c & t \\ a & b & c & d \end{bmatrix}.$$

Find the four conditions on a, b, c, d, r, s, t to get $A = LU$ with four pivots.

9 Solve the triangular system $Lc = b$ to find c. Then solve $Ux = c$ to find x:

$$L = \begin{bmatrix} 1 & 0 \\ 4 & 1 \end{bmatrix} \quad \text{and} \quad U = \begin{bmatrix} 2 & 4 \\ 0 & 1 \end{bmatrix} \quad \text{and} \quad b = \begin{bmatrix} 2 \\ 11 \end{bmatrix}.$$

For safety multiply LU and solve $Ax = b$ as usual. Circle c when you see it.

10 Solve $Lc = b$ to find c. Then solve $Ux = c$ to find x. **What was** A?

$$L = \begin{bmatrix} 1 & 0 & 0 \\ 1 & 1 & 0 \\ 1 & 1 & 1 \end{bmatrix} \quad \text{and} \quad U = \begin{bmatrix} 1 & 1 & 1 \\ 0 & 1 & 1 \\ 0 & 0 & 1 \end{bmatrix} \quad \text{and} \quad b = \begin{bmatrix} 4 \\ 5 \\ 6 \end{bmatrix}.$$

11 (a) When you apply the usual elimination steps to L, what matrix do you reach?

$$L = \begin{bmatrix} 1 & 0 & 0 \\ \ell_{21} & 1 & 0 \\ \ell_{31} & \ell_{32} & 1 \end{bmatrix}.$$

(b) When you apply the same steps to I, what matrix do you get?

(c) When you apply the same steps to LU, what matrix do you get?

12 If $A = LDU$ and also $A = L_1D_1U_1$ with all factors invertible, then $L = L_1$ and $D = D_1$ and $U = U_1$. *"The three factors are unique."*

Derive the equation $L_1^{-1}LD = D_1U_1U^{-1}$. Are the two sides triangular or diagonal? Deduce $L = L_1$ and $U = U_1$ (they all have diagonal 1's). Then $D = D_1$.

13 *Tridiagonal matrices* have zero entries except on the main diagonal and the two adjacent diagonals. Factor these into $A = LU$ and $A = LDL^{\mathrm{T}}$:

$$A = \begin{bmatrix} 1 & 1 & 0 \\ 1 & 2 & 1 \\ 0 & 1 & 2 \end{bmatrix} \quad \text{and} \quad A = \begin{bmatrix} a & a & 0 \\ a & a+b & b \\ 0 & b & b+c \end{bmatrix}.$$

14 If A and B have nonzeros in the positions marked by x, which zeros (marked by 0) *stay zero* in their factors L and U?

$$A = \begin{bmatrix} x & x & x & x \\ x & x & x & 0 \\ 0 & x & x & x \\ 0 & 0 & x & x \end{bmatrix} \qquad B = \begin{bmatrix} x & x & x & 0 \\ x & x & 0 & x \\ x & 0 & x & x \\ 0 & x & x & x \end{bmatrix}.$$

15 *Easy but important.* If A has pivots $5, 9, 3$ with no row exchanges, what are the pivots for the upper left 2 by 2 submatrix A_2 (without row 3 and column 3 of A)?

16 The second proof of $A = LU$ removes which three rank 1 matrices adding to A?

$$A = \begin{bmatrix} 1 & 2 & 0 \\ 1 & 5 & 1 \\ 0 & 6 & 4 \end{bmatrix} = \boldsymbol{\ell}_1\boldsymbol{u}_1 + \boldsymbol{\ell}_2\boldsymbol{u}_2 + \boldsymbol{\ell}_3\boldsymbol{u}_3 = LU! \quad \text{Columns multiply rows}$$

17 Multiply $L^{\mathrm{T}}L$ and LL^{T} by columns times rows when the 3 by 3 lower triangular L has six 1's.

2.4 Permutations and Transposes

1 A **permutation matrix** P has the same rows as I (in any order). There are $n\,!$ different orders.

2 Then P times x puts the components x_1 to x_n in that new order. And $\boldsymbol{P^{\mathrm{T}}}$ **equals** $\boldsymbol{P^{-1}}$.

3 **Columns of A are rows of A^{T}. The transposes of Ax and AB are $x^{\mathrm{T}}A^{\mathrm{T}}$ and $B^{\mathrm{T}}A^{\mathrm{T}}$.**

4 The idea behind A^{T} is that $\boldsymbol{Ax \cdot y = x \cdot A^{\mathrm{T}}y}$ because $(Ax)^{\mathrm{T}}y = x^{\mathrm{T}}A^{\mathrm{T}}y = x^{\mathrm{T}}(A^{\mathrm{T}}y)$.

5 A **symmetric matrix** has $\boldsymbol{S^{\mathrm{T}} = S}$. The products $A^{\mathrm{T}}A$ and AA^{T} are always symmetric.

Permutations

Permutation matrices have a 1 in every row and a 1 in every column. All other entries are zero. When this matrix P multiplies a vector, it changes the order of its components:

P has the rows of I
Px = Circular shift of x
$1,2,3$ to $3,1,2$

$$Px = \begin{bmatrix} 0 & 0 & 1 \\ 1 & 0 & 0 \\ 0 & 1 & 0 \end{bmatrix} \begin{bmatrix} x_1 \\ x_2 \\ x_3 \end{bmatrix} = \begin{bmatrix} \boldsymbol{x_3} \\ \boldsymbol{x_1} \\ \boldsymbol{x_2} \end{bmatrix}$$

For this example, P shifts x_1 to second position. If we repeat, P^2 shifts x_1 to third position. Then $P^3 = I$ and we recover the order x_1, x_2, x_3. P, P^2, and I are permutations.

Shuffling a deck of cards would be a permutation of size $n = 52$. Persi Diaconis proved that a new deck is well shuffled after 7 random permutations.

There are $3! = 6$ permutation matrices of size 3, and $4! = 24$ permutations of size 4. Here are specific tasks for these 4 by 4 permutations, when they multiply a vector $\boldsymbol{x}$:

Px reverses the order to x_4, x_3, x_2, x_1
$\boldsymbol{P^2 = I}$

$$\begin{bmatrix} 0 & 0 & 0 & 1 \\ 0 & 0 & 1 & 0 \\ 0 & 1 & 0 & 0 \\ 1 & 0 & 0 & 0 \end{bmatrix}$$

x_4 stays fixed
Input x_1, x_2, x_3, x_4
output x_3, x_1, x_2, x_4
$\boldsymbol{P^3 = I}$

$$\begin{bmatrix} 0 & 0 & 1 & 0 \\ 1 & 0 & 0 & 0 \\ 0 & 1 & 0 & 0 \\ 0 & 0 & 0 & 1 \end{bmatrix}$$

Circular shift
Input x_1, x_2, x_3, x_4
output x_4, x_1, x_2, x_3
$\boldsymbol{P^4 = I}$

$$\begin{bmatrix} 0 & 0 & 0 & 1 \\ 1 & 0 & 0 & 0 \\ 0 & 1 & 0 & 0 \\ 0 & 0 & 1 & 0 \end{bmatrix}$$

Evens before odds
x_0, x_2 before x_1, x_3
Fast Fourier Transform
Number from 0 to 3

$$\begin{bmatrix} 1 & 0 & 0 & 0 \\ 0 & 0 & 1 & 0 \\ 0 & 1 & 0 & 0 \\ 0 & 0 & 0 & 1 \end{bmatrix}$$

Half of the $\boldsymbol{n}$ **! permutations of size** n are "even" and half are "odd". An even permutation needs an even number of simple row exchanges to reach the matrix I. The last example (1 exchange) was **odd**. The first example (exchange 1 and 4, exchange 2 and 3) was **even**.

The rows of any P are
the columns of $P^{-1} = P^{\mathrm{T}}$
P^{T} = "transpose of P"

$$P^{\mathrm{T}}P = \begin{bmatrix} 0 & 1 & 0 \\ 0 & 0 & 1 \\ 1 & 0 & 0 \end{bmatrix} \begin{bmatrix} 0 & 0 & 1 \\ 1 & 0 & 0 \\ 0 & 1 & 0 \end{bmatrix} = \begin{bmatrix} 1 & & \\ & 1 & \\ & & 1 \end{bmatrix} = I$$

Properties of Permutation Matrices

1. The n 1's appear in n different rows and n different columns of P.
2. The columns of P are *orthogonal*: dot products between columns are all zero.
3. The product P_1P_2 of permutations is a permutation. So is the inverse of P.
4. If A is invertible, there is a permutation P to order its rows in advance, so that **elimination on PA meets no zeros in the pivot positions**. Then $PA = LU$.

The $PA = LU$ Factorization: Row Exchanges from P

An example will show how elimination can often succeed, even when a zero appears in the pivot position. Suppose elimination starts with 1 as the first pivot. Subtracting 2 times row 1 produces 0 as an unacceptable second pivot:

$$\boldsymbol{A} = \begin{bmatrix} 1 & 2 & a \\ 2 & 4 & b \\ 3 & 7 & c \end{bmatrix} \xrightarrow{\boldsymbol{E}} \begin{bmatrix} 1 & 2 & a \\ 0 & \mathbf{0} & b-2a \\ 0 & \mathbf{1} & c-3a \end{bmatrix} \xrightarrow{\boldsymbol{P}} \begin{bmatrix} 1 & 2 & a \\ 0 & \mathbf{1} & c-3a \\ 0 & \mathbf{0} & b-2a \end{bmatrix} = \boldsymbol{U}$$

In spite of that zero, this A can be invertible. To rescue elimination, P exchanged *row* 2 *with row* 3. That brings 1 into the second pivot as shown. So we can continue.

A is invertible if and only if $b - 2a$ is not zero in the third pivot. If $b = 2a$, then row 2 of A equals 2 (row 1 of A). In that case, A is surely not invertible.

We can exchange rows 2 and 3 *first* to get PA. Then LU factorization becomes $\boldsymbol{PA = LU}$. The matrix PA sails through elimination without seeing that zero pivot.

$$\underset{\boldsymbol{P}}{\begin{bmatrix} 1 & 0 & 0 \\ 0 & 0 & 1 \\ 0 & 1 & 0 \end{bmatrix}} \underset{\boldsymbol{A}}{\begin{bmatrix} 1 & 2 & a \\ 2 & 4 & b \\ 3 & 7 & c \end{bmatrix}} = \underset{\boldsymbol{PA}}{\begin{bmatrix} 1 & 2 & a \\ 3 & 7 & c \\ 2 & 4 & b \end{bmatrix}} = \underset{\boldsymbol{L}}{\begin{bmatrix} 1 & 0 & 0 \\ 3 & 1 & 0 \\ 2 & 0 & 1 \end{bmatrix}} \underset{\boldsymbol{U}}{\begin{bmatrix} 1 & 2 & a \\ 0 & 1 & c-3a \\ 0 & 0 & b-2a \end{bmatrix}}$$

In principle we might need several row exchanges. Then the overall permutation P includes them all, and still produces $PA = LU$. Daniel Drucker showed me a neat way to keep track of P, by adding a special **1 2 3** column to the matrix A. That column tracks the original row numbers, as rows are exchanged. Here the final permutation P is **1 3 2**.

$$\begin{bmatrix} 1 & 2 & a & \mathbf{1} \\ 2 & 4 & b & \mathbf{2} \\ 3 & 7 & c & \mathbf{3} \end{bmatrix} \rightarrow \begin{bmatrix} 1 & 2 & a & \mathbf{1} \\ 0 & \mathbf{0} & b-2a & \mathbf{2} \\ 0 & \mathbf{1} & c-3a & \mathbf{3} \end{bmatrix} \boldsymbol{\rightarrow} \begin{bmatrix} 1 & 2 & a & \mathbf{1} \\ 0 & \mathbf{1} & c-3a & \mathbf{3} \\ 0 & \mathbf{0} & b-2a & \mathbf{2} \end{bmatrix}. \ \boldsymbol{P_{132}} \text{ is } \begin{bmatrix} \mathbf{1} & 0 & 0 \\ 0 & 0 & \mathbf{1} \\ 0 & \mathbf{1} & 0 \end{bmatrix}$$

"Partial Pivoting" to Reduce Roundoff Errors

Every good code for elimination allows for extra row exchanges that add safety. **Small pivots are unsafe!** The code does not blindly accept a small pivot when a larger number appears below it (in the same column). The computation is more stable if we exchange those rows, to produce the *largest possible number in the pivot.*

This example has first pivot equal to 1, but column 1 offers larger numbers 2 and 3. **The code will choose the largest number 3 as the first pivot: exchange rows 1 and 3.** The order of rows is tracked by the last column $*$ which is not part of the matrix.

$$\underset{\boldsymbol{A}\qquad *}{\begin{bmatrix} 1 & 2 & a & \mathbf{1} \\ 2 & 4 & b & \mathbf{2} \\ 3 & 7 & c & \mathbf{3} \end{bmatrix}} \xrightarrow{\boldsymbol{P}} \begin{bmatrix} 3 & 7 & c & \mathbf{3} \\ 2 & 4 & b & \mathbf{2} \\ 1 & 2 & a & \mathbf{1} \end{bmatrix} \to \begin{bmatrix} 3 & 7 & c & \mathbf{3} \\ 0 & -\frac{2}{3} & b-\frac{2}{3}c & \mathbf{2} \\ 0 & -\frac{1}{3} & a-\frac{1}{3}c & \mathbf{1} \end{bmatrix} \to \underset{\boldsymbol{U}\qquad\qquad *}{\begin{bmatrix} 3 & 7 & c & \mathbf{3} \\ 0 & -\frac{2}{3} & b-\frac{2}{3}c & \mathbf{2} \\ 0 & 0 & a-\frac{1}{2}b & \mathbf{1} \end{bmatrix}}$$

This example has $P = \begin{bmatrix} 0 & 0 & \mathbf{1} \\ 0 & \mathbf{1} & 0 \\ \mathbf{1} & 0 & 0 \end{bmatrix}$ and $L = \begin{bmatrix} 1 & 0 & 0 \\ \mathbf{\frac{2}{3}} & 1 & 0 \\ \mathbf{\frac{1}{3}} & \mathbf{\frac{1}{2}} & 1 \end{bmatrix}$ with $PA = LU$.

All entries of L are ≤ 1 when we make each pivot larger than all the numbers below it.

PAQ Has Row Permutation P and Column Permutation Q

Start with a 3 by 3 matrix A. Reorder its rows $1, 2, 3$ by P in the new order $2, 3, 1$,:

$$\boldsymbol{P} = \begin{bmatrix} 0 & \mathbf{1} & 0 \\ 0 & 0 & \mathbf{1} \\ \mathbf{1} & 0 & 0 \end{bmatrix} \qquad \boldsymbol{PA} = \begin{bmatrix} a_{21} & a_{22} & a_{23} \\ a_{31} & a_{32} & a_{33} \\ a_{11} & a_{12} & a_{13} \end{bmatrix} \tag{1}$$

The column indices are still in their original order $1, 2, 3$. Now reorder those columns by Q in the order $3, 2, 1$. For column permutations Q follows PA to give PAQ:

$$\boldsymbol{Q} = \begin{bmatrix} 0 & 0 & \mathbf{1} \\ 0 & \mathbf{1} & 0 \\ \mathbf{1} & 0 & 0 \end{bmatrix} \qquad \boldsymbol{PAQ} = \begin{bmatrix} a_{23} & a_{22} & a_{21} \\ a_{33} & a_{32} & a_{31} \\ a_{13} & a_{12} & a_{11} \end{bmatrix} \tag{2}$$

Please notice PAQ. It has row subscripts $2, 3, 1$ and column subscripts $3, 2, 1$. We had 6 row orders P to choose from and 6 column orders Q. That makes **36 different matrices $\boldsymbol{PAQ}$.**

Question: Are those all the possible permutations of the 9 *numbers in* A? *Absolutely not.* Nine numbers have $9! = 362,880$ possible orders. Our matrices PAQ are very special, because the first index 2 or 3 or 1 is constant on each row. And the second index 3 or 2 or 1 is constant down each column. For the column spaces, does $\mathbf{C}(PAQ)$ equal $\mathbf{C}(A)$?

The Transpose of A

We need one more matrix, and fortunately it is much simpler than the inverse. It is the ***"transpose"*** of A, which is denoted by $\boldsymbol{A}^{\mathrm{T}}$. **The columns of $\boldsymbol{A}^{\mathrm{T}}$ are the rows of $\boldsymbol{A}$.**

When A is an m by n matrix, the transpose is n by m. Then 2 by 3 becomes 3 by 2.

$$\textbf{Transpose} \qquad \text{If} \quad \boldsymbol{A} = \begin{bmatrix} \mathbf{1} & \mathbf{2} & \mathbf{3} \\ 0 & 0 & 4 \end{bmatrix} \quad \text{then} \quad \boldsymbol{A}^{\mathrm{T}} = \begin{bmatrix} \mathbf{1} & 0 \\ \mathbf{2} & 0 \\ \mathbf{3} & 4 \end{bmatrix}.$$

You can write the rows of A into the columns of A^{T}. Or you can write the columns of A into the rows of A^{T}. The matrix "flips over" its main diagonal. The entry in row i, column j of A^{T} comes from row j, column i of the original A:

$$\textbf{Rows of } \boldsymbol{A} \textbf{ become columns of } \boldsymbol{A}^{\mathrm{T}} \qquad \boxed{(\boldsymbol{A}^{\mathrm{T}})_{ij} = \boldsymbol{A}_{ji}}$$

The transpose of a lower triangular matrix is upper triangular. The transpose of A^{T} is A.

Note MATLAB's symbol for A^{T} is A'. Typing $[1 \ \ 2 \ \ 3]$ gives a row vector and the column vector is $\boldsymbol{v} = [1 \ \ 2 \ \ 3]'$. The matrix M with second column $\boldsymbol{w} = [\ 4 \ 5 \ 6\ ]'$ is $M = [\ \boldsymbol{v} \ \ \boldsymbol{w}\ ]$. Quicker to enter by rows and transpose: $M = [1 \ \ 2 \ \ 3; \ \ 4 \ \ 5 \ \ 6]'$.

The rules for transposes are very direct. We can transpose $A + B$ to get $(A + B)^{\mathrm{T}}$. Or we can transpose A and B separately, and then add $A^{\mathrm{T}} + B^{\mathrm{T}}$—with the same result.

The serious questions are about the transposes of a product AB and an inverse A^{-1}:

Sum	The transpose of	$A+B$	is	$A^{\mathrm{T}} + B^{\mathrm{T}}$.	(1)
Product	The transpose of	AB	is	$(\boldsymbol{AB})^{\mathrm{T}} = \boldsymbol{B}^{\mathrm{T}}\boldsymbol{A}^{\mathrm{T}}$.	(2)
Inverse	The transpose of	A^{-1}	is	$(A^{-1})^{\mathrm{T}} = (A^{\mathrm{T}})^{-1}$.	(3)

Notice especially how $B^{\mathrm{T}}A^{\mathrm{T}}$ comes in reverse order. For inverses, this reverse order was quick to check: $B^{-1}A^{-1}$ times AB produces I because $A^{-1}A = B^{-1}B = I$. To understand $(AB)^{\mathrm{T}} = B^{\mathrm{T}}A^{\mathrm{T}}$, start with $(A\boldsymbol{x})^{\mathrm{T}} = \boldsymbol{x}^{\mathrm{T}}A^{\mathrm{T}}$ when B is just a vector:

$A\boldsymbol{x}$ ***combines the columns of*** A ***while*** $\boldsymbol{x}^{\mathrm{T}}A^{\mathrm{T}}$ ***combines the rows of*** A^{T}.

It is the same combination of the same vectors! In A they are columns, in A^{T} they are rows. So the transpose of the column $A\boldsymbol{x}$ is the row $\boldsymbol{x}^{\mathrm{T}}A^{\mathrm{T}}$. That fits our formula $(A\boldsymbol{x})^{\mathrm{T}} = \boldsymbol{x}^{\mathrm{T}}A^{\mathrm{T}}$. Now we can prove $(AB)^{\mathrm{T}} = B^{\mathrm{T}}A^{\mathrm{T}}$, when B has several columns.

If B has two columns $\boldsymbol{x}_1$ and $\boldsymbol{x}_2$, apply the same idea to each column. The columns of AB are $A\boldsymbol{x}_1$ and $A\boldsymbol{x}_2$. Their transposes $\boldsymbol{x}^{\mathrm{T}}A^{\mathrm{T}}$ appear correctly in the rows of $B^{\mathrm{T}}A^{\mathrm{T}}$:

$$\text{Transposing } AB = \begin{bmatrix} A\boldsymbol{x}_1 & A\boldsymbol{x}_2 & \cdots \end{bmatrix} \text{ gives } \begin{bmatrix} \boldsymbol{x}_1^{\mathrm{T}}A^{\mathrm{T}} \\ \boldsymbol{x}_2^{\mathrm{T}}A^{\mathrm{T}} \\ \vdots \end{bmatrix} \text{ which is } B^{\mathrm{T}}A^{\mathrm{T}}. \qquad (6)$$

The right answer $B^{\mathrm{T}}A^{\mathrm{T}}$ comes out a row at a time. Here are numbers in $(AB)^{\mathrm{T}} = B^{\mathrm{T}}A^{\mathrm{T}}$:

$$AB = \begin{bmatrix} 1 & 0 \\ 1 & 1 \end{bmatrix}\begin{bmatrix} 5 & 0 \\ 4 & 1 \end{bmatrix} = \begin{bmatrix} \mathbf{5} & \mathbf{0} \\ \mathbf{9} & \mathbf{1} \end{bmatrix} \text{ and } B^{\mathrm{T}}A^{\mathrm{T}} = \begin{bmatrix} 5 & 4 \\ 0 & 1 \end{bmatrix}\begin{bmatrix} 1 & 1 \\ 0 & 1 \end{bmatrix} = \begin{bmatrix} \mathbf{5} & \mathbf{9} \\ \mathbf{0} & \mathbf{1} \end{bmatrix}.$$

The reverse order rule extends to three or more factors: $(ABC)^{\mathrm{T}}$ equals $C^{\mathrm{T}}B^{\mathrm{T}}A^{\mathrm{T}}$.

Now apply this product rule by transposing both sides of $A^{-1}A = I$. On one side, I^{T} is I. We confirm the rule that $(A^{-1})^{\mathrm{T}}$ *is the inverse of* A^{T}.

$$\textbf{Transpose of inverse} \qquad A^{-1}A = I \quad \text{is transposed to} \quad A^{\mathrm{T}}(A^{-1})^{\mathrm{T}} = I. \qquad (7)$$

Similarly $AA^{-1} = I$ leads to $(A^{-1})^{\mathrm{T}}A^{\mathrm{T}} = I$. We can invert the transpose or we can transpose the inverse. Notice especially: A^{T} ***is invertible exactly when*** A ***is invertible.***

$$\text{The inverse of } A = \begin{bmatrix} 1 & 0 \\ 6 & 1 \end{bmatrix} \text{ is } A^{-1} = \begin{bmatrix} 1 & 0 \\ -6 & 1 \end{bmatrix}. \text{ The transpose is } A^{\mathrm{T}} = \begin{bmatrix} 1 & 6 \\ 0 & 1 \end{bmatrix}.$$

$$(\boldsymbol{A}^{-1})^{\mathrm{T}} \quad \textit{and} \quad (\boldsymbol{A}^{\mathrm{T}})^{-1} \quad \textit{are both equal to} \quad \begin{bmatrix} 1 & -6 \\ 0 & 1 \end{bmatrix}.$$

The Meaning of Inner Products

The dot product (inner product) of $\boldsymbol{x}$ and $\boldsymbol{y}$ is the sum of numbers $x_i y_i$. Now we have a better way to write $\boldsymbol{x} \cdot \boldsymbol{y}$, without using that unprofessional dot. Use matrix notation $\boldsymbol{x}^{\mathrm{T}}\boldsymbol{y}$:

^{T} is inside ***The dot product or inner product is*** $\boldsymbol{x}^{\mathrm{T}}\boldsymbol{y}$ $\quad (1 \times n)(n \times 1) = \mathbf{1} \times \mathbf{1}$

^{T} is outside ***The rank one product or outer product is*** $\boldsymbol{x}\boldsymbol{y}^{\mathrm{T}}$ $\quad (n \times 1)(1 \times n) = \boldsymbol{n} \times \boldsymbol{n}$

$\boldsymbol{x}^{\mathrm{T}}\boldsymbol{y}$ is a number, $\boldsymbol{x}\boldsymbol{y}^{\mathrm{T}}$ is a matrix. Quantum mechanics would write those as $< \boldsymbol{x}|\boldsymbol{y} >$ (inner) and $|\boldsymbol{x}><\boldsymbol{y}|$ (outer). Possibly our universe is governed by linear algebra? Here are three more examples where the inner product has meaning:

From mechanics	Work = (Movements) (Forces) = $\boldsymbol{x}^{\mathrm{T}}\boldsymbol{f}$
From circuits	Heat loss = (Voltage drops) (Currents) = $\boldsymbol{e}^{\mathrm{T}}\boldsymbol{y}$
From economics	Income = (Quantities) (Prices) = $\boldsymbol{q}^{\mathrm{T}}\boldsymbol{p}$

We are really close to the heart of applied mathematics, and there is one more point to emphasize. It is the deeper connection between inner products and the transpose of A.

We defined A^{T} by flipping the matrix across its main diagonal. That's not mathematics. There is a better definition. A^{T} *is the matrix that makes these inner products equal*:

$$(\boldsymbol{A}\boldsymbol{x})^{\mathrm{T}}\boldsymbol{y} = \boldsymbol{x}^{\mathrm{T}}(\boldsymbol{A}^{\mathrm{T}}\boldsymbol{y}) \quad \textbf{Inner product of } \boldsymbol{A}\boldsymbol{x} \textbf{ with } \boldsymbol{y} = \textbf{Inner product of } \boldsymbol{x} \textbf{ with } \boldsymbol{A}^{\mathrm{T}}\boldsymbol{y}$$

Example 1 Start with $A = \begin{bmatrix} -1 & 1 & 0 \\ 0 & -1 & 1 \end{bmatrix}$ $\quad x = \begin{bmatrix} x_1 \\ x_2 \\ x_3 \end{bmatrix}$ $\quad y = \begin{bmatrix} y_1 \\ y_2 \end{bmatrix}$

On one side we have Ax multiplying y to produce $(x_2 - x_1)\, y_1 + (x_3 - x_2)\, y_2$
That is the same as $x_1\,(-y_1) + x_2\,(y_1 - y_2) + x_3\,(y_2)$. Now x is multiplying $A^{\mathrm{T}}y$.

$$A^{\mathrm{T}}y = \begin{bmatrix} -1 & 0 \\ 1 & -1 \\ 0 & 1 \end{bmatrix} \begin{bmatrix} y_1 \\ y_2 \end{bmatrix} = \begin{bmatrix} -y_1 \\ y_1 - y_2 \\ y_2 \end{bmatrix} \qquad \begin{matrix} (Ax)^{\mathrm{T}}y \\ = \\ x^{\mathrm{T}}(A^{\mathrm{T}}y) \end{matrix}$$

Example 2 Will you allow a little calculus? It is important or I wouldn't leave linear algebra. (This is linear algebra for functions.) **Change the matrix to a derivative**: $A = d/dt$. The transpose of d/dt comes from the rule that $(Ax)^{\mathrm{T}}y = x^{\mathrm{T}}(A^{\mathrm{T}}y)$.

First, the dot product $x^{\mathrm{T}}y$ changes from $x_1y_1 + \cdots + x_ny_n$ to an *integral of* $x(t)y(t)$.

Inner product of functions x and y
$$x^{\mathrm{T}}y = \int_{-\infty}^{\infty} x(t)\, y(t)\, dt \ \text{ by definition} \tag{8}$$

Transpose rule for functions $(Ax)^{\mathrm{T}}y = x^{\mathrm{T}}(A^{\mathrm{T}}y)$
$$\int_{-\infty}^{\infty} \frac{dx}{dt}\, y(t)\, dt = \int_{-\infty}^{\infty} x(t) \left(-\frac{dy}{dt}\right) dt \tag{9}$$

I hope you recognize "***integration by parts***". The derivative A moves from the first function $x(t)$ to the second function $y(t)$. During that move, a minus sign appears. This tells us that ***the transpose of*** $A = d/dt$ ***is*** $A^{\mathrm{T}} = -A = -d/dt$.

The derivative is *anti-symmetric*. Symmetric matrices have $A^{\mathrm{T}} = A$, anti-symmetric matrices have $A^{\mathrm{T}} = -A$. In some way, the 2 by 3 difference matrix in Example 1 followed this pattern. The 3 by 2 matrix A^{T} was *minus* a difference matrix. It produced $y_1 - y_2$ in the middle component of $A^{\mathrm{T}}y$ instead of the difference $y_2 - y_1$.

Integration by parts is deceptively important and not just a trick. I left out boundary conditions by assuming that $x(-\infty)$ and $x(+\infty)$ are zero.

Symmetric Matrices

For a *symmetric matrix*, transposing A to A^{T} produces no change. In this case A^{T} equals A. Its (j, i) entry across the main diagonal equals its (i, j) entry. In my opinion, these are the most important matrices of all. We give symmetric matrices the special letter S.

A symmetric matrix has $S^{\mathrm{T}} = S$. This means that every $s_{ji} = s_{ij}$.

Symmetric matrices $\quad S = \begin{bmatrix} 1 & 2 \\ 2 & 5 \end{bmatrix} = S^{\mathrm{T}}$ and $D = \begin{bmatrix} 1 & 0 \\ 0 & 10 \end{bmatrix} = D^{\mathrm{T}}$.

The inverse of a symmetric matrix is a symmetric matrix. The transpose of S^{-1} is $(S^{-1})^{\mathrm{T}} = (S^{\mathrm{T}})^{-1} = S^{-1}$. When S is invertible, this says that S^{-1} is symmetric .

Symmetric inverses $\quad S^{-1} = \begin{bmatrix} 5 & -2 \\ -2 & 1 \end{bmatrix}$ and $D^{-1} = \begin{bmatrix} 1 & 0 \\ 0 & 0.1 \end{bmatrix}$.

Next we will produce a symmetric matrix S by ***multiplying any matrix A by*** $\boldsymbol{A}^{\mathrm{T}}$.

Symmetric Products $A^{\mathrm{T}}A$ and AA^{T} and LDL^{T}

Choose any matrix A, probably rectangular. Multiply A^{T} times A. Then the product $S = A^{\mathrm{T}}A$ is automatically a square symmetric matrix:

$$\textit{The transpose of}\quad A^{\mathrm{T}}A \quad \textit{is} \quad A^{\mathrm{T}}(A^{\mathrm{T}})^{\mathrm{T}} \quad \textit{which is} \quad A^{\mathrm{T}}A \ \textit{again.} \tag{10}$$

The matrix AA^{T} is also symmetric. (The shapes of A and A^{T} allow multiplication.) But AA^{T} *is a different matrix from* $A^{\mathrm{T}}A$. In our experience, most scientific problems that start with a rectangular matrix A end up with $A^{\mathrm{T}}A$ or AA^{T} or both. As in least squares.

Example 3 Multiply $\boldsymbol{A} = \begin{bmatrix} -1 & 1 & 0 \\ 0 & -1 & 1 \end{bmatrix}$ times $\boldsymbol{A}^{\mathrm{T}} = \begin{bmatrix} -1 & 0 \\ 1 & -1 \\ 0 & 1 \end{bmatrix}$ in both orders.

$$\boldsymbol{AA}^{\mathrm{T}} = \begin{bmatrix} 2 & -1 \\ -1 & 2 \end{bmatrix} \text{ and } \boldsymbol{A}^{\mathrm{T}}\boldsymbol{A} = \begin{bmatrix} 1 & -1 & 0 \\ -1 & 2 & -1 \\ 0 & -1 & 1 \end{bmatrix} \text{ are both symmetric matrices.}$$

The product AA^{T} is m by m. In the opposite order, $A^{\mathrm{T}}A$ is n by n. Both are symmetric, with positive diagonal (*why*?). But even if $m = n$, it is very likely that $\boldsymbol{A}^{\mathrm{T}}\boldsymbol{A} \neq \boldsymbol{AA}^{\mathrm{T}}$.

Symmetric matrices in elimination $\quad S^{\mathrm{T}} = S$ makes elimination twice as fast. We can work with half the matrix (plus the diagonal). ***The symmetry is in the triple product*** $\boldsymbol{S = LDL^{\mathrm{T}}}$. The diagonal matrix D of pivots can be divided out, to leave $\boldsymbol{U = L^{\mathrm{T}}}$.

$$\begin{bmatrix} 1 & 2 \\ 2 & 7 \end{bmatrix} = \begin{bmatrix} 1 & 0 \\ 2 & 1 \end{bmatrix}\begin{bmatrix} 1 & 2 \\ 0 & 3 \end{bmatrix} \qquad \begin{array}{l} LU \text{ misses the symmetry of } S \\ \text{Divide the pivots } 1, 3 \text{ out of } U \end{array}$$

$$\begin{bmatrix} 1 & 2 \\ 2 & 7 \end{bmatrix} = \begin{bmatrix} \mathbf{1} & \mathbf{0} \\ \mathbf{2} & \mathbf{1} \end{bmatrix}\begin{bmatrix} 1 & 0 \\ 0 & 3 \end{bmatrix}\begin{bmatrix} \mathbf{1} & \mathbf{2} \\ 0 & \mathbf{1} \end{bmatrix} \qquad \begin{array}{l} \boldsymbol{S = LDL^{\mathrm{T}}} \textbf{ captures the symmetry} \\ \textit{Pivots in } \boldsymbol{D}\textit{. Now } \boldsymbol{U} \textit{ is the transpose of } \boldsymbol{L} \end{array}$$

Example 4 For a rectangular A, this ***saddle-point matrix S*** is symmetric and important:

Block matrix from least squares $\quad S = \begin{bmatrix} \boldsymbol{I} & \boldsymbol{A} \\ \boldsymbol{A}^{\mathrm{T}} & \mathbf{0} \end{bmatrix} = \boldsymbol{S}^{\mathrm{T}}$ has size $m + n$.

Apply block elimination to find a **block factorization** $\boldsymbol{S = LDL^{\mathrm{T}}}$. Then test invertibility:

Block elimination Subtract $\boldsymbol{A}^{\mathrm{T}}$**(row 1)** $\quad \boldsymbol{S} = \begin{bmatrix} I & A \\ A^{\mathrm{T}} & 0 \end{bmatrix}$ goes to $\begin{bmatrix} I & A \\ 0 & -\boldsymbol{A}^{\mathrm{T}}\boldsymbol{A} \end{bmatrix}$. This is $\boldsymbol{U}$.

The block pivot matrix $\boldsymbol{D}$ contains I and $-A^{\mathrm{T}}A$. Then L and L^{T} contain A^{T} and A:

Block factorization $\quad S = LDL^{\mathrm{T}} = \begin{bmatrix} I & 0 \\ \boldsymbol{A}^{\mathrm{T}} & I \end{bmatrix}\begin{bmatrix} I & 0 \\ 0 & -\boldsymbol{A}^{\mathrm{T}}\boldsymbol{A} \end{bmatrix}\begin{bmatrix} I & \boldsymbol{A} \\ 0 & I \end{bmatrix}$.

S is invertible $\iff$ $\boldsymbol{A}^{\mathrm{T}}\boldsymbol{A}$ ***is invertible*** $\iff$ $\boldsymbol{Ax \neq 0}$ ***whenever*** $\boldsymbol{x \neq 0}$

Problem Set 2.4

Questions 1–7 are about the rules for transpose matrices.

1 Find A^{T} and A^{-1} and $(A^{-1})^{\mathrm{T}}$ and $(A^{\mathrm{T}})^{-1}$ for $A = \begin{bmatrix} 1 & 0 \\ 9 & 3 \end{bmatrix}$ and $\begin{bmatrix} 1 & c \\ c & 0 \end{bmatrix}$.

2 Verify that $(AB)^{\mathrm{T}}$ equals $B^{\mathrm{T}}A^{\mathrm{T}}$ but those are different from $A^{\mathrm{T}}B^{\mathrm{T}}$:

$$A = \begin{bmatrix} 1 & 0 \\ 2 & 1 \end{bmatrix} \qquad B = \begin{bmatrix} 1 & 3 \\ 0 & 1 \end{bmatrix} \qquad AB = \begin{bmatrix} 1 & 3 \\ 2 & 7 \end{bmatrix}.$$

Show also that AA^{T} is different from $A^{\mathrm{T}}A$. But both of those matrices are _____ .

3 (a) The matrix $((AB)^{-1})^{\mathrm{T}}$ comes from $(A^{-1})^{\mathrm{T}}$ and $(B^{-1})^{\mathrm{T}}$. *In what order*?

(b) If U is upper triangular then $(U^{-1})^{\mathrm{T}}$ is _____ triangular.

4 Show that $A^2 = 0$ is possible but $A^{\mathrm{T}}A = 0$ is not possible (unless A = zero matrix).

5 (a) Compute the number $\boldsymbol{x}^{\mathrm{T}}A\boldsymbol{y} = \begin{bmatrix} 0 & 1 \end{bmatrix} \begin{bmatrix} 1 & 2 & 3 \\ 4 & 5 & 6 \end{bmatrix} \begin{bmatrix} 0 \\ 1 \\ 0 \end{bmatrix} =$ _____ .

(b) This is the row $\boldsymbol{x}^{\mathrm{T}}A =$ _____ times the column $\boldsymbol{y} = (0, 1, 0)$.

(c) This is the row $\boldsymbol{x}^{\mathrm{T}} = \begin{bmatrix} 0 & 1 \end{bmatrix}$ times the column $A\boldsymbol{y} =$ _____ .

6 The transpose of a block matrix $M = \begin{bmatrix} \mathbf{A} & \mathbf{B} \\ \mathbf{C} & \mathbf{D} \end{bmatrix}$ is $M^{\mathrm{T}} =$ _____ if A, D are square. Under what conditions on A, B, C, D is this block matrix symmetric?

7 True or false:

(a) The block matrix $\begin{bmatrix} \mathbf{0} & \mathbf{A} \\ \mathbf{A} & \mathbf{0} \end{bmatrix}$ is automatically symmetric.

(b) If A and B are symmetric then their product AB is symmetric.

(c) If A is not symmetric then A^{-1} is not symmetric.

(d) When A, B, C are symmetric, the transpose of ABC is CBA.

Questions 8–15 are about permutation matrices.

8 Why are there $n!$ permutation matrices of order n?

9 If P_1 and P_2 are permutation matrices, so is P_1P_2. This still has the rows of I in some order. Give examples with $P_1P_2 \neq P_2P_1$ and $P_3P_4 = P_4P_3$.

10 There are 12 "*even*" permutations of $(1, 2, 3, 4)$, with an *even number of exchanges*. Two of them are $(1, 2, 3, 4)$ with zero exchanges and $(4, 3, 2, 1)$ with 2 exchanges. List the other ten. Instead of writing each 4 by 4 matrix, just order the numbers.

11 If P has 1's on the antidiagonal from $(1, n)$ to $(n, 1)$, describe PAP. Note $P = P^{\mathrm{T}}$.

12 Explain why the dot product of $\boldsymbol{x}$ and $\boldsymbol{y}$ equals the dot product of $P\boldsymbol{x}$ and $P\boldsymbol{y}$. Then $(P\boldsymbol{x})^{\mathrm{T}}(P\boldsymbol{y}) = \boldsymbol{x}^{\mathrm{T}}\boldsymbol{y}$ tells us that $\boldsymbol{P}^{\mathrm{T}}\boldsymbol{P} = \boldsymbol{I}$ for any permutation. With $\boldsymbol{x} = (1,2,3)$ and $\boldsymbol{y} = (1,4,2)$ choose P to show that $P\boldsymbol{x}\cdot\boldsymbol{y}$ is not always $\boldsymbol{x}\cdot P\boldsymbol{y}$.

13 Which permutation makes PA upper triangular? Which permutations make P_1AP_2 lower triangular? ***Multiplying*** A ***on the right by*** P_2 ***exchanges the*** _____ ***of*** A.

$$A = \begin{bmatrix} 0 & 0 & 6 \\ 1 & 2 & 3 \\ 0 & 4 & 5 \end{bmatrix}.$$

14 Find a 3 by 3 permutation matrix with $P^3 = I$ (but not $P = I$). Why can't P simply exchange two rows? Find a 4 by 4 permutation $\widehat{P}$ with $\widehat{P}^4 \neq I$.

15 All row exchange matrices are symmetric: $P^{\mathrm{T}} = P$. Then $P^{\mathrm{T}}P = I$ becomes $P^2 = I$. Other permutation matrices may or may not be symmetric.

(a) If P sends row 1 to row 4, then P^{T} sends row _____ to row _____.
When $P^{\mathrm{T}} = P$ the row exchanges come in pairs with no overlap.

(b) Find a 4 by 4 example with $P^{\mathrm{T}} = P$ that moves all four rows.

Questions 16–18 are about symmetric matrices and their factorizations.

16 If $A = A^{\mathrm{T}}$ and $B = B^{\mathrm{T}}$, which of these matrices are certainly symmetric?

(a) $A^2 - B^2$ (b) $(A+B)(A-B)$ (c) ABA (d) $ABAB$.

17 (a) How many entries of S can be chosen independently, if $S = S^{\mathrm{T}}$ is 5 by 5?

(b) How do L and D (still 5 by 5) give the same number of choices in LDL^{T}?

(c) How many entries can be chosen if A is *skew-symmetric*? ($A^{\mathrm{T}} = -A$).

(d) Why does $A^{\mathrm{T}}A$ have no negative numbers on its diagonal?

18 Factor these symmetric matrices into $S = LDL^{\mathrm{T}}$. The pivot matrix D is diagonal:

$$S = \begin{bmatrix} 1 & 3 \\ 3 & 2 \end{bmatrix} \quad\text{and}\quad S = \begin{bmatrix} 1 & b \\ b & c \end{bmatrix} \quad\text{and}\quad S = \begin{bmatrix} 2 & -1 & 0 \\ -1 & 2 & -1 \\ 0 & -1 & 2 \end{bmatrix}.$$

19 Find the $PA = LU$ factorizations (and check them) for

$$A = \begin{bmatrix} 0 & 1 & 1 \\ 1 & 0 & 1 \\ 2 & 3 & 4 \end{bmatrix} \quad\text{and}\quad A = \begin{bmatrix} 1 & 2 & 0 \\ 2 & 4 & 1 \\ 1 & 1 & 1 \end{bmatrix}.$$

20 Find a 4 by 4 permutation matrix (call it A) that needs 3 row exchanges to reach the end of elimination. For this matrix, what are its factors P, L, and U?

21 Prove that the identity matrix cannot be the product of three row exchanges (or five). It can be the product of two exchanges (or four).

22 If every row of a 4 by 4 matrix contains the numbers $0, 1, 2, 3$ in some order, can the matrix be symmetric?

23 Start with 9 entries in a 3 by 3 matrix A. Prove that no reordering of rows and reordering of columns can produce A^{T}. (Watch the diagonal entries.)

24 Wires go between Boston, Chicago, and Seattle. Those cities are at voltages x_B, x_C, x_S. With unit resistances between cities, the currents between cities are in $\boldsymbol{y}$:

$$\boldsymbol{y} = A\boldsymbol{x} \quad \text{is} \quad \begin{bmatrix} y_{BC} \\ y_{CS} \\ y_{BS} \end{bmatrix} = \begin{bmatrix} 1 & -1 & 0 \\ 0 & 1 & -1 \\ 1 & 0 & -1 \end{bmatrix} \begin{bmatrix} x_B \\ x_C \\ x_S \end{bmatrix}.$$

(a) Find the total currents $A^{\mathrm{T}}\boldsymbol{y}$ out of the three cities.

(b) Verify that $(A\boldsymbol{x})^{\mathrm{T}}\boldsymbol{y}$ agrees with $\boldsymbol{x}^{\mathrm{T}}(A^{\mathrm{T}}\boldsymbol{y})$—six terms in both.

25 The matrix P that multiplies (x, y, z) to give (z, x, y) is also a rotation matrix. Find P and P^3. The rotation axis $\boldsymbol{a} = (1, 1, 1)$ doesn't move, it equals $P\boldsymbol{a}$. What is the angle of rotation from $\boldsymbol{v} = (2, 3, -5)$ to $P\boldsymbol{v} = (-5, 2, 3)$?

26 Here is a new factorization $A = LS =$ *triangular times symmetric*:

Start from $A = LDU$. Then A equals $L\,(U^{\mathrm{T}})^{-1}$ times $S = U^{\mathrm{T}}DU$.

Why is $L\,(U^{\mathrm{T}})^{-1}$ triangular? Why is $U^{\mathrm{T}}DU$ symmetric?

27 In algebra, a *group* of matrices includes AB and A^{-1} if it includes A and B. "*Products and inverses stay in the group.*" Which of these sets are groups?
Lower triangular matrices L with 1's on the diagonal, symmetric matrices S, positive matrices M, diagonal invertible matrices D, permutation matrices P, orthogonal matrices with $Q^{\mathrm{T}} = Q^{-1}$. ***Invent two more matrix groups.***

Challenge Problems

28 If you take powers of a permutation matrix, why is some P^k eventually equal to I? Find a 5 by 5 permutation P so that the smallest power to equal I is P^6.

29 (a) Write down any 3 by 3 matrix M. Split M into $S + A$ where $S = S^{\mathrm{T}}$ is symmetric and $A = -A^{\mathrm{T}}$ is anti-symmetric.

(b) Find formulas for S and A involving M and M^{T}. We want $M = S + A$.

30 Suppose Q^{T} *equals* Q^{-1}, so $Q^{\mathrm{T}}Q = I$. Then Q is called an **orthogonal matrix**.

(a) Show that the columns $\boldsymbol{q}_1, \ldots, \boldsymbol{q}_n$ are unit vectors: $\|\boldsymbol{q}_i\|^2 = 1$.

(b) Show that every two columns of Q are perpendicular: $\boldsymbol{q}_1^{\mathrm{T}}\boldsymbol{q}_2 = 0$.

(c) Find a 2 by 2 example with first entry $q_{11} = \cos\theta$.

2.5 Derivatives and Finite Difference Matrices

This section aims to connect key points of calculus and linear algebra. The matrices imitate derivatives (as nearly as possible). Derivatives tell us what is happening at one point x of space or at one moment t in time:

$\dfrac{dy}{dx} > 0$: The graph of y **goes upward**	$\dfrac{du}{dt} > 0$: The function u is increasing
$\dfrac{d^2y}{dx^2} > 0$: The graph of y **bends upward**	$\dfrac{d^2u}{dt^2} > 0$: The growth rate $\dfrac{du}{dt}$ is increasing

It would be hard to write down ideas about $y(x)$ and $u(t)$ more important than those. Increasing or decreasing is different from bending up (accelerating) or bending down.

The parabola $y = x^2$ has derivatives $dy/dx = 2x$ and $d^2y/dx^2 = 2$. Its graph goes *downward* (decreasing) when x is negative and *upward* when x is positive. Everywhere the graph of x^2 is *bending up*. Its second derivative is positive $(+2)$ so its first derivative $2x$ is increasing. Increasing the slope makes the graph bend upwards.

Now come differences instead of derivatives: **y at two points and u at two times. The two points are x and $x + h$.** Previously we looked at derivatives (one point only). Now we look at differences between those points. Here are three important statements:

1. The difference $\Delta y = y(x+h) - y(x)$ is approximately **h times the slope** $\dfrac{dy}{dx}$ at x

2. A better approximation to Δy adds on $\dfrac{1}{2}h^2$ **times the second derivative** $\dfrac{d^2y}{dx^2}$ at x

3. The exact Δy is $y(x+h) - y(x) =$ **the integral** of $\dfrac{dy}{dx}$ from x to $x + h$

Statements **1** and **2** are connecting the change $y(x+h) - y(x)$ to the derivatives of y *at one point*. The first derivative gives "first order" approximation—good but not great. By including the second derivative (times $\frac{1}{2}h^2$) we get a "second order" approximation—much better. Compare those options for $y = x^2$ with $dy/dx = 2x$ and $d^2y/dx^2 = 2$. The exact $\Delta y = (x+h)^2 - x^2$ is $x^2 + 2hx + h^2 - x^2$.

$$\Delta y = \underset{\text{first order}}{2hx} + \underset{\text{second order}}{h^2} = \underset{\text{tangent line}}{h\frac{dy}{dx}} + \underset{\text{parabola}}{\frac{1}{2}h^2\frac{d^2y}{dx^2}}$$

The derivative dy/dx multiplied the step length $h = \Delta x$. That first approximation follows the **tangent line** to the graph. The second derivative multiplies $\frac{1}{2}h^2$ to give a valuable correction. The second order approximation produces the **tangent parabola**.

For the special example $y = x^2$, the tangent parabola is the actual parabola. If we want to estimate the error from the tangent parabola, we can go onward to the function $y = x^3$. The graphs of $\boldsymbol{y = x^2}$ and $\boldsymbol{y = x^3}$ from $x = 1$ to $x = 1 + h = 2$ show how the tangent parabola improves on the tangent line.

The formula for $y(x+h) - y(x)$ that includes **all** the derivatives of y is the **Taylor Series**. It starts with h times dy/dx and $\frac{1}{2}h^2$ times the second derivative d^2y/dx^2. The next term is $\frac{1}{6}h^3$ times the third derivative. Every term is $h^n/n!$ times the nth derivative.

Here are two popular ways to write the Taylor series. First, we use derivatives at $x = 0$ to predict $y(x)$ at other points x. Second, we use derivatives at a point x to predict $y(x+h)$ at a different point $x + h$.

Taylor series for $\boldsymbol{y(x)}$ $\quad y(0) + x\dfrac{dy}{dx}(0) + \dfrac{1}{2}x^2\dfrac{d^2y}{dx^2}(0) + \cdots + \dfrac{1}{n!}x^n\dfrac{d^ny}{dx^n}(0) + \cdots$

Taylor series for $\boldsymbol{y(x+h)}$ $\quad y(x) + hy'(x) + \dfrac{1}{2}h^2y''(x) + \cdots + \dfrac{1}{n!}h^n\dfrac{d^ny}{dx^n}(x) + \cdots$

These are infinite series! For famous functions like $y(x) = e^x$, the series converges at all points. (That makes e^x an entire **analytic function**.) For the function $y(x) = 1/(1-x)$, the series $1 + x + x^2 + \cdots$ will only converge for $|x| < 1$. So $1/(1-x)$ is analytic up to that point but $x = 1$ is a "singularity" called a "pole" of $1/(1-x)$.

Turning the Formulas Around : Derivatives from Differences

Our starting point is still the important approximation by a tangent parabola:

$$\boldsymbol{y(x+h) \approx y(x) + h\frac{dy}{dx}(x) + \frac{1}{2}\,h^2\,\frac{d^2y}{dx^2}(x)} \tag{1}$$

Our questions are now in the opposite direction. If we know $y(x)$ and $y(x+h)$, how do we estimate dy/dx? **If we also know $\boldsymbol{y(x-h)}$, how do we estimate $\boldsymbol{dy/dx}$** (improved!) **and also $\boldsymbol{d^2y/dx^2}$?** With three values $y(x-h), y(x), y(x+h)$ we can get the most important approximations in scientific computing. We must do this, and it is not difficult. Write the approximation (1) at $x + h$ and then write it again at $x - h$ (change h to $-h$):

$$y(x+h) - y(x) \approx \quad \boldsymbol{h}\frac{dy}{dx} + \frac{1}{2}\,h^2\,\frac{d^2y}{dx^2} \tag{2a}$$

$$y(x-h) - y(x) \approx \boldsymbol{-h}\frac{dy}{dx} + \frac{1}{2}\,h^2\,\frac{d^2y}{dx^2} \tag{2b}$$

Subtract (2b) from (2a) and divide by $2h$. This gives an improved approximation to dy/dx:

$$\textbf{Centered difference} \qquad \boldsymbol{\frac{y(x+h) - y(x-h)}{2h} \approx \frac{dy}{dx}} \tag{3}$$

Add (2b) to (2a) and divide by h^2 for an excellent approximation to the second derivative!

$$\textbf{Second difference} \qquad \frac{\boldsymbol{y(x+h) - 2y(x) + y(x-h)}}{\boldsymbol{h^2}} \approx \frac{\boldsymbol{d^2y}}{\boldsymbol{dx^2}} \tag{4}$$

Those are the two formulas we need for the big step we can now take—approximating a differential equation. Many important differential equations are "second order". They include first and second derivatives as in

$$\frac{d^2y}{dx^2} + a\,\frac{dy}{dx} + by = f(x) \tag{5}$$

At a typical point x, a good finite difference approximation is

$$\frac{y(x+h) - 2y(x) + y(x-h)}{h^2} + a\,\frac{y(x+h) - y(x-h)}{2h} + by(x) = f(x) \tag{6}$$

That step went from derivatives in (5) to differences in (6), using the approximations (3) and (4). You can compare (3) with these uncentered differences, only **first order accurate**:

$$\begin{array}{lll} \textbf{Forward difference} & \dfrac{dy}{dx} = \dfrac{y(x+h) - y(x)}{h} + \boldsymbol{O(h)} \text{ error} \\[2ex] \textbf{Backward difference} & \dfrac{dy}{dx} = \dfrac{y(x) - y(x-h)}{h} + \boldsymbol{O(h)} \text{ error} \end{array} \tag{7}$$

When we average forward and backward in (7), we get centered in (3). The leading $O(h)$ error term cancels out. In that way the centered difference has second order accuracy.

Second Difference Matrices K, T, B

This section introduces three important second difference matrices $\boldsymbol{K, T, B}$. They all have the $1, -2, 1$ pattern that equation (4) found in approximating d^2y/dx^2 (bending of a graph and acceleration of motion).

The actual numbers in those matrices will *reverse signs* to $-1, 2, -1$. Now we are approximating $-d^2y/dx^2$. By extending the pattern from a second difference at *one point* to second differences at *many points*, this section takes the step to matrices—crucially important.

The big step is to approximate a differential equation by N difference equations at N meshpoints. We normally use equally spaced points $x_1 = h, x_2 = 2h, \ldots, x_N = Nh$. The spacing h is the **meshwidth**. The **boundary points** are $x_0 = 0$ and $x_{N+1} = 1$. Thus $(N+1)h$ equals the total length 1. The meshwidth is $\boldsymbol{h = 1/(N+1)}$. That number h will appear in first differences, and h^2 appears in second differences.

Start with the equation $-d^2u/dx^2 = f(x)$. Suppose the boundary points $x = 0$ and $x = 1$ are both **fixed at zero**. This means $\boldsymbol{u(0) = 0}$ and also $\boldsymbol{u(1) = 0}$. For the difference equation those boundary conditions translate into $u_0 = 0$ and $u_{N+1} = 0$. The unknown numbers $u_1, u_2, \ldots, u_N$ will be **approximations** to the values $u(h), u(2h), \ldots, u(Nh)$ of the exact solution of the differential equation between $x = 0$ and $x = 1$.

Our unknown is the vector $U = (u_1, \ldots, u_N)$. We need N difference equations, one at each meshpoint. We plan to express those N equations as **one matrix equation**. Here is that matrix equation $(1/h^2)KU = F$ for $N = 4$ and $h = \frac{1}{5}$:

$$-\frac{d^2u}{dx^2} = f(x) \quad \text{becomes } KU/h^2 = F \qquad \frac{1}{h^2}\begin{bmatrix} 2 & -1 & 0 & 0 \\ -1 & 2 & -1 & 0 \\ 0 & -1 & 2 & -1 \\ 0 & 0 & -1 & 2 \end{bmatrix}\begin{bmatrix} u_1 \\ u_2 \\ u_3 \\ u_4 \end{bmatrix} = \begin{bmatrix} f(h) \\ f(2h) \\ f(3h) \\ f(4h) \end{bmatrix}.$$

Equation (4) said that d^2u/dx^2 is close to $u(x+h) - 2u(x) + u(x-h)$ *divided by* h^2. This matrix K gives a natural approximation to $-d^2u/dx^2$ with **fixed-fixed boundary conditions $u_0 = 0$ and $u_{N+1} = 0$.**

The key point is : $N = 4$ equations went into an N by N matrix. The usual *row times column rule* for multiplying KU reproduces our N difference equations. Row 1 times U is $\boldsymbol{2u_1 - u_2}$ = dot product of the row $(2, -1, 0, 0)$ and the column (u_1, u_2, u_3, u_4). For this first row—and also for the last row—the matrix K has built in the fixed boundary conditions $u_0 = 0$ and $u_5 = 0$. A typical row is $(-u_1 + 2u_2 - u_3)/h^2 = f(h)$.

The division by h^2 makes K/h^2 a **second difference matrix**, replacing $\boldsymbol{-d^2u/dx^2}$.

Properties of K

At this point I would like to ask : *What are the important patterns in the matrix K* ? When I ask my class that question, I usually get three answers—sometimes four. Let me put those answers in the order that many students have given. Here is the matrix K for $N = 4$.

Second difference matrix
Fixed-fixed endpoints

$$\boldsymbol{K_4} = \begin{bmatrix} \mathbf{2} & -1 & 0 & 0 \\ -1 & \mathbf{2} & -1 & 0 \\ 0 & -1 & \mathbf{2} & -1 \\ 0 & 0 & -1 & \mathbf{2} \end{bmatrix} \qquad (8)$$

1. **K is symmetric.** The number in row i, column j is also in row j, column i. Thus $K_{ij} = K_{ji}$, on opposite sides of the main diagonal. "K transpose equals K." So the matrix K is symmetric across the diagonal.

   The simplest way to express symmetry will be $\boldsymbol{K^{\mathrm{T}} = K}$. That transpose puts the rows of any matrix A into the columns of A^{T}. In our case K^{T} stays the same as K.

$$A = \begin{bmatrix} 1 & 2 \\ 3 & 4 \\ 5 & 6 \end{bmatrix} \text{ has } A^{\mathrm{T}} = \begin{bmatrix} 1 & 3 & 5 \\ 2 & 4 & 6 \end{bmatrix} \text{ and } (A^{\mathrm{T}})^{\mathrm{T}} = A \text{ and } (A^{\mathrm{T}}A)^{\mathrm{T}} = A^{\mathrm{T}}A.$$

2. ***K* is banded**. All the nonzeros in K lie in a "band" around the main diagonal. The band has only three diagonals, so ***K* is a tridiagonal matrix**. This means that $KU/h^2 = F$ can be quickly solved. Elimination is easy and fast.

   A matrix with a narrow band is **sparse—mostly zeros**. We don't see this well for $N = 4$ (small matrix). But we see it clearly for $N = 100$. In that case K has $100 + 99 + 99$ nonzeros out of $(100)^2 = 10000$ entries : *less than* 3% *nonzeros*. And $N = 100$ is by no means a large matrix.

3. ***K* has constant diagonals.** A diagonal of -1's, then a diagonal of 2's, then a diagonal of -1's. The matrix is **"shift-invariant"**. This is because the underlying differential equation had a constant coefficient -1. The approximation to $-d^2u/dx^2$ has the same numbers $-1, 2, -1$ at every x. (Strictly speaking the shift-invariance fails at the boundaries—always an awkward problem.)

   Right away that constant-diagonal property wakes up Fourier. The whole world of Fourier transforms is linked to constant-diagonal matrices. We also call them **filters** or **convolution matrices**. The mathematical name is **Toeplitz matrix**:

   This K is certainly one of the most important matrices in scientific computing.

   The command $K =$ toeplitz$[2, -1, 0, 0]$ **constructs** $K = K^{\mathrm{T}}$ **from its first row.**

That completes three properties of K that are directly visible: **symmetry, sparsity,** and **shift-invariance**. All the matrices K_n have a fourth property that is equally important. But this fact is not directly obvious:

4. ***K* is invertible**

   It has an inverse matrix $\boldsymbol{K^{-1}}$

   Then $\boldsymbol{K^{-1}K = I}$ **and** $\boldsymbol{KK^{-1} = I}$

$$K_4^{-1} = \frac{1}{5}\begin{bmatrix} 4 & 3 & 2 & 1 \\ 3 & 6 & 4 & 2 \\ 2 & 4 & 6 & 3 \\ 1 & 2 & 3 & 4 \end{bmatrix}. \tag{9}$$

This inverse matrix K^{-1} is also symmetric, but it is not tridiagonal. It is a dense matrix (in fact no entries are zero). Numerically we don't need to know K^{-1}, but *we do need to know that the inverse exists.*

Invertibility is not easy to decide from a quick look at the matrix. Theoretically, we can compute the determinant. (Here it was 5.) Since we have to divide by it, *the determinant must not be zero*. Then the matrix has an inverse—but computing the determinant or K^{-1} is almost never done in practice! It is a very poor way to find the solution vector $K^{-1}F$.

What we actually do is to go ahead with the elimination steps that solve $KU = F$. The nonzero pivots on the main diagonal of the triangular matrix show that K is invertible.

The backslash command $\boldsymbol{U = K\backslash F}$ produces the fastest solution in MATLAB.

5. I am going to add one more property: **The symmetric matrices $\boldsymbol{K_n}$ are positive definite**. A goal of Chapter 6 is to explain what this means. We plan to contrast K and T (positive definite) with the matrix B (positive **semidefinite**).

All depends on the difference between "positive" and "nonnegative": >0 and ≥ 0.

Pivots
- An invertible matrix has n *nonzero* pivots
- A positive *definite* symmetric matrix has n *positive* pivots
- A positive *semidefinite* symmetric matrix has n *nonnegative* pivots

The Free-Fixed Matrices T_n

To complete this section, we will describe T_n and B_n. They are variations on K_n. Those variations come from "changing the boundary conditions". We are thinking of an elastic bar that is fixed at two ends or only one end or entirely free. This model problem will be described after we see the matrices.

The second difference matrices T_n are symmetric and tridiagonal like K_n. But the $(1,1)$ entry in T_n is changed from 2 to 1. The first row $1,-1$ represents a **free boundary condition** $du/dx = 0$, whose meaning we will soon understand.

Free-fixed boundary conditions

Still positive definite

$$\boldsymbol{T_4} = \begin{bmatrix} \mathbf{1} & -1 & 0 & 0 \\ -1 & 2 & -1 & 0 \\ 0 & -1 & 2 & -1 \\ 0 & 0 & -1 & \mathbf{2} \end{bmatrix} \tag{10}$$

T is no longer Toeplitz, because its main diagonal is not constant. But it does have a simpler factorization than K (because every pivot of T equals 1).

$$\boldsymbol{T} = \begin{bmatrix} 1 & & & \\ -1 & 1 & & \\ 0 & -1 & 1 & \\ 0 & 0 & -1 & 1 \end{bmatrix} \begin{bmatrix} 1 & -1 & 0 & 0 \\ & 1 & -1 & 0 \\ & & 1 & -1 \\ & & & 1 \end{bmatrix} = \boldsymbol{LU} \tag{11}$$

T comes from multiplying a backward difference L times its transpose: $U = L^{\mathrm{T}}$. Notice that U is *minus* a forward difference. That gives the minus sign we need, because T is *minus* a second difference.

That minus sign produces a positive definite matrix $T = LL^{\mathrm{T}}$! Please notice that L, U, and $T = LU$ all have beautiful inverses.

$$\boldsymbol{U^{-1}} = \begin{bmatrix} 1 & 1 & 1 & 1 \\ & 1 & 1 & 1 \\ & & 1 & 1 \\ & & & 1 \end{bmatrix} \quad \boldsymbol{L^{-1}} = \begin{bmatrix} 1 & & & \\ 1 & 1 & & \\ 1 & 1 & 1 & \\ 1 & 1 & 1 & 1 \end{bmatrix} \quad \boldsymbol{T^{-1}} = U^{-1}L^{-1} = \begin{bmatrix} 4 & 3 & 2 & 1 \\ 3 & 3 & 2 & 1 \\ 2 & 2 & 2 & 1 \\ 1 & 1 & 1 & 1 \end{bmatrix}. \tag{12}$$

The Free-Free Matrices B_n are Singular

For a square matrix B, "singular" means "not invertible". One test is *determinant = zero*. The inverse of B includes a division by its determinant D, so we need $D \neq 0$. A better test (since computing D is not efficient) is to find a nonzero vector $\boldsymbol{x}$ such that $B\boldsymbol{x} = \mathbf{0}$.

> If B multiplies a nonzero vector x to produce $Bx = 0$, then B cannot be invertible.
> The inverse would give $B^{-1}Bx = x$. But this would mean $B^{-1}\mathbf{0} \neq \mathbf{0}$. Not possible.

Example The free-free matrix B_3 has 1 (not 2) in its $(1,1)$ and $(3,3)$ corners:

$$B_3 = \begin{bmatrix} \mathbf{1} & -1 & 0 \\ -1 & 2 & -1 \\ 0 & -1 & \mathbf{1} \end{bmatrix} \quad \text{has } B_3x = \begin{bmatrix} 1 & -1 & 0 \\ -1 & 2 & -1 \\ 0 & -1 & 1 \end{bmatrix} \begin{bmatrix} \mathbf{1} \\ \mathbf{1} \\ \mathbf{1} \end{bmatrix} = \begin{bmatrix} \mathbf{0} \\ \mathbf{0} \\ \mathbf{0} \end{bmatrix}. \tag{13}$$

The all-ones vector $x = (1,1,1)$ solves $B_3x = 0$. That vector x is in the **nullspace** of B_3. Any constant vector $x = (c,c,c)$ will be in that nullspace, with $B_3x = \mathbf{0}$. This is because **all the rows of B_3 add to zero**. Such a matrix cannot be invertible.

To see why B_3 is singular when K_3 and T_3 are invertible, look at these models. They have three masses connected by springs. The differences are at the ends (the boundaries). For K_3, both ends are connected to **fixed supports**. For B_3, both of the ends are **free: no supports**. In between, T_3 is *free* at the top and *fixed* at the bottom.

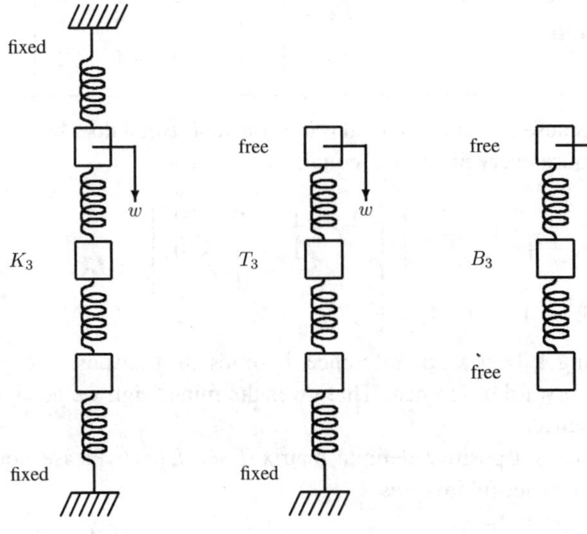

Figure 2.3: K is **fixed-fixed**, T is **free-fixed**, B is **free-free** (singular, no inverse).

If you add a weight W in the middle box, the B_3 springs cannot adjust because they have no support. But the K_3 and T_3 springs will adjust to balance that weight:

$$\begin{bmatrix} 2 & -1 & \\ -1 & 2 & -1 \\ & -1 & 2 \end{bmatrix} \begin{bmatrix} W/2 \\ W \\ W/2 \end{bmatrix} = \begin{bmatrix} 0 \\ W \\ 0 \end{bmatrix} \quad \text{and} \quad \begin{bmatrix} 1 & -1 & \\ -1 & 2 & -1 \\ & -1 & 2 \end{bmatrix} \begin{bmatrix} 2W \\ 2W \\ W \end{bmatrix} = \begin{bmatrix} 0 \\ W \\ 0 \end{bmatrix}. \tag{14}$$

Masses and springs lead to matrices. Elastic bars lead to differential equations.

Problem Set 2.5

1 Find the first order and second order terms in Δy (involving h and h^2) for the function $y(x) = x^3$ with $\Delta y = (x+h)^3 - x^3$. Multiply out $(x+h)^3$ and show that your answers match $h\,dy/dx$ and $\frac{1}{2}\,h^2\,d^2y/dx^2$.

2 For the function $y = e^x$, what is the slope of the tangent line at the point $x = 0$, $y = 1$? What is the equation $y = \cdots$ for that tangent line?

3 Continuing with $y = e^x$, what is the second derivative at $x = 0$ and what is the equation for the tangent parabola? This is the second order approximation to e^h: $y(h) = y(0) + h\,dy/dx + \frac{1}{2}\,h^2\,d^2y/dx^2$.

4 Now find all the derivatives d^ny/dx^n of $y(x) = e^x$ at $x = 0$. Write the **Taylor Series**!

$$y(h) = e^h = y(0) + h\frac{dy}{dx}(0) + \cdots + \frac{h^n}{n!}\frac{d^ny}{dx^n}(0) + \cdots$$

I believe that this is the most important infinite series in mathematics.

5 For $y(x) = \sin x$, what order of accuracy is the approximation $\sin h \approx h$?

6 Euler's great formula $\boldsymbol{e^{ix} = \cos x + i\,\sin x}$ can come by comparing Taylor series! What are all the derivatives of $y = e^{ix}$ at $x = 0$?

$$\boldsymbol{e^{ix}} = 1 + ix + \frac{1}{2}(ix)^2 + \frac{1}{6}(ix)^3 + \cdots = \left(1 - \frac{1}{2}x^2 + \cdots\right) + i\left(x - \frac{1}{6}x^3 + \cdots\right)$$

7 For $y(x) = x^3$ and $h = 1/10$, compute these approximations to dy/dx at $x = 0$:

Centered $\dfrac{y(h) - y(-h)}{2h}$ **Forward** $\dfrac{y(h) - y(0)}{h}$ **Backward** $\dfrac{y(0) - y(-h)}{h}$

8 We know that $y(h) \approx y(0) + h\,dy/dx\,(0) + \frac{1}{2}h^2\,d^2y/dx^2\,(0)$. Substitute into the three formulas in the previous problem. Up to this second order h^2 term, which formula is exact? What are the errors in the other two formulas?

9 The derivative of $y = e^x$ at $x = 0$ is $dy/dx = 1$. Find the centered-forward-backward approximations to $dy/dx\,(0)$ using $h = 1$. Which is closest to 1?

10 For $y = e^x$ and $h = 1$ how close is $(e - 2 + e^{-1})/1^2$ to the true second derivative of e^x at $x = 0$?

11 When a first difference has coefficients $1, -1$ and a second difference has coefficients $1, -2, 1$, what four numbers do you think appear in a third difference? (You would divide by h^3 to approximate d^3y/dx^3. Third differences are pretty rare, but try yours on the functions $y = x^3$ and $y = x^4$.)

12 The transpose of the centered difference Δ_0 is $-\Delta_0$ (*antisymmetric*). That is like the minus sign in integration by parts, when $f(x)g(x)$ drops to zero at $\pm\infty$:

$$\textbf{Integration by parts} \qquad \int_{-\infty}^{\infty} f(x)\,\frac{dg}{dx}\,dx = -\int_{-\infty}^{\infty} \frac{df}{dx}\,g(x)\,dx\,.$$

Verify the **summation by parts** $\displaystyle\sum_{-\infty}^{\infty} f_i\,(g_{i+1} - g_{i-1}) = -\sum_{-\infty}^{\infty} (f_{i+1} - f_{i-1})\,g_i$.

13 Show that four samples of u can give fourth-order accuracy for du/dx at the center:

$$\frac{-u_2 + 8u_1 - 8u_{-1} + u_{-2}}{12h} = \frac{du}{dx} + bh^4\,\frac{d^5u}{dx^5} + \cdots \qquad \begin{array}{ll}\text{Test} & u = x^2\\ \text{and} & u = x^4\end{array}$$

14 The inverses of K_3 and K_4 have $\dfrac{1}{\text{determinant}} = \dfrac{1}{4}$ and $\dfrac{1}{5}$:

$$K_3^{-1} = \frac{1}{4}\begin{bmatrix} 3 & 2 & 1\\ 2 & 4 & 2\\ 1 & 2 & 3\end{bmatrix} \quad \text{and} \quad K_4^{-1} = \frac{1}{5}\begin{bmatrix} 4 & 3 & 2 & 1\\ 3 & 6 & 4 & 2\\ 2 & 4 & 6 & 3\\ 1 & 2 & 3 & 4\end{bmatrix}.$$

First guess the determinant of $K = K_5$. *Challenge*: Guess the inverse of K_5. Then compute $\mathsf{det}(K) * \mathsf{inv}(K)$—any software is allowed.

15 The matrices B, T, K come from A_0, A_1, A_2 with $0, 1, 2$ boundary conditions.

$$A_0 = \begin{bmatrix} -1 & 1 & 0 & 0\\ 0 & -1 & 1 & 0\\ 0 & 0 & -1 & 1\end{bmatrix} \qquad \begin{array}{l} A_0 \text{ is a } \textbf{forward difference matrix}\\ B = A_0^{\mathrm{T}} A_0 \text{ is a second difference matrix}\end{array}$$

Which column of A_0 would you remove to produce A_1 with $T = A_1^{\mathrm{T}} A_1$? Which column would you remove next to produce A_2 with $K = A_2^{\mathrm{T}} A_2$?

16 Show that $D_4 D_4^{\mathrm{T}}$ gives a second difference. Is D_4 invertible?

$$D_4 = \begin{bmatrix} 1 & -1 & 0 & 0\\ 0 & 1 & -1 & 0\\ 0 & 0 & 1 & -1\\ -1 & 0 & 0 & 1\end{bmatrix} \quad \text{times } D_4^{\mathrm{T}} = \begin{bmatrix} 1 & 0 & 0 & -1\\ -1 & 1 & 0 & 0\\ 0 & -1 & 1 & 0\\ 0 & 0 & -1 & 1\end{bmatrix} \quad \text{equals } C_4$$

17 Solve $-u'' = \cos 4\pi x$ with fixed-fixed conditions $u(0) = u(1) = 0$. Use K_4 and K_8 to compute $u_1, \ldots, u_n$ and plot on the same graph with $u(x)$:

$$u_{i+1} - 2u_i + u_{i-1} = h^2 \cos 4\pi i h \quad \text{with} \quad u_0 = u_{n+1} = 0\,.$$

18 The centered first difference matrix might not be invertible!

$$\text{Test} \qquad \Delta_3 = \begin{bmatrix} 0 & 1 & 0\\ -1 & 0 & 1\\ 0 & -1 & 0\end{bmatrix} \quad \text{and} \quad \Delta_4 = \begin{bmatrix} 0 & 1 & 0 & 0\\ -1 & 0 & 1 & 0\\ 0 & -1 & 0 & 1\\ 0 & 0 & -1 & 0\end{bmatrix}$$

Thoughts on Chapter 2

I think of Chapter 2 as useful and practical. It solves $A\boldsymbol{x} = \boldsymbol{b}$ for a square invertible matrix. The properties of A^{-1} are important to know: $(AB)^{-1} = B^{-1}A^{-1}$ and $\boldsymbol{x} = A^{-1}\boldsymbol{b}$. But we seldom compute inverse matrices. It is expensive and unnecessary to find A^{-1} and multiply A^{-1} times $\boldsymbol{b}$.

The solution is much easier for an **upper triangular matrix** $\boldsymbol{U}$. All entries below the main diagonal are zero. Then the nth equation of $U\boldsymbol{x} = \boldsymbol{c}$ is just $u_{nn}x_n = c_n$. Dividing by u_{nn} produces x_n. Now the equation above is also easy: It tells us x_{n-1}. The whole problem $U\boldsymbol{x} = \boldsymbol{c}$ is solved from last equation to first equation: this is **back substitution**.

The goal now is to change $A\boldsymbol{x} = \boldsymbol{b}$ into $U\boldsymbol{x} = \boldsymbol{c}$. In principle this is easy to do. *For equations* $i = 2$ *to* $i = n$, *subtract* a_{i1}/a_{11} *times equation* 1 *from equation* i. That leaves $n-1$ equations with unknowns x_2 to x_n. Repeat on those equations, to find $n-2$ equations with unknowns x_3 to x_n. After $n^3/3$ steps, you have $U\boldsymbol{x} = \boldsymbol{c}$ with an upper triangular matrix U—ready for back substitution: $\boldsymbol{x} = U^{-1}\boldsymbol{c} = A^{-1}\boldsymbol{b}$.

The beautiful formula is $\boldsymbol{A = LU}$. L is the **lower triangular** matrix of multipliers (like $\ell_{i1} = a_{i1}/a_{11}$ in the first column). U is the **upper triangular** matrix that appears when elimination produces $U\boldsymbol{x} = \boldsymbol{c}$.

The only unhappy note is when a diagonal entry u_{ii} is zero or small—in which case a later number ℓ_{ij} will be infinite or large. Then we exchange equation i with equation j below it, to make the pivot u_{ii} as large as possible.

The Matrices in Chapter 2

We won't see much more of $A = LU$ or $PA = LU$. The equations $A\boldsymbol{x} = \boldsymbol{b}$ that they solve are the nice ones—and mathematics is not so much about "nice ones". The fun is to work harder and climb higher.

What to do if A is not invertible? What to do if A is not square?

What to do if A has very many columns and/or very many rows?

Those are real questions and the answers will fill this book. What we carry forward from Chapter 2 is the memory of useful matrices—and of ideas which Chapter 3 will organize and extend. Here are the "matrix ideas" from Chapter 2:

Inverse matrix	Singular matrix	Finite difference matrix
Upper triangular $\boldsymbol{U}$	Lower triangular $\boldsymbol{L}$	Centered, Forward, Backward
Elimination matrix $\boldsymbol{E}$	Permutation matrix $\boldsymbol{P}$	Second difference matrix
Transpose matrix $\boldsymbol{A}^{\mathrm{T}}$	Symmetric matrix $\boldsymbol{S = S}^{\mathrm{T}}$	K, T, B
Factorization $\boldsymbol{A = LU}$	Factorization $\boldsymbol{S = LDL}^{\mathrm{T}}$	Fixed ends or free ends

Invertible n by n matrices have column space $\mathbf{R}^n$ and row space $\mathbf{R}^n$. The columns of A are independent and the rows of A are independent. Chapter 3 will allow every matrix, and it will need new ideas.

3 The Four Fundamental Subspaces

Section 3.1 opens with a pure algebra question. *How do we define a "vector space"?* Looking at $\mathbf{R}^3$, the key operations are $\boldsymbol{v} + \boldsymbol{w}$ and $c\boldsymbol{v}$. They are connected by simple laws like $c(\boldsymbol{v} + \boldsymbol{w}) = c\boldsymbol{v} + c\boldsymbol{w}$. We must be able to add $\boldsymbol{v} + \boldsymbol{w}$, and multiply by c. Section 3.1 will give eight rules that the vectors $\boldsymbol{v}$ and the scalars c must satisfy.

Notice that $\boldsymbol{v}$ and $\boldsymbol{w}$ could be matrices! We can add matrices and we can add functions (and multiply by c). So we can have **matrix spaces** and **function spaces**. And inside $\mathbf{R}^n$, we could allow only vectors $\boldsymbol{x}$ that satisfy $A\boldsymbol{x} = \mathbf{0}$. That produces the "**nullspace of A**". All combinations of solutions to $A\boldsymbol{x} = \mathbf{0}$ are also solutions: *The nullspace is a subspace.*

Linear algebra gives us a way to solve $A\boldsymbol{x} = \mathbf{0}$. The best system is *elimination*: Simplify $A\boldsymbol{x} = \mathbf{0}$ to $R\boldsymbol{x} = \mathbf{0}$. Then $A = CR$ tells us how each dependent column of A is a combination of the first r independent columns in C. **One column is a combination of other columns!** Right there we have a special solution to $A\boldsymbol{x} = \mathbf{0}$.

Finally comes the idea of a **basis**: A set of vectors that perfectly describes the space. Their combinations give one and only one way to produce every vector in the space. The r independent columns of A are a basis for $\mathbf{C}(A)$. The $n - r$ special solutions to $A\boldsymbol{x} = \mathbf{0}$ are a basis for $\mathbf{N}(A)$. Those subspaces $\mathbf{C}$ and $\mathbf{N}$ have "*dimensions*" r and $n - r$.

Chapter 2 was about square invertible matrices. All four of the matrices in $PA = LU$ had full rank $r = m = n$. The column space and row space of A were the full space $\mathbf{R}^n$. The nullspaces of A and A^{T} contained only the **zero vector**.

Chapter 3 moves to a higher level! It may be the most important chapter in the book. Every m by n matrix is allowed, and there will surely be nonzero solutions to $A\boldsymbol{x} = \mathbf{0}$ if $n > m$. Notice that this nullspace of A is not like the column space and row space. **The nullspace starts with equations $A\boldsymbol{x} = \mathbf{0}$, not with columns or rows from A.**

The prime goal of this chapter is the *"Fundamental Theorem of Linear Algebra"*. In Section 3.5 this connects the four subspaces and their dimensions $\boldsymbol{r}, \boldsymbol{r}, \boldsymbol{n} - \boldsymbol{r}, \boldsymbol{m} - \boldsymbol{r}$.

Column space of A, row space of A, nullspace of A, nullspace of A^{T}.

I hope the picture that goes with it makes the Fundamental Theorem easy to remember.

3.1 Vector Spaces and Subspaces

1 Requirement: All linear combinations $\boldsymbol{cv} + \boldsymbol{dw}$ must stay in the vector space.

2 The row space of A is "spanned" by the rows of A. The columns of A span $\mathbf{C}(A)$.

3 Matrices M_1 to M_N and functions f_1 to f_N span matrix spaces and function spaces.

Start with the vector spaces $\mathbf{R}^1, \mathbf{R}^2, \mathbf{R}^3, \ldots$ The space $\mathbf{R}^n$ contains all column vectors $\boldsymbol{v}$ of length n. The components v_1 to v_n are real numbers. When complex numbers like $v_1 = 2 + 3i$ are allowed, the spaces become $\mathbf{C}^1, \mathbf{C}^2, \mathbf{C}^3, \ldots$. We know how to add vectors $\boldsymbol{v}$ and $\boldsymbol{w}$ in $\mathbf{R}^n$. We know how to multiply a vector by a number c or d to get $c\boldsymbol{v}$ or $d\boldsymbol{w}$. So we can find **linear combinations** $c\boldsymbol{v} + d\boldsymbol{w}$ in the vector space $\mathbf{R}^n$.

This operation of "linear combinations" is fundamental for any vector space. It must satisfy eight rules. Those eight rules are listed at the start of Problem Set 3.1 — they begin with $\boldsymbol{v} + \boldsymbol{w} = \boldsymbol{w} + \boldsymbol{v}$ and they are easy to check in $\mathbf{R}^n$. They don't need to be memorized!

One important requirement: All linear combinations $c\boldsymbol{v}+d\boldsymbol{w}$ **stay in the vector space**. The set of all positive vectors $(v_1, \ldots, v_n)$ with every $v_i > 0$ is **not** a vector space. The set of solutions to $A\boldsymbol{x} = (1, 1, \ldots, 1)$ is **not** a vector space. A line in $\mathbf{R}^n$ is not a vector space unless it goes through the center point $(0, 0, \ldots, 0)$.

If the line does go through $\mathbf{0}$, we can multiply points on the line by any number c. And we can add points on the line—without leaving the line. That line through $\mathbf{0}$ in $\mathbf{R}^n$ shows the **idea of a subspace: A vector space inside another vector space**.

Examples of Vector Spaces

This book is mainly about the vector spaces $\mathbf{R}^n$ and their subspaces like lines and planes. The space $\mathbf{Z}$ that only contains the zero vector $\mathbf{0} = (0, 0, \ldots, 0)$ counts as a subspace! Combinations $c\,\mathbf{0} + d\,\mathbf{0}$ are still $\mathbf{0}$ (inside the subspace). $\mathbf{Z}$ is the smallest vector space. We often see $\mathbf{Z}$ as the nullspace of an invertible matrix: If the only solution to $A\boldsymbol{x} = \mathbf{0}$ is the zero vector $\boldsymbol{x} = \mathbf{0}$, then the columns are independent and the nullspace of A is $\mathbf{Z}$.

We can certainly accept vector spaces of matrices. The space $\mathbf{R}^{3 \times 3}$ contains **all 3 by 3 matrices**. We can take combinations $cA + dB$ of those matrices. They easily satisfy the eight rules. One subspace would be the 3 by 3 matrices with all 9 entries equal—a "line of matrices". $\mathbf{S}$ = all symmetric 3 by 3 matrices is also a subspace: $A + B$ stays symmetric. But the invertible matrices are *not* a subspace. Why not?

We can also accept vector spaces of **functions**. The line of functions $\boldsymbol{y} = \boldsymbol{c}\boldsymbol{e}^{\boldsymbol{x}}$ (any c) would be a "line in function space". That line contains all the solutions to the differential equation $dy/dx = y$. Another function space contains all quadratics $\boldsymbol{y} = \boldsymbol{a} + \boldsymbol{b}\boldsymbol{x} + \boldsymbol{c}\boldsymbol{x}^{\mathbf{2}}$. Those are the solutions to $d^3y/dx^3 = 0$. You see how linear differential equations replace linear algebraic equations $A\boldsymbol{x} = \mathbf{0}$ when we move into function space.

In some way the space of all 3 by 3 matrices is essentially the same space as $\mathbf{R}^{\mathbf{9}}$. The space of functions $f(x) = a + bx + cx^2$ is essentially the same as $\mathbf{R}^{\mathbf{3}}$.

Linear combinations of the matrices and functions are safely in those spaces. This book will stay almost entirely with ordinary column vectors and not functions.

To repeat: The word "space" means that all linear combinations of the vectors or matrices or functions stay *inside the space*.

Subspaces of Vector Spaces

At different times, we will ask you to think of matrices and functions as vectors. But at all times, the vectors that we need most are ordinary column vectors. They are vectors with n components—but *maybe not all* of the vectors with n components. There are important vector spaces ***inside*** $\mathbf{R}^n$. Those are ***subspaces*** of $\mathbf{R}^n$.

Start with the usual three-dimensional space $\mathbf{R}^3$. Choose a plane through the origin $(0,0,0)$. ***That plane is a vector space in its own right.*** If we add two vectors in the plane, their sum is in the plane. If we multiply an in-plane vector by 2 or -5, it stays in the plane. A plane in three-dimensional space is not $\mathbf{R}^2$ (even if it looks like $\mathbf{R}^2$). The vectors have three components and they belong to $\mathbf{R}^3$. The plane is a vector space ***inside*** $\mathbf{R}^3$.

This illustrates one of the most fundamental ideas in linear algebra. The plane going through $(0,0,0)$ is a ***subspace*** of the full vector space $\mathbf{R}^3$.

DEFINITION A ***subspace*** of a vector space is a set of vectors (including $\mathbf{0}$) that satisfies two requirements: ***If v and w are vectors in the subspace and c is any scalar, then***

(i) $v + w$ is in the subspace	**(ii)** cv is in the subspace

In other words, the set of vectors is "closed" under addition $v + w$ and multiplication cv (and dw). Those operations leave us in the subspace. We can also subtract, because $-w$ is in the subspace and its sum with v is $v - w$. ***All linear combinations stay in the subspace.***

These operations follow the rules of the host space, so the eight required conditions are automatic. We just have to check that $cv + dw$ stays in the subspace.

First fact: ***Every subspace contains the zero vector***. The plane in $\mathbf{R}^3$ has to go through $(0,0,0)$. We mention this separately, for extra emphasis, but it follows directly from rule **(ii)**. Choose $c = 0$, and that rule requires $0v$ to be in the subspace.

Planes that don't contain the origin fail those tests. Those planes are not subspaces.

Lines through the origin are also subspaces. When we multiply by 5, or add two vectors on the line, then we do stay on the line. But the line must go through $(0,0,0)$.

Another subspace is all of $\mathbf{R}^3$. The whole space is a subspace (*of itself*). Here is a list of all the possible subspaces of $\mathbf{R}^3$:

($\mathbf{L}$) Any **line** through $(0,0,0)$	($\mathbf{R}^3$) The **whole space**
($\mathbf{P}$) Any **plane** through $(0,0,0)$	($\mathbf{Z}$) The **single vector** $(0,0,0)$

If we try to keep only *part* of a plane or line, the requirements for a subspace don't hold. Look at these examples in $\mathbf{R}^2$—they are not subspaces.

Example 1 Keep only the vectors (x, y) whose components are positive or zero (this is a quarter-plane). The vector $(2, 3)$ is included but $(-2, -3)$ is not. So rule (**ii**) is violated when we try to multiply by $c = -1$. ***The quarter-plane is not a subspace.***

Example 2 Include also the vectors whose components are both negative. Now we have two quarter-planes. Requirement (**ii**) is satisfied; we can multiply by any c. But rule (**i**) now fails. The sum of $\boldsymbol{v} = (2, 3)$ and $\boldsymbol{w} = (-3, -2)$ is $(-1, 1)$, which is outside both quarter-planes. ***Two quarter-planes don't make a subspace.***

Rules (**i**) and (**ii**) involve vector addition $\boldsymbol{v} + \boldsymbol{w}$ and multiplication by scalars c and d. The rules can be combined into a single requirement—*the rule for subspaces*:

A subspace containing v and w must contain all linear combinations $cv + dw$.

Example 3 Inside the vector space $\mathbf{M}$ of all 2 by 2 matrices, here are two subspaces:

$$(\mathbf{U})\text{ All upper triangular matrices } \begin{bmatrix} a & b \\ 0 & d \end{bmatrix} \quad (\mathbf{D})\text{ All diagonal matrices } \begin{bmatrix} a & 0 \\ 0 & d \end{bmatrix}.$$

Add any upper triangular matrices in $\mathbf{U}$, and the sum is in $\mathbf{U}$. Add diagonal matrices, and the sum is diagonal. In this case $\mathbf{D}$ is also a subspace of $\mathbf{U}$! Of course the zero matrix is in these subspaces, when a, b, and d all equal zero. $\mathbf{Z}$ is always a subspace with one vector.

Multiples of the identity matrix also form a subspace of M. Those matrices cI form a "line of matrices" inside $\mathbf{M}$. It is also inside $\mathbf{U}$ and $\mathbf{D}$.

Is the matrix I a subspace by itself? Certainly not. Only the zero matrix is. Your mind will invent more subspaces of 2 by 2 matrices—write them down for Problem 5.

The Column Space of A

The most important subspaces are tied directly to a matrix A. We are trying to solve $A\boldsymbol{x} = \boldsymbol{b}$. We want to describe the good right sides $\boldsymbol{b}$—the vectors that can be written as A times some vector $\boldsymbol{x}$. Those $\boldsymbol{b}$'s form the "*column space*" of A.

Remember that $A\boldsymbol{x}$ is a combination of the columns of A. To get every possible $\boldsymbol{b}$, we use every possible $\boldsymbol{x}$. Start with the columns of A and ***take all their linear combinations. This produces the column space of A.*** **It is a vector space made up of column vectors.**

DEFINITION The ***column space*** consists of ***all linear combinations of the columns***. Those combinations are all possible vectors $A\boldsymbol{x}$. They fill the column space $\mathbf{C}(\boldsymbol{A})$.

This column space is crucial to the whole book, and here is why. ***To solve $A\boldsymbol{x} = \boldsymbol{b}$ is to express $\boldsymbol{b}$ as a combination of the columns***. The right side $\boldsymbol{b}$ has to be *in the column space* produced by A, or $A\boldsymbol{x} = \boldsymbol{b}$ has no solution!

The equations $A\boldsymbol{x} = \boldsymbol{b}$ are solvable if and only if $\boldsymbol{b}$ is in the column space of A.

When $\boldsymbol{b}$ is in the column space $\mathbf{C}(A)$, it is a combination of the columns of A. The coefficients in that combination will solve $A\boldsymbol{x} = \boldsymbol{b}$. The word "space" is justified by taking *all combinations* of the columns. This includes the zero vector in $\mathbf{R}^m$.

Caution: The columns of A do not form a subspace ! The invertible matrices do not form a subspace. The singular matrices do not form a subspace. You have to include all linear combinations. The columns of A **"span"** a subspace when *we take their combinations.*

The Row Space of A

The rows of A are the columns of A^{T}, the n by m transpose matrix. Since we prefer to work with column vectors, we welcome A^{T}—it contains rows of A.

The row space of A is the column space $\mathbf{C}(A^{\mathrm{T}})$ of the transpose matrix A^{T}

This row space is a subspace of $\mathbf{R}^n$. It contains m column vectors from A^{T} and all their combinations. The equations $A^{\mathrm{T}}\boldsymbol{y} = \boldsymbol{c}$ are solvable exactly when the vector $\boldsymbol{c}$ is in the subspace $\mathbf{C}(A^{\mathrm{T}}) =$ *row space of* A.

Chapter 1 explained why $\mathbf{C}(A)$ and $\mathbf{C}(A^{\mathrm{T}})$ both contain r independent vectors and no more. Then $r =$ **rank of A** $=$ **rank of A^{T}**. A new proof is in Section 3.5.

Example The row space of the rank 1 matrix $A = \boldsymbol{u}\boldsymbol{v}^{\mathrm{T}}$ is the line of all column vectors $c\boldsymbol{v}$. This is because every column of $A^{\mathrm{T}} = \boldsymbol{v}\boldsymbol{u}^{\mathrm{T}}$ is a multiple of $\boldsymbol{v}$. One vector $\boldsymbol{v}$ spans the row space of A, one vector $\boldsymbol{u}$ spans the column space. Those spaces are lines through $\mathbf{0}$.

The Columns of A Span the Vector Space $\mathbf{C}(A)$

One useful new word: *"Span"*. Suppose we start with a set $\boldsymbol{S}$ of vectors in $\mathbf{R}^m$. If $\boldsymbol{S}$ contains only N vectors, it is certainly not a subspace. But if we include *all combinations of the vectors in* $\boldsymbol{S}$, then we have a vector space $\mathbf{V}$. In this case **the set $\boldsymbol{S}$ spans $\mathbf{V}$**. In fact $\mathbf{V}$ is the smallest vector space containing $\boldsymbol{S}$ (because we are forced to include all combinations to produce a vector space).

This is exactly what we did for the columns of A. Those n columns span the column space $\mathbf{C}(A)$ = all combinations of the columns. Independence is not required by the word *span*. In the same way, the m columns of A^{T} span the row space $\mathbf{C}(A^{\mathrm{T}})$.

Test question. Show that the invertible 2 by 2 matrices span $\mathbf{R}^{2\times 2} =$ **all 2 by 2's**.

Examples If the n by n matrix A is invertible, then its columns span $\mathbf{R}^n$.

The invertible 3 by 3 matrices span the whole matrix space $\mathbf{R}^{3\times 3}$.

The singular (**not invertible**) 3 by 3 matrices also span $\mathbf{R}^{3\times 3}$.

Next comes the nullspace $\mathbf{N}(A)$ and that requires new ideas. We start with $A\boldsymbol{x} = \mathbf{0}$ (*m equations and not n vectors*). The solutions $\boldsymbol{x}$ to those equations give the nullspace. It is a vector space because $A\boldsymbol{x} = \mathbf{0}$ and $A\boldsymbol{y} = \mathbf{0}$ lead to $A(c\boldsymbol{x} + d\boldsymbol{y}) = \mathbf{0}$. But we have to work to find those solutions $\boldsymbol{x}$ and $\boldsymbol{y}$.

Test question: When do ten vectors span $\mathbf{R}^5$? This is very possible. But not any ten.

Problem Set 3.1

The first problems 1–7 are about vector spaces in general. The vectors in those spaces are not necessarily column vectors. In the definition of a *vector space*, vector addition $\boldsymbol{x}+\boldsymbol{y}$ and scalar multiplication $c\boldsymbol{x}$ must obey the following eight rules:

(1) $\boldsymbol{x}+\boldsymbol{y}=\boldsymbol{y}+\boldsymbol{x}$

(2) $\boldsymbol{x}+(\boldsymbol{y}+\boldsymbol{z})=(\boldsymbol{x}+\boldsymbol{y})+\boldsymbol{z}$

(3) There is a unique "zero vector" such that $\boldsymbol{x}+\mathbf{0}=\boldsymbol{x}$ for all $\boldsymbol{x}$

(4) For each $\boldsymbol{x}$ there is a unique vector $-\boldsymbol{x}$ such that $\boldsymbol{x}+(-\boldsymbol{x})=\mathbf{0}$

(5) 1 times $\boldsymbol{x}$ equals $\boldsymbol{x}$

(6) $(c_1c_2)\boldsymbol{x}=c_1(c_2\boldsymbol{x})$ — rules (1) to (4) are about $\boldsymbol{x}+\boldsymbol{y}$

(7) $c(\boldsymbol{x}+\boldsymbol{y})=c\boldsymbol{x}+c\boldsymbol{y}$ — rules (5) to (6) are about $c\boldsymbol{x}$

(8) $(c_1+c_2)\boldsymbol{x}=c_1\boldsymbol{x}+c_2\boldsymbol{x}$. — rules **(7) to (8) connect them**

---

1 Suppose $(x_1,x_2)+(y_1,y_2)$ is defined to be (x_1+y_2,x_2+y_1). With the usual multiplication $c\boldsymbol{x}=(cx_1,cx_2)$, which of the eight conditions are not satisfied?

2 Suppose the multiplication $c\boldsymbol{x}$ is defined to produce $(cx_1,0)$ instead of (cx_1,cx_2). With the usual addition in $\mathbf{R}^2$, are the eight conditions satisfied?

3 (a) Which rules are broken if we keep only the positive numbers $x>0$ in $\mathbf{R}^1$? Every c must be allowed. This half-line is not a subspace.

(b) The positive numbers with $\boldsymbol{x}+\boldsymbol{y}$ and $c\boldsymbol{x}$ redefined to equal the usual xy and x^c *do satisfy the eight rules.* Test rule 7 when $c=3,x=2,y=1$. (Then $\boldsymbol{x}+\boldsymbol{y}=2$ and $c\boldsymbol{x}=8$.) Which number acts as the "zero vector" in this space?

4 The matrix $A=\begin{bmatrix}2&-2\\2&-2\end{bmatrix}$ is a "vector" in the space $\mathbf{M}$ of all 2 by 2 matrices. Write down the zero vector in this space, the vector $\frac{1}{2}A$, and the vector $-A$. What matrices are in the smallest subspace containing A (the subspace spanned by A)?

5 (a) Describe a subspace of $\mathbf{M}$ that contains $A=\begin{bmatrix}1&0\\0&0\end{bmatrix}$ but not $B=\begin{bmatrix}0&0\\0&-1\end{bmatrix}$.

(b) If a subspace of $\mathbf{M}$ does contain A and B, must it contain the identity matrix?

(c) Describe a subspace of $\mathbf{M}$ that contains no nonzero diagonal matrices.

6 The functions $\boldsymbol{f}(x) = x^2$ and $\boldsymbol{g}(x) = 5x$ are "vectors" in $\mathbf{F}$. This is the vector space of all real functions. (The functions are defined for $-\infty < x < \infty$.) The combination $3\boldsymbol{f}(x) - 4\boldsymbol{g}(x)$ is the function $\boldsymbol{h}(x) =$ ____.

7 Which rule is broken if multiplying $\boldsymbol{f}(x)$ by c gives the function $\boldsymbol{f}(cx)$? Keep the usual addition $\boldsymbol{f}(x) + \boldsymbol{g}(x)$.

Questions 8–15 are about the "subspace requirements": $\boldsymbol{x} + \boldsymbol{y}$ and $c\boldsymbol{x}$ (and then all linear combinations $c\boldsymbol{x} + d\boldsymbol{y}$) stay in the subspace.

8 One subspace requirement can be met while the other fails. Show this by finding

(a) A set of vectors in $\mathbf{R}^2$ for which $\boldsymbol{x} + \boldsymbol{y}$ stays in the set but $\frac{1}{2}\boldsymbol{x}$ may be outside.

(b) A set of vectors in $\mathbf{R}^2$ (other than two quarter-planes) for which every $c\boldsymbol{x}$ stays in the set but $\boldsymbol{x} + \boldsymbol{y}$ may be outside.

9 Which of these subsets of $\mathbf{R}^3$ are actually subspaces ? They all span subspaces !

(a) The plane of vectors (b_1, b_2, b_3) with $b_1 = b_2$.

(b) The plane of vectors with $b_1 = 1$.

(c) The vectors with $b_1 b_2 b_3 = 0$.

(d) All linear combinations of $\boldsymbol{v} = (1, 4, 0)$ and $\boldsymbol{w} = (2, 2, 2)$.

(e) All vectors that satisfy $b_1 + b_2 + b_3 = 0$.

(f) All vectors with $b_1 \le b_2 \le b_3$.

10 Describe the smallest subspace of the matrix space $\mathbf{M}$ that contains

(a) $\begin{bmatrix} 1 & 0 \\ 0 & 0 \end{bmatrix}$ and $\begin{bmatrix} 0 & 1 \\ 0 & 0 \end{bmatrix}$ (b) $\begin{bmatrix} 1 & 1 \\ 0 & 0 \end{bmatrix}$ (c) $\begin{bmatrix} 1 & 1 \\ 1 & 1 \end{bmatrix}$ and $\begin{bmatrix} 1 & 0 \\ 0 & 1 \end{bmatrix}$.

11 Let P be the plane in $\mathbf{R}^3$ with equation $x + y - 2z = 4$. The origin $(0, 0, 0)$ is not in P! Find two vectors in P and check that their sum is not in P.

12 Let $\mathbf{P}_0$ be the plane through $(0, 0, 0)$ parallel to the previous plane P. What is the equation for $\mathbf{P}_0$? Find two vectors in $\mathbf{P}_0$ and check that their sum is in $\mathbf{P}_0$.

13 Suppose $\mathbf{P}$ is a plane through $(0, 0, 0)$ and $\mathbf{L}$ is a line through $(0, 0, 0)$. The smallest vector space containing both $\mathbf{P}$ and $\mathbf{L}$ is either ____ or ____.

14 (a) Show that the set of *invertible* matrices in $\mathbf{M}$ is not a subspace.

(b) Show that the set of *singular* matrices in $\mathbf{M}$ is not a subspace.

15 True or false (check addition in each case by an example):

(a) The symmetric matrices in $\mathbf{M}$ (with $A^{\mathrm{T}} = A$) form a subspace.

(b) The skew-symmetric matrices in $\mathbf{M}$ (with $A^{\mathrm{T}} = -A$) form a subspace.

(c) The unsymmetric matrices in $\mathbf{M}$ (with $A^{\mathrm{T}} \neq A$) **span** a subspace.

Questions 16–26 are about column spaces $\mathbf{C}(A)$ and the equation $A\boldsymbol{x} = \boldsymbol{b}$.

16 Describe the column spaces (lines or planes) of these particular matrices:

$$A = \begin{bmatrix} 1 & 2 \\ 0 & 0 \\ 0 & 0 \end{bmatrix} \quad \text{and} \quad B = \begin{bmatrix} 1 & 0 \\ 0 & 2 \\ 0 & 0 \end{bmatrix} \quad \text{and} \quad C = \begin{bmatrix} 1 & 0 \\ 2 & 0 \\ 0 & 0 \end{bmatrix}.$$

17 For which right sides (find a condition on b_1, b_2, b_3) are these systems solvable?

$$\text{(a)} \begin{bmatrix} 1 & 4 & 2 \\ 2 & 8 & 4 \\ -1 & -4 & -2 \end{bmatrix} \begin{bmatrix} x_1 \\ x_2 \\ x_3 \end{bmatrix} = \begin{bmatrix} b_1 \\ b_2 \\ b_3 \end{bmatrix} \qquad \text{(b)} \begin{bmatrix} 1 & 4 \\ 2 & 9 \\ -1 & -4 \end{bmatrix} \begin{bmatrix} x_1 \\ x_2 \end{bmatrix} = \begin{bmatrix} b_1 \\ b_2 \\ b_3 \end{bmatrix}.$$

18 Adding row 1 of A to row 2 produces B. Adding column 1 to column 2 produces C. A combination of the columns of (B or C ?) is also a combination of the columns of A. Which two matrices have the same column _____ ?

$$A = \begin{bmatrix} 1 & 2 \\ 2 & 4 \end{bmatrix} \quad \text{and} \quad B = \begin{bmatrix} 1 & 2 \\ 3 & 6 \end{bmatrix} \quad \text{and} \quad C = \begin{bmatrix} 1 & 3 \\ 2 & 6 \end{bmatrix}.$$

19 For which (b_1, b_2, b_3) do these systems have a solution ? Then $\boldsymbol{b}$ is in $\mathbf{C}(A)$.

$$\begin{bmatrix} 1 & 1 & 1 \\ 0 & 1 & 1 \\ 0 & 0 & 1 \end{bmatrix} \begin{bmatrix} x_1 \\ x_2 \\ x_3 \end{bmatrix} = \begin{bmatrix} b_1 \\ b_2 \\ b_3 \end{bmatrix} \quad \text{and} \quad \begin{bmatrix} 1 & 1 & 1 \\ 0 & 1 & 1 \\ 0 & 0 & 0 \end{bmatrix} \begin{bmatrix} x_1 \\ x_2 \\ x_3 \end{bmatrix} = \begin{bmatrix} b_1 \\ b_2 \\ b_3 \end{bmatrix}$$

$$\text{and} \quad \begin{bmatrix} 1 & 1 & 1 \\ 0 & 0 & 1 \\ 0 & 0 & 1 \end{bmatrix} \begin{bmatrix} x_1 \\ x_2 \\ x_3 \end{bmatrix} = \begin{bmatrix} b_1 \\ b_2 \\ b_3 \end{bmatrix}.$$

20 (Recommended) If we add an extra column $\boldsymbol{b}$ to a matrix A, then the column space gets larger unless _____. Give an example where the column space gets larger and an example where it doesn't. Why is $A\boldsymbol{x} = \boldsymbol{b}$ solvable exactly when the column space *doesn't* get larger ? Then that space is the same for A and $\begin{bmatrix} A & \boldsymbol{b} \end{bmatrix}$.

21 The columns of AB are combinations of the columns of A. This means: *The column space of AB is contained in* (possibly equal to) *the column space of A.* Give an example where the column spaces of A and AB are not equal.

22 Suppose $Ax = b$ and $Ay = b^*$ are both solvable. Then $Az = b + b^*$ is solvable. What is z? This translates into: If b and b^* are in the column space $\mathbf{C}(A)$, *then* $b + b^*$ is in $\mathbf{C}(A)$. That is a requirement for a vector space.

23 If A is any 5 by 5 invertible matrix, then its column space is _____. Why?

24 True or false (with a counterexample if false):

(a) The vectors b that are not in the column space $\mathbf{C}(A)$ form a subspace.

(b) If $\mathbf{C}(A)$ contains only the zero vector, then A is the zero matrix.

(c) The column space of $2A$ equals the column space of A.

(d) The column space of $A - I$ equals the column space of A (test this).

25 Construct a 3 by 3 matrix whose column space contains $(1,1,0)$ and $(1,0,1)$ but not $(1,1,1)$. Construct a 3 by 3 matrix whose column space is only a line.

26 If the 9 by 12 system $Ax = b$ is solvable for every b, then $\mathbf{C}(A) =$ _____.

Challenge Problems

27 Suppose $\mathbf{S}$ and $\mathbf{T}$ are two subspaces of a vector space $\mathbf{V}$.

(a) **Definition**: The **sum** $\mathbf{S} + \mathbf{T}$ contains all sums $s + t$ of a vector s in $\mathbf{S}$ and a vector t in $\mathbf{T}$. Show that $\mathbf{S} + \mathbf{T}$ satisfies the requirements for a vector space. Addition and scalar multiplication stay inside $\mathbf{S} + \mathbf{T}$.

(b) If $\mathbf{S}$ and $\mathbf{T}$ are lines in $\mathbf{R}^m$, what is the difference between $\mathbf{S} + \mathbf{T}$ and $\mathbf{S} \cup \mathbf{T}$? That union contains all vectors from $\mathbf{S}$ or $\mathbf{T}$ or both. Explain this statement: **The span of $\mathbf{S} \cup \mathbf{T}$ is $\mathbf{S} + \mathbf{T}$**. (Section 3.4 returns to this word "span".)

28 If $\mathbf{S}$ is the column space of A and $\mathbf{T}$ is $\mathbf{C}(B)$, then $\mathbf{S} + \mathbf{T}$ is the column space of what matrix M? The columns of A and B and M are all in $\mathbf{R}^m$. I don't think $A + B$ is always a correct M. We want the columns of M to span $\mathbf{S} + \mathbf{T}$.

29 Show that the matrices A and $\begin{bmatrix} A & AB \end{bmatrix}$ (with extra columns) have the same column space. But find a square matrix with $\mathbf{C}(A^2)$ smaller than $\mathbf{C}(A)$. Important point:

An n by n matrix has $\mathbf{C}(A) = \mathbf{R}^n$ exactly when A is an _____ matrix.

30 Find another independent solution (after $y = e^x$) to the second order differential equation $d^2y/dx^2 = y$. Find two independent solutions to $d^2y/dx^2 = -y$.

Then the 2-dimensional solution space contains all linear combinations $y =$ _____.

31 Suppose V and W are two subspaces of $\mathbf{R}^n$. Their "intersection" $V \cap W$ contains the vectors that are in both subspaces. (Notice that the zero vector is in V and W.) Show that $V \cap W$ is a *subspace* by testing the requirement: If x and y are in $V \cap W$, why is $cx + dy$ in $V \cap W$?

3.2 Computing the Nullspace by Elimination: $A = CR$

1 The **nullspace** $\mathbf{N}(A)$ in $\mathbf{R}^n$ contains all solutions $\boldsymbol{x}$ to $A\boldsymbol{x} = \mathbf{0}$. This includes $\boldsymbol{x} = \mathbf{0}$.

2 Elimination from A to R_0 to R does not change the nullspace: $\mathbf{N}(A) = \mathbf{N}(R_0) = \mathbf{N}(R)$.

3 **The reduced row echelon form $\boldsymbol{R_0} = \mathbf{rref}(\boldsymbol{A})$** has $\boldsymbol{I}$ in r columns and $\boldsymbol{F}$ in $n - r$ columns.

4 If column j is dependent on previous columns, $A\boldsymbol{x} = \mathbf{0}$ has a "*special solution*" with $x_j = 1$.

5 **The $n - r$ special solutions to $A\boldsymbol{x} = 0$ contain $-\boldsymbol{F}$ and $\boldsymbol{I}$ (page 97).**

6 Every short wide matrix with $m < n$ has nonzero solutions to $A\boldsymbol{x} = \mathbf{0}$ in its nullspace.

This section finds all solutions to $A\boldsymbol{x} = \mathbf{0}$. When A is a square invertible matrix (in this case its rank is $r = n$), the only solution is $\boldsymbol{x} = \mathbf{0}$. Then the nullspace only contains the zero vector: the columns of A are independent. But in general A has r independent columns (r = rank). *The other $n - r$ columns of A are combinations of those independent columns.* We will find $n - r$ vectors in the nullspace of A—special solutions to $A\boldsymbol{x} = \mathbf{0}$.

With square invertible matrices, Chapter 2 simplified A to an upper triangular U. For matrices of all shapes, elimination will simplify $A\boldsymbol{x} = \mathbf{0}$ to an **"echelon form" $\boldsymbol{Rx} = 0$**. (Actually $R = I$ when A is invertible, so elimination is now going further than L and U: as far as it can.) We begin with two examples of R, to show where we are going.

Here is a matrix R of rank $r = 2$. It has $n = 4$ columns so we look for $n - r = 4 - 2 = 2$ independent solutions to $R\boldsymbol{x} = \mathbf{0}$. The nullspace $\mathbf{N}(R)$ will have dimension 2.

Example 1 $\quad \boldsymbol{R} = [\, \boldsymbol{I} \quad \boldsymbol{F} \,] P = \begin{bmatrix} \mathbf{1} & \mathbf{0} & 3 & 5 \\ \mathbf{0} & \mathbf{1} & 4 & 6 \end{bmatrix} \qquad R\boldsymbol{x} = \mathbf{0}$ is $\quad \begin{matrix} x_1 + 3x_3 + 5x_4 = 0 \\ x_2 + 4x_3 + 6x_4 = 0 \end{matrix}$

Two "special solutions" are easy to find, when x_3 and x_4 are 1 and 0 or 0 and 1.

Set $x_3 = \mathbf{1}$ and $x_4 = \mathbf{0}$. Equation 1 gives $x_1 = \mathbf{-3}$. Equation 2 gives $x_2 = \mathbf{-4}$.

Set $x_3 = \mathbf{0}$ and $x_4 = \mathbf{1}$. Equation 1 gives $x_3 = \mathbf{-5}$. Equation 2 gives $x_2 = \mathbf{-6}$.

These two special solutions $\boldsymbol{s}_1 = (\mathbf{-3}, \mathbf{-4}, \mathbf{1}, \mathbf{0})$ and $\boldsymbol{s}_2 = (\mathbf{-5}, \mathbf{-6}, \mathbf{0}, \mathbf{1})$ are in the nullspace of R. They give $R\boldsymbol{s}_1 = \mathbf{0}$ and $R\boldsymbol{s}_2 = \mathbf{0}$. Any combination of those two solutions will also be in the nullspace. The matrix R times the vector $\boldsymbol{x} = c_1\boldsymbol{s}_1 + c_2\boldsymbol{s}_2$ produces zero. Soon we will call those special solutions $\boldsymbol{s}_1$ and $\boldsymbol{s}_2$ a **basis for the nullspace**.

That matrix R was easy to work with. *Its first two columns contained the identity matrix* (P was just I). R is an example of a matrix in "reduced row echelon form". We will give one more example to show a variation R_0 that is still in reduced row echelon form and still simple. The subscript in R_0 indicates that there can also be rows of zeros.

Example 2 $R_0 = \begin{bmatrix} \mathbf{1} & 7 & \mathbf{0} & 8 \\ \mathbf{0} & 0 & \mathbf{1} & 9 \\ \mathbf{0} & 0 & \mathbf{0} & 0 \end{bmatrix}$ $\quad R_0\boldsymbol{x} = \mathbf{0}$ is $\quad \begin{aligned} x_1 + 7x_2 + 0x_3 + 8x_4 &= 0 \\ x_3 + 9x_4 &= 0 \\ 0 &= 0 \end{aligned}$

Now the identity matrix is inside columns 1 and 3. And row 3 is all zero. This still counts as a reduced row echelon form—elimination can't make it simpler. *The 1's in the identity matrix are still the first nonzeros in their rows.* The word "echelon" refers to the "staircase" of 1's. Any zero rows always come last in R_0.

The special solutions still have 1 and 0 for the **"free variables"**—which are $\boldsymbol{x_2}$ and $\boldsymbol{x_4}$.

Set $\boldsymbol{x_2} = \mathbf{1}$ and $\boldsymbol{x_4} = \mathbf{0}$. Equation 1 gives $\boldsymbol{x_1} = \mathbf{-7}$. Equation 2 gives $\boldsymbol{x_3} = \mathbf{0}$.

Set $\boldsymbol{x_2} = \mathbf{0}$ and $\boldsymbol{x_4} = \mathbf{1}$. Equation 1 gives $\boldsymbol{x_1} = \mathbf{-8}$. Equation 2 gives $\boldsymbol{x_3} = \mathbf{-9}$.

Those special solutions are now $\boldsymbol{s}_1 = (\mathbf{-7, 1, 0, 0})$ and $\boldsymbol{s}_2 = (\mathbf{-8, 0, -9, 1})$. For the free variables x_2 and x_4, we freely choose $1, 0$ and then $0, 1$. Then the equations $R_0\boldsymbol{x} = \mathbf{0}$ tell us x_1 and x_3. If we want to keep I in columns 1 and 2, we need a permutation P.

Here is the plan for this section of the book. We start with any m by n matrix A. We apply elimination (to be explained). That changes A into its reduced row echelon form $\boldsymbol{R_0} = \mathbf{rref}\,(\boldsymbol{A})$. Our two examples showed the simplest form $R_0 = R$, and then the most general form when R_0 may have zero rows. **Removing all zero rows of $\boldsymbol{R_0}$ leaves $\boldsymbol{R}$.**

$r, m, n = 2, 2, 4$	**Simplest case**	$\boldsymbol{R} = \begin{bmatrix} \boldsymbol{I} & \boldsymbol{F} \end{bmatrix}$	as in $\begin{bmatrix} \mathbf{1} & 0 & 3 & 5 \\ 0 & \mathbf{1} & 4 & 6 \end{bmatrix}$
$r, m, n = 2, 3, 4$	**General case**	$\boldsymbol{R_0} = \begin{bmatrix} \boldsymbol{I} & \boldsymbol{F} \\ \mathbf{0} & \mathbf{0} \end{bmatrix} \boldsymbol{P}$	as in $\begin{bmatrix} \mathbf{1} & 7 & 0 & 8 \\ 0 & 0 & \mathbf{1} & 9 \\ 0 & 0 & 0 & 0 \end{bmatrix}$

I and F have r rows. The reduced matrix R_0 and the original A have m rows. So R_0 has $m - r$ rows of zeros. When we remove those zero rows, we have $\boldsymbol{R} = \begin{bmatrix} \boldsymbol{I} & \boldsymbol{F} \end{bmatrix} \boldsymbol{P}$.

Summary In Chapter 2, A was square and invertible. Elimination on $A\boldsymbol{x} = \boldsymbol{b}$ stopped at $U\boldsymbol{x} = \boldsymbol{c}$ (upper triangular U). Then back-substitution solved for $\boldsymbol{x}$.

Here we allow any matrix A of rank r. Elimination continues until we reach an r by r identity matrix I, as in Chapter 1. (We didn't know about elimination then, but we somehow produced the first r independent columns of A in C.) Now we go from A to C and R.

Key question for each new column : Is it dependent on the previous columns ? In this section, $A\boldsymbol{x} = \mathbf{0}$ becomes $R\boldsymbol{x} = \mathbf{0}$ and we find the nullspace. In Section 3.3, $A\boldsymbol{x} = \boldsymbol{b}$ becomes $R\boldsymbol{x} = \boldsymbol{d}$ and we find all solutions. Our examples showed R after elimination has done all it can to simplify A. We will soon see those steps to R.

Appendix 9 summarizes **rref** (A) as the key step to $A = CR$ and solving $A\boldsymbol{x} = \mathbf{0}$.

Elimination from A to rref(A) : Reduced Row Echelon Form

How does elimination work ? In any order, we may execute these three different steps :

1. Subtract a multiple of one row from another row (above or below !)
2. Multiply a row by any nonzero number
3. Exchange any rows.

Let me stay with Examples 1 and 2, the simplest case R and the general case R_0. Here is a 2 by 4 matrix A that elimination reduces to our 2 by 4 example $R = \begin{bmatrix} I & F \end{bmatrix}$.

$$A = \begin{bmatrix} 1 & 2 & 11 & 17 \\ 3 & 7 & 37 & 57 \end{bmatrix} \rightarrow \begin{bmatrix} 1 & 2 & 11 & 17 \\ 0 & 1 & 4 & 6 \end{bmatrix} \rightarrow \begin{bmatrix} 1 & 0 & 3 & 5 \\ 0 & 1 & 4 & 6 \end{bmatrix} = \boldsymbol{R}$$

Elimination starts with column 1. *It subtracts* 3 *times row* 1 *from row* 2. That produces the zero in the middle matrix. Now column 1 is set (the corner pivot was $A_{11} = 1$ which is what we want). Moving to column 2, **we subtract 2 times the new row 2 from row 1.** Upward elimination produces the second zero in R. Now R starts with the r by r identity matrix I. The rank is $r = 2$ and elimination on this matrix A is complete.

What did elimination actually do ? **It inverted the leading 2 by 2 matrix** $\boldsymbol{W} = \begin{bmatrix} \mathbf{1} & \mathbf{2} \\ \mathbf{3} & \mathbf{7} \end{bmatrix}$. The matrix W at the start of A became I at the start of R :

Multiply $\boldsymbol{W^{-1}A = W^{-1}} \begin{bmatrix} \boldsymbol{W} & \boldsymbol{H} \end{bmatrix}$ **to produce** $\boldsymbol{R} = \begin{bmatrix} \boldsymbol{I} & \boldsymbol{W^{-1}H} \end{bmatrix} = \begin{bmatrix} \boldsymbol{I} & \boldsymbol{F} \end{bmatrix}$.

We always knew that the dependent columns of A (in H) would be some combination of the independent columns (in W). Now we see that $\boldsymbol{H = WF}$. The matrix F is telling us how to combine the independent columns of A to produce the dependent columns. We can understand the echelon form R and the role of F !

Dependent columns $\quad \boldsymbol{H} = \begin{bmatrix} 11 & 17 \\ 37 & 57 \end{bmatrix} = \boldsymbol{WF} = \begin{bmatrix} 1 & 2 \\ 3 & 7 \end{bmatrix}$ times $\begin{bmatrix} 3 & 5 \\ 4 & 6 \end{bmatrix}$.

However you compute R from A, you always reach the same R. R is completely determined by A (even if there are different elimination steps that lead from A to R).

> **1** The first r independent columns of A locate the columns of R containing I
> The permutation P puts those columns into their correct places in A.
>
> **2** The remaining columns F in R are determined by the equation $H = WF$:
> **(Dependent columns of A) = (Independent columns of A)** times $\boldsymbol{F}$
>
> **3** The last $m - r$ rows of R_0 are rows of zeros.

Example 2 Come back to the matrix A that leads to our second reduced echelon form R_0. Both A and R_0 are 3 by 4 matrices of rank $r = 2$. Watch each step of elimination :

$$\boldsymbol{A} = \begin{bmatrix} 1 & 7 & 3 & 35 \\ 2 & 14 & 6 & 70 \\ 2 & 14 & 9 & 97 \end{bmatrix} \rightarrow \begin{bmatrix} 1 & 7 & 3 & 35 \\ \mathbf{0} & \mathbf{0} & \mathbf{0} & \mathbf{0} \\ \mathbf{0} & \mathbf{0} & \mathbf{3} & \mathbf{27} \end{bmatrix} \rightarrow \begin{bmatrix} \mathbf{1} & \mathbf{7} & \mathbf{0} & \mathbf{8} \\ 0 & 0 & 0 & 0 \\ 0 & 0 & 3 & 27 \end{bmatrix} \rightarrow \begin{bmatrix} 1 & 7 & 0 & 8 \\ \mathbf{0} & \mathbf{0} & \mathbf{1} & \mathbf{9} \\ 0 & 0 & 0 & 0 \end{bmatrix} = \boldsymbol{R_0}$$

Elimination Column by Column : The Steps from A to R_0

After those examples, here is an algorithm ! Elimination goes a column at a time, from left to right. After k columns, that part of the matrix is in its own rref form and we are ready for column $k+1$. This new column has an upper part $\boldsymbol{u}$ and a lower part $\boldsymbol{\ell}$.

$$\textbf{First } k+1 \textbf{ columns} \qquad \begin{bmatrix} I_k & F_k \\ 0 & 0 \end{bmatrix} P_k \text{ followed by } \begin{bmatrix} \boldsymbol{u} \\ \boldsymbol{\ell} \end{bmatrix}.$$

The big question is : **Does this new column $k+1$ join with I_k or F_k ?**

If $\boldsymbol{\ell}$ is all zeros, the new column is **dependent** on the first k columns. Then $\boldsymbol{u}$ joins with F_k to produce F_{k+1} in the next step to column $k+2$.

If $\boldsymbol{\ell}$ is not all zero, the new column is **independent** of the first k columns. Pick any nonzero in $\boldsymbol{\ell}$ (preferably the largest) as the pivot. Move that row of A up into row $k+1$. Then subtract multiples of that pivot row to zero out (by standard elimination) all the rest of column $k+1$. (That step is expected to change the columns after $k+1$.) Column $k+1$ joins with I_k to produce I_{k+1}. We are ready for column $k+2$.

At the end of elimination, we have a most desirable list of column numbers. They tell us the **first r independent columns of A**. Those are the columns of C. They led to the identity matrix $I_{r \text{ by } r}$ in the row factor R of $A = CR$.

Example 2 showed the three allowed row operations in elimination from A to R_0 :

1) Subtract a multiple of one row from another row (**below or above**)
2) Divide a row like $\begin{bmatrix} 0 & 0 & 3 & 27 \end{bmatrix}$ by its first nonzero entry (*to reach pivot* $= 1$)
3) Exchange rows (*to move all zero rows to the bottom of* R_0)

A different series of steps could reach the same R_0. But that result $R_0 = \text{rref}(A)$ can't change. The pieces of R_0 are all fully determined by the original matrix A.

$\boldsymbol{R_0}$ had a zero row in Example 2 because A has rank $r = 2$

$\boldsymbol{I}$ is in columns 1 and 3 of R_0 because those are the first independent columns of A

$\boldsymbol{F}$ in columns 2, 4 of R **combines columns 1, 3 of $\boldsymbol{A}$** to give its dependent columns **2, 4**

$$\boldsymbol{C} \textbf{ times } \boldsymbol{F} = \begin{bmatrix} 1 & 3 \\ 2 & 6 \\ 2 & 9 \end{bmatrix} \begin{bmatrix} 7 & 8 \\ 0 & 9 \end{bmatrix} = \begin{bmatrix} 7 & 35 \\ 14 & 70 \\ 14 & 97 \end{bmatrix} = \begin{matrix} \textbf{dependent} \\ \textbf{columns} \\ \textbf{2 and 4 of } \boldsymbol{A} \end{matrix}$$

We could never have seen in Chapter 1 that $(35, 70, 97)$ combines columns 1 and 3 of A.

Please remember how the matrix R shows us the nullspace of A. To solve $A\boldsymbol{x} = \boldsymbol{0}$ we just have to solve $R\boldsymbol{x} = \boldsymbol{0}$. This is easy because of the identity matrix inside R.

The Matrix Factorization $A = CR$ and the Nullspace

This is our chance to complete Chapter 1. That chapter introduced the factorization $A = CR$ by small examples: We learned the meaning of independent columns, but we had no systematic way to find them. Now we have a way: *Apply elimination to reduce A to R_0.* ***Then I in R_0 locates the matrix C of independent columns in A.*** And removing any zero rows from R_0 produces **the row matrix R in $A = CR$.**

We find two special solutions s_1 and s_2—one solution for every column of F in R.

Example 2 again

$$\boldsymbol{Rs_1 = 0} \qquad \begin{bmatrix} 1 & 7 & 0 & 8 \\ 0 & 0 & 1 & 9 \end{bmatrix} \begin{bmatrix} \mathbf{-7} \\ \mathbf{1} \\ 0 \\ \mathbf{0} \end{bmatrix} = \begin{bmatrix} 0 \\ 0 \end{bmatrix} \qquad \text{Put } \mathbf{1} \text{ and } \mathbf{0} \text{ in positions 2 and 4}$$

$$\boldsymbol{Rs_2 = 0} \qquad \begin{bmatrix} 1 & 7 & 0 & 8 \\ 0 & 0 & 1 & 9 \end{bmatrix} \begin{bmatrix} \mathbf{-8} \\ \mathbf{0} \\ \mathbf{-9} \\ \mathbf{1} \end{bmatrix} = \begin{bmatrix} 0 \\ 0 \end{bmatrix} \qquad \text{Put } \mathbf{0} \text{ and } \mathbf{1} \text{ in positions 2 and 4}$$

I think s_1 and s_2 are easiest to see using the matrices $-F$ and I and P^{T}.

The two special solutions to $\begin{bmatrix} I & F \end{bmatrix} \boldsymbol{x} = \mathbf{0}$ are the columns of $\begin{bmatrix} -F \\ I \end{bmatrix}$ in Example 1
The special solutions to $\begin{bmatrix} I & F \end{bmatrix} P\boldsymbol{x} = \mathbf{0}$ are the columns of $P^{\mathrm{T}} \begin{bmatrix} -F \\ I \end{bmatrix}$ in Example 2

The first one is easy because the permutation is $P = I$. The second one is correct because PP^{T} is the identity matrix for any permutation matrix P:

$$\boldsymbol{Rx = 0} \qquad \begin{bmatrix} I & F \end{bmatrix} P \text{ times } P^{\mathrm{T}} \begin{bmatrix} -F \\ I \end{bmatrix} \text{ reduces to } \begin{bmatrix} I & F \end{bmatrix} \begin{bmatrix} -F \\ I \end{bmatrix} = \begin{bmatrix} 0 \end{bmatrix}$$

Review Suppose the m by n matrix A has rank r. To find the $n - r$ special solutions to $A\boldsymbol{x} = \mathbf{0}$, compute the reduced row echelon form R_0 of A. Remove the $m - r$ zero rows of R_0 to produce $R = \begin{bmatrix} I & F \end{bmatrix} P$ and $A = CR$. Then the special solutions to $A\boldsymbol{x} = \mathbf{0}$ are the $n - r$ columns of $P^{\mathrm{T}} \begin{bmatrix} -F \\ I \end{bmatrix}$. **Please see the last page of Chapter 3.**

Example 3 **Elimination on A gives R_0 and R. Then R reveals the nullspace of A.**

$$\boldsymbol{A} = \begin{bmatrix} 1 & 2 & 1 \\ 2 & 4 & 5 \\ 3 & 6 & 9 \end{bmatrix} \to \begin{bmatrix} 1 & 2 & 1 \\ 0 & 0 & 3 \\ 0 & 0 & 6 \end{bmatrix} \to \begin{bmatrix} 1 & 2 & 0 \\ 0 & 0 & 1 \\ 0 & 0 & 0 \end{bmatrix} = \boldsymbol{R_0} \text{ with rank 2}$$

Then $R = \begin{bmatrix} 1 & 2 & 0 \\ 0 & 0 & 1 \end{bmatrix}$ and the independent columns of A and R_0 and R are 1 and 3.

To solve $A\boldsymbol{x} = \mathbf{0}$ and $R\boldsymbol{x} = \mathbf{0}$, set $x_2 = 1$. Solve for $x_1 = -2$ and $x_3 = 0$. Special solution $\boldsymbol{s} = (-2, 1, 0)$. All solutions $\boldsymbol{x} = (-2c, c, 0)$. And here is $A = CR$.

$$\boldsymbol{A} = \begin{bmatrix} 1 & 2 & 1 \\ 2 & 4 & 5 \\ 3 & 6 & 9 \end{bmatrix} = \boldsymbol{CR} = \begin{bmatrix} 1 & 1 \\ 2 & 5 \\ 3 & 9 \end{bmatrix} \begin{bmatrix} 1 & 2 & 0 \\ 0 & 0 & 1 \end{bmatrix} = \begin{matrix} \text{(column basis)} & \text{(row basis)} \\ \text{in } C & \text{in } R \end{matrix}$$

Can you write each row of A as a combination of the rows of R? *Yes.* Just use $A = CR$. The three rows of A are the three rows of C times the two rows of R.

For many matrices, the only solution to $A\boldsymbol{x} = \mathbf{0}$ is $\boldsymbol{x} = \mathbf{0}$. The columns of A are independent. The nullspace $\mathbf{N}(A)$ contains only the zero vector: *no special solutions.* The only combination of the columns that produces $A\boldsymbol{x} = \mathbf{0}$ is the zero combination $\boldsymbol{x} = \mathbf{0}$.

This case of a zero nullspace $\mathbf{Z}$ is of the greatest importance. It says that the columns of A are **independent**. No combination of columns gives the zero vector (except $\boldsymbol{x} = \mathbf{0}$). But this can't happen if $n > m$. We can't have n independent columns in $\mathbf{R}^m$.

Important** **Suppose** A **has more columns than rows. With** $n > m$ **there is at least one free variable. The system** $A\boldsymbol{x} = \mathbf{0}$ **has at least one nonzero solution.

Suppose $A\boldsymbol{x} = \mathbf{0}$ has more unknowns than equations ($n > m$). There must be **at least $n - m$ free columns. $A\boldsymbol{x} = \mathbf{0}$ must have nonzero solutions** in $\mathbf{N}(A)$.

The nullspace is a subspace. Its "dimension" is the number of free variables. This central idea—the ***dimension of a subspace***—is explained in Section 3.5 of this chapter.

Example 4 Find the nullspaces of A, B, M and the two special solutions to $M\boldsymbol{x} = \mathbf{0}$.

$$\boldsymbol{A} = \begin{bmatrix} 1 & 2 \\ 3 & 8 \end{bmatrix} \quad \boldsymbol{B} = \begin{bmatrix} A \\ 2A \end{bmatrix} = \begin{bmatrix} 1 & 2 \\ 3 & 8 \\ 2 & 4 \\ 6 & 16 \end{bmatrix} \quad \boldsymbol{M} = \begin{bmatrix} A & 2A \end{bmatrix} = \begin{bmatrix} 1 & 2 & 2 & 4 \\ 3 & 8 & 6 & 16 \end{bmatrix}.$$

Solution The equation $A\boldsymbol{x} = \mathbf{0}$ has only the zero solution $\boldsymbol{x} = \mathbf{0}$. *The nullspace is* $\mathbf{Z}$. It contains only the single point $\boldsymbol{x} = \mathbf{0}$ in $\mathbf{R}^2$. This fact comes from elimination:

$$A\boldsymbol{x} = \begin{bmatrix} 1 & 2 \\ 3 & 8 \end{bmatrix} \rightarrow \begin{bmatrix} 1 & 2 \\ 0 & 2 \end{bmatrix} \rightarrow \begin{bmatrix} 1 & 0 \\ 0 & 1 \end{bmatrix} = R = I \quad \textbf{No free variables, no } \boldsymbol{F}$$

A is invertible. There are no special solutions. Both columns of this matrix have pivots.

The rectangular matrix B has the same nullspace $\mathbf{Z}$. The first two equations in $B\boldsymbol{x} = \mathbf{0}$ again require $\boldsymbol{x} = \mathbf{0}$. The last two equations would also force $\boldsymbol{x} = \mathbf{0}$. When we add extra equations (giving extra rows), the nullspace certainly cannot become larger. Extra rows impose more conditions on the vectors $\boldsymbol{x}$ in the nullspace.

The rectangular matrix M is different. It has extra columns instead of extra rows. The solution vector $\boldsymbol{x}$ has *four* components. Elimination will produce pivots in the first two columns of M. **The last two columns of M are "free". They don't have pivots.**

$$M=\begin{bmatrix}1&2&2&4\\3&8&6&16\end{bmatrix}\qquad \boldsymbol{R}=\begin{bmatrix}\mathbf{1}&\mathbf{0}&2&0\\\mathbf{0}&\mathbf{1}&0&2\end{bmatrix}=[\boldsymbol{I}\ \ \boldsymbol{F}]$$

For the free variables x_3 and x_4, we make special choices of ones and zeros. First $x_3=1$, $x_4=0$ and second $x_3=0$, $x_4=1$. The pivot variables x_1 and x_2 are determined by the equation $Rx=\mathbf{0}$. We get two special solutions in the nullspace of M. This is also the nullspace of R: *elimination doesn't change solutions.*

Special solutions to $Mx=0$

$$\boldsymbol{R}=\begin{bmatrix}1&0&2&0\\0&1&0&2\end{bmatrix}\qquad s_1=\begin{bmatrix}-2\\0\\1\\0\end{bmatrix}\ \text{and}\ s_2=\begin{bmatrix}0\\-2\\0\\1\end{bmatrix}\begin{matrix}\leftarrow \textbf{2 pivot}\\ \leftarrow \textbf{variables}\\ \leftarrow \textbf{2 free}\\ \leftarrow \textbf{variables}\end{matrix}$$

$$\boldsymbol{R}s_1=\mathbf{0}\quad \boldsymbol{R}s_2=\mathbf{0}$$

Block Elimination in Three Steps : Final Thoughts

The special value of matrix notation is to show the big picture. So far we have described elimination as it is usually executed, a column at a time. But if we work with blocks of the original A, then **block elimination** can be described in three quick steps. A has rank r.

Step 1 Exchange columns of A by P_C and exchange rows of A by P_R to put the r independent columns first and r independent rows first in P_RAP_C.

$$P_RAP_C=\begin{bmatrix}\boldsymbol{W}&\boldsymbol{H}\\\boldsymbol{J}&\boldsymbol{K}\end{bmatrix}\qquad C=\begin{bmatrix}W\\J\end{bmatrix}\ \text{and}\ B=\begin{bmatrix}W&H\end{bmatrix}\ \text{have full rank } r$$

Step 2 Multiply the r top rows by W^{-1} to produce $W^{-1}B=\begin{bmatrix}I&W^{-1}H\end{bmatrix}=\begin{bmatrix}I&F\end{bmatrix}$

Step 3 Subtract $J\begin{bmatrix}I&W^{-1}H\end{bmatrix}$ from the $m-r$ lower rows $\begin{bmatrix}J&K\end{bmatrix}$ to produce $\begin{bmatrix}0&0\end{bmatrix}$

The result of Steps 1, 2, 3 is the reduced row echelon form $\boldsymbol{R_0}$

$$\boldsymbol{P_RAP_C}=\begin{bmatrix}W&H\\J&K\end{bmatrix}\rightarrow\begin{bmatrix}I&W^{-1}H\\J&K\end{bmatrix}\rightarrow\begin{bmatrix}\boldsymbol{I}&\boldsymbol{W^{-1}H}\\\mathbf{0}&\mathbf{0}\end{bmatrix}=\boldsymbol{R_0}\tag{1}$$

There are two facts that need explanation. They led to Step 2 and Step 3 :

1. The r by r matrix W is invertible **2. The blocks satisfy $JW^{-1}H=K$.**

1. For the invertibility of W, we look back to the factorization $A=CR$. Focusing on the r independent rows of A that go into B, this is $B=WR$. Since B and R have rank r and W is r by r, W must have rank r and must be invertible.
2. We know that the first r rows $\begin{bmatrix}I&W^{-1}H\end{bmatrix}$ are linearly independent. Since A has rank r, the lower rows $\begin{bmatrix}J&K\end{bmatrix}$ must be combinations of those upper rows. The combinations must be given by J to get the first r columns correct: $JI=J$. Then J times $W^{-1}H$ must equal K to make the last columns correct.

The conclusion is that $$\boldsymbol{P_RAP_C}=\begin{bmatrix}\boldsymbol{W}\\\boldsymbol{J}\end{bmatrix}\boldsymbol{W^{-1}}\begin{bmatrix}\boldsymbol{W}&\boldsymbol{H}\end{bmatrix}=\boldsymbol{CW^{-1}B}$$

We need that middle factor W^{-1} because the columns C and the rows B both contain W.

Problem Set 3.2

1 Why do A and $R = EA$ have the same nullspace? We know that E is invertible.

2 Find the row reduced form R and the rank r of A and B (*those depend on* c). Which are the pivot columns? Find the special solutions to $A\boldsymbol{x} = \mathbf{0}$ and $B\boldsymbol{x} = \mathbf{0}$.

Find special solutions $A = \begin{bmatrix} 1 & 2 & 1 \\ 3 & 6 & 3 \\ 4 & 8 & c \end{bmatrix}$ and $B = \begin{bmatrix} c & c \\ c & c \end{bmatrix}$.

3 Create a 2 by 4 matrix R whose special solutions to $R\boldsymbol{x} = \mathbf{0}$ are $\boldsymbol{s}_1$ and $\boldsymbol{s}_2$:

$$\boldsymbol{s}_1 = \begin{bmatrix} -3 \\ \mathbf{1} \\ 0 \\ \mathbf{0} \end{bmatrix} \quad \text{and} \quad \boldsymbol{s}_2 = \begin{bmatrix} -2 \\ \mathbf{0} \\ -6 \\ \mathbf{1} \end{bmatrix}$$

pivot columns 1 and 3
free variables x_2 and x_4
x_2 and x_4 are $1, 0$ and $0, 1$

Describe all 2 by 4 matrices with this nullspace $\mathbf{N}(A)$ spanned by $\boldsymbol{s}_1$ and $\boldsymbol{s}_2$.

4 Reduce A and B to their echelon forms R. Factor A and B into C times R.

$$\text{(a) } A = \begin{bmatrix} 1 & 2 & 2 & 4 & 6 \\ 1 & 2 & 3 & 6 & 9 \\ 0 & 0 & 1 & 2 & 3 \end{bmatrix} \qquad \text{(b) } B = \begin{bmatrix} 2 & 4 & 2 \\ 0 & 4 & 4 \\ 0 & 8 & 8 \end{bmatrix}.$$

5 For the matrix A in Problem 4, find a special solution to $R\boldsymbol{x} = \mathbf{0}$ for each free variable. Set the free variable to 1. Set the other free variables to zero. Solve $R\boldsymbol{x} = \mathbf{0}$.

6 True or false (with reason if true or example to show it is false):

(a) A square matrix has no free variables (its columns are independent).

(b) An invertible matrix has no free variables.

(c) An m by n matrix has no more than n pivot variables.

(d) An m by n matrix has no more than m pivot variables.

7 C is an r by r invertible matrix and $A = \begin{bmatrix} C & C \end{bmatrix}$. Factor A into CR and find r independent solutions to $A\boldsymbol{x} = \mathbf{0}$: a basis for the nullspace of A.

8 Put as many 1's as possible in a 4 by 8 *reduced* echelon matrix R so that the free columns (dependent on previous columns) are (a) $2, 4, 5, 6$ or (b) $1, 3, 6, 7, 8$.

9 Suppose column 4 of a 3 by 5 matrix is all zero. Then x_4 is certainly a _____ variable. The special solution for this variable is the vector $\boldsymbol{x} =$ _____.

10 Suppose the first and last columns of a 3 by 4 matrix are the same (not zero). Then _____ is a free variable. Find the special solution for this free variable.

11 The nullspace of a 5 by 5 matrix contains only $\boldsymbol{x} = \mathbf{0}$ when the matrix has _____ pivots. In that case the column space is $\mathbf{R}^5$. Explain why.

12 Suppose an m by n matrix has r pivots. The number of special solutions is _____ by the Counting Theorem. The nullspace contains only $\boldsymbol{x} = \mathbf{0}$ when $r =$ _____. The column space is all of $\mathbf{R}^m$ when the rank is $r =$ _____.

13 (Recommended) The plane $x - 3y - z = 12$ is parallel to $x - 3y - z = 0$. One particular point on this plane is $(12, 0, 0)$. All points on the plane have the form

$$\begin{bmatrix} x \\ y \\ z \end{bmatrix} = \begin{bmatrix} \ \\ 0 \\ 0 \end{bmatrix} + y \begin{bmatrix} \ \\ 1 \\ 0 \end{bmatrix} + z \begin{bmatrix} \ \\ 0 \\ 1 \end{bmatrix}.$$

14 Suppose column 1 + column 3 + column 5 = $\mathbf{0}$ in a 4 by 5 matrix with four pivots. Which column has no pivot? What is the special solution? Describe $\mathbf{N}(A)$.

Questions 15–20 ask for matrices (if possible) with specific properties.

15 Construct a matrix for which $\mathbf{N}(A) =$ all combinations of $(2, 2, 1, 0)$ and $(3, 1, 0, 1)$.

16 Construct A so that $\mathbf{N}(A) =$ all multiples of $(4, 3, 2, 1)$. Its rank is _____.

17 Construct a matrix whose column space contains $(1, 1, 5)$ and $(0, 3, 1)$ and whose nullspace contains $(1, 1, 2)$.

18 Construct a matrix whose column space contains $(1, 1, 0)$ and $(0, 1, 1)$ and whose nullspace contains $(1, 0, 1)$.

19 Construct a 2 by 2 matrix whose nullspace equals its column space. This is possible.

20 Why does no 3 by 3 matrix have a nullspace that equals its column space?

21 If $AB = 0$ then the column space of B is contained in the _____ of A. Why?

22 The reduced form R of a 3 by 3 matrix with randomly chosen entries is almost sure to be _____. What R is virtually certain if the random A is 4 by 3?

23 If $\mathbf{N}(A) =$ all multiples of $\boldsymbol{x} = (2, 1, 0, 1)$, what is R and what is its rank?

24 If the special solutions to $R\boldsymbol{x} = \mathbf{0}$ are in the columns of these nullspace matrices N, go backward to find the nonzero rows of the reduced matrices R:

$$N = \begin{bmatrix} 2 & 3 \\ 1 & 0 \\ 0 & 1 \end{bmatrix} \quad \text{and} \quad N = \begin{bmatrix} 0 \\ 0 \\ 1 \end{bmatrix} \quad \text{and} \quad N = \begin{bmatrix} \ \\ \ \\ \ \end{bmatrix} \quad \text{(empty 3 by 1).}$$

25 (a) What are the five 2 by 2 reduced matrices R whose entries are all 0's and 1's?

(b) What are the eight 1 by 3 matrices containing only 0's and 1's? Are all eight of them reduced echelon matrices R?

26 If A is 4 by 4 and invertible, describe the nullspace of the 4 by 8 matrix $B = [A \ \ A]$. Explain why A and $-A$ always have the same reduced echelon form R.

27 How is the nullspace $\mathbf{N}(C)$ related to the spaces $\mathbf{N}(A)$ and $\mathbf{N}(B)$, if $C = \begin{bmatrix} A \\ B \end{bmatrix}$?

28 Find the reduced R_0 and R for each of these matrices:

$$A = \begin{bmatrix} 0 & 0 & 0 \\ 2 & 4 & 6 \end{bmatrix} \qquad B = \begin{bmatrix} A & A \\ 0 & A \end{bmatrix} \qquad C = \begin{bmatrix} A & A \\ A & 0 \end{bmatrix}$$

29 Suppose the 2 pivot variables come *last* instead of first. Describe the reduced matrix R (3 columns) and the nullspace matrix N containing the special solutions.

30 If A has r pivot columns, how do you know that A^{T} has r pivot columns? Give a 3 by 3 example with different pivot column numbers for A and A^{T}.

31 Fill out these matrices so that they have rank 1:

$$A = \begin{bmatrix} a & b & c \\ d & & \\ g & & \end{bmatrix} \quad \text{and} \quad B = \begin{bmatrix} & 9 & \\ 1 & & \\ 2 & 6 & -3 \end{bmatrix} \quad \text{and} \quad M = \begin{bmatrix} a & b \\ c & \end{bmatrix}.$$

32 If A is a rank one matrix, the second row of R is _____. Do an example.

33 *If A has rank r, then it has an r by r submatrix S that is invertible.* Remove $m - r$ rows and $n - r$ columns to find an invertible submatrix S inside A, B, and C. You could keep the pivot rows and pivot columns:

$$A = \begin{bmatrix} 1 & 2 & 3 \\ 1 & 2 & 4 \end{bmatrix} \qquad B = \begin{bmatrix} 1 & 2 & 3 \\ 2 & 4 & 6 \end{bmatrix} \qquad C = \begin{bmatrix} 0 & 1 & 0 \\ 0 & 0 & 0 \\ 0 & 0 & 1 \end{bmatrix}.$$

34 Suppose A and B have the *same* reduced row echelon form R.

(a) Show that A and B have the same nullspace and the same row space.

(b) We know $E_1 A = R$ and $E_2 B = R$. So A equals an _____ matrix times B.

35 Kirchhoff's Current Law $A^{\mathrm{T}}\boldsymbol{y} = \mathbf{0}$ says that *current in* = *current out* at every node. At node 1 this is $y_3 = y_1 + y_4$ (arrows show the positive direction of currents). Reduce A^{T} to R (3 rows) and find three special solutions in the nullspace of A^{T}.

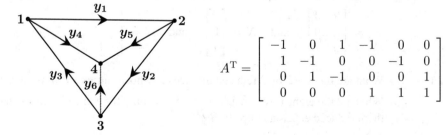

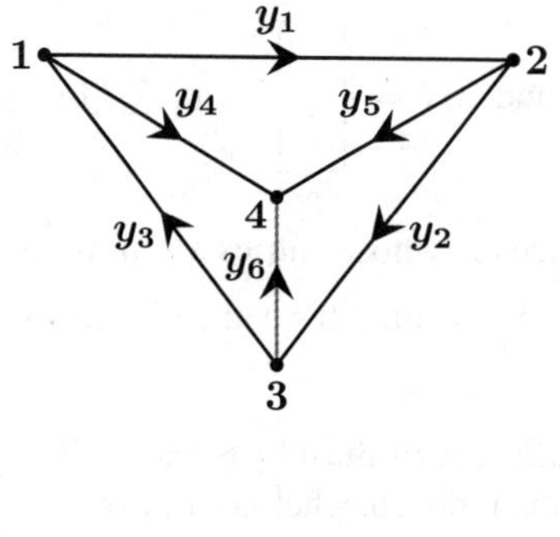

$$A^{\mathrm{T}} = \begin{bmatrix} -1 & 0 & 1 & -1 & 0 & 0 \\ 1 & -1 & 0 & 0 & -1 & 0 \\ 0 & 1 & -1 & 0 & 0 & 1 \\ 0 & 0 & 0 & 1 & 1 & 1 \end{bmatrix}$$

36 C contains the r pivot columns of A. Find the r pivot columns of C^{T} (r by m). Transpose back to find an r by r invertible submatrix S inside A:

$$\text{For } A=\begin{bmatrix}1&2&3\\2&4&6\\2&4&7\end{bmatrix} \quad \text{find } C \text{ (3 by 2) and then the invertible } S \text{ (2 by 2).}$$

37 Why is the column space $\mathbf{C}(AB)$ a subspace of $\mathbf{C}(A)$? Then rank$(AB) \leq$ rank(A).

38 Suppose column j of B is a combination of previous columns of B. Show that column j of AB is the same combination of previous columns of AB. Then AB cannot have new pivot columns, so $\mathbf{rank}(\boldsymbol{AB}) \leq \mathbf{rank}(\boldsymbol{B})$.

39 *(Important)* Suppose A and B are n by n matrices, and $AB = I$. Prove from $\operatorname{rank}(AB) \leq \operatorname{rank}(A)$ that the rank of A is n. So A is invertible and B must be its inverse. Therefore $BA = I$ *(which is not so obvious!)*.

40 If A is 2 by 3 and B is 3 by 2 and $AB = I$, show from its rank that $BA \neq I$. Give an example of A and B with $AB = I$. For $m < n$, a right inverse is not a left inverse.

41 What is the nullspace matrix N (containing the special solutions) for A, B, C ?

$$\text{2 by 2 blocks} \quad A=\begin{bmatrix}I & I\end{bmatrix} \quad \text{and} \quad B=\begin{bmatrix}I & I\\0 & 0\end{bmatrix} \quad \text{and} \quad C=\begin{bmatrix}I & I & I\end{bmatrix}.$$

42 Suppose A is an m by n matrix of rank r. Its reduced echelon form (including any zero rows) is R_0. Describe exactly the matrix Z (its shape and all its entries) that comes from *transposing the reduced row echelon form of* R_0^{T} : $\boldsymbol{Z} = (\text{rref}(\boldsymbol{A}^{\mathrm{T}}))^{\mathrm{T}}$.

43 (Recommended) Suppose $R_0=\begin{bmatrix}I & F\\0 & 0\end{bmatrix}$ is m by n of rank r. Pivot columns first:

(a) What are the shapes of those four blocks, based on m and n and r ?

(b) Find a *right inverse* B with $R_0B = I$ if $r = m$. The zero blocks are gone.

(c) Find a *left inverse* C with $CR_0 = I$ if $r = n$. The F and 0 column is gone.

(d) What is the reduced row echelon form of R_0^{T} (with shapes)?

(e) What is the reduced row echelon form of $R_0^{\mathrm{T}} R_0$ (with shapes)?

44 Suppose you allow elementary *column* operations on A as well as elementary row operations (which get to R_0). What is the "row-and-column reduced form" for an m by n matrix A of rank r ?

45 Find the factorizations $A = CR = CW^{-1}B$ for

$$A=\begin{bmatrix}1&4&7\\2&5&8\\3&6&9\end{bmatrix}$$

46 What multiple of block row 1 will equal block row 2 of this matrix ?

$$\begin{bmatrix}W\\J\end{bmatrix}\begin{bmatrix}W^{-1}\end{bmatrix}\begin{bmatrix}W & H\end{bmatrix}=\begin{bmatrix}W & H\\J & JW^{-1}H\end{bmatrix}$$

3.3 The Complete Solution to $Ax = b$

1 Complete solution to $A\boldsymbol{x} = \boldsymbol{b}$: $\boldsymbol{x} =$ (one particular solution $\boldsymbol{x}_p$) $+$ (any $\boldsymbol{x}_n$ in the nullspace).

2 Elimination on $A\boldsymbol{x} = \boldsymbol{b}$ leads to $R_0\boldsymbol{x} = \boldsymbol{d}$: Solvable when zero rows of R_0 have zeros in $\boldsymbol{d}$.

3 When $R_0\boldsymbol{x} = \boldsymbol{d}$ is solvable, one very particular solution $\boldsymbol{x}_p$ has all free variables equal to zero.

4 A has **full column rank** $\boldsymbol{r = n}$ when its nullspace $\mathbf{N}(A) =$ zero vector: *no free variables.*

5 A has **full row rank** $\boldsymbol{r = m}$ when its column space $\mathbf{C}(A)$ is $\mathbf{R}^m$: $A\boldsymbol{x} = \boldsymbol{b}$ is always solvable.

The last section totally solved $A\boldsymbol{x} = \mathbf{0}$. Elimination converted the problem to $R_0\boldsymbol{x} = \mathbf{0}$. The free variables were given special values (one and zero). Then the pivot variables were found by back substitution. We paid no attention to the right side $\boldsymbol{b}$ because it stayed at zero. Then zero rows in R_0 were no problem.

Now $\boldsymbol{b}$ is not zero. Row operations on the left side must act also on the right side. $A\boldsymbol{x} = \boldsymbol{b}$ is reduced to a simpler system $R_0\boldsymbol{x} = \boldsymbol{d}$ with the same solutions (if any). One way to organize that is to ***add b as an extra column of the matrix***. I will *"augment"* A with the right side $(b_1, b_2, b_3) = (1, 6, 7)$ to produce the **augmented matrix** $\begin{bmatrix} A & \boldsymbol{b} \end{bmatrix}$:

$$\begin{bmatrix} 1 & 3 & 0 & 2 \\ 0 & 0 & 1 & 4 \\ 1 & 3 & 1 & 6 \end{bmatrix} \begin{bmatrix} x_1 \\ x_2 \\ x_3 \\ x_4 \end{bmatrix} = \begin{bmatrix} \mathbf{1} \\ \mathbf{6} \\ \mathbf{7} \end{bmatrix} \quad \begin{matrix}\text{has the} \\ \text{augmented} \\ \text{matrix}\end{matrix} \quad \begin{bmatrix} 1 & 3 & 0 & 2 & \mathbf{1} \\ 0 & 0 & 1 & 4 & \mathbf{6} \\ 1 & 3 & 1 & 6 & \mathbf{7} \end{bmatrix} = \begin{bmatrix} A & \boldsymbol{b} \end{bmatrix}.$$

When we apply the usual elimination steps to A, reaching R_0, $\boldsymbol{b}$ turns into $\boldsymbol{d}$.

In this example we subtract row 1 and row 2 from row 3. Elimination produces a *row of zeros in* R_0, and it changes $\boldsymbol{b} = (1, 6, 7)$ to a new right side $\boldsymbol{d} = (\mathbf{1}, \mathbf{6}, \mathbf{0})$:

$$\begin{bmatrix} 1 & 3 & 0 & 2 \\ 0 & 0 & 1 & 4 \\ 0 & 0 & 0 & 0 \end{bmatrix} \begin{bmatrix} x_1 \\ x_2 \\ x_3 \\ x_4 \end{bmatrix} = \begin{bmatrix} \mathbf{1} \\ \mathbf{6} \\ \mathbf{0} \end{bmatrix} \quad \begin{matrix}\text{has the} \\ \text{augmented} \\ \text{matrix}\end{matrix} \quad \boxed{\begin{bmatrix} 1 & 3 & 0 & 2 & \mathbf{1} \\ 0 & 0 & 1 & 4 & \mathbf{6} \\ 0 & 0 & 0 & 0 & \mathbf{0} \end{bmatrix} = \begin{bmatrix} R_0 & \boldsymbol{d} \end{bmatrix}}.$$

That very last row is crucial. The third equation has become $0 = 0$. So the equations can be solved. In the original matrix A, the first row plus the second row equals the third row. To solve $A\boldsymbol{x} = \boldsymbol{b}$, we need $b_1 + b_2 = b_3$ on the right side too. The all-important property of $\boldsymbol{b}$ was $\mathbf{1} + \mathbf{6} = \mathbf{7}$. That led to $0 = 0$ in the third equation. This was essential.

Here are the same augmented matrices for a general $\boldsymbol{b} = (b_1, b_2, b_3)$:

$$\begin{bmatrix} A & \boldsymbol{b} \end{bmatrix} = \begin{bmatrix} 1 & 3 & 0 & 2 & \boldsymbol{b}_1 \\ 0 & 0 & 1 & 4 & \boldsymbol{b}_2 \\ 1 & 3 & 1 & 6 & \boldsymbol{b}_3 \end{bmatrix} \longrightarrow \begin{bmatrix} 1 & 3 & 0 & 2 & \boldsymbol{b}_1 \\ 0 & 0 & 1 & 4 & \boldsymbol{b}_2 \\ 0 & 0 & 0 & 0 & \boldsymbol{b}_3 - \boldsymbol{b}_1 - \boldsymbol{b}_2 \end{bmatrix} = \begin{bmatrix} R_0 & \boldsymbol{d} \end{bmatrix}$$

Now we get $0 = 0$ in the third equation only if $b_3 - b_1 - b_2 = 0$. This is $b_1 + b_2 = b_3$.

One Particular Solution $A\boldsymbol{x}_p = \boldsymbol{b}$

For an easy solution $\boldsymbol{x}_p$, *choose the free variables to be zero:* $x_2 = x_4 = 0$. Then the two nonzero equations give the two pivot variables $x_1 = 1$ and $x_3 = 6$. Our particular solution to $A\boldsymbol{x} = \boldsymbol{b}$ (and also $R_0\boldsymbol{x} = \boldsymbol{d}$) is $\boldsymbol{x}_p = (\mathbf{1}, 0, \mathbf{6}, 0)$. This particular solution is my favorite: ***free variables = zero, pivot variables from d.***

For a solution to exist, zero rows in R_0 must also be zero in $\boldsymbol{d}$. Since I is in the pivot rows and pivot columns of R_0, the pivot variables in $\boldsymbol{x}_{\text{particular}}$ come from $\boldsymbol{d}$:

$$R_0\boldsymbol{x}_p = \begin{bmatrix} \mathbf{1} & 3 & \mathbf{0} & 2 \\ \mathbf{0} & 0 & \mathbf{1} & 4 \\ 0 & 0 & 0 & 0 \end{bmatrix} \begin{bmatrix} \mathbf{1} \\ 0 \\ \mathbf{6} \\ 0 \end{bmatrix} = \begin{bmatrix} \mathbf{1} \\ \mathbf{6} \\ 0 \end{bmatrix} = \boldsymbol{d} \qquad \begin{array}{l} \textbf{Pivot variables } \mathbf{1, 6} \\ \textbf{Free variables } \mathbf{0, 0} \\ \boldsymbol{x}_{\textbf{particular}} = (\mathbf{1, 0, 6, 0}). \end{array}$$

Notice how we *choose* the free variables (as zero) and then *solve* for the pivot variables. After the row reduction to R_0, those steps are quick. When the free variables are zero, the pivot variables for $\boldsymbol{x}_p$ are already seen in the right side vector $\boldsymbol{d}$.

One $\boldsymbol{x}_{\textbf{particular}}$	***The particular solution solves***	$A\boldsymbol{x}_p = \boldsymbol{b}$
All $\boldsymbol{x}_{\textbf{nullspace}}$	***The $n - r$ special solutions solve***	$A\boldsymbol{x}_n = \mathbf{0}$

That particular solution is $(1, 0, 6, 0)$. The $4 - 2$ special (nullspace) solutions came in Section 3.2 from the two free columns of R_0, by reversing signs of $3, 2$, and 4. ***Please notice how the complete solution $\boldsymbol{x}_p + \boldsymbol{x}_n$ to $A\boldsymbol{x} = \boldsymbol{b}$ includes all $\boldsymbol{x}_n$:***

Complete solution
one $\boldsymbol{x}_p$+many $\boldsymbol{x}_n$
particular $\boldsymbol{x}_p$
nullspace vectors $\boldsymbol{x}_n$

$$\boldsymbol{x} = \boldsymbol{x}_p + \boldsymbol{x}_n = \begin{bmatrix} 1 \\ 0 \\ 6 \\ 0 \end{bmatrix} + x_2 \begin{bmatrix} -3 \\ 1 \\ 0 \\ 0 \end{bmatrix} + x_4 \begin{bmatrix} -2 \\ 0 \\ -4 \\ 1 \end{bmatrix}.$$

Question Suppose A is a square invertible matrix, $m = n = r$. What are $\boldsymbol{x}_p$ and $\boldsymbol{x}_n$?
Answer The particular solution is the one and *only* solution $\boldsymbol{x}_p = A^{-1}\boldsymbol{b}$. There are no special solutions or free variables. $R_0 = I$ has no zero rows. The only vector in the nullspace is $\boldsymbol{x}_n = \mathbf{0}$. The complete solution is $\boldsymbol{x} = \boldsymbol{x}_p + \boldsymbol{x}_n = A^{-1}\boldsymbol{b} + \mathbf{0}$.

We didn't mention the nullspace in Chapter 2, because A was invertible. A was reduced all the way to I. $\begin{bmatrix} A & \boldsymbol{b} \end{bmatrix}$ went to $\begin{bmatrix} I & A^{-1}\boldsymbol{b} \end{bmatrix}$. Then $A\boldsymbol{x} = \boldsymbol{b}$ became $\boldsymbol{x} = A^{-1}\boldsymbol{b}$ which is $\boldsymbol{d}$. This is a special case here, but square invertible matrices are the best. So they got their own chapter before this one.

For small examples we can reduce $\begin{bmatrix} A & \boldsymbol{b} \end{bmatrix}$ to $\begin{bmatrix} R_0 & \boldsymbol{d} \end{bmatrix}$. For a large matrix, MATLAB does it better. One particular solution (not necessarily ours) is $\boldsymbol{x} = A\backslash\boldsymbol{b}$ from backslash. Here is an example with *full column rank*. Both columns have pivots.

Example 1 Find the condition on (b_1, b_2, b_3) for $A\boldsymbol{x} = \boldsymbol{b}$ to be solvable, if

$$A = \begin{bmatrix} 1 & 1 \\ 1 & 2 \\ -2 & -3 \end{bmatrix} \quad \text{and} \quad \boldsymbol{b} = \begin{bmatrix} b_1 \\ b_2 \\ b_3 \end{bmatrix}.$$

Solution Elimination uses the augmented matrix $\begin{bmatrix} A & \boldsymbol{b} \end{bmatrix}$ with its extra column $\boldsymbol{b}$. Subtract row 1 of $\begin{bmatrix} A & \boldsymbol{b} \end{bmatrix}$ from row 2. Then add 2 times row 1 to row 3 to reach $\begin{bmatrix} R_0 & \boldsymbol{d} \end{bmatrix}$:

$$\begin{bmatrix} 1 & 1 & b_1 \\ 1 & 2 & b_2 \\ -2 & -3 & b_3 \end{bmatrix} \rightarrow \begin{bmatrix} 1 & 1 & b_1 \\ 0 & 1 & b_2 - b_1 \\ 0 & -1 & b_3 + 2b_1 \end{bmatrix} \rightarrow \begin{bmatrix} 1 & 0 & 2b_1 - b_2 \\ 0 & 1 & b_2 - b_1 \\ 0 & 0 & \boldsymbol{b_3 + b_1 + b_2} \end{bmatrix} = \begin{bmatrix} R_0 & \boldsymbol{d} \end{bmatrix}$$

Equation 3 is $0 = 0$ provided $b_3 + b_1 + b_2 = 0$. This is the condition to put $\boldsymbol{b}$ in the column space. Then $A\boldsymbol{x} = \boldsymbol{b}$ will be solvable. The rows of A add to the zero row. So for consistency (these are equations!) the entries of $\boldsymbol{b}$ must also add to zero.

This example has no free variables since $n - r = 2 - 2$. Therefore no special solutions. The nullspace solution is $\boldsymbol{x}_n = \boldsymbol{0}$. The particular solution to $A\boldsymbol{x} = \boldsymbol{b}$ and $R_0\boldsymbol{x} = \boldsymbol{d}$ is at the top of the final column $\boldsymbol{d}$:

$$\boldsymbol{r = n} \qquad \textbf{One solution to } A\boldsymbol{x} = \boldsymbol{b} \qquad \boldsymbol{x} = \boldsymbol{x}_p + \boldsymbol{x}_n = \begin{bmatrix} 2b_1 - b_2 \\ b_2 - b_1 \end{bmatrix} + \begin{bmatrix} 0 \\ 0 \end{bmatrix}.$$

If $b_3 + b_1 + b_2$ is not zero, there is no solution to $A\boldsymbol{x} = \boldsymbol{b}$ ($\boldsymbol{x}_p$ and $\boldsymbol{x}$ don't exist).

This example is typical of an extremely important case: A has *full column rank*. Every column has a pivot. *The rank is* $r = n$. The matrix is tall and thin ($m \geq n$). Row reduction puts $R = I$ at the top, when A is reduced to R_0 with rank n:

$$\textbf{Full column rank } \boldsymbol{r = n} \quad R_0 = \begin{bmatrix} I \\ 0 \end{bmatrix} = \begin{bmatrix} n \text{ by } n \text{ identity matrix} \\ m - n \text{ rows of zeros} \end{bmatrix} \tag{1}$$

This matrix A has independent columns. The nullspace of A is $\mathbf{Z} = \{\text{zero vector}\}$.

We will collect together the different ways of recognizing this type of matrix.

Every matrix A with **full column rank** ($\boldsymbol{r = n}$) has all these properties:
1. All columns of A are pivot columns (**independent**). No free variables.
2. The nullspace $\mathbf{N}(A)$ contains only the zero vector $\boldsymbol{x} = \boldsymbol{0}$.
3. If $A\boldsymbol{x} = \boldsymbol{b}$ has a solution (it might not) then it has only *one solution*.

$A\boldsymbol{x} = \boldsymbol{0}$ only happens when $\boldsymbol{x} = \boldsymbol{0}$. Later we will add one more fact to the list above: *The square matrix* $A^{\mathrm{T}}A$ *is invertible when the rank of* A *is* n. A may have many rows.

In this case the nullspace of A has shrunk to the zero vector. The solution to $A\boldsymbol{x} = \boldsymbol{b}$ is *unique* (if there is a solution). There will be $m - n$ zero rows in R_0. So there are $m - n$ conditions on $\boldsymbol{b}$ in order to have $0 = 0$ in those rows, and $\boldsymbol{b}$ in the column space. With full column rank, $A\boldsymbol{x} = \boldsymbol{b}$ will have **one solution** or **no solution**.

Full Row Rank and the Complete Solution

The other extreme case is full row rank. Now $\boldsymbol{A x}=\boldsymbol{b}$ has *one or infinitely many* solutions. In this case A must be *short and wide* ($m \leq n$). ***A matrix has full row rank if*** $\boldsymbol{r=m}$. "*The rows are independent.*" Every row has a pivot, and here is an example.

Example 2 This system $\boldsymbol{A x}=\boldsymbol{b}$ has $n=3$ unknowns but only $m=2$ equations:

Full row rank
$$\begin{aligned} x+\ y+z&=3 \\ x+2y-z&=4 \end{aligned} \qquad (\text{rank } \boldsymbol{r}=\boldsymbol{m}=\mathbf{2})$$

These are two planes in xyz space. The planes are not parallel so they intersect in a line. This line of solutions is exactly what elimination will find. ***The particular solution will be one point on the line. Adding the nullspace vectors*** $\boldsymbol{x}_n$ ***will move us along the line in Figure 3.1***. Then $\boldsymbol{x}=\boldsymbol{x}_p+$ all $\boldsymbol{x}_n$ gives the whole line of solutions.

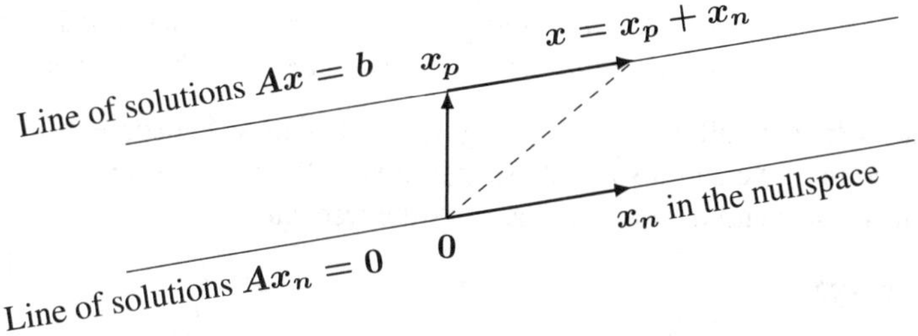

Figure 3.1: Complete solution $\boldsymbol{x}=$ *one* particular solution $\boldsymbol{x}_p+$ *all* nullspace solutions $\boldsymbol{x}_n$.

We find $\boldsymbol{x}_p$ and $\boldsymbol{x}_n$ by elimination downwards and then upwards on $\begin{bmatrix} A & \boldsymbol{b} \end{bmatrix}$.

$$\begin{bmatrix} 1 & 1 & 1 & \mathbf{3} \\ 1 & 2 & -1 & \mathbf{4} \end{bmatrix} \rightarrow \begin{bmatrix} 1 & 1 & 1 & \mathbf{3} \\ \mathbf{0} & 1 & -2 & \mathbf{1} \end{bmatrix} \rightarrow \begin{bmatrix} 1 & \mathbf{0} & 3 & \mathbf{2} \\ 0 & 1 & -2 & \mathbf{1} \end{bmatrix} = \begin{bmatrix} R & \boldsymbol{d} \end{bmatrix}.$$

The particular solution $(2,1,0)$ *has free variable* $x_3=0$. It comes directly from $\boldsymbol{d}$. *The special solution* $\boldsymbol{s}$ *has* $x_3=1$. Then $-x_1$ and $-x_2$ come from the free column of R.

It is wise to check that $\boldsymbol{x}_p$ and $\boldsymbol{s}$ satisfy the original equations $A\boldsymbol{x}_p=\boldsymbol{b}$ and $A\boldsymbol{s}=\mathbf{0}$:

$$\begin{aligned} 2+1 &= 3 \qquad & -3+2+1 &= 0 \\ 2+2 &= 4 \qquad & -3+4-1 &= 0 \end{aligned}$$

The nullspace solution $\boldsymbol{x}_n$ is any multiple of $\boldsymbol{s}$. It moves along the line of solutions, starting at $\boldsymbol{x}_{\textbf{particular}}$. ***Please notice again how to write the answer***:

Complete solution
Particular + nullspace
$$\boldsymbol{x}=\boldsymbol{x}_p+\boldsymbol{x}_n=\begin{bmatrix} 2 \\ 1 \\ 0 \end{bmatrix}+x_3\begin{bmatrix} -3 \\ 2 \\ 1 \end{bmatrix}.$$

This line of solutions is drawn in Figure 3.1. Any point on the line *could* have been chosen as the particular solution. We chose $\boldsymbol{x}_p$ as the point with $x_3=0$.

The particular solution is *not* multiplied by an arbitrary constant!

> Every matrix A with ***full row rank*** ($r{=}m$) has all these properties:
>
> **1.** All rows have pivots, and R_0 **has no zero rows**: $R_0 = R$.
>
> **2.** $Ax = b$ **has a solution for every right side** b.
>
> **3.** The column space of A is the whole space $\mathbf{R}^m$.
>
> **4.** If $m < n$ the equation $Ax = b$ is **underdetermined** (many solutions).

In this case with m pivots, ***the m rows are linearly independent***. So the columns of A^{T} are linearly independent. The nullspace of A^{T} contains only the zero vector.

The four possibilities for linear equations depend on the rank r

$r = m$	and	$r = n$	*Square and invertible*	$Ax = b$	has 1 solution
$r = m$	and	$r < n$	*Short and wide*	$Ax = b$	has ∞ solutions
$r < m$	and	$r = n$	*Tall and thin*	$Ax = b$	has 0 or 1 solution
$r < m$	and	$r < n$	*Not full rank*	$Ax = b$	has 0 or ∞ solutions

The reduced R_0 will fall in the same category as the matrix A. In case the pivot columns happen to come first, we can display these four possibilities. For $R_0 x = d$ and $Ax = b$ to be solvable, d must end in $m - r$ zeros. F is the free part of R_0.

Four types for R_0	$\begin{bmatrix} I \end{bmatrix}$	$\begin{bmatrix} I & F \end{bmatrix}$	$\begin{bmatrix} I \\ 0 \end{bmatrix}$	$\begin{bmatrix} I & F \\ 0 & 0 \end{bmatrix}$
Their ranks	$r = m = n$	$r = m < n$	$r = n < m$	$r < m, r < n$

Cases 1 and 2 have full row rank $r = m$. Cases 1 and 3 have full column rank $r = n$.

To end this important section of the book, here is a note about *computational linear algebra*. Linear equations $Ax = b$ are obviously fundamental. In practice, the steps of elimination are reordered for the sake of speed and numerical stability. We can solve systems of order 1000 on a laptop (allowing roundoff errors in single precision or double precision). Supercomputers can solve much larger systems. But there is a limit on the matrix size. What to do beyond that limit?

The surprising answer is **randomized linear algebra**. We *sample* the columns of A. We accept the errors involved. In practice matrices are not completely random, and the final results are remarkably good. Often the approximation to A is expressed in the 3-factor form $\boldsymbol{A \approx CUR}$. C comes from sampling the columns of A and R comes from the rows of A.

The smaller mixing matrix U is constructed as we go. With high probability, the approximate solution is surprisingly accurate.

Linear algebra is alive. The demands of computation (speed and accuracy) lead to new ideas. The same will be true for eigenvalues and singular values—later in this book.

■ WORKED EXAMPLES ■

3.3 A This question connects elimination (**pivot columns and back substitution**) to **column space-nullspace-rank-solvability** (the higher level picture). A has rank 2:

$$A\boldsymbol{x} = \boldsymbol{b} \quad \text{is} \quad \begin{aligned} x_1 + 2x_2 + 3x_3 + \ \ 5x_4 &= b_1 \\ 2x_1 + 4x_2 + 8x_3 + 12x_4 &= b_2 \\ 3x_1 + 6x_2 + 7x_3 + 13x_4 &= b_3 \end{aligned} \qquad A = \begin{bmatrix} 1 & 2 & 3 & 5 \\ 2 & 4 & 8 & 12 \\ 3 & 6 & 7 & 13 \end{bmatrix}$$

1. Reduce $[\,A \ \ \boldsymbol{b}\,]$ to $[\,U \ \ \boldsymbol{c}\,]$, so that $A\boldsymbol{x} = \boldsymbol{b}$ becomes a triangular system $U\boldsymbol{x} = \boldsymbol{c}$.
2. Find the condition on b_1, b_2, b_3 for $A\boldsymbol{x} = \boldsymbol{b}$ to have a solution.
3. Describe the column space of A. Which plane in $\mathbf{R}^3$?
4. Describe the nullspace of A. Which special solutions in $\mathbf{R}^4$?
5. Reduce $[\,U \ \ \boldsymbol{c}\,]$ to $[\,R_0 \ \ \boldsymbol{d}\,]$: Special solutions from R_0, particular solution from $\boldsymbol{d}$.
6. Find a particular solution to $A\boldsymbol{x} = (0, 6, -6)$ and then the complete solution.

Solution

1. The multipliers in elimination are 2 and 3 and -1. They take $[\,A \ \ \boldsymbol{b}\,]$ into $[\,U \ \ \boldsymbol{c}\,]$.

$$\begin{bmatrix} 1 & 2 & 3 & 5 & \mathbf{b_1} \\ 2 & 4 & 8 & 12 & \mathbf{b_2} \\ 3 & 6 & 7 & 13 & \mathbf{b_3} \end{bmatrix} \rightarrow \left[\begin{array}{cccc|l} 1 & 2 & 3 & 5 & \mathbf{b_1} \\ 0 & 0 & 2 & 2 & \mathbf{b_2 - 2b_1} \\ 0 & 0 & -2 & -2 & \mathbf{b_3 - 3b_1} \end{array}\right] \rightarrow \left[\begin{array}{cccc|l} 1 & 2 & 3 & 5 & \mathbf{b_1} \\ 0 & 0 & 2 & 2 & \mathbf{b_2 - 2b_1} \\ 0 & 0 & 0 & 0 & \mathbf{b_3 + b_2 - 5b_1} \end{array}\right]$$

2. The last equation shows the solvability condition $b_3 + b_2 - 5b_1 = 0$. Then $0 = 0$.
3. **First description**: The column space is the plane containing all combinations of A's pivot columns $(1, 2, 3)$ and $(3, 8, 7)$. The pivots are in columns 1 and 3. **Second description**: The column space contains all vectors with $b_3 + b_2 - 5b_1 = 0$. That makes $A\boldsymbol{x} = \boldsymbol{b}$ solvable, so $\boldsymbol{b}$ is in the column space. *All columns of A pass this test $b_3 + b_2 - 5b_1 = 0$. This is the equation for the plane in the first description*!
4. The special solutions have free variables $x_2 = 1, x_4 = 0$ and then $x_2 = 0, x_4 = 1$:

Special solutions to $A\boldsymbol{x} = \mathbf{0}$
Back substitution in $U\boldsymbol{x} = \mathbf{0}$
or change signs of 2, 2, 1 in R

$$\boldsymbol{s}_1 = \begin{bmatrix} -2 \\ \mathbf{1} \\ 0 \\ \mathbf{0} \end{bmatrix} \qquad \boldsymbol{s}_2 = \begin{bmatrix} -2 \\ \mathbf{0} \\ -1 \\ \mathbf{1} \end{bmatrix}$$

The nullspace $\mathbf{N}(A)$ in $\mathbf{R}^4$ contains all $\boldsymbol{x}_n = c_1\boldsymbol{s}_1 + c_2\boldsymbol{s}_2$.

5. In the reduced form R_0, the third column changes from $(3, 2, 0)$ in U to $(0, 1, 0)$. The right side $\boldsymbol{c} = (0, 6, 0)$ becomes $\boldsymbol{d} = (-9, 3, 0)$ showing $\mathbf{-9}$ **and 3 in** $\boldsymbol{x}_p$:

$$[\,U \ \ \boldsymbol{c}\,] = \begin{bmatrix} 1 & 2 & 3 & 5 & \mathbf{0} \\ 0 & 0 & 2 & 2 & \mathbf{6} \\ 0 & 0 & 0 & 0 & \mathbf{0} \end{bmatrix} \longrightarrow [\,R_0 \ \ \boldsymbol{d}\,] = \begin{bmatrix} 1 & 2 & 0 & 2 & \mathbf{-9} \\ 0 & 0 & 1 & 1 & \mathbf{3} \\ 0 & 0 & 0 & 0 & \mathbf{0} \end{bmatrix}$$

6. $\boldsymbol{x}_p = (\mathbf{-9}, 0, \mathbf{3}, 0)$ is the very particular solution with free variables $x_2 = x_4 = 0$.

The complete solution to $A\boldsymbol{x} = (0, 6, -6)$ is $\boldsymbol{x} = \boldsymbol{x}_p + \boldsymbol{x}_n = \boldsymbol{x}_p + c_1\boldsymbol{s}_1 + c_2\boldsymbol{s}_2$.

3.3 B Suppose you have this information about the solutions to $Ax = b$ for a specific b. What does that tell you about m and n and r (and A itself)? And possibly about b.

1. There is exactly one solution.
2. All solutions to $Ax = b$ have the form $x = \begin{bmatrix} 2 \\ 1 \end{bmatrix} + c\begin{bmatrix} 1 \\ 1 \end{bmatrix}$.
3. There are no solutions.
4. All solutions to $Ax = b$ have the form $x = \begin{bmatrix} 1 \\ 1 \\ 0 \end{bmatrix} + c\begin{bmatrix} 1 \\ 0 \\ 1 \end{bmatrix}$

Solution In case 1, with exactly one solution, A must have full column rank $r = n$. The nullspace of A contains only the zero vector. And b is in the column space.

In case 2, A must have $n = 2$ columns (and m is arbitrary). With $\begin{bmatrix} 1 \\ 1 \end{bmatrix}$ in the nullspace of A, column 2 is the *negative* of column 1. Also $A \neq 0$: the rank is 1. With $x = \begin{bmatrix} 2 \\ 1 \end{bmatrix}$ as a solution, $b = 2(\text{column } 1) + (\text{column } 2)$. My choice for x_p would be $(1, 0)$.

In case 3 we only know that b is not in the column space of A. The rank of A must be less than m. I guess we know $b \neq 0$, otherwise $x = 0$ would be a solution.

In case 4, A must have $n = 3$ columns. With $(1, 0, 1)$ in the nullspace of A, column 3 of A is $-(\text{column } 1)$. The rank is $3 - 1 = 2$ and b is column 1 + column 2.

3.3 C Find the complete solution $x = x_p + x_n$ by forward elimination on $[A \ \ b]$:

Solution Elimination leads us to $R_0 x = d$

$$\begin{bmatrix} 1 & 2 & 1 & 0 & \mathbf{4} \\ 2 & 4 & 4 & 8 & \mathbf{2} \\ 4 & 8 & 6 & 8 & \mathbf{10} \end{bmatrix} \longrightarrow \begin{bmatrix} 1 & 2 & 1 & 0 & \mathbf{4} \\ 0 & 0 & 2 & 8 & \mathbf{-6} \\ 0 & 0 & 0 & 0 & \mathbf{0} \end{bmatrix} \longrightarrow \begin{bmatrix} 1 & 2 & 0 & -4 & \mathbf{7} \\ 0 & 0 & 1 & 4 & \mathbf{-3} \\ 0 & 0 & 0 & 0 & \mathbf{0} \end{bmatrix}.$$

For the nullspace part x_n with $b = 0$, set the free variables x_2, x_4 to $1, 0$ and also $0, 1$:

Special solutions $s_1 = (-2, 1, 0, 0)$ and $s_2 = (4, 0, -4, 1)$ **Particular** $x_p = (7, 0, -3, 0)$

Then the complete solution to $Ax = b$ is $x_{\text{complete}} = x_p + c_1 s_1 + c_2 s_2$.

The rows of A produced the zero row from $2(\text{row } 1) + (\text{row } 2) - (\text{row } 3) = (0, 0, 0, 0)$. The same combination for $b = (4, 2, 10)$ gives $2(4) + (2) - (10) = 0$.

If a combination of the rows of A (on the left side) gives the zero row, then the same combination must give zero on the right side. Of course! *Otherwise no solution.*

We could say this again in different words: If every column of A is perpendicular to $y = (2, 1, -1)$, then any combination b of those columns must also be perpendicular to y. Otherwise b is not in the column space and $Ax = b$ is not solvable.

And again: If y is in the nullspace of A^T then y must be perpendicular to every b in the column space of A.

Problem Set 3.3

1 (Recommended) Execute the six steps of Worked Example **3.3 A** to describe the column space and nullspace of A and the complete solution to $A\boldsymbol{x} = \boldsymbol{b}$:

$$A = \begin{bmatrix} 2 & 4 & 6 & 4 \\ 2 & 5 & 7 & 6 \\ 2 & 3 & 5 & 2 \end{bmatrix} \qquad \boldsymbol{b} = \begin{bmatrix} b_1 \\ b_2 \\ b_3 \end{bmatrix} = \begin{bmatrix} 4 \\ 3 \\ 5 \end{bmatrix}$$

2 Carry out the same six steps for this matrix A with rank one. You will find *two* conditions on b_1, b_2, b_3 for $A\boldsymbol{x} = \boldsymbol{b}$ to be solvable. Together these two conditions put $\boldsymbol{b}$ into the ______ space (two planes give a line):

$$A = \begin{bmatrix} 1 \\ 3 \\ 2 \end{bmatrix} \begin{bmatrix} 2 & 1 & 3 \end{bmatrix} = \begin{bmatrix} 2 & 1 & 3 \\ 6 & 3 & 9 \\ 4 & 2 & 6 \end{bmatrix} \qquad \boldsymbol{b} = \begin{bmatrix} b_1 \\ b_2 \\ b_3 \end{bmatrix} = \begin{bmatrix} 10 \\ 30 \\ 20 \end{bmatrix}$$

Questions 3–15 are about the solution of $A\boldsymbol{x} = \boldsymbol{b}$. Follow the steps in the text to $\boldsymbol{x}_p$ and $\boldsymbol{x}_n$. Start from the augmented matrix with last column $\boldsymbol{b}$.

3 Write the complete solution as $\boldsymbol{x}_p$ plus any multiple of $\boldsymbol{s}$ in the nullspace:

$$\begin{aligned} x + 3y &= 7 \\ 2x + 6y &= 14 \end{aligned} \qquad \begin{aligned} x + 3y + 3z &= 1 \\ 2x + 6y + 9z &= 5 \\ -x - 3y + 3z &= 5 \end{aligned}$$

4 Find the complete solution $\boldsymbol{x} = \boldsymbol{x}_p +$ any $\boldsymbol{x}_n$ (also called the *general solution*) to

$$\begin{bmatrix} 1 & 3 & 1 & 2 \\ 2 & 6 & 4 & 8 \\ 0 & 0 & 2 & 4 \end{bmatrix} \begin{bmatrix} x \\ y \\ z \\ t \end{bmatrix} = \begin{bmatrix} 1 \\ 3 \\ 1 \end{bmatrix}.$$

5 Under what conditions on b_1, b_2, b_3 are these systems solvable? Include $\boldsymbol{b}$ as a fourth column in elimination. Find all solutions when that condition holds:

$$\begin{aligned} x + 2y - 2z &= b_1 \\ 2x + 5y - 4z &= b_2 \\ 4x + 9y - 8z &= b_3 \end{aligned} \qquad \begin{aligned} 2x + 2y &= b_1 \\ 4x + 4y &= b_2 \\ 8x + 8y &= b_3 \end{aligned}$$

6 What conditions on b_1, b_2, b_3, b_4 make each system solvable? Find $\boldsymbol{x}$ in that case:

$$\begin{bmatrix} 1 & 2 \\ 2 & 4 \\ 2 & 5 \\ 3 & 9 \end{bmatrix} \begin{bmatrix} x_1 \\ x_2 \end{bmatrix} = \begin{bmatrix} b_1 \\ b_2 \\ b_3 \\ b_4 \end{bmatrix} \qquad \begin{bmatrix} 1 & 2 & 3 \\ 2 & 4 & 6 \\ 2 & 5 & 7 \\ 3 & 9 & 12 \end{bmatrix} \begin{bmatrix} x_1 \\ x_2 \\ x_3 \end{bmatrix} = \begin{bmatrix} b_1 \\ b_2 \\ b_3 \\ b_4 \end{bmatrix}.$$

7 Show by elimination that (b_1, b_2, b_3) is in the column space if $b_3 - 2b_2 + 4b_1 = 0$. The rank is $\boldsymbol{r} = \mathbf{2}$. What combination of the rows of A gives the zero row?

$$A = \begin{bmatrix} 1 & 3 & 1 \\ 3 & 8 & 2 \\ 2 & 4 & 0 \end{bmatrix}.$$

8 Which vectors (b_1, b_2, b_3) are in the column space of A? Which combinations of the rows of A give zero?

$$\text{(a) } A = \begin{bmatrix} 1 & 2 & 1 \\ 2 & 6 & 3 \\ 0 & 2 & 5 \end{bmatrix} \qquad \text{(b) } A = \begin{bmatrix} 1 & 1 & 1 \\ 1 & 2 & 4 \\ 2 & 4 & 8 \end{bmatrix}.$$

9 Find the complete solution in the form $\boldsymbol{x}_p + \boldsymbol{x}_n$ to these full rank systems:

$$\text{(a) } x + y + z = 4 \qquad \text{(b) } \begin{matrix} x + y + z = 4 \\ x - y + z = 4. \end{matrix}$$

10 Construct a 2 by 3 system $A\boldsymbol{x} = \boldsymbol{b}$ with particular solution $\boldsymbol{x}_p = (2, 4, 0)$ and homogeneous solution $\boldsymbol{x}_n =$ any multiple of $(1, 1, 1)$.

11 Why can't a 1 by 3 system have $\boldsymbol{x}_p = (2, 4, 0)$ and $\boldsymbol{x}_n =$ any multiple of $(1, 1, 1)$?

12 (a) If $A\boldsymbol{x} = \boldsymbol{b}$ has two solutions $\boldsymbol{x}_1$ and $\boldsymbol{x}_2$, find two solutions to $A\boldsymbol{x} = \mathbf{0}$.

(b) Then find another solution to $A\boldsymbol{x} = \mathbf{0}$ and another solution to $A\boldsymbol{x} = \boldsymbol{b}$.

13 Explain why these are all false:

(a) The complete solution is any linear combination of $\boldsymbol{x}_p$ and $\boldsymbol{x}_n$.

(b) A system $A\boldsymbol{x} = \boldsymbol{b}$ has at most one particular solution. This is true if A is ____.

(c) The solution $\boldsymbol{x}_p$ with all free variables zero is the shortest solution (minimum length $\|\boldsymbol{x}\|$). Find a 2 by 2 counterexample.

(d) If A is invertible there is no solution $\boldsymbol{x}_n$ in the nullspace.

14 Suppose column 5 of U has no pivot. Then x_5 is a ____ variable. The zero vector (is) (is not) the only solution to $A\boldsymbol{x} = \mathbf{0}$. If $A\boldsymbol{x} = \boldsymbol{b}$ has a solution, then it has ____ solutions.

15 Suppose row 3 of U has no pivot. Then that row is ____. The equation $U\boldsymbol{x} = \boldsymbol{c}$ is only solvable provided ____. The equation $A\boldsymbol{x} = \boldsymbol{b}$ (*is*) (*is not*) (*might not be*) solvable.

16 The largest possible rank of a 3 by 5 matrix is ____. Then there is a pivot in every ____ of U and R. The solution to $Ax = b$ (*always exists*) (*is unique*). The column space of A is ____. An example is $A =$ ____.

17 The largest possible rank of a 6 by 4 matrix is ____. Then there is a pivot in every ____ of U and R. The solution to $Ax = b$ (*always exists*) (*is unique*). The nullspace of A is ____. An example is $A =$ ____.

18 Find by elimination the rank of A and also the rank of A^{T}:

$$A = \begin{bmatrix} 1 & 4 & 0 \\ 2 & 11 & 5 \\ -1 & 2 & 10 \end{bmatrix} \quad \text{and} \quad A = \begin{bmatrix} 1 & 0 & 1 \\ 1 & 1 & 2 \\ 1 & 1 & q \end{bmatrix} \text{ (rank depends on } q\text{)}.$$

19 If $Ax = b$ has infinitely many solutions, why is it impossible for $Ax = B$ (new right side) to have only one solution? Could $Ax = B$ have no solution ?

20 Choose the number q so that (if possible) the ranks are (a) 1 (b) 2 (c) 3 :

$$A = \begin{bmatrix} 6 & 4 & 2 \\ -3 & -2 & -1 \\ 9 & 6 & q \end{bmatrix} \quad \text{and} \quad B = \begin{bmatrix} 3 & 1 & 3 \\ q & 2 & q \end{bmatrix}.$$

21 Give examples of matrices A for which the number of solutions to $Ax = b$ is

(a) 0 or 1, depending on b (b) ∞, regardless of b

(c) 0 or ∞, depending on b (d) 1, regardless of b.

22 Write down all known relations between r and m and n if $Ax = b$ has

(a) no solution for some b (b) one solution for some b, no solution for other b

(c) infinitely many solutions for every b (d) exactly one solution for every b.

Questions 23–27 are about the reduced echelon matrices R_0 and R.

23 Divide rows by pivots. Then produce zeros *above* those pivots to reach R_0 and R.

$$U = \begin{bmatrix} 2 & 4 & 4 \\ 0 & 3 & 6 \\ 0 & 0 & 0 \end{bmatrix} \quad \text{and} \quad U = \begin{bmatrix} 2 & 4 & 4 \\ 0 & 3 & 6 \\ 0 & 0 & 5 \end{bmatrix} \quad \text{and} \quad U = \begin{bmatrix} 0 & 0 & 4 \\ 0 & 1 & 0 \end{bmatrix}.$$

24 If A is a triangular matrix, when is $R_0 = \text{rref}(A)$ equal to I ?

25 Apply elimination to $Ux = \mathbf{0}$ and $Ux = c$. Reach $R_0 x = \mathbf{0}$ and $R_0 x = d$:

$$\begin{bmatrix} U & \mathbf{0} \end{bmatrix} = \begin{bmatrix} 1 & 2 & 3 & \mathbf{0} \\ 0 & 0 & 4 & \mathbf{0} \end{bmatrix} \quad \text{and} \quad \begin{bmatrix} U & c \end{bmatrix} = \begin{bmatrix} 1 & 2 & 3 & \mathbf{5} \\ 0 & 0 & 4 & \mathbf{8} \end{bmatrix}.$$

Solve $R_0 x = \mathbf{0}$ to find x_n with $x_2 = 1$. Solve $R_0 x = d$ to find x_p with $x_2 = 0$.

26 Reduce $U\boldsymbol{x} = \boldsymbol{0}$ and $U\boldsymbol{x} = \boldsymbol{c}$ to $R_0\boldsymbol{x} = \boldsymbol{0}$ and $R_0\boldsymbol{x} = \boldsymbol{d}$. What are the solutions to $R_0\boldsymbol{x} = \boldsymbol{d}$?

$$\begin{bmatrix} U & 0 \end{bmatrix} = \begin{bmatrix} 3 & 0 & 6 & \mathbf{0} \\ 0 & 0 & 0 & \mathbf{0} \\ 0 & 0 & 2 & \mathbf{0} \end{bmatrix} \quad \text{and} \quad \begin{bmatrix} U & \boldsymbol{c} \end{bmatrix} = \begin{bmatrix} 3 & 0 & 6 & \mathbf{9} \\ 0 & 0 & 0 & \mathbf{4} \\ 0 & 0 & 2 & \mathbf{5} \end{bmatrix}.$$

27 Reduce to $U\boldsymbol{x} = \boldsymbol{c}$ (Gaussian elimination) and then $R_0\boldsymbol{x} = \boldsymbol{d}$. Find a particular solution $\boldsymbol{x}_p$ and all homogeneous solutions $\boldsymbol{x}_n$.

$$A\boldsymbol{x} = \begin{bmatrix} 1 & 0 & 2 & 3 \\ 1 & 3 & 2 & 0 \\ 2 & 0 & 4 & 9 \end{bmatrix} \begin{bmatrix} x_1 \\ x_2 \\ x_3 \\ x_4 \end{bmatrix} = \begin{bmatrix} 2 \\ 5 \\ 10 \end{bmatrix} = \boldsymbol{b}.$$

28 Find matrices A and B with the given property or explain why you can't:

(a) The only solution of $A\boldsymbol{x} = \begin{bmatrix} \mathbf{1} \\ \mathbf{2} \\ \mathbf{3} \end{bmatrix}$ is $\boldsymbol{x} = \begin{bmatrix} \mathbf{0} \\ \mathbf{1} \end{bmatrix}$.

(b) The only solution of $B\boldsymbol{x} = \begin{bmatrix} \mathbf{0} \\ \mathbf{1} \end{bmatrix}$ is $\boldsymbol{x} = \begin{bmatrix} \mathbf{1} \\ \mathbf{2} \\ \mathbf{3} \end{bmatrix}$.

29 Find the LU factorization of A and all solutions to $A\boldsymbol{x} = \boldsymbol{b}$:

$$A = \begin{bmatrix} 1 & 3 & 1 \\ 1 & 2 & 3 \\ 2 & 4 & 6 \\ 1 & 1 & 5 \end{bmatrix} \quad \text{and} \quad \boldsymbol{b} = \begin{bmatrix} 1 \\ 3 \\ 6 \\ 5 \end{bmatrix} \quad \text{and then} \quad \boldsymbol{b} = \begin{bmatrix} 1 \\ 0 \\ 0 \\ 0 \end{bmatrix}.$$

30 The complete solution to $A\boldsymbol{x} = \begin{bmatrix} \mathbf{1} \\ \mathbf{3} \end{bmatrix}$ is $\boldsymbol{x} = \begin{bmatrix} \mathbf{1} \\ \mathbf{0} \end{bmatrix} + c \begin{bmatrix} \mathbf{0} \\ \mathbf{1} \end{bmatrix}$. Find A.

31 (Recommended!) Suppose you know that the 3 by 4 matrix A has the vector $\boldsymbol{s} = (2, 3, 1, 0)$ as the only special solution to $A\boldsymbol{x} = \boldsymbol{0}$.

(a) What is the *rank* of A and the complete solution to $A\boldsymbol{x} = \boldsymbol{0}$?

(b) What is the exact row reduced echelon form R_0 of A?

(c) How do you know that $A\boldsymbol{x} = \boldsymbol{b}$ can be solved for all $\boldsymbol{b}$?

32 Suppose $A\boldsymbol{x} = \boldsymbol{b}$ and $C\boldsymbol{x} = \boldsymbol{b}$ have the same (complete) solutions for every $\boldsymbol{b}$. Is it true that A equals C?

33 Describe the column space of a reduced row echelon matrix R_0 with rank r. Removing any zero rows, describe the column space of R.

3.4 Independence, Basis, and Dimension

1 Independent vectors: The only zero combination $c_1\boldsymbol{v}_1 + \cdots + c_k\boldsymbol{v}_k = 0$ has all c's $= 0$.

2 The vectors $\boldsymbol{v}_1, \ldots, \boldsymbol{v}_k$ **span the space S** if **S** = all combinations of the $\boldsymbol{v}$'s.

3 The vectors $\boldsymbol{v}_1, \ldots, \boldsymbol{v}_k$ are a **basis for S** if (**1**) they are **independent** and (**2**) they **span S**.

4 The **dimension of a vector space S** is the number k of vectors in every basis for **S**.

This important section is about the true size of a subspace. There are n columns in an m by n matrix. But the true "dimension" of the column space is not necessarily n. The dimension of $\mathbf{C}(A)$ is measured by counting ***independent columns***. We will see again that **the true dimension of the column space is the rank** $\boldsymbol{r}$. This is partly review.

The idea of independence applies to any vectors $\boldsymbol{v}_1, \ldots, \boldsymbol{v}_n$ in any vector space. Most of this section concentrates on the subspaces that we know and use—especially the column space and the nullspace of A. In the last part we also study "vectors" that are not column vectors. They can be matrices and functions; they can be linearly independent or dependent. First come the key examples using column vectors.

The goal is to understand a ***basis*** : ***independent vectors that "span the space"***.

Every vector in the space is a unique combination of the basis vectors.

We are at the heart of our subject, and we cannot go on without a basis. The four essential ideas in this section (with first hints at their meaning) are:

1. Independent vectors	(*no extra vectors*)
2. Spanning a space	(*enough vectors to produce the rest*)
3. Basis for a space	(*not too many and not too few*)
4. Dimension of a space	(*the number of vectors in every basis*)

Linear Independence

Our first definition of independence is not so conventional, but you are ready for it.

DEFINITION The columns of A are ***linearly independent*** when the only solution to $A\boldsymbol{x} = \mathbf{0}$ is $\boldsymbol{x} = \mathbf{0}$. ***No other combination*** $A\boldsymbol{x}$ ***of the columns gives the zero vector.***

The columns are independent when the nullspace $\mathbf{N}(A)$ contains only the zero vector. Let me illustrate linear independence (and dependence) with three vectors in $\mathbf{R}^3$.

1. If three vectors in $\mathbf{R}^3$ are *not* in the same plane, those vectors are independent. No combination of $\boldsymbol{v}_1, \boldsymbol{v}_2, \boldsymbol{v}_3$ in Figure 3.2 gives zero except $0\boldsymbol{v}_1 + 0\boldsymbol{v}_2 + 0\boldsymbol{v}_3$.

2. If three vectors $\boldsymbol{w}_1, \boldsymbol{w}_2, \boldsymbol{w}_3$ are *in the same plane* in $\mathbf{R}^3$, they are dependent.

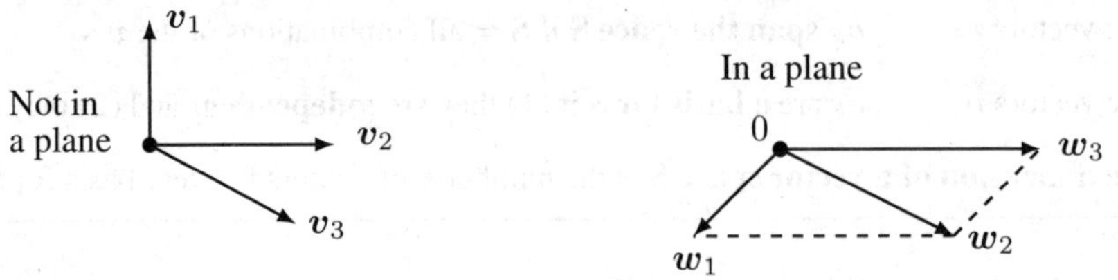

Figure 3.2: **Independent**: Only $0\boldsymbol{v}_1+0\boldsymbol{v}_2+0\boldsymbol{v}_3$ gives $\mathbf{0}$. **Dependent**: $\boldsymbol{w}_1-\boldsymbol{w}_2+\boldsymbol{w}_3=\mathbf{0}$.

This idea of independence applies to 7 vectors in 12-dimensional space. If they are the columns of A, and *independent*, the nullspace of the 12 by 7 matrix only contains $\boldsymbol{x} = \mathbf{0}$. None of the column vectors is a combination of the other six vectors.

Now we choose different words to express the same idea in any vector space.

DEFINITION The sequence of vectors $\boldsymbol{v}_1, \ldots, \boldsymbol{v}_n$ is ***linearly independent*** if the only combination that gives the zero vector is $0\boldsymbol{v}_1 + 0\boldsymbol{v}_2 + \cdots + 0\boldsymbol{v}_n$.

Linear independence

$$x_1\boldsymbol{v}_1 + x_2\boldsymbol{v}_2 + \cdots + x_n\boldsymbol{v}_n = \mathbf{0} \quad \text{only happens when all } x\text{'s are zero.} \tag{1}$$

If a combination gives $\mathbf{0}$, when the x's are not all zero, the vectors are *dependent*.

Correct language: "The sequence of vectors is linearly independent." *Acceptable shortcut*: "The vectors are independent." *Unacceptable*: "The matrix is independent."

A sequence of vectors is either dependent or independent. They can be combined to give the zero vector (with nonzero x's) or they can't. So the key question is: Which combinations of the vectors give zero? We begin with some small examples in $\mathbf{R}^2$:

(a) The vectors $(1, 0)$ and $(1, 0.00001)$ are independent.

(b) The vectors $(1, 1)$ and $(-1, -1)$ are *dependent*.

(c) The vectors $(1, 1)$ and $(0, 0)$ are *dependent* because of the zero vector.

(d) In $\mathbf{R}^2$, any three vectors (a, b) and (c, d) and (e, f) are *dependent*.

Dependent columns
$\boldsymbol{x} \neq \mathbf{0}$ **in the nullspace**

$$\begin{bmatrix} 1 & -1 \\ 1 & -1 \end{bmatrix} \begin{bmatrix} x_1 \\ x_2 \end{bmatrix} = \begin{bmatrix} 0 \\ 0 \end{bmatrix} \quad \text{for } x_1 = 1 \text{ and } x_2 = 1.$$

Three vectors in $\mathbf{R}^2$ cannot be independent! One way to see this: the matrix A with those three columns must have a free variable and then a special solution to $A\boldsymbol{x} = \mathbf{0}$.

Now move to three vectors in $\mathbf{R}^3$. If one of them is a multiple of another one, these vectors are dependent. But the complete test involves all three vectors at once. We put them in a matrix and try to solve $A\boldsymbol{x} = \mathbf{0}$.

Example 1 The columns of this A are dependent. $A\boldsymbol{x} = \mathbf{0}$ has a nonzero solution:

$$A\boldsymbol{x} = \begin{bmatrix} 1 & 0 & 3 \\ 2 & 1 & 5 \\ 1 & 0 & 3 \end{bmatrix} \begin{bmatrix} -3 \\ 1 \\ 1 \end{bmatrix} \quad \text{is} \quad -3 \begin{bmatrix} 1 \\ 2 \\ 1 \end{bmatrix} + 1 \begin{bmatrix} 0 \\ 1 \\ 0 \end{bmatrix} + 1 \begin{bmatrix} 3 \\ 5 \\ 3 \end{bmatrix} = \begin{bmatrix} 0 \\ 0 \\ 0 \end{bmatrix}.$$

The rank is only $r = 2$. *Independent columns produce full column rank* $r = n = 3$. For a *square matrix*, dependent columns imply dependent rows and vice versa.

Question How to find that solution to $A\boldsymbol{x} = \mathbf{0}$? The systematic way is elimination.

$$A = \begin{bmatrix} 1 & 0 & 3 \\ 2 & 1 & 5 \\ 1 & 0 & 3 \end{bmatrix} \text{ reduces to } R_0 = \begin{bmatrix} 1 & 0 & 3 \\ 0 & 1 & -1 \\ 0 & 0 & 0 \end{bmatrix}. \quad \text{Then } F = \begin{bmatrix} 3 \\ -1 \end{bmatrix} \text{ and } \boldsymbol{x} = \begin{bmatrix} -3 \\ 1 \\ 1 \end{bmatrix}.$$

Full column rank The columns of A are independent exactly when the rank is $\boldsymbol{r = n}$. There are n pivots and no free variables and $A = C$. Only $\boldsymbol{x} = \mathbf{0}$ is in the nullspace.

One case is of special importance because it is clear from the start. Suppose seven columns have five components each ($\boldsymbol{m} = \mathbf{5}$ **is less than** $\boldsymbol{n} = \mathbf{7}$). Then the columns *must be dependent*. Any seven vectors from $\mathbf{R}^5$ are dependent. The rank of A cannot be larger than 5. There cannot be more than five pivots in five rows. $A\boldsymbol{x} = \mathbf{0}$ has at least $7 - 5 = 2$ free variables, so it has nonzero solutions—which means that the columns are dependent.

Any set of n vectors in $\mathbf{R}^m$ must be linearly dependent if $n > m$.

This type of matrix has more columns than rows—it is short and wide. The columns are certainly dependent if $n > m$, because $A\boldsymbol{x} = \mathbf{0}$ has a nonzero solution.

The columns might be dependent or might be independent if $n \leq m$. Elimination will reveal the r pivot columns. *It is those r pivot columns that are independent in C.*

Note Another way to describe linear dependence is this: "*One vector is a combination of the other vectors.*" That sounds clear. Why don't we say this from the start? Our definition was longer: "*Some combination gives the zero vector, other than the trivial combination with every $x = 0$.*" We must rule out the easy way to get the zero vector.

The point is, our definition doesn't pick out one particular vector as guilty. All columns of A are treated the same. We look at $A\boldsymbol{x} = \mathbf{0}$, and it has a nonzero solution or it hasn't.

Vectors that Span a Subspace

The first subspace in this book was the column space. Starting with columns $\boldsymbol{v}_1, \ldots, \boldsymbol{v}_n$, $\mathbf{C}(A)$ includes all combinations $x_1\boldsymbol{v}_1 + \cdots + x_n\boldsymbol{v}_n$. *The column space consists of all combinations $A\boldsymbol{x}$ of the columns.* The single word "span" describes $\mathbf{C}(A)$.

The columns of a matrix span its column space. They might be dependent.

Example 2 Describe the column space and the row space of A.

$$m = 3 \quad n = 2 \qquad A = \begin{bmatrix} 1 & 4 \\ 2 & 7 \\ 3 & 5 \end{bmatrix} \text{ and } A^{\mathrm{T}} = \begin{bmatrix} 1 & 2 & 3 \\ 4 & 7 & 5 \end{bmatrix}.$$

The column space of A is the plane in $\mathbf{R}^3$ spanned by the two columns of A. *The row space of A is spanned by the three rows of A* (which are columns of A^{T}). This row space is all of $\mathbf{R}^2$. Remember: The rows are in $\mathbf{R}^n$ spanning the row space. The columns are in $\mathbf{R}^m$ spanning the column space. Same numbers, different vectors, different spaces.

A Basis for a Vector Space

Two vectors can't span all of $\mathbf{R}^3$, even if they are independent. Four vectors can't be independent, even if they span $\mathbf{R}^3$. We want ***enough independent vectors to span the space*** (and not more). A "***basis***" is just right.

DEFINITION A ***basis*** for a vector space is a sequence of vectors with two properties:

The basis vectors are linearly independent and they span the space.

This combination of properties is fundamental to linear algebra. Every vector $\boldsymbol{v}$ in the space is a combination of the basis vectors, because they span the space. More than that, the combination that produces $\boldsymbol{v}$ is *unique*, because the basis vectors $\boldsymbol{v}_1, \ldots, \boldsymbol{v}_n$ are independent:

There is one and only one way to write $\boldsymbol{v}$ as a combination of the basis vectors.

Reason : Suppose $\boldsymbol{v} = a_1\boldsymbol{v}_1 + \cdots + a_n\boldsymbol{v}_n$ and also $\boldsymbol{v} = b_1\boldsymbol{v}_1 + \cdots + b_n\boldsymbol{v}_n$. By subtraction $(a_1 - b_1)\boldsymbol{v}_1 + \cdots + (a_n - b_n)\boldsymbol{v}_n$ is the zero vector. From the independence of the $\boldsymbol{v}$'s, each $a_i - b_i = 0$. Hence $a_i = b_i$, and there are not two ways to produce $\boldsymbol{v}$.

Example 3 The columns of $I = \begin{bmatrix} 1 & 0 \\ 0 & 1 \end{bmatrix}$ produce the "standard basis" for $\mathbf{R}^2$.

$$\text{The basis vectors} \quad \boldsymbol{i} = \begin{bmatrix} 1 \\ 0 \end{bmatrix} \text{ and } \boldsymbol{j} = \begin{bmatrix} 0 \\ 1 \end{bmatrix} \text{ are independent. They span } \mathbf{R}^2.$$

Everybody thinks of this basis first. The vector $\boldsymbol{i}$ goes across and $\boldsymbol{j}$ goes straight up. The columns of the n by n identity matrix give the "**standard basis**" for $\mathbf{R}^n$.

Now we find many other bases (infinitely many). The basis is not unique!

Example 4 (Important) The columns of every invertible n by n matrix give a basis for $\mathbf{R}^n$:

Invertible matrix
Independent columns $A = \begin{bmatrix} 1 & 0 & 0 \\ 1 & 1 & 0 \\ 1 & 1 & 1 \end{bmatrix}$
Column space $= \mathbf{R}^3$

Singular matrix
Dependent columns $B = \begin{bmatrix} 1 & 0 & 1 \\ 1 & 1 & 2 \\ 1 & 1 & 2 \end{bmatrix}$.
Column space $\neq \mathbf{R}^3$

The only solution to $A\boldsymbol{x} = \mathbf{0}$ is $\boldsymbol{x} = A^{-1}\mathbf{0} = \mathbf{0}$. The columns are independent. They span the whole space $\mathbf{R}^n$—because every vector $\boldsymbol{b}$ is a combination of the columns. $A\boldsymbol{x} = \boldsymbol{b}$ can always be solved by $\boldsymbol{x} = A^{-1}\boldsymbol{b}$. Do you see how everything comes together for invertible matrices? Here it is in one sentence:

> The vectors $\boldsymbol{v}_1, \ldots, \boldsymbol{v}_n$ are a ***basis for*** $\mathbf{R}^n$ exactly when they are ***the columns of an n by n invertible matrix***. $\mathbf{R}^n$ has infinitely many different bases.

When the columns are dependent, we keep only the *pivot columns*—the first two columns of B above. Those two columns are independent and they span the column space.

> **Every set of independent vectors can be extended to a basis.**
>
> **Every spanning set of vectors can be reduced to a basis.**

Example 5 This matrix is not invertible. Its columns are not a basis for anything!

One pivot column
One pivot row $(r = 1)$

$$A = \begin{bmatrix} 2 & 4 \\ 3 & 6 \end{bmatrix} \text{ reduces to } R_0 = \begin{bmatrix} 1 & 2 \\ 0 & 0 \end{bmatrix}.$$

Example 6 Find bases for the column and row spaces of this rank two matrix:

$$R_0 = \begin{bmatrix} 1 & 2 & 0 & 3 \\ 0 & 0 & 1 & 4 \\ 0 & 0 & 0 & 0 \end{bmatrix}.$$

Columns 1 and 3 are the pivot columns. They are a basis for the column space of R_0. The column space is the "xy plane" inside xyz space $\mathbf{R}^3$. That plane is not $\mathbf{R}^2$, it is a subspace of $\mathbf{R}^3$. Columns 2 and 3 are also a basis for the same column space. Which pair of columns of R_0 is *not* a basis for its column space?

The row space is a subspace of $\mathbf{R}^4$. The simplest basis for that row space is the two nonzero rows of R_0. **The zero vector is never in a basis.**

> **Question** Given five vectors in $\mathbf{R}^7$, ***how do you find a basis for the space they span?***

First answer Make them the **rows of** $\boldsymbol{A}$, and eliminate to find the nonzero rows in R.

Second answer Put the five vectors into the **columns of** $\boldsymbol{A}$. Eliminate to find the pivot columns. Those pivot columns in C are a basis for the column space.

Could another basis have more vectors, or fewer? This is a crucial question with a good answer: *No.* ***All bases for a vector space contain the same*** *number* ***of vectors.***

> ***The number of vectors in any and every basis is the "dimension" of the space.***

Dimension of a Vector Space

We have to prove what was just stated. There are many choices for the basis vectors, but **the number of basis vectors doesn't change.**

> If $v_1, \ldots, v_m$ and $w_1, \ldots, w_n$ are both bases for the same vector space, then $m = n$.

Proof Suppose that there are more w's than v's. From $n > m$ we want to reach a contradiction. The v's are a basis, so each w must be a combination of the v's. If w_1 equals $a_{11}v_1 + \cdots + a_{m1}v_m$, this is the first column of a matrix multiplication $VA = W$:

Each w is a combination of the v's

$$W = \begin{bmatrix} w_1 & w_2 & \ldots & w_n \end{bmatrix} = \begin{bmatrix} v_1 & \ldots & v_m \end{bmatrix} \begin{bmatrix} a_{11} & & a_{1n} \\ \vdots & & \vdots \\ a_{m1} & & a_{mn} \end{bmatrix} = VA.$$

We don't know each a_{ij}, but we know the shape of A (it is m by n). The second vector w_2 is also a combination of the v's. The coefficients in that combination fill the second column of A. The key is that A has a row for every v and a column for every w. A is a *short wide matrix*, since we assumed $n > m$. So $A\boldsymbol{x} = \mathbf{0}$ ***has a nonzero solution***.

$A\boldsymbol{x} = 0$ gives $VA\boldsymbol{x} = \mathbf{0}$ which is $W\boldsymbol{x} = \mathbf{0}$. *A combination of the w's gives zero*! Then the w's could not be a basis—our assumption $n > m$ is **not possible** for two bases.

If $m > n$ we exchange the v's and w's and repeat the same steps. The only way to avoid a contradiction is to have $m = n$. This completes the proof that $m = n$.

The number of basis vectors is the dimension. So the dimension of $\mathbf{R}^n$ is n. We now define the important word ***dimension***.

> **DEFINITION** The ***dimension of a space*** is the ***number of vectors in every basis***.

The dimension matches our intuition. The line through $v = (1, 5, 2)$ has dimension one. It is a subspace with this one vector v in its basis. Perpendicular to that line is the plane $x + 5y + 2z = 0$. This plane has dimension 2. To prove it, we find a basis $(-5, 1, 0)$ and $(-2, 0, 1)$. The dimension is 2 because the basis contains two vectors.

The plane is the nullspace of the matrix $A = \begin{bmatrix} 1 & 5 & 2 \end{bmatrix}$, which has two free variables. Our basis vectors $(-5, 1, 0)$ and $(-2, 0, 1)$ are the "special solutions" to $A\boldsymbol{x} = \mathbf{0}$. The $n - r$ special solutions always give *a basis for the nullspace*: dimension $n - r$.

Note about the language of linear algebra We would never say "the rank of a space" or "the dimension of a basis" or "the basis of a matrix". Those terms have no meaning. It is the ***dimension of the column space*** that equals the ***rank of the matrix.***

The row space has dimension r	**The column space has dimension r**
The nullspace has dimension $n - r$	$\mathbf{N}(A^{\mathrm{T}})$ **has dimension $m - r$**

Bases for Matrix Spaces and Function Spaces

The words "independence" and "basis" and "dimension" are not limited to column vectors. We can ask whether three matrices A_1, A_2, A_3 are independent. When they are 3 by 4 matrices, some combination might give the zero matrix. We can also ask the dimension of the full 3 by 4 matrix space. (That dimension is 12. Twelve matrices in every basis.)

In differential equations, $d^2y/dx^2 = y$ has a space of solutions. One basis is $y = e^x$ and $y = e^{-x}$. Counting the basis functions gives the dimension 2 for this solution space. (The dimension is 2 because the linear equation starts with the second derivative.)

Matrix spaces The vector space $\mathbf{M}$ contains all 2 by 2 matrices. Its dimension is 4.

$$\textbf{One basis is}\quad A_1, A_2, A_3, A_4 = \begin{bmatrix} \mathbf{1} & 0 \\ 0 & 0 \end{bmatrix}, \begin{bmatrix} 0 & \mathbf{1} \\ 0 & 0 \end{bmatrix}, \begin{bmatrix} 0 & 0 \\ \mathbf{1} & 0 \end{bmatrix}, \begin{bmatrix} 0 & 0 \\ 0 & \mathbf{1} \end{bmatrix}.$$

Those matrices are linearly independent. We are not looking at their columns, but at the whole matrix. Combinations of those four matrices can produce any matrix in $\mathbf{M}$.

$$\begin{array}{l}\textbf{Every } A \textbf{ combines} \\ \textbf{the basis matrices}\end{array} \qquad c_1A_1 + c_2A_2 + c_3A_3 + c_4A_4 = \begin{bmatrix} c_1 & c_2 \\ c_3 & c_4 \end{bmatrix} = A.$$

A is zero only if the c's are all zero—this proves independence of A_1, A_2, A_3, A_4.

The three matrices A_1, A_2, A_4 are a basis for a subspace—the upper triangular matrices. Its dimension is 3. A_1 and A_4 are a basis for the diagonal matrices. What is a basis for the symmetric matrices? Keep A_1 and A_4, and throw in $A_2 + A_3$.

The dimension of the whole n by n matrix space is n^2.

The dimension of the subspace of *upper triangular* matrices is $\frac{1}{2}n^2 + \frac{1}{2}n$.

The dimension of the subspace of *diagonal* matrices is n.

The dimension of the subspace of *symmetric* matrices is $\frac{1}{2}n^2 + \frac{1}{2}n$ (why ?).

Function spaces The equations $d^2y/dx^2 = 0$ and $d^2y/dx^2 = -y$ and $d^2y/dx^2 = y$ involve the second derivative. In calculus we solve to find the functions $y(x)$:

$$\begin{array}{ll} y'' = 0 & \text{is solved by any linear function } y = cx + d \\ y'' = -y & \text{is solved by any combination } y = c\sin x + d\cos x \\ y'' = y & \text{is solved by any combination } y = ce^x + de^{-x}. \end{array}$$

That solution space for $y'' = -y$ has two basis functions: $\sin x$ and $\cos x$. The space for $y'' = 0$ has x and 1. It is the "nullspace" of the second derivative! The dimension is 2 in each case (these are second-order equations).

The solutions of $y'' = 2$ don't form a subspace—the right side $b = 2$ is not zero. A particular solution is $y(x) = x^2$. The complete solution is $y(x) = x^2 + cx + d$. All those functions satisfy $y'' = 2$. Notice the *particular solution plus any function $cx + d$ in the nullspace*. A linear differential equation is like a linear matrix equation $A\boldsymbol{x} = \boldsymbol{b}$.

We end here with the space $\mathbf{Z}$ that contains only the zero vector. The dimension of this space is *zero*. ***The empty set*** (containing no vectors) ***is a basis for*** $\mathbf{Z}$. We can never allow the zero vector into a basis, because then linear independence is lost.

The key words in this section were **independence, span, basis, dimension.**

1. The columns of A are ***independent*** if $\boldsymbol{x} = \mathbf{0}$ is the only solution to $A\boldsymbol{x} = \mathbf{0}$.
2. The vectors $\boldsymbol{v}_1, \ldots, \boldsymbol{v}_r$ ***span*** a space if their combinations fill that space.
3. ***A basis consists of linearly independent vectors that span the space.*** Every vector in the space is a *unique* combination of the basis vectors.
4. All bases for a space have the same number of vectors. This number of vectors in a basis is the ***dimension*** of the space.
5. The pivot columns are one basis for the column space. The dimension of $\mathbf{C}(A)$ is r.

■ WORKED EXAMPLES ■

3.4 A (***Important example***) Suppose $\boldsymbol{v}_1, \ldots, \boldsymbol{v}_n$ is a basis for $\mathbf{R}^n$ and the n by n matrix A is invertible. **Show that $A\boldsymbol{v}_1, \ldots, A\boldsymbol{v}_n$ is also a basis for $\mathbf{R}^n$.**

Solution In *matrix language*: Put the basis vectors $\boldsymbol{v}_1, \ldots, \boldsymbol{v}_n$ in the columns of an invertible (!) matrix V. Then $A\boldsymbol{v}_1, \ldots, A\boldsymbol{v}_n$ are the columns of AV. Since A is invertible, so is AV. Its columns give a basis.

In *vector language*: Suppose $c_1 A\boldsymbol{v}_1 + \cdots + c_n A\boldsymbol{v}_n = \mathbf{0}$. This is $A\boldsymbol{v} = \mathbf{0}$ with $\boldsymbol{v} = c_1\boldsymbol{v}_1 + \cdots + c_n\boldsymbol{v}_n$. Multiply by A^{-1} to reach $\boldsymbol{v} = \mathbf{0}$. By linear independence of the $\boldsymbol{v}$'s, all $c_i = 0$. This shows that the $A\boldsymbol{v}$'s are independent.

To show that the $A\boldsymbol{v}$'s span $\mathbf{R}^n$, solve $c_1 A\boldsymbol{v}_1 + \cdots + c_n A\boldsymbol{v}_n = \boldsymbol{b}$ which is the same as $c_1\boldsymbol{v}_1 + \cdots + c_n\boldsymbol{v}_n = A^{-1}\boldsymbol{b}$. Since the $\boldsymbol{v}$'s are a basis, this must be solvable.

3.4 B Start with the vectors $\boldsymbol{v}_1 = (1, 2, 0)$ and $\boldsymbol{v}_2 = (2, 3, 0)$. **(a)** Are they linearly independent? **(b)** Are they a basis for any space $\boldsymbol{V}$? **(c)** What is the dimension of $\mathbf{V}$? **(d)** Which matrices A have $\mathbf{V}$ as their column space? **(e)** Which matrices have $\mathbf{V}$ as their nullspace? **(f)** Describe all vectors $\boldsymbol{v}_3$ that complete a basis $\boldsymbol{v}_1, \boldsymbol{v}_2, \boldsymbol{v}_3$ for $\mathbf{R}^3$.

Solution

(a) $\boldsymbol{v}_1$ and $\boldsymbol{v}_2$ are independent—the only combination to give $\mathbf{0}$ is $0\boldsymbol{v}_1 + 0\boldsymbol{v}_2$.

(b) Yes, they are a basis for the space $\mathbf{V}$ that they span: All vectors $(x, y, 0)$.

(c) The dimension of $\mathbf{V}$ is 2 since the basis contains two vectors.

(d) This $\mathbf{V}$ is the column space of any 3 by n matrix A of rank 2, if every column is a combination of $\boldsymbol{v}_1$ and $\boldsymbol{v}_2$. In particular A could just have columns $\boldsymbol{v}_1$ and $\boldsymbol{v}_2$.

(e) This $\mathbf{V}$ is the nullspace of any m by 3 matrix B of rank 1, if every row is a multiple of $(0, 0, 1)$. In particular take $B = [0\ \ 0\ \ 1]$. Then $B\boldsymbol{v}_1 = \mathbf{0}$ and $B\boldsymbol{v}_2 = \mathbf{0}$.

(f) Any third vector $\boldsymbol{v}_3 = (a, b, c)$ will complete a basis for $\mathbf{R}^3$ provided $c \neq 0$.

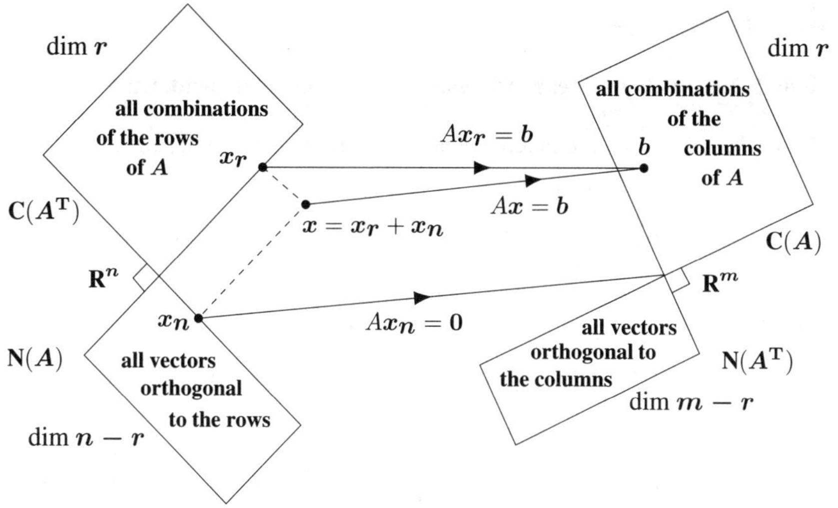

Figure 3.3: $Ax_r = b$ **is in the column space of** A **and** $Ax_n = 0$**. The complete solution to** $Ax = b$ **is** $x =$ **one** $x_r +$ **any** x_n**.**

Change of basis Start with independent vectors w_1, w_2, w_3. Take combinations of those vectors to produce v_1, v_2, v_3. Write the combinations in matrix form as $V = WB$:

$$\begin{aligned} v_1 &= w_1 + w_2 \\ v_2 &= w_1 + 2w_2 + w_3 \\ v_3 &= w_2 + cw_3 \end{aligned} \quad \text{which is} \quad \begin{bmatrix} v_1 & v_2 & v_3 \end{bmatrix} = \begin{bmatrix} w_1 & w_2 & w_3 \end{bmatrix} \begin{bmatrix} 1 & 1 & 0 \\ 1 & 2 & 1 \\ 0 & 1 & c \end{bmatrix}$$

What is the test on B to see if $V = WB$ has independent columns? If $c \neq 1$ show that v_1, v_2, v_3 are linearly independent. If $c = 1$ show that the v's are linearly *dependent.*

Solution For independent columns, *the nullspace of V must contain only the zero vector.* $Vx = 0$ requires $x = (0, 0, 0)$.

If $c = 1$ in our problem, we can see *dependence* in two ways. First, $v_1 + v_3$ will be the same as v_2. Then $v_1 - v_2 + v_3 = \mathbf{0}$—which says that the v's are dependent.

The other way is to look at the nullspace of B. If $c = 1$, the vector $x = (1, -1, 1)$ is in that nullspace, and $Bx = \mathbf{0}$. Then certainly $WBx = \mathbf{0}$ which is the same as $Vx = 0$. So the v's are dependent: $v_1 - v_2 + v_3 = \mathbf{0}$. If $c \neq 1$ then B is invertible.

The general rule is "independent v's from independent w's when B is invertible". And if these vectors are in $\mathbf{R}^3$, they are not only independent—they are a basis for $\mathbf{R}^3$. ***"Basis of v's from basis of w's when the change of basis matrix B is invertible."***

Problem Set 3.4

Questions 1–10 are about linear independence and linear dependence.

1 Show that $\boldsymbol{v}_1, \boldsymbol{v}_2, \boldsymbol{v}_3$ are independent but $\boldsymbol{v}_1, \boldsymbol{v}_2, \boldsymbol{v}_3, \boldsymbol{v}_4$ are dependent:

$$\boldsymbol{v}_1 = \begin{bmatrix} 1 \\ 0 \\ 0 \end{bmatrix} \quad \boldsymbol{v}_2 = \begin{bmatrix} 1 \\ 1 \\ 0 \end{bmatrix} \quad \boldsymbol{v}_3 = \begin{bmatrix} 1 \\ 1 \\ 1 \end{bmatrix} \quad \boldsymbol{v}_4 = \begin{bmatrix} 2 \\ 3 \\ 4 \end{bmatrix}.$$

Solve $c_1\boldsymbol{v}_1 + c_2\boldsymbol{v}_2 + c_3\boldsymbol{v}_3 + c_4\boldsymbol{v}_4 = \mathbf{0}$ or $A\boldsymbol{x} = \mathbf{0}$. The $\boldsymbol{v}$'s go in the columns of A.

2 (Recommended) Find the largest possible number of independent vectors among

$$\boldsymbol{v}_1 = \begin{bmatrix} 1 \\ -1 \\ 0 \\ 0 \end{bmatrix} \quad \boldsymbol{v}_2 = \begin{bmatrix} 1 \\ 0 \\ -1 \\ 0 \end{bmatrix} \quad \boldsymbol{v}_3 = \begin{bmatrix} 1 \\ 0 \\ 0 \\ -1 \end{bmatrix} \quad \boldsymbol{v}_4 = \begin{bmatrix} 0 \\ 1 \\ -1 \\ 0 \end{bmatrix} \quad \boldsymbol{v}_5 = \begin{bmatrix} 0 \\ 1 \\ 0 \\ -1 \end{bmatrix} \quad \boldsymbol{v}_6 = \begin{bmatrix} 0 \\ 0 \\ 1 \\ -1 \end{bmatrix}$$

3 Prove that if $a = 0$ or $d = 0$ or $f = 0$ (3 cases), the columns of U are dependent:

$$U = \begin{bmatrix} a & b & c \\ 0 & d & e \\ 0 & 0 & f \end{bmatrix}.$$

4 If a, d, f in Question 3 are nonzero, show that the only solution to $U\boldsymbol{x} = \mathbf{0}$ is $\boldsymbol{x} = \mathbf{0}$. An upper triangular U with no diagonal zeros has independent columns.

5 Decide the dependence or independence of

(a) the vectors $(1, 3, 2)$ and $(2, 1, 3)$ and $(3, 2, 1)$

(b) the vectors $(1, -3, 2)$ and $(2, 1, -3)$ and $(-3, 2, 1)$.

6 Choose three independent columns of U. Then make two other choices. Do the same for A.

$$U = \begin{bmatrix} 2 & 3 & 4 & 1 \\ 0 & 6 & 7 & 0 \\ 0 & 0 & 0 & 9 \\ 0 & 0 & 0 & 0 \end{bmatrix} \quad \text{and} \quad A = \begin{bmatrix} 2 & 3 & 4 & 1 \\ 0 & 6 & 7 & 0 \\ 0 & 0 & 0 & 9 \\ 4 & 6 & 8 & 2 \end{bmatrix}.$$

7 If $\boldsymbol{w}_1, \boldsymbol{w}_2, \boldsymbol{w}_3$ are independent vectors, show that the differences $\boldsymbol{v}_1 = \boldsymbol{w}_2 - \boldsymbol{w}_3$ and $\boldsymbol{v}_2 = \boldsymbol{w}_1 - \boldsymbol{w}_3$ and $\boldsymbol{v}_3 = \boldsymbol{w}_1 - \boldsymbol{w}_2$ are *dependent*. Find a combination of the $\boldsymbol{v}$'s that gives zero. Which matrix A in $[\, \boldsymbol{v}_1 \ \boldsymbol{v}_2 \ \boldsymbol{v}_3 \,] = [\, \boldsymbol{w}_1 \ \boldsymbol{w}_2 \ \boldsymbol{w}_3 \,] A$ is singular?

8 If $\boldsymbol{w}_1, \boldsymbol{w}_2, \boldsymbol{w}_3$ are independent vectors, show that the sums $\boldsymbol{v}_1 = \boldsymbol{w}_2 + \boldsymbol{w}_3$ and $\boldsymbol{v}_2 = \boldsymbol{w}_1 + \boldsymbol{w}_3$ and $\boldsymbol{v}_3 = \boldsymbol{w}_1 + \boldsymbol{w}_2$ are *independent*. (Write $c_1\boldsymbol{v}_1 + c_2\boldsymbol{v}_2 + c_3\boldsymbol{v}_3 = \mathbf{0}$ in terms of the $\boldsymbol{w}$'s. Find and solve equations for the c's, to show they are zero.)

9 Suppose $\boldsymbol{v}_1, \boldsymbol{v}_2, \boldsymbol{v}_3, \boldsymbol{v}_4$ are vectors in $\mathbf{R}^3$.

(a) These four vectors are dependent because ______.

(b) The two vectors $\boldsymbol{v}_1$ and $\boldsymbol{v}_2$ will be dependent if ______.

(c) The vectors $\boldsymbol{v}_1$ and $(0,0,0)$ are dependent because ______.

10 Find two independent vectors on the plane $x+2y-3z-t=0$ in $\mathbf{R}^4$. Then find three independent vectors. Why not four? This plane is the nullspace of what matrix?

Questions 11–14 are about the space *spanned* by a set of vectors. Take all linear combinations of the vectors.

11 Describe the subspace of $\mathbf{R}^3$ (is it a line or plane or $\mathbf{R}^3$?) spanned by

(a) the two vectors $(1,1,-1)$ and $(-1,-1,1)$

(b) the three vectors $(0,1,1)$ and $(1,1,0)$ and $(0,0,0)$

(c) all vectors in $\mathbf{R}^3$ with whole number components

(d) all vectors with positive components.

12 The vector $\boldsymbol{b}$ is in the subspace spanned by the columns of A when ______ has a solution. The vector $\boldsymbol{c}$ is in the row space of A when ______ has a solution.

True or false: If the zero vector is in the row space, the rows are dependent.

13 Find the dimensions of these 4 spaces. Which two of the spaces are the same? (a) column space of A, (b) column space of U, (c) row space of A, (d) row space of U:

$$A=\begin{bmatrix}1&1&0\\1&3&1\\3&1&-1\end{bmatrix} \quad \text{and} \quad U=\begin{bmatrix}1&1&0\\0&2&1\\0&0&0\end{bmatrix}.$$

14 $\boldsymbol{v}+\boldsymbol{w}$ and $\boldsymbol{v}-\boldsymbol{w}$ are combinations of $\boldsymbol{v}$ and $\boldsymbol{w}$. Write $\boldsymbol{v}$ and $\boldsymbol{w}$ as combinations of $\boldsymbol{v}+\boldsymbol{w}$ and $\boldsymbol{v}-\boldsymbol{w}$. The two pairs of vectors ______ the same space. When are they a basis for the same space?

Questions 15–25 are about the requirements for a basis.

15 If $\boldsymbol{v}_1, \ldots, \boldsymbol{v}_n$ are linearly independent, the space they span has dimension ______. These vectors are a ______ for that space. If the vectors are the columns of an m by n matrix, then m is ______ than n. If $m=n$, that matrix is ______.

16 Find a basis for each of these subspaces of $\mathbf{R}^4$:

(a) All vectors whose components are equal.

(b) All vectors whose components add to zero.

(c) All vectors that are perpendicular to $(1,1,0,0)$ and $(1,0,1,1)$.

(d) The column space and the nullspace of I (4 by 4).

17 Find three different bases for the column space of $U = \begin{bmatrix} 1 & 0 & 1 & 0 & 1 \\ 0 & 1 & 0 & 1 & 0 \end{bmatrix}$. Then find two different bases for the row space of U.

18 Suppose $\boldsymbol{v}_1, \boldsymbol{v}_2, \ldots, \boldsymbol{v}_6$ are six vectors in $\mathbf{R}^4$.

(a) Those vectors (do)(do not)(might not) span $\mathbf{R}^4$.

(b) Those vectors (are)(are not)(might be) linearly independent.

(c) Any four of those vectors (are)(are not)(might be) a basis for $\mathbf{R}^4$.

19 The columns of A are n vectors from $\mathbf{R}^m$. If they are linearly independent, what is the rank of A? If they span $\mathbf{R}^m$, what is the rank? If they are a basis for $\mathbf{R}^m$, what then? *Looking ahead*: The rank r counts the number of _____ columns.

20 Find a basis for the plane $x-2y+3z=0$ in $\mathbf{R}^3$. Then find a basis for the intersection of that plane with the xy plane. Then find a basis for all vectors perpendicular to the plane.

21 Suppose the columns of a 5 by 5 matrix A are a basis for $\mathbf{R}^5$.

(a) The equation $A\boldsymbol{x} = \mathbf{0}$ has only the solution $\boldsymbol{x} = \mathbf{0}$ because _____.

(b) If $\boldsymbol{b}$ is in $\mathbf{R}^5$ then $A\boldsymbol{x} = \boldsymbol{b}$ is solvable because the basis vectors _____ $\mathbf{R}^5$.

Conclusion: A is invertible. Its rank is 5. Its rows are also a basis for $\mathbf{R}^5$.

22 Suppose $\mathbf{S}$ is a 5-dimensional subspace of $\mathbf{R}^6$. True or false (example if false):

(a) Every basis for $\mathbf{S}$ can be extended to a basis for $\mathbf{R}^6$ by adding one more vector.

(b) Every basis for $\mathbf{R}^6$ can be reduced to a basis for $\mathbf{S}$ by removing one vector.

23 U comes from A by subtracting row 1 from row 3:

$$A = \begin{bmatrix} 1 & 3 & 2 \\ 0 & 1 & 1 \\ 1 & 3 & 2 \end{bmatrix} \quad \text{and} \quad U = \begin{bmatrix} 1 & 3 & 2 \\ 0 & 1 & 1 \\ 0 & 0 & 0 \end{bmatrix}.$$

Find bases for the two column spaces. Find bases for the two row spaces. Find bases for the two nullspaces. Which spaces stay fixed in elimination?

24 True or false (give a good reason):

(a) If the columns of a matrix are dependent, so are the rows.

(b) The column space of a 2 by 2 matrix is the same as its row space.

(c) The column space of a 2 by 2 matrix has the same dimension as its row space.

(d) The columns of a matrix are a basis for the column space.

25 Suppose $\boldsymbol{v}_1, \ldots, \boldsymbol{v}_k$ span $\mathbf{R}^n$. How would you reduce this set to a basis? Suppose $\boldsymbol{v}_1, \ldots, \boldsymbol{v}_j$ are independent in $\mathbf{R}^n$. How would you add vectors to reach a basis?

26 For which numbers c and d do these matrices have rank 2?

$$A = \begin{bmatrix} 1 & 2 & 5 & 0 & 5 \\ 0 & 0 & c & 2 & 2 \\ 0 & 0 & 0 & d & 2 \end{bmatrix} \quad \text{and} \quad B = \begin{bmatrix} c & d \\ d & c \end{bmatrix}.$$

Questions 27–31 are about spaces where the "vectors" are matrices.

27 Find a basis (and the dimension) for each of these subspaces of 3 by 3 matrices:

(a) All diagonal matrices.

(b) All symmetric matrices ($A^{\mathrm{T}} = A$).

(c) All skew-symmetric matrices ($A^{\mathrm{T}} = -A$).

28 Construct six linearly independent 3 by 3 echelon matrices $U_1, \ldots, U_6$.

29 Find a basis for the space of all 2 by 3 matrices whose columns add to zero. Find a basis for the subspace whose rows also add to zero.

30 What subspace of 3 by 3 matrices is spanned (take all combinations) by

(a) the invertible matrices?

(b) the rank one matrices?

(c) the identity matrix?

31 Find a basis for the space of 2 by 3 matrices whose nullspace contains $(2, 1, 1)$.

Questions 32–36 are about spaces where the "vectors" are functions.

32 (a) Find all functions that satisfy $\frac{dy}{dx} = 0$.

(b) Choose a particular function that satisfies $\frac{dy}{dx} = 3$.

(c) Find all functions that satisfy $\frac{dy}{dx} = 3$.

33 The cosine space $\mathbf{F}_3$ contains all combinations $y(x) = A\cos x + B\cos 2x + C\cos 3x$. Find a basis for the subspace with $y(0) = 0$.

34 Find a basis for the space of functions that satisfy

(a) $\frac{dy}{dx} - 2y = 0$

(b) $\frac{dy}{dx} - \frac{y}{x} = 0$.

35 Suppose $y_1(x), y_2(x), y_3(x)$ are three different functions of x. The vector space they span could have dimension 1, 2, or 3. Give an example of y_1, y_2, y_3 to show each possibility.

36 Find a basis for the space of polynomials $p(x)$ of degree ≤ 3. Find a basis for the subspace with $p(1) = 0$.

37 Find a basis for the space $\mathbf{S}$ of vectors (a, b, c, d) with $a + c + d = 0$ and also for the space $\mathbf{T}$ with $a + b = 0$ and $c = 2d$. What is the dimension of the intersection $\mathbf{S} \cap \mathbf{T}$?

38 Suppose A is 5 by 4 with rank 4. Show that $A\boldsymbol{x} = \boldsymbol{b}$ has no solution when the 5 by 5 matrix $[\,A\ \ \boldsymbol{b}\,]$ is invertible. Show that $A\boldsymbol{x} = \boldsymbol{b}$ is solvable when $[\,A\ \ \boldsymbol{b}\,]$ is singular.

39 Find bases for all solutions to $d^2y/dx^2 = y(x)$ and then $d^2y/dx^2 = -y(x)$.

Challenge Problems

40 Write the 3 by 3 identity matrix as a combination of the other five permutation matrices! Then show that those five matrices are linearly independent. This is a basis for the subspace of 3 by 3 matrices with row and column sums all equal.

41 Choose $\boldsymbol{x} = (x_1, x_2, x_3, x_4)$ in $\mathbf{R}^4$. It has 24 rearrangements like (x_2, x_1, x_3, x_4) and (x_4, x_3, x_1, x_2). Those 24 vectors, including $\boldsymbol{x}$ itself, span a subspace **S**. Find specific vectors $\boldsymbol{x}$ so that the dimension of **S** is: (a) zero, (b) one, (c) three, (d) four.

42 Intersections and sums have $\dim(\mathbf{V}) + \dim(\mathbf{W}) = \dim(\mathbf{V} \cap \mathbf{W}) + \dim(\mathbf{V}+\mathbf{W})$. Start with a basis $\boldsymbol{u}_1, \ldots, \boldsymbol{u}_r$ for the intersection $\mathbf{V} \cap \mathbf{W}$. Extend with $\boldsymbol{v}_1, \ldots, \boldsymbol{v}_s$ to a basis for **V**, and separately $\boldsymbol{w}_1, \ldots, \boldsymbol{w}_t$ to a basis for **W**. Prove that the $\boldsymbol{u}$'s, $\boldsymbol{v}$'s and $\boldsymbol{w}$'s together are ***independent***: dimensions $(\boldsymbol{r}+\boldsymbol{s})+(\boldsymbol{r}+\boldsymbol{t}) = (\boldsymbol{r})+(\boldsymbol{r}+\boldsymbol{s}+\boldsymbol{t})$.

43 Inside $\mathbf{R}^n$, suppose dimension (**V**) + dimension (**W**) $> n$. Show that some nonzero vector is in both **V** and **W**. Hint: Put a basis for V and a basis for W into the columns of a matrix A, and solve $A\boldsymbol{x} = \mathbf{0}$.

44 Suppose A is 10 by 10 and $A^2 = 0$ (zero matrix). So A multiplies each column of A to give the zero vector. Then the column space of A is contained in the _____. If A has rank r, those subspaces have dimension $r \leq 10 - r$. So the rank is $\boldsymbol{r} \leq \mathbf{5}$.

3.5 Dimensions of the Four Subspaces

1 The column space $\mathbf{C}(A)$ and the row space $\mathbf{C}(A^{\mathrm{T}})$ both have *dimension* r (the rank of A).

2 The nullspace $\mathbf{N}(A)$ has *dimension* $n - r$. The left nullspace $\mathbf{N}(A^{\mathrm{T}})$ has *dimension* $m - r$.

3 Elimination from A to R_0 changes $\mathbf{C}(A)$ and $\mathbf{N}(A^{\mathrm{T}})$ (but their dimensions don't change).

The main theorem in this chapter connects ***rank*** and ***dimension***. The ***rank*** of a matrix counts independent columns. The ***dimension*** of a subspace is the number of vectors in a basis. We can count pivots or basis vectors. *The rank of A reveals the dimensions of all four fundamental subspaces.* Here are the subspaces, including the new one.

Two subspaces come directly from A, and the other two come from A^{T}.

	Four Fundamental Subspaces	**Dimensions**
1.	The ***row space*** is $\mathbf{C}(A^{\mathrm{T}})$, a subspace of $\mathbf{R}^n$.	r
2.	The ***column space*** is $\mathbf{C}(A)$, a subspace of $\mathbf{R}^m$.	r
3.	The ***nullspace*** is $\mathbf{N}(A)$, a subspace of $\mathbf{R}^n$.	$n - r$
4.	The ***left nullspace*** is $\mathbf{N}(A^{\mathrm{T}})$, a subspace of $\mathbf{R}^m$.	$m - r$

We know $\mathbf{C}(A)$ and $\mathbf{N}(A)$ pretty well. Now $\mathbf{C}(A^{\mathrm{T}})$ and $\mathbf{N}(A^{\mathrm{T}})$ come forward. The row space contains all combinations of the rows. *This row space is the column space of* A^{T}.

For the left nullspace we solve $A^{\mathrm{T}}\boldsymbol{y} = \mathbf{0}$—that system is n by m. In Example 2 this produces one of the great equations of applied mathematics—**Kirchhoff's Current Law**. The currents flow around a network, and they can't pile up at the nodes. The matrix A is the **incidence matrix of a graph**. Its four subspaces come from nodes and edges and loops and trees. Those subspaces are connected in an absolutely beautiful way.

Part 1 of the Fundamental Theorem finds the dimensions of the four subspaces. One fact stands out: ***The row space and column space have the same dimension*** r. This number r is the rank of A (Chapter 1). The other important fact involves the two nullspaces:

$\mathbf{N}(A)$ ***and*** $\mathbf{N}(A^{\mathrm{T}})$ ***have dimensions*** $n - r$ ***and*** $m - r$***, to make up the full*** n ***and*** m.

Part 2 of the Fundamental Theorem will describe how the four subspaces fit together: Nullspace perpendicular to row space, and $\mathbf{N}(A^{\mathrm{T}})$ perpendicular to $\mathbf{C}(A)$. That completes the "right way" to understand $A\boldsymbol{x} = \boldsymbol{b}$. Stay with it—you are doing real mathematics.

The Four Subspaces for R_0

Suppose A is reduced to its row echelon form R_0. For that special form, the four subspaces are easy to identify. We will find a basis for each subspace and check its dimension. Then we watch how the subspaces change (two of them don't change!) as we look back at A. The main point is that ***the four dimensions are the same for A and*** $\boldsymbol{R_0}$.

For A and R, one of the four subspaces can have different dimensions—because zero rows are removed in R, which changes m.

As a specific 3 by 5 example, look at the four subspaces for this echelon matrix R_0:

$$\begin{matrix} m = 3 \\ n = 5 \\ r = 2 \end{matrix} \qquad R_0 = \begin{bmatrix} \mathbf{1} & 3 & 5 & 0 & 7 \\ 0 & 0 & 0 & \mathbf{1} & 2 \\ 0 & 0 & 0 & 0 & 0 \end{bmatrix} \qquad \begin{matrix} \textbf{pivot rows 1 and 2} \\ \\ \textbf{pivot columns 1 and 4} \end{matrix}$$

The rank of this matrix is $\boldsymbol{r = 2}$ (*two pivots*). Take the four subspaces in order.

1. The ***row space*** has dimension 2, matching the rank.

Reason: The first two rows are a basis. The row space contains combinations of all three rows, but the third row (the zero row) adds nothing to the row space.

The pivot rows 1 and 2 are independent. That is obvious for this example, and it is always true. If we look only at the pivot columns, we see the r by r identity matrix. There is no way to combine its rows to give the zero row (except by the combination with all coefficients zero). So the r pivot rows (the rows of R) are a basis for the row space.

The dimension of the row space is the rank r. The nonzero rows of $\boldsymbol{R_0}$ ***form a basis.***

2. The ***column space*** of R_0 also has dimension $r = 2$.

Reason: The pivot columns 1 and 4 form a basis. They are independent because they contain the r by r identity matrix. No combination of those pivot columns can give the zero column (except the combination with all coefficients zero). And they also span the column space. Every other (free) column is a combination of the pivot columns. Actually the combinations we need are the three special solutions !

Column 2 is 3 (column 1). The special solution is $(-3, 1, 0, 0, 0)$.

Column 3 is 5 (column 1). The special solution is $(-5, 0, 1, 0, 0,)$.

Column 5 is 7 (column 1) + 2 (column 4). That solution is $(-7, 0, 0, -2, 1)$.

The pivot columns are independent, and they span $\mathbf{C}(R_0)$, so they are a basis for $\mathbf{C}(R_0)$.

The dimension of the column space is the rank r. The pivot columns form a basis.

3. The ***nullspace*** of R_0 has dimension $\boldsymbol{n-r=5-2}$. The 3 free variables give **3 special solutions** to $R_0\boldsymbol{x}=\mathbf{0}$. Set the free variables to 1 and 0 and 0.

$$s_2=\begin{bmatrix}\mathbf{-3}\\ \mathbf{1}\\ 0\\ 0\\ 0\end{bmatrix}\quad s_3=\begin{bmatrix}\mathbf{-5}\\ 0\\ \mathbf{1}\\ 0\\ 0\end{bmatrix}\quad s_5=\begin{bmatrix}\mathbf{-7}\\ 0\\ 0\\ \mathbf{-2}\\ \mathbf{1}\end{bmatrix}$$

$R_0\boldsymbol{x}=\mathbf{0}$ has the complete solution $\boldsymbol{x}=x_2\boldsymbol{s}_2+x_3\boldsymbol{s}_3+x_5\boldsymbol{s}_5$. The nullspace has dimension 3.

Reason : There is a special solution for each free variable. With n variables and r pivots, that leaves $n-r$ free variables and special solutions. The special solutions are independent, because you can see the identity matrix in rows 2, 3, 5.

The nullspace* N(A) *has dimension $n-r$. The special solutions form a basis.

4. The ***nullspace of $R_0^{\mathrm T}$ (left nullspace of R_0) has dimension $m-r=3-2$.***

Reason : R_0 has r independent rows and $\boldsymbol{m-r}$ **zero rows**. Then $R_0^{\mathrm T}$ has r independent columns and $\boldsymbol{m-r}$ **zero columns**. So $\boldsymbol{y}$ in the nullspace of $R_0^{\mathrm T}$ can have nonzeros in its last $m-r$ entries. The example has $m-r=1$ zero column in $R_0^{\mathrm T}$ and 1 nonzero in $\boldsymbol{y}$.

$$\boldsymbol{R_0^{\mathrm T}y}=\begin{bmatrix}\mathbf{1}&0&0\\ \mathbf{3}&0&0\\ \mathbf{5}&0&0\\ 0&\mathbf{1}&0\\ \mathbf{7}&\mathbf{2}&0\end{bmatrix}\begin{bmatrix}y_1\\ y_2\\ y_3\end{bmatrix}=\begin{bmatrix}0\\0\\0\\0\\0\end{bmatrix}\quad\text{is solved by } \boldsymbol{y}=\begin{bmatrix}0\\0\\y_3\end{bmatrix}. \tag{1}$$

Because of zero rows in R_0 and zero columns in $R_0^{\mathrm T}$, it is easy to see the dimension (and even a basis) for this fourth fundamental subspace :

If R_0 has $m-r$ zero rows, its left nullspace has dimension $m-r$.

Why is this a "*left* nullspace"? Because we can transpose $R_0^{\mathrm T}\boldsymbol{y}=\mathbf{0}$ to $\boldsymbol{y}^{\mathrm T}R_0=\mathbf{0}^{\mathrm T}$. Now $\boldsymbol{y}^{\mathrm T}$ is a row vector to the *left* of R. This subspace came fourth, and some linear algebra books omit it—but that misses the beauty of the whole subject.

***In* $\mathbf{R}^n$ *the row space and nullspace have dimensions* r and $n-r$ (adding to n).**

***In* $\mathbf{R}^m$ *the column space and left nullspace have dimensions* r and $m-r$ (total m).**

We have a job still to do. ***The four subspace dimensions for A are the same as for R_0.*** The job is to explain why. A is now any matrix that reduces to $R_0=\mathrm{rref}(A)$.

This A reduces to R_0 $\quad A=\begin{bmatrix}1&3&5&0&7\\ 0&0&0&1&2\\ \mathbf{1}&\mathbf{3}&\mathbf{5}&\mathbf{1}&\mathbf{9}\end{bmatrix}$ $\quad$ **Same row space as R_0**
Different column space
But same dimension !

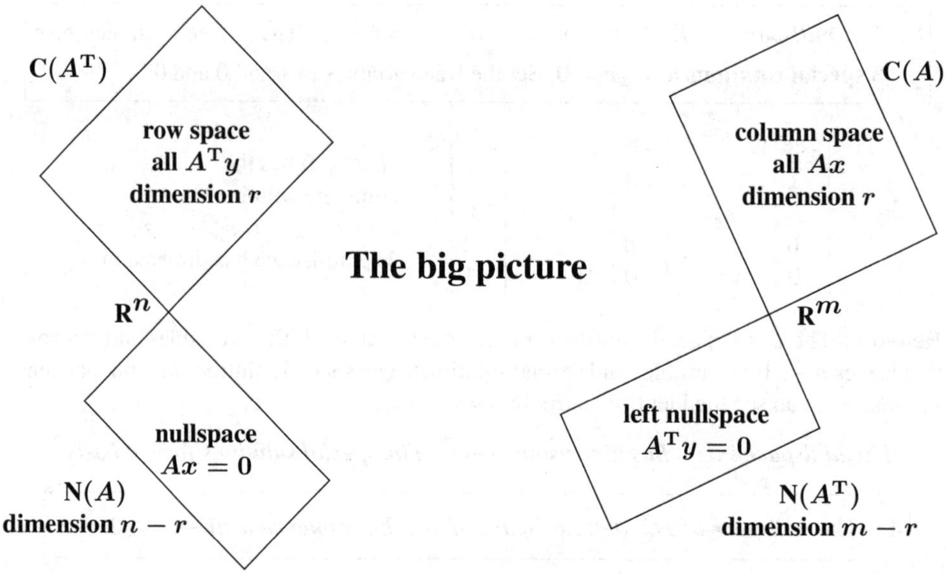

Figure 3.4: The dimensions of the Four Fundamental Subspaces (for R_0 and for A).

The Four Subspaces for A

1 ***A has the same row space as R_0 and R.* Same dimension r and same basis.**

Reason: Every row of A is a combination of the rows of R_0. Also every row of R_0 is a combination of the rows of A. **Elimination changes rows, but not row spaces.**

Since A has the same row space as R_0, the first r rows of R_0 are still a basis. Or we could choose r suitable rows of the original A. They might not always be the *first* r rows of A, because those could be dependent. The good r rows of A are the ones that end up as pivot rows in R_0 and R.

2 ***The column space of A has dimension r. The column rank equals the row rank.***

The number of independent columns = the number of independent rows.

Wrong reason: "A and R_0 have the same column space." This is false. *The columns of R_0 often end in zeros.* The columns of A don't often end in zeros. Then $\mathbf{C}(A)$ is not $\mathbf{C}(R_0)$.

Right reason: The ***same combinations*** of the columns are zero (or not) for A and R_0. Dependent in $A \Leftrightarrow$ dependent in R_0. Say that another way: $A\boldsymbol{x} = \mathbf{0}$ *exactly when* $R_0\boldsymbol{x} = \mathbf{0}$. The column spaces are different, but their *dimensions* are the same—equal to the rank r.

Conclusion The r pivot columns of A are a basis for *its* column space $\mathbf{C}(A)$.

3 ***A has the same nullspace as R_0.* Same dimension $n - r$ and same basis.**

Reason: Elimination doesn't change the solutions to $A\boldsymbol{x} = \mathbf{0}$. The special solutions are a basis for this nullspace (as we always knew). There are $n-r$ free variables, so the nullspace dimension is $n - r$. This is the **Counting Theorem**: $\boldsymbol{r + (n - r)}$ **equals** $\boldsymbol{n}$.

(dimension of column space)+(dimension of nullspace) = dimension of $\mathbf{R}^n$.

4 ***The left nullspace of A*** (the nullspace of A^{T}) ***has dimension $m - r$.***

Reason: A^{T} is just as good a matrix as A. When we know the dimensions for every A, we also know them for A^{T}. Its column space was proved to have dimension r. Since A^{T} is n by m, the "whole space" is now $\mathbf{R}^m$. The counting rule for A was $r + (n-r) = n$. **The counting rule for A^{T} is** $r + (m - r) = m$. We have all details of a big theorem:

***Fundamental Theorem of Linear Algebra*, Part 1**

The column space and row space both have dimension r.

The nullspaces have dimensions $n - r$ and $m - r$.

By concentrating on *spaces* of vectors, not on individual numbers or vectors, we get these clean rules. You will soon take them for granted—eventually they begin to look obvious. But if you write down an 11 by 17 matrix with 187 nonzero entries, I don't think most people would see why these facts are true:

Two key facts

$$\text{dimension of } \mathbf{C}(A) = \text{dimension of } \mathbf{C}(A^{\mathrm{T}}) = \text{rank of } A$$
$$\text{dimension of } \mathbf{C}(A) + \text{dimension of } \mathbf{N}(A) = 17.$$

Every vector $A\boldsymbol{x} = \boldsymbol{b}$ in the column space comes from exactly one $\boldsymbol{x}$ in the row space! (If we also have $A\boldsymbol{y} = \boldsymbol{b}$ then $A(\boldsymbol{x} - \boldsymbol{y}) = \boldsymbol{b} - \boldsymbol{b} = \mathbf{0}$. So $\boldsymbol{x} - \boldsymbol{y}$ is in the nullspace as well as the row space, which forces $\boldsymbol{x} = \boldsymbol{y}$.) From its row space to its column space, *A is like an r by r invertible matrix.*

It is the nullspaces that force us to define a **"pseudoinverse of *A*"** in Section 4.5.

Example 1 $A = \begin{bmatrix} 1 & 2 & 3 \\ 2 & 4 & 6 \end{bmatrix}$ has $m = 2$ with $n = 3$. The rank is $r = 1$.

The row space is the line through $(1, 2, 3)$. The nullspace is the plane $x_1 + 2x_2 + 3x_3 = 0$. The line and plane dimensions still add to $1 + 2 = 3$. The column space and left nullspace are **perpendicular lines in $\mathbf{R}^2$**. Dimensions $1 + 1 = 2$.

$$\text{Column space} = \text{line through } \begin{bmatrix} 1 \\ 2 \end{bmatrix} \qquad \text{Left nullspace} = \text{line through } \begin{bmatrix} 2 \\ -1 \end{bmatrix}.$$

Final point:* *The y's in the left nullspace combine the rows of A to give the zero row.

Example 2 You have nearly finished three chapters with made-up equations, and this can't continue forever. Here is a better example of five equations (one equation for every edge in Figure 3.4). The five equations have four unknowns (one for every node). **The important matrix in $Ax = b$ is an incidence matrix.** It has 1 and -1 on every row.

Differences $Ax = b$
across edges $1, 2, 3, 4, 5$
between nodes $1, 2, 3, 4$
$m = 5$ and $n = 4$

$$\boxed{\begin{array}{rrrrl} -x_1 & +x_2 & & & = b_1 \\ -x_1 & & +x_3 & & = b_2 \\ & -x_2 & +x_3 & & = b_3 \\ & -x_2 & & +x_4 & = b_4 \\ & & -x_3 & +x_4 & = b_5 \end{array}} \qquad (2)$$

If you understand the four fundamental subspaces for this matrix (*the column spaces and the nullspaces for A and A^{T}*) you have captured a central idea of linear algebra.

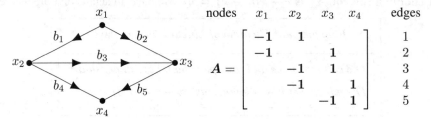

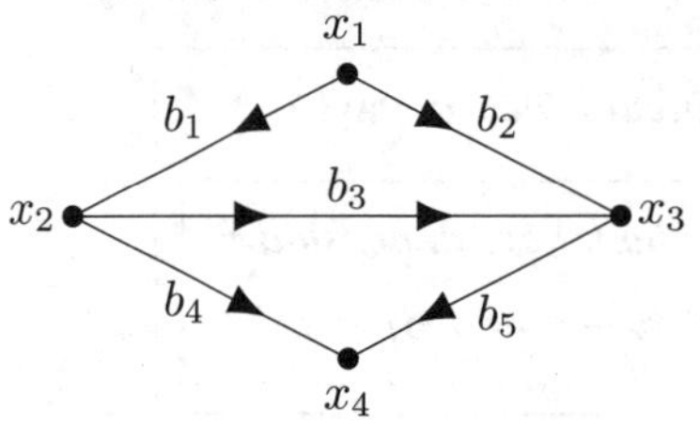

$$\begin{array}{c} \text{nodes} \\ \\ \boldsymbol{A} = \end{array} \begin{array}{c} x_1 \quad x_2 \quad x_3 \quad x_4 \\ \begin{bmatrix} -1 & 1 & & \\ -1 & & 1 & \\ & -1 & 1 & \\ & -1 & & 1 \\ & & -1 & 1 \end{bmatrix} \end{array} \begin{array}{c} \text{edges} \\ 1 \\ 2 \\ 3 \\ 4 \\ 5 \end{array}$$

Figure 3.5: A **"graph"** with 5 edges and 4 nodes. ***A* is its 5 by 4 incidence matrix.**

The nullspace $\mathbf{N}(A)$ To find the nullspace we set $\boldsymbol{b} = \mathbf{0}$. Then the first equation says $x_1 = x_2$. The second equation is $x_3 = x_1$. Equation 4 is $x_2 = x_4$. *All four unknowns x_1, x_2, x_3, x_4 have the same value c.* The vectors $\boldsymbol{x} = (c, c, c, c)$ fill the nullspace of A.

That nullspace is a line in $\mathbf{R}^4$. The special solution $\boldsymbol{x} = (1, 1, 1, 1)$ is a basis for $\mathbf{N}(A)$. The dimension of $\mathbf{N}(A)$ is 1 (one vector in the basis). *The rank of A must be* 3, *since* $n - r = 4 - 3 = 1$. We now know the dimensions of all four subspaces.

The column space $\mathbf{C}(A)$ There must be $r = 3$ independent columns. The fast way is to look at the first 3 columns of A. The systematic way is to find $R_0 = \mathrm{rref}(A)$.

$$\begin{array}{c}\textbf{Columns} \\ \mathbf{1, 2, 3} \\ \textbf{of } \boldsymbol{A}\end{array} \quad \begin{array}{rrr} -1 & 1 & 0 \\ -1 & 0 & 1 \\ 0 & -1 & 1 \\ 0 & -1 & 0 \\ 0 & 0 & -1 \end{array} \qquad \boldsymbol{R_0} = \begin{array}{c}\textbf{reduced row} \\ \textbf{echelon form} \\ \textbf{of } \boldsymbol{A}\end{array} = \begin{bmatrix} 1 & 0 & 0 & -1 \\ 0 & 1 & 0 & -1 \\ 0 & 0 & 1 & -1 \\ 0 & 0 & 0 & 0 \\ 0 & 0 & 0 & 0 \end{bmatrix}$$

From R_0 we see again the special solution $\boldsymbol{x} = (1, 1, 1, 1)$. The first 3 columns are basic, the fourth column is free. To produce a basis for $\mathbf{C}(A)$ and not $\mathbf{C}(R_0)$, we must go back to **columns $1, 2, 3$ of $\boldsymbol{A}$**. The column space has dimension $r = 3$.

The row space $\mathbf{C}(A^{\mathrm{T}})$ The dimension must again be $r = 3$. But the first 3 rows of A are *not independent*: row 3 = row 2 − row 1. So row 3 became zero in elimination, and row 3 was exchanged with row 4. *The first three independent rows are rows* $1, 2, 4$. Those three rows are a basis (one possible basis) for the row space.

Edges $1, 2, 3$ form a **loop** in the graph:	Dependent rows $1, 2, 3$.
Edges $1, 2, 4$ form a **tree. Trees have no loops**!	Independent rows $1, 2, 4$.

The left nullspace $\mathbf{N}(A^{\mathrm{T}})$ Now we solve $A^{\mathrm{T}}\boldsymbol{y} = \mathbf{0}$. Combinations of the rows give zero. We already noticed that row 3 = row 2 − row 1, so one solution is $\boldsymbol{y} = (1, -1, 1, 0, 0)$. I would say: *That* $\boldsymbol{y}$ *comes from following the upper loop in the graph.* Another $\boldsymbol{y}$ comes from going around the lower loop and it is $\boldsymbol{y} = (0, 0, \mathbf{-1}, \mathbf{1}, \mathbf{-1})$: *row* 3 = *row* 4 − *row* 5. Those two $\boldsymbol{y}$'s are independent, they solve $A^{\mathrm{T}}\boldsymbol{y} = \mathbf{0}$, and the dimension of $\mathbf{N}(A^{\mathrm{T}})$ is $\boldsymbol{m} - \boldsymbol{r} = 5 - 3 = \mathbf{2}$. So we have a basis for the left nullspace.

You may ask how "loops" and "trees" got into this problem. That didn't have to happen. We could have used elimination to solve $A^{\mathrm{T}}\boldsymbol{y} = \mathbf{0}$. The 4 by 5 matrix A^{T} would have three pivot columns $1, 2, 4$ and two free columns $3, 5$. There are two special solutions and the nullspace of A^{T} has dimension two: $m - r = 5 - 3 = 2$. But **loops** and **trees** identify *dependent rows* and *independent rows* in a beautiful way for every incidence matrix.

The equations $A\boldsymbol{x} = \boldsymbol{b}$ give "voltages" x_1, x_2, x_3, x_4 at the four nodes. The equations $A^{\mathrm{T}}\boldsymbol{y} = \mathbf{0}$ give "currents" y_1, y_2, y_3, y_4, y_5 on the five edges. These two equations are **Kirchhoff's Voltage Law** and **Kirchhoff's Current Law**. Those laws apply to an electrical network. But the ideas behind the words apply all over engineering and science and economics and business. Linear algebra connects the laws to the four subspaces.

Graphs are *the most important model in discrete applied mathematics.* You see graphs everywhere: roads, pipelines, blood flow, the brain, the Web, the economy of a country or the world. We can understand their matrices A and A^{T}. Here is a summary.

The incidence matrix A comes from a connected graph with n nodes and m edges. The row space and column space have dimensions $r = n - 1$. The nullspaces of A and A^{T} have dimensions 1 and $m - n + 1$:

$\mathbf{N}(A)$ The constant vectors $(c, c, \ldots, c)$ make up the nullspace of A : dim $= 1$.

$\mathbf{C}(A^{\mathrm{T}})$ The edges of any tree give r independent rows of A : $r = n - 1$.

$\mathbf{C}(A)$ ***Voltage Law***: The components of $A\boldsymbol{x}$ add to zero around all loops: dim $= n - 1$.

$\mathbf{N}(A^{\mathrm{T}})$ ***Current Law***: $A^{\mathrm{T}}\boldsymbol{y} =$ (**flow in**)−(**flow out**) $= \mathbf{0}$ is solved by loop currents.

There are $\boldsymbol{m - r = m - n + 1}$ ***independent small loops in the graph.***

For every graph in a plane, linear algebra yields ***Euler's formula***: Theorem 1 in topology!

$$\textit{(nodes)} - \textit{(edges)} + \textit{(small loops)} = (n) - (m) + (m - n + 1) = 1$$

Rank Two Matrices = Rank One plus Rank One

Rank one matrices have the form $\boldsymbol{u}\boldsymbol{v}^{\mathrm{T}}$. Here is a matrix A of rank $r = 2$. We can't see r immediately from A. So we reduce the matrix by row operations to R_0. **R_0 has the same row space as A.** Throw away its zero row to find R—also with the same row space.

$$\textbf{Rank two} \qquad A = \begin{bmatrix} 1 & 0 & 3 \\ 1 & 1 & 7 \\ 4 & 2 & 20 \end{bmatrix} = \begin{bmatrix} 1 & 0 \\ 1 & 1 \\ 4 & 2 \end{bmatrix} \begin{bmatrix} 1 & 0 & 3 \\ 0 & 1 & 4 \end{bmatrix} = CR \qquad (3)$$

Now look at columns. The pivot columns of R are clearly $(1,0)$ and $(0,1)$. Then the pivot columns of A are also in columns 1 and 2: $\boldsymbol{u}_1 = (1,1,4)$ and $\boldsymbol{u}_2 = (0,1,2)$. Notice that C has those same first two columns! That was guaranteed since multiplying by two columns of the identity matrix (in R) won't change the pivot columns $\boldsymbol{u}_1$ and $\boldsymbol{u}_2$.

When you put in letters for the columns and rows, you see **rank 2 = rank 1 + rank 1**.

$$\begin{matrix}\textbf{Matrix } A \\ \textbf{Rank two}\end{matrix} \qquad A = \begin{bmatrix} & & \\ \boldsymbol{u}_1 & \boldsymbol{u}_2 & \boldsymbol{u}_3 \\ & & \end{bmatrix} \begin{bmatrix} \boldsymbol{v}_1^{\mathrm{T}} \\ \boldsymbol{v}_2^{\mathrm{T}} \\ \textbf{zero row} \end{bmatrix} = \boldsymbol{u}_1\boldsymbol{v}_1^{\mathrm{T}} + \boldsymbol{u}_2\boldsymbol{v}_2^{\mathrm{T}}$$

Columns of C times rows of R. *Every rank r matrix is a sum of r rank one matrices*

■ WORKED EXAMPLES ■

3.5 A **Put four 1's into a 5 by 6 matrix of zeros**, keeping the dimension of its *row space* as small as possible. Describe all the ways to make the dimension of its *column space* as small as possible. Then describe all the ways to make the dimension of its *nullspace* as small as possible. How to make the *sum of the dimensions of all four subspaces small*?

Solution The rank is 1 if the four 1's go into the same row, or into the same column. They can also go into *two rows and two columns* (so $a_{ii} = a_{ij} = a_{ji} = a_{jj} = 1$). Since the column space and row space always have the same dimensions, this answers the first two questions: Dimension 1.

The nullspace has its smallest possible dimension $6 - 4 = 2$ when the rank is $r = 4$. To achieve rank 4, the 1's must go into four different rows and four different columns.

You can't do anything about the sum $r + (n-r) + r + (m-r) = \boldsymbol{n} + \boldsymbol{m}$. It will be $6 + 5 = 11$ no matter how the 1's are placed. The sum is 11 even if there aren't any 1's...

If all the other entries of A are 2's instead of 0's, how do these answers change?

3.5 B All the rows of AB are combinations of the rows of B. So the row space of AB is contained in (possibly equal to) the row space of B. **Rank $(AB) \le$ rank (B).**

All columns of AB are combinations of the columns of A. So the column space of AB is contained in (possibly equal to) the column space of A. **Rank $(AB) \le$ rank (A).**

If we multiply A by an *invertible* matrix B, the rank will not change. The rank can't drop, because when we multiply by the inverse matrix the rank can't jump back up.

Appendix 1 collects the key facts about the ranks of matrices.

Problem Set 3.5

1 (a) If a 7 by 9 matrix has rank 5, what are the dimensions of the four subspaces? What is the sum of all four dimensions?

(b) If a 3 by 4 matrix has rank 3, what are its column space and left nullspace?

2 Find bases and dimensions for the four subspaces associated with A and B:

$$A = \begin{bmatrix} 1 & 2 & 4 \\ 2 & 4 & 8 \end{bmatrix} \quad \text{and} \quad B = \begin{bmatrix} 1 & 2 & 4 \\ 2 & 5 & 8 \end{bmatrix}.$$

3 Find a basis for each of the four subspaces associated with A:

$$A = \begin{bmatrix} 0 & 1 & 2 & 3 & 4 \\ 0 & 1 & 2 & 4 & 6 \\ 0 & 0 & 0 & 1 & 2 \end{bmatrix} = \begin{bmatrix} 1 & 0 & 0 \\ 1 & 1 & 0 \\ 0 & 1 & 1 \end{bmatrix} \begin{bmatrix} 0 & 1 & 2 & 3 & 4 \\ 0 & 0 & 0 & 1 & 2 \\ 0 & 0 & 0 & 0 & 0 \end{bmatrix}.$$

4 Construct a matrix with the required property or explain why this is impossible:

(a) Column space contains $\begin{bmatrix} 1 \\ 1 \\ 0 \end{bmatrix}, \begin{bmatrix} 0 \\ 0 \\ 1 \end{bmatrix}$, row space contains $\begin{bmatrix} 1 \\ 2 \end{bmatrix}, \begin{bmatrix} 2 \\ 5 \end{bmatrix}$.

(b) Column space has basis $\begin{bmatrix} 1 \\ 1 \\ 3 \end{bmatrix}$, nullspace has basis $\begin{bmatrix} 3 \\ 1 \\ 1 \end{bmatrix}$.

(c) Dimension of nullspace $= 1 +$ dimension of left nullspace.

(d) Nullspace contains $\begin{bmatrix} 1 \\ 3 \end{bmatrix}$, column space contains $\begin{bmatrix} 3 \\ 1 \end{bmatrix}$.

(e) Row space $=$ column space, nullspace $\neq$ left nullspace.

5 If $\mathbf{V}$ is the subspace spanned by $(1,1,1)$ and $(2,1,0)$, find a matrix A that has $\mathbf{V}$ as its row space. Find a matrix B that has $\mathbf{V}$ as its nullspace. Multiply AB.

6 Without using elimination, find dimensions and bases for the four subspaces for

$$A = \begin{bmatrix} 0 & 3 & 3 & 3 \\ 0 & 0 & 0 & 0 \\ 0 & 1 & 0 & 1 \end{bmatrix} \quad \text{and} \quad B = \begin{bmatrix} 1 \\ 4 \\ 5 \end{bmatrix}.$$

7 Suppose the 3 by 3 matrix A is invertible. Write down bases for the four subspaces for A, and also for the 3 by 6 matrix $B = [\, A \;\; A \,]$. (*The basis for $\boldsymbol{Z}$ is empty.*)

8 What are the dimensions of the four subspaces for A, B, and C, if I is the 3 by 3 identity matrix and 0 is the 3 by 2 zero matrix?

$$A = \begin{bmatrix} I & 0 \end{bmatrix} \quad \text{and} \quad B = \begin{bmatrix} I & I \\ 0^{\mathrm{T}} & 0^{\mathrm{T}} \end{bmatrix} \quad \text{and} \quad C = \begin{bmatrix} 0 \end{bmatrix}.$$

9 Which subspaces are the same for these matrices of different sizes?

$$\text{(a)} \ \begin{bmatrix} A \end{bmatrix} \text{ and } \begin{bmatrix} A \\ A \end{bmatrix} \qquad \text{(b)} \ \begin{bmatrix} A \\ A \end{bmatrix} \text{ and } \begin{bmatrix} A & A \\ A & A \end{bmatrix}.$$

Prove that all three of those matrices have the *same rank* r.

10 If the entries of a 3 by 3 matrix are chosen randomly between 0 and 1, what are the most likely dimensions of the four subspaces? What if the random matrix is 3 by 5?

11 (Important) A is an m by n matrix of rank r. Suppose there are right sides $\boldsymbol{b}$ for which $A\boldsymbol{x} = \boldsymbol{b}$ has *no solution*.

(a) What are all inequalities ($<$ or $\leq$) that must be true between m, n, and r?

(b) How do you know that $A^{\mathrm{T}}\boldsymbol{y} = \boldsymbol{0}$ has solutions other than $\boldsymbol{y} = \boldsymbol{0}$?

12 Construct a matrix with $(1, 0, 1)$ and $(1, 2, 0)$ as a basis for its row space and its column space. Why can't this be a basis for the row space and nullspace?

13 True or false (with a reason or a counterexample):

(a) If $m = n$ then the row space of A equals the column space.

(b) The matrices A and $-A$ share the same four subspaces.

(c) If A and B share the same four subspaces then A is a multiple of B.

14 Without computing A, find bases for its four fundamental subspaces:

$$A = \begin{bmatrix} 1 & 0 & 0 \\ 6 & 1 & 0 \\ 9 & 8 & 1 \end{bmatrix} \begin{bmatrix} 1 & 2 & 3 & 4 \\ 0 & 1 & 2 & 3 \\ 0 & 0 & 1 & 2 \end{bmatrix}.$$

15 If you exchange the first two rows of A, which of the four subspaces stay the same? If $\boldsymbol{v} = (1, 2, 3, 4)$ is in the left nullspace of A, write down a vector in the left nullspace of the new matrix after the row exchange.

16 *Explain why $\boldsymbol{v} = (1, 0, -1)$ cannot be a row of A and also in the nullspace of A.*

17 Describe the four subspaces of $\mathbf{R}^3$ associated with

$$A = \begin{bmatrix} 0 & 1 & 0 \\ 0 & 0 & 1 \\ 0 & 0 & 0 \end{bmatrix} \quad \text{and} \quad I + A = \begin{bmatrix} 1 & 1 & 0 \\ 0 & 1 & 1 \\ 0 & 0 & 1 \end{bmatrix}.$$

18 Can tic-tac-toe be completed (5 ones and 4 zeros in A) so that rank $(A) = 2$ but neither side passed up a winning move?

19 (Left nullspace) Add the extra column $\boldsymbol{b}$ and reduce A to echelon form:

$$\begin{bmatrix} A & \boldsymbol{b} \end{bmatrix} = \begin{bmatrix} 1 & 2 & 3 & b_1 \\ 4 & 5 & 6 & b_2 \\ 7 & 8 & 9 & b_3 \end{bmatrix} \rightarrow \begin{bmatrix} 1 & 2 & 3 & b_1 \\ 0 & -3 & -6 & b_2 - 4b_1 \\ 0 & 0 & 0 & b_3 - 2b_2 + b_1 \end{bmatrix}.$$

A combination of the rows of A has produced the zero row. What combination is it? (Look at $b_3 - 2b_2 + b_1$ on the right side.) Which vectors are in the nullspace of A^{T} and which vectors are in the nullspace of A?

20 (**Patience needed**) Describe the row operations that reduce a matrix A to its echelon form R_0.

21 Suppose A is the sum of two matrices of rank one: $A = \boldsymbol{u}\boldsymbol{v}^{\mathrm{T}} + \boldsymbol{w}\boldsymbol{z}^{\mathrm{T}}$.

(a) Which vectors span the column space of A?

(b) Which vectors span the row space of A?

(c) The rank is less than 2 if ______ or if ______.

(d) Compute A and its rank if $\boldsymbol{u} = \boldsymbol{z} = (1,0,0)$ and $\boldsymbol{v} = \boldsymbol{w} = (0,0,1)$.

22 Construct $A = \boldsymbol{u}\boldsymbol{v}^{\mathrm{T}} + \boldsymbol{w}\boldsymbol{z}^{\mathrm{T}}$ whose column space has basis $(1,2,4), (2,2,1)$ and whose row space has basis $(1,0), (1,1)$. Write A as (3 by 2) times (2 by 2).

23 Without multiplying matrices, find bases for the row and column spaces of A:

$$A = \begin{bmatrix} 1 & 2 \\ 4 & 5 \\ 2 & 7 \end{bmatrix} \begin{bmatrix} 3 & 0 & 3 \\ 1 & 1 & 2 \end{bmatrix}.$$

How do you know from these shapes that A cannot be invertible?

24 (Important) $A^{\mathrm{T}}\boldsymbol{y} = \boldsymbol{d}$ is solvable when $\boldsymbol{d}$ is in which of the four subspaces? The solution $\boldsymbol{y}$ is unique when the ______ contains only the zero vector.

25 True or false (with a reason or a counterexample):

(a) A and A^{T} have the same number of pivots.

(b) A and A^{T} have the same left nullspace.

(c) If the row space equals the column space then $A^{\mathrm{T}} = A$.

(d) If $A^{\mathrm{T}} = -A$ then the row space of A equals the column space.

26 If a, b, c are given with $a \neq 0$, how would you choose d so that $\begin{bmatrix} a & b \\ c & d \end{bmatrix}$ has rank 1? Find a basis for the row space and nullspace. Show they are perpendicular!

Challenge Problems

27 Find the ranks of the 8 by 8 checkerboard matrix B and the chess matrix C:

$$B=\begin{bmatrix}1&0&1&0&1&0&1&0\\0&1&0&1&0&1&0&1\\1&0&1&0&1&0&1&0\\\cdot&\cdot&\cdot&\cdot&\cdot&\cdot&\cdot&\cdot\\0&1&0&1&0&1&0&1\end{bmatrix}\quad\text{and}\quad C=\begin{bmatrix}r&n&b&q&k&b&n&r\\p&p&p&p&p&p&p&p\\&&&\text{four zero rows}&&&&\\p&p&p&p&p&p&p&p\\r&n&b&q&k&b&n&r\end{bmatrix}$$

The numbers r, n, b, q, k, p are all different. Find bases for the row space and left nullspace of B and C. Find a basis for the nullspace of C.

28 If $A = \boldsymbol{u}\boldsymbol{v}^{\mathrm{T}}$ is a 2 by 2 matrix of rank 1, redraw Figure 3.5 to show clearly the Four Fundamental Subspaces. If B produces those same four subspaces, what is the exact relation of B to A?

29 $\mathbf{M}$ is the space of 3 by 3 matrices. Multiply every matrix X in $\mathbf{M}$ by A:

$$A=\begin{bmatrix}1&0&-1\\-1&1&0\\0&-1&1\end{bmatrix}.\quad\text{Notice: } A\begin{bmatrix}1\\1\\1\end{bmatrix}=\begin{bmatrix}0\\0\\0\end{bmatrix}.$$

(a) Which matrices X lead to $AX =$ zero matrix?

(b) Which matrices have the form AX for some matrix X?

(a) finds the "nullspace" of that operation AX and (b) finds the "column space". What are the dimensions of those two subspaces of $\mathbf{M}$? Why do they add to 9?

30 Suppose the m by n matrices A and B have *the same four subspaces*. If they are both in row reduced echelon form, is it true that $\boldsymbol{F}$ **must equal** $\boldsymbol{G}$?

$$A=\begin{bmatrix}I&F\\0&0\end{bmatrix}\qquad B=\begin{bmatrix}I&G\\0&0\end{bmatrix}.$$

31 Find the **incidence matrix** and its rank and one vector in each subspace for this complete graph—all six edges included.

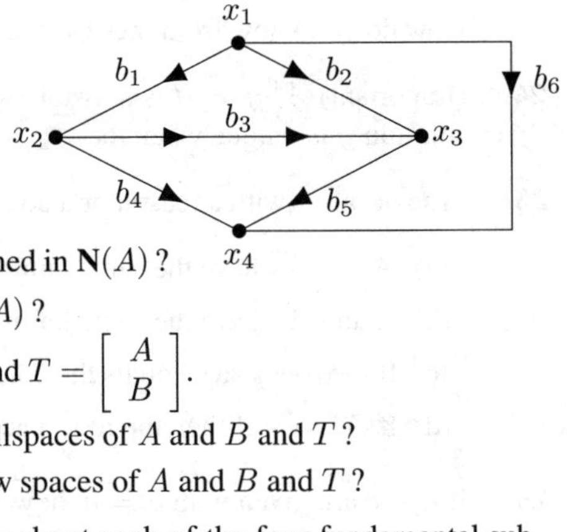

32 (Review) (a) Is $\mathbf{N}(AB)$ or $\mathbf{N}(BA)$ contained in $\mathbf{N}(A)$?

(b) Is $\mathbf{C}(AB)$ or $\mathbf{C}(BA)$ contained in $\mathbf{C}(A)$?

33 Suppose A is m by n and B is M by n and $T=\begin{bmatrix}A\\B\end{bmatrix}$.

(a) What are the relations between the nullspaces of A and B and T?

(b) What are the relations between the row spaces of A and B and T?

34 Suppose A is m by n. What can you say about each of the four fundamental subspaces for the matrices A and $W=\begin{bmatrix}A & A\end{bmatrix}$?

35 If A and B are n by n, is it always true that rank $\begin{bmatrix}A\\B\end{bmatrix}=$ rank $\begin{bmatrix}A & B\end{bmatrix}$?

Thoughts on Chapter 3 : The Big Picture of Elimination

This page explains elimination at the vector level and subspace level, when A is reduced to R. You know the steps and I won't repeat them. Elimination starts with the first pivot. It moves a column at a time (left to right) and a row at a time (top to bottom) for $\boldsymbol{U}$. Continuing elimination **upward** produces $\boldsymbol{R}_0$ and $\boldsymbol{R}$. Elimination answers two questions :

Question 1 Is this column a combination of previous columns?

If the column contains a pivot, the answer is no. Pivot columns are "independent" of previous columns. If column 4 has no pivot, it is a combination of columns $1, 2, 3$.

Question 2 Is this row a combination of previous rows?

If the row contains a pivot, the answer is no. Pivot rows are independent of previous rows, and their first nonzero is 1 from I. Rows that are all zero in R_0 were not and are not independent. Those zero rows disappear in R. That matrix is r by n.

It is amazing to me that one pass through the matrix answers both questions 1 and 2. Elimination acts on the rows but the result tells us about the columns ! The identity matrix in R locates the first r independent columns in A. Then the free columns F in R tell us *the combinations of those independent columns that produce the dependent columns in* A. This is easy to miss without seeing the factorization $A = CR$.

$\boldsymbol{R}$ tells us the special solutions to $\boldsymbol{Ax} = \mathbf{0}$. We could reach R from A by different row exchanges and elimination steps, but it will always be the same R. (This is because the special solutions are fully decided by A. The formula comes before Problem Set 3.2.) In the language coming soon, R reveals a "basis" for three of the fundamental subspaces :

The **column space** of A—choose the $\boldsymbol{r}$ **columns of $\boldsymbol{A}$** that produce pivots in R.

The **row space** of A—choose the $\boldsymbol{r}$ **rows of $\boldsymbol{R}$** as a basis.

The **nullspace** of A—choose the $\boldsymbol{n} - \boldsymbol{r}$ **special solutions** to $Rx = \mathbf{0}$ (and $Ax = \mathbf{0}$).

For the **left nullspace** $\mathbf{N}(A^{\mathrm{T}})$, we look at the elimination step $EA = R_0$. The last $m - r$ rows of R_0 are zero. The **last $\boldsymbol{m} - \boldsymbol{r}$ rows of $\boldsymbol{E}$** are a basis for the left nullspace ! In reducing the extended matrix $[A \;\; I]$ to $[R_0 \;\; E]$, the matrix E keeps a record of elimination that is otherwise lost.

Suppose we fix C and B (m by r and r by n, both rank r). Choose any invertible r by r mixing matrix M. **All the matrices $\boldsymbol{CMB}$** (and only those) have the **same four fundamental subspaces**.

Note This is the first textbook to express the result of elimination in its matrix form $\boldsymbol{A} = \boldsymbol{CR} = \boldsymbol{C}\begin{bmatrix}\boldsymbol{I} & \boldsymbol{F}\end{bmatrix}\boldsymbol{P}$. Elimination reveals $\boldsymbol{C}$ and $\boldsymbol{F}$ and $\boldsymbol{P}$ and $\boldsymbol{A} = \begin{bmatrix}\boldsymbol{C} & \boldsymbol{CF}\end{bmatrix}\boldsymbol{P}$.

$$\boldsymbol{A} = \big[\textbf{Independent columns in } \boldsymbol{C} \quad \textbf{Dependent columns in } \boldsymbol{CF}\big] \textbf{ Permute columns}$$

Row Operations on an m by n Matrix A: Review

(i)	Subtract a multiple of one row from another row
(ii)	Exchange two rows
(iii)	Multiply a row by any nonzero constant

The important point is: **Those row operations are reversible (invertible).**

(i) Add back the multiple of one row to the other row

(ii) Exchange the rows again

(iii) Divide the row by that nonzero constant

Total effect of those row operations: **An m by m invertible matrix E multiplies A.**

The nullspace is not changed by E: $A\boldsymbol{x} = \mathbf{0} \Rightarrow EA\boldsymbol{x} = \mathbf{0} \Rightarrow A\boldsymbol{x} = \mathbf{0}$.
A and EA have different rows but **the same row space and nullspace.**

Reduced Row Echelon Form: E can produce $EA = R_0 = \begin{bmatrix} I & F \\ 0 & 0 \end{bmatrix} P = \textbf{rref}(A)$

The identity I is r by r, F is r by $n-r$, P puts the n columns in correct order.

Factorization: $A = CR =$ [First r independent columns] $\begin{bmatrix} I & F \end{bmatrix} P$
$\boldsymbol{A} = \begin{bmatrix} \boldsymbol{C} & \boldsymbol{CF} \end{bmatrix} \boldsymbol{P} = \begin{bmatrix} \textbf{Independent cols} & \textbf{Dependent cols} \end{bmatrix}$ **Reorder columns**

Nullspace of A: Each column of F leads to one of the $n-r$ "special solutions" to $A\boldsymbol{x} = \mathbf{0}$:

Special solution
page 88, Example I

$$\boldsymbol{s}_k = P^{\mathrm{T}} \begin{bmatrix} -\text{ column } k \text{ of } F(r \text{ by } n-r) \\ +\text{ column } k \text{ of } I(n-r \text{ by } n-r) \end{bmatrix}$$

That permutation P^{T} puts the n components of the solution $\boldsymbol{s}_k$ in the right order.

Example Special solution $\boldsymbol{s}_1$ to $A\boldsymbol{x} = \mathbf{0}$ and $R\boldsymbol{x} = \mathbf{0}$ with $P = I$ and rank $r = 3$

$$R = \begin{bmatrix} 1 & 0 & 0 & 3 \\ 0 & 1 & 0 & 4 \\ 0 & 0 & 1 & 5 \end{bmatrix} \qquad \boldsymbol{s}_1 = \begin{bmatrix} -3 \\ -4 \\ -5 \\ 1 \end{bmatrix} \qquad R\boldsymbol{s}_1 = \mathbf{0} \qquad A\boldsymbol{s}_1 = CR\boldsymbol{s}_1 = \mathbf{0}$$

4 Orthogonality

Two vectors are orthogonal when their dot product is zero: $\boldsymbol{v} \cdot \boldsymbol{w} = \boldsymbol{v}^{\mathrm{T}}\boldsymbol{w} = 0$. This chapter moves to **orthogonal subspaces** and **orthogonal bases** and **orthogonal matrices**. The vectors in two subspaces, and the vectors in a basis, and the column vectors in Q, all pairs will be orthogonal. Think of $a^2 + b^2 = c^2$ for a *right triangle* with sides $\boldsymbol{v}$ and $\boldsymbol{w}$.

Orthogonal vectors	$\boldsymbol{v}^{\mathrm{T}}\boldsymbol{w} = 0$	and	$\\|\boldsymbol{v}\\|^2 + \\|\boldsymbol{w}\\|^2 = \\|\boldsymbol{v} + \boldsymbol{w}\\|^2$.

The right side is $(\boldsymbol{v} + \boldsymbol{w})^{\mathrm{T}}(\boldsymbol{v} + \boldsymbol{w})$. This equals $\boldsymbol{v}^{\mathrm{T}}\boldsymbol{v} + \boldsymbol{w}^{\mathrm{T}}\boldsymbol{w}$ when $\boldsymbol{v}^{\mathrm{T}}\boldsymbol{w} = \boldsymbol{w}^{\mathrm{T}}\boldsymbol{v} = 0$.

Subspaces entered Chapter 3 to throw light on $A\boldsymbol{x} = \boldsymbol{b}$. Right away we needed the column space and the nullspace. Then the light turned onto A^{T}, uncovering two more subspaces. Those four fundamental subspaces reveal what a matrix really does.

A matrix multiplies a vector: *A times* $\boldsymbol{x}$. At the first level this is only numbers. At the second level $A\boldsymbol{x}$ is a combination of column vectors. The third level shows subspaces. But I don't think you have seen the whole picture until you study Figure 4.1. **Those fundamental subspaces are orthogonal:**

The nullspace $\mathbf{N}(A)$ contains all vectors orthogonal to the row space $\mathbf{C}(A^{\mathrm{T}})$.
The nullspace $\mathbf{N}(A^{\mathrm{T}})$ contains all vectors orthogonal to the column space $\mathbf{C}(A)$.

$A\boldsymbol{x} = \mathbf{0}$ makes $\boldsymbol{x}$ orthogonal to each row. $A^{\mathrm{T}}\boldsymbol{y} = \mathbf{0}$ makes $\boldsymbol{y}$ orthogonal to each column.

A key idea in this chapter is **projection**: If $\boldsymbol{b}$ is outside the column space of A, find the closest point $\boldsymbol{p}$ that is inside. The line from $\boldsymbol{b}$ to $\boldsymbol{p}$ shows the error $\boldsymbol{e}$. That line is perpendicular to the column space. The *least squares equation* $A^{\mathrm{T}}A\widehat{\boldsymbol{x}} = A^{\mathrm{T}}\boldsymbol{b}$ produces the closest $\boldsymbol{p} = A\widehat{\boldsymbol{x}}$ and smallest possible error $\boldsymbol{e}$. It gives the best $\widehat{\boldsymbol{x}}$ when $A\boldsymbol{x} = \boldsymbol{b}$ is unsolvable. **That best $\widehat{\boldsymbol{x}}$ makes $||A\widehat{\boldsymbol{x}} - \boldsymbol{b}||^2$ as small as possible: Least squares.**

The equation $A^{\mathrm{T}}A\widehat{\boldsymbol{x}} = A^{\mathrm{T}}\boldsymbol{b}$ is easy when $A^{\mathrm{T}}A = I$. Then A has orthonormal columns: *perpendicular unit vectors.* That won't happen by accident but we can make it happen. Orthogonalizing the columns $\boldsymbol{a}_1$ to $\boldsymbol{a}_n$ in 4.4 produces $\boldsymbol{q}_1$ to $\boldsymbol{q}_n$ with $\boldsymbol{q}_i^{\mathrm{T}}\boldsymbol{q}_j = 0$. **The matrix Q has $Q^{\mathrm{T}}Q = I$ and $QR = A$. This R is upper triangular.**

Orthogonal matrices are perfect for computations. Q_1 times Q_2 is still orthogonal. In many ways $A = QR$ is better than $A = LU$, and this chapter shows why.

4.1 Orthogonality of Vectors and Subspaces

1 Orthogonal vectors have $\boldsymbol{v}^{\mathrm{T}}\boldsymbol{w}=0$. Then $||\boldsymbol{v}||^2+||\boldsymbol{w}||^2=||\boldsymbol{v}+\boldsymbol{w}||^2$ as in $a^2+b^2=c^2$.

2 Subspaces **V** and **W** are orthogonal when $\boldsymbol{v}^{\mathrm{T}}\boldsymbol{w}=0$ for every $\boldsymbol{v}$ in **V** and every $\boldsymbol{w}$ in **W**.

3 The row space of A is orthogonal to the nullspace. The column space is orthogonal to $\mathbf{N}(A^{\mathrm{T}})$.

4 The dimensions add to $r+(n-r)=n$ and $r+(m-r)=m$: Orthogonal complements.

5 If n vectors in $\mathbf{R}^n$ are independent, they span $\mathbf{R}^n$. If n vectors span $\mathbf{R}^n$, they are independent.

Chapter 1 connected dot products $\boldsymbol{v}^{\mathrm{T}}\boldsymbol{w}$ to the angle between $\boldsymbol{v}$ and $\boldsymbol{w}$. For 90° angles we have $\boldsymbol{v}^{\mathrm{T}}\boldsymbol{w}=0$. The vectors $\boldsymbol{v}, \boldsymbol{w}$ and $\boldsymbol{v}+\boldsymbol{w}$ produce a right triangle. The side lengths a, b, c or $||\boldsymbol{v}||, ||\boldsymbol{w}||, ||\boldsymbol{v}+\boldsymbol{w}||$ satisfy the famous rule $a^2+b^2=c^2$ of Pythagoras:

$$\textbf{Orthogonal vectors} \quad \boldsymbol{v}^{\mathrm{T}}\boldsymbol{w}=0 \text{ and } ||\boldsymbol{v}||^2+||\boldsymbol{w}||^2=||\boldsymbol{v}+\boldsymbol{w}||^2$$

The right side is $(\boldsymbol{v}+\boldsymbol{w})^{\mathrm{T}}(\boldsymbol{v}+\boldsymbol{w})=\boldsymbol{v}^{\mathrm{T}}\boldsymbol{v}+\boldsymbol{w}^{\mathrm{T}}\boldsymbol{w}+\boldsymbol{v}^{\mathrm{T}}\boldsymbol{w}+\boldsymbol{w}^{\mathrm{T}}\boldsymbol{v}=\boldsymbol{v}^{\mathrm{T}}\boldsymbol{v}+\boldsymbol{w}^{\mathrm{T}}\boldsymbol{w}+0+0$. These vectors can be in a plane or in n-dimensional space $\mathbf{R}^n$. We still see a right triangle, and orthogonal means perpendicular.

A line (1-dimensional subspace) is perpendicular to a plane (2-dimensional subspace) when *every vector on the line* is perpendicular to *every vector in the plane*. Now we are in 3 or more dimensions. The test for perpendicular subspaces is passed by the row space and nullspace of any matrix A:

The **nullspace of A** is orthogonal to the **row space of A**. Look at $A\boldsymbol{x}=\mathbf{0}$.

$$A\boldsymbol{x}=\begin{bmatrix}\textbf{row 1 of } A\\ \vdots\\ \textbf{row } m \textbf{ of } A\end{bmatrix}\begin{bmatrix}\\ \boldsymbol{x}\\ \\ \end{bmatrix}=\begin{bmatrix}0\\ \vdots\\ 0\end{bmatrix}\quad\begin{matrix}\longleftarrow\ \textbf{(row 1)}\cdot\boldsymbol{x}\textbf{ is zero}\\ \\ \longleftarrow\ \textbf{(row } m\textbf{)}\cdot\boldsymbol{x}\textbf{ is zero}\end{matrix}\tag{1}$$

This says: Every row has a zero dot product with $\boldsymbol{x}$. Then every combination of the rows is perpendicular to $\boldsymbol{x}$. The whole row space $\mathbf{C}(A^{\mathrm{T}})$ is orthogonal to the whole nullspace.

Remember our convention—to stay with column vectors. The rows of A become the **columns of A^{T}**. The vectors in the row space become combinations $A^{\mathrm{T}}\boldsymbol{y}$ of those columns. Then here is the official short proof of perpendicular subspaces:

$$\boldsymbol{x}\textbf{ in nullspace orthogonal to }A^{\mathrm{T}}\boldsymbol{y}\textbf{ in row space}\quad \boldsymbol{x}^{\mathrm{T}}(A^{\mathrm{T}}\boldsymbol{y})=(A\boldsymbol{x})^{\mathrm{T}}\boldsymbol{y}=\mathbf{0}^{\mathrm{T}}\boldsymbol{y}=0.\tag{2}$$

We like the first proof in (1). You can see each row of A multiplying $\boldsymbol{x}$ to give zero. Then all linear combinations of the rows multiply $\boldsymbol{x}$ to give zero.

Important There is another pair of fundamental subspaces: The **column space** $\mathbf{C}(A)$ and $\mathbf{N}(A^{\mathrm{T}})$ = **nullspace of** $\boldsymbol{A}^{\mathbf{T}}$. Those are perpendicular inside m-dimensional space $\mathbf{R}^m$.

We could just apply the original proof to A^{T} instead of A. The row space of A^{T} is the column space of A. Then row space perpendicular to nullspace (applied to A^{T}) becomes **column space** $\mathbf{C}(A)$ **perpendicular to** $\mathbf{N}(A^{\mathrm{T}})$.

Example 1 The two rows of A are perpendicular to $\boldsymbol{x}$ in the nullspace of A:

row space
nullspace
$$A\boldsymbol{x} = \begin{bmatrix} 1 & -2 & 1 \\ 1 & 0 & -1 \end{bmatrix}\begin{bmatrix} 1 \\ 1 \\ 1 \end{bmatrix} = \begin{bmatrix} 0 \\ 0 \end{bmatrix} \quad \begin{matrix} \text{Dot product } 1-2+1=0 \\ \text{Dot product } 1+0-1=0 \end{matrix}$$

And the three columns of A (rows of A^{T}) are perpendicular to $\boldsymbol{y} = \mathbf{0}$ in $\mathbf{N}(A^{\mathrm{T}})$:

column space
left nullspace
$$A^{\mathrm{T}}\boldsymbol{y} = \begin{bmatrix} 1 & 1 \\ -2 & 0 \\ 1 & -1 \end{bmatrix}\begin{bmatrix} 0 \\ 0 \end{bmatrix} = \begin{bmatrix} 0 \\ 0 \\ 0 \end{bmatrix}$$

That was an extreme case with $\mathbf{C}(A)$ = all of $\mathbf{R}^2$ and nullspace of A^{T} = zero vector.

The row space – nullspace part had dimensions $2+1=3$. Those subspaces accounted for all vectors in $\mathbf{R}^3 = \mathbf{R}^n$. The column space – left nullspace part had dimensions $2+0=2$. Those subspaces accounted for all vectors in $\mathbf{R}^2 = \mathbf{R}^m$.

There is a very important restriction on the dimensions of any two orthogonal subspaces:

If $\boldsymbol{V}$ and $\boldsymbol{W}$ are orthogonal subspaces in $\mathbf{R}^n$ then $\dim \boldsymbol{V} + \dim \boldsymbol{W} \le n$.

That seems so clear and natural. But look what it tells us about a wall of your room and the floor of your room. Those are 2-dimensional subspaces in $\mathbf{R}^3$. **They can't be orthogonal subspaces! $2+2$ is more than 3**. Some vector must lie in both the wall and the floor. Yes, the line where wall meets floor is in both subspaces.

Two orthogonal subspaces that account for the whole space have a special name: **orthogonal complements. The orthogonal complement $\boldsymbol{V}^{\perp}$ of $\boldsymbol{V}$ contains all vectors orthogonal to $\boldsymbol{V}$.** So the two pairs of subspaces in the big picture of linear algebra are actually orthogonal complements:

Orthogonal	Row space and nullspace	$r + (n-r) = n$
complements	Column space and left nullspace	$r + (m-r) = m$

Any vector $\boldsymbol{x}$ in $\mathbf{R}^n$ is the sum $\boldsymbol{x} = \boldsymbol{x}_{\mathbf{row}} + \boldsymbol{x}_{\mathbf{null}}$ of its row space component and its nullspace component. Any vector $\boldsymbol{y}$ in $\mathbf{R}^m$ is the sum $\boldsymbol{y} = \boldsymbol{y}_{\mathbf{col}} + \boldsymbol{y}_{\mathbf{null}}$ of its column space component and its component in $\mathbf{N}(A^{\mathrm{T}})$.

> ***Fundamental Theorem of Linear Algebra*, Part 2**
>
> **N**(A) ***is the orthogonal complement of the row space*** **C**(A^{T}) **(in R^n).**
>
> **N**(A^{T}) ***is the orthogonal complement of the column space*** **C**(A) **(in R^m).**

Part 1 gave the dimensions of the subspaces. Part 2 gives the 90° angles between them. Every $\boldsymbol{x}$ can be split into a *row space component* $\boldsymbol{x}_r$ and a *nullspace component* $\boldsymbol{x}_n$. When A multiplies $\boldsymbol{x} = \boldsymbol{x}_r + \boldsymbol{x}_n$, Figure 4.1 shows $A\boldsymbol{x}_n = \mathbf{0}$ and $A\boldsymbol{x}_r = A\boldsymbol{x}$.

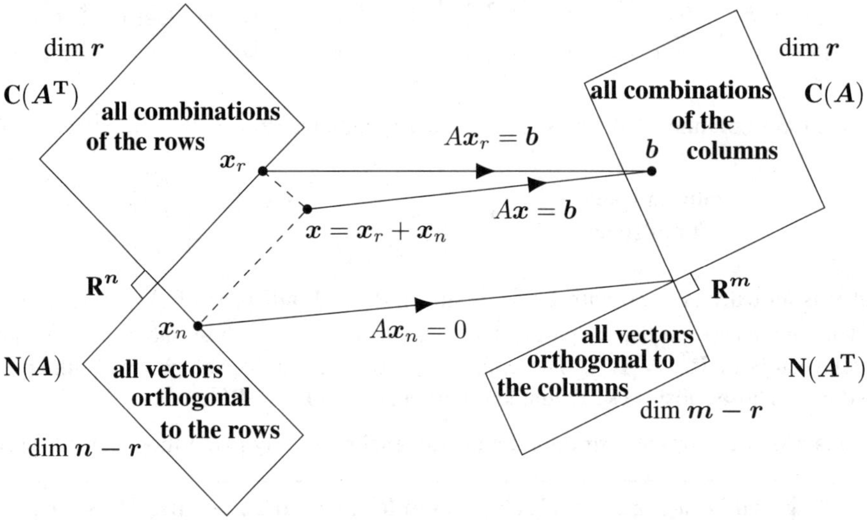

Figure 4.1: $A\boldsymbol{x_r} = \boldsymbol{b}$ is in the column space and $A\boldsymbol{x_n} = \mathbf{0}$. The complete solution to $A\boldsymbol{x} = \boldsymbol{b}$ is $\boldsymbol{x} =$ **one** $\boldsymbol{x_r}+$ **any** $\boldsymbol{x_n}$. Then $||x||^2 = ||\boldsymbol{x_r}||^2 + ||\boldsymbol{x_n}||^2$ and the **minimum norm solution** to $A\boldsymbol{x} = \boldsymbol{b}$ is $\boldsymbol{x} = \boldsymbol{x_r}$ in the row space plus $\boldsymbol{x}_n = \mathbf{0}$ from **N**(A).

Every vector $A\boldsymbol{x}$ is in the column space ! Multiplying by A cannot do anything else. More than that: *Every vector* $\boldsymbol{b}$ *in the column space comes from exactly one vector* $\boldsymbol{x}_r$ *in the row space*. Proof: If $A\boldsymbol{x}_r = A\boldsymbol{x}'_r$, the difference $\boldsymbol{x}_r - \boldsymbol{x}'_r$ is in the nullspace. It is also in the row space, where $\boldsymbol{x}_r$ and $\boldsymbol{x}'_r$ came from. This difference must be the zero vector, because the nullspace and row space are perpendicular. Therefore $\boldsymbol{x}_r = \boldsymbol{x}'_r$.

There is an r by r invertible matrix hiding inside A, if we throw away the two nullspaces. ***A is invertible from row space to column space***. The **pseudoinverse A^+** in Section 4.5 will invert that part of A : column space back to row space.

Example 2 **Every matrix of rank r has an r by r invertible submatrix**. A has rank 2 :

$$\boldsymbol{A} = \begin{bmatrix} \mathbf{1} & 2 & \mathbf{3} & 4 & 5 \\ 1 & 2 & 4 & 5 & 6 \\ \mathbf{1} & 2 & \mathbf{4} & 5 & 6 \end{bmatrix} \quad \text{contains} \quad \begin{bmatrix} \mathbf{1} & \mathbf{3} \\ \mathbf{1} & \mathbf{4} \end{bmatrix} \quad \text{in the pivot rows and pivot columns.}$$

Proof for any A: Its submatrix C has r independent columns (Chapter 1). Then C and C^{T} have rank r. So C^{T} has r independent columns—in other words C has r independent rows. This locates an r by r invertible submatrix of A.

Let me repeat: The only vector in two orthogonal subspaces is the zero vector.

Combining Bases from Subspaces

A basis contains *linearly independent* vectors *that span the space.* Normally we have to check both properties of a basis. When the count is right, one property implies the other: Every vector is a combination of the basis vectors **in exactly one way**.

Any n independent vectors in $\mathbf{R}^n$ must span $\mathbf{R}^n$. So they are a basis.
Any n vectors that span $\mathbf{R}^n$ must be independent. So they are a basis.

This is true in any vector space, but we care most about $\mathbf{R}^n$. When the vectors go into the columns of an n by n *square* matrix A, here are the same two facts:

If the n columns of A are independent, they span $\mathbf{R}^n$. So $A\boldsymbol{x} = \boldsymbol{b}$ is solvable.
If the n columns span $\mathbf{R}^n$, they are independent. So $A\boldsymbol{x} = \boldsymbol{b}$ has only one solution.
If $\boldsymbol{AB} = \boldsymbol{I}$ for square matrices A and B, then also $\boldsymbol{BA} = \boldsymbol{I}$.

Uniqueness implies existence and existence implies uniqueness. ***Then A is invertible***. If there are no free variables, the solution $\boldsymbol{x}$ is unique. There must be n pivot columns. Then back substitution solves $A\boldsymbol{x} = \boldsymbol{b}$ (the solution exists).

Starting in the opposite direction, suppose that $A\boldsymbol{x} = \boldsymbol{b}$ can be solved for every $\boldsymbol{b}$ (*existence of solutions*). Then elimination produced no zero rows. There are n pivots and no free variables. The nullspace contains only $\boldsymbol{x} = \mathbf{0}$ (*uniqueness of solutions*).

With bases for the row space and the nullspace, we have $r + (n - r) = n$ vectors. This is the right number. Those n vectors are independent.[1] *Therefore they span* $\mathbf{R}^n$.

Each $\boldsymbol{x}$ is the sum $\boldsymbol{x}_r + \boldsymbol{x}_n$ of a row space vector $\boldsymbol{x}_r$ and a nullspace vector $\boldsymbol{x}_n$. The splitting $\boldsymbol{x}_r + \boldsymbol{x}_n$ in Figure 4.1 shows the key point of orthogonal complements—the dimensions add to n and all vectors are fully accounted for.

Example 3 For $A = \begin{bmatrix} 1 & 2 \\ 3 & 6 \end{bmatrix}$ split $\boldsymbol{x} = \begin{bmatrix} 4 \\ 3 \end{bmatrix}$ into $\boldsymbol{x}_r + \boldsymbol{x}_n = \begin{bmatrix} 2 \\ 4 \end{bmatrix} + \begin{bmatrix} 2 \\ -1 \end{bmatrix}$.

The vector $(2, 4)$ is in the row space. The orthogonal vector $(2, -1)$ is in the nullspace. The next section will compute this splitting by a projection matrix P.

Example 4 Suppose S is a six-dimensional subspace of nine-dimensional space R^9.

(a) What are the possible dimensions of subspaces orthogonal to **S**? **$0, 1, 2, 3$**

(b) What are the possible dimensions of the orthogonal complement $\mathbf{S}^\perp$ of **S**? **3**

(c) What is the smallest possible size of a matrix A that has row space **S**? **6 by 9**

(d) What is the smallest possible size of a matrix B that has nullspace $\mathbf{S}^\perp$? **6 by 9**

---

[1] If a combination of all n vectors gives $\boldsymbol{x}_r + \boldsymbol{x}_n = \mathbf{0}$, then $\boldsymbol{x}_r = -\boldsymbol{x}_n$ is in both subspaces. So $\boldsymbol{x}_r = \boldsymbol{x}_n = \mathbf{0}$. All coefficients of the row space basis and of the nullspace basis must be zero. This proves independence of the n vectors together.

Problem Set 4.1

1 Construct any 2 by 3 matrix of rank one. Copy Figure 4.1 and put one vector in each subspace (and put two in the nullspace). Which vectors are orthogonal?

2 Redraw Figure 4.1 for a 3 by 2 matrix of rank $r = 2$. Which subspace is $\boldsymbol{Z}$ (zero vector only)? The nullspace part of any vector $\boldsymbol{x}$ in $\mathbf{R}^2$ is $\boldsymbol{x}_n =$ _____.

3 Construct a matrix with the required property or say why that is impossible:

(a) Column space contains $\begin{bmatrix} 1 \\ 2 \\ -3 \end{bmatrix}$ and $\begin{bmatrix} 2 \\ -3 \\ 5 \end{bmatrix}$, nullspace contains $\begin{bmatrix} 1 \\ 1 \\ 1 \end{bmatrix}$

(b) Row space contains $\begin{bmatrix} 1 \\ 2 \\ -3 \end{bmatrix}$ and $\begin{bmatrix} 2 \\ -3 \\ 5 \end{bmatrix}$, nullspace contains $\begin{bmatrix} 1 \\ 1 \\ 1 \end{bmatrix}$

(c) $A\boldsymbol{x} = \begin{bmatrix} 1 \\ 1 \\ 1 \end{bmatrix}$ has a solution and $A^{\mathrm{T}} \begin{bmatrix} 1 \\ 0 \\ 0 \end{bmatrix} = \begin{bmatrix} 0 \\ 0 \\ 0 \end{bmatrix}$

(d) Every row is orthogonal to every column (A is not the zero matrix)

(e) Columns add up to a column of zeros, rows add to a row of 1's.

4 If $AB = 0$ then the columns of B are in the _____ of A. The rows of A are in the _____ of B. With $AB = 0$, why can't A and B be 3 by 3 matrices of rank 2?

5 (a) If $A\boldsymbol{x} = \boldsymbol{b}$ has a solution and $A^{\mathrm{T}}\boldsymbol{y} = \boldsymbol{0}$, *is* $(\boldsymbol{y}^{\mathrm{T}}\boldsymbol{x} = 0)$ *or* $(\boldsymbol{y}^{\mathrm{T}}\boldsymbol{b} = 0)$?

(b) If $A^{\mathrm{T}}\boldsymbol{y} = (1, 1, 1)$ has a solution and $A\boldsymbol{x} = \boldsymbol{0}$, then _____.

6 Any system of equations $A\boldsymbol{x} = \boldsymbol{b}$ with *no solution* can be combined into $\mathbf{0} = \mathbf{1}$.

$$\begin{aligned} x + 2y + 2z &= 5 \\ 2x + 2y + 3z &= 5 \\ 3x + 4y + 5z &= 9 \end{aligned}$$

Find numbers y_1, y_2, y_3 to multiply the equations so they add to $0 = 1$. You have found a vector $\boldsymbol{y}$ in which subspace? Its dot product $\boldsymbol{y}^{\mathrm{T}}\boldsymbol{b}$ is 1, so no solution $\boldsymbol{x}$.

7 Every system $A\boldsymbol{x} = \boldsymbol{b}$ with no solution is like the one in Problem 6. There are numbers $y_1, \ldots, y_m$ that multiply the m equations so they add up to $0 = 1$. This is called **Fredholm's Alternative: If $\boldsymbol{b}$ is not in $\mathbf{C}(A)$, then part of $\boldsymbol{b}$ is in $\mathbf{N}(A^{\mathrm{T}})$.**

Exactly one problem has a solution: $A\boldsymbol{x} = \boldsymbol{b}$ **OR** $A^{\mathrm{T}}\boldsymbol{y} = \mathbf{0}$ with $\boldsymbol{y}^{\mathrm{T}}\boldsymbol{b} = 1$.

Multiply the equations $x_1 - x_2 = 1$ and $x_2 - x_3 = 1$ and $x_1 - x_3 = 1$ by numbers y_1, y_2, y_3 chosen so the equations add up to $0 = 1$. The equations are *unsolvable*.

8 In Figure 4.1, how do we know that $A\boldsymbol{x}_r$ is equal to $A\boldsymbol{x}$? How do we know that this vector $A\boldsymbol{x}$ is in the column space? If $A = \begin{bmatrix} 1 & 1 \\ 1 & 1 \end{bmatrix}$ and $\boldsymbol{x} = \begin{bmatrix} 1 \\ 0 \end{bmatrix}$ what is $\boldsymbol{x}_r$?

9 **If $A^{\mathrm{T}}Ax = \mathbf{0}$ then $Ax = \mathbf{0}$.** Reason: Ax is in the nullspace of A^{T} and also in the _____ of A and those spaces are _____. *Conclusion*: $Ax = \mathbf{0}$ and therefore *$A^{\mathrm{T}}A$ has the same nullspace as A. This key fact will be repeated when we need it.*

10 Suppose A is a symmetric matrix ($A^{\mathrm{T}} = A$).

(a) Why is its column space perpendicular to its nullspace?

(b) If $Ax = \mathbf{0}$ and $Az = 5z$, which subspaces contain these "eigenvectors" x and z? **Symmetric matrices have perpendicular eigenvectors $x^{\mathrm{T}}z = 0$.**

11 Draw Figure 4.1 to show each subspace correctly for $A = \begin{bmatrix} 1 & 2 \\ 3 & 6 \end{bmatrix}$ and $B = \begin{bmatrix} 1 & 0 \\ 3 & 0 \end{bmatrix}$.

12 Find x_r and x_n and draw Figure 4.1 properly if $A = \begin{bmatrix} 1 & -1 \\ 0 & 0 \end{bmatrix}$ and $x = \begin{bmatrix} 2 \\ 0 \end{bmatrix}$.

Questions 13–23 are about orthogonal subspaces.

13 Put bases for the subspaces $\mathbf{V}$ and $\mathbf{W}$ into the columns of matrices V and W. Explain why the test for orthogonal subspaces can be written $V^{\mathrm{T}}W =$ zero matrix. This matches $v^{\mathrm{T}}w = 0$ for orthogonal vectors.

14 The floor $\mathbf{V}$ and the wall $\mathbf{W}$ are not orthogonal subspaces, because they share a nonzero vector (along the line where they meet). No planes $\mathbf{V}$ and $\mathbf{W}$ in $\mathbf{R}^3$ can be orthogonal! Find a vector in the column spaces of both matrices:

$$A = \begin{bmatrix} 1 & 2 \\ 1 & 3 \\ 1 & 2 \end{bmatrix} \quad \text{and} \quad B = \begin{bmatrix} 5 & 4 \\ 6 & 3 \\ 5 & 1 \end{bmatrix}$$

This will be a vector Ax and also $B\widehat{x}$. Think 3 by 4 with the matrix $[\,A \;\; B\,]$.

15 Extend Problem 14 to a p-dimensional subspace $\mathbf{V}$ and a q-dimensional subspace $\mathbf{W}$ of $\mathbf{R}^n$. What inequality on $p+q$ guarantees that $\mathbf{V}$ intersects $\mathbf{W}$ in a nonzero vector? These subspaces cannot be orthogonal.

16 Prove that every y in $\mathbf{N}(A^{\mathrm{T}})$ is perpendicular to every Ax in the column space, using the matrix shorthand of equation (2). Start from $A^{\mathrm{T}}y = \mathbf{0}$.

17 If $\mathbf{S}$ is the subspace of $\mathbf{R}^3$ containing only the zero vector, what is $\mathbf{S}^{\perp}$? If $\mathbf{S}$ is spanned by $(1,1,1)$, what is $\mathbf{S}^{\perp}$? If $\mathbf{S}$ is spanned by $(1,1,1)$ and $(1,1,-1)$, what is a basis for $\mathbf{S}^{\perp}$?

18 Suppose $\mathbf{S}$ only contains two vectors $(1,5,1)$ and $(2,2,2)$ (*not a subspace*). Then $\mathbf{S}^{\perp}$ is the nullspace of the matrix $A =$ _____. $\mathbf{S}^{\perp}$ is a subspace even if $\mathbf{S}$ is not.

19 Suppose $\mathbf{L}$ is a one-dimensional subspace (a line) in $\mathbf{R}^3$. Its orthogonal complement $\mathbf{L}^{\perp}$ is the _____ perpendicular to $\mathbf{L}$. Then $(\mathbf{L}^{\perp})^{\perp}$ is a _____ perpendicular to $\mathbf{L}^{\perp}$. In fact $(\mathbf{L}^{\perp})^{\perp}$ is the same as _____.

20 Suppose $\mathbf{V}$ is the whole space $\mathbf{R}^4$. Then $\mathbf{V}^\perp$ contains only the vector _____. Then $(\mathbf{V}^\perp)^\perp$ is _____. So $(\mathbf{V}^\perp)^\perp$ is the same as _____.

21 Suppose $\mathbf{S}$ is spanned by the vectors $(1,2,2,3)$ and $(1,3,3,2)$. Find two vectors that span $\mathbf{S}^\perp$. This is the same as solving $A\boldsymbol{x} = \mathbf{0}$ for which A?

22 If $\mathbf{P}$ is the plane of vectors in $\mathbf{R}^4$ satisfying $x_1 + x_2 + x_3 + x_4 = 0$, write a basis for $\mathbf{P}^\perp$. Construct a matrix that has $\mathbf{P}$ as its nullspace.

23 If a subspace $\mathbf{S}$ is contained in a subspace $\mathbf{V}$, explain why $\mathbf{S}^\perp$ contains $\mathbf{V}^\perp$.

Questions 24-28 are about perpendicular columns and rows.

24 Suppose an n by n matrix is invertible: $AA^{-1} = I$. Then the first column of A^{-1} is orthogonal to the space spanned by which rows of A?

25 Find $A^{\mathrm{T}}A$ if the columns of A are unit vectors, perpendicular to each other.

26 Construct a 3 by 3 matrix A with no zero entries whose columns are mutually perpendicular. Compute $A^{\mathrm{T}}A$. Why is it a diagonal matrix?

27 The lines $3x + y = b_1$ and $6x + 2y = b_2$ are _____. They are the same line if _____. In that case (b_1, b_2) is perpendicular to the vector _____. The nullspace of the matrix is the line $3x + y =$ _____. One particular vector in that nullspace is _____.

28 Why is each of these statements false?

(a) $(1,1,1)$ is perpendicular to $(1,1,-2)$ so the planes $x + y + z = 0$ and $x + y - 2z = 0$ are orthogonal subspaces.

(b) The subspace spanned by $(1,1,0,0,0)$ and $(0,0,0,1,1)$ is the orthogonal complement of the subspace spanned by $(1,-1,0,0,0)$ and $(2,-2,3,4,-4)$.

(c) Two subspaces that meet only in the zero vector are orthogonal.

29 Find a matrix A with $\boldsymbol{v} = (1,2,3)$ in the row space and column space. Find B with $\boldsymbol{v}$ in the nullspace and column space. Which pairs of subspaces can't share $\boldsymbol{v}$?

30 Suppose A is 3 by 4 and B is 4 by 5 and $AB = 0$. So $\mathbf{N}(A)$ contains $\mathbf{C}(B)$. Prove from the dimensions of $\mathbf{N}(A)$ and $\mathbf{C}(B)$ that $\operatorname{rank}(A) + \operatorname{rank}(B) \le 4$.

31 Suppose the command $\mathbf{N} = \mathsf{null}(A)$ will produce a basis for the nullspace of A. Then the command $\mathbf{B} = \mathsf{null}(N^{\mathrm{T}})$ will produce a basis for the _____ of A.

32 What are the conditions for nonzero vectors $\boldsymbol{r}, \boldsymbol{n}, \boldsymbol{c}, \boldsymbol{l}$ in $\mathbf{R}^2$ to be bases for the four fundamental subspaces $\mathbf{C}(A^{\mathrm{T}}), \mathbf{N}(A), \mathbf{C}(A), \mathbf{N}(A^{\mathrm{T}})$ of a 2 by 2 matrix?

33 When can the 8 vectors $\boldsymbol{r}_1, \boldsymbol{r}_2, \boldsymbol{n}_1, \boldsymbol{n}_2, \boldsymbol{c}_1, \boldsymbol{c}_2, \boldsymbol{l}_1, \boldsymbol{l}_2$ in $\mathbf{R}^4$ be bases for the four fundamental subspaces of a 4 by 4 matrix? What is one possible matrix A?

4.2 Projections onto Lines and Subspaces

1 The projection of $\boldsymbol{b}$ onto the line through $\boldsymbol{a}$ is the **closest point to $\boldsymbol{b}$**: $\boldsymbol{p} = \boldsymbol{a}(\boldsymbol{a}^{\mathrm{T}}\boldsymbol{b} / \boldsymbol{a}^{\mathrm{T}}\boldsymbol{a})$.

2 The error $\boldsymbol{e} = \boldsymbol{b} - \boldsymbol{p}$ is perpendicular to $\boldsymbol{a}$: Right triangle $\boldsymbol{b}\,\boldsymbol{p}\,\boldsymbol{e}$ has $||\boldsymbol{p}||^2 + ||\boldsymbol{e}||^2 = ||\boldsymbol{b}||^2$.

3 The **projection** of $\boldsymbol{b}$ onto a subspace $\boldsymbol{S}$ is the closest vector $\boldsymbol{p}$ in $\boldsymbol{S}$; $\boldsymbol{b} - \boldsymbol{p}$ is orthogonal to $\boldsymbol{S}$.

4 $A^{\mathrm{T}}A$ is invertible (and symmetric) when A has independent columns: $\mathbf{N}(A^{\mathrm{T}}A) = \mathbf{N}(A)$.

5 Then the projection of $\boldsymbol{b}$ onto the column space of $\mathbf{C}(A)$ is the vector $\boldsymbol{p} = A(A^{\mathrm{T}}A)^{-1}A^{\mathrm{T}}\boldsymbol{b}$.

6 The **projection matrix** onto $\mathbf{C}(A)$ is $\boxed{P = A(A^{\mathrm{T}}A)^{-1}A^{\mathrm{T}}.}$ It has $\boldsymbol{p} = \boldsymbol{P}\boldsymbol{b}$ and $\boldsymbol{P}^2 = \boldsymbol{P} = \boldsymbol{P}^{\mathrm{T}}$.

Projections on a subspace S are easy to visualize: Each vector $\boldsymbol{b}$ goes to the closest point $\boldsymbol{p}$ in S. Then the error vector $\boldsymbol{e} = \boldsymbol{b} - \boldsymbol{p}$ is perpendicular to the subspace. If A has independent columns, the projection of $\boldsymbol{b}$ onto $\mathbf{C}(A)$ is $\boldsymbol{p} = A(A^{\mathrm{T}}A)^{-1}A^{\mathrm{T}}\boldsymbol{b}$. That **projection matrix** $\boldsymbol{P} = \boldsymbol{A}(\boldsymbol{A}^{\mathrm{T}}\boldsymbol{A})^{-1}\boldsymbol{A}^{\mathrm{T}}$ is symmetric. Its special property is $\boldsymbol{P}^2 = \boldsymbol{P}$—a second projection changes nothing, because $\boldsymbol{p}$ projects to $\boldsymbol{p}$.

For specially nice subspaces you can see their projection matrices P:

1 What are the projections of $\boldsymbol{b} = (2, 3, 4)$ onto the z axis and the xy plane?

2 What matrices P_1 and P_2 produce those projections onto a line and a plane?

When $\boldsymbol{b}$ is projected onto a line, ***its projection p is the part of b along that line.*** If $\boldsymbol{b}$ is projected onto a plane, $\boldsymbol{p}$ is the part in that plane. ***The projection p is*** Pb.

The projection onto the z axis we call $\boldsymbol{p}_1$. The projection p_2 drops straight down to the xy plane. Start with $\boldsymbol{b} = (\mathbf{2}, \mathbf{3}, \mathbf{4})$. The projection across gives $\boldsymbol{p}_1 = (\mathbf{0}, \mathbf{0}, \mathbf{4})$. The projection down gives $\boldsymbol{p}_2 = (\mathbf{2}, \mathbf{3}, \mathbf{0})$. Those are the parts of $\boldsymbol{b}$ along the z axis and in the xy plane (Figure 4.2).

The projection matrices P_1 and P_2 are 3 by 3. They multiply $\boldsymbol{b}$ with 3 components to produce $\boldsymbol{p}$ with 3 components. Projection onto a line comes from a *rank one* matrix. Projection onto a plane comes from a *rank two* matrix:

Projection matrix
Onto the z axis: $P_1 = \begin{bmatrix} 0 & 0 & 0 \\ 0 & 0 & 0 \\ 0 & 0 & \mathbf{1} \end{bmatrix}$ Onto the xy plane: $P_2 = \begin{bmatrix} \mathbf{1} & 0 & 0 \\ 0 & \mathbf{1} & 0 \\ 0 & 0 & 0 \end{bmatrix}$.

P_1 picks out the z component of every vector. P_2 picks out the x and y components. To find the projections $\boldsymbol{p}_1$ and $\boldsymbol{p}_2$ of $\boldsymbol{b}$, multiply $\boldsymbol{b}$ by P_1 and P_2 (small $\boldsymbol{p}$ for the vector, capital P for the matrix that produces it):

$$\boldsymbol{p}_1 = P_1\boldsymbol{b} = \begin{bmatrix} 0 & 0 & 0 \\ 0 & 0 & 0 \\ 0 & 0 & 1 \end{bmatrix} \begin{bmatrix} x \\ y \\ z \end{bmatrix} = \begin{bmatrix} \mathbf{0} \\ \mathbf{0} \\ \boldsymbol{z} \end{bmatrix} \qquad \boldsymbol{p}_2 = P_2\boldsymbol{b} = \begin{bmatrix} 1 & 0 & 0 \\ 0 & 1 & 0 \\ 0 & 0 & 0 \end{bmatrix} \begin{bmatrix} x \\ y \\ z \end{bmatrix} = \begin{bmatrix} \boldsymbol{x} \\ \boldsymbol{y} \\ \mathbf{0} \end{bmatrix}.$$

In this case the projections p_1 and p_2 are perpendicular. The xy plane and the z axis are ***orthogonal subspaces***, like the floor of a room and the line straight upward.

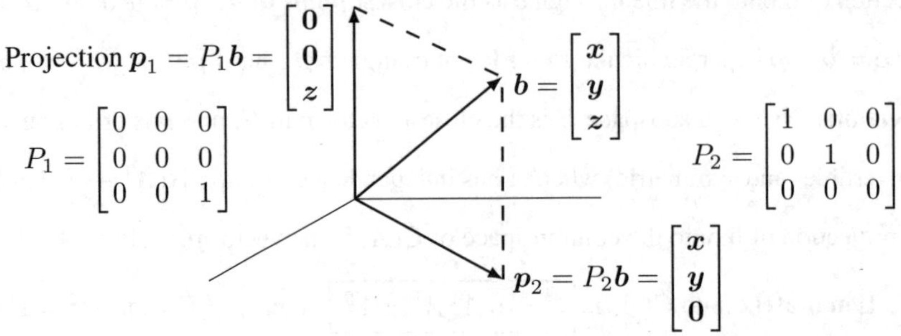

Figure 4.2: Projections $\boldsymbol{p}_1 = P_1\boldsymbol{b}$ and $\boldsymbol{p}_2 = P_2\boldsymbol{b}$ with $P_1 + P_2 = I$ and $P_1P_2 = 0$.

More than just orthogonal, the line and plane are orthogonal ***complements***. Their dimensions add to $1 + 2 = 3$. Every vector $\boldsymbol{b}$ in the whole space is the sum of its parts in the two subspaces. The projections $(0, 0, z)$ and $(x, y, 0)$ are exactly those two parts of $\boldsymbol{b}$:

$$\text{The vectors give } \boldsymbol{p}_1 + \boldsymbol{p}_2 = \boldsymbol{b}. \qquad \text{The matrices give } P_1 + P_2 = I \text{ and } P_1P_2 = 0. \tag{1}$$

This is perfect. Our goal is reached—for this example. We have the same goal for any line and any plane and any n-dimensional subspace in m dimensions. The object is to find the part $\boldsymbol{p}$ in each subspace, and the projection matrix P that produces that part $\boldsymbol{p} = P\boldsymbol{b}$. Every subspace of $\mathbf{R}^m$ has its own m by m projection matrix P.

The best description of a subspace is a basis. We put the basis vectors into the columns of A. ***Now we are projecting onto the column space of A***! Certainly the z axis is the column space of the 3 by 1 matrix A_1. The xy plane is the column space of A_2. That plane is *also* the column space of A_3 (a subspace has many bases). So $\boldsymbol{p}_2 = \boldsymbol{p}_3$ and $P_2 = P_3$.

$$A_1 = \begin{bmatrix} 0 \\ 0 \\ 1 \end{bmatrix} \quad \text{and} \quad A_2 = \begin{bmatrix} 1 & 0 \\ 0 & 1 \\ 0 & 0 \end{bmatrix} \quad \text{and} \quad A_3 = \begin{bmatrix} 1 & 2 \\ 2 & 3 \\ 0 & 0 \end{bmatrix}.$$

Our problem is ***to project any b onto the column space of any m by n matrix***. Start with a line (dimension $n = 1$). The matrix A will have only one column. Call it $\boldsymbol{a}$.

Projection Onto a Line

A line goes through the origin in the direction of $\boldsymbol{a} = (a_1, \ldots, a_m)$. Along that line, we want the point $\boldsymbol{p}$ closest to $\boldsymbol{b} = (b_1, \ldots, b_m)$. The key to projection is orthogonality: ***The line from b to p is perpendicular to the vector a***. This is the dotted line marked $\boldsymbol{e} = \boldsymbol{b} - \boldsymbol{p}$ for the error on the left side of Figure 4.3. We now compute $\boldsymbol{p}$ by algebra.

The projection $\boldsymbol{p}$ will be some multiple of $\boldsymbol{a}$. Call it $\boldsymbol{p} = \widehat{\boldsymbol{x}}\boldsymbol{a}$ = "x hat" times $\boldsymbol{a}$. Computing this number $\widehat{\boldsymbol{x}}$ will give the vector $\boldsymbol{p}$. Then from the formula for $\boldsymbol{p}$, we will read off the projection matrix P. These three steps will lead to all projection matrices: **First find $\widehat{\boldsymbol{x}}$, then find the vector $\boldsymbol{p}$, then find the matrix $\boldsymbol{P}$.**

The dashed line $\boldsymbol{b} - \boldsymbol{p}$ is the "error" $\boldsymbol{e} = \boldsymbol{b} - \widehat{\boldsymbol{x}}\boldsymbol{a}$. It is perpendicular to $\boldsymbol{a}$—this will determine $\widehat{\boldsymbol{x}}$. **Use the fact that $\boldsymbol{b} - \widehat{\boldsymbol{x}}\boldsymbol{a}$ is perpendicular to $\boldsymbol{a}$: their dot product is zero.**

Projecting $\boldsymbol{b}$ onto $\boldsymbol{a}$ with error $\boldsymbol{e} = \boldsymbol{b} - \widehat{\boldsymbol{x}}\boldsymbol{a}$ $\boldsymbol{a} \cdot (\boldsymbol{b} - \widehat{\boldsymbol{x}}\boldsymbol{a}) = 0$ or $\boldsymbol{a} \cdot \boldsymbol{b} - \widehat{\boldsymbol{x}}\boldsymbol{a} \cdot \boldsymbol{a} = 0$	$\widehat{\boldsymbol{x}} = \dfrac{\boldsymbol{a} \cdot \boldsymbol{b}}{\boldsymbol{a} \cdot \boldsymbol{a}} = \dfrac{\boldsymbol{a}^{\mathrm{T}}\boldsymbol{b}}{\boldsymbol{a}^{\mathrm{T}}\boldsymbol{a}}.$	(2)

The multiplication $\boldsymbol{a}^{\mathrm{T}}\boldsymbol{b}$ is the same as $\boldsymbol{a} \cdot \boldsymbol{b}$. Using the transpose is better, because it applies also to matrices. Our formula $\widehat{\boldsymbol{x}} = \boldsymbol{a}^{\mathrm{T}}\boldsymbol{b}/\boldsymbol{a}^{\mathrm{T}}\boldsymbol{a}$ gives the projection $\boldsymbol{p} = \widehat{\boldsymbol{x}}\boldsymbol{a}$.

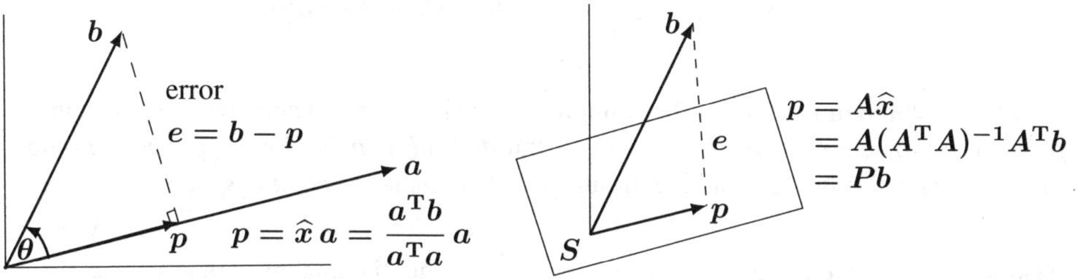

Figure 4.3: The projection $\boldsymbol{p}$ of a vector $\boldsymbol{b}$ onto a line and onto $\boldsymbol{S}$ = column space of A.

> **The projection of $\boldsymbol{b}$ onto the line through $\boldsymbol{a}$ is the vector $\boldsymbol{p} = \widehat{\boldsymbol{x}}\boldsymbol{a} = \dfrac{\boldsymbol{a}^{\mathrm{T}}\boldsymbol{b}}{\boldsymbol{a}^{\mathrm{T}}\boldsymbol{a}}\boldsymbol{a}$.**
>
> Special case 1: If $\boldsymbol{b} = \boldsymbol{a}$ then $\widehat{\boldsymbol{x}} = 1$. The projection of $\boldsymbol{a}$ onto $\boldsymbol{a}$ is itself. $P\boldsymbol{a} = \boldsymbol{a}$.
>
> Special case 2: If $\boldsymbol{b}$ is perpendicular to $\boldsymbol{a}$ then $\boldsymbol{a}^{\mathrm{T}}\boldsymbol{b} = 0$. The projection is $\boldsymbol{p} = \boldsymbol{0}$.

Example 1 Project $\boldsymbol{b} = \begin{bmatrix} 1 \\ 1 \\ 1 \end{bmatrix}$ onto $\boldsymbol{a} = \begin{bmatrix} 1 \\ 2 \\ 2 \end{bmatrix}$ to find $\boldsymbol{p} = \widehat{\boldsymbol{x}}\boldsymbol{a}$ in Figure 4.3.

Solution The number $\widehat{\boldsymbol{x}}$ is the ratio of $\boldsymbol{a}^{\mathrm{T}}\boldsymbol{b} = 5$ to $\boldsymbol{a}^{\mathrm{T}}\boldsymbol{a} = 9$. So the projection is $\boldsymbol{p} = \dfrac{5}{9}\boldsymbol{a}$.

The error vector between $\boldsymbol{b}$ and $\boldsymbol{p}$ is $\boldsymbol{e} = \boldsymbol{b} - \boldsymbol{p}$. Those vectors $\boldsymbol{p}$ and $\boldsymbol{e}$ will add to $\boldsymbol{b} = (\boldsymbol{1}, \boldsymbol{1}, \boldsymbol{1})$:

$$\boldsymbol{p} = \frac{5}{9}\boldsymbol{a} = \left(\frac{5}{9}, \frac{10}{9}, \frac{10}{9}\right) \quad \text{and} \quad \boldsymbol{e} = \boldsymbol{b} - \boldsymbol{p} = \left(\frac{4}{9}, -\frac{1}{9}, -\frac{1}{9}\right).$$

The error $\boldsymbol{e}$ should be perpendicular to $\boldsymbol{a} = (1, 2, 2)$ and it is: $\boldsymbol{e}^{\mathrm{T}}\boldsymbol{a} = \frac{4}{9} - \frac{2}{9} - \frac{2}{9} = 0$.

Look at the right triangle of $\boldsymbol{b}$, $\boldsymbol{p}$, and $\boldsymbol{e}$. The vector $\boldsymbol{b}$ is split into two parts—its component along the line is $\boldsymbol{p}$, its perpendicular part is $\boldsymbol{e}$. Those two sides $\boldsymbol{p}$ and $\boldsymbol{e}$ have length $\|\boldsymbol{p}\| = \|\boldsymbol{b}\|\cos\theta$ and $\|\boldsymbol{e}\| = \|\boldsymbol{b}\|\sin\theta$. Trigonometry matches the dot product:

$$\boldsymbol{p} = \frac{\boldsymbol{a}^{\mathrm{T}}\boldsymbol{b}}{\boldsymbol{a}^{\mathrm{T}}\boldsymbol{a}}\boldsymbol{a} \quad \text{has length} \quad \boxed{\|\boldsymbol{p}\|} = \frac{\|\boldsymbol{a}\|\,\|\boldsymbol{b}\|\cos\theta}{\|\boldsymbol{a}\|^2}\|\boldsymbol{a}\| = \boxed{\|\boldsymbol{b}\|\cos\theta.} \tag{3}$$

The dot product is a lot simpler than getting involved with $\cos\theta$ and the length of $\boldsymbol{b}$. The example has square roots in $\cos\theta = 5/3\sqrt{3}$ and $\|\boldsymbol{b}\| = \sqrt{3}$. There are no square roots in the projection $\boldsymbol{p} = 5\boldsymbol{a}/9$. The good way to $5/9$ is $\boldsymbol{a}^{\mathrm{T}}\boldsymbol{b}/\boldsymbol{a}^{\mathrm{T}}\boldsymbol{a}$.

Now comes the ***projection matrix***. In the formula for $\boldsymbol{p}$, what matrix is multiplying $\boldsymbol{b}$? You can see the matrix better if the number $\widehat{\boldsymbol{x}}$ is on the right side of $\boldsymbol{a}$:

Projection matrix P	$\boldsymbol{p} = \boldsymbol{a}\widehat{\boldsymbol{x}} = \boldsymbol{a}\dfrac{\boldsymbol{a}^{\mathrm{T}}\boldsymbol{b}}{\boldsymbol{a}^{\mathrm{T}}\boldsymbol{a}} = P\boldsymbol{b}$ ***when the matrix is*** $P = \dfrac{\boldsymbol{a}\boldsymbol{a}^{\mathrm{T}}}{\boldsymbol{a}^{\mathrm{T}}\boldsymbol{a}}$.

P is a column times a row! The column is $\boldsymbol{a}$, the row is $\boldsymbol{a}^{\mathrm{T}}$. Then divide by the number $\boldsymbol{a}^{\mathrm{T}}\boldsymbol{a}$. The projection matrix P is m by m, but ***its rank is one***. We are projecting onto a one-dimensional subspace, the line through $\boldsymbol{a}$. *That line is the column space of P.*

Example 2 Find the projection matrix $P = \dfrac{\boldsymbol{a}\boldsymbol{a}^{\mathrm{T}}}{\boldsymbol{a}^{\mathrm{T}}\boldsymbol{a}}$ onto the line through $\boldsymbol{a} = \begin{bmatrix}\mathbf{1}\\\mathbf{2}\\\mathbf{2}\end{bmatrix}$.

Solution Multiply column $\boldsymbol{a}$ times row $\boldsymbol{a}^{\mathrm{T}}$ and divide by $\boldsymbol{a}^{\mathrm{T}}\boldsymbol{a} = 9$:

$$\textbf{Projection matrix} \qquad P = \frac{\boldsymbol{a}\boldsymbol{a}^{\mathrm{T}}}{\boldsymbol{a}^{\mathrm{T}}\boldsymbol{a}} = \frac{1}{9}\begin{bmatrix}1\\2\\2\end{bmatrix}\begin{bmatrix}1 & 2 & 2\end{bmatrix} = \frac{1}{9}\begin{bmatrix}1 & 2 & 2\\2 & 4 & 4\\2 & 4 & 4\end{bmatrix}.$$

This matrix projects *any* vector $\boldsymbol{b}$ onto $\boldsymbol{a}$. Check $\boldsymbol{p} = P\boldsymbol{b}$ for $\boldsymbol{b} = (1,1,1)$ in Example 1:

$$\boldsymbol{p} = P\boldsymbol{b} = \frac{1}{9}\begin{bmatrix}1 & 2 & 2\\2 & 4 & 4\\2 & 4 & 4\end{bmatrix}\begin{bmatrix}1\\1\\1\end{bmatrix} = \frac{1}{9}\begin{bmatrix}\mathbf{5}\\\mathbf{10}\\\mathbf{10}\end{bmatrix} \quad \text{with error } \boldsymbol{e} = \boldsymbol{b} - \boldsymbol{p} = \begin{bmatrix}1\\1\\1\end{bmatrix} - \frac{1}{9}\begin{bmatrix}5\\10\\10\end{bmatrix} = \frac{\mathbf{1}}{\mathbf{9}}\begin{bmatrix}\mathbf{4}\\\mathbf{-1}\\\mathbf{-1}\end{bmatrix}.$$

If the vector $\boldsymbol{a}$ is doubled, the matrix P stays the same! It still projects onto the same line. If the matrix is squared, P^2 equals P. ***Projecting a second time doesn't change anything*, so $\boldsymbol{P^2 = P}$.** The diagonal entries of P add up to $\frac{1}{9}(1+4+4) = 1$. And $\boldsymbol{e}^{\mathrm{T}}\boldsymbol{p} = 0$.

The matrix $I - P$ should be a projection too. It produces the other side $\boldsymbol{e}$ of the triangle.

When P projects onto one subspace, $I - P$ projects onto the perpendicular subspace. Then $I - P$ projects onto the plane perpendicular to $\boldsymbol{a}$—the orthogonal complement.

Now we move beyond projection onto a line. Projecting onto an n-dimensional subspace of $\mathbf{R}^m$ takes more effort. The crucial formulas will be collected in equations (5)–(6)–(7). Basically you need to remember those three equations.

Projection Onto a Subspace

Start with n vectors $\boldsymbol{a}_1, \ldots, \boldsymbol{a}_n$ in $\mathbf{R}^m$. Assume that these $\boldsymbol{a}$'s are linearly independent.

Problem: Find the combination $\boldsymbol{p} = \widehat{x}_1\boldsymbol{a}_1 + \cdots + \widehat{x}_n\boldsymbol{a}_n$ closest to a given vector $\boldsymbol{b}$. We are projecting each $\boldsymbol{b}$ in $\mathbf{R}^m$ onto the n-dimensional subspace spanned by the $\boldsymbol{a}$'s.

With $n = 1$ (one vector $\boldsymbol{a}_1$) this is projection onto a line. The line is the column space of A, which has just one column. In general the matrix A has n columns $\boldsymbol{a}_1, \ldots, \boldsymbol{a}_n$.

The combinations in $\mathbf{R}^m$ are the vectors $A\boldsymbol{x}$ in the column space. We are looking for the particular combination $\boldsymbol{p} = A\widehat{\boldsymbol{x}}$ (***the projection***) that is closest to $\boldsymbol{b}$. The hat over $\widehat{\boldsymbol{x}}$ indicates the *best* choice $\widehat{\boldsymbol{x}}$, to give the closest vector in the column space. That choice is $\widehat{x} = \boldsymbol{a}^{\mathrm{T}}\boldsymbol{b}/\boldsymbol{a}^{\mathrm{T}}\boldsymbol{a}$ when $n = 1$. For $n > 1$, the best $\widehat{\boldsymbol{x}} = (\widehat{x}_1, \ldots, \widehat{x}_n)$ is to be found now.

We compute projections onto n-dimensional subspaces in three steps as before: ***Find the vector $\widehat{\boldsymbol{x}}$ in (S), the projection $\boldsymbol{p} = A\widehat{\boldsymbol{x}}$ in (C), and the projection matrix P in (T).***

The key is in the geometry! The dotted line in Figure 4.3 went from $\boldsymbol{b}$ to the nearest point $A\widehat{\boldsymbol{x}}$ in the subspace. ***This error vector $\boldsymbol{b} - A\widehat{\boldsymbol{x}}$ is perpendicular to the subspace.*** The error $\boldsymbol{b} - A\widehat{\boldsymbol{x}}$ makes a right angle with all the vectors $\boldsymbol{a}_1, \ldots, \boldsymbol{a}_n$ in the subspace. Those n right angles give n equations for $\widehat{\boldsymbol{x}}$:

$$\begin{matrix} \boldsymbol{a}_1^{\mathrm{T}}(\boldsymbol{b} - A\widehat{\boldsymbol{x}}) = 0 \\ \vdots \\ \boldsymbol{a}_n^{\mathrm{T}}(\boldsymbol{b} - A\widehat{\boldsymbol{x}}) = 0 \end{matrix} \quad \text{or} \quad \begin{bmatrix} - \boldsymbol{a}_1^{\mathrm{T}} - \\ \vdots \\ - \boldsymbol{a}_n^{\mathrm{T}} - \end{bmatrix} \begin{bmatrix} \\ \boldsymbol{b} - A\widehat{\boldsymbol{x}} \\ \\ \end{bmatrix} = \begin{bmatrix} \\ \mathbf{0} \\ \\ \end{bmatrix} \quad \text{or} \quad A^{\mathrm{T}}A\widehat{\boldsymbol{x}} = A^{\mathrm{T}}\boldsymbol{b}. \tag{4}$$

The matrix with those rows $\boldsymbol{a}_i^{\mathrm{T}}$ is A^{T}. The n equations are exactly $A^{\mathrm{T}}(\boldsymbol{b} - A\widehat{\boldsymbol{x}}) = \mathbf{0}$.

Rewrite $A^{\mathrm{T}}(\boldsymbol{b} - A\widehat{\boldsymbol{x}}) = \mathbf{0}$ in its famous form $A^{\mathrm{T}}A\widehat{\boldsymbol{x}} = A^{\mathrm{T}}\boldsymbol{b}$. This is the equation for $\widehat{\boldsymbol{x}}$, and the coefficient matrix is $A^{\mathrm{T}}A$. Now we can find $\widehat{\boldsymbol{x}}$ and $\boldsymbol{p}$ and P, in that order.

The next three equations (5), (6), (7) are the keys to projection.

The combination $\boldsymbol{p} = \widehat{x}_1\boldsymbol{a}_1 + \cdots + \widehat{x}_n\boldsymbol{a}_n = A\widehat{\boldsymbol{x}}$ that is closest to $\boldsymbol{b}$ comes from $\widehat{\boldsymbol{x}}$:

$$\textbf{Find } \widehat{\boldsymbol{x}}\ (n \times 1) \quad A^{\mathrm{T}}(\boldsymbol{b} - A\widehat{\boldsymbol{x}}) = \mathbf{0} \quad \text{or} \quad A^{\mathrm{T}}A\widehat{\boldsymbol{x}} = A^{\mathrm{T}}\boldsymbol{b}. \tag{5}$$

This symmetric matrix $A^{\mathrm{T}}A$ is n by n. It is invertible if the $\boldsymbol{a}$'s are independent. The solution is $\widehat{\boldsymbol{x}} = (A^{\mathrm{T}}A)^{-1}A^{\mathrm{T}}\boldsymbol{b}$. The ***projection*** of $\boldsymbol{b}$ onto the subspace is $\boldsymbol{p}$:

$$\textbf{Find } \boldsymbol{p}\ (m \times 1) \qquad \boldsymbol{p} = A\widehat{\boldsymbol{x}} = A(A^{\mathrm{T}}A)^{-1}A^{\mathrm{T}}\boldsymbol{b}. \tag{6}$$

The ***projection matrix P*** is multiplying $\boldsymbol{b}$ in (6). It has four A's:

$$\textbf{Find } \boldsymbol{P}\ (m \times m) \qquad \boldsymbol{P} = \boldsymbol{A}(\boldsymbol{A}^{\mathrm{T}}\boldsymbol{A})^{-1}\boldsymbol{A}^{\mathrm{T}}. \tag{7}$$

Compare with projection onto a line, when A has only one column: $A^{\mathrm{T}}A = \boldsymbol{a}^{\mathrm{T}}\boldsymbol{a}$ is 1 by 1.

$$\textbf{For } n=1 \qquad \widehat{\boldsymbol{x}}=\frac{\boldsymbol{a}^{\mathrm{T}}\boldsymbol{b}}{\boldsymbol{a}^{\mathrm{T}}\boldsymbol{a}} \quad\text{and}\quad \boldsymbol{p}=\boldsymbol{a}\frac{\boldsymbol{a}^{\mathrm{T}}\boldsymbol{b}}{\boldsymbol{a}^{\mathrm{T}}\boldsymbol{a}} \quad\text{and}\quad P=\frac{\boldsymbol{a}\boldsymbol{a}^{\mathrm{T}}}{\boldsymbol{a}^{\mathrm{T}}\boldsymbol{a}}. \tag{8}$$

Those formulas are identical with (5) and (6) and (7). The number $\boldsymbol{a}^{\mathrm{T}}\boldsymbol{a}$ becomes the matrix $A^{\mathrm{T}}A$. When it is a number, we divide by it. When it is a matrix, we invert it. The linear independence of the columns $\boldsymbol{a}_1,\ldots,\boldsymbol{a}_n$ guarantees that $A^{\mathrm{T}}A$ is invertible.

The key step was $A^{\mathrm{T}}(\boldsymbol{b}-A\widehat{\boldsymbol{x}})=\mathbf{0}$. We used geometry ($\boldsymbol{e}$ is orthogonal to each $\boldsymbol{a}$). *Linear algebra gives this "normal equation" too*, in a very quick and beautiful way:

1. Our subspace $\mathbf{C}(A)$ is the column space of A.

2. The error $\boldsymbol{e}=\boldsymbol{b}-A\widehat{\boldsymbol{x}}$ is in the perpendicular subspace $\mathbf{N}(A^{\mathrm{T}})$.

This means $A^{\mathrm{T}}(\boldsymbol{b}-A\widehat{\boldsymbol{x}})=\mathbf{0}$. The left nullspace $\mathbf{N}(A^{\mathrm{T}})$ is important in projections. That nullspace contains the error vector $\boldsymbol{e}=\boldsymbol{b}-A\widehat{\boldsymbol{x}}$. The vector $\boldsymbol{b}$ is split into the projection $\boldsymbol{p}$ and the error $\boldsymbol{e}=\boldsymbol{b}-\boldsymbol{p}$. *Projection produces a right triangle with sides* $\boldsymbol{p}$, $\boldsymbol{e}$, *and* $\boldsymbol{b}$.

Example 3 If $A=\begin{bmatrix}\mathbf{1}&\mathbf{0}\\\mathbf{1}&\mathbf{1}\\\mathbf{1}&\mathbf{2}\end{bmatrix}$ and $\boldsymbol{b}=\begin{bmatrix}\mathbf{6}\\\mathbf{0}\\\mathbf{0}\end{bmatrix}$ find $\widehat{\boldsymbol{x}}$ and $\boldsymbol{p}$ and P.

Solution Compute the square matrix $A^{\mathrm{T}}A$ and also the vector $A^{\mathrm{T}}\boldsymbol{b}$:

$$A^{\mathrm{T}}A=\begin{bmatrix}1&1&1\\0&1&2\end{bmatrix}\begin{bmatrix}1&0\\1&1\\1&2\end{bmatrix}=\begin{bmatrix}3&3\\3&5\end{bmatrix} \quad\text{and}\quad A^{\mathrm{T}}\boldsymbol{b}=\begin{bmatrix}1&1&1\\0&1&2\end{bmatrix}\begin{bmatrix}6\\0\\0\end{bmatrix}=\begin{bmatrix}6\\0\end{bmatrix}.$$

Now solve the normal equation $A^{\mathrm{T}}A\widehat{\boldsymbol{x}}=A^{\mathrm{T}}\boldsymbol{b}$ to find $\widehat{\boldsymbol{x}}$:

$$\begin{bmatrix}3&3\\3&5\end{bmatrix}\begin{bmatrix}\widehat{x}_1\\\widehat{x}_2\end{bmatrix}=\begin{bmatrix}6\\0\end{bmatrix} \quad\text{gives}\quad \widehat{\boldsymbol{x}}=\begin{bmatrix}\widehat{x}_1\\\widehat{x}_2\end{bmatrix}=\begin{bmatrix}\mathbf{5}\\\mathbf{-3}\end{bmatrix}. \tag{9}$$

The combination $\boldsymbol{p}=A\widehat{\boldsymbol{x}}$ is the projection of $\boldsymbol{b}$ onto the column space of A:

$$\boldsymbol{p}=\mathbf{5}\begin{bmatrix}1\\1\\1\end{bmatrix}-\mathbf{3}\begin{bmatrix}0\\1\\2\end{bmatrix}=\begin{bmatrix}5\\2\\-1\end{bmatrix}. \quad\text{The error is}\quad \boldsymbol{e}=\boldsymbol{b}-\boldsymbol{p}=\begin{bmatrix}1\\-2\\1\end{bmatrix}. \tag{10}$$

Two checks on the calculation. First, the error $\boldsymbol{e}=(1,-2,1)$ is perpendicular to both columns $(1,1,1)$ and $(0,1,2)$. Second, the matrix P times $\boldsymbol{b}=(6,0,0)$ correctly gives $\boldsymbol{p}=(5,2,-1)$. That solves the problem for one particular $\boldsymbol{b}$, as soon as we find P.

The projection matrix is $P=A(A^{\mathrm{T}}A)^{-1}A^{\mathrm{T}}$. The determinant of $A^{\mathrm{T}}A$ is $15-9=6$; then its 2 by 2 inverse is easy. Multiply A times $(A^{\mathrm{T}}A)^{-1}$ times A^{T} to reach P:

$$(A^{\mathrm{T}}A)^{-1}=\frac{1}{6}\begin{bmatrix}5&-3\\-3&3\end{bmatrix} \quad\text{and}\quad P=\frac{1}{6}\begin{bmatrix}5&2&-1\\2&2&2\\-1&2&5\end{bmatrix}. \tag{11}$$

We must have $P^2=P$, because a second projection doesn't change the first projection.

Warning The matrix $P = A(A^{\mathrm{T}}A)^{-1}A^{\mathrm{T}}$ is deceptive. You might try to split $(A^{\mathrm{T}}A)^{-1}$ into A^{-1} times $(A^{\mathrm{T}})^{-1}$. If you make that mistake, and substitute it into P, you will find $P = AA^{-1}(A^{\mathrm{T}})^{-1}A^{\mathrm{T}}$. Apparently everything cancels to produce $P = I$. We want to say why this is wrong.

The matrix A is rectangular. It has no inverse matrix. We cannot split $(A^{\mathrm{T}}A)^{-1}$ into A^{-1} times $(A^{\mathrm{T}})^{-1}$ because there is no A^{-1} in the first place, when $m > n$.

In our experience, a problem that involves a rectangular matrix almost always leads to $A^{\mathrm{T}}A$. When A (m by n) has independent columns, $A^{\mathrm{T}}A$ is invertible. This fact is so crucial that we state it and prove it again (Section 3.5 proved it first).

$A^{\mathrm{T}}A$ is invertible if and only if A has linearly independent columns.

Proof $A^{\mathrm{T}}A$ is a square matrix (n by n). For every matrix A, we will now show that $A^{\mathrm{T}}A$ ***has the same nullspace as*** A. When the columns of A are linearly independent, its nullspace contains only the zero vector. Then $A^{\mathrm{T}}A$, with this same nullspace, is invertible.

Let A be any matrix. If $\boldsymbol{x}$ is in its nullspace, then $A\boldsymbol{x} = \mathbf{0}$. Multiplying by A^{T} gives $A^{\mathrm{T}}A\boldsymbol{x} = \mathbf{0}$. So $\boldsymbol{x}$ is also in the nullspace of $A^{\mathrm{T}}A$.

Now start with the nullspace of $A^{\mathrm{T}}A$. **From** $A^{\mathrm{T}}A\boldsymbol{x} = \mathbf{0}$ **we must prove** $A\boldsymbol{x} = \mathbf{0}$. We can't multiply by $(A^{\mathrm{T}})^{-1}$, which generally doesn't exist. Just multiply by $\boldsymbol{x}^{\mathrm{T}}$:

$$(\boldsymbol{x}^{\mathrm{T}})A^{\mathrm{T}}A\boldsymbol{x} = 0 \quad \text{or} \quad (A\boldsymbol{x})^{\mathrm{T}}(A\boldsymbol{x}) = 0 \quad \text{or} \quad \|A\boldsymbol{x}\|^2 = 0. \tag{12}$$

We have shown: If $A^{\mathrm{T}}A\boldsymbol{x} = \mathbf{0}$ then $A\boldsymbol{x}$ has length zero. Therefore $A\boldsymbol{x} = \mathbf{0}$. Every vector $\boldsymbol{x}$ in one nullspace is in the other nullspace. If $A^{\mathrm{T}}A$ has dependent columns, so has A. If $A^{\mathrm{T}}A$ has independent columns, so has A. This is the good case: $A^{\mathrm{T}}A$ is invertible.

When A has independent columns, $A^{\mathrm{T}}A$ is square, symmetric, and invertible.

To repeat for emphasis: $A^{\mathrm{T}}A$ is (n by m) times (m by n). Then $A^{\mathrm{T}}A$ is square (n by n). It is symmetric, because its transpose is $(A^{\mathrm{T}}A)^{\mathrm{T}} = A^{\mathrm{T}}(A^{\mathrm{T}})^{\mathrm{T}}$ which equals $A^{\mathrm{T}}A$. We just proved that $A^{\mathrm{T}}A$ is invertible—provided A has independent columns. Watch the difference between dependent and independent columns:

$$\underset{A^{\mathrm{T}}}{\begin{bmatrix}1&1&0\\2&2&0\end{bmatrix}}\underset{\substack{A\\ \text{dependent}}}{\begin{bmatrix}1&2\\1&2\\0&0\end{bmatrix}}=\underset{\substack{A^{\mathrm{T}}A\\ \text{singular}}}{\begin{bmatrix}2&4\\4&8\end{bmatrix}} \qquad \underset{A^{\mathrm{T}}}{\begin{bmatrix}1&1&0\\2&2&1\end{bmatrix}}\underset{\substack{A\\ \text{indep.}}}{\begin{bmatrix}1&2\\1&2\\0&1\end{bmatrix}}=\underset{\substack{A^{\mathrm{T}}A\\ \text{invertible}}}{\begin{bmatrix}2&4\\4&9\end{bmatrix}}$$

Very brief summary To find the projection $\boldsymbol{p} = \widehat{x}_1\boldsymbol{a}_1 + \cdots + \widehat{x}_n\boldsymbol{a}_n$, solve $A^{\mathrm{T}}A\widehat{\boldsymbol{x}} = A^{\mathrm{T}}\boldsymbol{b}$. This gives $\widehat{\boldsymbol{x}}$. The projection is $\boldsymbol{p} = A\widehat{\boldsymbol{x}}$ and the error is $\boldsymbol{e} = \boldsymbol{b} - \boldsymbol{p} = \boldsymbol{b} - A\widehat{\boldsymbol{x}}$. The projection matrix $P = A(A^{\mathrm{T}}A)^{-1}A^{\mathrm{T}}$ gives $\boldsymbol{p} = P\boldsymbol{b}$. P is invertible only if $P = I$.

This matrix satisfies $P^2 = P$. ***The distance from b to the subspace* $\mathbf{C}(A)$ *is* $\|e\|$.**

■ REVIEW OF THE KEY IDEAS ■

1. The projection of $\boldsymbol{b}$ onto the line through $\boldsymbol{a}$ is $\boldsymbol{p} = \boldsymbol{a}\widehat{x} = \boldsymbol{a}(\boldsymbol{a}^{\mathrm{T}}\boldsymbol{b}/\boldsymbol{a}^{\mathrm{T}}\boldsymbol{a})$.
2. The rank one projection matrix $P = \boldsymbol{a}\boldsymbol{a}^{\mathrm{T}}/\boldsymbol{a}^{\mathrm{T}}\boldsymbol{a}$ multiplies $\boldsymbol{b}$ to produce $\boldsymbol{p}$.
3. Projecting $\boldsymbol{b}$ onto a subspace leaves $\boldsymbol{e} = \boldsymbol{b} - \boldsymbol{p}$ perpendicular to the subspace.
4. When A has full rank n, the equation $A^{\mathrm{T}}A\widehat{\boldsymbol{x}} = A^{\mathrm{T}}\boldsymbol{b}$ leads to $\widehat{\boldsymbol{x}}$ and $\boldsymbol{p} = A\widehat{\boldsymbol{x}}$.
5. The projection matrix $P = A(A^{\mathrm{T}}A)^{-1}A^{\mathrm{T}}$ has $P^{\mathrm{T}} = P$ and $P^2 = P$ and $P\boldsymbol{b} = \boldsymbol{p}$.

■ WORKED EXAMPLES ■

4.2 A Project the vector $\boldsymbol{b} = (3, 4, 4)$ onto the line through $\boldsymbol{a} = (2, 2, 1)$ and then onto the plane that also contains $\boldsymbol{a}^* = (1, 0, 0)$. Check that the first error vector $\boldsymbol{b} - \boldsymbol{p}$ is perpendicular to $\boldsymbol{a}$, and the second error vector $\boldsymbol{e}^* = \boldsymbol{b} - \boldsymbol{p}^*$ is also perpendicular to $\boldsymbol{a}^*$.

Find the 3 by 3 projection matrix P onto that plane of $\boldsymbol{a}$ and $\boldsymbol{a}^*$. Find a vector whose *projection onto the plane is* $\boldsymbol{p} = \boldsymbol{0}$.

Solution The projection of $\boldsymbol{b} = (3, 4, 4)$ onto the line through $\boldsymbol{a} = (2, 2, 1)$ is $\boldsymbol{p} = 2\boldsymbol{a}$:

Onto a line $$\boldsymbol{p} = \frac{\boldsymbol{a}^{\mathrm{T}}\boldsymbol{b}}{\boldsymbol{a}^{\mathrm{T}}\boldsymbol{a}}\boldsymbol{a} = \frac{18}{9}(2, 2, 1) = (4, 4, 2) = 2\boldsymbol{a}.$$

The error vector $\boldsymbol{e} = \boldsymbol{b} - \boldsymbol{p} = (-1, 0, 2)$ is perpendicular to $\boldsymbol{a} = (2, 2, 1)$. So $\boldsymbol{p}$ is correct.

The plane of $\boldsymbol{a} = (2, 2, 1)$ and $\boldsymbol{a}^* = (1, 0, 0)$ is the column space of $A = [\boldsymbol{a}\ \boldsymbol{a}^*]$:

$$A = \begin{bmatrix} 2 & 1 \\ 2 & 0 \\ 1 & 0 \end{bmatrix} \quad A^{\mathrm{T}}A = \begin{bmatrix} 9 & 2 \\ 2 & 1 \end{bmatrix} \quad (A^{\mathrm{T}}A)^{-1} = \frac{1}{5}\begin{bmatrix} 1 & -2 \\ -2 & 9 \end{bmatrix} \quad P = \begin{bmatrix} 1 & 0 & 0 \\ 0 & .8 & .4 \\ 0 & .4 & .2 \end{bmatrix}$$

Now $\boldsymbol{p}^* = P\boldsymbol{b} = (3, 4.8, 2.4)$. The error $\boldsymbol{e}^* = \boldsymbol{b} - \boldsymbol{p}^* = (0, -.8, 1.6)$ is perpendicular to $\boldsymbol{a}$ and $\boldsymbol{a}^*$. This $\boldsymbol{e}^*$ is in the nullspace of P. *Its projection is zero*! Note $\boldsymbol{P^2 = P = P^{\mathrm{T}}}$.

4.2 B Suppose your pulse is measured at 70 beats per minute, then at $x = 80$, then at $x = 120$. Those three equations $Ax = \boldsymbol{b}$ in one unknown have $A^{\mathrm{T}} = [1\ 1\ 1]$ and $\boldsymbol{b} = (70, 80, 120)$. ***The best $\widehat{x}$ is the _____ of 70, 80, 120***. Use calculus and projection:

1. Minimize $E = (x - 70)^2 + (x - 80)^2 + (x - 120)^2$ by solving $dE/dx = 0$.
2. Project $\boldsymbol{b} = (70, 80, 120)$ onto $\boldsymbol{a} = (1, 1, 1)$ to find $\widehat{x} =$ **average** $= 90$.

4.2 C Suppose you know the average $\widehat{x}_{\text{old}}$ of $b_1, b_2, \ldots, b_{999}$. When b_{1000} arrives, check that the new average is a combination of $\widehat{x}_{\text{old}}$ and the mismatch $b_{1000} - \widehat{x}_{\text{old}}$:

$$\widehat{x}_{\text{new}} = \frac{b_1 + \cdots + b_{1000}}{1000} = \frac{b_1 + \cdots + b_{999}}{999} + \frac{1}{1000}\left(b_{1000} - \frac{b_1 + \cdots + b_{999}}{999}\right).$$

This is a "**Kalman filter**" $\widehat{x}_{\text{new}} = \widehat{x}_{\text{old}} + \frac{1}{1000}(b_{1000} - \widehat{x}_{\text{old}})$ with gain matrix $\frac{1}{1000}$.

Problem Set 4.2

0 A projection satisfies $P^2 = P$. The extra symmetry condition $P^{\mathrm{T}} = P$ makes it an **"orthogonal projection"**. Examples will show the meaning of those words.

(a) Verify $P^2 = P$ and $(I-P)^2 = I-P$ for $P = \begin{bmatrix} 0 & 1 \\ 0 & 1 \end{bmatrix}$ and $I-P = \begin{bmatrix} 1 & -1 \\ 0 & 0 \end{bmatrix}$.

(b) Find vectors $\boldsymbol{v}$ and $\boldsymbol{w}$ in the column spaces of P and $I-P$. Are those spaces orthogonal?

(c) Show that if $P^2 = P$ and $P^{\mathrm{T}} = P$ then the dot product of $P\boldsymbol{v}$ with $(I-P)\boldsymbol{w}$ is zero: $\boldsymbol{v}^{\mathrm{T}} P^{\mathrm{T}} (I-P)\boldsymbol{w} = 0$. Now the projection is orthogonal.

Summary For the projection $\boldsymbol{p} = P\boldsymbol{b}$ to be perpendicular to the error $\boldsymbol{e} = \boldsymbol{b} - \boldsymbol{p}$, we need $\boldsymbol{P^2 = P = P^{\mathrm{T}}}$.

Questions 1–9 ask for projections $\boldsymbol{p}$ onto lines. Also errors $\boldsymbol{e} = \boldsymbol{b} - \boldsymbol{p}$ and matrices $\boldsymbol{P}$.

1 Project the vector $\boldsymbol{b}$ onto the line through $\boldsymbol{a}$. Check that $\boldsymbol{e}$ is perpendicular to $\boldsymbol{a}$:

$$\text{(a) } \boldsymbol{b} = \begin{bmatrix} 1 \\ 2 \\ 2 \end{bmatrix} \text{ and } \boldsymbol{a} = \begin{bmatrix} 1 \\ 1 \\ 1 \end{bmatrix} \qquad \text{(b) } \boldsymbol{b} = \begin{bmatrix} 1 \\ 3 \\ 1 \end{bmatrix} \text{ and } \boldsymbol{a} = \begin{bmatrix} -1 \\ -3 \\ -1 \end{bmatrix}.$$

2 *Draw* the projection of $\boldsymbol{b}$ onto $\boldsymbol{a}$ and also compute it from $\boldsymbol{p} = \widehat{x}\boldsymbol{a}$:

$$\text{(a) } \boldsymbol{b} = \begin{bmatrix} \cos\theta \\ \sin\theta \end{bmatrix} \text{ and } \boldsymbol{a} = \begin{bmatrix} 1 \\ 0 \end{bmatrix} \qquad \text{(b) } \boldsymbol{b} = \begin{bmatrix} 1 \\ 1 \end{bmatrix} \text{ and } \boldsymbol{a} = \begin{bmatrix} 1 \\ -1 \end{bmatrix}.$$

3 In Problem 1, find the projection matrix $P = \boldsymbol{a}\boldsymbol{a}^{\mathrm{T}}/\boldsymbol{a}^{\mathrm{T}}\boldsymbol{a}$ onto the line through each vector $\boldsymbol{a}$. Verify in both cases that $P^2 = P$. Multiply $P\boldsymbol{b}$ in each case to compute the projection $\boldsymbol{p}$.

4 Construct the projection matrices P_1 and P_2 onto the lines through the $\boldsymbol{a}$'s in Problem 2. Is it true that $(P_1 + P_2)^2 = P_1 + P_2$? This *would* be true if $P_1 P_2 = 0$.

5 Compute the projection matrices $\boldsymbol{a}\boldsymbol{a}^{\mathrm{T}}/\boldsymbol{a}^{\mathrm{T}}\boldsymbol{a}$ onto the lines through $\boldsymbol{a}_1 = (-1, 2, 2)$ and $\boldsymbol{a}_2 = (2, 2, -1)$. Multiply those projection matrices and explain why their product $P_1 P_2$ is what it is.

6 Project $\boldsymbol{b} = (1, 0, 0)$ onto the lines through $\boldsymbol{a}_1$ and $\boldsymbol{a}_2$ in Problem 5 and also onto $\boldsymbol{a}_3 = (2, -1, 2)$. Add up the three projections $\boldsymbol{p}_1 + \boldsymbol{p}_2 + \boldsymbol{p}_3$.

7 Continuing Problems 5–6, find the projection matrix P_3 onto $\boldsymbol{a}_3 = (2, -1, 2)$. Verify that $P_1 + P_2 + P_3 = I$. This is because the basis $\boldsymbol{a}_1, \boldsymbol{a}_2, \boldsymbol{a}_3$ is orthogonal!

8 Project the vector $\boldsymbol{b} = (1, 1)$ onto the lines through $\boldsymbol{a}_1 = (1, 0)$ and $\boldsymbol{a}_2 = (1, 2)$. Draw the projections $\boldsymbol{p}_1$ and $\boldsymbol{p}_2$ and add $\boldsymbol{p}_1 + \boldsymbol{p}_2$. The projections do not add to $\boldsymbol{b}$ because the $\boldsymbol{a}$'s are not orthogonal.

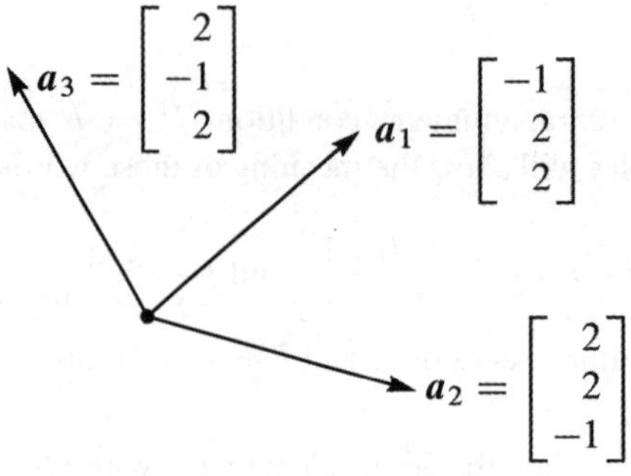

Questions 5–6–7: orthogonal $\boldsymbol{a}$'s

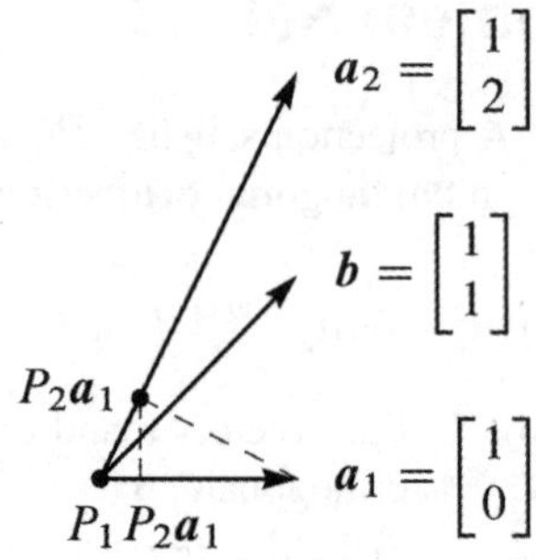

Questions 8–9–10: not orthogonal

9 In Problem 8, the projection of $\boldsymbol{b}$ onto the *plane* of $\boldsymbol{a}_1$ and $\boldsymbol{a}_2$ will equal $\boldsymbol{b}$. Find $P = A(A^{\mathrm{T}}A)^{-1}A^{\mathrm{T}}$ for $A = \begin{bmatrix} \boldsymbol{a}_1 & \boldsymbol{a}_2 \end{bmatrix} = \left[\begin{smallmatrix} 1 & 1 \\ 0 & 2 \end{smallmatrix}\right] =$ invertible matrix.

10 Project $\boldsymbol{a}_1 = (1, 0)$ onto $\boldsymbol{a}_2 = (1, 2)$. Then project the result back onto $\boldsymbol{a}_1$. Draw these projections and multiply the projection matrices $P_1 P_2$: Is this a projection?

Questions 11–21 ask for projections, and projection matrices, onto subspaces.

11 If P (m by n) projects $\mathbf{R}^m$ onto a subspace $\mathbf{S}$, what are the four fundamental subspaces for the matrix P?

12 Project $\boldsymbol{b}$ onto the column space of A by solving $A^{\mathrm{T}}A\widehat{\boldsymbol{x}} = A^{\mathrm{T}}\boldsymbol{b}$ and $\boldsymbol{p} = A\widehat{\boldsymbol{x}}$:

$$\text{(a)}\ A = \begin{bmatrix} 1 & 1 \\ 0 & 1 \\ 0 & 0 \end{bmatrix} \text{ and } \boldsymbol{b} = \begin{bmatrix} 2 \\ 3 \\ 4 \end{bmatrix} \qquad \text{(b)}\quad A = \begin{bmatrix} 1 & 1 \\ 1 & 1 \\ 0 & 1 \end{bmatrix} \text{ and } \boldsymbol{b} = \begin{bmatrix} 4 \\ 4 \\ 6 \end{bmatrix}.$$

Find $\boldsymbol{e} = \boldsymbol{b} - \boldsymbol{p}$. It should be perpendicular to the columns of A.

13 Compute the projection matrices P_1 and P_2 onto the column spaces in Problem 12. Verify that $P_1\boldsymbol{b}$ gives the first projection $\boldsymbol{p}_1$. Also verify $P_2^2 = P_2$.

14 (Quick and Recommended) Suppose A is the 4 by 4 identity matrix with its last column removed. A is 4 by 3. Project $\boldsymbol{b} = (1, 2, 3, 4)$ onto the column space of A. What shape is the projection matrix P and what is P?

15 Suppose $\boldsymbol{b}$ equals 2 times the first column of A. What is the projection of $\boldsymbol{b}$ onto the column space of A? Is $P = I$ for sure in this case? Compute $\boldsymbol{p}$ and P when $\boldsymbol{b} = (0, 2, 4)$ and the columns of A are $(0, 1, 2)$ and $(1, 2, 0)$.

16 If A is doubled, then $P = 2A(4A^{\mathrm{T}}A)^{-1}2A^{\mathrm{T}}$. This is the same as $A(A^{\mathrm{T}}A)^{-1}A^{\mathrm{T}}$. The column space of $2A$ is the same as _____. Is $\widehat{\boldsymbol{x}}$ the same for A and $2A$?

17 What linear combination of $(1,2,-1)$ and $(1,0,1)$ is closest to $\boldsymbol{b}=(2,1,1)$?

18 (*Important*) If $P^2 = P$ show that $(I-P)^2 = I-P$.When P projects onto the column space of A, $I-P$ projects onto the ______.

19 (a) If P is the 2 by 2 projection matrix onto the line through $(1,1)$, then $I-P$ is the projection matrix onto ______.

(b) If P is the 3 by 3 projection matrix onto the line through $(1,1,1)$, then $I-P$ is the projection matrix onto ______.

20 To find the projection matrix onto the plane $x-y-2z=0$, choose two vectors in that plane and make them the columns of A. The plane will be the column space of A! Then compute $P=A(A^{\mathrm{T}}A)^{-1}A^{\mathrm{T}}$.

21 To find the projection matrix P onto the same plane $x-y-2z=0$, write down a vector $\boldsymbol{e}$ that is perpendicular to that plane. Compute the projection $Q=\boldsymbol{e}\boldsymbol{e}^{\mathrm{T}}/\boldsymbol{e}^{\mathrm{T}}\boldsymbol{e}$ and then $P=I-Q$.

Questions 22–27 show that projection matrices satisfy $P^2=P$ and $P^{\mathrm{T}}=P$.

22 Multiply the matrix $P=A(A^{\mathrm{T}}A)^{-1}A^{\mathrm{T}}$ by itself. Cancel to prove that $P^2=P$. Explain why $P(P\boldsymbol{b})=$ projection of $P\boldsymbol{b}$ always equals $P\boldsymbol{b}$.

23 Prove that $P=A(A^{\mathrm{T}}A)^{-1}A^{\mathrm{T}}$ is symmetric by computing P^{T}. Remember that the inverse of a symmetric matrix is symmetric.

24 If A is square and invertible, the warning against splitting $(A^{\mathrm{T}}A)^{-1}$ does not apply. It becomes true that $AA^{-1}(A^{\mathrm{T}})^{-1}A^{\mathrm{T}}=I$. *When A is invertible, why is $P=I$? **What is the error** $\boldsymbol{e}$?*

25 The nullspace of A^{T} is ______ to the column space $\mathbf{C}(A)$. So if $A^{\mathrm{T}}\boldsymbol{b}=\mathbf{0}$, the projection of $\boldsymbol{b}$ onto $\mathbf{C}(A)$ should be $\boldsymbol{p}=$ ______. Check that $A(A^{\mathrm{T}}A)^{-1}A^{\mathrm{T}}\boldsymbol{b}=\boldsymbol{p}$.

26 The projection matrix P onto an n-dimensional subspace of $\mathbf{R}^m$ has rank $r=n$. ***Reason:*** The projections $P\boldsymbol{b}$ fill the subspace S. So S is the ______ of P.

27 If an m by m matrix has $A^2=A$ and its rank is m, how do you know that $A=I$?

28 The important fact that ends the section is this: ***If*** $A^{\mathrm{T}}A\boldsymbol{x}=\mathbf{0}$ ***then*** $A\boldsymbol{x}=\mathbf{0}$. *New Proof*: The vector $A\boldsymbol{x}$ is in the nullspace of ______. $A\boldsymbol{x}$ is always in the column space of ______. To be in both of those perpendicular spaces, $A\boldsymbol{x}$ must be zero.

29 Use $P^{\mathrm{T}}=P$ and $P^2=P$ to prove that the length squared of column 2 of P always equals the diagonal entry P_{22}. Give an example.

30 If B has rank m (full row rank, independent rows) show that BB^{T} is invertible.

Challenge Problems

31 (a) Find the projection matrix P_C onto the column space of this matrix A:

$$A = \begin{bmatrix} 3 & 6 & 6 \\ 4 & 8 & 8 \end{bmatrix}$$

(b) Find the 3 by 3 projection matrix P_R onto the row space of A. Multiply $B = P_C A P_R$. Your answer B should be a little surprising—can you explain it?

32 In $\mathbf{R}^m$, suppose I give you $\boldsymbol{b}$ and also a combination $\boldsymbol{p}$ of $\boldsymbol{a}_1, \ldots, \boldsymbol{a}_n$. How would you test to see if $\boldsymbol{p}$ is the projection of $\boldsymbol{b}$ onto the subspace spanned by the $\boldsymbol{a}$'s?

33 Suppose P_1 is the projection matrix onto the 1-dimensional subspace spanned by the first column of A. Suppose P_2 is the projection matrix onto the 2-dimensional column space of A. After thinking a little, compute the product $P_2 P_1$.

$$A = \begin{bmatrix} 1 & 0 \\ 2 & 1 \\ 0 & 1 \end{bmatrix}.$$

34 Suppose P_1 and P_2 are projection matrices ($P_i^2 = P_i = P_i^{\mathrm{T}}$). Prove this fact:

$$P_1 P_2 \text{ is a projection matrix if and only if } P_1 P_2 = P_2 P_1.$$

4.3 Least Squares Approximations

1 Solving $\boxed{A^{\mathrm{T}}A\,\widehat{\boldsymbol{x}} = A^{\mathrm{T}}\boldsymbol{b}}$ gives the projection $\boldsymbol{p} = A\widehat{\boldsymbol{x}}$ of $\boldsymbol{b}$ onto the column space of A.

2 When $A\boldsymbol{x} = \boldsymbol{b}$ has no solution, $\widehat{\boldsymbol{x}}$ is the "least-squares solution": $||A\widehat{\boldsymbol{x}} - \boldsymbol{b}||^2 =$ minimum.

3 Setting derivatives of $E = ||A\boldsymbol{x} - \boldsymbol{b}||^2$ to zero $\left(\dfrac{\partial E}{\partial x_i} = 0\right)$ also produces $A^{\mathrm{T}}A\widehat{\boldsymbol{x}} = A^{\mathrm{T}}\boldsymbol{b}$.

4 To fit points $(t_1, b_1), \ldots, (t_m, b_m)$ by a straight line, A has columns $(1, \ldots, 1)$ and $(t_1, \ldots, t_m)$.

5 In that case $A^{\mathrm{T}}A$ is the 2 by 2 matrix $\begin{bmatrix} m & \Sigma\, t_i \\ \Sigma\, t_i & \Sigma\, t_i^2 \end{bmatrix}$ and $A^{\mathrm{T}}\boldsymbol{b}$ is the vector $\begin{bmatrix} \Sigma\, b_i \\ \Sigma\, t_i b_i \end{bmatrix}$.

It often happens that $A\boldsymbol{x} = \boldsymbol{b}$ has no solution. The usual reason is: *too many equations.* The matrix A has more rows than columns. There are more equations than unknowns (m is greater than n). The n columns span a small part of m-dimensional space. Unless all measurements are perfect, $\boldsymbol{b}$ is outside that column space of A. Elimination reaches an impossible equation and stops. But we can't just stop when measurements include noise!

To repeat: We cannot always get the error $\boldsymbol{e} = \boldsymbol{b} - A\boldsymbol{x}$ down to zero. When $\boldsymbol{e}$ is zero, $\boldsymbol{x}$ is an exact solution to $A\boldsymbol{x} = \boldsymbol{b}$. *When the length of $\boldsymbol{e}$ is as small as possible, so that $||\boldsymbol{b} - A\widehat{\boldsymbol{x}}||^2$ reaches its minimum, then $\widehat{\boldsymbol{x}}$ is a least squares solution.* Our goal in this section is to compute $\widehat{\boldsymbol{x}}$ and use it. These are real problems and they need an answer.

The previous section emphasized $\boldsymbol{p}$ (the projection). This section emphasizes $\widehat{\boldsymbol{x}}$ (the least squares solution). They are connected by $\boldsymbol{p} = A\widehat{\boldsymbol{x}}$. The fundamental equation is still $A^{\mathrm{T}}A\widehat{\boldsymbol{x}} = A^{\mathrm{T}}\boldsymbol{b}$. Here is a short unofficial way to reach this "*normal equation*":

When $A\boldsymbol{x} = \boldsymbol{b}$ has no solution, multiply by A^{T} and solve $A^{\mathrm{T}}A\widehat{\boldsymbol{x}} = A^{\mathrm{T}}\boldsymbol{b}$.

Example 1 A crucial application of least squares is fitting a straight line to m points. Start with three points: *Find the closest line to the points* $(\mathbf{0}, \mathbf{6})$, $(\mathbf{1}, \mathbf{0})$, *and* $(\mathbf{2}, \mathbf{0})$.

No straight line $b = C + Dt$ goes through those three points. We are asking for two numbers C and D that satisfy three equations: $n = 2$ and $m = 3$. Here are the three equations at $t = 0, 1, 2$ to match the given values $b = 6, 0, 0$:

$\boldsymbol{t = 0}$	The first point is on the line $b = C + Dt$ if	$\boldsymbol{C + D \cdot 0 = 6}$
$\boldsymbol{t = 1}$	The second point is on the line $b = C + Dt$ if	$\boldsymbol{C + D \cdot 1 = 0}$
$\boldsymbol{t = 2}$	The third point is on the line $b = C + Dt$ if	$\boldsymbol{C + D \cdot 2 = 0}$.

This 3 by 2 system has *no solution*: $\boldsymbol{b} = (6, 0, 0)$ is not a combination of the columns $(1, 1, 1)$ and $(0, 1, 2)$. Read off A, $\boldsymbol{x}$, and $\boldsymbol{b}$ from those equations:

$$A = \begin{bmatrix} 1 & 0 \\ 1 & 1 \\ 1 & 2 \end{bmatrix} \quad \boldsymbol{x} = \begin{bmatrix} C \\ D \end{bmatrix} \quad \boldsymbol{b} = \begin{bmatrix} 6 \\ 0 \\ 0 \end{bmatrix} \quad A\boldsymbol{x} = \boldsymbol{b} \text{ is } \textit{not} \text{ solvable.}$$

The same numbers were in Example 3 in the last section. We computed $\widehat{\boldsymbol{x}} = (5, -3)$. **Those numbers are the best C and D, so $5 - 3t$ will be the best line for the 3 points.** We must connect projections to least squares, by explaining why $A^{\mathrm{T}}A\widehat{\boldsymbol{x}} = A^{\mathrm{T}}\boldsymbol{b}$.

In practical problems, there could easily be $m = 100$ points instead of $m = 3$. They don't exactly match any straight line $C + Dt$. Our numbers $6, 0, 0$ exaggerate the error so you can see e_1, e_2, and e_3 in Figure 4.6.

Minimizing the Error

How do we make the error $\boldsymbol{e} = \boldsymbol{b} - A\boldsymbol{x}$ as small as possible? This is an important question with a beautiful answer. The best $\boldsymbol{x}$ (called $\widehat{\boldsymbol{x}}$) can be found by geometry (the error $\boldsymbol{e}$ meets the column space of A at 90°). The key comes from algebra: $A^{\mathrm{T}}A\widehat{\boldsymbol{x}} = A^{\mathrm{T}}\boldsymbol{b}$. Calculus gives the same $\widehat{\boldsymbol{x}}$: the derivative of the error $||A\boldsymbol{x} - \boldsymbol{b}||^2$ is zero at $\widehat{\boldsymbol{x}}$.

By geometry Every $A\boldsymbol{x}$ lies in the plane of the columns $(1, 1, 1)$ and $(0, 1, 2)$. In that plane, we look for the point closest to $\boldsymbol{b}$. *The nearest point is the projection* $\boldsymbol{p}$.

The best choice for $A\widehat{\boldsymbol{x}}$ is $\boldsymbol{p}$. The smallest possible error is $\boldsymbol{e} = \boldsymbol{b} - \boldsymbol{p}$, perpendicular to the columns of A. *The three points at heights* (p_1, p_2, p_3) *do lie on a line*, because $\boldsymbol{p}$ is in the column space of A. In fitting a straight line, $\widehat{\boldsymbol{x}} = (C, D)$ is the best choice.

By algebra Every vector $\boldsymbol{b}$ splits into two parts. The part in the column space is $\boldsymbol{p}$. The perpendicular part is $\boldsymbol{e}$. There is an equation we cannot solve ($A\boldsymbol{x} = \boldsymbol{b}$). There is an equation $A\widehat{\boldsymbol{x}} = \boldsymbol{p}$ we can and do solve (by removing $\boldsymbol{e}$ and solving $A^{\mathrm{T}}A\widehat{\boldsymbol{x}} = A^{\mathrm{T}}\boldsymbol{b}$):

$$A\boldsymbol{x} = \boldsymbol{b} = \boldsymbol{p} + \boldsymbol{e} \quad \text{is impossible} \qquad A\widehat{\boldsymbol{x}} = \boldsymbol{p} \quad \text{is solvable} \qquad \widehat{\boldsymbol{x}} \text{ is } (A^{\mathrm{T}}A)^{-1}A^{\mathrm{T}}\boldsymbol{b}. \quad (1)$$

The solution to $A\widehat{\boldsymbol{x}} = \boldsymbol{p}$ leaves the least possible error (which is $\boldsymbol{e}$):

Squared error for any $\boldsymbol{x}$ $$||A\boldsymbol{x} - \boldsymbol{b}||^2 = ||A\boldsymbol{x} - \boldsymbol{p}||^2 + ||\boldsymbol{e}||^2. \quad (2)$$

This is the law $c^2 = a^2 + b^2$ for a right triangle. The vector $A\boldsymbol{x} - \boldsymbol{p}$ in the column space is perpendicular to $\boldsymbol{e}$ in the left nullspace. We reduce $A\boldsymbol{x} - \boldsymbol{p}$ to **zero** by choosing $\boldsymbol{x} = \widehat{\boldsymbol{x}}$. That leaves the smallest possible error $\boldsymbol{e} = (e_1, e_2, e_3)$ which we can't reduce.

Notice what "smallest" means. The *squared length* of $A\boldsymbol{x} - \boldsymbol{b}$ is minimized:

The least squares solution $\widehat{\boldsymbol{x}}$ makes $E = ||A\boldsymbol{x} - \boldsymbol{b}||^2$ as small as possible.

Figure 4.6a shows the closest line. It misses by distances $e_1, e_2, e_3 = 1, -2, 1$. *Those are vertical distances.* That least squares line minimizes $E = e_1^2 + e_2^2 + e_3^2$.

Figure 4.6b shows the same problem in 3-dimensional space ($\boldsymbol{b}\,\boldsymbol{p}\,\boldsymbol{e}$ space). The vector $\boldsymbol{b}$ is not in the column space of A. That is why we could not solve $A\boldsymbol{x} = \boldsymbol{b}$. No line goes through the three points. The smallest possible error is the perpendicular vector $\boldsymbol{e}$. This is $\boldsymbol{e} = \boldsymbol{b} - A\widehat{\boldsymbol{x}}$, the vector of errors $(1, -2, 1)$ in the three equations. Those are the distances from the best line. Behind both figures is the fundamental equation $A^{\mathrm{T}}A\widehat{\boldsymbol{x}} = A^{\mathrm{T}}\boldsymbol{b}$.

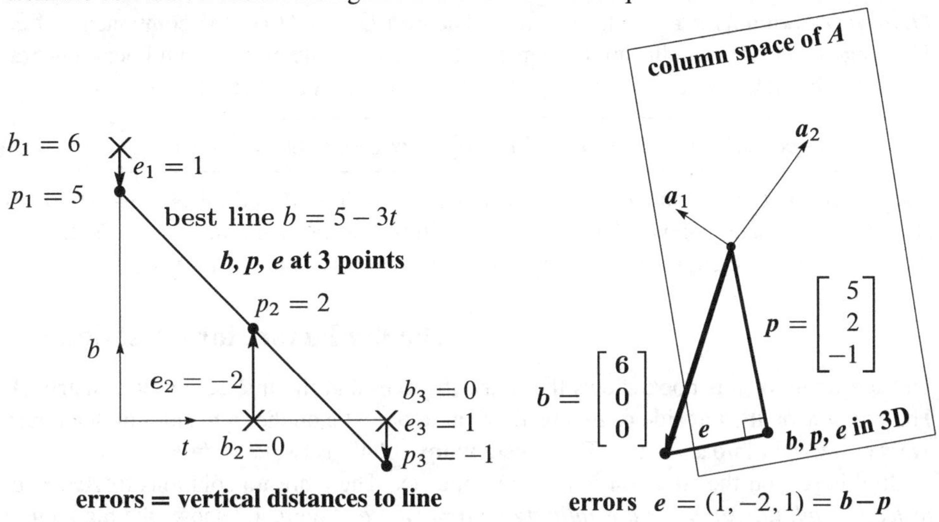

Figure 4.6: **Best line and projection: Two pictures, same problem.** The line has heights $\boldsymbol{p} = (5, 2, -1)$ with errors $\boldsymbol{e} = (1, -2, 1)$. The equations $A^{\mathrm{T}}A\widehat{\boldsymbol{x}} = A^{\mathrm{T}}\boldsymbol{b}$ give $\widehat{\boldsymbol{x}} = (5, -3)$. Same answer! The best line is $b = 5 - 3t$ and the closest point is $\boldsymbol{p} = 5\boldsymbol{a}_1 - 3\boldsymbol{a}_2$.

Notice that the errors $1, -2, 1$ add to zero. *Reason*: The error $\boldsymbol{e} = (e_1, e_2, e_3)$ is perpendicular to the first column $(1, 1, 1)$ in A. The dot product gives $e_1 + e_2 + e_3 = 0$.

By calculus Most functions are minimized by calculus ! The graph bottoms out and the derivative in every direction is zero. Here the error $\boldsymbol{E}$ to be minimized is a *sum of squares* $e_1^2 + e_2^2 + e_3^2$ (the squares of the errors in three equations):

$$\boldsymbol{E} = \|A\boldsymbol{x} - \boldsymbol{b}\|^2 = (C + D \cdot 0 - 6)^2 + (C + D \cdot 1)^2 + (C + D \cdot 2)^2. \qquad (3)$$

The unknowns are C and D. With two unknowns there are *two derivatives*—both zero at the minimum. They are "partial derivatives" because $\partial E/\partial C$ treats D as constant and $\partial E/\partial D$ treats C as constant. Here are the derivatives of E in (3):

$$\boldsymbol{\partial E/\partial C} = 2(C + D \cdot 0 - 6) \quad + 2(C + D \cdot 1) \quad + 2(C + D \cdot 2) \quad = 0$$

$$\boldsymbol{\partial E/\partial D} = 2(C + D \cdot 0 - 6)(\mathbf{0}) + 2(C + D \cdot 1)(\mathbf{1}) + 2(C + D \cdot 2)(\mathbf{2}) = 0.$$

$\partial E/\partial D$ contains the extra factors $\mathbf{0, 1, 2}$ from the chain rule. (The last derivative from $(C + 2D)^2$ was 2 times $C + 2D$ times that extra 2.) Those factors are not in $\partial E/\partial C$.

It is no accident that those factors 1, 1, 1 and 0, 1, 2 in the derivatives of $||Ax - b||^2$ are the columns of A. Now cancel 2 from every term and collect all C's and all D's:

$$\begin{array}{ll}\text{The } C \text{ derivative is zero:} & 3C + 3D = 6 \\ \text{The } D \text{ derivative is zero:} & 3C + 5D = 0\end{array} \qquad \textbf{This matrix} \begin{bmatrix} 3 & 3 \\ 3 & 5 \end{bmatrix} \textbf{ is } A^{\mathrm{T}}A \tag{4}$$

Those two equations are $A^{\mathrm{T}}A\widehat{x} = A^{\mathrm{T}}b$. The best C and D are the components of $\widehat{x}$. The equations from calculus are the same as the "normal equations" from linear algebra. These are the equations to minimize $||Ax - b||^2 = x^{\mathrm{T}}A^{\mathrm{T}}Ax - 2x^{\mathrm{T}}A^{\mathrm{T}}b + b^{\mathrm{T}}b$:

$$\textit{The partial derivatives of } \ ||Ax - b||^2 \ \textit{ are zero when } A^{\mathrm{T}}A\widehat{x} = A^{\mathrm{T}}b.$$

The solution is $C = 5$ and $D = -3$. Therefore $b = 5 - 3t$ is the best line—it comes closest to the three points. At $t = 0, 1, 2$ this line goes through $p = 5, 2, -1$. It could not go through $b = 6, 0, 0$. The errors are $1, -2, 1$. This is the vector e!

The Big Picture for Least Squares

The key figure of this book shows the four subspaces and the true action of a matrix A. The vector x on the left side of Figure 4.1 went to $b = Ax$ on the right side. In that figure every x was split into $x_r + x_n$. There were **many** solutions to $Ax = b$.

In this section the situation is just the opposite. There are **no** solutions to $Ax = b$. *Instead of splitting up* x *we are splitting up* $b = p + e$. Figure 4.7 shows the big picture for least squares. Instead of $Ax = b$ we solve $A\widehat{x} = p$. The error $e = b - p$ is unavoidable.

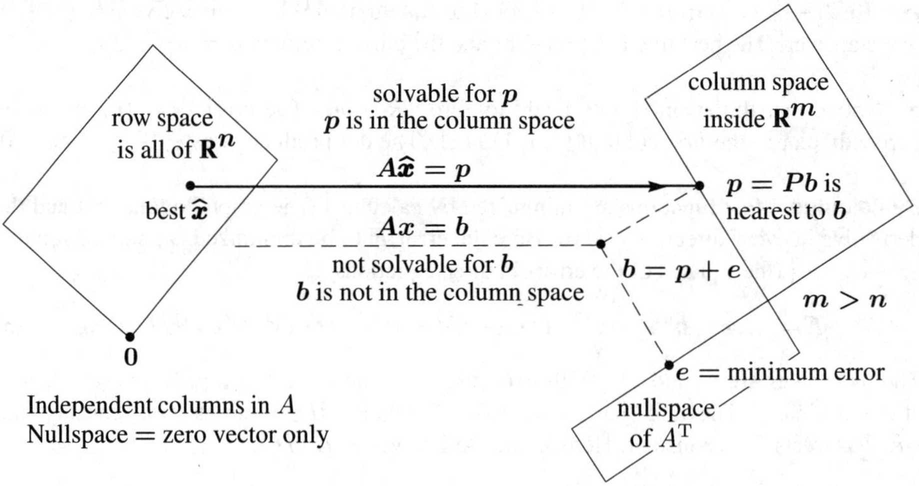

Figure 4.7: The projection $p = A\widehat{x}$ is closest to b, so $\widehat{x}$ minimizes $E = ||b - Ax||^2$.

Notice how the nullspace $\mathbf{N}(A)$ is very small—just one point. With independent columns, the only solution to $Ax = \mathbf{0}$ is $x = \mathbf{0}$. Then $A^{\mathrm{T}}A$ is invertible. The equation $A^{\mathrm{T}}A\widehat{x} = A^{\mathrm{T}}b$ fully determines the best vector $\widehat{x}$. The error $e = b - p$ has $A^{\mathrm{T}}e = \mathbf{0}$.

Chapter 7 will have the complete picture—all four subspaces included. Every $\boldsymbol{x}$ splits into $\boldsymbol{x}_r + \boldsymbol{x}_n$, and every $\boldsymbol{b}$ splits into $\boldsymbol{p} + \boldsymbol{e}$. The best solution is $\widehat{\boldsymbol{x}} = \widehat{\boldsymbol{x}}_r$ in the row space. We can't help $\boldsymbol{e}$ and we don't want $\boldsymbol{x}_n$ from the nullspace. This leaves $A\widehat{\boldsymbol{x}} = \boldsymbol{p}$.

Fitting a Straight Line

Fitting a line is the clearest application of least squares. It starts with $m > 2$ points, hopefully near a straight line. At times $t_1, \ldots, t_m$ those m points are at heights $b_1, \ldots, b_m$. The best line $C + Dt$ misses the points by vertical distances $e_1, \ldots, e_m$. No line is perfect, and the least squares line minimizes $E = e_1^2 + \cdots + e_m^2$.

The first example in this section had three points in Figure 4.6. Now we allow m points (and m can be large). The two components of $\widehat{\boldsymbol{x}}$ are still C and D.

A line goes through the m points when we exactly solve $A\boldsymbol{x} = \boldsymbol{b}$. Generally we can't do it. Two unknowns C and D determine a line, so A has only $n = 2$ columns. To fit the m points, we are trying to solve m equations (and we only have two unknowns!).

$$A\boldsymbol{x} = \boldsymbol{b} \quad \text{is} \quad \begin{aligned} C + Dt_1 &= b_1 \\ C + Dt_2 &= b_2 \\ &\vdots \\ C + Dt_m &= b_m \end{aligned} \qquad \text{with} \qquad A = \begin{bmatrix} 1 & t_1 \\ 1 & t_2 \\ \vdots & \vdots \\ 1 & t_m \end{bmatrix}. \tag{5}$$

The column space is so thin that almost certainly $\boldsymbol{b}$ is outside of it. When $\boldsymbol{b}$ happens to lie in the column space, the points happen to lie on a line. In that case $\boldsymbol{b} = \boldsymbol{p}$. Then $A\boldsymbol{x} = \boldsymbol{b}$ is solvable and the errors are $\boldsymbol{e} = (0, \ldots, 0)$.

The closest line $C + Dt$ has heights $p_1, \ldots, p_m$ with errors $e_1, \ldots, e_m$.

Solve $A^{\mathrm{T}}A\widehat{\boldsymbol{x}} = A^{\mathrm{T}}\boldsymbol{b}$ for $\widehat{\boldsymbol{x}} = (C, D)$. The errors are $e_i = b_i - C - Dt_i$.

Fitting points by a straight line is so important that we give the two equations $A^{\mathrm{T}}A\widehat{\boldsymbol{x}} = A^{\mathrm{T}}\boldsymbol{b}$, once and for all. The two columns of A are independent (unless all times t_i are the same). So we turn to least squares and solve $A^{\mathrm{T}}A\widehat{\boldsymbol{x}} = A^{\mathrm{T}}\boldsymbol{b}$.

$$\textbf{Dot-product matrix} \quad A^{\mathrm{T}}A = \begin{bmatrix} 1 & \cdots & 1 \\ t_1 & \cdots & t_m \end{bmatrix} \begin{bmatrix} 1 & t_1 \\ \vdots & \vdots \\ 1 & t_m \end{bmatrix} = \begin{bmatrix} m & \sum t_i \\ \sum t_i & \sum t_i^2 \end{bmatrix}. \tag{6}$$

On the right side of the normal equation is the 2 by 1 vector $A^{\mathrm{T}}\boldsymbol{b}$:

$$A^{\mathrm{T}}A\begin{bmatrix} C \\ D \end{bmatrix} = A^{\mathrm{T}}A\widehat{\boldsymbol{x}} = A^{\mathrm{T}}\boldsymbol{b} = \begin{bmatrix} 1 & \cdots & 1 \\ t_1 & \cdots & t_m \end{bmatrix} \begin{bmatrix} b_1 \\ \vdots \\ b_m \end{bmatrix} = \begin{bmatrix} \sum b_i \\ \sum t_i b_i \end{bmatrix}. \tag{7}$$

In a specific problem, these numbers are given. The best $\widehat{\boldsymbol{x}} = (C, D)$ is $(A^{\mathrm{T}}A)^{-1}A^{\mathrm{T}}\boldsymbol{b}$.

The line $C + Dt$ minimizes $e_1^2 + \cdots + e_m^2 = \|A\boldsymbol{x} - \boldsymbol{b}\|^2$ when $A^{\mathrm{T}}A\widehat{\boldsymbol{x}} = A^{\mathrm{T}}\boldsymbol{b}$:

$$A^{\mathrm{T}}A\widehat{\boldsymbol{x}} = A^{\mathrm{T}}\boldsymbol{b} \qquad \begin{bmatrix} m & \sum t_i \\ \sum t_i & \sum t_i^2 \end{bmatrix}\begin{bmatrix} C \\ D \end{bmatrix} = \begin{bmatrix} \sum b_i \\ \sum t_i b_i \end{bmatrix}. \tag{8}$$

The vertical errors at the m points on the line are the components of $\boldsymbol{e} = \boldsymbol{b} - \boldsymbol{p}$. This error vector (the *residual* $\boldsymbol{b} - A\widehat{\boldsymbol{x}}$) is perpendicular to the columns of A (geometry). The error is in the nullspace of A^{T} (linear algebra). The best $\widehat{\boldsymbol{x}} = (C, D)$ minimizes the total error E, the sum of squares (calculus):

$$E(\boldsymbol{x}) = \|A\boldsymbol{x} - \boldsymbol{b}\|^2 = (C + Dt_1 - b_1)^2 + \cdots + (C + Dt_m - b_m)^2.$$

Calculus sets the derivatives $\partial E/\partial C$ and $\partial E/\partial D$ to zero, and recovers $A^{\mathrm{T}}A\widehat{\boldsymbol{x}} = A^{\mathrm{T}}\boldsymbol{b}$.

Other least squares problems have more than two unknowns. Fitting by the best parabola has $n = 3$ coefficients C, D, E (see below). In general we are fitting m data points by n parameters $x_1, \ldots, x_n$. The matrix A has n columns and $n < m$. The derivatives of $\|A\boldsymbol{x} - \boldsymbol{b}\|^2$ give the n equations $A^{\mathrm{T}}A\widehat{\boldsymbol{x}} = A^{\mathrm{T}}\boldsymbol{b}$. **The derivative of a square is linear.** This is one reason why the method of least squares is so popular.

Example 2 A has *orthogonal columns* when the measurement times t_i add to zero.

Suppose $b = 1, 2, 4$ at times $t = -2, 0, 2$. *Those times add to zero*. The columns of A have *zero dot product*: $(1, 1, 1)$ is orthogonal to $(-2, 0, 2)$:

$$\begin{aligned} C + D(-2) &= 1 \\ C + \ D(0) &= 2 \\ C + \ D(2) &= 4 \end{aligned} \qquad \text{or} \qquad A\boldsymbol{x} = \begin{bmatrix} 1 & -2 \\ 1 & 0 \\ 1 & 2 \end{bmatrix}\begin{bmatrix} C \\ D \end{bmatrix} = \begin{bmatrix} 1 \\ 2 \\ 4 \end{bmatrix}.$$

When the columns of A are orthogonal, $\boldsymbol{A^{\mathrm{T}}A}$ **will be a diagonal matrix**:

$$A^{\mathrm{T}}A\widehat{\boldsymbol{x}} = A^{\mathrm{T}}\boldsymbol{b} \quad \text{is} \quad \begin{bmatrix} 3 & 0 \\ \mathbf{0} & 8 \end{bmatrix}\begin{bmatrix} C \\ D \end{bmatrix} = \begin{bmatrix} 7 \\ 6 \end{bmatrix}. \tag{9}$$

Main point: Since $A^{\mathrm{T}}A$ *is diagonal*, we can solve separately for $C = \frac{7}{3}$ and $D = \frac{6}{8}$. The zeros in $A^{\mathrm{T}}A$ are dot products of perpendicular columns in A. The diagonal matrix $A^{\mathrm{T}}A$, with entries $m = 3$ and $t_1^2 + t_2^2 + t_3^2 = 8$, is virtually as simple as the identity matrix.

Orthogonal columns are so helpful that it can be worth *shifting the times by subtracting the average time* $\widehat{t} = (t_1 + \cdots + t_m)/m$. If the original times were $1, 3, 5$ then their average is $\widehat{t} = 3$. The shifted times $T = t - \widehat{t} = t - 3$ add up to zero!

$$\begin{aligned} T_1 &= 1 - 3 = -2 \\ T_2 &= 3 - 3 = \ \ 0 \\ T_3 &= 5 - 3 = \ \ 2 \end{aligned} \qquad A_{\text{new}} = \begin{bmatrix} 1 & T_1 \\ 1 & T_2 \\ 1 & T_3 \end{bmatrix} \qquad A_{\text{new}}^{\mathrm{T}}A_{\text{new}} = \begin{bmatrix} 3 & \mathbf{0} \\ \mathbf{0} & 8 \end{bmatrix}.$$

Now C and D come from the easy equation (9). Then the best straight line uses $C + DT$ which is $C + D(t - \widehat{t}\,) = C + D(t - 3)$. Problem 30 even gives a formula for C and D.

That was a perfect example of the "Gram-Schmidt idea" coming in the next section: *Make the columns orthogonal in advance.* Then $A_{\text{new}}^{\mathrm{T}} A_{\text{new}}$ is diagonal and $\widehat{\boldsymbol{x}}_{\text{new}}$ is easy.

Dependent Columns in A: What is $\widehat{\boldsymbol{x}}$?

From the start, this chapter has assumed independent columns in A. Then $A^{\mathrm{T}}A$ is invertible. Then $A^{\mathrm{T}}A\widehat{\boldsymbol{x}} = A^{\mathrm{T}}\boldsymbol{b}$ produces the least squares solution $\widehat{\boldsymbol{x}}$ to $A\boldsymbol{x} = \boldsymbol{b}$.

Which $\widehat{\boldsymbol{x}}$ is best if A has *dependent columns*? Here is a specific example.

$$\begin{bmatrix} 1 & 1 \\ 1 & 1 \end{bmatrix} \begin{bmatrix} x_1 \\ x_2 \end{bmatrix} = \begin{bmatrix} 3 \\ 1 \end{bmatrix} = \boldsymbol{b} \qquad \begin{bmatrix} 1 & 1 \\ 1 & 1 \end{bmatrix} \begin{bmatrix} \widehat{x}_1 \\ \widehat{x}_2 \end{bmatrix} = \begin{bmatrix} 2 \\ 2 \end{bmatrix} = \boldsymbol{p}$$

$$A\boldsymbol{x} = \boldsymbol{b} \qquad\qquad A\widehat{\boldsymbol{x}} = \boldsymbol{p}$$

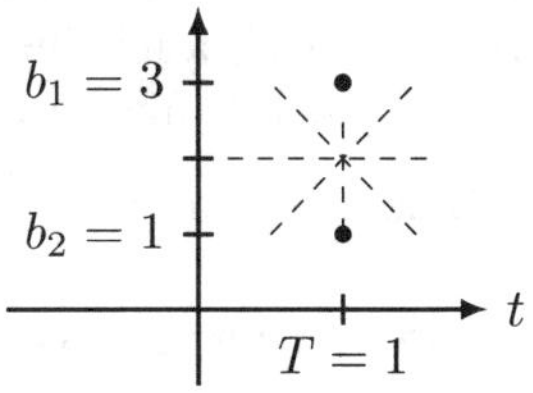

The measurements $b_1 = 3$ and $b_2 = 1$ are **at the same time T**. A straight line $C + Dt$ cannot go through both points. I think we are right to project $\boldsymbol{b} = (3, 1)$ to $\boldsymbol{p} = (2, 2)$ in the column space of A. That changes the equation $A\boldsymbol{x} = \boldsymbol{b}$ to the equation $A\widehat{\boldsymbol{x}} = \boldsymbol{p}$. An equation with no solution has become an equation with infinitely many solutions. The problem is that A has dependent columns and $(1, -1)$ is in its nullspace.

Which solution $\widehat{\boldsymbol{x}}$ should we choose? All the dashed lines in the figure have the same two errors 1 and -1 at time T. Those errors $(1, -1) = \boldsymbol{e} = \boldsymbol{b} - \boldsymbol{p}$ are as small as possible. But this doesn't tell us which dashed line is best.

My instinct is to go for the horizontal line at height 2. If the equation for the best line is $b = C + Dt$, then my choice will have $\widehat{x}_1 = C = 2$ and $\widehat{x}_2 = D = 0$. But what if the line had been written as $b = ct + d$? This is equally correct (just reversing C and D). Now the horizontal line has $\widehat{x}_1 = c = 0$ and $\widehat{x}_2 = d = 2$. I don't see any way out.

In Section 4.5, the *"pseudoinverse" of* A finds the **shortest solution to $A\widehat{\boldsymbol{x}} = \boldsymbol{p}$**. Here, that shortest solution will be $\boldsymbol{x}^{+} = (1, 1)$. This is the particular solution in the row space of A, and $\boldsymbol{x}^{+}$ has length $\sqrt{2}$. (Both solutions $\widehat{\boldsymbol{x}} = (2, 0)$ and $(0, 2)$ have length 2.) We are choosing the nullspace component of the solution $\boldsymbol{x}^{+}$ to be zero.

If A has independent columns, the $\boldsymbol{x}^{+}$ pseudoinverse is our usual left inverse $(A^{\mathrm{T}}A)^{-1}A^{\mathrm{T}}$. When I write it that way, the pseudoinverse sounds like the best way to choose $\widehat{\boldsymbol{x}}$.

Comment MATLAB experiments with singular matrices produced either **Inf** or **NaN** (Not a Number) or $\mathbf{10^{16}}$ (a bad number). There is a warning in every case! I believe that **Inf** and **NaN** and $\mathbf{10^{16}}$ come from the possibilities $\mathbf{0}\boldsymbol{x} = \boldsymbol{b}$ and $\mathbf{0}\boldsymbol{x} = \mathbf{0}$ and $\mathbf{10^{-16}}\,\boldsymbol{x} = \mathbf{1}$.

Those are three small examples of three big difficulties: singular with no solution, singular with many solutions, and very very close to singular.

Fitting by a Parabola

If we throw a ball, it would be crazy to fit the path by a straight line. A parabola $b = C + Dt + Et^2$ allows the ball to go up and come down again (b is the height at time t). The actual path is not a perfect parabola, but the whole theory of projectiles starts with that approximation.

When Galileo dropped a stone from the Leaning Tower of Pisa, it accelerated. The distance contains a quadratic term $\frac{1}{2}gt^2$. (Galileo's point was that the stone's mass is not involved.) Without that t^2 term we could never send a satellite into its orbit. But even with a nonlinear function like t^2, the unknowns C, D, E still appear linearly! Fitting points by the best parabola is still a problem in linear algebra.

Problem Try to fit heights $b_1, \ldots, b_m$ at times $t_1, \ldots, t_m$ by a parabola $C + Dt + Et^2$.

Solution With $m > 3$ points, the m equations for an exact fit are generally unsolvable:

$$\begin{aligned} C + Dt_1 + Et_1^2 &= b_1 \\ &\vdots \\ C + Dt_m + Et_m^2 &= b_m \end{aligned} \qquad \begin{array}{l} \text{is } A\boldsymbol{x} = \boldsymbol{b} \text{ with} \\ \text{the } m \text{ by 3 matrix} \end{array} \qquad A = \begin{bmatrix} 1 & t_1 & t_1^2 \\ \vdots & \vdots & \vdots \\ 1 & t_m & t_m^2 \end{bmatrix}. \qquad (10)$$

Least squares The closest parabola $C + Dt + Et^2$ chooses $\widehat{\boldsymbol{x}} = (C, D, E)$ to satisfy the three normal equations $A^{\mathrm{T}}A\widehat{\boldsymbol{x}} = A^{\mathrm{T}}\boldsymbol{b}$.

May I ask you to convert this to a problem of projection? The column space of A has dimension ______. The projection of $\boldsymbol{b}$ is $\boldsymbol{p} = A\widehat{\boldsymbol{x}}$, which combines the three columns using the coefficients C, D, E. The error at the first data point is $e_1 = b_1 - C - Dt_1 - Et_1^2$. If you prefer to minimize by calculus, take the partial derivatives of $e_1^2 + \cdots + e_m^2$ with respect to ______, ______, ______. These three derivatives will be zero when $\widehat{\boldsymbol{x}} = (C, D, E)$ solves the 3 by 3 system of equations $A^{\mathrm{T}}A\widehat{\boldsymbol{x}} = A^{\mathrm{T}}\boldsymbol{b}$.

Fourier series is least squares in infinite dimensions—approximating functions instead of vectors. The function to be minimized changes from a sum of squared errors to an integral of the squared error.

Example 3 For a parabola $b = C + Dt + Et^2$ to go through the three heights $b = 6, 0, 0$ when $t = 0, 1, 2$, the equations for C, D, E are

$$\begin{aligned} C + D \cdot 0 + E \cdot 0^2 &= 6 \\ C + D \cdot 1 + E \cdot 1^2 &= 0 \\ C + D \cdot 2 + E \cdot 2^2 &= 0. \end{aligned} \qquad (11)$$

This is $A\boldsymbol{x} = \boldsymbol{b}$. We can solve it exactly. Three data points give three equations and a square matrix. The solution is $\boldsymbol{x} = (C, D, E) = (\mathbf{6}, \mathbf{-9}, \mathbf{3})$. The parabola through the three points is $\boldsymbol{b} = \mathbf{6} - \mathbf{9}\boldsymbol{t} + \mathbf{3}\boldsymbol{t}^{\mathbf{2}}$.

■ WORKED EXAMPLES ■

4.3 A Start with nine measurements b_1 to b_9, *all zero*, at times $t = 1, \ldots, 9$. The tenth measurement $b_{10} = 40$ is an outlier. Find the **best horizontal line** $y = C$ to fit the ten points $(1,0), (2,0), \ldots, (9,0), (10,40)$ using three options for the error E:

(1) Least *squares* $E_2 = e_1^2 + \cdots + e_{10}^2$ (then the normal equation for C is linear)

(2) Least *maximum* error $E_\infty = |e_{\max}|$ **(3)** Least *sum* of errors $E_1 = |e_1| + \cdots + |e_{10}|$.

Solution **(1)** The least squares fit to $0, 0, \ldots, 0, 40$ by a horizontal line is $\boldsymbol{C = 4}$:

$$A = \text{column of 1's} \quad A^{\mathrm{T}}A = 10 \quad A^{\mathrm{T}}\boldsymbol{b} = \text{sum of } b_i = 40. \quad \text{So } 10\,C = 40.$$

(2) The least maximum error requires $\boldsymbol{C = 20}$, halfway between 0 and 40.

(3) The least sum requires $\boldsymbol{C = 0}$ (!!). The sum of errors $9|C| + |40 - C|$ would increase if C moves up from zero.

The least sum comes from the *median* measurement (the median of $0, \ldots, 0, 40$ is zero). Many statisticians feel that the least squares solution is too heavily influenced by outliers like $b_{10} = 40$, and they prefer least sum. But the equations become *nonlinear*.

Now find the least squares line $C + Dt$ through those ten points $(1,0)$ to $(10,40)$:

$$A^{\mathrm{T}}A = \begin{bmatrix} 10 & \sum t_i \\ \sum t_i & \sum t_i^2 \end{bmatrix} = \begin{bmatrix} 10 & 55 \\ 55 & 385 \end{bmatrix} \qquad A^{\mathrm{T}}\boldsymbol{b} = \begin{bmatrix} \sum b_i \\ \sum t_i b_i \end{bmatrix} = \begin{bmatrix} 40 \\ 400 \end{bmatrix}$$

Those come from equation (8). Then $A^{\mathrm{T}}A\widehat{\boldsymbol{x}} = A^{\mathrm{T}}\boldsymbol{b}$ gives $C = -8$ and $D = 24/11$.

What happens to C and D if you multiply $\boldsymbol{b} = (0, 0, \ldots, 40)$ by 3 and then add 30 to get $\boldsymbol{b}_{\text{new}} = (30, 30, \ldots, 150)$? Linearity allows us to rescale $\boldsymbol{b}$. Multiplying $\boldsymbol{b}$ by 3 will multiply C and D by 3. Adding 30 to all b_i will add 30 to C.

4.3 B Find the parabola $C + Dt + Et^2$ that comes closest (least squares error) to the values $\boldsymbol{b} = (0, 0, 1, 0, 0)$ at the times $t = -2, -1, 0, 1, 2$. First write down the five equations $A\boldsymbol{x} = \boldsymbol{b}$ in three unknowns $\boldsymbol{x} = (C, D, E)$ for a parabola to go through the five points. No solution because no such parabola exists. Solve $A^{\mathrm{T}}A\widehat{\boldsymbol{x}} = A^{\mathrm{T}}\boldsymbol{b}$.

I would predict $D = 0$. Why should the best parabola be symmetric around $t = 0$? In $A^{\mathrm{T}}A\widehat{\boldsymbol{x}} = A^{\mathrm{T}}\boldsymbol{b}$, equation 2 for D should uncouple from equations 1 and 3.

Solution The five equations $A\boldsymbol{x} = \boldsymbol{b}$ have a rectangular *Vandermonde matrix* A:

$$\begin{aligned} C + D\,(-2) + E\,(-2)^2 &= 0 \\ C + D\,(-1) + E\,(-1)^2 &= 0 \\ C + D\quad (0) + E\quad (0)^2 &= 1 \\ C + D\quad (1) + E\quad (1)^2 &= 0 \\ C + D\quad (2) + E\quad (2)^2 &= 0 \end{aligned} \qquad A = \begin{bmatrix} 1 & -2 & 4 \\ 1 & -1 & 1 \\ 1 & 0 & 0 \\ 1 & 1 & 1 \\ 1 & 2 & 4 \end{bmatrix} \qquad A^{\mathrm{T}}A = \begin{bmatrix} 5 & \mathbf{0} & 10 \\ \mathbf{0} & 10 & \mathbf{0} \\ 10 & \mathbf{0} & 34 \end{bmatrix}$$

Those zeros in $A^{\mathrm{T}}A$ mean that column 2 of A is orthogonal to columns 1 and 3. We see this directly in A (the times $-2,-1,0,1,2$ are symmetric). The best C, D, E in the parabola $C+Dt+Et^2$ come from $A^{\mathrm{T}}A\widehat{\boldsymbol{x}}=A^{\mathrm{T}}\boldsymbol{b}$, and D is uncoupled from C and E:

$$\begin{bmatrix}5&0&10\\0&10&0\\10&0&34\end{bmatrix}\begin{bmatrix}C\\D\\E\end{bmatrix}=\begin{bmatrix}1\\0\\0\end{bmatrix}\quad\text{leads to}\quad\begin{aligned}C&=34/70\\D&=0\ \text{ as predicted}\\E&=-10/70\end{aligned}$$

Problem Set 4.3

Problems 1–11 use four data points $\boldsymbol{b}=(0,8,8,20)$ to bring out the key ideas.

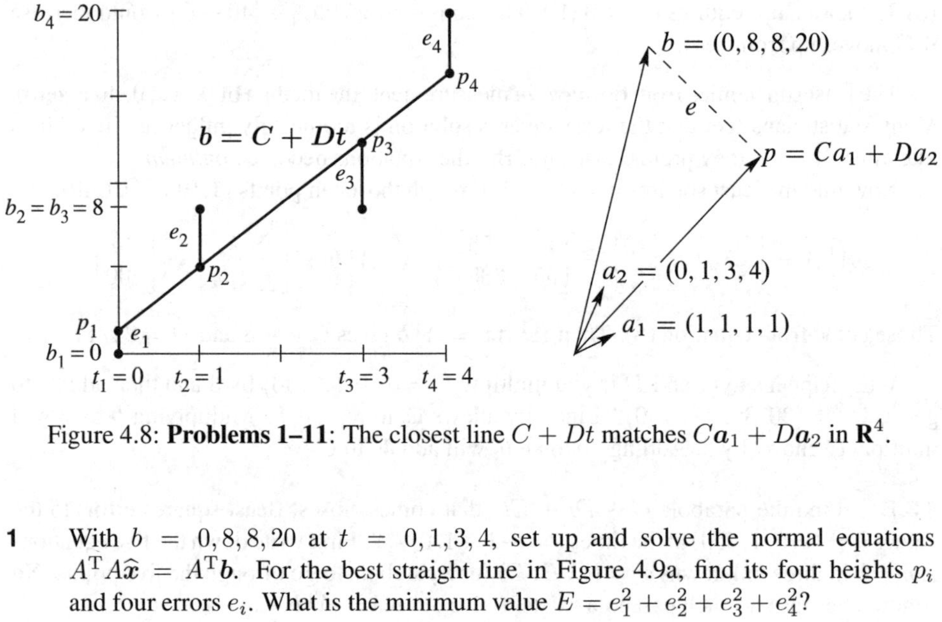

Figure 4.8: **Problems 1–11**: The closest line $C+Dt$ matches $C\boldsymbol{a}_1+D\boldsymbol{a}_2$ in $\mathbf{R}^4$.

1 With $b=0,8,8,20$ at $t=0,1,3,4$, set up and solve the normal equations $A^{\mathrm{T}}A\widehat{\boldsymbol{x}}=A^{\mathrm{T}}\boldsymbol{b}$. For the best straight line in Figure 4.9a, find its four heights p_i and four errors e_i. What is the minimum value $E=e_1^2+e_2^2+e_3^2+e_4^2$?

2 (Line $C+Dt$ does go through p's) With $b=0,8,8,20$ at times $t=0,1,3,4$, write down the four equations $A\boldsymbol{x}=\boldsymbol{b}$ (unsolvable). Change the measurements to $p=1,5,13,17$ and find an exact solution to $A\widehat{\boldsymbol{x}}=\boldsymbol{p}$.

3 Check that $\boldsymbol{e}=\boldsymbol{b}-\boldsymbol{p}=(-1,3,-5,3)$ is perpendicular to both columns of the same matrix A. What is the shortest distance $\|\boldsymbol{e}\|$ from $\boldsymbol{b}$ to the column space of A?

4 (By calculus) Write down $E=\|A\boldsymbol{x}-\boldsymbol{b}\|^2$ as a sum of four squares—the last one is $(C+4D-20)^2$. Find the derivative equations $\partial E/\partial C=0$ and $\partial E/\partial D=0$. Divide by 2 to obtain the normal equations $A^{\mathrm{T}}A\widehat{\boldsymbol{x}}=A^{\mathrm{T}}\boldsymbol{b}$.

5 Find the height C of the best *horizontal line* to fit $\boldsymbol{b} = (0, 8, 8, 20)$. An exact fit would solve the unsolvable equations $C = 0$, $C = 8$, $C = 8$, $C = 20$. Find the 4 by 1 matrix A in these equations and solve $A^{\mathrm{T}}A\widehat{x} = A^{\mathrm{T}}\boldsymbol{b}$. Draw the horizontal line at height $\widehat{x} = C$ and the four errors in $\boldsymbol{e}$.

6 Project $\boldsymbol{b} = (0, 8, 8, 20)$ onto the line through $\boldsymbol{a} = (1, 1, 1, 1)$. Find $\widehat{x} = \boldsymbol{a}^{\mathrm{T}}\boldsymbol{b}/\boldsymbol{a}^{\mathrm{T}}\boldsymbol{a}$ and the projection $\boldsymbol{p} = \widehat{x}\boldsymbol{a}$. Check that $\boldsymbol{e} = \boldsymbol{b} - \boldsymbol{p}$ is perpendicular to $\boldsymbol{a}$, and find the shortest distance $\|\boldsymbol{e}\|$ from $\boldsymbol{b}$ to the line through $\boldsymbol{a}$.

7 Find the closest line $b = Dt$, *through the origin*, to the same four points. An exact fit would solve $D \cdot 0 = 0$, $D \cdot 1 = 8$, $D \cdot 3 = 8$, $D \cdot 4 = 20$. Find the 4 by 1 matrix and solve $A^{\mathrm{T}}A\widehat{x} = A^{\mathrm{T}}\boldsymbol{b}$. Redraw Figure 4.9a showing the best line $b = Dt$ and the e's.

8 Project $\boldsymbol{b} = (0, 8, 8, 20)$ onto the line through $\boldsymbol{a} = (0, 1, 3, 4)$. Find $\widehat{x} = D$ and $\boldsymbol{p} = \widehat{x}\boldsymbol{a}$. The best C in Problems 5–6 and the best D in Problems 7–8 do *not* agree with the best (C, D) in Problems 1–4. That is because $(1, 1, 1, 1)$ and $(0, 1, 3, 4)$ are _____ perpendicular.

9 For the closest parabola $b = C + Dt + Et^2$ to the same four points, write down the unsolvable equations $A\boldsymbol{x} = \boldsymbol{b}$ in three unknowns $\boldsymbol{x} = (C, D, E)$. Set up the three normal equations $A^{\mathrm{T}}A\widehat{\boldsymbol{x}} = A^{\mathrm{T}}\boldsymbol{b}$ (solution not required). In Figure 4.9a you are now fitting a parabola to 4 points—what is happening in Figure 4.9b?

10 For the closest cubic $b = C + Dt + Et^2 + Ft^3$ to the same four points, write down the four equations $A\boldsymbol{x} = \boldsymbol{b}$. Solve them by elimination. In Figure 4.9a this cubic now goes exactly through the points. What are $\boldsymbol{p}$ and $\boldsymbol{e}$?

11 The average of the four times is $\widehat{t} = \frac{1}{4}(0 + 1 + 3 + 4) = 2$. The average of the four b's is $\widehat{b} = \frac{1}{4}(0 + 8 + 8 + 20) = 9$.

(a) Verify that the best line goes through the center point $(\widehat{t}, \widehat{b}) = (2, 9)$.

(b) Explain why $C + D\widehat{t} = \widehat{b}$ comes from the first equation in $A^{\mathrm{T}}A\widehat{\boldsymbol{x}} = A^{\mathrm{T}}\boldsymbol{b}$.

Questions 12–16 introduce basic ideas of statistics—the foundation for least squares.

12 (Recommended) This problem projects $\boldsymbol{b} = (b_1, \ldots, b_m)$ onto the line through $\boldsymbol{a} = (1, \ldots, 1)$. We solve m equations $\boldsymbol{a}x = \boldsymbol{b}$ in 1 unknown (by least squares).

(a) Solve $\boldsymbol{a}^{\mathrm{T}}\boldsymbol{a}\widehat{x} = \boldsymbol{a}^{\mathrm{T}}\boldsymbol{b}$ to show that $\widehat{x}$ is the *mean* (the average) of the $\boldsymbol{b}$'s.

(b) Find $\boldsymbol{e} = \boldsymbol{b} - \boldsymbol{a}\widehat{x}$ and the *variance* $\|\boldsymbol{e}\|^2$ and the *standard deviation* $\|\boldsymbol{e}\|$.

(c) The horizontal line $\widehat{\boldsymbol{b}} = 3$ is closest to $\boldsymbol{b} = (1, 2, 6)$. Check that $\boldsymbol{p} = (3, 3, 3)$ is perpendicular to $\boldsymbol{e}$ and find the 3 by 3 projection matrix P.

13 First assumption behind least squares: $A\boldsymbol{x} = \boldsymbol{b}-$ ***(noise e with mean zero)***. Multiply the error vectors $\boldsymbol{e} = \boldsymbol{b} - A\boldsymbol{x}$ by $(A^{\mathrm{T}}A)^{-1}A^{\mathrm{T}}$ to get $\widehat{\boldsymbol{x}} - \boldsymbol{x}$ on the right. The estimation errors $\widehat{\boldsymbol{x}} - \boldsymbol{x}$ also average to zero. The estimate $\widehat{\boldsymbol{x}}$ is *unbiased*.

14 Second assumption behind least squares: The m errors e_i are independent with variance σ^2, so the average of $(\boldsymbol{b}-A\boldsymbol{x})(\boldsymbol{b}-A\boldsymbol{x})^{\mathrm{T}}$ is $\sigma^2 I$. Multiply on the left by $(A^{\mathrm{T}}A)^{-1}A^{\mathrm{T}}$ and on the right by $A(A^{\mathrm{T}}A)^{-1}$ to show that the average matrix $(\widehat{\boldsymbol{x}}-\boldsymbol{x})(\widehat{\boldsymbol{x}}-\boldsymbol{x})^{\mathrm{T}}$ is $\sigma^2(A^{\mathrm{T}}A)^{-1}$. This is the ***covariance matrix*** W in Section 10.2.

15 A doctor takes 4 readings of your heart rate. The best solution to $x=b_1,\ldots,x=b_4$ is the average $\widehat{x}$ of $b_1,\ldots,b_4$. The matrix A is a column of 1's. Problem 14 gives the expected error $(\widehat{x}-x)^2$ as $\sigma^2(A^{\mathrm{T}}A)^{-1}=$ _____. ***By averaging, the variance drops from*** σ^2 ***to*** $\sigma^2/4$.

16 If you know the average $\widehat{x}_9$ of 9 numbers $b_1,\ldots,b_9$, how can you quickly find the average $\widehat{x}_{10}$ with one more number b_{10}? The idea of *recursive* least squares is to avoid adding 10 numbers. What number multiplies $\widehat{x}_9$ in computing $\widehat{x}_{10}$?

$$\widehat{x}_{10}=\tfrac{1}{10}b_{10}+\_\_\_\_\_\_\widehat{x}_9=\tfrac{1}{10}(b_1+\cdots+b_{10})\quad\text{as in Worked Example 4.2 C.}$$

Questions 17–24 give more practice with $\widehat{\boldsymbol{x}}$ and $\boldsymbol{p}$ and $\boldsymbol{e}$.

17 Write down three equations for the line $b=C+Dt$ to go through $b=7$ at $t=-1$, $b=7$ at $t=1$, and $b=21$ at $t=2$. Find the least squares solution $\widehat{\boldsymbol{x}}=(C,D)$ and draw the closest line.

18 Find the projection $\boldsymbol{p}=A\widehat{\boldsymbol{x}}$ in Problem 17. This gives the three heights of the closest line. Show that the error vector is $\boldsymbol{e}=(2,-6,4)$. Why is $P\boldsymbol{e}=\mathbf{0}$?

19 Suppose the measurements at $t=-1,1,2$ are the errors $2,-6,4$ in Problem 18. Compute $\widehat{\boldsymbol{x}}$ and the closest line to these new measurements. Explain the answer: $\boldsymbol{b}=(2,-6,4)$ is perpendicular to _____ so the projection is $\boldsymbol{p}=\mathbf{0}$.

20 Suppose the measurements at $t=-1,1,2$ are $\boldsymbol{b}=(5,13,17)$. Compute $\widehat{\boldsymbol{x}}$ and the closest line and $\boldsymbol{e}$. The error is $\boldsymbol{e}=\mathbf{0}$ because this $\boldsymbol{b}$ is _____.

21 Which of the four subspaces contains the error vector $\boldsymbol{e}$? Which contains $\boldsymbol{p}$? Which contains $\widehat{\boldsymbol{x}}$? What is the nullspace of A?

22 Find the best line $C+Dt$ to fit $b=4,2,-1,0,0$ at times $t=-2,-1,0,1,2$.

23 Is the error vector $\boldsymbol{e}$ orthogonal to $\boldsymbol{b}$ or $\boldsymbol{p}$ or $\boldsymbol{e}$ or $\widehat{\boldsymbol{x}}$? Show that $\|\boldsymbol{e}\|^2$ equals $\boldsymbol{e}^{\mathrm{T}}\boldsymbol{b}$ which equals $\boldsymbol{b}^{\mathrm{T}}\boldsymbol{b}-\boldsymbol{p}^{\mathrm{T}}\boldsymbol{b}$. This is the smallest total error E.

24 The partial derivatives of $\|A\boldsymbol{x}\|^2$ with respect to $x_1,\ldots,x_n$ fill the vector $2A^{\mathrm{T}}A\boldsymbol{x}$. The derivatives of $2\boldsymbol{b}^{\mathrm{T}}A\boldsymbol{x}$ fill the vector $2A^{\mathrm{T}}\boldsymbol{b}$. So the derivatives of $\|A\boldsymbol{x}-\boldsymbol{b}\|^2$ are zero when _____.

Challenge Problems

25 *What condition on $(t_1, b_1), (t_2, b_2), (t_3, b_3)$ puts those three points onto a straight line?* A column space answer is: (b_1, b_2, b_3) must be a combination of $(1, 1, 1)$ and (t_1, t_2, t_3). Try to reach a specific equation connecting the t's and b's. I should have thought of this question sooner!

26 Find the *plane* that gives the best fit to the 4 values $\boldsymbol{b} = (0, 1, 3, 4)$ at the corners $(1, 0)$ and $(0, 1)$ and $(-1, 0)$ and $(0, -1)$ of a square. The equations $C + Dx + Ey = b$ at those 4 points are $A\boldsymbol{x} = \boldsymbol{b}$ with 3 unknowns $\boldsymbol{x} = (C, D, E)$. What is A? At the center $(0, 0)$ of the square, show that $C + Dx + Ey =$ average of the b's.

27 (Distance between lines) The points $P = (x, x, x)$ and $Q = (y, 3y, -1)$ are on two lines in space that don't meet. Choose x and y to minimize the squared distance $\|P - Q\|^2$. The line connecting the closest P and Q is perpendicular to ____.

28 Suppose the columns of A are not independent. How could you find a matrix B so that $P = B(B^{\mathrm{T}}B)^{-1}B^{\mathrm{T}}$ does give the projection onto the column space of A? (The usual formula will fail when $A^{\mathrm{T}}A$ is not invertible.)

29 Usually there will be exactly one hyperplane in $\mathbf{R}^n$ that contains the n given points $\boldsymbol{x} = \mathbf{0}, \boldsymbol{a}_1, \ldots, \boldsymbol{a}_{n-1}$. (Example for $n = 3$: There will be one plane containing $0, \boldsymbol{a}_1, \boldsymbol{a}_2$ unless ____.) What is the test to have exactly one plane in $\mathbf{R}^n$?

30 Example 2 shifted the times t_i to make them add to zero. We subtracted away the average time $\widehat{t} = (t_1 + \cdots + t_m)/m$ to get $T_i = t_i - \widehat{t}$. Those T_i add to zero.

With the columns $(1, \ldots, 1)$ and $(T_1, \ldots, T_m)$ now orthogonal, $A^{\mathrm{T}}A$ is diagonal. Its entries are m and $T_1^2 + \cdots + T_m^2$. Show that the best C and D have direct formulas:

$$\boldsymbol{T} \textbf{ is } \boldsymbol{t} - \widehat{\boldsymbol{t}} \qquad C = \frac{b_1 + \cdots + b_m}{m} \quad \text{and} \quad D = \frac{b_1T_1 + \cdots + b_mT_m}{T_1^2 + \cdots + T_m^2}.$$

The best line is $C + DT$ ***or*** $C + D(t - \widehat{t})$. The time shift that makes $A^{\mathrm{T}}A$ diagonal is an example of the Gram-Schmidt process: ***orthogonalize the columns of A in advance***. This is the subject of the next Section 4.4.

■ REVIEW OF THE KEY IDEAS ■

1. The least squares solution $\widehat{\boldsymbol{x}}$ minimizes $\|A\boldsymbol{x} - \boldsymbol{b}\|^2 = \boldsymbol{x}^{\mathrm{T}}A^{\mathrm{T}}A\boldsymbol{x} - 2\boldsymbol{x}^{\mathrm{T}}A^{\mathrm{T}}\boldsymbol{b} + \boldsymbol{b}^{\mathrm{T}}\boldsymbol{b}$. This is E, the sum of squares of the errors e_1 to e_m in the m equations $(m > n)$.

2. The best $\widehat{\boldsymbol{x}}$ comes from the normal equations $A^{\mathrm{T}}A\widehat{\boldsymbol{x}} = A^{\mathrm{T}}\boldsymbol{b}$. E is a minimum.

3. To fit m points by a line $b = C + Dt$, the normal equations give C and D.

4. The heights of the best line are $\boldsymbol{p} = (p_1, \ldots, p_m)$. The vertical distances to the data points are the errors $\boldsymbol{e} = (e_1, \ldots, e_m)$. A key equation is $A^{\mathrm{T}}\boldsymbol{e} = \mathbf{0}$.

5. If we try to fit m points by a combination of $n < m$ functions, the m equations $A\boldsymbol{x} = \boldsymbol{b}$ are generally unsolvable. The n equations $A^{\mathrm{T}}A\widehat{\boldsymbol{x}} = A^{\mathrm{T}}\boldsymbol{b}$ give the least squares solution—the combination with smallest MSE (**mean square error**).

4.4 Orthonormal Bases and Gram-Schmidt

1 The columns $q_1, \ldots, q_n$ are **orthonormal** if $\boldsymbol{q}_i^{\mathrm{T}}\boldsymbol{q}_j = \begin{cases} 0 \text{ for } i \neq j \\ 1 \text{ for } i = j \end{cases}$. Then $\boxed{\boldsymbol{Q}^{\mathrm{T}}\boldsymbol{Q} = \boldsymbol{I}}$.

2 If Q is also square, then $QQ^{\mathrm{T}} = I$ and $\boxed{\boldsymbol{Q}^{\mathrm{T}} = \boldsymbol{Q}^{-1}}$. Now Q is an **"orthogonal matrix"**.

3 The least squares solution to $Q\boldsymbol{x} = \boldsymbol{b}$ is $\widehat{\boldsymbol{x}} = Q^{\mathrm{T}}\boldsymbol{b}$. Projection of $\boldsymbol{b}$: $\boldsymbol{p} = QQ^{\mathrm{T}}\boldsymbol{b} = P\boldsymbol{b}$.

4 The **Gram-Schmidt** process takes independent $\boldsymbol{a}_i$ to orthonormal $\boldsymbol{q}_i$. Start with $\boldsymbol{q}_1 = \boldsymbol{a}_1 / \|\boldsymbol{a}_1\|$.

5 $\boldsymbol{q}_i$ is $(\boldsymbol{a}_i -$ its projection $\boldsymbol{p}_i) / \|\boldsymbol{a}_i - \boldsymbol{p}_i\|$; projection $\boldsymbol{p}_i = (\boldsymbol{a}_i^{\mathrm{T}}\boldsymbol{q}_1)\boldsymbol{q}_1 + \cdots + (\boldsymbol{a}_i^{\mathrm{T}}\boldsymbol{q}_{i-1})\boldsymbol{q}_{i-1}$.

6 Each $\boldsymbol{a}_i$ will be a combination of $\boldsymbol{q}_1$ to $\boldsymbol{q}_i$. Then $\boldsymbol{A} = \boldsymbol{QR}$: **orthogonal Q and triangular R.**

This section has two goals, **why** and **how**. The first is to see why orthogonal columns in A are good. Dot products are zero, so $A^{\mathrm{T}}A$ will be diagonal. It becomes so easy to find $\widehat{\boldsymbol{x}}$ and $\boldsymbol{p} = A\widehat{\boldsymbol{x}}$. ***The second goal is to construct orthogonal vectors $\boldsymbol{q}_i$.*** You will see how Gram-Schmidt combines the original basis vectors $\boldsymbol{a}_i$ to produce right angles for the $\boldsymbol{q}_i$. Those $\boldsymbol{a}_i$ are the columns of A, probably *not* orthogonal. ***The orthonormal basis vectors $\boldsymbol{q}_i$ will be the columns of a new matrix Q.***

From Chapter 3, a basis consists of independent vectors that span the space. The basis vectors could meet at any angle (except 0° and 180°). But every time we visualize axes, they are perpendicular. *In our imagination, the coordinate axes are practically always orthogonal.* This simplifies the picture and it greatly simplifies the computations.

The vectors $\boldsymbol{q}_1, \ldots, \boldsymbol{q}_n$ are ***orthogonal*** when their dot products $\boldsymbol{q}_i \cdot \boldsymbol{q}_j$ are zero. More exactly $\boldsymbol{q}_i^{\mathrm{T}}\boldsymbol{q}_j = 0$ whenever $i \neq j$. With one more step—just *divide each vector by its length*—the vectors become ***orthogonal unit vectors***. Their lengths are all 1 (normal). Then the basis is called ***orthonormal***.

DEFINITION The n vectors $\boldsymbol{q}_1, \ldots, \boldsymbol{q}_n$ are ***orthonormal*** if

$$\boldsymbol{q}_i^{\mathrm{T}}\boldsymbol{q}_j = \begin{cases} 0 & \text{when } i \neq j \quad (\textit{orthogonal} \text{ vectors}) \\ 1 & \text{when } i = j \quad (\textit{unit} \text{ vectors: } \|\boldsymbol{q}_i\| = 1) \end{cases}$$

A matrix Q with orthonormal columns has $Q^{\mathrm{T}}Q = I$. Typically $m > n$.

The matrix Q is easy to work with because $\boldsymbol{Q}^{\mathrm{T}}\boldsymbol{Q} = \boldsymbol{I}$. This repeats in matrix language that the columns $\boldsymbol{q}_1, \ldots, \boldsymbol{q}_n$ are orthonormal. Q is not required to be square: $QQ^{\mathrm{T}} \neq I$.

A matrix Q with orthonormal columns satisfies $\boldsymbol{Q^{\mathrm{T}}Q = I}$ **:**

$$\boldsymbol{Q^{\mathrm{T}}Q} = \begin{bmatrix} -\boldsymbol{q}_1^{\mathrm{T}}- \\ -\boldsymbol{q}_2^{\mathrm{T}}- \\ \\ -\boldsymbol{q}_n^{\mathrm{T}}- \end{bmatrix} \begin{bmatrix} | & | & & | \\ \boldsymbol{q}_1 & \boldsymbol{q}_2 & & \boldsymbol{q}_n \\ | & | & & | \end{bmatrix} = \begin{bmatrix} 1 & 0 & \cdots & 0 \\ 0 & 1 & \cdots & 0 \\ \vdots & \vdots & \ddots & \vdots \\ 0 & 0 & \cdots & 1 \end{bmatrix} = \boldsymbol{I}\,. \quad (1)$$

When row i of Q^{T} multiplies column j of Q, the dot product is $\boldsymbol{q}_i^{\mathrm{T}}\boldsymbol{q}_j$. Off the diagonal $(i \neq j)$ that dot product is zero by orthogonality. On the diagonal $(i = j)$ the unit vectors give $\boldsymbol{q}_i^{\mathrm{T}}\boldsymbol{q}_i = \|\boldsymbol{q}_i\|^2 = 1$. Often Q is rectangular $(m > n)$. Sometimes $m = n$.

When Q is square, $\boldsymbol{Q^{\mathrm{T}}Q = I}$ ***means that*** $\boldsymbol{Q^{\mathrm{T}} = Q^{-1}}$***: transpose = inverse.***

If the columns are only orthogonal (not unit vectors), dot products still give a diagonal matrix (not the identity matrix). This diagonal matrix is almost as good as I. The important thing is orthogonality—then it is easy to produce unit vectors.

To repeat: $Q^{\mathrm{T}}Q = I$ even when Q is rectangular. Then Q^{T} is a 1-sided inverse from the left. For square matrices we also have $QQ^{\mathrm{T}} = I$. Then Q^{T} is the two-sided inverse of Q. The rows of a square Q are orthonormal like the columns. ***The inverse is the transpose*** Q^{T}. In this square case we call Q an ***orthogonal matrix***.[1]

Here are three examples of orthogonal matrices—rotation and permutation and reflection. Those are important matrices ! The quickest test is to check $Q^{\mathrm{T}}Q = I$.

Example 1 **(Rotation)** Q rotates every vector in the plane by the angle θ:

$$Q = \begin{bmatrix} \cos\theta & -\sin\theta \\ \sin\theta & \cos\theta \end{bmatrix} \quad \text{and} \quad Q^{\mathrm{T}} = Q^{-1} = \begin{bmatrix} \cos\theta & \sin\theta \\ -\sin\theta & \cos\theta \end{bmatrix}.$$

The columns of Q are orthogonal (take their dot product). They are unit vectors because $\sin^2\theta + \cos^2\theta = 1$. Those columns give an ***orthonormal basis*** for the plane $\mathbf{R}^2$.

The standard basis vectors $\boldsymbol{i}$ and $\boldsymbol{j}$ are rotated through θ in Figure 4.10a. Then Q^{-1} **rotates vectors back through** $\boldsymbol{-\theta}$. It agrees with Q^{T}, because the cosine of $-\theta$ equals the cosine of θ, and $\sin(-\theta) = -\sin\theta$. We have $Q^{\mathrm{T}}Q = I$ and $QQ^{\mathrm{T}} = I$.

Example 2 **(Permutation)** These matrices change the order to (y, z, x) and (y, x):

$$\begin{bmatrix} 0 & 1 & 0 \\ 0 & 0 & 1 \\ 1 & 0 & 0 \end{bmatrix} \begin{bmatrix} x \\ y \\ z \end{bmatrix} = \begin{bmatrix} y \\ z \\ x \end{bmatrix} \quad \text{and} \quad \begin{bmatrix} 0 & 1 \\ 1 & 0 \end{bmatrix} \begin{bmatrix} x \\ y \end{bmatrix} = \begin{bmatrix} y \\ x \end{bmatrix}.$$

---

[1] "Orthonormal matrix" would have been a better name for Q, but it's not used. Any matrix with orthonormal columns has the letter Q. But we only call it an **orthogonal matrix when it is square**.

All columns of these Q's are unit vectors (their lengths are obviously 1). They are also orthogonal (the 1's appear in different places). *The inverse of a permutation matrix is its transpose*: $Q^{-1} = Q^{\mathrm{T}}$. The inverse puts the components back into their original order:

$$\textbf{Inverse = transpose:}\quad \begin{bmatrix} 0 & 0 & 1 \\ 1 & 0 & 0 \\ 0 & 1 & 0 \end{bmatrix} \begin{bmatrix} y \\ z \\ x \end{bmatrix} = \begin{bmatrix} x \\ y \\ z \end{bmatrix} \quad \text{and} \quad \begin{bmatrix} 0 & 1 \\ 1 & 0 \end{bmatrix} \begin{bmatrix} y \\ x \end{bmatrix} = \begin{bmatrix} x \\ y \end{bmatrix}.$$

Every permutation matrix is an orthogonal matrix.

Example 3 **(Reflection)** If $\boldsymbol{u}$ is any unit vector, set $Q = I - 2\boldsymbol{u}\boldsymbol{u}^{\mathrm{T}}$. Notice that $\boldsymbol{u}\boldsymbol{u}^{\mathrm{T}}$ is a matrix while $\boldsymbol{u}^{\mathrm{T}}\boldsymbol{u}$ is the number $\|\boldsymbol{u}\|^2 = 1$. Then Q^{T} and Q^{-1} both equal Q:

$$\boxed{Q^{\mathrm{T}} = I - 2\boldsymbol{u}\boldsymbol{u}^{\mathrm{T}} = Q} \quad \text{and} \quad Q^{\mathrm{T}}Q = I - 4\boldsymbol{u}\boldsymbol{u}^{\mathrm{T}} + 4\boldsymbol{u}\boldsymbol{u}^{\mathrm{T}}\boldsymbol{u}\boldsymbol{u}^{\mathrm{T}} = I. \qquad (2)$$

Reflection matrices $I - 2\boldsymbol{u}\boldsymbol{u}^{\mathrm{T}}$ are symmetric and also orthogonal. If you square them, you get the identity matrix: $Q^2 = Q^{\mathrm{T}}Q = I$. Reflecting twice through a mirror brings back the original, like $(-1)^2 = 1$. Notice $\boldsymbol{u}^{\mathrm{T}}\boldsymbol{u} = 1$ inside $4\boldsymbol{u}\boldsymbol{u}^{\mathrm{T}}\boldsymbol{u}\boldsymbol{u}^{\mathrm{T}}$ in equation (2).

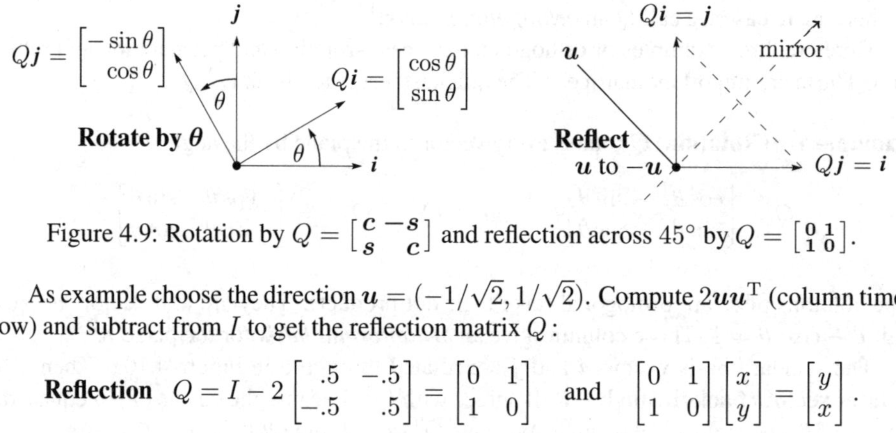

Figure 4.9: Rotation by $Q = \begin{bmatrix} c & -s \\ s & c \end{bmatrix}$ and reflection across 45° by $Q = \begin{bmatrix} 0 & 1 \\ 1 & 0 \end{bmatrix}$.

As example choose the direction $\boldsymbol{u} = (-1/\sqrt{2}, 1/\sqrt{2})$. Compute $2\boldsymbol{u}\boldsymbol{u}^{\mathrm{T}}$ (column times row) and subtract from I to get the reflection matrix Q:

$$\textbf{Reflection}\quad Q = I - 2\begin{bmatrix} .5 & -.5 \\ -.5 & .5 \end{bmatrix} = \begin{bmatrix} 0 & 1 \\ 1 & 0 \end{bmatrix} \quad \text{and} \quad \begin{bmatrix} 0 & 1 \\ 1 & 0 \end{bmatrix} \begin{bmatrix} x \\ y \end{bmatrix} = \begin{bmatrix} y \\ x \end{bmatrix}.$$

When (x, y) goes to (y, x), a vector like $(3, 3)$ doesn't move. It is on the mirror line.

Rotations preserve the length of every vector. So do reflections. So do permutations. So does multiplication by any orthogonal matrix Q—***lengths and angles don't change***.

Proof $\|Q\boldsymbol{x}\|^2$ equals $\|\boldsymbol{x}\|^2$ because $(Q\boldsymbol{x})^{\mathrm{T}}(Q\boldsymbol{x}) = \boldsymbol{x}^{\mathrm{T}}Q^{\mathrm{T}}Q\boldsymbol{x} = \boldsymbol{x}^{\mathrm{T}}I\boldsymbol{x} = \boldsymbol{x}^{\mathrm{T}}\boldsymbol{x}$.

If Q has orthonormal columns ($Q^{\mathrm{T}}Q = I$), it leaves lengths unchanged:

$$\textbf{Same length for } Q\boldsymbol{x} \qquad \|Q\boldsymbol{x}\| = \|\boldsymbol{x}\| \text{ for every vector } \boldsymbol{x}. \qquad (3)$$

***Q* also preserves dot products**: $(Q\boldsymbol{x})^{\mathrm{T}}(Q\boldsymbol{y}) = \boldsymbol{x}^{\mathrm{T}}Q^{\mathrm{T}}Q\boldsymbol{y} = \boldsymbol{x}^{\mathrm{T}}\boldsymbol{y}$. Just use $Q^{\mathrm{T}}Q = I$!

Projections QQ^{T} Using Orthonormal Bases: Q Replaces A

Orthogonal matrices are excellent for computations—numbers can never grow too large when lengths of vectors are fixed. Stable computer codes use Q's as much as possible.

For projections onto subspaces, all formulas involve $A^{\mathrm{T}}A$. The entries of $A^{\mathrm{T}}A$ are the dot products $\boldsymbol{a}_i^{\mathrm{T}}\boldsymbol{a}_j$ of the basis vectors $\boldsymbol{a}_1, \ldots, \boldsymbol{a}_n$.

Suppose the basis vectors are actually orthonormal. The $\boldsymbol{a}$'s become the $\boldsymbol{q}$'s. Then $A^{\mathrm{T}}A$ *simplifies to* $Q^{\mathrm{T}}Q = I$. Look at the improvements in $\widehat{\boldsymbol{x}}$ and $\boldsymbol{p}$ and P.

$$\widehat{\boldsymbol{x}} = Q^{\mathrm{T}}\boldsymbol{b} \quad \text{and} \quad \boldsymbol{p} = Q\widehat{\boldsymbol{x}} \quad \text{and} \quad P = QQ^{\mathrm{T}}. \tag{4}$$

The least squares solution of $Q\boldsymbol{x} = \boldsymbol{b}$ ***is*** $\widehat{\boldsymbol{x}} = Q^{\mathrm{T}}\boldsymbol{b}$. ***The projection matrix is*** QQ^{T}.

There are no matrices to invert. This is the point of an orthonormal basis. The best $\widehat{\boldsymbol{x}} = Q^{\mathrm{T}}\boldsymbol{b}$ just has dot products of $\boldsymbol{q}_1, \ldots, \boldsymbol{q}_n$ with $\boldsymbol{b}$. We have 1-dimensional projections! The "coupling matrix" or "correlation matrix" $A^{\mathrm{T}}A$ is now $Q^{\mathrm{T}}Q = I$. There is no coupling. When A is Q, with orthonormal columns, here is $\boldsymbol{p} = Q\widehat{\boldsymbol{x}} = QQ^{\mathrm{T}}\boldsymbol{b}$:

Projection onto q's

$$\boldsymbol{p} = \begin{bmatrix} | & & | \\ \boldsymbol{q}_1 & \cdots & \boldsymbol{q}_n \\ | & & | \end{bmatrix} \begin{bmatrix} \boldsymbol{q}_1^{\mathrm{T}}\boldsymbol{b} \\ \vdots \\ \boldsymbol{q}_n^{\mathrm{T}}\boldsymbol{b} \end{bmatrix} = \boldsymbol{q}_1(\boldsymbol{q}_1^{\mathrm{T}}\boldsymbol{b}) + \cdots + \boldsymbol{q}_n(\boldsymbol{q}_n^{\mathrm{T}}\boldsymbol{b}). \tag{5}$$

Important case: When Q is square and $m = n$, the subspace is the whole space. Then $Q^{\mathrm{T}} = Q^{-1}$ and $\widehat{\boldsymbol{x}} = Q^{\mathrm{T}}\boldsymbol{b}$ is the same as $\boldsymbol{x} = Q^{-1}\boldsymbol{b}$. The solution is exact! The projection of $\boldsymbol{b}$ onto the whole space is $\boldsymbol{b}$ itself. In this case $\boldsymbol{p} = \boldsymbol{b}$ and $P = QQ^{\mathrm{T}} = I$.

You may think that projection onto the whole space is not worth mentioning. But when $\boldsymbol{p} = \boldsymbol{b}$, our formula assembles $\boldsymbol{b}$ out of its 1-dimensional projections. If $\boldsymbol{q}_1, \ldots, \boldsymbol{q}_n$ is an orthonormal basis for the whole space, then Q is square. Every $\boldsymbol{b} = QQ^{\mathrm{T}}\boldsymbol{b}$ *is the sum of its components along the* $\boldsymbol{q}$'s:

$$\boldsymbol{b} = \boldsymbol{q}_1(\boldsymbol{q}_1^{\mathrm{T}}\boldsymbol{b}) + \boldsymbol{q}_2(\boldsymbol{q}_2^{\mathrm{T}}\boldsymbol{b}) + \cdots + \boldsymbol{q}_n(\boldsymbol{q}_n^{\mathrm{T}}\boldsymbol{b}). \tag{6}$$

Transforms $QQ^{\mathrm{T}} = I$ is the foundation of Fourier series and all the great "transforms" of applied mathematics. They break vectors $\boldsymbol{b}$ or functions $f(x)$ into perpendicular pieces. Then by adding the pieces in (6), the inverse transform puts $\boldsymbol{b}$ and $f(x)$ back together.

Example 4 The columns of this orthogonal Q are orthonormal vectors $\boldsymbol{q}_1, \boldsymbol{q}_2, \boldsymbol{q}_3$:

$$m = n = \mathbf{3} \qquad Q = \frac{1}{3}\begin{bmatrix} -1 & 2 & 2 \\ 2 & -1 & 2 \\ 2 & 2 & -1 \end{bmatrix} \quad \text{has} \quad \boldsymbol{Q^{\mathrm{T}}Q = QQ^{\mathrm{T}} = I}.$$

The separate projections of $\boldsymbol{b} = (0, 0, 1)$ onto $\boldsymbol{q}_1$ and $\boldsymbol{q}_2$ and $\boldsymbol{q}_3$ are $\boldsymbol{p}_1$ and $\boldsymbol{p}_2$ and $\boldsymbol{p}_3$:

$$\boldsymbol{q}_1(\boldsymbol{q}_1^{\mathrm{T}}\boldsymbol{b}) = \tfrac{2}{3}\boldsymbol{q}_1 \quad \text{and} \quad \boldsymbol{q}_2(\boldsymbol{q}_2^{\mathrm{T}}\boldsymbol{b}) = \tfrac{2}{3}\boldsymbol{q}_2 \quad \text{and} \quad \boldsymbol{q}_3(\boldsymbol{q}_3^{\mathrm{T}}\boldsymbol{b}) = -\tfrac{1}{3}\boldsymbol{q}_3.$$

The sum of the first two is the projection of $\boldsymbol{b}$ onto the *plane* of $\boldsymbol{q}_1$ and $\boldsymbol{q}_2$. The sum of all three is the projection of $\boldsymbol{b}$ onto the *whole space*—which is $\boldsymbol{p}_1 + \boldsymbol{p}_2 + \boldsymbol{p}_3 = \boldsymbol{b}$ itself:

Reconstruct $\boldsymbol{b}$
$\boldsymbol{b} = \boldsymbol{p}_1 + \boldsymbol{p}_2 + \boldsymbol{p}_3$

$$\frac{2}{3}\boldsymbol{q}_1 + \frac{2}{3}\boldsymbol{q}_2 - \frac{1}{3}\boldsymbol{q}_3 = \frac{1}{9}\begin{bmatrix} -2+4-2 \\ 4-2-2 \\ 4+4+1 \end{bmatrix} = \begin{bmatrix} 0 \\ 0 \\ 1 \end{bmatrix} = \boldsymbol{b}.$$

The Gram-Schmidt Process

The point of this section is that "orthogonal is good". Projections and least squares always involve $A^{\mathrm{T}}A$. When this matrix becomes $Q^{\mathrm{T}}Q = I$, the inverse is no problem. The one-dimensional projections are uncoupled. The best $\widehat{\boldsymbol{x}}$ is $Q^{\mathrm{T}}\boldsymbol{b}$ (just n separate dot products). For this to be true, we had to say "*If* the vectors are orthonormal". ***Now we explain the "Gram-Schmidt way" to create orthonormal vectors.***

Start with three **independent** vectors $\boldsymbol{a}, \boldsymbol{b}, \boldsymbol{c}$. We intend to construct three **orthogonal** vectors $\boldsymbol{A}, \boldsymbol{B}, \boldsymbol{C}$. Then (at the end or as we go) we divide $\boldsymbol{A}, \boldsymbol{B}, \boldsymbol{C}$ by their lengths. That produces three orthonormal vectors $\boldsymbol{q}_1 = \boldsymbol{A}/\|\boldsymbol{A}\|$, $\boldsymbol{q}_2 = \boldsymbol{B}/\|\boldsymbol{B}\|$, $\boldsymbol{q}_3 = \boldsymbol{C}/\|\boldsymbol{C}\|$.

Gram-Schmidt Begin by choosing $\boldsymbol{A} = \boldsymbol{a}$. This first direction is accepted as it comes. The next direction $\boldsymbol{B}$ must be perpendicular to $\boldsymbol{A}$. ***Start with $\boldsymbol{b}$ and subtract its projection along $\boldsymbol{A}$.*** This leaves the perpendicular part, which is the orthogonal vector $\boldsymbol{B}$:

First Gram-Schmidt step
$\boldsymbol{B}$ is perpendicular to $\boldsymbol{A}$

$$\boxed{\boldsymbol{B} = \boldsymbol{b} - \frac{\boldsymbol{A}^{\mathrm{T}}\boldsymbol{b}}{\boldsymbol{A}^{\mathrm{T}}\boldsymbol{A}}\boldsymbol{A}.} \tag{7}$$

$\boldsymbol{A}$ and $\boldsymbol{B}$ are orthogonal in Figure 4.11. Multiply equation (7) by $\boldsymbol{A}^{\mathrm{T}}$ to verify that $\boldsymbol{A}^{\mathrm{T}}\boldsymbol{B} = \boldsymbol{A}^{\mathrm{T}}\boldsymbol{b} - \boldsymbol{A}^{\mathrm{T}}\boldsymbol{b} = \boldsymbol{0}$. This vector $\boldsymbol{B}$ is what we have called the error vector $\boldsymbol{e}$, perpendicular to $\boldsymbol{A}$. Notice that $\boldsymbol{B}$ in equation (7) is not zero (otherwise $\boldsymbol{a}$ and $\boldsymbol{b}$ would be dependent). The directions $\boldsymbol{A}$ and $\boldsymbol{B}$ are now set.

The third direction starts with $\boldsymbol{c}$. This is not a combination of $\boldsymbol{A}$ and $\boldsymbol{B}$ (because $\boldsymbol{c}$ is not a combination of $\boldsymbol{a}$ and $\boldsymbol{b}$). But most likely $\boldsymbol{c}$ is not perpendicular to $\boldsymbol{A}$ and $\boldsymbol{B}$. So subtract off its components in those two directions to get a perpendicular direction $\boldsymbol{C}$:

Next Gram-Schmidt step
$\boldsymbol{C}$ is perpendicular to $\boldsymbol{A}$ and $\boldsymbol{B}$

$$\boxed{\boldsymbol{C} = \boldsymbol{c} - \frac{\boldsymbol{A}^{\mathrm{T}}\boldsymbol{c}}{\boldsymbol{A}^{\mathrm{T}}\boldsymbol{A}}\boldsymbol{A} - \frac{\boldsymbol{B}^{\mathrm{T}}\boldsymbol{c}}{\boldsymbol{B}^{\mathrm{T}}\boldsymbol{B}}\boldsymbol{B}.} \tag{8}$$

This is the one and only idea of the Gram-Schmidt process. ***Subtract from every new vector its projections in the directions already set.*** That idea is repeated at every step.[2] If we had a fourth vector $\boldsymbol{d}$, we would subtract three projections onto $\boldsymbol{A}, \boldsymbol{B}, \boldsymbol{C}$ to get $\boldsymbol{D}$. At the end, *or immediately when each one is found*, divide the orthogonal vectors $\boldsymbol{A}, \boldsymbol{B}, \boldsymbol{C}, \boldsymbol{D}$ by their lengths. The resulting vectors $\boldsymbol{q}_1, \boldsymbol{q}_2, \boldsymbol{q}_3, \boldsymbol{q}_4$ are orthonormal.

---

[2] I think Gram had the idea. I don't really know where Schmidt came in.

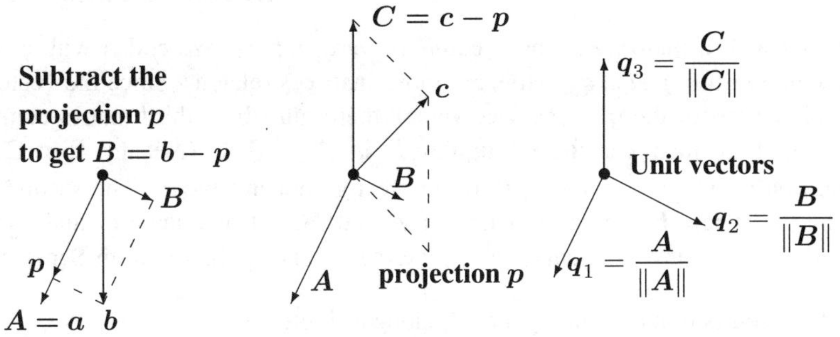

Figure 4.10: First project $\boldsymbol{b}$ onto the line through $\boldsymbol{a}$ and find the orthogonal $\boldsymbol{B} = \boldsymbol{b} - \boldsymbol{p}$. Then project $\boldsymbol{c}$ to $\boldsymbol{p}$ on the $\boldsymbol{AB}$ plane and find $\boldsymbol{C} = \boldsymbol{c} - \boldsymbol{p}$. Divide by $\|\boldsymbol{A}\|$, $\|\boldsymbol{B}\|$, $\|\boldsymbol{C}\|$.

Example of **Gram-Schmidt** Suppose the independent non-orthogonal vectors $\boldsymbol{a}, \boldsymbol{b}, \boldsymbol{c}$ are

$$\boldsymbol{a} = \begin{bmatrix} \mathbf{1} \\ \mathbf{-1} \\ \mathbf{0} \end{bmatrix} \quad \text{and} \quad \boldsymbol{b} = \begin{bmatrix} 2 \\ 0 \\ -2 \end{bmatrix} \quad \text{and} \quad \boldsymbol{c} = \begin{bmatrix} 3 \\ -3 \\ 3 \end{bmatrix}.$$

Then $\boldsymbol{A} = \boldsymbol{a}$ has $\boldsymbol{A}^{\mathrm{T}}\boldsymbol{A} = 2$ and $A^{\mathrm{T}}\boldsymbol{b} = 2$. Subtract from $\boldsymbol{b}$ its projection $\boldsymbol{p}$ along $\boldsymbol{A}$:

First step $$\boldsymbol{B} = \boldsymbol{b} - \frac{\boldsymbol{A}^{\mathrm{T}}\boldsymbol{b}}{\boldsymbol{A}^{\mathrm{T}}\boldsymbol{A}}\boldsymbol{A} = \boldsymbol{b} - \frac{2}{2}\boldsymbol{A} = \begin{bmatrix} \mathbf{1} \\ \mathbf{1} \\ \mathbf{-2} \end{bmatrix}.$$

Check: $\boldsymbol{A}^{\mathrm{T}}\boldsymbol{B} = 0$ as required. Now subtract the projections of $\boldsymbol{c}$ on $\boldsymbol{A}$ and $\boldsymbol{B}$ to get $\boldsymbol{C}$: **Two** projections to subtract for $\boldsymbol{C}$.

Next step $$\boldsymbol{C} = \boldsymbol{c} - \frac{\boldsymbol{A}^{\mathrm{T}}\boldsymbol{c}}{\boldsymbol{A}^{\mathrm{T}}\boldsymbol{A}}\boldsymbol{A} - \frac{\boldsymbol{B}^{\mathrm{T}}\boldsymbol{c}}{\boldsymbol{B}^{\mathrm{T}}\boldsymbol{B}}\boldsymbol{B} = \boldsymbol{c} - \frac{6}{2}\boldsymbol{A} + \frac{6}{6}\boldsymbol{B} = \begin{bmatrix} \mathbf{1} \\ \mathbf{1} \\ \mathbf{1} \end{bmatrix}.$$

Check: $\boldsymbol{C} = (1, 1, 1)$ is perpendicular to both $\boldsymbol{A}$ and $\boldsymbol{B}$. Finally convert $\boldsymbol{A}, \boldsymbol{B}, \boldsymbol{C}$ to unit vectors (length 1, orthonormal). The lengths of $\boldsymbol{A}$, $\boldsymbol{B}$, $\boldsymbol{C}$ are $\sqrt{2}$ and $\sqrt{6}$ and $\sqrt{3}$. Divide by those lengths, for an orthonormal basis $\boldsymbol{q}_1, \boldsymbol{q}_2, \boldsymbol{q}_3$.

$$q_1 = \frac{1}{\sqrt{2}}\begin{bmatrix} 1 \\ -1 \\ 0 \end{bmatrix} \quad \text{and} \quad q_2 = \frac{1}{\sqrt{6}}\begin{bmatrix} 1 \\ 1 \\ -2 \end{bmatrix} \quad \text{and} \quad q_3 = \frac{1}{\sqrt{3}}\begin{bmatrix} 1 \\ 1 \\ 1 \end{bmatrix}.$$

Usually $\boldsymbol{A}, \boldsymbol{B}, \boldsymbol{C}$ contain fractions. Almost always $\boldsymbol{q}_1, \boldsymbol{q}_2, \boldsymbol{q}_3$ will contain square roots. That is the price we pay for unit vectors and $Q^{\mathrm{T}}Q = I$.

The Factorization $A = QR$

We started with a matrix A, whose columns were $\boldsymbol{a}, \boldsymbol{b}, \boldsymbol{c}$. We ended with a matrix Q, whose columns are $\boldsymbol{q}_1, \boldsymbol{q}_2, \boldsymbol{q}_3$. How are those matrices related? Since the vectors $\boldsymbol{a}, \boldsymbol{b}, \boldsymbol{c}$ are combinations of the $\boldsymbol{q}$'s (and vice versa), there must be a third matrix connecting A to Q. This third matrix is **the triangular $\boldsymbol{R}$ in $\boldsymbol{A} = \boldsymbol{QR}$.** (Not the R of Chapter 1.)

The first step was $\boldsymbol{q}_1 = \boldsymbol{a}/\|\boldsymbol{a}\|$ (other vectors not involved). The second step was equation (7), where $\boldsymbol{b}$ is a combination of $\boldsymbol{A}$ and $\boldsymbol{B}$. At that stage $\boldsymbol{C}$ and $\boldsymbol{q}_3$ were not involved. This non-involvement of later vectors is the key point of Gram-Schmidt:

- The vectors $\boldsymbol{a}$ and $\boldsymbol{A}$ and $\boldsymbol{q}_1$ are all along a single line.
- The vectors $\boldsymbol{a}, \boldsymbol{b}$ and $\boldsymbol{A}, \boldsymbol{B}$ and $\boldsymbol{q}_1, \boldsymbol{q}_2$ are all in the same plane.
- The vectors $\boldsymbol{a}, \boldsymbol{b}, \boldsymbol{c}$ and $\boldsymbol{A}, \boldsymbol{B}, \boldsymbol{C}$ and $\boldsymbol{q}_1, \boldsymbol{q}_2, \boldsymbol{q}_3$ are in one subspace (dimension 3).

At every step $\boldsymbol{a}_1, \ldots, \boldsymbol{a}_k$ are combinations of $\boldsymbol{q}_1, \ldots, \boldsymbol{q}_k$. Later $\boldsymbol{q}$'s are not involved. The connecting matrix R is ***triangular***, and we have $A = \boldsymbol{Q}$ **times** $\boldsymbol{R}$.

$$\begin{bmatrix} & & \\ \boldsymbol{a} & \boldsymbol{b} & \boldsymbol{c} \\ & & \end{bmatrix} = \begin{bmatrix} & & \\ \boldsymbol{q}_1 & \boldsymbol{q}_2 & \boldsymbol{q}_3 \\ & & \end{bmatrix} \begin{bmatrix} \boldsymbol{q}_1^{\mathrm{T}}\boldsymbol{a} & \boldsymbol{q}_1^{\mathrm{T}}\boldsymbol{b} & \boldsymbol{q}_1^{\mathrm{T}}\boldsymbol{c} \\ & \boldsymbol{q}_2^{\mathrm{T}}\boldsymbol{b} & \boldsymbol{q}_2^{\mathrm{T}}\boldsymbol{c} \\ & & \boldsymbol{q}_3^{\mathrm{T}}\boldsymbol{c} \end{bmatrix} \quad \text{or} \quad \boldsymbol{A} = \boldsymbol{QR}. \tag{9}$$

$A = QR$ is Gram-Schmidt in a nutshell. Multiply by Q^{T} to see $\boldsymbol{R} = \boldsymbol{Q}^{\mathrm{T}}\boldsymbol{A}$ in (9).

(Gram-Schmidt) From independent vectors $\boldsymbol{a}_1, \ldots, \boldsymbol{a}_n$, Gram-Schmidt constructs orthonormal vectors $\boldsymbol{q}_1, \ldots, \boldsymbol{q}_n$. The matrices with these columns satisfy $\boldsymbol{A} = \boldsymbol{QR}$. Then $\boldsymbol{R} = \boldsymbol{Q}^{\mathrm{T}}\boldsymbol{A}$ ***is upper triangular*** because later $\boldsymbol{q}$'s are orthogonal to earlier $\boldsymbol{a}$'s.

Here are the original $\boldsymbol{a}$'s and the final $\boldsymbol{q}$'s from the example. The i, j entry of $R = Q^{\mathrm{T}}A$ is row i of Q^{T} times column j of A. The dot products $\boldsymbol{q}_i^{\mathrm{T}}\boldsymbol{a}_j$ go into R. **Then $\boldsymbol{A} = \boldsymbol{QR}$:**

$$A = \begin{bmatrix} 1 & 2 & 3 \\ -1 & 0 & -3 \\ 0 & -2 & 3 \end{bmatrix} = \begin{bmatrix} 1/\sqrt{2} & 1/\sqrt{6} & 1/\sqrt{3} \\ -1/\sqrt{2} & 1/\sqrt{6} & 1/\sqrt{3} \\ 0 & -2/\sqrt{6} & 1/\sqrt{3} \end{bmatrix} \begin{bmatrix} \boldsymbol{\sqrt{2}} & \sqrt{2} & \sqrt{18} \\ \boldsymbol{0} & \boldsymbol{\sqrt{6}} & -\sqrt{6} \\ \boldsymbol{0} & \boldsymbol{0} & \boldsymbol{\sqrt{3}} \end{bmatrix} = QR.$$

Look closely at Q and R. The lengths of $\boldsymbol{A}, \boldsymbol{B}, \boldsymbol{C}$ are $\sqrt{2}, \sqrt{6}, \sqrt{3}$ on the diagonal of R. The columns of Q are orthonormal. Because of the square roots, QR might look harder than LU. Both factorizations are absolutely central to calculations in linear algebra.

Any m by n matrix A with independent columns can be factored into $A = QR$. The m by n matrix Q has orthonormal columns, and the square matrix R is upper triangular with positive diagonal. We must not forget why this is useful for least squares: $\boldsymbol{A}^{\mathrm{T}}\boldsymbol{A} = (\boldsymbol{QR})^{\mathrm{T}}\boldsymbol{QR} = \boldsymbol{R}^{\mathrm{T}}\boldsymbol{Q}^{\mathrm{T}}\boldsymbol{QR} = \boldsymbol{R}^{\mathrm{T}}\boldsymbol{R}$. Then $A^{\mathrm{T}}A\widehat{\boldsymbol{x}} = A^{\mathrm{T}}\boldsymbol{b}$ simplifies to

$$R^{\mathrm{T}}R\widehat{\boldsymbol{x}} = R^{\mathrm{T}}Q^{\mathrm{T}}\boldsymbol{b}. \text{ Finally } \boldsymbol{R\widehat{x}} = \boldsymbol{Q}^{\mathrm{T}}\boldsymbol{b} \textbf{ with triangular } \boldsymbol{R}.$$

$$\textbf{Least squares} \qquad R^{\mathrm{T}}R\widehat{\boldsymbol{x}} = R^{\mathrm{T}}Q^{\mathrm{T}}\boldsymbol{b} \text{ or } R\widehat{\boldsymbol{x}} = Q^{\mathrm{T}}\boldsymbol{b} \text{ or } \widehat{\boldsymbol{x}} = R^{-1}Q^{\mathrm{T}}\boldsymbol{b} \qquad (10)$$

Instead of solving $A\boldsymbol{x} = \boldsymbol{b}$, which is impossible, we solve $R\widehat{\boldsymbol{x}} = Q^{\mathrm{T}}\boldsymbol{b}$ by back substitution—which is very fast. The real cost is the mn^2 multiplications for Gram-Schmidt, which are needed to construct the orthogonal Q and the triangular R with $A = QR$.

Below is an informal code. It executes equations (11) for $j = 1$ then $j = 2$ and eventually $j = n$. The important lines 4-5 subtract from $\boldsymbol{v} = \boldsymbol{a}_j$ its projection onto each $\boldsymbol{q}_i, i < j$. The last line of that code normalizes $\boldsymbol{v}$ (divides by $r_{jj} = ||\boldsymbol{v}||$) to get the unit vector $\boldsymbol{q}_j$:

$$\boldsymbol{r_{kj}} = \sum_{i=1}^{m} q_{ik}v_{ij} \text{ and } \boldsymbol{v_{ij}} = v_{ij} - q_{ik}r_{kj} \text{ and } \boldsymbol{r_{jj}} = \left(\sum_{i=1}^{m} v_{ij}^2\right)^{1/2} \text{ and } \boldsymbol{q_{ij}} = \frac{v_{ij}}{r_{jj}}. \quad (11)$$

Starting from $\boldsymbol{a}, \boldsymbol{b}, \boldsymbol{c} = \boldsymbol{a}_1, \boldsymbol{a}_2, \boldsymbol{a}_3$ this code will construct $\boldsymbol{q}_1$, then $\boldsymbol{B}$, $\boldsymbol{q}_2$, then $\boldsymbol{C}$, $\boldsymbol{q}_3$:

$$\boldsymbol{q}_1 = \boldsymbol{a}_1/\|\boldsymbol{a}_1\| \qquad \boldsymbol{B} = \boldsymbol{a}_2 - (\boldsymbol{q}_1^{\mathrm{T}}\boldsymbol{a}_2)\boldsymbol{q}_1 \qquad \boldsymbol{q}_2 = \boldsymbol{B}/\|\boldsymbol{B}\|$$

$$\boldsymbol{C}^* = \boldsymbol{a}_3 - (\boldsymbol{q}_1^{\mathrm{T}}\boldsymbol{a}_3)\boldsymbol{q}_1 \qquad \boldsymbol{C} = \boldsymbol{C}^* - (\boldsymbol{q}_2^{\mathrm{T}}\boldsymbol{C}^*)\boldsymbol{q}_2 \qquad \boldsymbol{q}_3 = \boldsymbol{C}/\|\boldsymbol{C}\|$$

Equation (11) subtracts **one projection at a time** as in $\boldsymbol{C}^*$ and $\boldsymbol{C}$. That change is called ***modified Gram-Schmidt***. This code is numerically more stable than equation (8) which subtracts all projections at once. We are using MATLAB notation.

```
for j = 1:n                          % modified Gram-Schmidt in n loops
  v = A(:, j);                       % v begins as column j of the original A
  for i = 1:j-1                      % columns q_1 to q_{j-1} are already settled in Q
    R(i, j) = Q(:, i)' * v;          % compute R_ij = q_i^T a_j which is q_i^T v
    v = v - R(i, j)*Q(:, i);         % subtract the projection (q_i^T v)q_i
  end                                % v is now perpendicular to all of q_1, ..., q_{j-1}
  R(j, j) = norm(v);                 % the diagonal entries R_jj in R are lengths
  Q(:, j) = v/R(j, j);               % divide v by its length to get the next q_j
end                                  % for j = 1 : n produces all of the q_j
```

To recover column j of A, undo the last step and the middle steps of the code:

$$R(j,j)\boldsymbol{q_j} = (\boldsymbol{v} \textbf{\textit{ minus its projections}}) = (\text{column } j \text{ of } A) - \sum_{i=1}^{j-1} R(i,j)\boldsymbol{q_i}\,. \qquad (12)$$

Moving the sum to the far left, equation (12) is column j in the multiplication $QR = A$.

Confession Good software like LAPACK, used in good systems like MATLAB and **Julia** and **Python**, might not use this Gram-Schmidt code. There is now another way. "Householder reflections" act on A to produce the upper triangular R. This happens one column at a time in the same way that elimination produces the upper triangular U in LU.

If A is tridiagonal we can simplify even more to use 2 by 2 rotations. The result is always $A = QR$ and the MATLAB command to orthogonalize A is $[Q, R] = \mathsf{qr}(A)$. Gram-Schmidt is still the good process to understand, and the most stable codes look for and use the **largest vector** $\boldsymbol{a}_j$ at each new step.

■ REVIEW OF THE KEY IDEAS ■

1. If the orthonormal vectors $\boldsymbol{q}_1, \ldots, \boldsymbol{q}_n$ are the columns of Q, then $\boldsymbol{q}_i^{\mathrm{T}} \boldsymbol{q}_j = 0$ and $\boldsymbol{q}_i^{\mathrm{T}} \boldsymbol{q}_i = 1$ translate into the matrix multiplication $Q^{\mathrm{T}} Q = I$.
2. If Q is square (an ***orthogonal matrix***) then $Q^{\mathrm{T}} = Q^{-1}$: ***transpose*** $=$ ***inverse***.
3. The length of $Q\boldsymbol{x}$ equals the length of $\boldsymbol{x}$: $\|Q\boldsymbol{x}\| = \|\boldsymbol{x}\|$.
4. The projection onto the column space of Q spanned by the $\boldsymbol{q}$'s is $P = QQ^{\mathrm{T}}$.
5. If Q is square then $P = QQ^{\mathrm{T}} = I$ and every $\boldsymbol{b} = \boldsymbol{q}_1(\boldsymbol{q}_1^{\mathrm{T}}\boldsymbol{b}) + \cdots + \boldsymbol{q}_n(\boldsymbol{q}_n^{\mathrm{T}}\boldsymbol{b})$.
6. Gram-Schmidt produces orthonormal vectors $\boldsymbol{q}_1, \boldsymbol{q}_2, \boldsymbol{q}_3$ from independent $\boldsymbol{a}, \boldsymbol{b}, \boldsymbol{c}$. In matrix form this is the factorization $\boldsymbol{A} =$ **(orthogonal $\boldsymbol{Q}$)(triangular $\boldsymbol{R}$)**.

■ WORKED EXAMPLES ■

4.4 A Add two more columns with all entries 1 or -1, so the columns of this 4 by 4 "Hadamard matrix" are orthogonal. How do you turn H_4 into an *orthogonal matrix* Q?

$$H_2 = \begin{bmatrix} 1 & 1 \\ 1 & -1 \end{bmatrix} \qquad H_4 = \begin{bmatrix} 1 & 1 & x & x \\ 1 & -1 & x & x \\ 1 & 1 & x & x \\ 1 & -1 & x & x \end{bmatrix} \qquad \text{and} \qquad Q_4 = \begin{bmatrix} & & & \\ & & & \\ & & & \\ & & & \end{bmatrix}$$

The block matrix $H_8 = \begin{bmatrix} H_4 & H_4 \\ H_4 & -H_4 \end{bmatrix}$ is the next Hadamard matrix with 1's and -1's. What is the product $H_8^{\mathrm{T}} H_8$?

The projection of $\boldsymbol{b} = (6, 0, 0, 2)$ onto the first column of H_4 is $\boldsymbol{p}_1 = (2, 2, 2, 2)$. The projection onto the second column is $\boldsymbol{p}_2 = (1, -1, 1, -1)$. What is the projection $\boldsymbol{p}_{1,2}$ of $\boldsymbol{b}$ onto the 2-dimensional space spanned by the first two columns?

Solution H_4 can be built from H_2 just as H_8 is built from H_4:

$$H_4 = \begin{bmatrix} H_2 & H_2 \\ H_2 & -H_2 \end{bmatrix} = \begin{bmatrix} 1 & 1 & 1 & 1 \\ 1 & -1 & 1 & -1 \\ 1 & 1 & -1 & -1 \\ 1 & -1 & -1 & 1 \end{bmatrix} \text{ has orthogonal columns.}$$

Then $Q = H/2$ has orthonormal columns. Dividing by 2 gives unit vectors in Q. A 5 by 5 Hadamard matrix is impossible because the dot product of columns would have five 1's and/or -1's and could not add to zero. H_8 has orthogonal columns of length $\sqrt{8}$.

$$H_8^{\mathrm T} H_8 = \begin{bmatrix} H^{\mathrm T} & H^{\mathrm T} \\ H^{\mathrm T} & -H^{\mathrm T} \end{bmatrix} \begin{bmatrix} H & H \\ H & -H \end{bmatrix} = \begin{bmatrix} 2H^{\mathrm T}H & 0 \\ 0 & 2H^{\mathrm T}H \end{bmatrix} = \begin{bmatrix} 8I & 0 \\ 0 & 8I \end{bmatrix}. \; Q_8 = \frac{H_8}{\sqrt{8}}$$

4.4 B What is the key point of orthogonal columns? Answer: $Q^{\mathrm T}Q$ **is diagonal** and easy to invert. **We can project onto lines and just add**. The axes are orthogonal.

Orthogonal Matrices Q and Orthogonal Projections P

$\boldsymbol{A}$ has n independent columns in $\mathbf{R}^m$	$\boldsymbol{Q}$ has n orthonormal columns in $\mathbf{R}^m$
$(m \times n)\quad \boldsymbol{A} = \begin{bmatrix} \boldsymbol{a}_1 & \cdots & \boldsymbol{a}_n \end{bmatrix}$	$\boldsymbol{Q} = \begin{bmatrix} \boldsymbol{q}_1 & \cdots & \boldsymbol{q}_n \end{bmatrix}$
$A^{\mathrm T}A$ is symmetric and invertible	$Q^{\mathrm T}Q = I$
Projection matrix P onto the column space	$\boldsymbol{Pb}$ = nearest point to $\boldsymbol{b}$
$\boldsymbol{P} = \boldsymbol{A}(\boldsymbol{A}^{\mathrm T}\boldsymbol{A})^{-1}\boldsymbol{A}^{\mathrm T}$	$\boldsymbol{P} = \boldsymbol{Q}\boldsymbol{Q}^{\mathrm T}$
Least squares solution to $Ax = b$	**Least squares solution to $Qx = b$**
$\boldsymbol{A}^{\mathrm T}\boldsymbol{A}\widehat{\boldsymbol{x}} = \boldsymbol{A}^{\mathrm T}\boldsymbol{b}$	$\widehat{\boldsymbol{x}} = \boldsymbol{Q}^{\mathrm T}\boldsymbol{b}$
Gram-Schmidt orthogonalization $\boldsymbol{A} = \boldsymbol{QR}$	**Independent columns a_j** **Orthonormal columns q_j**

$$\begin{bmatrix} \boldsymbol{a}_1 & \cdot & \cdot & \boldsymbol{a}_n \end{bmatrix} = \begin{bmatrix} \boldsymbol{q}_1 & \cdot & \cdot & \boldsymbol{q}_n \end{bmatrix} \begin{bmatrix} r_{11} & r_{12} & \cdot & r_{1n} \\ 0 & r_{22} & \cdot & r_{2n} \\ 0 & 0 & \cdot & \cdot \\ 0 & 0 & 0 & r_{nn} \end{bmatrix}$$

Pseudoinverse A^+ of $A = QR$	**Pseudoinverse Q^+ of Q**
$\boldsymbol{A}^+ = \boldsymbol{R}^+\boldsymbol{Q}^+ = \boldsymbol{R}^{-1}\boldsymbol{Q}^{\mathrm T}$	$\boldsymbol{Q}^+ = \boldsymbol{Q}^{\mathrm T}$
Pseudoinverse A^+ of $A = CR$ in 4.5	$\boldsymbol{A}^+ = \boldsymbol{R}^+\boldsymbol{C}^+ = \boldsymbol{R}^{\mathrm T}(\boldsymbol{C}^{\mathrm T}\boldsymbol{A}\boldsymbol{R}^{\mathrm T})^{-1}\boldsymbol{C}^{\mathrm T}$

Problem Set 4.4

Problems 1–12 are about orthogonal vectors and orthogonal matrices.

1 Are these pairs of vectors orthonormal or only orthogonal or only independent?

$$\text{(a)}\ \begin{bmatrix}1\\0\end{bmatrix} \text{ and } \begin{bmatrix}-1\\1\end{bmatrix} \qquad \text{(b)}\ \begin{bmatrix}.6\\.8\end{bmatrix} \text{ and } \begin{bmatrix}.4\\-.3\end{bmatrix} \qquad \text{(c)}\ \begin{bmatrix}\cos\theta\\ \sin\theta\end{bmatrix} \text{ and } \begin{bmatrix}-\sin\theta\\ \cos\theta\end{bmatrix}.$$

Change the second vector when necessary to produce orthonormal vectors.

2 The vectors $(2,2,-1)$ and $(-1,2,2)$ are orthogonal. Divide them by their lengths to find orthonormal vectors $\boldsymbol{q}_1$ and $\boldsymbol{q}_2$. Put those into the columns of Q and multiply $Q^{\mathrm{T}}Q$ and QQ^{T}.

3 (a) If A has three orthogonal columns each of length 4, what is $A^{\mathrm{T}}A$?

(b) If A has three orthogonal columns of lengths $1,2,3$, what is $A^{\mathrm{T}}A$?

4 Give an example of each of the following:

(a) A matrix Q that has orthonormal columns but $QQ^{\mathrm{T}} \neq I$.

(b) Two orthogonal vectors that are not linearly independent.

(c) An orthonormal basis for $\mathbf{R}^3$, including the vector $\boldsymbol{q}_1 = (1,1,1)/\sqrt{3}$.

5 Find two orthogonal vectors in the plane $x + y + 2z = 0$. Make them orthonormal.

6 If Q_1 and Q_2 are orthogonal matrices, show that their product Q_1Q_2 is also an orthogonal matrix. (Use $Q^{\mathrm{T}}Q = I$.)

7 If Q has orthonormal columns, what is the least squares solution $\widehat{\boldsymbol{x}}$ to $Q\boldsymbol{x} = \boldsymbol{b}$?

8 If $\boldsymbol{q}_1$ and $\boldsymbol{q}_2$ are orthonormal vectors in $\mathbf{R}^5$, what combination ____ $\boldsymbol{q}_1 +$ ____ $\boldsymbol{q}_2$ is closest to a given vector $\boldsymbol{b}$?

9 (a) Compute $P = QQ^{\mathrm{T}}$ when $\boldsymbol{q}_1 = (.8,.6,0)$ and $\boldsymbol{q}_2 = (-.6,.8,0)$. Verify that $P^2 = P$.

(b) Prove that always $(QQ^{\mathrm{T}})^2 = QQ^{\mathrm{T}}$ by using $Q^{\mathrm{T}}Q = I$. Then $P = QQ^{\mathrm{T}}$ is the projection matrix onto the column space of Q.

10 Orthonormal vectors are automatically linearly independent.

(a) Vector proof: When $c_1\boldsymbol{q}_1 + c_2\boldsymbol{q}_2 + c_3\boldsymbol{q}_3 = \mathbf{0}$, what dot product leads to $c_1 = 0$? Similarly $c_2 = 0$ and $c_3 = 0$. Thus the $\boldsymbol{q}$'s are independent.

(b) Matrix proof: Show that $Q\boldsymbol{x} = \mathbf{0}$ leads to $\boldsymbol{x} = \mathbf{0}$. Since Q may be rectangular, you can use Q^{T} but not Q^{-1}.

11 (a) Gram-Schmidt: Find orthonormal vectors $\boldsymbol{q}_1$ and $\boldsymbol{q}_2$ in the plane spanned by $\boldsymbol{a} = (1,3,4,5,7)$ and $\boldsymbol{b} = (-6,6,8,0,8)$.

(b) Which vector in this plane is closest to $(1,0,0,0,0)$?

12 If $\boldsymbol{a}_1, \boldsymbol{a}_2, \boldsymbol{a}_3$ is an orthogonal basis for $\mathbf{R}^3$, show that $x_1 = \boldsymbol{a}_1^{\mathrm{T}}\boldsymbol{b}/\boldsymbol{a}_1^{\mathrm{T}}\boldsymbol{a}_1$.

$$\boldsymbol{b} = x_1\boldsymbol{a}_1 + x_2\boldsymbol{a}_2 + x_3\boldsymbol{a}_3 \qquad \text{or} \qquad \begin{bmatrix} & & \\ \boldsymbol{a}_1 & \boldsymbol{a}_2 & \boldsymbol{a}_3 \\ & & \end{bmatrix}\begin{bmatrix} x_1 \\ x_2 \\ x_3 \end{bmatrix} = \boldsymbol{b}.$$

13 What multiple of $\boldsymbol{a} = \left[\begin{smallmatrix}1\\1\end{smallmatrix}\right]$ should be subtracted from $\boldsymbol{b} = \left[\begin{smallmatrix}4\\0\end{smallmatrix}\right]$ to make the result $\boldsymbol{B}$ orthogonal to $\boldsymbol{a}$? Sketch a figure to show $\boldsymbol{a}$, $\boldsymbol{b}$, and $\boldsymbol{B}$.

14 Complete the Gram-Schmidt process in Problem 13 by computing $\boldsymbol{q}_1 = \boldsymbol{a}/\|\boldsymbol{a}\|$ and $\boldsymbol{q}_2 = \boldsymbol{B}/\|\boldsymbol{B}\|$ and factoring into QR:

$$\begin{bmatrix} 1 & 4 \\ 1 & 0 \end{bmatrix} = \begin{bmatrix} \boldsymbol{q}_1 & \boldsymbol{q}_2 \end{bmatrix}\begin{bmatrix} \|\boldsymbol{a}\| & ? \\ 0 & \|\boldsymbol{B}\| \end{bmatrix}.$$

15 (a) Find orthonormal vectors $\boldsymbol{q}_1, \boldsymbol{q}_2, \boldsymbol{q}_3$ such that $\boldsymbol{q}_1, \boldsymbol{q}_2$ span the column space of

$$A = \begin{bmatrix} 1 & 1 \\ 2 & -1 \\ -2 & 4 \end{bmatrix}.$$

(b) Which of the four fundamental subspaces contains $\boldsymbol{q}_3$?

(c) Solve $A\boldsymbol{x} = (1, 2, 7)$ by least squares.

16 What multiple of $\boldsymbol{a} = (4, 5, 2, 2)$ is closest to $\boldsymbol{b} = (1, 2, 0, 0)$? Find orthonormal vectors $\boldsymbol{q}_1$ and $\boldsymbol{q}_2$ in the plane of $\boldsymbol{a}$ and $\boldsymbol{b}$.

17 Find the projection of $\boldsymbol{b}$ onto the line through $\boldsymbol{a}$:

$$\boldsymbol{a} = \begin{bmatrix} 1 \\ 1 \\ 1 \end{bmatrix} \quad \text{and} \quad \boldsymbol{b} = \begin{bmatrix} 1 \\ 3 \\ 5 \end{bmatrix} \quad \text{and} \quad \boldsymbol{p} = ? \quad \text{and} \quad \boldsymbol{e} = \boldsymbol{b} - \boldsymbol{p} = ?$$

Compute the orthonormal vectors $\boldsymbol{q}_1 = \boldsymbol{a}/\|\boldsymbol{a}\|$ and $\boldsymbol{q}_2 = \boldsymbol{e}/\|\boldsymbol{e}\|$.

18 (Recommended) Find orthogonal vectors $\boldsymbol{A}, \boldsymbol{B}, \boldsymbol{C}$ by Gram-Schmidt from $\boldsymbol{a}, \boldsymbol{b}, \boldsymbol{c}$:

$$\boldsymbol{a} = (1, -1, 0, 0) \qquad \boldsymbol{b} = (0, 1, -1, 0) \qquad \boldsymbol{c} = (0, 0, 1, -1).$$

$\boldsymbol{A}, \boldsymbol{B}, \boldsymbol{C}$ and $\boldsymbol{a}, \boldsymbol{b}, \boldsymbol{c}$ are bases for the vectors perpendicular to $\boldsymbol{d} = (1, 1, 1, 1)$.

19 If $A = QR$ then $A^{\mathrm{T}}A = R^{\mathrm{T}}R =$ _____ triangular times _____ triangular. *Gram-Schmidt on A corresponds to elimination on $A^{\mathrm{T}}A$.* The pivots for $A^{\mathrm{T}}A$ must be the squares of diagonal entries of R. Find Q and R by Gram-Schmidt for this A:

$$A = \begin{bmatrix} -1 & 1 \\ 2 & 1 \\ 2 & 4 \end{bmatrix} \quad \text{and} \quad A^{\mathrm{T}}A = \begin{bmatrix} 9 & 9 \\ 9 & 18 \end{bmatrix} = \begin{bmatrix} 1 & 0 \\ 1 & 1 \end{bmatrix}\begin{bmatrix} 9 & \\ & 9 \end{bmatrix}\begin{bmatrix} 1 & 1 \\ 0 & 1 \end{bmatrix}.$$

20 True or false (give an example in either case):

(a) Q^{-1} is an orthogonal matrix when Q is an orthogonal matrix.

(b) If Q (3 by 2) has orthonormal columns then $\|Q\boldsymbol{x}\|$ always equals $\|\boldsymbol{x}\|$.

21 Find an orthonormal basis for the column space of A and project $\boldsymbol{b}$ onto $\mathbf{C}(A)$.

$$A = \begin{bmatrix} 1 & -2 \\ 1 & 0 \\ 1 & 1 \\ 1 & 3 \end{bmatrix} \quad \text{and} \quad \boldsymbol{b} = \begin{bmatrix} -4 \\ -3 \\ 3 \\ 0 \end{bmatrix}.$$

22 Find orthogonal vectors $\boldsymbol{A}, \boldsymbol{B}, \boldsymbol{C}$ by Gram-Schmidt from

$$\boldsymbol{a} = \begin{bmatrix} 1 \\ 1 \\ 2 \end{bmatrix} \quad \text{and} \quad \boldsymbol{b} = \begin{bmatrix} 1 \\ -1 \\ 0 \end{bmatrix} \quad \text{and} \quad \boldsymbol{c} = \begin{bmatrix} 1 \\ 0 \\ 4 \end{bmatrix}.$$

23 Find $\boldsymbol{q}_1, \boldsymbol{q}_2, \boldsymbol{q}_3$ (orthonormal) as combinations of $\boldsymbol{a}, \boldsymbol{b}, \boldsymbol{c}$ (independent columns). Then write A as QR:

$$A = \begin{bmatrix} 1 & 2 & 4 \\ 0 & 0 & 5 \\ 0 & 3 & 6 \end{bmatrix}.$$

24 (a) Find a basis for the subspace $\boldsymbol{S}$ in $\mathbf{R}^4$ spanned by all solutions of

$$x_1 + x_2 + x_3 - x_4 = 0.$$

(b) Find a basis for the orthogonal complement $\boldsymbol{S}^\perp$.

(c) Find $\boldsymbol{b}_1$ in $\boldsymbol{S}$ and $\boldsymbol{b}_2$ in $\boldsymbol{S}^\perp$ so that $\boldsymbol{b}_1 + \boldsymbol{b}_2 = \boldsymbol{b} = (1, 1, 1, 1)$.

25 If $ad - bc > 0$, the entries in $A = QR$ are

$$\begin{bmatrix} a & b \\ c & d \end{bmatrix} = \frac{\begin{bmatrix} a & -c \\ c & a \end{bmatrix}}{\sqrt{a^2 + c^2}} \frac{\begin{bmatrix} a^2 + c^2 & ab + cd \\ 0 & ad - bc \end{bmatrix}}{\sqrt{a^2 + c^2}}.$$

Write $A = QR$ when $a, b, c, d = 2, 1, 1, 1$ and also $1, 1, 1, 1$. Which entry of R becomes zero when the columns are dependent and Gram-Schmidt breaks down?

Problems 26–29 use the QR code in equation (11). It executes Gram-Schmidt.

26 Show why $\boldsymbol{C}$ (found via $\boldsymbol{C}^*$ in the steps after (11)) is equal to $\boldsymbol{C}$ in equation (8).

27 Equation (8) subtracts from $\boldsymbol{c}$ its components along $\boldsymbol{A}$ and $\boldsymbol{B}$. Why not subtract the components along $\boldsymbol{a}$ and along $\boldsymbol{b}$?

28 Where are the mn^2 multiplications in equation (11)?

29 Apply the MATLAB qr code to $\boldsymbol{a} = (2, 2, -1)$, $\boldsymbol{b} = (0, -3, 3)$, $\boldsymbol{c} = (1, 0, 0)$. What are the $\boldsymbol{q}$'s?

Problems 30–35 involve orthogonal matrices that are special.

30 The first four ***wavelets*** are in the columns of this wavelet matrix W:

$$W = \frac{1}{2}\begin{bmatrix} 1 & 1 & \sqrt{2} & 0 \\ 1 & 1 & -\sqrt{2} & 0 \\ 1 & -1 & 0 & \sqrt{2} \\ 1 & -1 & 0 & -\sqrt{2} \end{bmatrix}.$$

What is special about the columns? Find the inverse wavelet transform W^{-1}.

31 Choose c so that Q is an orthogonal matrix:

$$Q = c\begin{bmatrix} 1 & -1 & -1 & -1 \\ -1 & 1 & -1 & -1 \\ -1 & -1 & 1 & -1 \\ -1 & -1 & -1 & 1 \end{bmatrix}.$$

Project $\boldsymbol{b} = (1,1,1,1)$ onto the first column. Then project $\boldsymbol{b}$ onto the plane of the first two columns.

32 If $\boldsymbol{u}$ is a unit vector, then $Q = I - 2\boldsymbol{u}\boldsymbol{u}^{\mathrm{T}}$ is a reflection matrix (Example 3). Find Q_1 from $\boldsymbol{u} = (0,1)$ and Q_2 from $\boldsymbol{u} = (0, \sqrt{2}/2, \sqrt{2}/2)$. Draw the reflections when Q_1 and Q_2 multiply the vectors $(1,2)$ and $(1,1,1)$.

33 Find all matrices that are both orthogonal and lower triangular.

34 $Q = I - 2\boldsymbol{u}\boldsymbol{u}^{\mathrm{T}}$ is a reflection matrix when $\boldsymbol{u}^{\mathrm{T}}\boldsymbol{u} = 1$. Two reflections give $Q^2 = I$.

(a) Show that $Q\boldsymbol{u} = -\boldsymbol{u}$. The mirror is perpendicular to $\boldsymbol{u}$.

(b) Find $Q\boldsymbol{v}$ when $\boldsymbol{u}^{\mathrm{T}}\boldsymbol{v} = 0$. The mirror contains $\boldsymbol{v}$. It reflects to itself.

Challenge Problems

35 (MATLAB) Factor $[Q, R] = \mathbf{qr}(A)$ for $A = \mathbf{eye}(4) - \mathbf{diag}([1\ \ 1\ \ 1], -1)$. You are orthogonalizing the columns $(1,-1,0,0)$ and $(0,1,-1,0)$ and $(0,0,1,-1)$ and $(0,0,0,1)$ of A. Can you scale the orthogonal columns of Q to get nice integer components?

36 If A is m by n with rank n, $\mathbf{qr}(A)$ produces a *square* Q and zeros below R:

The factors from MATLAB are $(m$ by $m)(m$ by $n)$ $\qquad A = [Q_1\ \ Q_2]\begin{bmatrix} R \\ 0 \end{bmatrix}.$

The n columns of Q_1 are an orthonormal basis for which fundamental subspace? The $m-n$ columns of Q_2 are an orthonormal basis for which fundamental subspace?

37 We know that $P = QQ^{\mathrm{T}}$ is the projection onto the column space of $Q(m$ by $n)$. Now add another column $\boldsymbol{a}$ to produce $A = [Q\ \ \boldsymbol{a}]$. Gram-Schmidt replaces $\boldsymbol{a}$ by what vector $\boldsymbol{q}$? Start with $\boldsymbol{a}$, subtract ______, divide by ______ to find $\boldsymbol{q}$.

4.5 The Pseudoinverse of a Matrix

Rank $r = m = n = 2$	$r = m < n = 3$	$r = n < m = 3$
$\boldsymbol{I} = \begin{bmatrix} 1 & 0 \\ 0 & 1 \end{bmatrix}$	$\boldsymbol{I_L} = \begin{bmatrix} 1 & 0 & 0 \\ 0 & 1 & 0 \end{bmatrix}$	$\boldsymbol{I_R} = \begin{bmatrix} 1 & 0 \\ 0 & 1 \\ 0 & 0 \end{bmatrix}$
$\boldsymbol{I}$ = **inverse of** $\boldsymbol{I}$	$\boldsymbol{I_L}$ = **left inverse of** $\boldsymbol{I_R}$	$(\boldsymbol{I_L})(\boldsymbol{I_R}) = \boldsymbol{I}$
$(\boldsymbol{I})(\boldsymbol{I}) = \boldsymbol{I}$	$\boldsymbol{I_R}$ = **right inverse of** $\boldsymbol{I_L}$	$(\boldsymbol{I_R})(\boldsymbol{I_L}) \neq \boldsymbol{I}$

Only the first matrix I is truly invertible. I_L and I_R have one-sided inverses but not true inverses. **Every matrix A has a pseudoinverse A^+. That word includes inverses and left inverses and right inverses.** *We will start with a summary of one-sided inverses.*

A has a left inverse : $A^+A = I$	**A has a right inverse : $AA^+ = I$**
A has independent columns	**A has independent rows**
A can be tall and thin : **rank $r = n$**	A can be short and wide : **rank $r = m$**
$A\boldsymbol{x} = \boldsymbol{b}$ has no solution or 1 solution	$A\boldsymbol{x} = \boldsymbol{b}$ has 1 solution or many solutions
The nullspace $\mathbf{N}(A) =$ **zero vector**	The left nullspace $\mathbf{N}(A^{\mathrm{T}}) =$ **zero vector**
$A^{\mathrm{T}}A$ is $n \times n$ and invertible	**AA^{T} is $m \times m$ and invertible**
The left inverse is $A^+ = (A^{\mathrm{T}}A)^{-1}A^{\mathrm{T}}$	**The right inverse is $A^+ = A^{\mathrm{T}}(AA^{\mathrm{T}})^{-1}$**

The left column with $A^+A = I$ describes the matrices in this least squares chapter. The rank is $r = n < m$ and the m equations $A\boldsymbol{x} = \boldsymbol{b}$ may be **unsolvable** : too many equations. Then $\widehat{\boldsymbol{x}} = A^+\boldsymbol{b}$ is the vector that solves $A^{\mathrm{T}}A\widehat{\boldsymbol{x}} = A^{\mathrm{T}}\boldsymbol{b}$ and makes the error $\boldsymbol{e} = \boldsymbol{b} - A\widehat{\boldsymbol{x}}$ as small as possible : this is **least squares.**

The right column with $AA^+ = I$ describes the opposite problem. Now the rank is $r = m < n$ and $A\boldsymbol{x} = \boldsymbol{b}$ has **infinitely many solutions.** In this case $\boldsymbol{x}^+ = A^+\boldsymbol{b}$ is the **minimum length solution** to $A\boldsymbol{x} = \boldsymbol{b}$. That solution is in the row space of A and all other solutions $\boldsymbol{x} = \boldsymbol{x}^+ + \boldsymbol{x}_n$ have a nullspace component $\boldsymbol{x}_n$—which increases the length.

What would you expect when the rank of A has $r < m$ **and** $r < n$? There is no one-sided inverse. The pseudoinverse A^+ (chosen next) will give the best $\boldsymbol{x}^+ = A^+\boldsymbol{b}$. **That vector $\boldsymbol{x}^+$ minimizes the error** $||A\boldsymbol{x} - \boldsymbol{b}||^2$. Also, among all the vectors $\boldsymbol{x} = \boldsymbol{x}^+ + \boldsymbol{x}_{\text{nullspace}}$ that give the same $A\boldsymbol{x}$ and same error, the vector $\boldsymbol{x}^+$ has **minimum length** :

$$\boldsymbol{x}^+ \textbf{ is the "minimum norm least squares solution" to } A\boldsymbol{x} = \boldsymbol{b}.$$

Among all the vectors that minimize $||A\boldsymbol{x} - \boldsymbol{b}||^2$, $\boldsymbol{x}^+ = A^+\boldsymbol{b}$ also minimizes $||\boldsymbol{x}||^2$.

Key idea : $\boldsymbol{x}^+$ **has nullspace component** $= \mathbf{0}$. If A has dependent columns, we need A^+. So this section is optional but the topic fits here. A^+ is explained using the four fundamental subspaces. It can be computed from $A = CR$ or $A = QR$ or the SVD.

The Pseudoinverse A^+ $(n \times m)$ of a Matrix A $(m \times n)$

1. Every vector $\boldsymbol{y}$ in m dimensions has two perpendicular parts $\boldsymbol{y} = \boldsymbol{b} + \boldsymbol{z}$

2. $\boldsymbol{b}$ is in the **column space of** $\boldsymbol{A}$ and $\boldsymbol{z}$ is in the **nullspace of** $\boldsymbol{A}^{\mathrm{T}}$: $A^{\mathrm{T}}\boldsymbol{z}=\mathbf{0}$ and $\boldsymbol{A}^+\boldsymbol{z}=\mathbf{0}$

3. $A\boldsymbol{x} = \boldsymbol{b}$ for one vector $\boldsymbol{x}$ in the row space of A. **Invert this part to find** $A^+\boldsymbol{b} = \boldsymbol{x}$

The pseudoinverse of A has $A^+\boldsymbol{b} = \boldsymbol{x}$ $\quad A^+\boldsymbol{z} = \mathbf{0}$ $\quad A^+\boldsymbol{y} = A^+\boldsymbol{b} + A^+\boldsymbol{z} = \boldsymbol{x}$

Here is the big picture for A^+. Notice that A^+ shares the same four subspaces as A^{T}. The difference is that A^+A brings $\boldsymbol{x}$ in the row space **back to the same** $\boldsymbol{x}$. In fact A^+A is exactly the n by n projection matrix P_{row} onto the row space of A. And symmetrically, AA^+ is exactly the m by m projection matrix P_{column} onto the column space of A. $\boldsymbol{A}^+$ **inverts** $\boldsymbol{A}$ **when it can**: column space to row space.

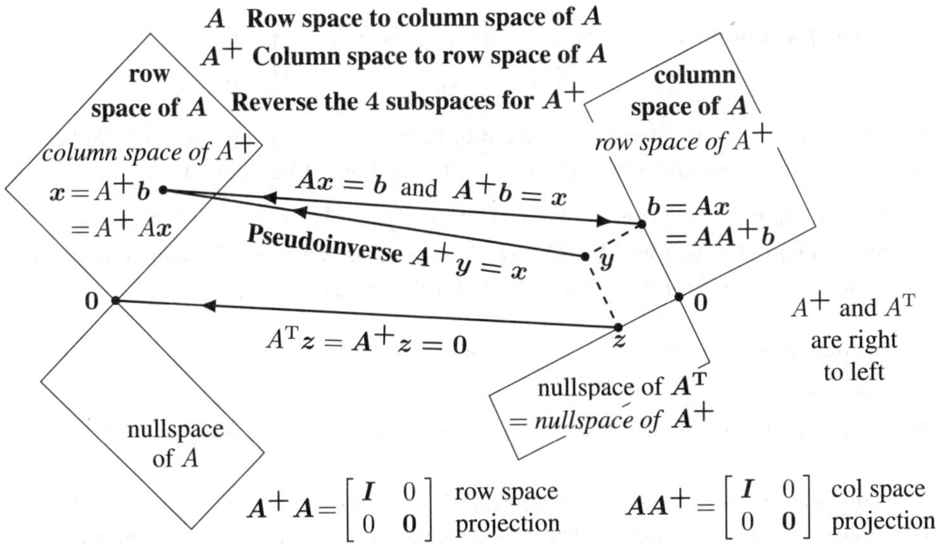

Figure 4.11: $\boldsymbol{b} = A\boldsymbol{x}$ in the column space of A goes back to $\boldsymbol{x} = \boldsymbol{A}^+\boldsymbol{b}$ in the row space.

Any $\boldsymbol{y} = \boldsymbol{b} + \boldsymbol{z}$ also goes back to $\boldsymbol{x} = \boldsymbol{A}^+\boldsymbol{b}$ because $\boldsymbol{A}^+\boldsymbol{z} = \mathbf{0}$. Then $(\boldsymbol{A}^+\boldsymbol{A})^2 = \boldsymbol{A}^+\boldsymbol{A}$.

A^+ was proposed by Moore and again by Penrose (who won the 2020 Nobel Prize in Physics). **It is the matrix that produces symmetric projections P_{row} and P_{column}**:

$P_{\text{row}} = \boldsymbol{A}^+\boldsymbol{A} = (\boldsymbol{A}^+\boldsymbol{A})^2 = (\boldsymbol{A}^+\boldsymbol{A})^{\mathrm{T}}$ $\qquad P_{\text{column}} = \boldsymbol{A}\boldsymbol{A}^+ = (\boldsymbol{A}\boldsymbol{A}^+)^2 = (\boldsymbol{A}\boldsymbol{A}^+)^{\mathrm{T}}$

Those tests for A^+ will be useful when we compute A^+ from $A = CR$ or $A = QR$.

Example 1 $A = \begin{bmatrix} 2 & 0 \\ 0 & 0 \end{bmatrix}$ $\quad \boldsymbol{y} = \begin{bmatrix} y_1 \\ 0 \end{bmatrix} + \begin{bmatrix} 0 \\ y_2 \end{bmatrix} = \boldsymbol{b} + \boldsymbol{z}$ $\quad A^+ = \begin{bmatrix} 1/2 & 0 \\ 0 & 0 \end{bmatrix}$

$\boldsymbol{b} = \begin{bmatrix} y_1 \\ 0 \end{bmatrix}$ is in the column space of A $\quad A\boldsymbol{x} = \boldsymbol{b}$ is $\begin{bmatrix} 2 & 0 \\ 0 & 0 \end{bmatrix}\begin{bmatrix} y_1/2 \\ 0 \end{bmatrix} = \begin{bmatrix} y_1 \\ 0 \end{bmatrix}$

$\boldsymbol{z} = \begin{bmatrix} 0 \\ y_2 \end{bmatrix}$ is in the nullspace of A^{T} $\quad A^{\mathrm{T}}\boldsymbol{z} = \mathbf{0}$ is $\begin{bmatrix} 2 & 0 \\ 0 & 0 \end{bmatrix}\begin{bmatrix} 0 \\ y_2 \end{bmatrix} = \begin{bmatrix} 0 \\ 0 \end{bmatrix}$

$A^+\boldsymbol{y} = A^+\boldsymbol{b} + A^+\boldsymbol{z} = \boldsymbol{x} + \mathbf{0} = \begin{bmatrix} \boldsymbol{y_1/2} \\ \mathbf{0} \end{bmatrix}$. Then $\boldsymbol{A^+} = \begin{bmatrix} \mathbf{1/2} & \mathbf{0} \\ \mathbf{0} & \mathbf{0} \end{bmatrix}$ and $A^+A = \begin{bmatrix} \mathbf{1} & \mathbf{0} \\ \mathbf{0} & \mathbf{0} \end{bmatrix}$

This extends to an easy and useful case: a **diagonal matrix D, square or rectangular:**

$$\textbf{The pseudoinverse of } D = \begin{bmatrix} \mathbf{2} & 0 & 0 \\ 0 & \mathbf{3} & 0 \\ 0 & 0 & \mathbf{0} \end{bmatrix} \textbf{ is } D^+ = \begin{bmatrix} \mathbf{1/2} & 0 & 0 \\ 0 & \mathbf{1/3} & 0 \\ 0 & 0 & \mathbf{0} \end{bmatrix}$$

We invert where we can. **But $\mathbf{1/0}$ is taken to be zero**! If the diagonal matrix D has additional zero rows (or zero columns) then D^+ has additional zero columns (or zero rows).

That example extends from a diagonal D to any matrix $A = BDC$, if B and C are invertible. **Then $\boldsymbol{A^+}$ will be $\boldsymbol{C^{-1}D^+B^{-1}}$.** Again, we can invert B and C but we can't invert the zeros in D. So D^+ replaces D^{-1} and A^+ replaces A^{-1}.

$$\text{The pseudoinverse of } A = \begin{bmatrix} 1 \\ 1 \end{bmatrix} \text{ is its left inverse } A^+ = \frac{1}{2}\begin{bmatrix} 1 & 1 \end{bmatrix}$$

$$\text{The pseudoinverse of } A = \begin{bmatrix} 1 & 1 \end{bmatrix} \text{ is its right inverse } A^+ = \frac{1}{2}\begin{bmatrix} 1 \\ 1 \end{bmatrix}$$

Many matrices don't have an inverse matrix A^{-1} with $A^{-1}A = I$ and $AA^{-1} = I$. That honor is limited to square matrices with independent columns and independent rows. For a true A^{-1}, only the zero vector is in the nullspace of A and the nullspace of A^{T}.

This chapter worked with tall thin matrices $(\boldsymbol{m} > \boldsymbol{n})$ with independent columns. The matrix has rank $r = n$. The zero vector is alone in the nullspace $\mathbf{N}(A)$. Then $\boldsymbol{A}^{\mathrm{T}}\boldsymbol{A}$ is invertible (n by n). In this case **$\boldsymbol{A^+} = (\boldsymbol{A}^{\mathrm{T}}\boldsymbol{A})^{-1}\boldsymbol{A}^{\mathrm{T}}$ is a left-inverse of $\boldsymbol{A}$.**

$$\textbf{We have } \boldsymbol{A^+A} = (\boldsymbol{A}^{\mathrm{T}}\boldsymbol{A})^{-1}\boldsymbol{A}^{\mathrm{T}}\boldsymbol{A} = \boldsymbol{I} \text{ but we do not have } AA^+ = I. \quad (1)$$

The next possibility is a short wide matrix $(m < n)$ with independent rows. The rank of A is m and the zero vector is alone in the left nullspace $\mathbf{N}(A^{\mathrm{T}})$. Then $\boldsymbol{A}\boldsymbol{A}^{\mathrm{T}}$ is invertible (m by m). In this case $\boldsymbol{A^+} = \boldsymbol{A}^{\mathrm{T}}(\boldsymbol{A}\boldsymbol{A}^{\mathrm{T}})^{-1}$ is a **right-inverse of $\boldsymbol{A}$.**

$$\textbf{Now we have } \boldsymbol{AA^+} = \boldsymbol{A}\boldsymbol{A}^{\mathrm{T}}(\boldsymbol{A}\boldsymbol{A}^{\mathrm{T}})^{-1} = \boldsymbol{I} \text{ but we do not have } A^+A = I. \quad (2)$$

The Important Action of A is Row Space to Column Space

Here is an important point about the row space and column space of any matrix A. We know that they have the same dimension r (the rank). That is part of the important point but not all of it. The other part is that **two vectors $\boldsymbol{x}$ and $\boldsymbol{X}$ in the row space cannot go to the same vector $A\boldsymbol{x} = A\boldsymbol{X}$ in the column space.** Every vector $\boldsymbol{b} = A\boldsymbol{x}$ in the column space **comes from one and only one vector $\boldsymbol{x}$ in the row space : Then $A^+\boldsymbol{b} = \boldsymbol{x}$.**

We can easily prove that useful fact. Suppose $A\boldsymbol{x} = \boldsymbol{b}$ and also $A\boldsymbol{X} = \boldsymbol{b}$ with $\boldsymbol{x}$ and $\boldsymbol{X}$ both in the row space. Then the difference $\boldsymbol{x} - \boldsymbol{X}$ is in the nullspace of A, because $A(\boldsymbol{x} - \boldsymbol{X}) = \boldsymbol{b} - \boldsymbol{b} = \boldsymbol{0}$. But $\boldsymbol{x} - \boldsymbol{X}$ is also in the row space (subtraction leaves us in the space). Since the nullspace is always orthogonal to the row space, $\boldsymbol{x} - \boldsymbol{X}$ in both spaces is orthogonal to itself. Then $\boldsymbol{x} - \boldsymbol{X}$ **must be the zero vector.**

We can express the same idea in other nice ways. Suppose the vectors $\boldsymbol{x}_1$ to $\boldsymbol{x}_r$ are a basis for the row space of A. Then the vectors $A\boldsymbol{x}_1$ to $A\boldsymbol{x}_r$ are a basis for the column space of A. It helps that both spaces have dimension r—the great fact from Chapter 1. Actually the neat factorization $A = CR$ that we learned in Chapter 1 (and completed in Chapter 3 by elimination) will now pay off. $A = CR$ tells us the pseudoinverse A^+.

> The matrix $\boldsymbol{C}$ is m by r with $\boldsymbol{r}$ **independent columns** coming directly from A. $\boldsymbol{R}$ is r by n with $\boldsymbol{r}$ **independent rows** coming from the reduced echelon form of A. Those led us to $A = CR$. **Their pseudoinverses now lead us to $\boldsymbol{A}^+ = \boldsymbol{R}^+\boldsymbol{C}^+$.**

The Pseudoinverse $A^+ = R^+C^+$ of $A = CR$

At this point we use the *one-sided inverses* of C and R. Again, C has r independent columns and R has r independent rows. Then $C^{\mathrm{T}}C$ and RR^{T} are invertible r by r matrices. The pseudoinverses of C and R are one-sided inverses : $C^+C = I$ and $RR^+ = I$.

$$\boldsymbol{C}^+ = (\boldsymbol{C}^{\mathrm{T}}\boldsymbol{C})^{-1}\boldsymbol{C}^{\mathrm{T}} \quad \text{and} \quad \boldsymbol{R}^+ = \boldsymbol{R}^{\mathrm{T}}(\boldsymbol{R}\boldsymbol{R}^{\mathrm{T}})^{-1}. \tag{3}$$

By good fortune that left inverse of C and right inverse of R are in the order R^+C^+ to tell us the pseudoinverse of $A = CR$. This is the key fact ! But see Problem 10.

$$\textbf{Pseudoinverse} \quad \boldsymbol{A}^+ = \boldsymbol{R}^+\boldsymbol{C}^+ = \boldsymbol{R}^{\mathrm{T}}(\boldsymbol{R}\boldsymbol{R}^{\mathrm{T}})^{-1}(\boldsymbol{C}^{\mathrm{T}}\boldsymbol{C})^{-1}\boldsymbol{C}^{\mathrm{T}} = \boldsymbol{R}^{\mathrm{T}}(\boldsymbol{C}^{\mathrm{T}}\boldsymbol{C}\boldsymbol{R}\boldsymbol{R}^{\mathrm{T}})^{-1}\boldsymbol{C}^{\mathrm{T}} = \boldsymbol{R}^{\mathrm{T}}(\boldsymbol{C}^{\mathrm{T}}\boldsymbol{A}\boldsymbol{R}^{\mathrm{T}})^{-1}\boldsymbol{C}^{\mathrm{T}} \tag{4}$$

We feel bound to give outline proofs for these various facts that fit together so well. The key points are that $C^{\mathrm{T}}C$ and RR^{T} are invertible when C has independent columns and R has independent rows. Here we give examples that apply formula (4) for A^+.

That discussion will continue on this book's website **math.mit.edu/linearalgebra**.

Remark We have explained A^+ based on the four fundamental subspaces of A. Moore and Penrose took a different approach. Penrose required these four conditions on the matrix A^+—and those conditions identify the same pseudoinverse:

$$\boxed{AA^+A = A \qquad A^+AA^+ = A^+ \qquad (A^+A)^{\mathrm{T}} = A^+A \qquad (AA^+)^{\mathrm{T}} = AA^+} \tag{5}$$

Then A^+A and AA^+ are projections onto the row space and column space of A.

Remark A lot of effort goes into choosing basis vectors with specially good properties. Among the very best properties is **orthogonality**: the basis vectors are **perpendicular**. The main point of Chapter 7 will be finding the famous **"singular vectors"** of A. Those are orthogonal unit vectors $\boldsymbol{v}_1$ to $\boldsymbol{v}_r$ in the row space *and* $A\boldsymbol{v}_1$ to $A\boldsymbol{v}_r$ are orthogonal in the column space. In applied linear algebra, that is perfection!

The same A^+ comes from $A = CR$ and $A = QR$ and the SVD: $A = U\Sigma V^{\mathrm{T}}$.

Example of A^+ from A = Incidence Matrix of a Graph

The m by n incidence matrix in Section 3.5 has a row for every edge in the graph and a column for every node. The ith row has two nonzeros, -1 and 1, to identify the leaving node and entering node for the ith edge. Flows can go with or against the arrows, which just set a positive direction on each edge.

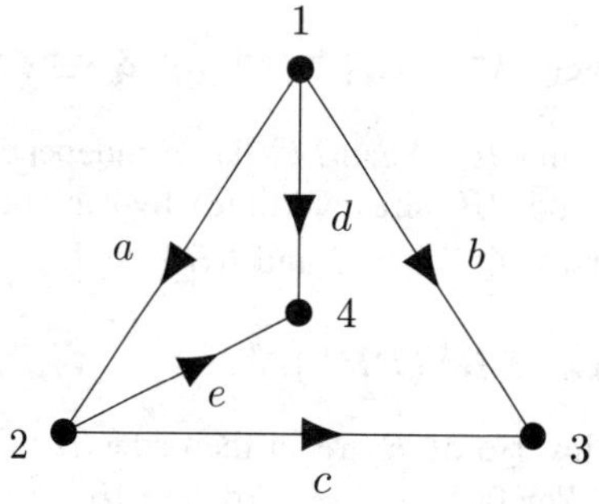

$$A = \begin{bmatrix} -1 & 1 & 0 & 0 \\ -1 & 0 & 1 & 0 \\ 0 & -1 & 1 & 0 \\ -1 & 0 & 0 & 1 \\ 0 & -1 & 0 & 1 \end{bmatrix} \begin{matrix} \text{edge } a \\ b \\ c \\ d \\ e \end{matrix}$$

Column 4 of $A = -$ Columns $1+2+3$,
so column 4 of R is $-1, -1, -1$

This matrix A has $\boldsymbol{m = 5, n = 4}$ and rank $\boldsymbol{r = 3}$. Its first three columns are independent:

$$\boldsymbol{A = CR} = \begin{bmatrix} -1 & 1 & 0 \\ -1 & 0 & 1 \\ 0 & -1 & 1 \\ -1 & 0 & 0 \\ 0 & -1 & 0 \end{bmatrix} \begin{bmatrix} \mathbf{1} & 0 & 0 & -\mathbf{1} \\ 0 & \mathbf{1} & 0 & -\mathbf{1} \\ 0 & 0 & \mathbf{1} & -\mathbf{1} \end{bmatrix} \tag{6}$$

Recall the basic facts about linear equations $Ax = b$. There is a solution when b is a combination of the columns of A. The *minimum norm solution* x_0 is the only solution in the row space of A (so x_0 is a combination of the rows). Other solutions $x = x_0 + x_n$ with larger norm $||x||^2 = ||x_0||^2 + ||x_n||^2$ include a nullspace part with $Ax_n = \mathbf{0}$.

Every A is invertible from its row space to its column space: dimension r to r.

x_0 in row space of A	$\xrightarrow{Ax_0 = b_0}$	$\xleftarrow{A^+b_0 = x_0}$	b_0 in column space of A
x_n in nullspace of A	$\xrightarrow[Ax_n = 0]{}$	$\xleftarrow[A^+b_n = 0]{}$	b_n in nullspace of A^{T}

How to compute A^+? **The recommended formula is $A^+ = V\Sigma^+U^{\mathrm{T}}$** from the singular value decomposition $A = U\Sigma V^{\mathrm{T}}$. Here U and V are square orthogonal matrices with $U^{-1} = U^{\mathrm{T}} = U^+$ and $V^{-1} = V^{\mathrm{T}} = V^+$. **The m by n diagonal matrix Σ has an n by m diagonal pseudoinverse Σ^+** (replace each $\sigma_i > 0$ in Σ by $1/\sigma_i$ in Σ^+). *Each diagonal zero in Σ remains a diagonal zero in Σ^+.*

When A is a small matrix of integers (like the incidence matrix), there is another way to compute A^+. Instead of the SVD, formula (6) showed $A = CR$. **Then $A^+ = R^+C^+$** in (4) involves **fractions but not irrational numbers (those come from the SVD).**

$$\textbf{Pseudoinverse of } A = CR \qquad A^+ = R^{\mathrm{T}}(C^{\mathrm{T}}AR^{\mathrm{T}})^{-1}C^{\mathrm{T}} \tag{7}$$

A^+ and A^{T} have the same row space and the same column space. The difference is: **A^+A is a symmetric projection matrix onto that row space.** We can check the four Penrose conditions in equation (5), and we can check $(A^+A)^2 = A^+A$ from (7):

$$\boxed{(A^+A)^2 = R^{\mathrm{T}}(C^{\mathrm{T}}AR^{\mathrm{T}})^{-1}C^{\mathrm{T}}AR^{\mathrm{T}}(C^{\mathrm{T}}AR^{\mathrm{T}})^{-1}C^{\mathrm{T}}A = A^+A} \tag{8}$$

Similarly $(AA^+)^2$ equals AA^+. And if A were invertible we would see that $A^+ = A^{-1}$. For the graph incidence matrix above, the computation of A^+ in formula (7) is exact:

$$C^{\mathrm{T}}AR^{\mathrm{T}} = \begin{bmatrix} 4 & 0 & 0 \\ 0 & 4 & 0 \\ -1 & -1 & 2 \end{bmatrix} \text{ and } A^+ = \frac{1}{8}\begin{bmatrix} -2 & -2 & 0 & -2 & 0 \\ 2 & 0 & -2 & 0 & -2 \\ 0 & 3 & 3 & -1 & -1 \\ 0 & -1 & -1 & 3 & 3 \end{bmatrix}. \tag{9}$$

The four subspaces for A^{T} are exactly the four subspaces for A^+. *Check the nullspaces of A^{T} and A^+.* Then the row spaces and the column spaces must agree by orthogonality.

Each row of A adds to zero	$\rightarrow$	Each column of A^+ adds to zero
Row 2 of A = Row 1 + Row 3	$\rightarrow$	Column 2 of A^+ = Column 1 + Column 3
Row 1 of A = Row 4 – Row 5	$\rightarrow$	Column 1 of A^+ = Column 4 – Column 5

The equation $A^{\mathrm{T}}y = \mathbf{0}$ is Kirchhoff's Current Law (flow in = flow out at each node). This is one of the great equations of circuit theory and computational engineering.

Problem Set 4.5

1 Find the pseudoinverses A^+ of these matrices:

$$A_1 = \begin{bmatrix} 1 & 1 \\ 1 & 1 \end{bmatrix} \qquad A_2 = \begin{bmatrix} 2 & 0 \\ 0 & 4 \\ 0 & 0 \end{bmatrix} \qquad A_3 = \begin{bmatrix} 2 & 0 & 0 \\ 0 & 4 & 0 \end{bmatrix}$$

2 If A has rank 1, then it has the simple form $A = \boldsymbol{c}\boldsymbol{r}^{\mathrm{T}}$ = one column times one row. Show that $A^+ = \boldsymbol{r}\boldsymbol{c}^{\mathrm{T}}/(\boldsymbol{r}^{\mathrm{T}}\boldsymbol{r})\,(\boldsymbol{c}^{\mathrm{T}}\boldsymbol{c})$ satisfies the four Penrose conditions in eqn (5).

3 If you add a row of zeros to a matrix A, show that you add a column of zeros to A^+. Hint: Do the four Penrose tests stay correct with one extra row of zeros in A and column of zeros in A^+?

4 What are the row space $\mathbf{C}(A^{\mathrm{T}})$ and column space $\mathbf{C}(A)$ for this matrix A? Find A^+ with the same four subspaces as A^{T}:

$$A = \begin{bmatrix} 1 & 0 & 0 \\ 0 & 1 & 0 \end{bmatrix}$$

5 Find the pseudoinverses of $A = \begin{bmatrix} 0 & 0 & 2 \\ 1 & 0 & 0 \end{bmatrix}$ and A^{T}. Explain $(A^{\mathrm{T}})^+ = (A^+)^{\mathrm{T}}$.

6 If P is a projection matrix ($P^{\mathrm{T}} = P = P^2$) show that $\boldsymbol{P^+ = P}$ satisfies Penrose in equation (5).

7 For any matrix $A = CR$, equation (8) verified that $(A^+A)^2 = A^+A$. Verify similarly that $(AA^+)^2 = AA^+$.

8 For any matrix, multiply $A^+ = R^{\mathrm{T}}(C^{\mathrm{T}}AR^{\mathrm{T}})^{-1}C^{\mathrm{T}}$ times $A = CR$ and simplify to show that $\boldsymbol{A^+A}$ **and** $\boldsymbol{AA^+}$ **are symmetric**. Combined with $(A^+A)^2 = A^+A$ from Problem 7, this verifies that $\boldsymbol{A^+A}$ **and** $\boldsymbol{AA^+}$ **are symmetric projections.**

9 Suppose you remove node 4 and the two edges into it from the graph in this section (leaving a triangle graph). Write down the incidence matrix A and find A^+. Can you factor $A = CR$ and recompute A^+ from equation (7)?

10 For $A = \begin{bmatrix} 1 & 0 \end{bmatrix}$ and $B = \begin{bmatrix} 1 \\ 1 \end{bmatrix}$ show that $\boldsymbol{(AB)^+}$ **is not** $\boldsymbol{B^+A^+}$ but $(BA)^+ = A^+B^+$. The first factor needs independent columns, the second factor needs independent rows.

11 *Why does the pseudoinverse of A^+ always equal A?*

Thoughts on the Victory of Orthogonality

If I look back at the linear algebra in this book, orthogonal vectors and matrices have won. You could say that they deserved to win. Their success goes all the way back to Chapter 1 on lengths and dot products. Let me recall some of their victories and add new ones.

1 The length of $Q\boldsymbol{x}$ equals the length of $\boldsymbol{x}$: $(Q\boldsymbol{x})^{\mathrm{T}}(Q\boldsymbol{x}) = \boldsymbol{x}^{\mathrm{T}}Q^{\mathrm{T}}Q\boldsymbol{x} = \boldsymbol{x}^{\mathrm{T}}\boldsymbol{x} = ||\boldsymbol{x}||^2$.

2 The dot product $(Q\boldsymbol{x})^{\mathrm{T}}(Q\boldsymbol{y})$ equals the dot product $\boldsymbol{x}^{\mathrm{T}}\boldsymbol{y}$: $\boldsymbol{x}^{\mathrm{T}}Q^{\mathrm{T}}Q\boldsymbol{y}=\boldsymbol{x}^{\mathrm{T}}\boldsymbol{y}$.

3 The projection matrix onto the column space of Q (m by n) is $P = QQ^{\mathrm{T}} = (QQ^{\mathrm{T}})^2$.

4 The least squares solution to $Q\boldsymbol{x} = \boldsymbol{b}$ ($m > n$) is $\widehat{\boldsymbol{x}} = Q^{\mathrm{T}}\boldsymbol{b} = Q^{+}\boldsymbol{b}$ (pseudoinverse).

If we add that Q is square, then $Q^{\mathrm{T}} = Q^{-1}$. More is coming in Chapters 5, 6, 7.

5 All powers Q^N and products $Q_1 Q_2 \ldots Q_N$ are orthogonal: $|$ determinant $| = 1$.

6 The eigenvectors $\boldsymbol{q}$ of a symmetric matrix S can be chosen orthonormal.

7 The singular vectors of every matrix are orthonormal.

That list shows something important. The success of orthogonal matrices is tied to the **sum of squares definition of length:** $||\boldsymbol{x}||^2 = \boldsymbol{x}^{\mathrm{T}}\boldsymbol{x}$. In least squares, the derivative of $||A\boldsymbol{x} - \boldsymbol{b}||^2$ leads to a symmetric matrix $S = A^{\mathrm{T}}A$. Then S is diagonalized by an orthogonal matrix Q (the eigenvectors will come in Chapter 6).

Every A is diagonalized by two orthogonal matrices Q_1 and Q_2 (the singular vectors in Chapter 7). And here is more about orthogonal matrices: $\boldsymbol{A} = \boldsymbol{QS}$ and $\boldsymbol{A} = \boldsymbol{QR}$.

8 Every invertible matrix equals an orthogonal $\boldsymbol{Q}$ times a symmetric $\boldsymbol{S}$.

Example $$\boldsymbol{A} = \begin{bmatrix} 3 & 0 \\ 4 & 5 \end{bmatrix} = \frac{1}{\sqrt{5}}\begin{bmatrix} 2 & -1 \\ 1 & 2 \end{bmatrix}\sqrt{5}\begin{bmatrix} 2 & 1 \\ 1 & 2 \end{bmatrix} = \boldsymbol{QS}$$

9 Every invertible matrix equals an orthogonal $\boldsymbol{Q}$ times an upper triangular $\boldsymbol{R}$.

Example Q and R come from the Gram-Schmidt algorithm in Section 4.4:

$$\boldsymbol{A} = \begin{bmatrix} 3 & 0 \\ 4 & 5 \end{bmatrix} = \frac{1}{5}\begin{bmatrix} 3 & -4 \\ 4 & 3 \end{bmatrix}\begin{bmatrix} 5 & 4 \\ 0 & 3 \end{bmatrix} = \boldsymbol{QR}$$

5 Determinants

The determinant of a square matrix is an astonishing number. If $\det A = 0$, this signals that the column vectors are dependent and A is not invertible. If $\det A$ is not zero, **the formula for A^{-1} will have a division by the determinant**. Section 5.1 will find 3 by 3 determinants and then the "cofactor formulas" for $\det A$ and the inverse matrix.

There are **three useful formulas** for $\det A$. After elimination we can multiply the pivots from the triangular matrix U in $PA = LU$. Always $\det P = 1$ or -1 and $\det L = 1$. So the **product of pivots in U** gives $\det A$ or $-\det A$. This is usually the fastest way.

A second formula uses determinants of size $n - 1$: the **"cofactors"** of A. They give the best formula for A^{-1}. *The i, j entry of A^{-1} is the j, i cofactor divided by* $\det A$.

The big formula for $\det A$ adds up $n\,!$ terms—one for every path down the matrix. Each path chooses one entry from every row and column, and we multiply the n entries on the path. Reverse the sign if the path has an odd permutation of column numbers. A 3 by 3 matrix has six paths and the big formula has six terms—one for every 3 by 3 permutation.

Section 5.2 begins with key properties like $\det AB = (\det A)(\det B)$. To prove that great fact, we begin with three simple properties that lead to every determinant. From those three, everything else follows—including the famous Cramer's Rule to solve linear equations $A\boldsymbol{x} = \boldsymbol{b}$.

Section 5.3 moves to geometry: **The volume of a tilted box in n dimensions**. The column vectors in the edge matrix E give n edges going out from the corner at $(0, \ldots, 0)$. Then the volume of the box is $|\det E|$. This connects geometry to linear algebra.

This chapter could easily become full of formulas! When an n by n matrix multiplies all points in an n-dimensional shape, it produces another shape. A cube goes to a tilted box. A circle goes to an ellipse, and a sphere goes to an ellipsoid. **The ratio of the output volume to the input volume is always $|\det A|$.**

Determinants could tell us everything, if only they were not so hard to compute.

5.1 3 by 3 Determinants and Cofactors

1 The **determinant** of $A = \begin{bmatrix} a & b \\ c & d \end{bmatrix}$ is $\boldsymbol{ad - bc}$. The singular matrix $\begin{bmatrix} a & 2a \\ c & 2c \end{bmatrix}$ has $\mathbf{det} = \mathbf{0}$.

2 **Row exchange reverses signs** $PA = \begin{bmatrix} 0 & 1 \\ 1 & 0 \end{bmatrix} \begin{bmatrix} a & b \\ c & d \end{bmatrix} = \begin{bmatrix} c & d \\ a & b \end{bmatrix}$ has $\mathbf{det}\, \boldsymbol{PA} = \boldsymbol{bc} - \boldsymbol{ad} = -\,\mathbf{det}\, \boldsymbol{A}$.

3 The determinant of $\begin{bmatrix} xa + yA & xb + yB \\ c & d \end{bmatrix}$ is $\boldsymbol{x(ad - bc) + y(Ad - Bc)}$. **Det is linear in row 1 by itself.**

4 3 by 3 determinants have $3! = 6$ terms in equation (1). They are separated into 3 terms using cofactors in equation (4). Those cofactors divided by $\det A$ produce A^{-1} in equation (7).

Determinants lead to beautiful formulas. But often they are not practical for computing. This section will focus on 3 by 3 matrices, to see how determinants produce the matrix A^{-1}. (This formula for A^{-1} was not seen in Chapters 1 to 4.) First come 2 by 2 matrices:

$$\det \begin{bmatrix} 1 & 0 \\ 0 & 1 \end{bmatrix} = \mathbf{1} \quad \det \begin{bmatrix} 0 & 1 \\ 1 & 0 \end{bmatrix} = \mathbf{-1} \quad \det \begin{bmatrix} a & b \\ c & d \end{bmatrix} = \boldsymbol{ad - bc} \quad \det \begin{bmatrix} c & d \\ a & b \end{bmatrix} = \boldsymbol{bc - ad}$$

Those examples show an important rule. **Determinants reverse sign when two rows are exchanged** (or two columns are exchanged). This rule becomes a key to determinants of all sizes. *The matrix has no inverse* when its determinant is zero.

For 2 by 2 matrices, $\det A = 0$ means $a/c = b/d$. The columns are parallel. For n by n matrices, $\mathbf{det}\, \boldsymbol{A} = \mathbf{0}$ **means that the columns of $\boldsymbol{A}$ are not independent**. Then a combination of columns gives the zero vector: $A\boldsymbol{x} = \mathbf{0}$ with $\boldsymbol{x} \neq \mathbf{0}$. A is not invertible.

These properties and more will follow after we define the determinant.

3 by 3 Determinants

2 by 2 matrices are easy: $ad - bc$. 4 by 4 matrices are hard. Better to use elimination (or a laptop). For 3 by 3 matrices, the determinant has $3! = 6$ terms, and often you can compute it by hand. We will show how. Every permutation has determinant 1 or -1.

Start with the identity matrix ($\det I = 1$) and exchange two rows (then $\det = -1$). Exchanging again brings back $\det = +1$. You quickly have all six permutation matrices. Each row exchange will reverse the sign of the determinant ($+1$ to -1 or else -1 to $+1$):

$$\begin{bmatrix} 1 & & \\ & 1 & \\ & & 1 \end{bmatrix} \begin{bmatrix} & 1 & \\ 1 & & \\ & & 1 \end{bmatrix} \begin{bmatrix} & 1 & \\ & & 1 \\ 1 & & \end{bmatrix} \begin{bmatrix} & & 1 \\ & 1 & \\ 1 & & \end{bmatrix} \begin{bmatrix} & & 1 \\ 1 & & \\ & 1 & \end{bmatrix} \begin{bmatrix} 1 & & \\ & & 1 \\ & 1 & \end{bmatrix}$$

$$\mathbf{det} = \mathbf{+1} \qquad \mathbf{-1} \qquad \mathbf{+1} \qquad \mathbf{-1} \qquad \mathbf{+1} \qquad \mathbf{-1}$$

If I exchange two rows of A—say rows 1 and 2—**then its determinant changes sign.** This will carry over to all determinants: **Row exchange multiplies $\mathbf{det}\, \boldsymbol{A}$ by $\mathbf{-1}$.**

When you multiply a row by a number, this multiplies the determinant by that number. Suppose the three rows are $\boldsymbol{a}\,\boldsymbol{b}\,\boldsymbol{c}$ and $\boldsymbol{p}\,\boldsymbol{q}\,\boldsymbol{r}$ and $\boldsymbol{x}\,\boldsymbol{y}\,\boldsymbol{z}$. Those nine numbers multiply ± 1.

$$\begin{bmatrix} a & & \\ & q & \\ & & z \end{bmatrix} \begin{bmatrix} & b & \\ p & & \\ & & z \end{bmatrix} \begin{bmatrix} & b & \\ & & r \\ x & & \end{bmatrix} \begin{bmatrix} & & c \\ & q & \\ x & & \end{bmatrix} \begin{bmatrix} & & c \\ p & & \\ & y & \end{bmatrix} \begin{bmatrix} a & & \\ & & r \\ & y & \end{bmatrix}$$

$$\det = +\boldsymbol{aqz} \qquad -\boldsymbol{bpz} \qquad +\boldsymbol{brx} \qquad -\boldsymbol{cqx} \qquad +\boldsymbol{cpy} \qquad -\boldsymbol{ary}$$

Finally we use the most powerful property we have. **The determinant of A is linear in each row separately**. As equation (4) will show, we can add those six determinants. To remember the plus and minus signs, I follow the arrows in this picture of the matrix.

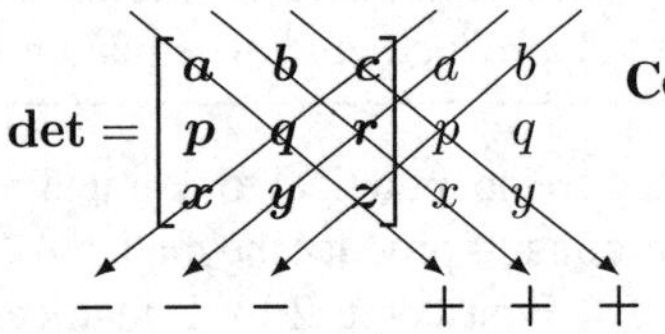

Combine those 6 simple determinants into det A =

$$+\,aqz + brx + cpy - ary - bpz - cqx \qquad (1)$$

Notice! Those six terms all have one entry from each row of the matrix. They also have one entry from each column. There are $6 = 3\,!$ terms in det A because there are six 3×3 permutation matrices. A 4×4 determinant will have $\mathbf{4\,!} = 4 \times 3 \times 2 \times 1 = \mathbf{24}$ **terms.**

Let me jump in with examples of sizes $1, 2, 3, 4$. This is pretty sudden but you can see their determinants. My favorite matrices have 2's on the main diagonal and -1's on the diagonals above and below. Here are sizes $n = 1, 2, 3$:

$$\boldsymbol{A_1} = \begin{bmatrix} \mathbf{2} \end{bmatrix} \qquad \boldsymbol{A_2} = \begin{bmatrix} \mathbf{2} & \mathbf{-1} \\ \mathbf{-1} & \mathbf{2} \end{bmatrix} \qquad \boldsymbol{A_3} = \begin{bmatrix} \mathbf{2} & \mathbf{-1} & 0 \\ \mathbf{-1} & \mathbf{2} & \mathbf{-1} \\ 0 & \mathbf{-1} & \mathbf{2} \end{bmatrix}$$

$$\det \boldsymbol{A_1} = \mathbf{2} \qquad \det \boldsymbol{A_2} = \mathbf{4 - 1 = 3} \qquad \det \boldsymbol{A_3} = \mathbf{8 - 2 - 2 = 4}$$

What are the six terms for det $A_3 = 4$? The diagonal gives $(2)\,(2)\,(2) = 8$. Five more:

Other five terms

$$\begin{bmatrix} 2 & & \\ & & -1 \\ & -1 & \end{bmatrix} \begin{bmatrix} & -1 & \\ -1 & & \\ & & 2 \end{bmatrix} \begin{bmatrix} & -1 & \\ & & -1 \\ 0 & & \end{bmatrix} \begin{bmatrix} & & 0 \\ -1 & & \\ & -1 & \end{bmatrix} \begin{bmatrix} & & 0 \\ & 2 & \\ 0 & & \end{bmatrix}$$

determinants **$-2, -2, 0, 0, 0$**

So the six terms add to $\mathbf{8 - 2 - 2 = 4}$ and this will be det A_3. Please watch the pattern: Each of the $n\,! = 3\,! = 6$ terms involves **one number from every row and every column.** When any of these numbers is zero, that whole term is zero. Computing det A is much easier when the matrix has many zeros.

Now we move to the 4×4 matrix A_4 with 24 terms. Here is the key simplification and it will teach us a lot. The 4×4 determinant comes from two 3×3 determinants!

$$A_4 = \left[\begin{array}{c|ccc} \mathbf{2} & \mathbf{-1} & & \\ \hline \mathbf{-1} & 2 & -1 & \\ & -1 & 2 & -1 \\ & & -1 & 2 \end{array}\right] \qquad \det A_4 = \mathbf{2}\det\begin{bmatrix} 2 & -1 & \\ -1 & 2 & -1 \\ & -1 & 2 \end{bmatrix} - (\mathbf{-1})\det\begin{bmatrix} -1 & -1 & 0 \\ 0 & 2 & -1 \\ 0 & -1 & 2 \end{bmatrix}$$

Those 3×3's from rows $2, 3, 4$ are the **cofactors of 2 and −1 in row 1**. (2)

Notice : 2 in the top row of A is multiplying a 3 by 3 from the other rows. One number from each row and each column ! The 2 came from row 1, column 1. So 2 multiplies the 3×3 determinant in rows and columns $2, 3, 4$. *That* 3×3 *matrix is* A_3 *with* $\det = 4$.

The final 3 by 3 matrix in equation (2) also uses rows $2, 3, 4$ of A. The bold -1 comes from row 1, column 2. So the 3 by 3 matrix must use *columns* $1, 3, 4$. The determinant of that 3 by 3 is -3 (please check this). Finally the overall determinant of A_4 is $\mathbf{8 - 3 = 5}$.

One more thing to learn about that last 3 by 3 from columns $1, 3, 4$. In equation (2) it is multiplied by the -1 from row 1, column 2 of A_4. But **equation (2) has another mysterious −1 !** *Where does that minus come from* ? It is there because row 1, column 2 is a "minus position". **1 + 2 is an odd number !** The sign pattern was already set for 2×2.

$$\det\begin{bmatrix} a & b \\ c & d \end{bmatrix} = \det\begin{bmatrix} a & \\ & d \end{bmatrix} + \det\begin{bmatrix} & b \\ c & \end{bmatrix} = a \text{ times } d \textbf{ minus } b \text{ times } c. \quad (3)$$

Finally A_1, A_2, A_3, A_4 have determinants $2, 3, 4, 5$. You can guess that $\det \boldsymbol{A_5} = \mathbf{6}$. *Same idea as before* $\det \boldsymbol{A_5} = \mathbf{2} \det \boldsymbol{A_4} - \det \boldsymbol{A_3} = 2(5) - 4 = \mathbf{6}$. We will explain.

Cofactors and a Formula for A^{-1}

For 3×3 determinants with 6 terms like aqz, I can explain the "cofactor formulas" for the determinant and the inverse matrix. *The idea is to reduce from* 3×3 *to* 2×2. Two of the six terms in $\det A$ start with a and with b and with c from row 1.

Cofactor formula
3 × 3 matrix

$$\boxed{\det A = a\,(\boldsymbol{qz - ry}) + b\,(\boldsymbol{rx - pz}) + c\,(\boldsymbol{py - qx}).} \quad (4)$$

We separated the factors a, b, c in row 1 from their "cofactors". **Each cofactor is a 2 by 2 determinant**. The factor a in the $1,1$ position takes its cofactor from row and column 2 and 3. Every row and column is used once ! **Notice that b in row 1, column 2 has cofactor $rx - pz$.** There we see the other rows $2, 3$ and columns $1, 3$. But the actual 2 by 2 determinant would be $\boldsymbol{pz - rx}$. The cofactor C_{12} reversed those $\pm$ signs : **1 + 2 is odd.**

Here is the definition of cofactors C_{ij} and the $(-1)^{i+j}$ rule for their plus and minus signs.

> For the i, j cofactor $\boldsymbol{C_{ij}}$, **remove row i and column j** from the matrix A.
>
> C_{ij} equals $\boldsymbol{(-1)^{i+j}}$ times the determinant of the remaining matrix (size $n - 1$).
>
> **The cofactor formula along row i is** $\det \boldsymbol{A} = \boldsymbol{a_{i1}C_{i1} + \cdots + a_{in}C_{in}}$ (5)

Key point: *The cofactor C_{ij} just collects all the terms in* $\det A$ *that are multiplied by* a_{ij}. Thus the cofactors of a 2 by 2 matrix involve 1 by 1 determinants $\boldsymbol{d, c, b, a}$ times $(-1)^{i+j}$.

$$\text{The cofactors of } A = \begin{bmatrix} a & b \\ c & d \end{bmatrix} \text{ are in the cofactor matrix } C = \begin{bmatrix} d & -c \\ -b & a \end{bmatrix}.$$

Now notice something wonderful. **If A multiplies C^{T} (not C) we get $\det A$ times I:**

$$AC^{\mathrm{T}} = \begin{bmatrix} ad-bc & 0 \\ 0 & ad-bc \end{bmatrix} = \begin{bmatrix} \det A & 0 \\ 0 & \det A \end{bmatrix} = (\det A)\, I. \tag{6}$$

Dividing by $\det A$, cofactors give our first and best formula for the inverse matrix A^{-1}.

$$\boxed{\textbf{Inverse matrix formula} \qquad A^{-1} = C^{\mathrm{T}} / \det A} \tag{7}$$

This formula shows why the determinant of an invertible matrix cannot be zero. We need to divide the matrix C^{T} by the number $\det A$. Every entry in the inverse matrix A^{-1} is a **ratio of two determinants** (size $n-1$ for the cofactor in C^{T} divided by size n for $\det A$).

The example below has determinant 1, so A^{-1} is exactly C^{T} = **adjugate matrix.** Notice how C_{32} removes row 3 and column 2. That leaves a 2 by 2 matrix with determinant 1. Since $(-1)^{2+3} = -\mathbf{1}$, this becomes -1 in C^{T}. It goes into row 2, column 3.

Example of A^{-1}
Determinant $= 1$
Cofactors in C^{T}

$$A^{-1} = \begin{bmatrix} 1 & 1 & 1 \\ 0 & 1 & 1 \\ 0 & 0 & 1 \end{bmatrix}^{-1} = \begin{bmatrix} 1 & -1 & 0 \\ 0 & 1 & -1 \\ 0 & 0 & 1 \end{bmatrix} = \frac{C^{\mathrm{T}}}{1} \tag{8}$$

The diagonal entries of AC^{T} are always $\det A$. That is the cofactor formula (5). Problem 18 will show why the off-diagonal entries of AC^{T} are always zero. Those numbers turn out to be determinants of matrices with two equal rows. Automatically zero.

A typical cofactor C_{31} removes row 3 and column 1. In our 3 by 3 example, that leaves a 2 by 2 matrix of 1's, with determinant $= 0$. This is the bold zero in A^{-1}: 1, 3 position. If we change to $2A$, the determinant is multiplied by $(2)\,(2)\,(2) = \mathbf{8}$. All cofactors in C are multiplied by $(2)\,(2) = \mathbf{4}$. Then $(\mathbf{2}A)^{-1} = \mathbf{4}C^{\mathrm{T}}/\mathbf{8} \det A$ is A^{-1} divided by 2. Right!

Example: The $-1, 2, -1$ Tridiagonal Matrix

The cofactor formula reduces n by n determinants to their cofactors of size $n-1$. The $-1, 2, -1$ matrix is a great example because row 1 has only two nonzeros 2 and -1. So we only need two cofactors of row 1—and they will lead us to A_{n-1} and A_{n-2}. Watch how this happens for the determinant of the $-1, 2, -1$ matrix A_n:

$$\det \begin{bmatrix} \mathbf{2} & \mathbf{-1} & 0 & 0 \\ -1 & 2 & -1 & 0 \\ 0 & -1 & 2 & -1 \\ \cdot & \cdot & \cdot & \cdot \end{bmatrix} = \mathbf{2}(-1)^{1+1} \det \begin{bmatrix} 2 & -1 & 0 \\ -1 & 2 & -1 \\ \cdot & \cdot & \cdot \end{bmatrix} \mathbf{-1}(-1)^{1+2} \det \begin{bmatrix} -1 & -1 & 0 \\ 0 & 2 & -1 \\ \cdot & \cdot & \cdot \end{bmatrix}$$

On the right hand side, the first matrix gives $\mathbf{2 \det A_{n-1}}$. The last term has $\mathbf{-1}$ from the 1, 2 entry of A_n. It also has $(-1)^{1+2} = -1$ from the sign $(-1)^{i+j}$ that goes with every cofactor. And that last cofactor itself has -1 in the top left entry, with makes **three -1's.**

That last cofactor of A_n is showing **its own cofactor of size** $\boldsymbol{n-2}$. That matrix is A_{n-2} !

$$\det \boldsymbol{A_n} = 2\det \boldsymbol{A_{n-1}} - \det \boldsymbol{A_{n-2}}. \tag{9}$$

Working directly we found $\det A_n = \mathbf{2, 3, 4, 5}$ for $n = 1, 2, 3, 4$. All those fit with equation (9), as in $\det A_4 = \mathbf{5} = \mathbf{2(4) - 3}$. But now we have found $\det \boldsymbol{A_n} = \boldsymbol{n+1}$ **for every** $\boldsymbol{n}$. This fits the formula (9) because $\boldsymbol{n+1 = 2(n) - (n-1)}$.

The cofactor formula is most useful when the matrix entries are mostly zero. Then we have few cofactors to find. And the entries of those cofactors are also mostly zero. For a dense matrix full of nonzeros, we look next for a better way to find its determinant. **The better way brings back elimination, to produce easy determinants of L and U.**

Problem Set 5.1

Questions 1–5 are about the rules for determinants.

1 If a 4 by 4 matrix has $\det A = \frac{1}{2}$, find $\det(2A)$ and $\det(-A)$ and $\det(A^2)$ and $\det(A^{-1})$.

2 If a 3 by 3 matrix has $\det A = -1$, find $\det\left(\frac{1}{2}A\right)$ and $\det(-A)$ and $\det(A^2)$ and $\det(A^{-1})$. What are those four answers if $\det A = 0$?

3 True or false, with a reason if true or a counterexample if false :

(a) The determinant of $I + A$ is $1 + \det A$.

(b) The determinant of $4A$ is 4 times the determinant of A.

(c) The determinant of $AB - BA$ is zero. Try an example with $A = \begin{bmatrix} 0 & 0 \\ 0 & 1 \end{bmatrix}$.

4 Exchanging rows reverses the sign of $\det A$. Which row exchanges show that these "reverse identity matrices" J_3 and J_4 have $|J_3| = -1$ but $|J_4| = +1$?

$$\det \begin{bmatrix} 0 & 0 & 1 \\ 0 & 1 & 0 \\ 1 & 0 & 0 \end{bmatrix} = \mathbf{-1} \quad \text{but} \quad \det \begin{bmatrix} 0 & 0 & 0 & 1 \\ 0 & 0 & 1 & 0 \\ 0 & 1 & 0 & 0 \\ 1 & 0 & 0 & 0 \end{bmatrix} = \mathbf{+1}.$$

5 For $n = 5, 6, 7$, count the row exchanges to permute the reverse identity J_n to the identity matrix I_n. Propose a rule for every $\det J_n$ and predict whether J_{101} has determinant $+1$ or -1.

6 Find the six terms in equation (1) like $+aqz$ (the main diagonal) and $-cqx$ (the anti-diagonal). Combine those six terms into the determinants of A, B, C.

$$A = \begin{bmatrix} 1 & -1 & 0 \\ -1 & 1 & -1 \\ 0 & -1 & 1 \end{bmatrix} \qquad B = \begin{bmatrix} 2 & 1 & 4 \\ 4 & 2 & 8 \\ 6 & 3 & 12 \end{bmatrix} \qquad C = \begin{bmatrix} 1 & 2 & 3 \\ 4 & 5 & 6 \\ 7 & 8 & 9 \end{bmatrix}$$

7 If you add row 1 $= \begin{bmatrix} a & b & c \end{bmatrix}$ to row 2 $= \begin{bmatrix} p & q & r \end{bmatrix}$ to get $\begin{bmatrix} p+a & q+b & r+c \end{bmatrix}$ in row 2, show from formula (1) for $\det A$ that the 3 by 3 determinant *does not change*. Here is another approach to the rule for adding two rows:

$$\det \begin{bmatrix} \text{row 1} \\ \textbf{row 1+row 2} \\ \text{row 3} \end{bmatrix} = \det \begin{bmatrix} \text{row 1} \\ \textbf{row 1} \\ \text{row 3} \end{bmatrix} + \det \begin{bmatrix} \text{row 1} \\ \textbf{row 2} \\ \text{row 3} \end{bmatrix} = \mathbf{0} + \det \begin{bmatrix} \text{row 1} \\ \text{row 2} \\ \text{row 3} \end{bmatrix}$$

8 Show that $\det A^{\mathrm{T}} = \det A$ because both of those 3 by 3 determinants come from the **same six terms like $\boldsymbol{brx}$**. This means $\det P^{\mathrm{T}} = \det P$ for every permutation P.

9 Do these matrices have determinant 0, 1, 2, or 3 ?

$$A = \begin{bmatrix} 0 & 0 & 1 \\ 1 & 0 & 0 \\ 0 & 1 & 0 \end{bmatrix} \quad B = \begin{bmatrix} 0 & 1 & 1 \\ 1 & 0 & 1 \\ 1 & 1 & 0 \end{bmatrix} \quad C = \begin{bmatrix} 1 & 1 & 1 \\ 1 & 1 & 1 \\ 1 & 1 & 1 \end{bmatrix} \quad D = \begin{bmatrix} 1 & 1 & 1 \\ 0 & 1 & 0 \\ 1 & 1 & 1 \end{bmatrix}.$$

10 If the entries in every row of A add to zero, solve $A\boldsymbol{x} = \mathbf{0}$ to prove $\det A = 0$. If those entries add to one, show that $\det(A - I) = 0$. Does this mean $\det A = 1$?

11 Why does $\det(P_1P_2) = (\det P_1)$ times $(\det P_2)$ for permutations? If P_1 needs 2 row exchanges and P_2 needs 3 row exchanges to reach I, why does P_1P_2 reach I from $2+3$ exchanges? Then their determinants will be $(-1)^5 = (-1)^2(-1)^3$.

12 Explain why half of all 5 by 5 permutations are even (with $\det P = 1$).

13 Find the determinants of a rank one matrix A and a skew-symmetric matrix B.

$$A = \begin{bmatrix} 1 \\ 2 \\ 3 \end{bmatrix} \begin{bmatrix} 1 & -4 & 5 \end{bmatrix} \quad \text{and} \quad B = \begin{bmatrix} 0 & 1 & 3 \\ -1 & 0 & 4 \\ -3 & -4 & 0 \end{bmatrix}.$$

14 If the i, j entry of A is i times j, show that $\det A = 0$. (Exception when $A = [\,1\,]$.) If the i, j entry of A is $i + j$, show that $\det A = 0$. (Exception when $n = 1$ or 2.)

15 Place **the smallest number of zeros** in a 4 by 4 matrix that will guarantee $\det A = 0$. Place as many zeros as possible while still allowing $\det A \neq 0$.

16 If all the cofactors are zero, how do you know that A has no inverse? If none of the cofactors are zero, is A sure to be invertible?

17 *Cofactor formula when two rows are equal.* Write out the 6 terms in $\det A$ when a 3 by 3 matrix has row 1 = row 2 = a, b, c. The determinant should be zero.

18 *Why is a matrix that has two equal rows always singular?* **Then $\det A = 0$.** If we combine the cofactors from one row with the numbers in another row, we will be computing $\det A^*$ when **A^* has equal rows.** Then $\det A^*$ equals 0. This is what produces the off-diagonal zeros in $AC^{\mathrm{T}} = (\det A)\, I$.

19 From the cofactor formula $AC^{\mathrm{T}} = (\det A)I$ show that $\det C = (\det A)^{n-1}$.

20 Suppose $\det A = 1$ and you know all the cofactors of A. How can you find A?

5.2 Computing and Using Determinants

1 Useful facts: $\det A^{\mathrm{T}} = \det A$ and $\mathbf{det}\, \boldsymbol{AB} = (\mathbf{det}\, \boldsymbol{A})\,(\mathbf{det}\, \boldsymbol{B})$ and $|\det Q| = \mathbf{1}$.

2 Elimination matrices have $\det E = 1$ so $\det EA = \det A$ is unchanged.

3 Cramer's Rule finds $\boldsymbol{x} = A^{-1}\boldsymbol{b}$ from ratios of determinants (a slow way).

4 $\det A = \pm$ product of the pivots in $A = LU$ (a much faster way).

5 The big formula for $\det A$ has $n!$ terms from $n!$ permutations (very slow if $n > 3$).

The determinant of a square matrix tells us a lot. First of all, an invertible matrix has $\det A \neq 0$. A singular matrix has $\det A = 0$. When we come to eigenvalues λ and eigenvectors $\boldsymbol{x}$ with $A\boldsymbol{x} = \lambda\boldsymbol{x}$, we will write that eigenvalue equation as $(\boldsymbol{A} - \boldsymbol{\lambda I})\boldsymbol{x} = \mathbf{0}$. This means that $A - \lambda I$ is singular and $\mathbf{det}(\boldsymbol{A} - \boldsymbol{\lambda I}) = \mathbf{0}$. We have an equation for λ.

Overall, the formulas are useful for small matrices and also for special matrices. And the properties of determinants can make those formulas simpler. If the matrix is triangular or diagonal, we just multiply the diagonal entries to find the determinant:

Triangular matrix
Diagonal matrix

$$\det\begin{bmatrix} a & b & c \\ & q & r \\ & & z \end{bmatrix} = \det\begin{bmatrix} a & & \\ & q & \\ & & z \end{bmatrix} = \boldsymbol{aqz} \qquad (1)$$

If we transpose A, the determinant formula gives the same result:

Transpose the matrix $$\mathbf{det}(\boldsymbol{A}^{\mathrm{T}}) = \mathbf{det}(\boldsymbol{A}) \qquad (2)$$

If we multiply AB, we just multiply determinants (this is a wonderful fact):

Multiply two matrices $$\mathbf{det}(\boldsymbol{AB}) = (\mathbf{det}\, \boldsymbol{A})\,(\mathbf{det}\, \boldsymbol{B}) \qquad (3)$$

A proof by algebra can get complicated. We will give a simple proof of (3) by geometry.

When we add matrices, we do not just add determinants! (See this from $I + I$. The determinant is 2^n not 2.) Here are two good consequences of equations (2) and (3):

Orthogonal matrices $\boldsymbol{Q}$ have determinant 1 or -1

We know that $Q^{\mathrm{T}}Q = I$. Then $(\det Q)^2 = (\det Q^{\mathrm{T}})\,(\det Q) = 1$. Therefore $\det Q$ is ± 1.

$$\boxed{\textbf{Invertible matrices have det } \boldsymbol{A} = \pm\ \textbf{(product of the pivots)}} \qquad (4)$$

If $A = LU$ then $\det A = (\det L)\,(\det U) = \det U$. Triangular U: *Multiply the pivots.*

If $PA = LU$ because of row exchanges, then $\det P = 1$ or -1. Permutation matrix!

Multiplying the pivots $U_{11}\; U_{22} \ldots U_{nn}$ on the diagonal reveals the determinant of A. **Elimination is how determinants are computed by virtually all computer systems for linear algebra**. The cost to find U in Chapter 2 was $n^3/3$ multiplications.

Proving those Properties (1) to (4)

The proofs of (1) to (4) can begin with three simple properties of determinants.

1. The n by n identity matrix has $\mathbf{det}\, \boldsymbol{I} = \mathbf{1}$.

2. Exchanging two rows of A reverses $\det A$ to $-\mathbf{det}\, \boldsymbol{A}$.

3. If row 1 of A is a combination $c\boldsymbol{v} + d\boldsymbol{w}$, then add 2 determinants:

$$\det \begin{bmatrix} c\boldsymbol{v} + d\boldsymbol{w} \\ \text{row } 2 \\ \cdot\cdot \\ \text{row } n \end{bmatrix} = c \det \begin{bmatrix} \boldsymbol{v} \\ \text{row } 2 \\ \cdot\cdot \\ \text{row } n \end{bmatrix} + d \det \begin{bmatrix} \boldsymbol{w} \\ \text{row } 2 \\ \cdot\cdot \\ \text{row } n \end{bmatrix} \tag{5}$$

In other words "the determinant is linear with respect to row 1." Then Property 2 leads us further: $\det A$ *is linear with respect to every row separately.* We can exchange rows and exchange back. So nothing is special about row 1.

Property 2 also leads to this simple fact. **If A has two equal rows, its determinant is zero**. Exchanging the rows has no effect on A. Then $\det A = -\det A$ forces $\det A = 0$.

Now use rule 3: **Subtracting d times row i from row j leaves $\det A$ unchanged.**

$$\det \begin{bmatrix} \text{row } 1 \\ \text{row } 2 - d \text{ row } 1 \\ \text{row } n \end{bmatrix} = \det \begin{bmatrix} \text{row } 1 \\ \text{row } 2 \\ \text{row } n \end{bmatrix} - d \begin{vmatrix} \textbf{row 1} \\ \textbf{row 1} \\ \textbf{row } \boldsymbol{n} \end{vmatrix} = \det A \tag{6}$$

This was "linearity in row 2 with row 1 fixed". It means (again) that our elimination steps from the original matrix A to an upper triangular U *do not change the determinant.* Then **elimination** (without row exchanges) **is the way to simplify A and its determinant**.

No row exchange
$$\boxed{\det \boldsymbol{A} = \det \boldsymbol{U} = U_{11}\, U_{22} \ldots U_{nn} = \textbf{product of the pivots}} \tag{7}$$

Here is a proof that $\det \boldsymbol{AB} = (\det \boldsymbol{A})\,(\det \boldsymbol{B})$. Assume $\det B \neq 0$. We will show that $\boldsymbol{D(A)} = \det \boldsymbol{AB} \,/\det \boldsymbol{B}$ has properties $\mathbf{1, 2, 3}$ above. Then $\boldsymbol{D(A)}$ must be the same as $\det A$. Here are those three properties **confirmed for $D(A)$**.

1. If $A = I$ then $D(I) = \det B \,/\, \det B$ correctly gives the answer $D(I) = 1$.

2. Exchanging rows of A will exchange rows of AB. Then $D(A)$ changes sign !

3. Multiplying row 1 of A by c will multiply $D(A)$ by c. If row 1 of A is $\boldsymbol{v} + \boldsymbol{w}$, then row 1 of AB is $\boldsymbol{v}B + \boldsymbol{w}B$. Equation (5) separates $\det(AB)$ into two determinants. So $D(A)$ also follows equation (5). From $\mathbf{1, 2, 3}$, $\mathbf{det}(\boldsymbol{AB})\,/\mathbf{det}\,\boldsymbol{B}$ must be $\mathbf{det}\,\boldsymbol{A}$.

Cramer's Rule to Solve $Ax = b$

A neat idea gives the first component x_1 of the solution vector $\boldsymbol{x}$ to $A\boldsymbol{x} = \boldsymbol{b}$. Replace the first column of I by $\boldsymbol{x}$. **This triangular M_1 has determinant x_1.** When you multiply by A, *the first column becomes $A\boldsymbol{x}$ which is $\boldsymbol{b}$.* The other columns of B_1 are copied from A.

$$\textbf{Key idea} \quad AM_1 = B_1 \qquad \begin{bmatrix} & & \\ & A & \\ & & \end{bmatrix} \begin{bmatrix} x_1 & 0 & 0 \\ x_2 & 1 & 0 \\ x_3 & 0 & 1 \end{bmatrix} = \begin{bmatrix} b_1 & a_{12} & a_{13} \\ b_2 & a_{22} & a_{23} \\ b_3 & a_{32} & a_{33} \end{bmatrix} = B_1. \tag{8}$$

We multiplied a column at a time. *Take determinants of the three matrices to find x_1*:

$$\textbf{Product rule} \qquad \boxed{(\det A)\,(x_1) = \det B_1 \quad \text{or} \quad x_1 = \det B_1/\det A.} \tag{9}$$

This is the first component of $\boldsymbol{x}$ in Cramer's Rule. Changing a column of A gave B_1. To find x_2 and B_2, put the vectors $\boldsymbol{x}$ and $\boldsymbol{b}$ into the *second columns* of I and A:

$$\textbf{Same idea} \quad AM_2 = B_2 \qquad \begin{bmatrix} \boldsymbol{a}_1 & \boldsymbol{a}_2 & \boldsymbol{a}_3 \end{bmatrix} \begin{bmatrix} 1 & x_1 & 0 \\ 0 & x_2 & 0 \\ 0 & x_3 & 1 \end{bmatrix} = \begin{bmatrix} \boldsymbol{a}_1 & \boldsymbol{b} & \boldsymbol{a}_3 \end{bmatrix} = B_2. \tag{10}$$

Take determinants to find $(\det A)(x_2) = \det B_2$. This gives $\boldsymbol{x_2 = (\det B_2)/(\det A)}$.

Example 1 Solving $3x_1 + 4x_2 = 2$ and $5x_1 + 6x_2 = 4$ needs three determinants:

$$\text{Put 2 and 4 into each } B \qquad \det A = \begin{vmatrix} 3 & 4 \\ 5 & 6 \end{vmatrix} \qquad \det B_1 = \begin{vmatrix} \mathbf{2} & 4 \\ \mathbf{4} & 6 \end{vmatrix} \qquad \det B_2 = \begin{vmatrix} 3 & \mathbf{2} \\ 5 & \mathbf{4} \end{vmatrix}$$

The determinants of B_1 and B_2 are -4 and 2. Those are divided by $\det A = -2$:

$$\textbf{Find } \boldsymbol{x} = A^{-1}\boldsymbol{b} \qquad x_1 = \frac{-4}{-2} = \mathbf{2} \qquad x_2 = \frac{2}{-2} = \mathbf{-1} \qquad \textbf{Check} \begin{bmatrix} 3 & 4 \\ 5 & 6 \end{bmatrix} \begin{bmatrix} \mathbf{2} \\ \mathbf{-1} \end{bmatrix} = \begin{bmatrix} 2 \\ 4 \end{bmatrix}$$

> **CRAMER'S RULE** If $\det A$ is not zero, $A\boldsymbol{x} = \boldsymbol{b}$ is solved by determinants:
>
> $$x_1 = \frac{\det B_1}{\det A} \qquad x_2 = \frac{\det B_2}{\det A} \qquad \ldots \qquad x_n = \frac{\det B_n}{\det A} \tag{11}$$
>
> ***The matrix B_j has the jth column of A replaced by the vector $\boldsymbol{b}$.***

To solve an n by n system, Cramer's Rule evaluates $n + 1$ determinants (of A and the n different B's). When each one is the sum of $n!$ terms—applying the "big formula" with all permutations—this makes a total of $(n+1)!$ terms. *It would be crazy to solve equations that way.* But we do finally have an explicit formula for the solution to $A\boldsymbol{x} = \boldsymbol{b}$.

Example 2 Cramer's Rule is not efficient for numbers but it is well suited to letters. For $n = 2$, find the two columns $\boldsymbol{x}$ and $\boldsymbol{y}$ of A^{-1} by solving $AA^{-1} = A\,[\boldsymbol{x} \;\; \boldsymbol{y}] = I$.

$$A\boldsymbol{x} = \begin{bmatrix} a & b \\ c & d \end{bmatrix}\begin{bmatrix} x_1 \\ x_2 \end{bmatrix} = \begin{bmatrix} \mathbf{1} \\ \mathbf{0} \end{bmatrix} \qquad A\boldsymbol{y} = \begin{bmatrix} a & b \\ c & d \end{bmatrix}\begin{bmatrix} y_1 \\ y_2 \end{bmatrix} = \begin{bmatrix} \mathbf{0} \\ \mathbf{1} \end{bmatrix}$$

Those share the same matrix A. We need $|A|$ and four determinants for $x_1,\ x_2,\ y_1,\ y_2$:

$$|A| = \begin{vmatrix} a & b \\ c & d \end{vmatrix} \quad \text{and} \quad \begin{vmatrix} \mathbf{1} & b \\ \mathbf{0} & d \end{vmatrix} \quad \begin{vmatrix} a & \mathbf{1} \\ c & \mathbf{0} \end{vmatrix} \quad \begin{vmatrix} \mathbf{0} & b \\ \mathbf{1} & d \end{vmatrix} \quad \begin{vmatrix} a & \mathbf{0} \\ c & \mathbf{1} \end{vmatrix}$$

The last four determinants are $d, -c, -b$, and a. (**They are the cofactors!**) Here is A^{-1}:

$$x_1 = \frac{d}{|A|},\ x_2 = \frac{-c}{|A|},\ y_1 = \frac{-b}{|A|},\ y_2 = \frac{a}{|A|} \text{ and } \boldsymbol{A^{-1}} = \frac{1}{\boldsymbol{ad-bc}}\begin{bmatrix} \boldsymbol{d} & \boldsymbol{-b} \\ \boldsymbol{-c} & \boldsymbol{a} \end{bmatrix}.$$

A^{-1} involves the cofactors of A. When the right side of $A\boldsymbol{x} = \boldsymbol{b}$ is a column of I, as in $AA^{-1} = I$, **the determinant of each B_j in Cramer's Rule is a cofactor of A.**

You can see those cofactors for $n = 3$. Solve $A\boldsymbol{x} = (1, 0, 0)$ to find column 1 of A^{-1}:

Determinants of B's are Cofactors of A
$$\begin{vmatrix} \mathbf{1} & a_{12} & a_{13} \\ \mathbf{0} & a_{22} & a_{23} \\ \mathbf{0} & a_{32} & a_{33} \end{vmatrix} \quad \begin{vmatrix} a_{11} & \mathbf{1} & a_{13} \\ a_{21} & \mathbf{0} & a_{23} \\ a_{31} & \mathbf{0} & a_{33} \end{vmatrix} \quad \begin{vmatrix} a_{11} & a_{12} & \mathbf{1} \\ a_{21} & a_{22} & \mathbf{0} \\ a_{31} & a_{32} & \mathbf{0} \end{vmatrix}$$

That determinant of B_1 is the cofactor $C_{11} = a_{22}a_{33} - a_{23}a_{32}$. Notice that the correct minus sign appears in $\det B_2 = -(a_{21}a_{33} - a_{23}a_{31})$. This cofactor C_{12} goes into column 1 of A^{-1}. When we divide by $\det A$, we have computed the same A^{-1} as before:

FORMULA FOR A^{-1} $\quad (A^{-1})_{ij} = \dfrac{C_{ji}}{\det A} \qquad A^{-1} = \dfrac{C^{\mathrm{T}}}{\det A}$ (12)

The Big Formula for the Determinant: $n!$ Terms

So far we have found two formulas for $\det A$. One was the **cofactor formula**, which reduces $\det A$ to a combination of smaller determinants. (This worked well when A had only 3 nonzero diagonals and lots of zeros.) The second was the **pivot formula**, coming from elimination and $PA = LU$. This worked well because L and U are triangular in Chapter 2. And $\det P = \pm 1$. Our **big formula** is not so great, because it has **all of the $n!$ terms in $\det A$.**

Every n by n permutation matrix P ($n!$ P's) produces a term D_P in the big formula. **Each matrix P has n 1's**: a single 1 in each row and each column. We multiply the n numbers $a_{1\alpha}a_{2\beta}\cdots a_{n\omega}$ in those positions, to find their contribution $\boldsymbol{D_P}$ to the overall determinant of A. And we remember to include the factor $\det P = \pm 1$ in $\boldsymbol{D_P}$. This accounts for an even or odd permutation. Now add up all the terms like $\boldsymbol{D_P} = \boldsymbol{ad}$ or $\boldsymbol{aqz}$.

Big Formula $\quad \det A$ = Sum of all D_P = $n!$ terms from $n!$ permutations P (13)

For $n = 2$ this formula is exactly $ad - bc$ (2 permutations give $D_P = ad$ and $-bc$). For $n = 3$ we have $3! = 6$ permutations and 6 terms in equation (13).

$$\det \boldsymbol{A} = +\boldsymbol{aqz} + \boldsymbol{brx} + \boldsymbol{cpy} - \boldsymbol{ary} - \boldsymbol{bpz} - \boldsymbol{cqx}.$$

Problem Set 5.2

1 If $\det A = 2$, what are $\det A^{-1}$ and $\det A^n$ and $\det A^{\mathrm{T}}$?

2 Compute the determinants of A, B, C, D. Are their columns independent?

$$A = \begin{bmatrix} 1 & 1 & 0 \\ 1 & 0 & 1 \\ 0 & 1 & 1 \end{bmatrix} \quad B = \begin{bmatrix} 1 & 2 & 3 \\ 4 & 5 & 6 \\ 7 & 8 & 9 \end{bmatrix} \quad C = \begin{bmatrix} A & 0 \\ 0 & A \end{bmatrix} \quad D = \begin{bmatrix} A & 0 \\ 0 & B \end{bmatrix}.$$

3 Show that $\det A = 0$, regardless of the five numbers marked by x's :

$$A = \begin{bmatrix} x & x & x \\ 0 & 0 & x \\ 0 & 0 & x \end{bmatrix}.$$

What are the cofactors of row 1?
What is the rank of A?
What are the 6 terms in $\det A$?

4 (a) If $D_n = \det(A^n)$, could D_n grow toward infinity even if all $|A_{ij}| < 1$?

(b) Could D_n approach zero even if all $|A_{ij}| > 1$?

Problems 5–9 are about Cramer's Rule for $\boldsymbol{x} = \boldsymbol{A^{-1}b}$.

5 Solve these linear equations by Cramer's Rule $x_j = \det B_j / \det A$:

(a) $\begin{aligned} 2x_1 + 5x_2 &= 1 \\ x_1 + 4x_2 &= 2 \end{aligned}$ (b) $\begin{aligned} 2x_1 + x_2 \quad\quad &= 1 \\ x_1 + 2x_2 + x_3 &= 0 \\ x_2 + 2x_3 &= 0. \end{aligned}$

6 Use Cramer's Rule to solve for y (only). Call the 3 by 3 determinant D:

(a) $\begin{aligned} ax + by &= 1 \\ cx + dy &= 0 \end{aligned}$ (b) $\begin{aligned} ax + by + cz &= 1 \\ dx + ey + fz &= 0 \\ gx + hy + iz &= 0. \end{aligned}$

7 Cramer's Rule breaks down when $\det A = 0$. Example (a) has no solution while (b) has infinitely many. What are the ratios $x_j = \det B_j / \det A$ in these two cases?

(a) $\begin{aligned} 2x_1 + 3x_2 &= 1 \\ 4x_1 + 6x_2 &= 1 \end{aligned}$ (parallel lines) (b) $\begin{aligned} 2x_1 + 3x_2 &= 1 \\ 4x_1 + 6x_2 &= 2 \end{aligned}$ (same line)

8 *Quick proof of Cramer's rule.* The determinant is a linear function of column 1. It is zero if two columns are equal. When $\boldsymbol{b} = A\boldsymbol{x} = x_1\boldsymbol{a}_1 + x_2\boldsymbol{a}_2 + x_3\boldsymbol{a}_3$ goes into the first column of A, we have the matrix B_1 and Cramer's Rule $x_1 = \det B_1 / \det A$:

$$|\boldsymbol{b} \quad \boldsymbol{a}_2 \quad \boldsymbol{a}_3| = |x_1\boldsymbol{a}_1 + x_2\boldsymbol{a}_2 + x_3\boldsymbol{a}_3 \quad \boldsymbol{a}_2 \quad \boldsymbol{a}_3| = x_1|\boldsymbol{a}_1 \quad \boldsymbol{a}_2 \quad \boldsymbol{a}_3| = x_1 \det A.$$

What steps lead to the middle equation?

9 If the right side $\boldsymbol{b}$ equals the first column of A, solve the 3 by 3 system $A\boldsymbol{x} = \boldsymbol{b}$. How does each determinant in Cramer's Rule lead to this solution $\boldsymbol{x}$?

10 Suppose E_n is the determinant of the n by n three-diagonal $1,1,1$ matrix. By cofactors of row 1 show that $\boldsymbol{E_n = E_{n-1} - E_{n-2}}$. Starting from $E_1 = 1$ and $E_2 = 0$ find $E_3, E_4, \ldots, E_8$. By noticing how the E's repeat, find E_{100}.

$$E_1 = |1| \quad E_2 = \begin{vmatrix} 1 & 1 \\ 1 & 1 \end{vmatrix} \quad E_3 = \begin{vmatrix} 1 & 1 & 0 \\ 1 & 1 & 1 \\ 0 & 1 & 1 \end{vmatrix} \quad E_4 = \begin{vmatrix} 1 & 1 & 0 & 0 \\ 1 & 1 & 1 & 0 \\ 0 & 1 & 1 & 1 \\ 0 & 0 & 1 & 1 \end{vmatrix}.$$

11 If a 3 by 3 matrix has entries $1,2,3,4,\ldots,9$, what is the maximum determinant? I would use a computer to decide. This problem does not seem easy.

12 True or false: The determinant of ABC is $(\det A)\,(\det B)\,(\det C)$

13 Reduce A to U and find $\det A =$ product of the pivots:

$$A = \begin{bmatrix} 1 & 1 & 1 \\ 1 & 2 & 2 \\ 1 & 2 & 3 \end{bmatrix} \qquad A = \begin{bmatrix} 1 & 2 & 3 \\ 2 & 2 & 3 \\ 3 & 3 & 3 \end{bmatrix}.$$

14 By applying elimination to produce an upper triangular matrix U, compute

$$\det \begin{bmatrix} 1 & 2 & 3 & 0 \\ 2 & 6 & 6 & 1 \\ -1 & 0 & 0 & 3 \\ 0 & 2 & 0 & 7 \end{bmatrix} \quad \text{and} \quad \det \begin{bmatrix} 1 & -1 & 0 & 0 \\ -1 & 2 & -1 & 0 \\ 0 & -1 & 2 & -1 \\ 0 & 0 & -1 & 2 \end{bmatrix}.$$

15 Use row operations to compute these determinants:

$$\det \begin{bmatrix} 101 & 201 & 301 \\ 102 & 202 & 302 \\ 103 & 203 & 303 \end{bmatrix} \quad \text{and} \quad \det \begin{bmatrix} 1 & a & a^2 \\ 1 & b & b^2 \\ 1 & c & c^2 \end{bmatrix} = \begin{array}{c} (b-a)(c-a)(c-b) \\ \text{“Vandermonde determinant”} \end{array}$$

16 In a 4 by 4 determinant, decide $\det P = 1$ or -1 for these big formula terms:

$$\pm a_{12}a_{21}a_{34}a_{43} \qquad \pm\, a_{13}a_{22}a_{31}a_{44} \qquad \pm\, a_{11}a_{22}a_{34}a_{43}$$

17 (a) If $a_{11} = a_{22} = a_{33} = 0$, how many of the six terms in $\det A$ will be zero?

(b) If $a_{11} = a_{22} = a_{33} = a_{44} = 0$, how many of the 24 products $a_{1j}a_{2k}a_{3l}a_{4m}$ are sure to be zero?

18 The big formula has $n!$ terms. But if an entry of A is zero, $(n-1)!$ terms disappear. If A has only *three nonzero diagonals* (in the center of A), how many terms are left?

For $n = 1,2,3,4$ that tridiagonal determinant of A has $1,2,3,5$ nonzero terms. Those are Fibonacci numbers! Show why a tridiagonal 5 by 5 determinant has $5+3=8$ nonzero terms (Fibonacci again). Use the cofactors of a_{11} and a_{12}.

19 The Big Formula has 24 terms if A is 4 by 4. How many terms include a_{13} and a_{22}? It is a challenge to write down all 24 terms in $\det(A)$.

5.3 Areas and Volumes by Determinants

1 A parallelogram in 2D starts from $(0,0)$ with sides $\boldsymbol{e}_1 = (a,b)$ and $\boldsymbol{e}_2 = (c,d)$.

2 Area of the parallelogram $= \left|\text{Determinant of the matrix } E = \begin{bmatrix} \boldsymbol{e}_1 & \boldsymbol{e}_2 \end{bmatrix}\right| = |ad - bc|$.

3 A tilted box in 3D starts with three edges $\boldsymbol{e}_1, \boldsymbol{e}_2, \boldsymbol{e}_3$ out from $(0,0,0)$.

4 Volume of a tilted box $= |$ Determinant of the 3×3 edge matrix $E\,|$.

Determinants lead to good formulas for areas and volumes. The regions have straight sides but they are not square. A typical region is a parallelogram—or half a parallelogram which is a triangle. The problem is: **Find the area**. For a triangle, the area is $\frac{1}{2}\,\boldsymbol{bh}$: half the base times the height. A parallelogram contains two triangles with equal area, so we omit the $\frac{1}{2}$. Then *parallelogram area = base times height.*

Those formulas are simple to remember. But they don't fit our problem, because we **are not given the base and height**. We only know **the positions of the corners**. For the triangle, suppose the corner points are $(0,0)$ and (a,b) and (c,d). For the parallelogram (twice as large) the fourth corner will be $(a+c, b+d)$. Knowing a, b, c, d, *what is the area*?

Triangle and parallelogram

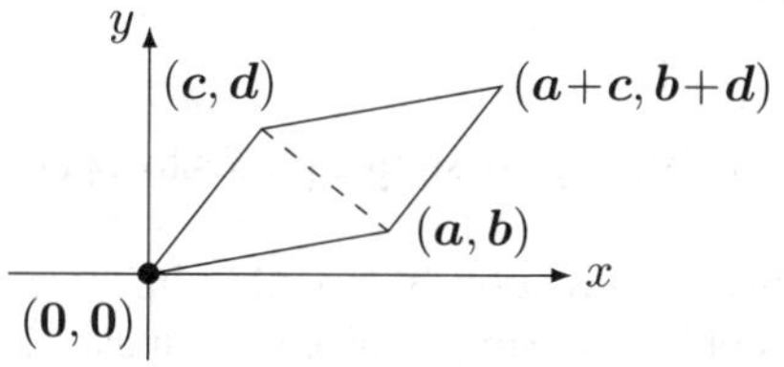

The base could be the lowest side and its length is $\sqrt{a^2+b^2}$. To find the height, we could create a line down from (c,d) that is perpendicular to the baseline. The length h of that perpendicular line would involve more square roots. But the area itself does not involve square roots! Its beautiful formula $ad - bc$ is simple for $a = d = 1, b = c = 0$ (a square).

$$\textbf{Area of parallelogram} = \left|\ \textbf{Determinant of matrix}\ \right| = \pm \begin{vmatrix} \boldsymbol{a} & \boldsymbol{c} \\ \boldsymbol{b} & \boldsymbol{d} \end{vmatrix} = |\boldsymbol{ad} - \boldsymbol{bc}|. \quad (1)$$

Our goal is to find that formula by linear algebra: no square roots or negative volumes. We took the absolute value $|ad - bc|$ to avoid a negative area.

We also have a more difficult goal. We need to move into 3 dimensions and eventually into n dimensions. We start with four corners $0,0,0$ and a,b,c and p,q,r and x,y,z of a box. (The box is tilted. Every side is a parallelogram. It will look lopsided.)

If we use the area formula (1) as a guide, we could guess the correct volume formula:

$$\textbf{Volume of box} = \left|\ \textbf{Determinant of matrix}\ \right| = \pm \begin{vmatrix} a & p & x \\ b & q & y \\ c & r & z \end{vmatrix} \qquad (2)$$

Our first effort stays in a plane. For this case we use geometry. Figure 5.1 shows how adding pieces to a parallelogram can produce a rectangle. When we subtract the areas of those six pieces, we arrive at the correct parallelogram area $\boldsymbol{ad - bc}$ (no square roots). The picture is not very elegant, but in two dimensions it succeeds.

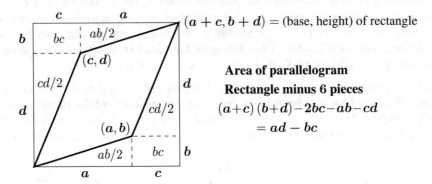

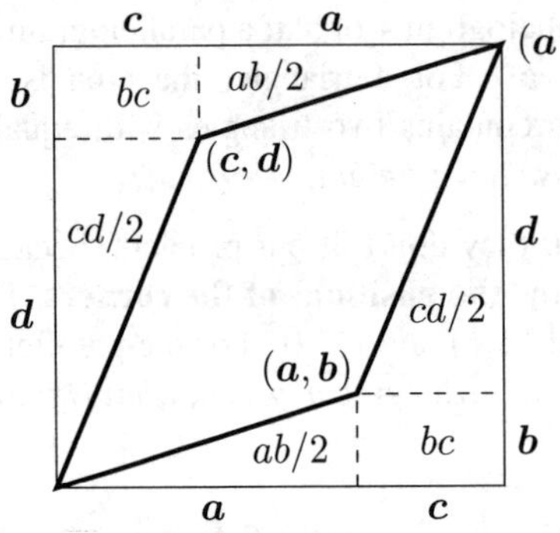

$(\boldsymbol{a+c, b+d})$ = (base, height) of rectangle

Area of parallelogram
Rectangle minus 6 pieces
$$(\boldsymbol{a+c})(\boldsymbol{b+d}) - \boldsymbol{2bc - ab - cd} = \boldsymbol{ad - bc}$$

Figure 5.1: Adding six simple pieces to a parallelogram produces a rectangle.

Would a similar construction be possible in three dimensions? Following Figure 5.1, I believe we could add simple pieces to make a tilted box into a rectangular box—but it doesn't look easy. And there is a much better way: *Use linear algebra.*

Areas and Volumes by Linear Algebra

In working on this problem, I came to an understanding. *If we do more algebra, then we need less geometry.* Very often, linear algebra comes down to factoring a matrix. We will look there for ideas.

A box in $\boldsymbol{n}$ dimensions has $\boldsymbol{n}$ edges $\boldsymbol{e_1, e_2, \ldots, e_n}$ going out from the origin. The parallelogram in two dimensions had two vectors $\boldsymbol{e}_1 = (a, b)$ and $\boldsymbol{e}_2 = (c, d)$. Those vectors $\boldsymbol{e}$ give two corners or n corners of the "box". In the 2-dimensional picture, the fourth corner was $\boldsymbol{e}_1 + \boldsymbol{e}_2$. In the n-dimensional picture, the other corners of the box would be sums of the $\boldsymbol{e}$'s. *The box is totally determined by the n edges in the matrix E*:

$$\textbf{Edge matrix} \quad \boldsymbol{E_2} = \begin{bmatrix} a & c \\ b & d \end{bmatrix} \quad \text{and} \quad \boldsymbol{E_n} = \begin{bmatrix} | & & | \\ e_1 & \cdots & e_n \\ | & & | \end{bmatrix} \quad \begin{matrix}\textbf{Columns} = \\ \textbf{Box edges}\end{matrix}$$

Our goal is to prove that the volume of the box is $|\det E|$. We considered three possible factorizations of E, to reach this goal. They are taken from Chapters 2 and 4 and 7.

The third factorization is called the Singular Value Decomposition of E: the SVD.

Lower times upper triangular	$E = LU$
Orthogonal times upper triangular R	$E = QR$
Orthogonal – Diagonal – Orthogonal	$E = U\Sigma V^{\mathrm{T}}$

Those factors of E are square matrices because E is square (n by n). Remember that the determinant of L is 1 (all ones on its diagonal). The determinant of any orthogonal matrix Q is 1 or -1. Then $\det \boldsymbol{L} = |\det \boldsymbol{Q}| = |\det \boldsymbol{U}| = |\det \boldsymbol{V}| = \mathbf{1}$. We will certainly depend on the multiplication rule $\det AB = (\det A)\,(\det B)$. Three options for $|\det E|$:

$$\boxed{\textbf{Box volume} = |\det \boldsymbol{E}| = |\det \boldsymbol{U}| = |\det \boldsymbol{R}| = |\det \boldsymbol{\Sigma}|.} \tag{3}$$

The problem is to connect the volume of the box to one of those determinants—to connect geometry to linear algebra. Start with the geometry of an orthogonal matrix Q.

Key idea for any region in $\mathbf{R}^n$: Multiplying all points by an orthogonal matrix Q **does not change the volume**. Let me understand this first and then use $\boldsymbol{E} = \boldsymbol{QR}$.

Multiply by any matrix: Straight lines stay straight, areas can change

Multiply by an orthogonal Q: $\boldsymbol{x}^{\mathrm{T}}\boldsymbol{x}$ and $\boldsymbol{x}^{\mathrm{T}}\boldsymbol{y}$ are the same as $(Q\boldsymbol{x})^{\mathrm{T}}(Q\boldsymbol{x})$ and $(Q\boldsymbol{x})^{\mathrm{T}}(Q\boldsymbol{y})$

Lengths and angles and box shapes and volumes are not changed by rotations $\boldsymbol{Q}$.

This remains true for curved regions. We divide them into many small cubes plus thin curved pieces. The total volume of those curved pieces can approach zero. The volumes of the cubes are not changed by Q. *The shapes for R and for $E = QR$ have the same volume.*

***R* is a triangular matrix.** Its box has a volume we can compute. For a parallelogram in the xy plane, **the base and height are the diagonal entries $\boldsymbol{u}$ and $\boldsymbol{w}$ of $\boldsymbol{R}$ in $\boldsymbol{E} = \boldsymbol{QR}$**:

$$\boldsymbol{R} = \begin{bmatrix} u & v \\ 0 & w \end{bmatrix}$$

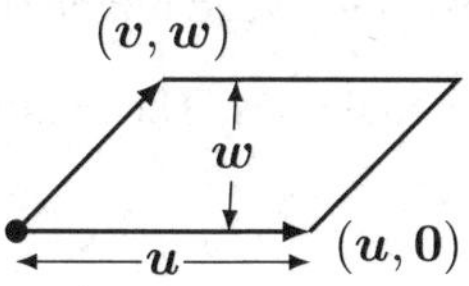

base $= u$, **height** $= w$

$|\textbf{area}| = |\boldsymbol{u} \text{ times } \boldsymbol{w}| = |\det \boldsymbol{R}|$

The key point is: The main diagonal of R shows the height in each new dimension. When we multiply those numbers on the diagonal of R, **we get the volume of the box and also the determinant of the triangular matrix $\boldsymbol{R}$.** The volume formula $|\det E|$ is now proved in all dimensions because $|\det \boldsymbol{Q}| = 1$ and $|\det \boldsymbol{E}| = |\det \boldsymbol{R}|$.

Final comment: The Singular Value Decomposition $E = U\Sigma V^{\mathrm{T}}$ has two orthogonal matrices U and V. The number $|\det E|$ is equal to $|\det \Sigma|$. And this matrix Σ is *diagonal.* Σ gives a perfectly normal *rectangular box* in $\mathbf{R}^n$. This SVD approach by $U\Sigma V^{\mathrm{T}}$ looks simpler than QR, where the triangular matrix R produced the tilted box above. But...

But that tilted figure shows a clear geometric meaning for the diagonal entries u and w : **base** and **height**. The geometry of the SVD will be seen in Chapter 7. It is beautifully clear for *ellipses* in n dimensions. But the singular values are not so clear for boxes. Σ gives the lengths of the axes of an *ellipse* but not the sides of a rectangular box. For a box with straight sides, $E = QR$ leads directly to volume of box $= |\det R|$.

Every shape has the growth factor $\det E$ when all points are multiplied by E.

Problem Set 5.3

1 For a 3-dimensional cube with four corners at $(0,0,0), (1,0,0), (0,1,0), (0,0,1)$, what is the edge matrix E ? For a tilted box with edges $\boldsymbol{e}_1, \boldsymbol{e}_2, \boldsymbol{e}_3$ going out from $(0,0,0)$, what are the 8 corners of the box ?

2 Suppose the edge vectors $\boldsymbol{e}_1, \boldsymbol{e}_2, \boldsymbol{e}_3$ in 3 dimensions are perpendicular unit vectors. Describe the box and find its volume.

3 If the edge vectors have lengths $||\boldsymbol{e}_1|| = 1, ||\boldsymbol{e}_2|| = 2, ||\boldsymbol{e}_3|| = 3$, guess the largest possible volume of the box. Describe the box that reaches this maximum volume.

4 Suppose an n by n matrix E has columns $\boldsymbol{e}_1, \boldsymbol{e}_2, \ldots, \boldsymbol{e}_n$. Problem 3 becomes

$$|\textbf{determinant of } \boldsymbol{E}\ | \leq \textbf{product of column lengths} = ||\boldsymbol{e}_1||\ ||\boldsymbol{e}_2|| \cdots ||\boldsymbol{e}_n||.$$

Hint: Why does the triangular R in $E = QR$ have diagonals $R_{ii} \leq ||\boldsymbol{e}_i||$?

5 (a) Find the area of the parallelogram with edges $\boldsymbol{v} = (3,2)$ and $\boldsymbol{w} = (1,4)$.

(b) Find the area of the triangle with sides $\boldsymbol{v}$, $\boldsymbol{w}$, and $\boldsymbol{v} + \boldsymbol{w}$. Draw it.

(c) Find the area of the triangle with sides $\boldsymbol{v}$, $\boldsymbol{w}$, and $\boldsymbol{w} - \boldsymbol{v}$. Draw it.

6 (a) The corners of a triangle are $(2,1)$ and $(3,4)$ and $(0,5)$. What is the area?

(b) Add a corner at $(-1,0)$ to make a lopsided region (four sides). Find the area.

7 This Hadamard matrix H has orthogonal rows. The box is a hypercube !

$$\text{Compute} \quad |\det H| = \begin{vmatrix} 1 & 1 & 1 & 1 \\ 1 & 1 & -1 & -1 \\ 1 & -1 & -1 & 1 \\ 1 & -1 & 1 & -1 \end{vmatrix} = \begin{array}{l} \text{volume of a hypercube in } \mathbf{R}^4 \\ \text{The sides have length 2} \end{array}$$

8 An n-dimensional cube has how many corners ? How many edges? How many $(n-1)$-dimensional faces? The cube in $\mathbf{R}^n$ whose edges are the rows of $2I$ has volume _____. A hypercube computer has parallel processors at the corners with connections along the edges.

9 The triangle with corners $(0,0), (1,0), (0,1)$ has area $\frac{1}{2}$. The pyramid in $\mathbf{R}^3$ with four corners $(0,0,0), (1,0,0), (0,1,0), (0,0,1)$ has volume _____. What is the volume of a pyramid in $\mathbf{R}^4$ with five corners at $(0,0,0,0)$ and the four columns of I ?

10 If the edge matrix E is an orthogonal matrix, the box has volume _____.
If the edge matrix E is a singular matrix, the box has volume _____.
If the volume in $\mathbf{R}^n$ is V, the box for $2E$ has volume _____.

11 Draw parallelograms for $\begin{bmatrix} 2 & 1 \\ 3 & 4 \end{bmatrix}$ and $\begin{bmatrix} 2 & 3 \\ 1 & 4 \end{bmatrix}$. Can you see any reason for equal areas?

12 Transposing the edge matrix $\begin{bmatrix} u & v \\ 0 & w \end{bmatrix}$ gives a matrix with the same determinant and a new parallelogram with the same area. Can you draw it and recompute its area?

13 Draw the parallelogram P with edges from $(0,0)$ to (a,b) and (c,d).

$$\text{Its area is det}\begin{bmatrix} a & c \\ b & d \end{bmatrix} = \det\begin{bmatrix} a & 0 \\ 0 & d \end{bmatrix} + \det\begin{bmatrix} 0 & c \\ b & 0 \end{bmatrix}.$$

Can you see how to produce the parallelogram P from the rectangle with sides a, d and the rectangle with sides b, c?

14 If the 5 by 5 matrix A contains five 1's and twenty 2's, what is the largest possible determinant?

15 Suppose a 2 by 2 matrix E has column vectors of length $||\boldsymbol{e}_1||$ and $||\boldsymbol{e}_2||$. Show that $|\det E| \leq ||\boldsymbol{e}_1||$ times $||\boldsymbol{e}_2||$. This is **Hadamard's inequality** (2 by 2 case). It was the key idea in Problem 4. Parallelogram area $\leq$ | side 1 | times | side 2 |.

Chapter 6

Eigenvalues and Eigenvectors

Eigenvalues and **eigenvectors** have new information about a square matrix—deeper than its rank or its column space. We look for **eigenvectors** $\boldsymbol{x}$ that don't change direction when they are multiplied by A. Then $\boldsymbol{A x = \lambda x}$ with **eigenvalue** $\boldsymbol{\lambda}$. (You could call λ the stretching factor.) Multiplying again gives $\boldsymbol{A^2 x = \lambda^2 x}$. We can go onwards to $\boldsymbol{A^{100} x = \lambda^{100} x}$. And we can combine two or more eigenvectors:

$$A(\boldsymbol{x}_1 + \boldsymbol{x}_2) = \lambda_1 \boldsymbol{x}_1 + \lambda_2 \boldsymbol{x}_2 \qquad A^2(c_1 \boldsymbol{x}_1 + c_2 \boldsymbol{x}_2) = c_1 \lambda_1^2 \boldsymbol{x}_1 + c_2 \lambda_2^2 \boldsymbol{x}_2$$

When we separate the input into eigenvectors, each eigenvector just goes its own way.

The eigenvalues are the growth factors in $\boldsymbol{A^n x = \lambda^n x}$. If all $|\lambda_i| < 1$ then A^n will eventually approach zero. If any $|\lambda_i| > 1$ then A^n eventually grows. If $\lambda = 1$ then $A^n \boldsymbol{x}$ never changes (a steady state). For the economy of a country or a company or a family, the size of λ is a critical number.

Properties of a matrix are reflected in the properties of the λ's and the $\boldsymbol{x}$'s. A symmetric matrix S has perpendicular eigenvectors—and all its eigenvalues are real numbers. The kings of linear algebra are **symmetric matrices with positive eigenvalues**. These "positive definite matrices" signal a minimum point for a function like the energy $f(\boldsymbol{x}) = \frac{1}{2}\boldsymbol{x}^{\mathrm{T}} S \boldsymbol{x}$. That is the n-dimensional form of the calculus test $d^2 f/dx^2 > 0$ for a minimum of $f(x)$.

This chapter ends by solving linear differential equations $d\boldsymbol{u}/dt = A\boldsymbol{u}$. The pieces of the solution are $\boldsymbol{u}(t) = e^{\lambda t}\boldsymbol{x}$ instead of $\boldsymbol{u}_n = \lambda^n \boldsymbol{x}$—exponentials instead of powers. The whole solution is $\boldsymbol{u}(t) = e^{At}\boldsymbol{u}(0)$. For linear differential equations with a constant matrix A, please use its eigenvectors.

Section 6.4 gives the rules for complex matrices—including the famous **Fourier matrix**.

6.1 Introduction to Eigenvalues : $Ax = \lambda x$

1 If $\boxed{Ax = \lambda x}$ then $x \neq \mathbf{0}$ is an **eigenvector** of A and the number λ is the **eigenvalue**.

2 Then $\boxed{A^n x = \lambda^n x}$ for every n and $(A + cI)x = (\lambda + c)x$ and $A^{-1}x = x/\lambda$ if $\lambda \neq 0$.

3 $(A - \lambda I)x = \mathbf{0} \Rightarrow$ **the determinant of $A - \lambda I$ is zero: this equation produces n λ's.**

4 Check λ's by $(\lambda_1)(\lambda_2)\cdots(\lambda_n) = \det A$ and $\lambda_1 + \cdots + \lambda_n =$ diagonal sum $a_{11} + \cdots + a_{nn}$.

5 Projections have $\lambda = 1$ and 0. Rotations have $\lambda = e^{i\theta}$ and $e^{-i\theta}$: *complex numbers*!

This chapter enters a new part of linear algebra. The first part was about $Ax = b$: linear equations for a steady state. Now the second part is about **change**. Time enters the picture—continuous time in a differential equation $du/dt = Au$ or time steps $k = 1, 2, 3, \ldots$ in $u_{k+1} = Au_k$. Those equations are NOT solved by elimination.

We want "eigenvectors" x that don't change direction when you multiply by A. The solution vector $u(t)$ or u_k stays in the direction of that fixed vector x. Then we only look for the number (changing with time) that multiplies x: a one-dimensional problem.

A good model comes from the powers $A, A^2, A^3, \ldots$ of a matrix. Suppose you need the hundredth power A^{100}. Its columns are very close to the **eigenvector** $x = (.6, .4)$:

$$A = \begin{bmatrix} .8 & .3 \\ .2 & .7 \end{bmatrix} \quad A^2 = \begin{bmatrix} .70 & .45 \\ .30 & .55 \end{bmatrix} \quad A^3 = \begin{bmatrix} .650 & .525 \\ .350 & .475 \end{bmatrix} \qquad A^{100} \approx \begin{bmatrix} .6000 & .6000 \\ .4000 & .4000 \end{bmatrix}$$

A^{100} was found by using the **eigenvalues 1 and 1/2** of this A, not by multiplying 100 matrices. Eigenvalues and eigenvectors are a new way to see into the heart of a matrix.

To explain eigenvalues, we first explain eigenvectors. Almost all vectors will change direction, when they are multiplied by A. ***Certain exceptional vectors x are in the same direction as Ax. Those are the "eigenvectors"***. Multiply an eigenvector by A, and the vector Ax is a number λ times the original x.

The basic equation is $Ax = \lambda x$. The number λ is an eigenvalue of A.

The eigenvalue λ tells whether the special vector x is stretched or shrunk or reversed—when it is multiplied by A. We may find $\lambda = 2$ or $\frac{1}{2}$ or -1. The eigenvalue λ could be zero ! Then $Ax = 0x$ means that this eigenvector x is in the nullspace of A.

If A is the identity matrix, every vector has $Ax = x$. All vectors are eigenvectors of I. All eigenvalues "lambda" are $\lambda = 1$. This is unusual to say the least. Most 2 by 2 matrices have *two* eigenvector directions and *two* eigenvalues. We will show that $\det(A - \lambda I) = 0$.

This section explains how to compute the x's and λ's. It can come early in the course. We only need the determinant $ad - bc$ of a 2 by 2 matrix. Example 1 uses $\det(A - \lambda I) = 0$ to find the eigenvalues $\lambda = 1$ and $\lambda = \frac{1}{2}$ for the matrix A that appears above.

Example 1 The key numbers in $\det(A-\lambda I)$ are $.8+.7=\mathbf{1.5}=\frac{\mathbf{3}}{\mathbf{2}}$ on the diagonal of A and $(.8)\,(.7)-(.3)\,(.2)=.5=$ **determinant of** $\boldsymbol{A}=\frac{\mathbf{1}}{\mathbf{2}}$:

$$\boldsymbol{A}=\begin{bmatrix}.8 & .3\\ .2 & .7\end{bmatrix}\quad \det\begin{bmatrix}.8-\lambda & .3\\ .2 & .7-\lambda\end{bmatrix}=\lambda^2-\frac{\mathbf{3}}{\mathbf{2}}\lambda+\frac{\mathbf{1}}{\mathbf{2}}=(\lambda-1)\left(\lambda-\frac{1}{2}\right).$$

I factored the quadratic into $\lambda-1$ times $\lambda-\frac{1}{2}$, to see the two eigenvalues $\boldsymbol{\lambda}=\mathbf{1}$ and $\frac{\mathbf{1}}{\mathbf{2}}$. **This tells us that $\boldsymbol{A-I}$ and $\boldsymbol{A-\frac{1}{2}I}$ (with zero determinant) are not invertible.** The eigenvectors $\boldsymbol{x}_1$ and $\boldsymbol{x}_2$ are in the nullspaces of $A-I$ and $A-\frac{1}{2}I$.

$(A-I)\boldsymbol{x}_1=0$ is $A\boldsymbol{x}_1=\boldsymbol{x}_1$. That first eigenvector is $\boldsymbol{x}_1=(.6,\ .4)$.
$(A-\frac{1}{2}I)\boldsymbol{x}_2=0$ is $A\boldsymbol{x}_2=\frac{1}{2}\boldsymbol{x}_2$. That second eigenvector is $\boldsymbol{x}_2=(\mathbf{1},\ \mathbf{-1})$:

$$\boldsymbol{x}_1=\begin{bmatrix}\mathbf{.6}\\ \mathbf{.4}\end{bmatrix}\quad\text{and}\quad A\boldsymbol{x}_1=\begin{bmatrix}.8 & .3\\ .2 & .7\end{bmatrix}\begin{bmatrix}.6\\ .4\end{bmatrix}=\begin{bmatrix}.6\\ .4\end{bmatrix}\qquad (A\boldsymbol{x}_1=\boldsymbol{x}_1 \text{ means that } \boldsymbol{\lambda_1=1})$$

$$\boldsymbol{x}_2=\begin{bmatrix}\mathbf{1}\\ \mathbf{-1}\end{bmatrix}\quad\text{and}\quad A\boldsymbol{x}_2=\begin{bmatrix}.8 & .3\\ .2 & .7\end{bmatrix}\begin{bmatrix}1\\ -1\end{bmatrix}=\begin{bmatrix}.5\\ -.5\end{bmatrix}\qquad (\text{this is } \tfrac{1}{2}\,\boldsymbol{x}_2 \text{ so } \boldsymbol{\lambda_2}=\tfrac{\mathbf{1}}{\mathbf{2}}).$$

From $A\boldsymbol{x}_1=\boldsymbol{x}_1$ we get $A^2\boldsymbol{x}_1=A\boldsymbol{x}_1=\boldsymbol{x}_1$. Every power of A will have $A^n\boldsymbol{x}_1=\boldsymbol{x}_1$. Multiplying $\boldsymbol{x}_2$ by A gave $\frac{1}{2}\boldsymbol{x}_2$, and if we multiply again we get $A^2\boldsymbol{x}_2=(\frac{1}{2})^2$ times $\boldsymbol{x}_2$.

The eigenvectors of A remain eigenvectors of $\boldsymbol{A^2}$. The eigenvalues $\boldsymbol{\lambda}$ are squared.

This pattern succeeds because the eigenvectors $\boldsymbol{x}_1,\boldsymbol{x}_2$ stay in their own directions. The eigenvalues of A^{100} are $1^{100}=1$ and $(\frac{1}{2})^{100}=$ very small number.

An eigenvector $\boldsymbol{x}$ of $\boldsymbol{A}$ is also an eigenvector of every $\boldsymbol{A^n}$. Then $\boldsymbol{A^n x=\lambda^n x}$.

Other vectors do change direction. But all other vectors are combinations of the two eigenvectors $\boldsymbol{x}_1$ and $\boldsymbol{x}_2$. The first column $(.8, .2)$ of A is the combination $(.6, .4)+(.2, -.2)$.

$$\textbf{Separate column 1 of } \boldsymbol{A} \text{ } \textbf{into eigenvectors}\qquad \begin{bmatrix}.8\\ .2\end{bmatrix}=\begin{bmatrix}.6\\ .4\end{bmatrix}+\begin{bmatrix}.2\\ -.2\end{bmatrix}=\boldsymbol{x}_1+.2\,\boldsymbol{x}_2 \tag{1}$$

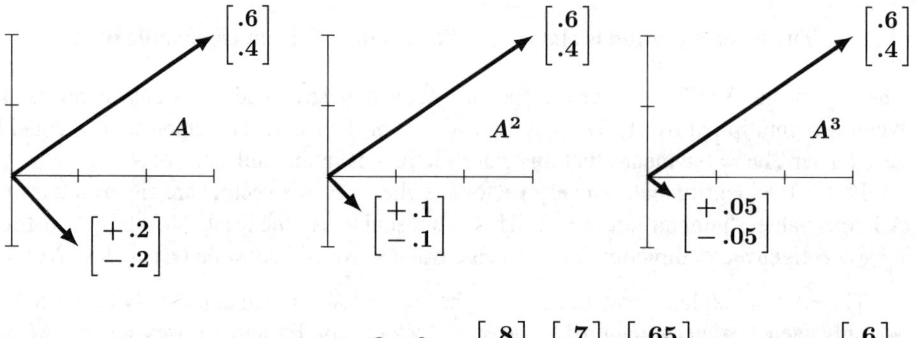

Figure 6.1: The first columns of A, A^2, A^3 are $\begin{bmatrix}.8\\ .2\end{bmatrix},\begin{bmatrix}.7\\ .3\end{bmatrix},\begin{bmatrix}.65\\ .35\end{bmatrix}$ approaching $\begin{bmatrix}.6\\ .4\end{bmatrix}$.

$$\textbf{Multiply each } x_i \textbf{ by } \lambda_i \quad \textbf{for column 1 of } A^2 \qquad A\begin{bmatrix}.8\\.2\end{bmatrix} \text{ is } x_1+\frac{1}{2}(.2)x_2=\begin{bmatrix}.6\\.4\end{bmatrix}+\begin{bmatrix}.1\\-.1\end{bmatrix}=\begin{bmatrix}.7\\.3\end{bmatrix} \quad (2)$$

Each eigenvector is multiplied by its eigenvalue, when we multiply by A. At every step x_1 is unchanged, because $\lambda_1 = 1$. But x_2 is multiplied 99 times by $\lambda_2 = \frac{1}{2}$:

$$\textbf{Column 1 of } A^{100} \qquad \boxed{A^{99}\begin{bmatrix}.8\\.2\end{bmatrix} \text{ is really } x_1 + (.2)\left(\tfrac{1}{2}\right)^{99} x_2 = \begin{bmatrix}.6\\.4\end{bmatrix} + \begin{bmatrix}\text{very}\\\text{small}\\\text{vector}\end{bmatrix}.} \quad (3)$$

This is the first column of A^{100}. The number we originally wrote as .6000 was not exact. We left out $(.2)$ times $(\frac{1}{2})^{99}$ which wouldn't show up for 30 decimal places.

The eigenvector x_1 is a "steady state" that doesn't change (because it has $\boldsymbol{\lambda_1 = 1}$). The eigenvector x_2 is a "decaying mode" that virtually disappears (because $\boldsymbol{\lambda_2 = .5}$). The higher the power of A, the more closely its columns approach the steady state.

This particular A is called a ***Markov matrix***. Its largest eigenvalue is $\lambda = 1$. Its eigenvector $x_1 = (.6, .4)$ is the *steady state*—which all columns of A^k will approach. *See Appendix* 8. *A giant Markov matrix is the key to Google's superfast web search.*

For projection matrices P, the column space projects to itself $(Px = x)$. The nullspace of P projects to zero $(Px = 0x)$. The eigenvalues of P are $\lambda = 1$ and $\lambda = 0$.

Example 2 **The projection matrix** $P = \begin{bmatrix}.5 & .5\\.5 & .5\end{bmatrix}$ has eigenvalues $\lambda = 1$ and $\lambda = 0$.

Its eigenvectors are $x_1 = (\mathbf{1}, \mathbf{1})$ and $x_2 = (\mathbf{1}, -\mathbf{1})$. Then $Px_1 = x_1$ (steady state) and $Px_2 = \mathbf{0}$ (nullspace). This example is a Markov matrix and a singular matrix (with $\lambda = 0$). Most important, P is a symmetric matrix. That makes its eigenvectors orthogonal.

1. **Markov matrix**: Each column of P adds to 1. Then $\lambda = 1$ is an eigenvalue.

2. ***P* is a singular matrix.** $\lambda = 0$ is an eigenvalue.

3. $\boldsymbol{P = P^{\mathrm{T}}}$ **is a symmetric matrix.** Perpendicular eigenvectors $\begin{bmatrix}1\\1\end{bmatrix}$ and $\begin{bmatrix}1\\-1\end{bmatrix}$.

The only eigenvalues of a projection matrix are 0 and 1. The eigenvectors for $\lambda = 0$ (which means $Px = 0x$) fill up the nullspace. The eigenvectors for $\lambda = 1$ (which means $Px = x$) fill up the column space. The nullspace is projected to zero. The column space projects onto itself. *The projection keeps the column space and destroys the nullspace*:

$$\textbf{Project each part} \quad v = \begin{bmatrix}1\\-1\end{bmatrix} + \begin{bmatrix}2\\2\end{bmatrix} = \begin{bmatrix}3\\1\end{bmatrix} \quad \textbf{projects onto} \quad Pv = \begin{bmatrix}\mathbf{0}\\\mathbf{0}\end{bmatrix} + \begin{bmatrix}\mathbf{2}\\\mathbf{2}\end{bmatrix}$$

Projections have $\lambda = 0$ and 1. Permutations have all $|\lambda| = 1$. The next matrix E is a reflection and at the same time a permutation. E also has special eigenvalues.

Example 3 **The exchange matrix $E = \begin{bmatrix} 0 & 1 \\ 1 & 0 \end{bmatrix}$ has eigenvalues 1 and -1.**

The eigenvector $(1, 1)$ is unchanged by E. The second eigenvector is $(1, -1)$—its signs are reversed by E. A matrix with no negative entries can still have a negative eigenvalue! The eigenvectors for E are the same as for P, because $E = 2P - I$:

$$\boldsymbol{E = 2P - I} \qquad \begin{bmatrix} 0 & 1 \\ 1 & 0 \end{bmatrix} = 2\begin{bmatrix} .5 & .5 \\ .5 & .5 \end{bmatrix} - \begin{bmatrix} 1 & 0 \\ 0 & 1 \end{bmatrix}. \tag{4}$$

***When a matrix is shifted by I, each λ is shifted by* 1**. No change in the eigenvectors.

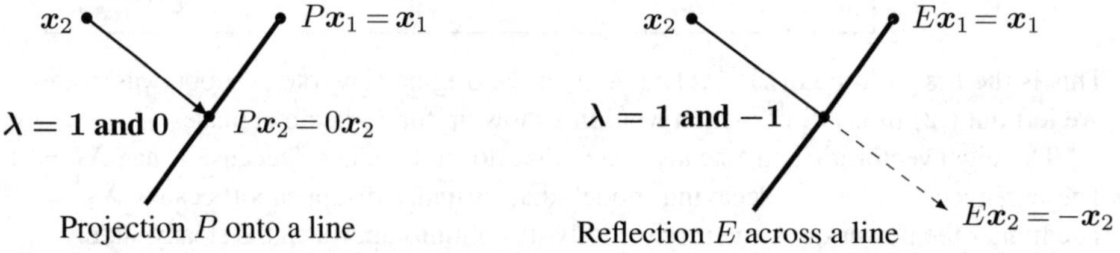

Figure 6.2: Projections P have eigenvalues 1 and 0. Exchanges E have $\lambda = 1$ and -1. A typical x changes direction, but an eigenvector $Ax = \lambda x$ stays on the line through x.

The Equation for the Eigenvalues: $\det(A - \lambda I) = 0$

For projection matrices we found λ's and x's by geometry: $Px = x$ and $Px = 0$. For other matrices we use determinants and linear algebra. *This is the key calculation*—almost every application starts by solving $\det(A - \lambda I) = 0$ and $Ax = \lambda x$.

First move λx to the left side. Write the equation $Ax = \lambda x$ as $(A - \lambda I)x = 0$. The matrix $A - \lambda I$ times the eigenvector x is the zero vector. ***The eigenvectors make up the nullspace of $A - \lambda I$.*** When we know an eigenvalue λ, we find an eigenvector by solving $(A - \lambda I)x = 0$.

Eigenvalues first. If $(A - \lambda I)x = 0$ has a nonzero solution, $A - \lambda I$ is not invertible. ***The determinant of $A - \lambda I$ must be zero.*** This is how to recognize an eigenvalue λ:

> **Eigenvalues** The number λ is an eigenvalue of A if and only if $A - \lambda I$ is singular.
>
> **Equation for the n eigenvalues of A** $\boxed{\det(A - \lambda I) = 0.}$ (5)

This "*characteristic polynomial*" $\det(A - \lambda I)$ involves only λ, not x. Since λ appears all along the main diagonal of $A - \lambda I$, the determinant in (5) includes $(-\lambda)^n$. Then equation (5) has n solutions λ_1 to λ_n and **A has n eigenvalues.**

An n by n matrix has n eigenvalues (repeated λ's are possible!) Each λ leads to x:

For each eigenvalue λ solve $(A - \lambda I)x = 0$ or $Ax = \lambda x$ to find an eigenvector x.

Example 4 $A = \begin{bmatrix} 1 & 2 \\ 2 & 4 \end{bmatrix}$ is already singular (zero determinant). Find its λ's and x's.

When A is singular, $\lambda = 0$ is one of the eigenvalues. The equation $Ax = 0x$ has solutions. They are the eigenvectors for $\lambda = 0$. But $\det(A - \lambda I) = 0$ is the way to find *all* λ's and x's. Always subtract λI from A:

$$\textit{Subtract } \lambda \textit{ along the diagonal to find} \quad A - \lambda I = \begin{bmatrix} 1-\lambda & 2 \\ 2 & 4-\lambda \end{bmatrix}. \tag{6}$$

Take the determinant "ad − bc" of this 2 by 2 matrix. From $1 - \lambda$ times $4 - \lambda$, the "ad" part is $\lambda^2 - 5\lambda + 4$. The "bc" part, not containing λ, is 2 times 2.

$$\det \begin{bmatrix} 1-\lambda & 2 \\ 2 & 4-\lambda \end{bmatrix} = (1-\lambda)(4-\lambda) - (2)(2) = \lambda^2 - 5\lambda. \tag{7}$$

Set this determinant $\lambda^2 - 5\lambda$ to zero. One solution is $\lambda = 0$ (as expected, since A is singular). Factoring into λ times $\lambda - 5$, the other root is $\lambda = 5$:

$$\boxed{\det(A - \lambda I) = \lambda^2 - 5\lambda = 0 \quad \text{yields the eigenvalues} \quad \boldsymbol{\lambda_1 = 0} \quad \text{and} \quad \boldsymbol{\lambda_2 = 5}.}$$

Now find the eigenvectors. Solve $(A - \lambda I)x = 0$ separately for $\lambda_1 = 0$ and $\lambda_2 = 5$:

$$(A - 0I)x = \begin{bmatrix} 1 & 2 \\ 2 & 4 \end{bmatrix} \begin{bmatrix} y \\ z \end{bmatrix} = \begin{bmatrix} 0 \\ 0 \end{bmatrix} \text{ yields an eigenvector } \boxed{\begin{bmatrix} y \\ z \end{bmatrix} = \begin{bmatrix} \mathbf{2} \\ \mathbf{-1} \end{bmatrix}} \text{ for } \boldsymbol{\lambda_1 = 0}$$

$$(A - 5I)x = \begin{bmatrix} -4 & 2 \\ 2 & -1 \end{bmatrix} \begin{bmatrix} y \\ z \end{bmatrix} = \begin{bmatrix} 0 \\ 0 \end{bmatrix} \text{ yields an eigenvector } \boxed{\begin{bmatrix} y \\ z \end{bmatrix} = \begin{bmatrix} \mathbf{1} \\ \mathbf{2} \end{bmatrix}} \text{ for } \boldsymbol{\lambda_2 = 5}$$

The matrices $A - 0I$ and $A - 5I$ are singular (because 0 and 5 are eigenvalues). The eigenvectors $(2, -1)$ and $(1, 2)$ are in the nullspaces: $(A - \lambda I)x = 0$ is $Ax = \lambda x$.

We need to emphasize: *There is nothing exceptional about $\lambda = 0$.* Like every other number, zero might be an eigenvalue and it might not. If A is singular, the eigenvectors for $\lambda = 0$ fill the nullspace: $Ax = 0x = \mathbf{0}$. If A is invertible, zero is not an eigenvalue.

In the example, the shifted matrix $A - 5I$ is singular and 5 is the other eigenvalue.

Summary To solve the eigenvalue problem for an n by n matrix, follow these steps:

1. ***Compute the determinant of*** $A - \lambda I$. With λ subtracted along the diagonal, this determinant starts with λ^n or $-\lambda^n$. It is a polynomial in λ of degree n.
2. ***Find the roots of this polynomial***, by solving $\det(A - \lambda I) = 0$. The n roots are the n eigenvalues of A. They make $A - \lambda I$ singular.
3. For each eigenvalue λ, ***solve*** $(A - \lambda I)x = \mathbf{0}$ ***to find an eigenvector*** x.

A note on the eigenvectors of 2 by 2 matrices. When $A - \lambda I$ is singular, both rows are multiples of a vector (a, b). *The eigenvector is any multiple of* $(b, -a)$. The example had

$\lambda = 0$: rows of $A - 0I$ in the direction $(1, 2)$; eigenvector in the direction $(2, -1)$

$\lambda = 5$: rows of $A - 5I$ in the direction $(-4, 2)$; eigenvector in the direction $(2, 4)$.

Previously we wrote that last eigenvector as $(1, 2)$. Both $(1, 2)$ and $(2, 4)$ are correct. There is a whole *line of eigenvectors*—any nonzero multiple of $\boldsymbol{x}$ is as good as $\boldsymbol{x}$. MATLAB's **eig**(A) divides by the length, to make each eigenvector $\boldsymbol{x}$ into a unit vector.

We must add a warning. Some 2 by 2 matrices have only *one* line of eigenvectors. This can only happen when two eigenvalues are equal. (On the other hand $A = I$ has equal eigenvalues and plenty of eigenvectors.) Without a full set of eigenvectors, we don't have a basis. We can't write every $\boldsymbol{v}$ as a combination of eigenvectors. In the language of the next section, *we can't diagonalize a matrix without n independent eigenvectors.*

Determinant and Trace

Bad news first: If you add a row of A to another row, or exchange rows, the eigenvalues usually change. *Elimination does not preserve the* λ's. The triangular U has *its* eigenvalues sitting along the diagonal—they are the pivots. But they are not the eigenvalues of A! Eigenvalues are changed when row 1 is added to row 2:

$$U = \begin{bmatrix} 1 & 3 \\ 0 & 0 \end{bmatrix} \quad \text{has } \lambda = 0 \text{ and } \lambda = 1; \quad A = \begin{bmatrix} 1 & 3 \\ 2 & 6 \end{bmatrix} \quad \text{has } \lambda = 0 \text{ and } \lambda = 7.$$

Good news second: The *product* λ_1 *times* λ_2 *and the sum* $\lambda_1 + \lambda_2$ *can be found quickly from the matrix.* For this A, the product is 0 times 7. That agrees with the determinant of A (which is 0). The sum of eigenvalues is $0 + 7$. That agrees with the sum down the main diagonal (the **trace** is $1 + 6$). These quick checks always work.

The product $(\boldsymbol{\lambda_1}) \cdots (\boldsymbol{\lambda_n})$ ***of the n eigenvalues*** *equals* ***the determinant of A***
$\boldsymbol{\lambda_1 + \lambda_2 + \cdots + \lambda_n}$ *equals* ***the sum of the n diagonal entries = (trace of A)***

Those checks are very useful. They are proved in Problems 16–17 and again in the next section. They don't remove the pain of computing λ's. But when the computation is wrong, they generally tell us so. To compute the correct λ's, go back to $\det(A - \lambda I) = 0$.

The trace and determinant *do* tell everything when the matrix is 2 by 2. We never want to get those wrong! Here trace = **3** and det = **2**, so they all have $\lambda = \mathbf{1}$ and **2**:

$$A = \begin{bmatrix} 1 & 9 \\ 0 & 2 \end{bmatrix} \text{ or } \begin{bmatrix} 3 & 1 \\ -2 & 0 \end{bmatrix} \text{ or } \begin{bmatrix} 7 & -3 \\ 10 & -4 \end{bmatrix}. \tag{8}$$

That first A is one of the best matrices for finding eigenvalues: *because it is triangular.*

Why do the eigenvalues 1 and 2 of that triangular matrix lie along its diagonal?

Imaginary Eigenvalues

One more bit of news (not too terrible). The eigenvalues might not be real numbers.

Example 5 *The* $90°$ *rotation* $Q = \begin{bmatrix} 0 & -1 \\ 1 & 0 \end{bmatrix}$ *has no real eigenvectors. Its eigenvalues are* $\lambda_1 = i$ *and* $\lambda_2 = -i$. *Then* $\lambda_1 + \lambda_2 =$ **trace** $= 0$ *and* $\lambda_1\lambda_2 =$ *determinant* $= 1$.

After a rotation, *no real vector* $Q\boldsymbol{x}$ *stays in the same direction as* $\boldsymbol{x}$ ($\boldsymbol{x} = \boldsymbol{0}$ is useless). There cannot be an eigenvector, unless we go to ***imaginary numbers***. Which we do.

To see how $i = \sqrt{-1}$ can help, look at Q^2 which is $-I$. If Q is rotation through $90°$, then Q^2 is rotation through $180°$. Its eigenvalues are -1 and -1. (Certainly $-I\boldsymbol{x} = -1\boldsymbol{x}$.) Squaring Q will square each λ, so we must have $\boldsymbol{\lambda^2 = -1}$. *The eigenvalues of the* $90°$ *rotation matrix* Q *are* $+i$ *and* $-i$, because $i^2 = -1$.

Those λ's come as usual from $\det(Q - \lambda I) = 0$. This equation gives $\lambda^2 + 1 = 0$. Its roots are i and $-i$. We meet the imaginary numbers i and $-i$ also in the eigenvectors:

Complex eigenvectors

$$\begin{bmatrix} 0 & -1 \\ 1 & 0 \end{bmatrix}\begin{bmatrix} 1 \\ i \end{bmatrix} = -i\begin{bmatrix} 1 \\ i \end{bmatrix} \quad \text{and} \quad \begin{bmatrix} 0 & -1 \\ 1 & 0 \end{bmatrix}\begin{bmatrix} i \\ 1 \end{bmatrix} = i\begin{bmatrix} i \\ 1 \end{bmatrix}.$$

Somehow these complex vectors $\boldsymbol{x}_1 = (1, i)$ and $\boldsymbol{x}_2 = (i, 1)$ keep their direction as they are rotated in complex space. Don't ask me how. This example makes the important point that real matrices can easily have complex eigenvalues and eigenvectors. The particular eigenvalues i and $-i$ also illustrate two properties of the special matrix Q.

1. Q is an orthogonal matrix so the absolute value of each λ is $|\lambda| = 1$.

2. Q is a skew-symmetric matrix so each λ is pure imaginary.

A symmetric matrix ($\boldsymbol{S}^{\mathrm{T}} = \boldsymbol{S}$) can be compared to a real number. A skew-symmetric matrix ($\boldsymbol{A}^{\mathrm{T}} = -\boldsymbol{A}$) can be compared to an imaginary number. An orthogonal matrix ($\boldsymbol{Q}^{\mathrm{T}}\boldsymbol{Q} = \boldsymbol{I}$) corresponds to a complex number with $|\lambda| = 1$. For the eigenvalues of S and A and Q, those are more than analogies—they are facts to be proved.

The eigenvectors for all these special matrices are perpendicular. Somehow $(i, 1)$ and $(1, i)$ are perpendicular (Section 6.4 will explain the dot product of complex vectors).

Eigenvalues of AB and $A+B$

The first guess about the eigenvalues of AB is not true. An eigenvalue λ of A times an eigenvalue β of B usually does *not* give an eigenvalue of AB:

False proof

$$AB\boldsymbol{x} = A\beta\boldsymbol{x} = \beta A\boldsymbol{x} = \beta\lambda\boldsymbol{x}. \tag{9}$$

It seems that β times λ is an eigenvalue. When $\boldsymbol{x}$ is an eigenvector for A and B, this proof is correct. ***The mistake is to expect that A and B automatically share the same eigenvector*** $\boldsymbol{x}$. Usually they don't. Eigenvectors of A are generally *not* eigenvectors of B.

A and B could have all zero eigenvalues while 1 is an eigenvalue of AB and $A+B$:

$$A = \begin{bmatrix} 0 & 1 \\ 0 & 0 \end{bmatrix} \quad \text{and} \quad B = \begin{bmatrix} 0 & 0 \\ 1 & 0 \end{bmatrix}; \quad \text{then} \quad AB = \begin{bmatrix} 1 & 0 \\ 0 & 0 \end{bmatrix} \quad \text{and} \quad A+B = \begin{bmatrix} 0 & 1 \\ 1 & 0 \end{bmatrix}.$$

For the same reason, the eigenvalues of $A+B$ are generally not $\lambda + \beta$. Here $A+B$ has eigenvalues 1 and -1 while λ and β are zero. (At least the trace of $A+B$ is zero.)

The false proof suggests what is true. Suppose $\boldsymbol{x}$ really is an eigenvector for both A and B. Then we do have $AB\boldsymbol{x} = \lambda\beta\boldsymbol{x}$ and $BA\boldsymbol{x} = \lambda\beta\boldsymbol{x}$. When all n eigenvectors are shared by A and B, we *can* multiply eigenvalues and we find $AB = BA$. That test is important in quantum mechanics—time out to mention this application of linear algebra:

> A and B share the same $\boldsymbol{n}$ *independent* eigenvectors if and only if $\boldsymbol{AB = BA}$.

Heisenberg's uncertainty principle In quantum mechanics, the position matrix P and the momentum matrix Q do not commute. In fact $QP - PQ = I$ (these are infinite matrices). To have $P\boldsymbol{x} = \boldsymbol{0}$ at the same time as $Q\boldsymbol{x} = \boldsymbol{0}$ would require $\boldsymbol{x} = I\boldsymbol{x} = \boldsymbol{0}$. If we knew the position exactly, we could not also know the momentum exactly. Problem 36 derives Heisenberg's uncertainty principle $\|P\boldsymbol{x}\|\,\|Q\boldsymbol{x}\| \geq \frac{1}{2}\|\boldsymbol{x}\|^2$.

■ REVIEW OF THE KEY IDEAS ■

1. $A\boldsymbol{x} = \lambda\boldsymbol{x}$ says that eigenvectors $\boldsymbol{x}$ keep the same direction when multiplied by A.

2. $A\boldsymbol{x} = \lambda\boldsymbol{x}$ also says that $\det(A - \lambda I) = 0$. This equation determines n eigenvalues.

3. The sum of the λ's equals the sum down the main diagonal of A (**the trace**). The product of the λ's equals the determinant of A.

4. Projections P, exchanges E, $90°$ rotations Q have eigenvalues $1, 0, -1, i, -i$. Singular matrices have $\lambda = 0$. Triangular matrices have λ's on their diagonal.

5. *Special properties of a matrix lead to special eigenvalues and eigenvectors.* That is a major theme of Chapter 6.

■ WORKED EXAMPLES ■

6.1 A Find the eigenvalues and eigenvectors of A and A^2 and A^{-1} and $A+4I$:

$$A = \begin{bmatrix} 2 & -1 \\ -1 & 2 \end{bmatrix} \quad \text{and} \quad A^2 = \begin{bmatrix} 5 & -4 \\ -4 & 5 \end{bmatrix}.$$

Check the trace $\lambda_1 + \lambda_2 = 4$ and the determinant λ_1 times $\lambda_2 = 3$.

Solution The eigenvalues of A come from $\det(A - \lambda I) = 0$:

$$A = \begin{bmatrix} 2 & -1 \\ -1 & 2 \end{bmatrix} \qquad \det(A - \lambda I) = \begin{vmatrix} \mathbf{2-\lambda} & \mathbf{-1} \\ \mathbf{-1} & \mathbf{2-\lambda} \end{vmatrix} = \lambda^2 - 4\lambda + 3 = 0.$$

This factors into $(\lambda - 1)(\lambda - 3) = 0$ so the eigenvalues of A are $\lambda_1 = \mathbf{1}$ and $\lambda_2 = \mathbf{3}$. For the trace, the sum $2+2$ agrees with $1+3$. The determinant 3 agrees with the product $\lambda_1\lambda_2$.

The eigenvectors come separately by solving $(A - \lambda I)\boldsymbol{x} = \mathbf{0}$ which is $A\boldsymbol{x} = \lambda\boldsymbol{x}$:

$$\boldsymbol{\lambda = 1}\text{:} \quad (A - I)\boldsymbol{x} = \begin{bmatrix} 1 & -1 \\ -1 & 1 \end{bmatrix}\begin{bmatrix} x \\ y \end{bmatrix} = \begin{bmatrix} 0 \\ 0 \end{bmatrix} \text{ gives the eigenvector } \boldsymbol{x}_1 = \begin{bmatrix} 1 \\ 1 \end{bmatrix}$$

$$\boldsymbol{\lambda = 3}\text{:} \quad (A - 3I)\boldsymbol{x} = \begin{bmatrix} -1 & -1 \\ -1 & -1 \end{bmatrix}\begin{bmatrix} x \\ y \end{bmatrix} = \begin{bmatrix} 0 \\ 0 \end{bmatrix} \text{ gives the eigenvector } \boldsymbol{x}_2 = \begin{bmatrix} 1 \\ -1 \end{bmatrix}$$

A^2 and A^{-1} and $A + 4I$ keep the *same eigenvectors as A itself.* Their eigenvalues are λ^2 and λ^{-1} and $\lambda + 4$:

$$A^2 \text{ has eigenvalues } 1^2 = 1 \text{ and } 3^2 = 9 \qquad A^{-1} \text{ has } \frac{1}{1} \text{ and } \frac{1}{3} \qquad A + 4I \text{ has } \begin{matrix} 1+4=5 \\ 3+4=7 \end{matrix}$$

Notes for later sections: A has *orthogonal eigenvectors* (Section 6.3 on symmetric matrices). A can be *diagonalized* since $\lambda_1 \neq \lambda_2$ (Section 6.2). A is *similar* to every 2 by 2 matrix with eigenvalues 1 and 3 (Section 6.2). A is a *positive definite matrix* (Section 6.3) since $A = A^{\mathrm{T}}$ and the λ's are positive.

6.1 B **How can you estimate the eigenvalues of any A?** Gershgorin gave this answer.

Every eigenvalue of A must be "near" at least one of the entries a_{ii} on the main diagonal. For λ to be "near a_{ii}" means that $|a_{ii} - \lambda|$ is no more than **the sum R_i of all other $|a_{ij}|$ in that row i of the matrix**. Then R_i is the radius of a circle centered at a_{ii}.

Every λ is in the circle around one or more diagonal entries a_{ii} : $|a_{ii} - \lambda| \leq R_i$.

Here is the reasoning. When λ is an eigenvalue, $A - \lambda I$ is not invertible. Then $A - \lambda I$ cannot be diagonally dominant (see Section 2.5). So at least one diagonal entry $a_{ii} - \lambda$ is *not larger* than the sum R_i of all other entries $|a_{ij}|$ (we take absolute values!) in row i.

Example Every eigenvalue λ of this A falls into one or both of the **Gershgorin circles**: The centers are a and d. The radii of the circles are $R_1 = |b|$ and $R_2 = |c|$.

$$A = \begin{bmatrix} a & b \\ c & d \end{bmatrix} \qquad \begin{matrix} \text{First circle:} & |\lambda - a| \leq |b| \\ \text{Second circle:} & |\lambda - d| \leq |c| \end{matrix}$$

Those are circles in the complex plane, since λ could certainly be a complex number.

6.1 C Find the eigenvalues and eigenvectors of this symmetric 3 by 3 matrix S:

Symmetric matrix
Singular matrix
Trace $1+2+1=4$

$$S = \begin{bmatrix} 1 & -1 & 0 \\ -1 & 2 & -1 \\ 0 & -1 & 1 \end{bmatrix}$$

Solution Since all rows of S add to zero, the vector $\boldsymbol{x} = (1,1,1)$ gives $S\boldsymbol{x} = \mathbf{0}$. This is an eigenvector for $\lambda = 0$. To find λ_2 and λ_3, compute the 3 by 3 determinant:

$$\det(S - \lambda I) = \begin{vmatrix} 1-\lambda & -1 & 0 \\ -1 & 2-\lambda & -1 \\ 0 & -1 & 1-\lambda \end{vmatrix} \begin{aligned} &= (1-\lambda)(2-\lambda)(1-\lambda) - 2(1-\lambda) \\ &= (1-\lambda)[(2-\lambda)(1-\lambda) - 2] \\ &= \mathbf{(1-\lambda)(-\lambda)(3-\lambda)}. \end{aligned}$$

Those three factors give $\lambda = 0, 1, 3$ as the solutions to $\det(S - \lambda I) = 0$. Each eigenvalue corresponds to an eigenvector (or a line of eigenvectors):

$$\boldsymbol{x}_1 = \begin{bmatrix} 1 \\ 1 \\ 1 \end{bmatrix} \quad S\boldsymbol{x}_1 = \mathbf{0}\boldsymbol{x}_1 \quad \boldsymbol{x}_2 = \begin{bmatrix} 1 \\ 0 \\ -1 \end{bmatrix} \quad S\boldsymbol{x}_2 = \mathbf{1}\boldsymbol{x}_2 \quad \boldsymbol{x}_3 = \begin{bmatrix} 1 \\ -2 \\ 1 \end{bmatrix} \quad S\boldsymbol{x}_3 = \mathbf{3}\boldsymbol{x}_3 .$$

I notice again that eigenvectors are perpendicular when S is symmetric. We were lucky to find $\lambda = 0, 1, 3$. For a larger matrix I would use **eig**(A), and never touch determinants.

The full command $[\boldsymbol{X}, \boldsymbol{E}] = $ **eig**(A) will produce unit eigenvectors in the columns of X.

Problem Set 6.1

1 The example at the start of the chapter has powers of this matrix A:

$$A = \begin{bmatrix} .8 & .3 \\ .2 & .7 \end{bmatrix} \quad \text{and} \quad A^2 = \begin{bmatrix} .70 & .45 \\ .30 & .55 \end{bmatrix} \quad \text{and} \quad A^\infty = \begin{bmatrix} .6 & .6 \\ .4 & .4 \end{bmatrix}.$$

Find the eigenvalues of these matrices. All powers have the same eigenvectors. Show from A how a row exchange can produce different eigenvalues.

2 Find the eigenvalues and eigenvectors of these two matrices:

$$A = \begin{bmatrix} 1 & 4 \\ 2 & 3 \end{bmatrix} \quad \text{and} \quad A + I = \begin{bmatrix} 2 & 4 \\ 2 & 4 \end{bmatrix}.$$

$A + I$ has the _____ eigenvectors as A. Its eigenvalues are _____ by 1.

3 Compute the eigenvalues and eigenvectors of A and A^{-1}. Check the trace !

$$A = \begin{bmatrix} 0 & 2 \\ 1 & 1 \end{bmatrix} \quad \text{and} \quad A^{-1} = \begin{bmatrix} -1/2 & 1 \\ 1/2 & 0 \end{bmatrix}.$$

A^{-1} has the _____ eigenvectors as A. When A has eigenvalues λ_1 and λ_2, its inverse has eigenvalues _____.

4 Compute the eigenvalues and eigenvectors of A and A^2:

$$A = \begin{bmatrix} -1 & 3 \\ 2 & 0 \end{bmatrix} \quad \text{and} \quad A^2 = \begin{bmatrix} 7 & -3 \\ -2 & 6 \end{bmatrix}.$$

A^2 has the same ______ as A. When A has eigenvalues λ_1 and λ_2, A^2 has eigenvalues ______. In this example, how do you see that $\lambda_1^2 + \lambda_2^2 = 13$?

5 Find the eigenvalues of A and B (easy for triangular matrices) and $A + B$:

$$A = \begin{bmatrix} 3 & 0 \\ 1 & 1 \end{bmatrix} \quad \text{and} \quad B = \begin{bmatrix} 1 & 1 \\ 0 & 3 \end{bmatrix} \quad \text{and} \quad A + B = \begin{bmatrix} 4 & 1 \\ 1 & 4 \end{bmatrix}.$$

Eigenvalues of $A + B$ (*are equal to*)(*are not equal to*) eigenvalues of A plus eigenvalues of B.

6 Find the eigenvalues of A and B and AB and BA:

$$A = \begin{bmatrix} 1 & 0 \\ 1 & 1 \end{bmatrix} \quad \text{and} \quad B = \begin{bmatrix} 1 & 2 \\ 0 & 1 \end{bmatrix} \quad \text{and} \quad AB = \begin{bmatrix} 1 & 2 \\ 1 & 3 \end{bmatrix} \quad \text{and} \quad BA = \begin{bmatrix} 3 & 2 \\ 1 & 1 \end{bmatrix}.$$

(a) Are the eigenvalues of AB equal to eigenvalues of A times eigenvalues of B?

(b) Are the eigenvalues of AB equal to the eigenvalues of BA?

7 Elimination produces $A = LU$. The eigenvalues of U are on its diagonal; they are the ______. The eigenvalues of L are on its diagonal; they are all ______. The eigenvalues of A are not the same as ______.

8 (a) If you know that $\boldsymbol{x}$ is an eigenvector, the way to find λ is to ______.

(b) If you know that λ is an eigenvalue, the way to find $\boldsymbol{x}$ is to ______.

9 What do you do to the equation $A\boldsymbol{x} = \lambda \boldsymbol{x}$, in order to prove (a), (b), and (c)?

(a) λ^2 is an eigenvalue of A^2, as in Problem 4.

(b) λ^{-1} is an eigenvalue of A^{-1}, as in Problem 3.

(c) $\lambda + 1$ is an eigenvalue of $A + I$, as in Problem 2.

10 Find the eigenvalues and eigenvectors for both of these Markov matrices A and A^∞. Explain from those answers why A^{100} is close to A^∞:

$$A = \begin{bmatrix} .6 & .2 \\ .4 & .8 \end{bmatrix} \quad \text{and} \quad A^\infty = \begin{bmatrix} 1/3 & 1/3 \\ 2/3 & 2/3 \end{bmatrix} = \text{limit of } A^n$$

11 Here is a strange fact about 2 by 2 matrices with eigenvalues $\lambda_1 \neq \lambda_2$: The columns of $A - \lambda_1 I$ are multiples of the eigenvector $\boldsymbol{x}_2$. Any idea why this should be?

12 Find three eigenvectors for this matrix P (projection matrices have $\lambda=1$ and 0):

$$\textbf{Projection matrix} \qquad P=\begin{bmatrix} .2 & .4 & 0 \\ .4 & .8 & 0 \\ 0 & 0 & 1 \end{bmatrix}.$$

If two eigenvectors share the same λ, so do all their linear combinations. Find an eigenvector of P with no zero components.

13 From the unit vector $\boldsymbol{u}=\left(\frac{1}{6},\frac{1}{6},\frac{3}{6},\frac{5}{6}\right)$ construct the rank one projection matrix $P=\boldsymbol{u}\boldsymbol{u}^{\mathrm{T}}$. This matrix has $P^2=P$ because $\boldsymbol{u}^{\mathrm{T}}\boldsymbol{u}=1$.

(a) $P\boldsymbol{u}=\boldsymbol{u}$ comes from $(\boldsymbol{u}\boldsymbol{u}^{\mathrm{T}})\boldsymbol{u}=\boldsymbol{u}(\_\_\_\_\_)$. Then $\lambda=1$.

(b) If $\boldsymbol{v}$ is perpendicular to $\boldsymbol{u}$ show that $P\boldsymbol{v}=\mathbf{0}$. Then $\lambda=0$.

(c) Find three independent eigenvectors of P all with eigenvalue $\lambda=0$.

14 Solve $\det(Q-\lambda I)=0$ by the quadratic formula to reach $\lambda=\cos\theta\pm i\sin\theta$:

$$Q=\begin{bmatrix} \cos\theta & -\sin\theta \\ \sin\theta & \cos\theta \end{bmatrix} \quad \text{rotates the } xy \text{ plane by the angle } \theta. \text{ No real } \lambda\text{'s.}$$

Find the eigenvectors of Q by solving $(Q-\lambda I)\boldsymbol{x}=\mathbf{0}$. Use $i^2=-1$.

15 Every permutation matrix leaves $\boldsymbol{x}=(1,1,\ldots,1)$ unchanged. Then $\lambda=1$. Find two more λ's (possibly complex) for these permutations, from $\det(P-\lambda I)=0$:

$$P=\begin{bmatrix} 0 & 1 & 0 \\ 0 & 0 & 1 \\ 1 & 0 & 0 \end{bmatrix} \quad \text{and} \quad P=\begin{bmatrix} 0 & 0 & 1 \\ 0 & 1 & 0 \\ 1 & 0 & 0 \end{bmatrix}. \qquad \textbf{All } |\boldsymbol{\lambda}|=\mathbf{1}$$

16 **The determinant of A equals the product $\lambda_1\lambda_2\cdots\lambda_n$.** Start with the polynomial $\det(A-\lambda I)$ separated into its n factors $\lambda_i-\lambda$ (always possible). Then set $\lambda=0$:

$$\det(A-\lambda I)=(\lambda_1-\lambda)(\lambda_2-\lambda)\cdots(\lambda_n-\lambda) \quad \text{so} \quad \det A=\_\_\_\_\_.$$

Check this rule for $\det A$ in Example 1 where the Markov matrix has $\lambda=1$ and $\frac{1}{2}$.

17 The sum of the diagonal entries (the **trace**) equals the sum of the eigenvalues:

$$A=\begin{bmatrix} a & b \\ c & d \end{bmatrix} \quad \text{has} \quad \det(A-\lambda I)=\lambda^2-(a+d)\lambda+ad-bc=0.$$

The quadratic formula gives the eigenvalues $\lambda=(a+d+\sqrt{})/2$ and $\lambda=\_\_\_\_\_$. Their sum is $\_\_\_\_\_$. If A has $\lambda_1=3$ and $\lambda_2=4$ then $\det(A-\lambda I)=\_\_\_\_\_$.

18 If A has $\lambda_1=4$ and $\lambda_2=5$ then $\det(A-\lambda I)=(\lambda-4)(\lambda-5)=\lambda^2-9\lambda+20$. Find three matrices that have trace $a+d=9$ and determinant 20 and $\lambda=4,5$.

19 A 3 by 3 matrix B is known to have eigenvalues $0, 1, 2$. This information is enough to find three of these (give the answers where possible):

(a) the rank of B
(b) the determinant of $B^{\mathrm{T}}B$
(c) the eigenvalues of $B^{\mathrm{T}}B$
(d) the eigenvalues of $(B^2 + I)^{-1}$.

20 Choose the last rows of A and C to give eigenvalues $4, 7$ and $1, 2, 3$:

Companion matrices $$A = \begin{bmatrix} 0 & 1 \\ * & * \end{bmatrix} \quad C = \begin{bmatrix} 0 & 1 & 0 \\ 0 & 0 & 1 \\ * & * & * \end{bmatrix}.$$

21 ***The eigenvalues of A equal the eigenvalues of A^{T}***. This is because $\det(A - \lambda I)$ equals $\det(A^{\mathrm{T}} - \lambda I)$. That is true because ______. Show by an example that the eigenvectors of A and A^{T} are *not* the same.

22 Construct any 3 by 3 Markov matrix M: positive entries down each column add to 1. Show that $M^{\mathrm{T}}(1, 1, 1) = (1, 1, 1)$. By Problem 21, $\lambda = 1$ is also an eigenvalue of M. Challenge: A 3 by 3 singular Markov matrix with trace $\frac{1}{2}$ has what λ's ?

23 Find three 2 by 2 matrices that have $\lambda_1 = \lambda_2 = 0$. The trace is zero and the determinant is zero. A might not be the zero matrix but check that $A^2 = 0$.

24 This matrix is singular with rank one. Find three λ's and three eigenvectors:

$$A = \begin{bmatrix} 1 \\ 2 \\ 1 \end{bmatrix} \begin{bmatrix} 2 & 1 & 2 \end{bmatrix} = \begin{bmatrix} 2 & 1 & 2 \\ 4 & 2 & 4 \\ 2 & 1 & 2 \end{bmatrix}.$$

25 Suppose A and B have the same eigenvalues $\lambda_1, \ldots, \lambda_n$ with the same independent eigenvectors $\boldsymbol{x}_1, \ldots, \boldsymbol{x}_n$. **Then $A = B$.** *Reason*: Any vector $\boldsymbol{x}$ is a combination $c_1\boldsymbol{x}_1 + \cdots + c_n\boldsymbol{x}_n$. What is $A\boldsymbol{x}$? What is $B\boldsymbol{x}$?

26 The block B has eigenvalues $1, 2$ and C has eigenvalues $3, 4$ and D has eigenvalues $5, 7$. Find the eigenvalues of the 4 by 4 matrix A:

$$A = \begin{bmatrix} B & C \\ 0 & D \end{bmatrix} = \begin{bmatrix} 0 & 1 & 3 & 0 \\ -2 & 3 & 0 & 4 \\ 0 & 0 & 6 & 1 \\ 0 & 0 & 1 & 6 \end{bmatrix}.$$

27 Find the rank and the four eigenvalues of A and C:

$$A = \begin{bmatrix} 1 & 1 & 1 & 1 \\ 1 & 1 & 1 & 1 \\ 1 & 1 & 1 & 1 \\ 1 & 1 & 1 & 1 \end{bmatrix} \quad \text{and} \quad C = \begin{bmatrix} 1 & 0 & 1 & 0 \\ 0 & 1 & 0 & 1 \\ 1 & 0 & 1 & 0 \\ 0 & 1 & 0 & 1 \end{bmatrix}.$$

28 Subtract I from A in Problem 27. Find the λ's and then the determinants of

$$B = A - I = \begin{bmatrix} 0 & 1 & 1 & 1 \\ 1 & 0 & 1 & 1 \\ 1 & 1 & 0 & 1 \\ 1 & 1 & 1 & 0 \end{bmatrix} \quad \text{and} \quad C = I - A = \begin{bmatrix} 0 & -1 & -1 & -1 \\ -1 & 0 & -1 & -1 \\ -1 & -1 & 0 & -1 \\ -1 & -1 & -1 & 0 \end{bmatrix}.$$

29 (Review) Find the eigenvalues of A, B, and C:

$$A = \begin{bmatrix} 1 & 2 & 3 \\ 0 & 4 & 5 \\ 0 & 0 & 6 \end{bmatrix} \quad \text{and} \quad B = \begin{bmatrix} 0 & 0 & 1 \\ 0 & 2 & 0 \\ 3 & 0 & 0 \end{bmatrix} \quad \text{and} \quad C = \begin{bmatrix} 2 & 2 & 2 \\ 2 & 2 & 2 \\ 2 & 2 & 2 \end{bmatrix}.$$

30 When $a + b = c + d$ show that $(1, 1)$ is an eigenvector and find both eigenvalues:

$$A = \begin{bmatrix} a & b \\ c & d \end{bmatrix}.$$

31 If we exchange rows 1 and 2 *and* columns 1 and 2, the eigenvalues don't change. Find eigenvectors of A and B for $\lambda = 11$. *Rank one gives* $\lambda_2 = \lambda_3 = 0$.

$$A = \begin{bmatrix} 1 & 2 & 1 \\ 3 & 6 & 3 \\ 4 & 8 & 4 \end{bmatrix} \quad \text{and} \quad B = PAP^{\mathrm{T}} = \begin{bmatrix} 6 & 3 & 3 \\ 2 & 1 & 1 \\ 8 & 4 & 4 \end{bmatrix}.$$

32 Suppose A has eigenvalues $0, 3, 5$ with independent eigenvectors $\boldsymbol{u}, \boldsymbol{v}, \boldsymbol{w}$.

(a) Give a basis for the nullspace and a basis for the column space.

(b) Find a particular solution to $A\boldsymbol{x} = \boldsymbol{v} + \boldsymbol{w}$. Find all solutions.

(c) $A\boldsymbol{x} = \boldsymbol{u}$ has no solution. If it did then ____ would be in the column space.

Challenge Problems

33 Show that $\boldsymbol{u}$ is an eigenvector of the rank one 2×2 matrix $A = \boldsymbol{u}\boldsymbol{v}^{\mathrm{T}}$. Find both eigenvalues of A. Check that $\lambda_1 + \lambda_2$ agrees with the trace $u_1 v_1 + u_2 v_2$.

34 Find the eigenvalues of this permutation matrix P from $\det(P - \lambda I) = 0$. Which vectors are not changed by the permutation? They are eigenvectors for $\lambda = 1$. Can you find three more eigenvectors?

$$P = \begin{bmatrix} 0 & 0 & 0 & 1 \\ 1 & 0 & 0 & 0 \\ 0 & 1 & 0 & 0 \\ 0 & 0 & 1 & 0 \end{bmatrix}.$$

35 There are six 3 by 3 permutation matrices P. What numbers can be the *determinants* of P? What numbers can be *pivots*? What numbers can be the *trace* of P? What *four numbers* can be eigenvalues of P, as in Problem 15?

36 (**Heisenberg's Uncertainty Principle**) $AB - BA = I$ can happen for infinite matrices with $A = A^{\mathrm{T}}$ and $B = -B^{\mathrm{T}}$. Then

$$x^{\mathrm{T}}x = x^{\mathrm{T}}ABx - x^{\mathrm{T}}BAx \leq 2\|Ax\|\,\|Bx\|.$$

Explain that last step by using the Schwarz inequality $|u^{\mathrm{T}}v| \leq ||u||\,||v||$. Then Heisenberg's inequality says that $\|Ax\|/\|x\|$ *times* $\|Bx\|/\|x\|$ is at least $\frac{1}{2}$. It is impossible to get the position error and momentum error both very small.

37 *Find a 2 by 2 rotation matrix* (other than I) *with* $A^3 = I$. Its eigenvalues must satisfy $\lambda^3 = 1$. They can be $e^{2\pi i/3}$ and $e^{-2\pi i/3}$. What are the trace and determinant of A?

38 (a) Find the eigenvalues and eigenvectors of A. They depend on c:

$$A = \begin{bmatrix} .4 & 1-c \\ .6 & c \end{bmatrix}.$$

(b) Show that A has just one line of eigenvectors when $c = 1.6$.

(c) This is a Markov matrix when $c = 0.8$. Then A^n approaches what matrix A^∞?

Eigshow in MATLAB

There is a MATLABdemo (just type **eigshow**), displaying the eigenvalue problem for a 2 by 2 matrix. It starts with the unit vector $x = (1, 0)$. *The mouse makes this vector move around the unit circle.* At the same time the screen shows Ax, in color and also moving. Possibly Ax is ahead of x. Possibly Ax is behind x. *Sometimes Ax is parallel to x.*

At that parallel moment, $Ax = \lambda x$ (at x_1 and x_2 in the second figure).

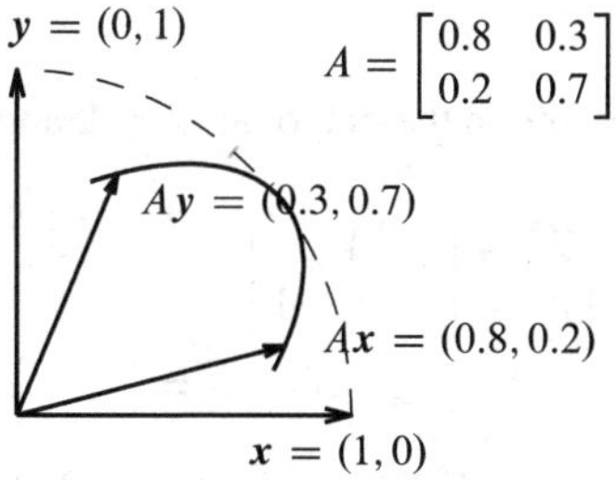

These are not eigenvectors

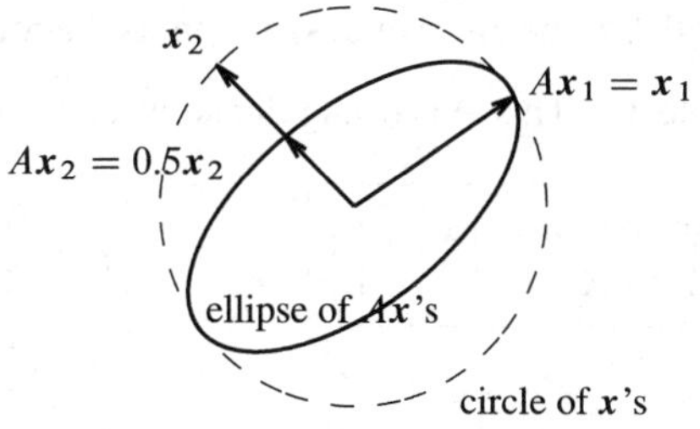

Ax lines up with x at eigenvectors

The eigenvalue λ is the length of Ax, when the unit eigenvector x lines up. The built-in choices for A illustrate three possibilities: $0, 1$, or 2 real vectors where Ax crosses x.

The axes of the ellipse are **singular vectors** in Section 7.1.

6.2 Diagonalizing a Matrix

1 The columns of $\boldsymbol{AX} = \boldsymbol{X\Lambda}$ are $A\boldsymbol{x}_k = \lambda_k \boldsymbol{x}_k$. The **eigenvalue matrix Λ** is diagonal.

2 n independent eigenvectors in X diagonalize A $\boxed{\boldsymbol{A = X\Lambda X^{-1}} \text{ and } \boldsymbol{A^k = X\Lambda^k X^{-1}}}$

3 Solve $\boldsymbol{u}_{k+1} = A\boldsymbol{u}_k$ by $\boldsymbol{u}_k = A^k\boldsymbol{u}_0 = X\Lambda^k X^{-1}\boldsymbol{u}_0 = \boxed{\boldsymbol{c_1(\lambda_1)^k x_1 + \cdots + c_n(\lambda_n)^k x_n}}$

4 **No equal eigenvalues** $\Rightarrow$ eigenvector matrix X is invertible and A can be diagonalized.
Repeated eigenvalue $\Rightarrow$ A *might* have too few independent eigenvectors. Then X^{-1} fails.

5 Every matrix $C = B^{-1}AB$ has the **same eigenvalues** as A. These C's are "**similar**" to A.

When $\boldsymbol{x}$ is an eigenvector, multiplication by A is just multiplication by a number λ: $A\boldsymbol{x} = \lambda\boldsymbol{x}$. All the difficulties of matrices are swept away, and we can follow the eigenvectors separately. The eigenvalues go into a *diagonal matrix*, and $A^{100} = X\Lambda^{100}X^{-1}$.

The point of this section is very direct. ***The matrix A turns into a diagonal matrix Λ when we use the eigenvectors properly***. This is the matrix form of our key idea. We start right off with that one essential computation. **The next page explains why $\boldsymbol{AX = X\Lambda}$.**

Diagonalization Suppose the n by n matrix A has n linearly independent eigenvectors $\boldsymbol{x}_1, \ldots, \boldsymbol{x}_n$. Put those $\boldsymbol{x}_i$ into the columns of an invertible ***eigenvector matrix X***. Then $X^{-1}AX$ is the diagonal ***eigenvalue matrix Λ***:

Eigenvector matrix X
Eigenvalue matrix Λ

$$\boldsymbol{X^{-1}AX = \Lambda} = \begin{bmatrix} \lambda_1 & & \\ & \ddots & \\ & & \lambda_n \end{bmatrix}. \qquad (1)$$

The matrix A is "diagonalized." We use capital lambda for the eigenvalue matrix, because the small λ's (the eigenvalues) are on its diagonal.

Example 1 This A is triangular so its eigenvalues are on the diagonal: $\lambda = 2$ and $\lambda = 6$.

$$\boxed{\textbf{Eigenvectors go into } \boldsymbol{X} \quad \begin{bmatrix} 1 \\ 0 \end{bmatrix} \begin{bmatrix} 1 \\ 1 \end{bmatrix}} \qquad \underset{\boldsymbol{X^{-1}}}{\begin{bmatrix} 1 & -1 \\ 0 & 1 \end{bmatrix}} \underset{\boldsymbol{A}}{\begin{bmatrix} \mathbf{2} & \mathbf{4} \\ 0 & \mathbf{6} \end{bmatrix}} \underset{\boldsymbol{X}}{\begin{bmatrix} 1 & 1 \\ 0 & 1 \end{bmatrix}} \underset{=}{=} \underset{\boldsymbol{\Lambda}}{\begin{bmatrix} \mathbf{2} & 0 \\ 0 & \mathbf{6} \end{bmatrix}}$$

In other words $A = X\Lambda X^{-1}$. Then watch $A^2 = X\Lambda X^{-1}X\Lambda X^{-1}$. So $\boldsymbol{A^2}$ **is** $\boldsymbol{X\Lambda^2 X^{-1}}$.

A^2 has the same eigenvectors in X. It has squared eigenvalues 4 and 36 in Λ^2.

Why is $AX = X\Lambda$? A multiplies its eigenvectors, which are the columns of X. The first column of AX is $A\boldsymbol{x}_1$. That is $\lambda_1\boldsymbol{x}_1$. Each column of X is multiplied by its eigenvalue:

$$\textbf{A times X} \qquad AX = A\begin{bmatrix} \\ \boldsymbol{x}_1 & \cdots & \boldsymbol{x}_n \\ \\ \end{bmatrix} = \begin{bmatrix} \\ \lambda_1\boldsymbol{x}_1 & \cdots & \lambda_n\boldsymbol{x}_n \\ \\ \end{bmatrix}.$$

This matrix AX splits into X times Λ (not ΛX, that would multiply rows of X):

$$\textbf{X times }\boldsymbol{\Lambda} \qquad \begin{bmatrix} \\ \lambda_1\boldsymbol{x}_1 & \cdots & \lambda_n\boldsymbol{x}_n \\ \\ \end{bmatrix} = \begin{bmatrix} \\ \boldsymbol{x}_1 & \cdots & \boldsymbol{x}_n \\ \\ \end{bmatrix}\begin{bmatrix} \lambda_1 & & \\ & \ddots & \\ & & \lambda_n \end{bmatrix} = X\Lambda.$$

Keep those matrices in the right order! Then λ_1 multiplies the first column $\boldsymbol{x}_1$, as shown. The diagonalization is complete, and we can write $AX = X\Lambda$ in two good ways:

$$\boxed{AX = X\Lambda \quad \text{is} \quad \boldsymbol{X^{-1}AX = \Lambda} \quad \text{or} \quad \boldsymbol{A = X\Lambda X^{-1}}.} \qquad (2)$$

The matrix X has an inverse, because its columns (the eigenvectors of A) were assumed to be linearly independent. *Without n independent eigenvectors, we can't diagonalize A.*

A and Λ have the same eigenvalues $\lambda_1, \ldots, \lambda_n$. The eigenvectors are different. The job of the original eigenvectors $\boldsymbol{x}_1, \ldots, \boldsymbol{x}_n$ was to diagonalize A. Those eigenvectors in X produce $A = X\Lambda X^{-1}$. You will soon see their simplicity and importance and meaning. The kth power will be $A^k = \boldsymbol{X\Lambda^k X^{-1}}$ which is easy to compute using $\boldsymbol{X^{-1}X = I}$:

$$\boldsymbol{A^k = (X\Lambda X^{-1})(X\Lambda X^{-1})\ldots(X\Lambda X^{-1}) = X\Lambda^k X^{-1}}.$$

$$\begin{matrix}\boldsymbol{A^k = X\Lambda^k X^{-1}} \\ \textbf{Example 1}\end{matrix} \qquad \begin{bmatrix} \mathbf{2} & \mathbf{4} \\ \mathbf{0} & \mathbf{6} \end{bmatrix}^{\boldsymbol{k}} = \begin{bmatrix} 1 & 1 \\ 0 & 1 \end{bmatrix}\begin{bmatrix} 2^k & \\ & 6^k \end{bmatrix}\begin{bmatrix} 1 & -1 \\ 0 & 1 \end{bmatrix} = \begin{bmatrix} \boldsymbol{2^k} & \boldsymbol{6^k - 2^k} \\ 0 & \boldsymbol{6^k} \end{bmatrix} = \boldsymbol{A^k}.$$

With $k = 1$ we get A. With $k = 0$ we get $A^0 = I$ (and $\lambda^0 = 1$). *With $k = -1$ we get A^{-1}.* When $k = 2$, you can see how $A^2 = [4\ \ 32;\ \ 0\ \ 36]$ fits that formula.

Here are four small remarks before we use Λ again in Example 2.

Remark 1 Suppose the eigenvalues are n different numbers (like 2 and 6). Then it is automatic that the n eigenvectors will be independent. The eigenvector matrix X will be *invertible.* ***Any matrix that has no repeated eigenvalues can be diagonalized.***

Remark 2 ***We can multiply eigenvectors by any nonzero constants.*** $A(c\boldsymbol{x}) = \lambda(c\boldsymbol{x})$ is still true. In Example 1, we can divide $\boldsymbol{x} = (1, 1)$ by $\sqrt{2}$ to produce a *unit vector.*

MATLAB and virtually all other codes produce eigenvectors of length $||\boldsymbol{x}|| = 1$.

Remark 3 The eigenvalues in Λ come in the same order as the eigenvectors in X. To reverse the order of 2 and 6 in Λ, put the eigenvector $(1,1)$ before $(1,0)$ in X:

$$\boldsymbol{X^{-1}AX = \Lambda_{\text{new}}} \qquad \begin{bmatrix} 0 & 1 \\ 1 & -1 \end{bmatrix} \begin{bmatrix} 2 & 4 \\ 0 & 6 \end{bmatrix} \begin{bmatrix} 1 & 1 \\ 1 & 0 \end{bmatrix} = \begin{bmatrix} \mathbf{6} & 0 \\ 0 & \mathbf{2} \end{bmatrix} = \boldsymbol{\Lambda_{\text{new}}}$$

To diagonalize A we *must* use an eigenvector matrix. From $X^{-1}AX = \Lambda$ we know that $AX = X\Lambda$. Suppose the first column of X is $\boldsymbol{x}$. Then the first columns of AX and $X\Lambda$ are $A\boldsymbol{x}$ and $\lambda_1 \boldsymbol{x}$. For those to be equal, $\boldsymbol{x}$ must be an eigenvector.

Remark 4 (repeated warning for repeated eigenvalues) Some matrices have too few eigenvectors. *Those matrices cannot be diagonalized.* Here are two examples:

$$\textbf{Not diagonalizable} \qquad A = \begin{bmatrix} 1 & -1 \\ 1 & -1 \end{bmatrix} \quad \textbf{and} \quad B = \begin{bmatrix} 0 & 1 \\ 0 & 0 \end{bmatrix}.$$

Their eigenvalues happen to be 0 and 0. Nothing is special about $\lambda = 0$, the problem is the repetition of λ. All eigenvectors of the first matrix are multiples of $(1,1)$:

$$\begin{array}{l}\textbf{Only one line}\\ \textbf{of eigenvectors}\end{array} \qquad A\boldsymbol{x} = 0\boldsymbol{x} \quad \text{means} \quad \begin{bmatrix} 1 & -1 \\ 1 & -1 \end{bmatrix} \begin{bmatrix} \boldsymbol{x} \end{bmatrix} = \begin{bmatrix} 0 \\ 0 \end{bmatrix} \quad \text{and} \quad \boldsymbol{x} = c \begin{bmatrix} 1 \\ 1 \end{bmatrix}.$$

There is no second eigenvector, so this unusual matrix A cannot be diagonalized.

Those matrices are the best examples to test any statement about eigenvectors. In many true-false questions, non-diagonalizable matrices lead to *false*.

Remember that there is no connection between invertibility and diagonalizability:

- ***Invertibility*** is concerned with the ***eigenvalues*** (we must have $\lambda \neq 0$).
- ***Diagonalizability*** is concerned with the ***eigenvectors*** (n independent $\boldsymbol{x}$'s).

Each eigenvalue has at least one eigenvector! $\boldsymbol{A - \lambda I}$ **is singular**. If $(A - \lambda I)\boldsymbol{x} = \mathbf{0}$ leads you to $\boldsymbol{x} = \mathbf{0}$, λ is *not* an eigenvalue. Look for a mistake in solving $\det(A - \lambda I) = 0$.

Fact: **Eigenvectors for n different λ's are independent. Then we can diagonalize A.**

Independent $\boldsymbol{x}$ from different λ Eigenvectors that correspond to distinct (all different) eigenvalues are linearly independent. An n by n matrix that has n different eigenvalues (**no repeated λ's**) must be diagonalizable.

Proof Suppose $c_1\boldsymbol{x}_1 + c_2\boldsymbol{x}_2 = \mathbf{0}$ $(n = 2)$. Multiply by A to find $c_1\lambda_1\boldsymbol{x}_1 + c_2\lambda_2\boldsymbol{x}_2 = \mathbf{0}$. Multiply by λ_2 to find $c_1\lambda_2\boldsymbol{x}_1 + c_2\lambda_2\boldsymbol{x}_2 = \mathbf{0}$. Now subtract one from the other:

$$\text{Subtraction leaves} \quad (\lambda_1 - \lambda_2)c_1\boldsymbol{x}_1 = \mathbf{0}. \quad \text{Therefore } c_1 = 0.$$

Since the λ's are different and $\boldsymbol{x}_1 \neq \mathbf{0}$, we are forced to the conclusion that $c_1 = 0$. Similarly $c_2 = 0$. Only the combination with $c_1 = c_2 = 0$ gives $c_1\boldsymbol{x}_1 + c_2\boldsymbol{x}_2 = \mathbf{0}$. So the eigenvectors $\boldsymbol{x}_1$ and $\boldsymbol{x}_2$ must be independent.

This proof extends directly to n eigenvectors. Suppose that $c_1\boldsymbol{x}_1 + \cdots + c_n\boldsymbol{x}_n = \mathbf{0}$. Multiply by A, multiply by λ_n, and subtract. This multiplies $\boldsymbol{x}_n$ by $\lambda_n - \lambda_n = 0$, and $\boldsymbol{x}_n$ is gone. Now multiply by A and by λ_{n-1} and subtract. This removes $\boldsymbol{x}_{n-1}$. Eventually only $\boldsymbol{x}_1$ is left:

$$\text{We reach} \quad (\lambda_1 - \lambda_2)\cdots(\lambda_1 - \lambda_n)c_1\boldsymbol{x}_1 = \mathbf{0} \quad \text{which forces} \quad c_1 = 0. \qquad (3)$$

Similarly every $c_i = 0$. **When the λ's are all different, the eigenvectors are independent.** A full set of eigenvectors can go into the columns of the invertible eigenvector matrix X.

Example 2 **Powers of A** The Markov matrix $A = \begin{bmatrix} .8 & .3 \\ .2 & .7 \end{bmatrix}$ in the last section had $\lambda_1 = 1$ and $\lambda_2 = .5$. Here is $A = X\Lambda X^{-1}$ with those eigenvalues in the diagonal Λ:

$$\textbf{Markov example} \qquad \boldsymbol{A} = \begin{bmatrix} .8 & .3 \\ .2 & .7 \end{bmatrix} = \begin{bmatrix} .6 & 1 \\ .4 & -1 \end{bmatrix} \begin{bmatrix} 1 & 0 \\ 0 & .5 \end{bmatrix} \begin{bmatrix} 1 & 1 \\ .4 & -.6 \end{bmatrix} = \boldsymbol{X\Lambda X^{-1}}.$$

The eigenvectors $(.6, .4)$ and $(1, -1)$ are in the columns of X. They are also the eigenvectors of A^2. Watch how A^2 has the same X, and ***the eigenvalue matrix of A^2 is Λ^2***:

$$\textbf{Same } \boldsymbol{X} \textbf{ for } \boldsymbol{A^2} \qquad \boxed{\boxed{\boldsymbol{A^2}} = X\Lambda X^{-1}X\Lambda X^{-1} = \boxed{\boldsymbol{X\Lambda^2 X^{-1}}.}} \qquad (4)$$

Just keep going, and you see why the high powers A^k approach a "steady state":

$$\textbf{Powers of } \boldsymbol{A} \qquad \boldsymbol{A^k} = \boldsymbol{X\Lambda^k X^{-1}} = \begin{bmatrix} .6 & 1 \\ .4 & -1 \end{bmatrix} \begin{bmatrix} 1^k & 0 \\ 0 & (.5)^k \end{bmatrix} \begin{bmatrix} 1 & 1 \\ .4 & -.6 \end{bmatrix}.$$

As k gets larger, $(.5)^k$ gets smaller. In the limit it disappears completely. That limit is A^∞:

$$\textbf{Limit } \boldsymbol{k \to \infty} \qquad \boldsymbol{A^\infty} = \begin{bmatrix} .6 & 1 \\ .4 & -1 \end{bmatrix} \begin{bmatrix} 1 & 0 \\ 0 & 0 \end{bmatrix} \begin{bmatrix} 1 & 1 \\ .4 & -.6 \end{bmatrix} = \begin{bmatrix} \mathbf{.6} & \mathbf{.6} \\ \mathbf{.4} & \mathbf{.4} \end{bmatrix}.$$

The limit has the eigenvector $\boldsymbol{x}_1$ in both columns. We saw this A^∞ on the very first page of Chapter 6. Now we see it coming from powers like $A^{100} = X\Lambda^{100}X^{-1}$.

Question	***When does $A^k \to$ zero matrix?***	**Answer**	***All*** $|\lambda| < 1$.

Similar Matrices : Same Eigenvalues

Suppose the eigenvalue matrix Λ is fixed. As we change the eigenvector matrix X, we get a whole family of different matrices $A = X\Lambda X^{-1}$—*all with the same eigenvalues in* Λ. All those matrices A (with the same Λ) are called **similar**.

This idea extends to matrices that can't be diagonalized. Again we choose one constant matrix C (not necessarily Λ). And we look at the whole family of matrices $\boldsymbol{A = BCB^{-1}}$, allowing all invertible matrices B. Again **those matrices A and C are called similar**.

We are using C instead of Λ because C might not be diagonal. We are using B instead of X because the columns of B might not be eigenvectors. We only require that B is invertible—the key fact about similar matrices stays true: **Similar matrices A and C have the same n eigenvalues.**

All the matrices $A = BCB^{-1}$ are "similar." They all share the eigenvalues of C.

Proof If $Cx = \lambda x$, then BCB^{-1} has the same eigenvalue λ with new eigenvector Bx:

$$\textit{Similar matrix, same } \lambda \qquad (BCB^{-1})(Bx) = BCx = B\lambda x = \lambda(Bx). \qquad (5)$$

A fixed matrix C produces a family of similar matrices BCB^{-1}, allowing all B. When C is the identity matrix, the "family" is very small. The only member is $BIB^{-1} = I$. The identity matrix is the only diagonalizable matrix with all eigenvalues $\lambda = 1$.

The family is larger when $\lambda = 1$ and 1 *with only one eigenvector* (not diagonalizable). The simplest C is the *Jordan form*—developed in Appendix 5. All the similar matrices have two parameters r and s, not both zero: always determinant $= 1$ and trace $= 2$.

$$C = \begin{bmatrix} \mathbf{1} & \mathbf{1} \\ 0 & \mathbf{1} \end{bmatrix} = \text{Jordan form gives } A = BCB^{-1} = \begin{bmatrix} 1-rs & r^2 \\ -s^2 & 1+rs \end{bmatrix}. \qquad (6)$$

For an important example I will take eigenvalues $\lambda = 1$ and 0 (not repeated!). Now the whole family is diagonalizable with the same eigenvalue matrix Λ. We get every 2 by 2 matrix with eigenvalues 1 and 0. The trace is 1, the determinant is zero, here they are:

$$\textbf{All similar} \quad \Lambda = \begin{bmatrix} \mathbf{1} & 0 \\ 0 & \mathbf{0} \end{bmatrix} \quad A = \begin{bmatrix} 1 & 1 \\ 0 & 0 \end{bmatrix} \text{ or } A = \begin{bmatrix} .5 & .5 \\ .5 & .5 \end{bmatrix} \text{ or any } A = \frac{xy^{\mathrm{T}}}{x^{\mathrm{T}}y}.$$

The family contains all 2×2 matrices with $A^2 = A$, including $A = \Lambda$ when $B = I$.

If A is m by n and B is n by m, AB and BA have the same nonzero eigenvalues.

Proof. Start with this identity between square matrices (easily checked). The first and third matrices are inverses. The "size matrix" shows the shapes of all blocks.

$$\begin{bmatrix} I & -A \\ 0 & I \end{bmatrix}\begin{bmatrix} AB & 0 \\ B & 0 \end{bmatrix}\begin{bmatrix} I & A \\ 0 & I \end{bmatrix} = \begin{bmatrix} 0 & 0 \\ B & BA \end{bmatrix} \qquad \begin{bmatrix} m \times m & m \times n \\ n \times m & n \times n \end{bmatrix}$$

This equation $D^{-1}ED = F$ says **F is similar to E**—they have the *same* $m+n$ eigenvalues.

$\begin{bmatrix} AB & 0 \\ B & 0 \end{bmatrix}$ has the m eigenvalues of AB plus n zeros $\qquad \begin{bmatrix} 0 & 0 \\ B & BA \end{bmatrix}$ has the n eigenvalues of BA plus m zeros

So AB and BA have the **same eigenvalues** except for $|n - m|$ zeros. Wow.

Fibonacci Numbers

We present a famous example, where eigenvalues tell how fast the Fibonacci numbers grow. ***Every new Fibonacci number is the sum of the two previous F's*:**

The sequence	$0, 1, 1, 2, 3, 5, 8, 13, \ldots$	***comes from***	$F_{k+2} = F_{k+1} + F_k$.

These numbers turn up in a fantastic variety of applications. Plants and trees grow in a spiral pattern, and a pear tree has 8 growths for every 3 turns. For a willow those numbers can be 13 and 5. The champion is a sunflower of Daniel O'Connell, which had 233 seeds in 144 loops. Those are the Fibonacci numbers F_{13} and F_{12}. Our problem is more basic.

Problem: Find the Fibonacci number F_{100} The slow way is to apply the rule $F_{k+2} = F_{k+1} + F_k$ one step at a time. By adding $F_6 = 8$ to $F_7 = 13$ we reach $F_8 = 21$. Eventually we come to F_{100}. Linear algebra gives a better way.

The key is to begin with a matrix equation $\mathbf{u}_{k+1} = A\mathbf{u}_k$. That is a *one-step* rule for vectors, while Fibonacci gave a two-step rule for scalars. We match those rules by putting two Fibonacci numbers F_{k+1} and F_k into a vector. Then you will see the matrix A.

$$\text{Let } \boldsymbol{u}_k = \begin{bmatrix} F_{k+1} \\ F_k \end{bmatrix}. \quad \text{The rule} \quad \begin{matrix} F_{k+2} = F_{k+1} + F_k \\ F_{k+1} = F_{k+1} \end{matrix} \quad \text{is} \quad \boldsymbol{u}_{k+1} = \begin{bmatrix} \mathbf{1} & \mathbf{1} \\ \mathbf{1} & \mathbf{0} \end{bmatrix} \boldsymbol{u}_k = \boldsymbol{A}\boldsymbol{u}_k. \tag{7}$$

Every step multiplies by $A = \begin{bmatrix} 1 & 1 \\ 1 & 0 \end{bmatrix}$. After 100 steps we reach $\boldsymbol{u}_{100} = A^{100}\boldsymbol{u}_0$:

$$\boldsymbol{u}_0 = \begin{bmatrix} 1 \\ 0 \end{bmatrix}, \quad \boldsymbol{u}_1 = \begin{bmatrix} 1 \\ 1 \end{bmatrix}, \quad \boldsymbol{u}_2 = \begin{bmatrix} 2 \\ 1 \end{bmatrix}, \quad \boldsymbol{u}_3 = \begin{bmatrix} 3 \\ 2 \end{bmatrix}, \quad \ldots, \quad \boldsymbol{u}_{100} = \begin{bmatrix} F_{101} \\ F_{100} \end{bmatrix}.$$

Powers of A are just right for eigenvalues! Subtract λ from the diagonal of A:

$$A - \lambda I = \begin{bmatrix} 1-\lambda & 1 \\ 1 & -\lambda \end{bmatrix} \quad \text{leads to} \quad \det(A - \lambda I) = \boldsymbol{\lambda^2 - \lambda - 1} \qquad \begin{matrix} a = +1 \\ b = -1 \\ c = -1 \end{matrix}$$

The equation $\lambda^2 - \lambda - 1 = 0$ is solved by the quadratic formula $\left(-b \pm \sqrt{b^2 - 4ac}\right)/2a$:

$$\textbf{Eigenvalues of } \boldsymbol{A} \qquad \lambda_1 = \frac{1+\sqrt{5}}{2} \approx 1.618 \quad \text{and} \quad \lambda_2 = \frac{1-\sqrt{5}}{2} \approx -.618$$

Solving $(A - \lambda I)\boldsymbol{x} = \mathbf{0}$ leads to eigenvectors $\boldsymbol{x}_1 = (\boldsymbol{\lambda_1}, \mathbf{1})$ and $\boldsymbol{x}_2 = (\boldsymbol{\lambda_2}, \mathbf{1})$. Step 2 finds the combination of those eigenvectors that gives the starting vector $\boldsymbol{u}_0 = (F_1, F_0) = (1, 0)$:

$$\begin{bmatrix} 1 \\ 0 \end{bmatrix} = \frac{1}{\lambda_1 - \lambda_2}\left(\begin{bmatrix} \lambda_1 \\ 1 \end{bmatrix} - \begin{bmatrix} \lambda_2 \\ 1 \end{bmatrix}\right) \quad \text{or} \quad \boldsymbol{u}_0 = \frac{\boldsymbol{x}_1 - \boldsymbol{x}_2}{\lambda_1 - \lambda_2}. \tag{8}$$

Step 3 multiplies $\boldsymbol{u}_0$ by A^{100} to find $\boldsymbol{u}_{100}$. The eigenvectors $\boldsymbol{x}_1$ and $\boldsymbol{x}_2$ stay separate! $\boldsymbol{x}_1$ is multiplied by $(\lambda_1)^{100}$ and $\boldsymbol{x}_2$ is multiplied by $(\lambda_2)^{100}$:

$$\textbf{100 steps from } \boldsymbol{u}_0 \qquad \boldsymbol{u}_{100} = \frac{(\lambda_1)^{100}\boldsymbol{x}_1 - (\lambda_2)^{100}\boldsymbol{x}_2}{\lambda_1 - \lambda_2} \tag{9}$$

We want F_{100} = second component of $\boldsymbol{u}_{100}$. The second components of $\boldsymbol{x}_1$ and $\boldsymbol{x}_2$ are 1. The difference between $\lambda_1 = (1+\sqrt{5})/2$ and $\lambda_2 = (1-\sqrt{5})/2$ is $\boldsymbol{\sqrt{5}}$. And $\boldsymbol{\lambda_2^{100} \approx 0}$.

$$\text{100th Fibonacci number} = \frac{\lambda_1^{100} - \lambda_2^{100}}{\lambda_1 - \lambda_2} = \text{nearest integer to } \frac{1}{\sqrt{5}}\left(\frac{1+\sqrt{5}}{2}\right)^{100}. \tag{10}$$

Every F_k is a whole number. The ratio F_{101}/F_{100} must be very close to the limiting ratio $\left(1+\sqrt{5}\right)/2$. The Greeks called this number the *"golden mean"*. For some reason a rectangle with sides 1.618 and 1 looks especially graceful.

Matrix Powers A^k

Fibonacci's example is a typical difference equation $u_{k+1} = Au_k$. ***Each step multiplies by A. The solution is*** $\boldsymbol{u_k = A^k u_0}$. We want to make clear how diagonalizing the matrix gives a quick way to compute A^k and find u_k in three steps.

The eigenvector matrix X produces $A = X\Lambda X^{-1}$. This is a factorization of the matrix, like $A = LU$ or $A = QR$. The new factorization is perfectly suited to computing powers, because ***every time*** $\boldsymbol{X^{-1}}$ ***multiplies X we get I***:

$$\textbf{Powers of } \boldsymbol{A} \qquad A^k u_0 = (X\Lambda X^{-1})\cdots(X\Lambda X^{-1})u_0 = X\Lambda^k X^{-1}u_0$$

I will split $X\Lambda^k X^{-1}u_0$ into Steps 1, 2, 3 that show how eigenvalues lead to u_k.

1. $\boxed{\text{Write } u_0 \text{ as a combination } Xc = c_1x_1 + \cdots + c_nx_n \text{ of eigenvectors.}}$ $\boldsymbol{c = X^{-1}u_0}$.

2. $\boxed{\text{Multiply each eigenvector } x_i \text{ by } (\lambda_i)^k.}$ Now we have $\boldsymbol{\Lambda^k X^{-1} u_0}$.

3. $\boxed{\text{Add up the pieces } \boldsymbol{c_i(\lambda_i)^k x_i} \text{ to find the solution } u_k = A^k u_0.}$ This is $\boldsymbol{X\Lambda^k X^{-1}u_0}$.

$$\boxed{\textbf{Solution for } u_{k+1} = Au_k \quad \boxed{u_k = A^k u_0 = c_1(\lambda_1)^k x_1 + \cdots + c_n(\lambda_n)^k x_n.}} \tag{11}$$

In matrix language A^k equals $(X\Lambda X^{-1})^k$ which is X times Λ^k times X^{-1}. In Step 1, the eigenvectors in X lead to the c's in the combination $u_0 = c_1x_1 + \cdots + c_nx_n$:

$$\textbf{Step 1 was} \qquad u_0 = \begin{bmatrix} & & \\ x_1 & \cdots & x_n \\ & & \end{bmatrix}\begin{bmatrix} c_1 \\ \vdots \\ c_n \end{bmatrix}. \quad \text{This says that } \boxed{u_0 = X\boldsymbol{c}.} \tag{12}$$

The coefficients in Step 1 are $\boldsymbol{c} = X^{-1}u_0$. Then Step 2 multiplies by Λ^k. The final result $u_k = \sum c_i(\lambda_i)^k x_i$ in Step 3 is the product of X and Λ^k and $X^{-1}u_0$:

$$A^k u_0 = X\Lambda^k X^{-1}u_0 = X\Lambda^k \boldsymbol{c} = \begin{bmatrix} & & \\ x_1 & \ldots & x_n \\ & & \end{bmatrix}\begin{bmatrix} (\lambda_1)^k & & \\ & \ddots & \\ & & (\lambda_n)^k \end{bmatrix}\begin{bmatrix} c_1 \\ \vdots \\ c_n \end{bmatrix}. \tag{13}$$

This result is exactly $\boldsymbol{u_k = c_1(\lambda_1)^k x_1 + \cdots + c_n(\lambda_n)^k x_n}$. It solves $u_{k+1} = Au_k$.

Example 3 Start from $u_0 = (1, 0)$. Compute $A^k u_0$ for this faster Fibonacci:

$$A = \begin{bmatrix} 1 & 2 \\ 1 & 0 \end{bmatrix} \quad \text{has} \quad \lambda_1 = \mathbf{2} \quad \text{and} \quad x_1 = \begin{bmatrix} 2 \\ 1 \end{bmatrix}, \quad \lambda_2 = \mathbf{-1} \quad \text{and} \quad x_2 = \begin{bmatrix} 1 \\ -1 \end{bmatrix}.$$

This matrix is like Fibonacci except the rule is changed to $F_{k+2} = F_{k+1} + \mathbf{2}F_k$. The new numbers start with $0, 1, 1, 3$. Then $\lambda_1 = 2$ makes these F_k grow faster.

Find $\boldsymbol{u}_k = \boldsymbol{A}^k\boldsymbol{u}_0$ in 3 steps $\quad \boldsymbol{u}_0 = c_1\boldsymbol{x}_1 + c_2\boldsymbol{x}_2$ and $\boldsymbol{u}_k = c_1(\lambda_1)^k\boldsymbol{x}_1 + c_2(\lambda_2)^k\boldsymbol{x}_2$

Step 1 $\quad \boldsymbol{u}_0 = \begin{bmatrix}1\\0\end{bmatrix} = \frac{1}{3}\begin{bmatrix}2\\1\end{bmatrix} + \frac{1}{3}\begin{bmatrix}1\\-1\end{bmatrix}$ so $c_1 = c_2 = \boldsymbol{\frac{1}{3}}$

Step 2 $\quad$ Multiply the two parts by $(\lambda_1)^k = \boldsymbol{2^k}$ and $(\lambda_2)^k = \boldsymbol{(-1)^k}$

Step 3 $\quad$ Combine eigenvectors $c_1(\lambda_1)^k\boldsymbol{x}_1$ and $c_2(\lambda_2)^k\boldsymbol{x}_2$ into $\boldsymbol{u}_k$:

$$\boldsymbol{u_k = A^k u_0} \qquad \boldsymbol{u}_k = \frac{1}{3}2^k\begin{bmatrix}2\\1\end{bmatrix} + \frac{1}{3}(-1)^k\begin{bmatrix}1\\-1\end{bmatrix} = \begin{bmatrix}F_{k+1}\\F_k\end{bmatrix}.$$

The new F_k is $\boldsymbol{(2^k - (-1)^k)/3}$, starting with $0, 1, 1, 3$. Now $\boldsymbol{F_{100} \approx 2^{100}/3}$.

Behind these numerical examples lies a fundamental idea: ***Follow the eigenvectors***. In Section 6.5 this is the crucial link from linear algebra to differential equations: $\boldsymbol{\lambda^k}$ **will become** $\boldsymbol{e^{\lambda t}}$. Chapter 8 sees the same idea as "transforming to an eigenvector basis." The best example of all is a ***Fourier series***, built from the **eigenvectors** $\boldsymbol{e^{ikx}}$.

Nondiagonalizable Matrices (Optional)

Suppose λ is an eigenvalue of A. We discover that fact in two ways:

1. **Eigenvectors** (geometric) $\quad$ There are nonzero solutions to $A\boldsymbol{x} = \lambda\boldsymbol{x}$.
2. **Eigenvalues** (algebraic) $\quad$ The determinant of $A - \lambda I$ is zero.

The number λ may be a simple eigenvalue or a multiple eigenvalue, and we want to know its ***multiplicity***. Most eigenvalues have multiplicity $M = 1$ (simple eigenvalues). Then there is a single line of eigenvectors, and $\det(A - \lambda I)$ does not have a double factor.

For exceptional matrices, an eigenvalue can be ***repeated***. Then there are two different ways to count its multiplicity. Always GM $\leq$ AM for each λ:

1. (Geometric Multiplicity = **GM**) $\quad$ Count the **independent eigenvectors** for λ. Then GM is the dimension of the nullspace of $A - \lambda I$.

2. (Algebraic Multiplicity = **AM**) $\quad$ AM counts the **repetitions of $\boldsymbol{\lambda}$** among the eigenvalues. Look at the n roots of $\det(A - \lambda I) = 0$.

If A has $\lambda = 4, 4, 4$, then that eigenvalue has AM $= 3$ and GM $= 1, 2$, or 3.

The following matrix A is the standard example of trouble. Its eigenvalue $\lambda = 0$ is repeated. It is a double eigenvalue (AM $= 2$) with only one eigenvector (GM $= 1$).

AM = 2
GM = 1 $\quad A = \begin{bmatrix}0 & 1\\0 & 0\end{bmatrix}$ has $\det(A - \lambda I) = \begin{vmatrix}-\lambda & 1\\0 & -\lambda\end{vmatrix} = \lambda^2.$ $\quad$ $\boldsymbol{\lambda = 0, 0}$ **but 1 eigenvector** $\begin{bmatrix}1\\0\end{bmatrix}$

There "should" be two eigenvectors, because $\lambda^2 = 0$ has a double root. The double factor λ^2 makes AM $= 2$. But there is only one eigenvector $\boldsymbol{x} = (1, 0)$ and GM $= 1$. ***This shortage of eigenvectors when* GM *is below* AM *means that A is not diagonalizable*.**

The next three matrices all have the same shortage of eigenvectors. Their repeated eigenvalue is $\lambda = 5$. Traces are 5 plus $5 = 10$ and determinants are 5 times $5 = 25$:

$$A = \begin{bmatrix} 5 & 1 \\ 0 & 5 \end{bmatrix} \quad \text{and} \quad A = \begin{bmatrix} 6 & -1 \\ 1 & 4 \end{bmatrix} \quad \text{and} \quad A = \begin{bmatrix} 7 & 2 \\ -2 & 3 \end{bmatrix}.$$

Those all have $\det(A - \lambda I) = (\lambda - 5)^2$. The algebraic multiplicity is AM $= 2$. But each $A - 5I$ has rank $r = 1$. The geometric multiplicity is GM $= 1$. There is only one line of eigenvectors for $\lambda = 5$, and these matrices are not diagonalizable.

■ REVIEW OF THE KEY IDEAS ■

1. If A has n independent eigenvectors $\boldsymbol{x}_1, \ldots, \boldsymbol{x}_n$, they go into the columns of X.

   A is diagonalized by X $\qquad X^{-1}AX = \Lambda \quad \text{and} \quad A = X\Lambda X^{-1}$.

2. The powers of A are $A^k = X\Lambda^k X^{-1}$. The eigenvectors in X are unchanged.

3. The eigenvalues of A^k are $(\lambda_1)^k, \ldots, (\lambda_n)^k$ in the matrix Λ^k.

4. The solution to $\boldsymbol{u}_{k+1} = A\boldsymbol{u}_k$ starting from $\boldsymbol{u}_0$ is $\boldsymbol{u}_k = A^k\boldsymbol{u}_0 = X\Lambda^k X^{-1}\boldsymbol{u}_0$:

$$\boxed{\boldsymbol{u}_k = c_1(\lambda_1)^k\boldsymbol{x}_1 + \cdots + c_n(\lambda_n)^k\boldsymbol{x}_n \text{ provided } \boldsymbol{u}_0 = c_1\boldsymbol{x}_1 + \cdots + c_n\boldsymbol{x}_n.}$$

5. A is diagonalizable if every eigenvalue has enough eigenvectors (GM $=$ AM).

■ WORKED EXAMPLES ■

6.2 A The **Lucas numbers** are like the Fibonacci numbers except they start with $L_1 = 1$ and $L_2 = 3$. Using the same rule $L_{k+2} = L_{k+1} + L_k$, the next Lucas numbers are $4, 7, 11, 18$. Show that the Lucas number L_{100} is $\lambda_1^{100} + \lambda_2^{100}$.

Solution $\boldsymbol{u}_{k+1} = \left[\begin{smallmatrix} 1 & 1 \\ 1 & 0 \end{smallmatrix}\right]\boldsymbol{u}_k$ is the same as for Fibonacci, because $L_{k+2} = L_{k+1} + L_k$ is the same rule (with different starting values). The equation becomes a 2 by 2 system:

$$\boxed{\text{Let } \boldsymbol{u}_k = \begin{bmatrix} L_{k+1} \\ L_k \end{bmatrix}. \quad \text{The rule} \quad \begin{aligned} L_{k+2} &= L_{k+1} + L_k \\ L_{k+1} &= L_{k+1} \end{aligned} \quad \text{is} \quad \boldsymbol{u}_{k+1} = \begin{bmatrix} 1 & 1 \\ 1 & 0 \end{bmatrix}\boldsymbol{u}_k.}$$

The eigenvalues and eigenvectors of $A = \left[\begin{smallmatrix} 1 & 1 \\ 1 & 0 \end{smallmatrix}\right]$ still come from $\lambda^2 = \lambda + 1$:

$$\lambda_1 = \frac{1 + \sqrt{5}}{2} \quad \text{and} \quad \boldsymbol{x}_1 = \begin{bmatrix} \lambda_1 \\ 1 \end{bmatrix} \qquad \lambda_2 = \frac{1 - \sqrt{5}}{2} \quad \text{and} \quad \boldsymbol{x}_2 = \begin{bmatrix} \lambda_2 \\ 1 \end{bmatrix}.$$

Now solve $c_1\boldsymbol{x}_1 + c_2\boldsymbol{x}_2 = \boldsymbol{u}_1 = (3,1)$. The solution is $c_1 = \lambda_1$ and $c_2 = \lambda_2$. Check:

$$\lambda_1\boldsymbol{x}_1 + \lambda_2\boldsymbol{x}_2 = \begin{bmatrix} \lambda_1^2 + \lambda_2^2 \\ \lambda_1 + \lambda_2 \end{bmatrix} = \begin{bmatrix} \text{trace of } A^2 \\ \text{trace of } A \end{bmatrix} = \begin{bmatrix} 3 \\ 1 \end{bmatrix} = \boldsymbol{u}_1$$

$\boldsymbol{u}_{100} = A^{99}\boldsymbol{u}_1$ tells us the Lucas numbers (L_{101}, L_{100}). The second components of the eigenvectors $\boldsymbol{x}_1$ and $\boldsymbol{x}_2$ are 1, so the second component of $\boldsymbol{u}_{100}$ is the answer we want:

Lucas number	$L_{100} = c_1\lambda_1^{99} + c_2\lambda_2^{99} = \boldsymbol{\lambda_1^{100} + \lambda_2^{100}}.$

Lucas starts faster than Fibonacci, and ends up larger by a factor near $\sqrt{5}$.

6.2 B Find the inverse and the eigenvalues and the determinant of this matrix A:

$$\boldsymbol{A = 5I - \text{ones}(4)} = \begin{bmatrix} 4 & -1 & -1 & -1 \\ -1 & 4 & -1 & -1 \\ -1 & -1 & 4 & -1 \\ -1 & -1 & -1 & 4 \end{bmatrix}.$$

Describe an orthogonal eigenvector matrix Q that diagonalizes $\boldsymbol{Q^{-1}AQ = \Lambda}$.

Solution What are the eigenvalues of the 4 by 4 **all-ones matrix**? Its rank is 1, so three eigenvalues are $\lambda = 0,0,0$. Its trace is 4, so the other eigenvalue is $\lambda = 4$. Subtract this all-ones matrix from $5I$ to get our matrix A above.

Subtract the eigenvalues $4,0,0,0$ from $5,5,5,5$. The eigenvalues of A are $1,5,5,5$.

The determinant of A is 125, the product of those four eigenvalues. The eigenvector for $\lambda = 1$ is $\boldsymbol{x} = (1,1,1,1)$ or (c,c,c,c). The other eigenvectors are perpendicular to $\boldsymbol{x}$ (since A is symmetric). The nicest eigenvector matrix Q is the **symmetric orthogonal Hadamard matrix $\boldsymbol{H}$**. The factor $\frac{1}{2}$ produces unit column vectors.

$$\textbf{Orthonormal eigenvectors} \quad Q = \boldsymbol{H} = \frac{1}{2}\begin{bmatrix} 1 & 1 & 1 & 1 \\ 1 & -1 & 1 & -1 \\ 1 & 1 & -1 & -1 \\ 1 & -1 & -1 & 1 \end{bmatrix} = \boldsymbol{H}^{\mathrm{T}} = \boldsymbol{H}^{-1}.$$

The eigenvalues of A^{-1} are $1, \frac{1}{5}, \frac{1}{5}, \frac{1}{5}$. The eigenvectors are the same as for A. So $A^{-1} = H\Lambda^{-1}H^{-1}$. The inverse matrix is surprisingly neat:

$$\boldsymbol{A^{-1}} = \frac{1}{5} * (\boldsymbol{I} + \textbf{ones}(4)) = \frac{1}{5}\begin{bmatrix} 2 & 1 & 1 & 1 \\ 1 & 2 & 1 & 1 \\ 1 & 1 & 2 & 1 \\ 1 & 1 & 1 & 2 \end{bmatrix}$$

$A = 5I - \text{ones}(4)$ is a rank-one change from $5I$. So A^{-1} is a rank-one change from $I/5$.

In a complete graph with 5 nodes, the determinant 125 counts the "spanning trees" (trees that touch all nodes). *Trees have no loops* (graphs and trees are on pages 134–135).

With 6 nodes, the matrix $\boldsymbol{6I - \textbf{ones}(5)}$ has the five eigenvalues $1,6,6,6,6$.

Problem Set 6.2

1 (a) Factor these two matrices into $A = X\Lambda X^{-1}$:

$$A = \begin{bmatrix} 1 & 2 \\ 0 & 3 \end{bmatrix} \quad \text{and} \quad A = \begin{bmatrix} 1 & 1 \\ 3 & 3 \end{bmatrix}.$$

(b) If $A = X\Lambda X^{-1}$ then $A^3 = (\quad)(\quad)(\quad)$ and $A^{-1} = (\quad)(\quad)(\quad)$.

2 If A has $\lambda_1 = 2$ with eigenvector $\boldsymbol{x}_1 = \left[\begin{smallmatrix} 1 \\ 0 \end{smallmatrix}\right]$ and $\lambda_2 = 5$ with $\boldsymbol{x}_2 = \left[\begin{smallmatrix} 1 \\ 1 \end{smallmatrix}\right]$, use $X\Lambda X^{-1}$ to find A. No other matrix has the same λ's and $\boldsymbol{x}$'s.

3 Suppose $A = X\Lambda X^{-1}$. What is the eigenvalue matrix for $A + 2I$? What is the eigenvector matrix? Check that $A + 2I = (\quad)(\quad)(\quad)^{-1}$.

4 True or false : If the columns of X (eigenvectors of A) are linearly independent, then

(a) A is invertible (b) A is diagonalizable

(c) X is invertible (d) X is diagonalizable.

5 If the eigenvectors of A are the columns of I, then A is a _____ matrix. If the eigenvector matrix X is triangular, then X^{-1} is triangular. Prove that A is also triangular.

6 Describe all matrices X that diagonalize this matrix A (find all eigenvectors of A):

$$A = \begin{bmatrix} 4 & 0 \\ 1 & 2 \end{bmatrix}. \qquad \text{Then describe all matrices that diagonalize } A^{-1}.$$

7 Write down the most general matrix that has eigenvectors $\left[\begin{smallmatrix} 1 \\ 1 \end{smallmatrix}\right]$ and $\left[\begin{smallmatrix} 1 \\ -1 \end{smallmatrix}\right]$.

Questions 8–10 are about Fibonacci and Gibonacci numbers.

8 Diagonalize the Fibonacci matrix by completing X^{-1}:

$$\begin{bmatrix} 1 & 1 \\ 1 & 0 \end{bmatrix} = \begin{bmatrix} \lambda_1 & \lambda_2 \\ 1 & 1 \end{bmatrix} \begin{bmatrix} \lambda_1 & 0 \\ 0 & \lambda_2 \end{bmatrix} \begin{bmatrix} & \\ & \end{bmatrix}.$$

Do the multiplication $X\Lambda^k X^{-1}\left[\begin{smallmatrix} 1 \\ 0 \end{smallmatrix}\right]$ to find its second component. This is the kth Fibonacci number $F_k = (\lambda_1^k - \lambda_2^k)/(\lambda_1 - \lambda_2)$.

9 Suppose G_{k+2} is the *average* of the two previous numbers G_{k+1} and G_k:

$$\begin{aligned} G_{k+2} &= \tfrac{1}{2}G_{k+1} + \tfrac{1}{2}G_k \\ G_{k+1} &= G_{k+1} \end{aligned} \quad \text{is} \quad \begin{bmatrix} G_{k+2} \\ G_{k+1} \end{bmatrix} = \begin{bmatrix} & A & \end{bmatrix} \begin{bmatrix} G_{k+1} \\ G_k \end{bmatrix}.$$

(a) Find the eigenvalues and eigenvectors of A.

(b) Find the limit as $n \to \infty$ of the matrices $A^n = X\Lambda^n X^{-1}$.

(c) If $G_0 = 0$ and $G_1 = 1$ show that the Gibonacci numbers approach $\frac{2}{3}$.

10 Prove that every third Fibonacci number in $0, 1, 1, 2, 3, \ldots$ is even.

Questions 11–14 are about diagonalizability.

11 True or false: If the eigenvalues of A are $2, 2, 5$ then the matrix is certainly

(a) invertible (b) diagonalizable (c) not diagonalizable.

12 True or false: If the only eigenvectors of A are multiples of $(1, 4)$ then A has

(a) no inverse (b) a repeated eigenvalue (c) no diagonalization $X\Lambda X^{-1}$.

13 Complete these matrices so that $\det A = 25$. Then check that $\lambda = 5$ is repeated—the trace is 10 so the determinant of $A - \lambda I$ is $(\lambda - 5)^2$. Find an eigenvector with $Ax = 5x$. These matrices are not diagonalizable—no second line of eigenvectors.

$$A = \begin{bmatrix} 8 & \\ & 2 \end{bmatrix} \quad \text{and} \quad A = \begin{bmatrix} 9 & 4 \\ & 1 \end{bmatrix} \quad \text{and} \quad A = \begin{bmatrix} 10 & 5 \\ -5 & \end{bmatrix}$$

14 The matrix $A = \begin{bmatrix} 3 & 1 \\ 0 & 3 \end{bmatrix}$ is not diagonalizable because the rank of $A - 3I$ is _____. Change one entry to make A diagonalizable. Which entries could you change?

Questions 15–19 are about powers of matrices.

15 $A^k = X\Lambda^k X^{-1}$ approaches the zero matrix as $k \to \infty$ if and only if every λ has absolute value less than _____. Which of these matrices has $A^k \to 0$?

$$A_1 = \begin{bmatrix} .6 & .9 \\ .4 & .1 \end{bmatrix} \quad \text{and} \quad A_2 = \begin{bmatrix} .6 & .9 \\ .1 & .6 \end{bmatrix}.$$

16 (Recommended) Find Λ and X to diagonalize A_1 in Problem 15. What is the limit of Λ^k as $k \to \infty$? What is the limit of $X\Lambda^k X^{-1}$? In the columns of this limiting matrix you see the _____.

17 Find Λ and X to diagonalize A_2 in Problem 15. What is $(A_2)^{10}\boldsymbol{u}_0$ for these $\boldsymbol{u}_0$?

$$\boldsymbol{u}_0 = \begin{bmatrix} 3 \\ 1 \end{bmatrix} \quad \text{and} \quad \boldsymbol{u}_0 = \begin{bmatrix} 3 \\ -1 \end{bmatrix} \quad \text{and} \quad \boldsymbol{u}_0 = \begin{bmatrix} 6 \\ 0 \end{bmatrix}.$$

18 Diagonalize A and compute $X\Lambda^k X^{-1}$ to prove this formula for A^k:

$$A = \begin{bmatrix} 2 & -1 \\ -1 & 2 \end{bmatrix} \quad \text{has} \quad A^k = \frac{1}{2}\begin{bmatrix} 1+3^k & 1-3^k \\ 1-3^k & 1+3^k \end{bmatrix}.$$

19 Diagonalize B and compute $X\Lambda^k X^{-1}$ to prove this formula for B^k:

$$B = \begin{bmatrix} 5 & 1 \\ 0 & 4 \end{bmatrix} \quad \text{has} \quad B^k = \begin{bmatrix} 5^k & 5^k - 4^k \\ 0 & 4^k \end{bmatrix}.$$

20 Suppose $A = X\Lambda X^{-1}$. Take determinants to prove $\det A = \det \Lambda = \lambda_1\lambda_2 \cdots \lambda_n$. This quick proof only works when A can be _____.

21 Show that trace XY = trace YX, by adding the diagonal entries of XY and YX:

$$X = \begin{bmatrix} a & b \\ c & d \end{bmatrix} \quad \text{and} \quad Y = \begin{bmatrix} q & r \\ s & t \end{bmatrix}.$$

Now choose Y to be ΛX^{-1}. Then $X\Lambda X^{-1}$ has the same trace as $\Lambda X^{-1}X = \Lambda$. This proves that *the trace of A equals the trace of Λ = the sum of the eigenvalues.* $\boldsymbol{AB - BA = I}$ **is impossible** since the left side has trace = _____.

22 If $A = X\Lambda X^{-1}$, diagonalize the block matrix $B = \begin{bmatrix} A & 0 \\ 0 & 2A \end{bmatrix}$. Find its eigenvalue and eigenvector (block) matrices.

23 Consider all 4 by 4 matrices A that are diagonalized by the same fixed eigenvector matrix X. Show that the A's form a subspace (cA and $A_1 + A_2$ have this same X). What is this subspace when $X = I$? What is its dimension?

24 Suppose $A^2 = A$. On the left side A multiplies each column of A. Which of our four subspaces contains eigenvectors with $\lambda = 1$? Which subspace contains eigenvectors with $\lambda = 0$? From the dimensions of those subspaces, A has a full set of independent eigenvectors. So a matrix with $A^2 = A$ can be diagonalized.

25 (Recommended) Suppose $A\boldsymbol{x} = \lambda\boldsymbol{x}$. If $\lambda = 0$ then $\boldsymbol{x}$ is in the nullspace. If $\lambda \neq 0$ then $\boldsymbol{x}$ is in the column space. Those spaces have dimensions $(n - r) + r = n$. So why doesn't every square matrix have n linearly independent eigenvectors?

26 The eigenvalues of A are 1 and 9. The eigenvalues of B are -1 and 9. Find a matrix square root $\sqrt{A} = X\sqrt{\Lambda}\,X^{-1}$. Why is there no real matrix $\sqrt{B}$?

$$A = \begin{bmatrix} 5 & 4 \\ 4 & 5 \end{bmatrix} \quad \text{and} \quad B = \begin{bmatrix} 4 & 5 \\ 5 & 4 \end{bmatrix}.$$

27 If A and B have the same λ's with the same independent eigenvectors, their factorizations into _____ are the same. So $\boldsymbol{A = B}$.

28 Suppose the same X diagonalizes both A and B. They have the *same eigenvectors* in $A = X\Lambda_1 X^{-1}$ and $B = X\Lambda_2 X^{-1}$. Prove that $AB = BA$.

29 (a) If $A = \begin{bmatrix} a & b \\ 0 & d \end{bmatrix}$ then the determinant of $A - \lambda I$ is $(\lambda - a)(\lambda - d)$. Check the "Cayley-Hamilton Theorem" that $(A - aI)(A - dI) =$ *zero matrix.*

(b) Test the Cayley-Hamilton Theorem on Fibonacci's $A = \begin{bmatrix} 1 & 1 \\ 1 & 0 \end{bmatrix}$. The theorem predicts that $A^2 - A - I = 0$, since the polynomial $\det(A - \lambda I)$ is $\lambda^2 - \lambda - 1$.

30 Substitute $A = X\Lambda X^{-1}$ into the product $(A - \lambda_1 I)(A - \lambda_2 I) \cdots (A - \lambda_n I)$ and explain why this produces the zero matrix. We are substituting the matrix A for the number λ in the polynomial $p(\lambda) = \det(A - \lambda I)$. The ***Cayley-Hamilton Theorem*** says that this product is always $p(A) =$ *zero matrix*, even if A is not diagonalizable.

31 If $A = \begin{bmatrix} 1 & 0 \\ 0 & 2 \end{bmatrix}$ and $AB = BA$, show that $B = \begin{bmatrix} \mathrm{a} & \mathrm{b} \\ \mathrm{c} & \mathrm{d} \end{bmatrix}$ is also a diagonal matrix. B has the same eigen_____ as A but different eigen_____ . These diagonal matrices B form a two-dimensional subspace of matrix space. $AB - BA = 0$ gives four equations for the unknowns **a, b, c, d**—find the rank of the 4 by 4 matrix.

32 The powers A^k blow up if any $|\lambda_i| > 1$ and D^k approaches zero if all $|\lambda_i| < 1$. Peter Lax gives these striking examples in his book *Linear Algebra*. Show $B^4 = I$ and $C^3 = -I$.

$$A = \begin{bmatrix} 3 & 2 \\ 1 & 4 \end{bmatrix} \qquad B = \begin{bmatrix} 3 & 2 \\ -5 & -3 \end{bmatrix} \quad C = \begin{bmatrix} 5 & 7 \\ -3 & -4 \end{bmatrix} \quad D = \begin{bmatrix} 5 & 6.9 \\ -3 & -4 \end{bmatrix}$$

$$\|\boldsymbol{A}^{1024}\| > \mathbf{10}^{700} \quad \boldsymbol{B}^{1024} = \boldsymbol{I} \qquad \boldsymbol{C}^{1024} = -\boldsymbol{C} \qquad \|\boldsymbol{D}^{1024}\| < \mathbf{10}^{-78}$$

Challenge Problems

33 The nth power of rotation through θ is rotation through $n\theta$:

$$A^n = \begin{bmatrix} \cos\theta & -\sin\theta \\ \sin\theta & \cos\theta \end{bmatrix}^n = \begin{bmatrix} \cos n\theta & -\sin n\theta \\ \sin n\theta & \cos n\theta \end{bmatrix}.$$

Prove that neat formula by diagonalizing $A = X\Lambda X^{-1}$. The eigenvectors (columns of X) are $(1, i)$ and $(i, 1)$. You need to know Euler's formula $e^{i\theta} = \cos\theta + i\sin\theta$.

34 The transpose of $A = X\Lambda X^{-1}$ is $A^{\mathrm{T}} = (X^{-1})^{\mathrm{T}}\Lambda X^{\mathrm{T}}$. The eigenvectors in $A^{\mathrm{T}}\boldsymbol{y} = \lambda\boldsymbol{y}$ are the columns of that matrix $(X^{-1})^{\mathrm{T}}$. They are often called ***left eigenvectors of A***, because $\boldsymbol{y}^{\mathrm{T}}A = \lambda\boldsymbol{y}^{\mathrm{T}}$. How do you multiply matrices to find this formula for A?

Sum of rank-1 matrices $A = X\Lambda X^{-1} = \lambda_1\boldsymbol{x}_1\boldsymbol{y}_1^{\mathrm{T}} + \cdots + \lambda_n\boldsymbol{x}_n\boldsymbol{y}_n^{\mathrm{T}}$.

35 The inverse of $A = \mathbf{eye}(n) + \mathbf{ones}(n)$ is $A^{-1} = \mathbf{eye}(n) + C * \mathbf{ones}(n)$. Multiply AA^{-1} to find that number C (depending on n).

36 Suppose A_1 and A_2 are n by n invertible matrices. What matrix B shows that $A_2A_1 = B(A_1A_2)B^{-1}$? Then A_2A_1 is similar to A_1A_2: *same eigenvalues.*

37 **When is a matrix $\boldsymbol{A}$ similar to its eigenvalue matrix $\boldsymbol{\Lambda}$?**

A and Λ always have the same eigenvalues. But similarity requires a matrix B with $A = B\Lambda B^{-1}$. Then B is the _____ matrix and A must have n independent _____.

38 (Pavel Grinfeld) Without writing down any calculations, can you find the eigenvalues of this matrix ? Can you find the power A^{2023} ?

$$A = \begin{bmatrix} 110 & 55 & -164 \\ 42 & 21 & -62 \\ 88 & 44 & -131 \end{bmatrix}.$$

6.3 Symmetric Positive Definite Matrices

1 A symmetric matrix $\boldsymbol{S}$ has $\boldsymbol{n}$ **real eigenvalues** $\boldsymbol{\lambda_i}$ and $\boldsymbol{n}$ **orthonormal eigenvectors** $\boldsymbol{q_i}$.

2 S is diagonalized by an orthogonal eigenvector matrix Q $\boxed{S = Q\Lambda Q^{-1} = Q\Lambda Q^{\mathrm{T}}}$

3A Positive definite $\boldsymbol{S}$: all $\boldsymbol{\lambda > 0}$ and all **pivots** $\boldsymbol{> 0}$ and all upper left **determinants** $\boldsymbol{> 0}$.

3B The **energy test** is $\boldsymbol{x}^{\mathrm{T}}S\boldsymbol{x}>0$ for all $\boldsymbol{x}\neq\boldsymbol{0}$. Then $\boldsymbol{S}=\boldsymbol{A}^{\mathrm{T}}\boldsymbol{A}$ with independent columns in A.

4 **Positive semidefinite** allows $\lambda = 0$, pivot $= 0$, determinant $= 0$, energy $\boldsymbol{x}^{\mathrm{T}}S\boldsymbol{x} = 0$, any A.

Symmetric matrices $\boldsymbol{S} = \boldsymbol{S}^{\mathrm{T}}$ deserve all the attention they get. Looking at their eigenvalues and eigenvectors, you see why they are special:

1 **All $\boldsymbol{n}$ eigenvalues $\boldsymbol{\lambda}$ of a symmetric matrix $\boldsymbol{S}$ are real numbers.**

2 **The $\boldsymbol{n}$ eigenvectors $\boldsymbol{q}$ can be chosen orthogonal (perpendicular to each other).**

The identity matrix $S = I$ is an extreme case. All its eigenvalues are $\lambda = 1$. Every nonzero vector $\boldsymbol{x}$ is an eigenvector: $I\boldsymbol{x} = 1\boldsymbol{x}$. This shows why we wrote "can be chosen" in Property 2 above. Repeated eigenvalues like $\lambda_1 = \lambda_2 = 1$ give a choice of eigenvectors. We can choose them to be orthogonal. We can rescale them to be **unit vectors** (length 1).

Then those eigenvectors $\boldsymbol{q}_1, \ldots, \boldsymbol{q}_n$ are not just orthogonal, they are **orthonormal. The eigenvector matrix for $\boldsymbol{SQ = Q\Lambda}$ has $\boldsymbol{Q^{\mathrm{T}}Q = I}$.**

$$\boldsymbol{q_i}^{\mathrm{T}}\boldsymbol{q_j} = \begin{cases} \mathbf{0} & i \neq j \\ \mathbf{1} & i = j \end{cases} \quad \text{leads to} \quad \begin{bmatrix} — \boldsymbol{q}_1^{\mathrm{T}} — \\ \vdots \\ — \boldsymbol{q}_n^{\mathrm{T}} — \end{bmatrix} \begin{bmatrix} | & & | \\ \boldsymbol{q}_1 & \cdots & \boldsymbol{q}_n \\ | & & | \end{bmatrix} = \begin{bmatrix} \mathbf{1} & 0 & \cdot & 0 \\ 0 & \mathbf{1} & 0 & \cdot \\ \cdot & 0 & \mathbf{1} & 0 \\ 0 & \cdot & 0 & \mathbf{1} \end{bmatrix}$$

We write Q instead of X for the eigenvector matrix of S, to emphasize that these eigenvectors are orthonormal: $\boldsymbol{Q^{\mathrm{T}}Q = I}$ and $\boldsymbol{Q^{\mathrm{T}} = Q^{-1}}$. This eigenvector matrix is an **orthogonal matrix**. The usual $A = X\Lambda X^{-1}$ becomes $S = Q\Lambda Q^{\mathrm{T}}$:

Spectral Theorem	**Every real symmetric matrix $\boldsymbol{S}$ has the form $\boldsymbol{S = Q\Lambda Q^{\mathrm{T}}}$.**

Every matrix of that form is symmetric: Transpose $Q\Lambda Q^{\mathrm{T}}$ to get $Q^{\mathrm{TT}}\Lambda^{\mathrm{T}}Q^{\mathrm{T}} = Q\Lambda Q^{\mathrm{T}}$.

Quick Proofs: Orthogonal Eigenvectors and Real Eigenvalues

Suppose first that $S\boldsymbol{x} = \lambda\boldsymbol{x}$ and $S\boldsymbol{y} = 0\boldsymbol{y}$. The symmetric matrix S has a nonzero eigenvalue λ and a zero eigenvalue. Then $\boldsymbol{y}$ is in the nullspace of S and $\boldsymbol{x}$ is in the column space of S ($\boldsymbol{x} = S\boldsymbol{x}/\lambda$ is a combination of the columns of S). *But S is symmetric: column space = row space!* Since the row space and nullspace are always orthogonal, we have proved that **the eigenvector $\boldsymbol{x}$ is orthogonal to the eigenvector $\boldsymbol{y}$.**

When that second eigenvalue is not zero, we have $S\boldsymbol{y} = \alpha\boldsymbol{y}$. In this case we look at the matrix $S - \alpha I$. Then $(S - \alpha I)\boldsymbol{y} = 0\boldsymbol{y}$ and $(S - \alpha I)\boldsymbol{x} = (\lambda - \alpha)\boldsymbol{x}$ with $\lambda - \alpha \neq 0$. Now $\boldsymbol{y}$ is in the nullspace and $\boldsymbol{x}$ is in the column space (= row space !) of $S - \alpha I$. So $\boldsymbol{y}^{\mathrm{T}}\boldsymbol{x} = 0$: *Orthogonal eigenvectors of S whenever the eigenvalues are different.*

Note on complex numbers That proof of orthogonality assumed real eigenvalues and eigenvectors of S. To prove this, suppose they could involve complex numbers. Multiply $S\boldsymbol{x} = \lambda\boldsymbol{x}$ by the complex conjugate vector $\overline{\boldsymbol{x}}^{\mathrm{T}}$ (every i changes to $-i$). That gives $\overline{\boldsymbol{x}}^{\mathrm{T}}S\boldsymbol{x} = \lambda\overline{\boldsymbol{x}}^{\mathrm{T}}\boldsymbol{x}$. When we show that $\overline{\boldsymbol{x}}^{\mathrm{T}}\boldsymbol{x}$ and $\overline{\boldsymbol{x}}^{\mathrm{T}}S\boldsymbol{x}$ are real, we see that **$\boldsymbol{\lambda}$ is real.**

I would like to leave complex numbers and complex matrices for the next section 6.4. The rules and the matrices are important. But positive definite matrices are so beautiful, and they connect to so many ideas in linear algebra, that they deserve to come first.

Positive Definite Matrices

We are working with real symmetric matrices $S = S^{\mathrm{T}}$. All their eigenvalues are real. Some of those symmetric matrices (*not all*) have an additional powerful property. Here is that important property, which puts S at the center of applied mathematics.

Test 1 **A positive definite matrix S has all positive eigenvalues**

We would like to check for positive eigenvalues without computing those numbers λ. You will see four more tests for positive definite matrices, after these five examples.

1 $S = \begin{bmatrix} 2 & 0 \\ 0 & 6 \end{bmatrix}$ is positive definite. Its eigenvalues 2 and 6 are both positive

2 $S = Q\begin{bmatrix} 2 & 0 \\ 0 & 6 \end{bmatrix}Q^{\mathrm{T}}$ is positive definite if $Q^{\mathrm{T}} = Q^{-1}$: same $\lambda = 2$ and 6

3 $S = C\begin{bmatrix} 2 & 0 \\ 0 & 6 \end{bmatrix}C^{\mathrm{T}}$ is positive definite if C is invertible (energy test)

4 $S = \begin{bmatrix} a & b \\ b & c \end{bmatrix}$ is positive definite when $\boldsymbol{a > 0}$ and $\boldsymbol{ac > b^2}$ (det test)

5 $S = \begin{bmatrix} 2 & 0 \\ 0 & 0 \end{bmatrix}$ is only **positive semidefinite**. It has all $\boldsymbol{\lambda \geq 0}$ but not $\lambda > 0$

Try Test 1 on these examples: **No, No, Yes**. The other four tests may give faster answers.

$$S = \begin{bmatrix} 1 & 2 \\ 2 & 1 \end{bmatrix} \qquad S = \boldsymbol{v}\boldsymbol{v}^{\mathrm{T}}\text{(rank 1)} \qquad S = \begin{bmatrix} 2 & 1 & 0 \\ 1 & 2 & 1 \\ 0 & 1 & 2 \end{bmatrix}$$

$$\lambda = 3 \text{ and } -1 \qquad \lambda = ||\boldsymbol{v}||^2 \text{ and } 0 \qquad \lambda = 2, 2-\sqrt{2}, 2+\sqrt{2}$$

The Energy-based Definition : Test 2

May I bring forward the most important idea about positive definite matrices ? This new approach doesn't directly involve eigenvalues, but it turns out to be a perfect test for $\lambda > 0$. This is a good definition of positive definite matrices : **Test 2 is the energy test.**

$$\boxed{\textbf{S is positive definite if the energy } \boldsymbol{x}^{\mathrm{T}} S\boldsymbol{x} \textbf{ is positive for all vectors } \boldsymbol{x} \neq \mathbf{0}} \quad (1)$$

Of course $S = I$ is positive definite : All $\lambda_i = 1$. The energy $\boldsymbol{x}^{\mathrm{T}} I\boldsymbol{x} = \boldsymbol{x}^{\mathrm{T}}\boldsymbol{x}$ is positive if $\boldsymbol{x} \neq 0$. Let me show you the energy in a 2 by 2 matrix. It depends on $\boldsymbol{x} = (x_1, x_2)$.

$$\textbf{Energy} \quad \boxed{\boldsymbol{x}^{\mathrm{T}} S\boldsymbol{x} = \begin{bmatrix} x_1 & x_2 \end{bmatrix} \begin{bmatrix} 2 & 4 \\ 4 & 9 \end{bmatrix} \begin{bmatrix} x_1 \\ x_2 \end{bmatrix} = \mathbf{2}\,\boldsymbol{x}_1^2 + \mathbf{8}\,\boldsymbol{x}_1\boldsymbol{x}_2 + \mathbf{9}\,\boldsymbol{x}_2^2}$$

Is this positive for every $\boldsymbol{x} = (x_1, x_2)$ except $(0, 0)$? *Yes, it is a sum of two squares* :

$$\boldsymbol{x}^{\mathrm{T}} S\boldsymbol{x} = 2x_1^2 + 8x_1x_2 + 9x_2^2 = 2\,(x_1 + 2x_2)^2 + x_2^2 = \text{ positive energy.}$$

We must connect positive energy $\boldsymbol{x}^{\mathrm{T}} S\boldsymbol{x} > 0$ to positive eigenvalues $\lambda > 0$:

$$\textbf{If } S\boldsymbol{x} = \boldsymbol{\lambda}\boldsymbol{x} \textbf{ then } \boldsymbol{x}^{\mathrm{T}} S\boldsymbol{x} = \boldsymbol{\lambda}\boldsymbol{x}^{\mathrm{T}}\boldsymbol{x}.\ \textbf{So } \boldsymbol{\lambda} > \mathbf{0} \textbf{ leads to energy } \boldsymbol{x}^{\mathrm{T}} S\boldsymbol{x} > \mathbf{0}.$$

That line tested $\boldsymbol{x}^{\mathrm{T}} S\boldsymbol{x}$ for each separate eigenvector $\boldsymbol{x}$. But more is true. If every eigenvector has positive energy, **then all nonzero vectors $\boldsymbol{x}$ have positive energy** :

If $\boldsymbol{x}^{\mathrm{T}} S\boldsymbol{x} > 0$ for the eigenvectors of S, then $\boldsymbol{x}^{\mathrm{T}} S\boldsymbol{x} > 0$ for every nonzero vector $\boldsymbol{x}$.

Here is the reason. Every $\boldsymbol{x}$ is a combination $c_1\boldsymbol{x}_1 + \cdots + c_n\boldsymbol{x}_n$ of the eigenvectors. Those eigenvectors can be chosen *orthogonal* because S is symmetric. We will now show : $\boldsymbol{x}^{\mathrm{T}} S\boldsymbol{x}$ is a positive combination of the energies $\lambda_k \boldsymbol{x}_k^{\mathrm{T}}\boldsymbol{x}_k > 0$ in the separate eigenvectors.

$$\begin{aligned} \boldsymbol{x}^{\mathrm{T}} S\boldsymbol{x} &= (c_1\boldsymbol{x}_1^{\mathrm{T}} + \cdots + c_n\boldsymbol{x}_n^{\mathrm{T}})\, S\,(c_1\boldsymbol{x}_1 + \cdots + c_n\boldsymbol{x}_n) \\ &= (c_1\boldsymbol{x}_1^{\mathrm{T}} + \cdots + c_n\boldsymbol{x}_n^{\mathrm{T}})\,(c_1\lambda_1\boldsymbol{x}_1 + \cdots + c_n\lambda_n\boldsymbol{x}_n) \\ &= \boldsymbol{c_1^2\lambda_1 x_1^{\mathrm{T}} x_1} + \cdots + \boldsymbol{c_n^2\lambda_n x_n^{\mathrm{T}} x_n} > \mathbf{0} \textbf{ if every } \boldsymbol{\lambda_i} > \mathbf{0}. \end{aligned}$$

From line 2 to line 3 we used the orthogonality of the eigenvectors of S : $\boldsymbol{x}_i^{\mathrm{T}}\boldsymbol{x}_j = 0$. Here is a typical use for the energy test, without knowing any eigenvalues or eigenvectors.

$$\boxed{\textbf{If } \boldsymbol{S_1} \textbf{ and } \boldsymbol{S_2} \textbf{ are symmetric positive definite, so is } \boldsymbol{S_1 + S_2}}$$

$$\textit{Proof by adding energies} : \quad \boldsymbol{x}^{\mathrm{T}}(S_1 + S_2)\,\boldsymbol{x} = \boldsymbol{x}^{\mathrm{T}} S_1\,\boldsymbol{x} + \boldsymbol{x}^{\mathrm{T}} S_2\,\boldsymbol{x} > \mathbf{0} + \mathbf{0}$$

The eigenvalues and eigenvectors of $S_1 + S_2$ are not easy to find. The energies just add.

Three More Equivalent Tests

So far we have tests **1** and **2**: positive eigenvalues and positive energy. That energy test quickly produces three more useful tests (and probably others, but we stop with three):

Test 3	$\boldsymbol{S = A^{\mathrm{T}}A}$ **for some matrix** $\boldsymbol{A}$ **with independent columns**
Test 4	**All the leading determinants** $\boldsymbol{D_1, D_2, \ldots, D_n}$ **of** $\boldsymbol{S}$ **are positive**
Test 5	**All the pivots of** $\boldsymbol{S}$ **(coming from elimination) are positive**

Test 3 applies to $S = A^{\mathrm{T}}A$. That matrix is symmetric: $(A^{\mathrm{T}}A)^{\mathrm{T}} = A^{\mathrm{T}}A^{\mathrm{TT}} = A^{\mathrm{T}}A$. Why must columns of A be independent in this test? Connect $A^{\mathrm{T}}A$ to the energy test:

$$\boldsymbol{S = A^{\mathrm{T}}A} \qquad \textbf{Energy} = \boldsymbol{x}^{\mathrm{T}}S\boldsymbol{x} = \boldsymbol{x}^{\mathrm{T}}A^{\mathrm{T}}A\boldsymbol{x} = (A\boldsymbol{x})^{\mathrm{T}}(A\boldsymbol{x}) = ||A\boldsymbol{x}||^2. \tag{2}$$

The energy is the **length squared** of the vector $A\boldsymbol{x}$. This energy is positive provided $A\boldsymbol{x}$ is not the zero vector. To assure $A\boldsymbol{x} \neq \boldsymbol{0}$ when $\boldsymbol{x} \neq \boldsymbol{0}$, the columns of A must be independent.

In this 2 by 3 example, A has *dependent columns*. Now $S = A^{\mathrm{T}}A$ has rank 2, not 3.

$$\boldsymbol{S = A^{\mathrm{T}}A} = \begin{bmatrix} 1 & 1 \\ 1 & 2 \\ 1 & 3 \end{bmatrix} \begin{bmatrix} 1 & 1 & 1 \\ 1 & 2 & 3 \end{bmatrix} = \begin{bmatrix} 2 & 3 & 4 \\ 3 & 5 & 7 \\ 4 & 7 & 10 \end{bmatrix}$$

S is **not** positive definite.
It is positive **semi**definite.
$\boldsymbol{x}^{\mathrm{T}}S\boldsymbol{x} = ||A\boldsymbol{x}||^2 \geq \boldsymbol{0}$

This A has column $1 +$ column $3 = 2$ (column 2). Then $\boldsymbol{x} = (1, -2, 1)$ has zero energy. It is an eigenvector of $A^{\mathrm{T}}A$ with $\lambda = 0$. This $S = A^{\mathrm{T}}A$ is only positive semidefinite, because the energy $\boldsymbol{x}^{\mathrm{T}}S\boldsymbol{x}$ and the eigenvalues of S touch zero.

Equation (2) says that $\boldsymbol{S = A^{\mathrm{T}}A}$ ***is at least semidefinite***: $\boldsymbol{x}^{\mathrm{T}}S\boldsymbol{x} = ||A\boldsymbol{x}||^2$ is never negative. ***Semidefinite allows energy / eigenvalues / determinants / pivots of S to be zero.***

Determinant Test and Pivot Test

The determinant test is the fastest for a small matrix. I will mark the four "leading determinants" D_1, D_2, D_3, D_4 in this 4 by 4 symmetric second difference matrix.

Test 4 $\quad S = \begin{bmatrix} 2 & -1 & & \\ -1 & 2 & -1 & \\ & -1 & 2 & -1 \\ & & -1 & 2 \end{bmatrix}$ has

1st determinant $D_1 = \boldsymbol{2} > 0$
2nd determinant $D_2 = \boldsymbol{3} > 0$
3rd determinant $D_3 = \boldsymbol{4} > 0$
4th determinant $D_4 = \boldsymbol{5} > 0$

The determinant test is here passed! The energy $\boldsymbol{x}^{\mathrm{T}}S\boldsymbol{x}$ must be positive. Eigenvalues too.

Leading determinants are closely related to pivots (the numbers on the diagonal after elimination). Here the first pivot is **2**. The second pivot $\boldsymbol{\frac{3}{2}}$ appears when $\frac{1}{2}$(row 1) is added to row 2. The third pivot $\boldsymbol{\frac{4}{3}}$ appears when $\frac{2}{3}$(new row 2) is added to row 3. Those fractions $\boldsymbol{\frac{2}{1}, \frac{3}{2}, \frac{4}{3}}$ are ratios of determinants! The last pivot is $\boldsymbol{\frac{5}{4}}$. Always a ratio.

The $\boldsymbol{k}$**th pivot equals the ratio** $\boldsymbol{\dfrac{D_k}{D_{k-1}}}$ **of the leading determinants (sizes** $\boldsymbol{k}$ **and** $\boldsymbol{k-1}$**)**

So Test 5 is passed (next page) exactly when Test 4 is passed. **The pivots are positive.**

Test 4 The leading determinants are all positive	**Test 5 The pivots are positive**

I can quickly connect these tests **4** and **5** to the third test $S = A^{\mathrm{T}}A$. In fact elimination on S produces an important choice of A. Remember that *elimination = triangular factorization* ($\boldsymbol{S} = \boldsymbol{LU}$). Up to now L has had 1's on its diagonal and U contained the pivots. But with symmetric matrices we can balance S as $\boldsymbol{LDL^{\mathrm{T}}}$:

$$\begin{bmatrix} 2 & -1 & 0 \\ -1 & 2 & -1 \\ 0 & -1 & 2 \end{bmatrix} = \begin{bmatrix} 1 & & \\ -\frac{1}{2} & 1 & \\ 0 & -\frac{2}{3} & 1 \end{bmatrix} \begin{bmatrix} \mathbf{2} & -1 & 0 \\ & \mathbf{\frac{3}{2}} & -1 \\ & & \mathbf{\frac{4}{3}} \end{bmatrix} \quad \textbf{This is } \boldsymbol{S = LU} \quad (3)$$

Put pivots into D for Test 5
$$\begin{bmatrix} 1 & & \\ -\frac{1}{2} & 1 & \\ 0 & -\frac{2}{3} & 1 \end{bmatrix} \begin{bmatrix} \mathbf{2} & & \\ & \mathbf{\frac{3}{2}} & \\ & & \mathbf{\frac{4}{3}} \end{bmatrix} \begin{bmatrix} 1 & -\frac{1}{2} & 0 \\ & 1 & -\frac{2}{3} \\ & & 1 \end{bmatrix} \quad \boldsymbol{S = LDL^{\mathrm{T}}} \quad (4)$$

Share those pivots between $\boldsymbol{A^{\mathrm{T}}}$ and $\boldsymbol{A}$ for Test 3
$$\begin{bmatrix} \sqrt{2} & & \\ -\sqrt{\frac{1}{2}} & \sqrt{\frac{3}{2}} & \\ 0 & -\sqrt{\frac{2}{3}} & \sqrt{\frac{4}{3}} \end{bmatrix} \begin{bmatrix} \sqrt{2} & -\sqrt{\frac{1}{2}} & 0 \\ & \sqrt{\frac{3}{2}} & -\sqrt{\frac{2}{3}} \\ & & \sqrt{\frac{4}{3}} \end{bmatrix} = \boldsymbol{A^{\mathrm{T}}A} \; (5)$$

I am sorry about those square roots—but the pattern $S = A^{\mathrm{T}}A$ is a success: $\boldsymbol{A = \sqrt{D}L^{\mathrm{T}}}$.

Elimination factors every positive definite $\boldsymbol{S}$ into $\boldsymbol{A^{\mathrm{T}}A}$ ($\boldsymbol{A}$ is upper triangular)

(5) is the Cholesky factorization $S = A^{\mathrm{T}}A$ with $\sqrt{\text{pivots}}$ on the main diagonal of A.

To apply the $S = A^{\mathrm{T}}A$ test when S is positive definite, we must find at least one possible A. There are many choices for A, including **(1) symmetric** and **(2) triangular**.

1 If $S = Q\Lambda Q^{\mathrm{T}}$, take square roots of those eigenvalues. $A = Q\sqrt{\Lambda}Q^{\mathrm{T}} = A^{\mathrm{T}}$ has $S = A^{\mathrm{T}}A$.

2 If $S = LU = LDL^{\mathrm{T}}$ with positive pivots in D, then $S = (L\sqrt{D})(\sqrt{D}L^{\mathrm{T}}) = A^{\mathrm{T}}A$.

> **Summary** The 5 tests for positive definiteness of $S = S^{\mathrm{T}}$ involve 5 different parts of linear algebra—**pivots, determinants, eigenvalues, $\boldsymbol{S = A^{\mathrm{T}}A}$, and energy**. Each test gives a complete answer by itself: positive definite or semidefinite or neither.
>
> **Positive energy $\boldsymbol{x^{\mathrm{T}}Sx > 0}$ is the best definition. It connects them all.**

When S is a symmetric positive definite 2 by 2 matrix, here are four of the tests:

$$S = \begin{bmatrix} a & b \\ b & c \end{bmatrix} \quad \begin{array}{ll} \textbf{determinants} & a > 0,\, ac - b^2 > 0 \\ \textbf{eigenvalues} & \lambda_1 > 0,\; \lambda_2 > 0 \end{array} \quad \begin{array}{ll} \textbf{pivots} & a > 0,\, (ac - b^2)/a > 0 \\ \textbf{energy} & ax^2 + 2bxy + cy^2 > 0 \end{array}$$

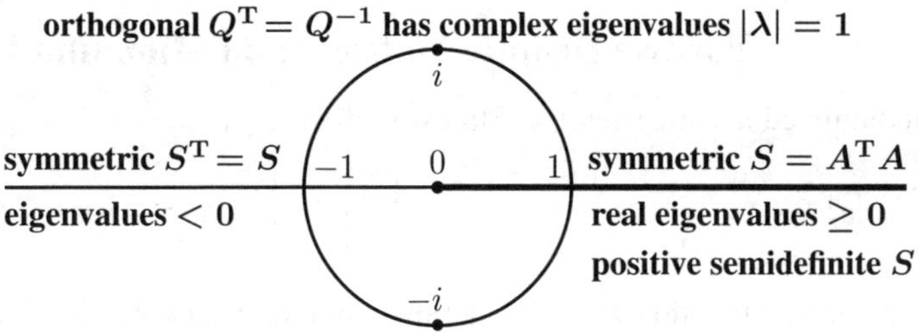

This page analyzes two examples: S is definite, T is semidefinite.

Review

$S = \begin{bmatrix} 9 & 3 \\ 3 & 3 \end{bmatrix}$ **Positive Definite**

Determinants **9** and **18**

Pivots **9** and **2** (invertible matrix)

Energy $9x^2 + 6xy + 3y^2 = \boldsymbol{(3x+y)^2 + 2y^2}$

Trace **12**, Det **18**, $\lambda^2 - 12\lambda + 18 = 0$

Eigenvalues $\lambda_1 = \boldsymbol{6 + 3\sqrt{2}}$ and $\lambda_2 = \boldsymbol{6 - 3\sqrt{2}}$

Eigenvectors $\begin{bmatrix} \boldsymbol{1} \\ \boldsymbol{\sqrt{2} - 1} \end{bmatrix}$ and $\begin{bmatrix} \boldsymbol{-1} \\ \boldsymbol{\sqrt{2} + 1} \end{bmatrix}$

$$\boldsymbol{S = LDL^{\mathrm{T}}} = \begin{bmatrix} 1 & 0 \\ \frac{1}{3} & 1 \end{bmatrix} \begin{bmatrix} \boldsymbol{9} & 0 \\ 0 & \boldsymbol{2} \end{bmatrix} \begin{bmatrix} 1 & \frac{1}{3} \\ 0 & 1 \end{bmatrix}$$

$$= \begin{bmatrix} 3 & 0 \\ 1 & \sqrt{2} \end{bmatrix} \begin{bmatrix} 3 & 1 \\ 0 & \sqrt{2} \end{bmatrix} = \boldsymbol{A^{\mathrm{T}}A}$$

The graph of **energy** $\boldsymbol{E}(x, y)$ is a **bowl**

$E(x, y)$ is a **strictly convex** function

Cross-section $E = 1$ is an **ellipse**

Axes of the ellipse along eigenvectors of S

Energy *E* has its minimum at a point

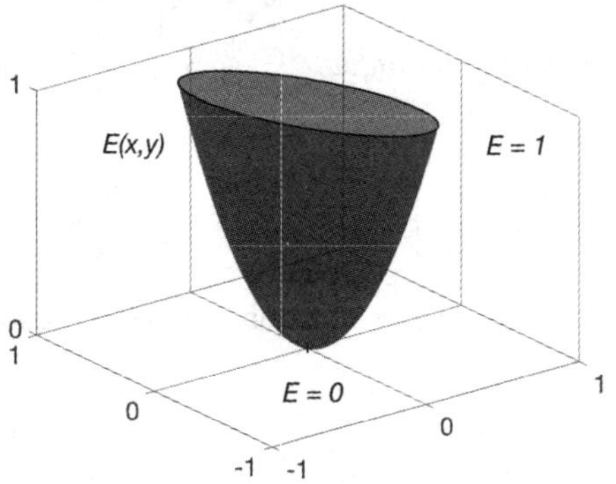

$T = \begin{bmatrix} 9 & 3 \\ 3 & 1 \end{bmatrix}$ **Positive Semidefinite**

Determinants **9** and **0**

Pivots **9** and **0** (no inverse)

Energy $9x^2 + 6xy + y^2 = \boldsymbol{(3x+y)^2} \geq 0$

Trace **10**, Det **0**, $\lambda^2 - 10\lambda = 0$

Eigenvalues $\lambda_1 = \boldsymbol{10}$ and $\lambda_2 = \boldsymbol{0}$

Eigenvectors $\begin{bmatrix} \boldsymbol{3} \\ \boldsymbol{1} \end{bmatrix}$ and $\begin{bmatrix} \boldsymbol{-1} \\ \boldsymbol{3} \end{bmatrix}$

$$\boldsymbol{T = LDL^{\mathrm{T}}} = \begin{bmatrix} 1 & 0 \\ \frac{1}{3} & 1 \end{bmatrix} \begin{bmatrix} \boldsymbol{9} & 0 \\ 0 & \boldsymbol{0} \end{bmatrix} \begin{bmatrix} 1 & \frac{1}{3} \\ 0 & 0 \end{bmatrix}$$

$$= \begin{bmatrix} 3 & 0 \\ 1 & 0 \end{bmatrix} \begin{bmatrix} 3 & 1 \\ 0 & 0 \end{bmatrix} = \boldsymbol{A^{\mathrm{T}}A}$$

The graph of **energy** $\boldsymbol{E}(x, y)$ is a **valley**

$E(x, y)$ is a **convex** function

Cross-section $E = 1$ is a **band** $3x + y = \pm 1$

Axes of the band along eigenvectors of T

Energy E has its minimum along a line

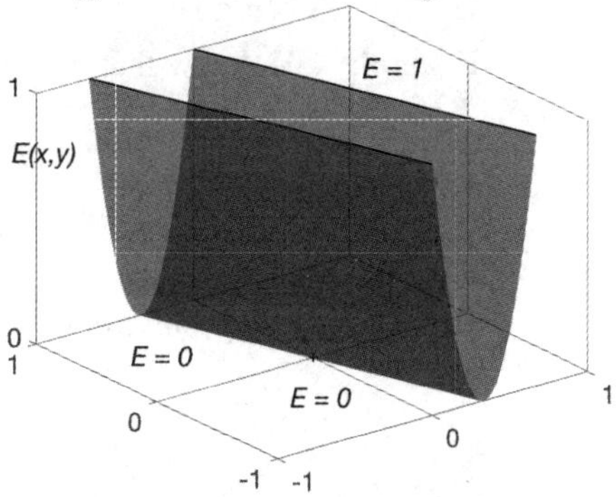

Positive Definite Matrices and Minimum Problems

This page is about functions of a vector $\boldsymbol{x}$. Start with the energy E:

$$\textbf{Energy } \boldsymbol{E} = \boldsymbol{x}^{\mathrm{T}} \boldsymbol{S} \boldsymbol{x} \qquad \begin{bmatrix} x & y \end{bmatrix} \begin{bmatrix} \mathbf{5} & \mathbf{4} \\ \mathbf{4} & \mathbf{5} \end{bmatrix} \begin{bmatrix} x \\ y \end{bmatrix} = \mathbf{5}\,x^2 + \mathbf{8}\,xy + \mathbf{5}\,y^2 > 0$$

The graph of that energy function $E(x, y)$ is **a bowl opening upwards**. The bottom point of the bowl has energy $E = 0$ when $x = y = 0$. This connects minimum problems in calculus with positive definite matrices in linear algebra.

For the best minimization problems, the function $f(\boldsymbol{x})$ is **strictly convex**—*the bowl curves upwards. The matrix of second derivatives (partial derivatives) is positive definite at all points.* We are in 3 or more dimensions, but linear algebra identifies the crucial properties of the second derivative matrix S.

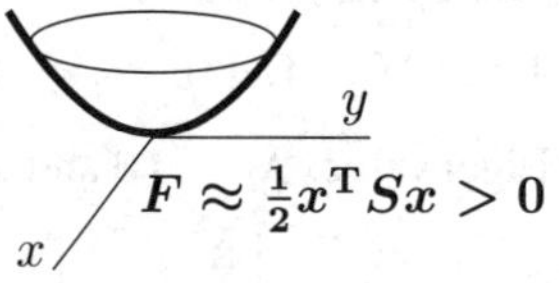

The second derivatives of the energy $\frac{1}{2}\,\boldsymbol{x}^{\mathrm{T}} S \boldsymbol{x}$ are in the matrix S

For an ordinary function $f(x)$ of one variable x, the test for a minimum at x_0 is famous:

$$\begin{matrix}\textbf{Minimum} \\ \textbf{at } \boldsymbol{x_0}\end{matrix} \qquad \begin{matrix}\textbf{First derivative} \\ \textbf{is zero}\end{matrix} \;\; \frac{d\boldsymbol{f}}{d\boldsymbol{x}} = 0 \qquad \begin{matrix}\textbf{Second derivative} \\ \textbf{is positive}\end{matrix} \;\; \frac{d^2\boldsymbol{f}}{d\boldsymbol{x}^2} > 0$$

For $f(x, y)$ with two variables, the second derivatives go into a matrix: positive definite!

$$\begin{matrix}\textbf{Minimum} \\ \textbf{at } \boldsymbol{x_0, y_0}\end{matrix} \;\; \frac{\partial f}{\partial x} = \frac{\partial f}{\partial y} = 0 \text{ and } \boldsymbol{S} = \begin{bmatrix} \partial^2 f/\partial x^2 & \partial^2 f/\partial x \partial y \\ \partial^2 f/\partial x \partial y & \partial^2 f/\partial y^2 \end{bmatrix} \begin{matrix}\textbf{is positive definite} \\ \textbf{at } \boldsymbol{x_0, y_0}\end{matrix}$$

The graph of $z = f(x, y)$ is flat at that point x_0, y_0 because $\partial f/\partial x = \partial f/\partial y = 0$. The graph goes upwards because the second derivative matrix $\boldsymbol{S}$ is positive definite. So we have a minimum point of the function $f(x, y)$. Similarly for $f(x, y, z)$.

Example We know that $E = x^2 + y^2$ has its minimum at $x = y = 0$. What about $\boldsymbol{f = e^{x^2+y^2}}$?

$$\begin{matrix}\partial f/\partial x = 2xe^{x^2+y^2} \\ \partial f/\partial y = 2ye^{x^2+y^2}\end{matrix} \qquad S = \begin{matrix}\text{second} \\ \text{derivatives}\end{matrix} \begin{bmatrix} f_{xx} & f_{xy} \\ f_{yx} & f_{yy} \end{bmatrix} = \begin{bmatrix} 2 + 4x^2 & 4xy \\ 4xy & 2 + 4y^2 \end{bmatrix} e^{x^2+y^2}$$

That matrix S is positive definite for every x and y! Such a function f is **strictly convex**.

Positive Semidefinite Matrices

Often we are at the edge of positive definiteness. The determinant is zero. The smallest eigenvalue is $\boldsymbol{\lambda} = \mathbf{0}$. The energy in its eigenvector is $\boldsymbol{x}^{\mathrm{T}} S \boldsymbol{x} = \boldsymbol{x}^{\mathrm{T}} 0 \boldsymbol{x} = \mathbf{0}$. These matrices on the edge are "*positive semidefinite*". Here are two examples (not invertible):

$$S = \begin{bmatrix} \mathbf{1} & \mathbf{2} \\ \mathbf{2} & \mathbf{4} \end{bmatrix} \text{ and } T = \begin{bmatrix} \mathbf{2} & \mathbf{-1} & \mathbf{-1} \\ \mathbf{-1} & \mathbf{2} & \mathbf{-1} \\ \mathbf{-1} & \mathbf{-1} & \mathbf{2} \end{bmatrix} \quad \begin{array}{l}\text{are positive semidefinite} \\ \text{but not positive definite}\end{array}$$

S has eigenvalues 5 and 0. Its trace is $1 + 4 = 5$. Its upper left determinants are 1 and 0. The rank of S is only 1. This matrix S factors into $A^{\mathrm{T}}A$ with *dependent columns* in A:

$$\begin{array}{l}\textbf{Dependent columns in } \boldsymbol{A} \\ \textbf{Positive semidefinite } \boldsymbol{S}\end{array} \qquad \begin{bmatrix} 1 & 2 \\ 2 & 4 \end{bmatrix} = \begin{bmatrix} 1 & 0 \\ 2 & 0 \end{bmatrix} \begin{bmatrix} 1 & 2 \\ 0 & 0 \end{bmatrix} = A^{\mathrm{T}}A.$$

If 4 is increased by any small number, the matrix S will become positive definite.

The cyclic T also has zero determinant. The eigenvector $\boldsymbol{x} = (1, 1, 1)$ has $T\boldsymbol{x} = \mathbf{0}$ and energy $\boldsymbol{x}^{\mathrm{T}} T \boldsymbol{x} = 0$. Vectors $\boldsymbol{x}$ in all other directions do give positive energy.

$$\begin{array}{l}\textbf{Second differences } \boldsymbol{T} \\ \textbf{from first differences } \boldsymbol{A} \\ \textbf{Columns add to } \mathbf{(0, 0, 0)}\end{array} \qquad \begin{bmatrix} 2 & -1 & -1 \\ -1 & 2 & -1 \\ -1 & -1 & 2 \end{bmatrix} = \begin{bmatrix} 1 & -1 & 0 \\ 0 & 1 & -1 \\ -1 & 0 & 1 \end{bmatrix} \begin{bmatrix} 1 & 0 & -1 \\ -1 & 1 & 0 \\ 0 & -1 & 1 \end{bmatrix}.$$

Positive semidefinite matrices have all $\lambda \geq 0$ and all $\boldsymbol{x}^{\mathrm{T}} S \boldsymbol{x} \geq 0$. Those weak inequalities ($\geq$ **instead of** $>$) include positive definite S along with the singular matrices at the edge.

If S is positive semidefinite, so is every matrix $A^{\mathrm{T}}SA$:

$$\textbf{If } \boldsymbol{x}^{\mathrm{T}} \boldsymbol{S} \boldsymbol{x} \geq \mathbf{0} \textbf{ for every vector } \boldsymbol{x}, \textbf{ then } (\boldsymbol{A}\boldsymbol{x})^{\mathrm{T}} \boldsymbol{S} (\boldsymbol{A}\boldsymbol{x}) \geq \mathbf{0} \textbf{ for every } \boldsymbol{x}.$$

We can tighten this proof to show when $A^{\mathrm{T}}SA$ is actually positive definite. But we have to guarantee that $A\boldsymbol{x}$ is not the zero vector—to be sure that $(A\boldsymbol{x})^{\mathrm{T}} S (A\boldsymbol{x})$ is not zero.

Suppose $\boldsymbol{x}^{\mathrm{T}} S \boldsymbol{x} > 0$ and $A\boldsymbol{x} \neq 0$ whenever $\boldsymbol{x}$ is not zero. Then $A^{\mathrm{T}}SA$ is positive definite.

Again we use the energy test. For every $\boldsymbol{x} \neq 0$ we have $A\boldsymbol{x} \neq \mathbf{0}$. The energy in $A\boldsymbol{x}$ is strictly positive: $(A\boldsymbol{x})^{\mathrm{T}} S (A\boldsymbol{x}) > 0$. The matrix $A^{\mathrm{T}}SA$ is called "**congruent**" to S.

$A^{\mathrm{T}}SA$ is important in applied mathematics. We want to be sure it is positive definite (not just semidefinite). Then the equations $A^{\mathrm{T}}SA\,\boldsymbol{x} = \boldsymbol{f}$ in engineering can be solved.

Here is an extension called the **Law of Inertia**.

If $S^{\mathrm{T}} = S$ has P positive eigenvalues and N negative eigenvalues and Z zero eigenvalues, then **the same is true for $\boldsymbol{A}^{\mathrm{T}}\boldsymbol{S}\boldsymbol{A}$**—provided A is invertible.

The Ellipse $ax^2 + 2bxy + cy^2 = 1$

Think of a tilted ellipse $\boldsymbol{x}^{\mathrm{T}} S \boldsymbol{x} = 1$. Its center is $(0,0)$, as in Figure 6.3a. Turn it to line up with the coordinate axes (X and Y axes). That is Figure 6.3b. These two pictures show the geometry behind the eigenvalues in Λ and the eigenvectors of S in Q. **The eigenvector matrix Q lines up with the ellipse.**

The tilted ellipse has $\boldsymbol{x}^{\mathrm{T}} S \boldsymbol{x} = 1$. The lined-up ellipse has $\boldsymbol{X}^{\mathrm{T}} \boldsymbol{\Lambda} \boldsymbol{X} = 1$ with $\boldsymbol{X} = Q^{\mathrm{T}} \boldsymbol{x}$.

Example 1 Find the axes of this tilted ellipse $\mathbf{5}\boldsymbol{x}^{\mathbf{2}} + \mathbf{8}\boldsymbol{xy} + \mathbf{5}\boldsymbol{y}^{\mathbf{2}} = \mathbf{1}$.

Solution Start with the positive definite matrix S that matches this equation:

$$\text{The equation is} \quad \begin{bmatrix} x & y \end{bmatrix} \begin{bmatrix} 5 & 4 \\ 4 & 5 \end{bmatrix} \begin{bmatrix} x \\ y \end{bmatrix} = 1. \qquad \text{The matrix is} \quad S = \begin{bmatrix} \mathbf{5} & \mathbf{4} \\ \mathbf{4} & \mathbf{5} \end{bmatrix}.$$

The eigenvectors are $\left[\begin{smallmatrix} 1 \\ 1 \end{smallmatrix}\right]$ and $\left[\begin{smallmatrix} 1 \\ -1 \end{smallmatrix}\right]$. Divide by $\sqrt{2}$ for unit vectors. Then $S = Q\Lambda Q^{\mathrm{T}}$:

$$\begin{array}{l} \textbf{Eigenvectors in } \boldsymbol{Q} \\ \textbf{Eigenvalues 9 and 1} \end{array} \qquad \begin{bmatrix} 5 & 4 \\ 4 & 5 \end{bmatrix} = \frac{1}{\sqrt{2}} \begin{bmatrix} 1 & 1 \\ 1 & -1 \end{bmatrix} \begin{bmatrix} \mathbf{9} & 0 \\ 0 & \mathbf{1} \end{bmatrix} \frac{1}{\sqrt{2}} \begin{bmatrix} 1 & 1 \\ 1 & -1 \end{bmatrix}.$$

Now multiply by $\begin{bmatrix} x & y \end{bmatrix}$ on the left and $\begin{bmatrix} x \\ y \end{bmatrix}$ on the right to get $\boldsymbol{x}^{\mathrm{T}} S \boldsymbol{x} = (\boldsymbol{x}^{\mathrm{T}} Q) \Lambda (Q^{\mathrm{T}} \boldsymbol{x})$:

$$\boldsymbol{x}^{\mathrm{T}} S \boldsymbol{x} = \textbf{ sum of squares } \quad 5x^2 + 8xy + 5y^2 = 9\left(\frac{x+y}{\sqrt{2}}\right)^2 + 1\left(\frac{x-y}{\sqrt{2}}\right)^2. \qquad (6)$$

The coefficients are the eigenvalues 9 and 1 from Λ. Inside the squares are the eigenvectors $\boldsymbol{q}_1 = (1,1)/\sqrt{2}$ and $\boldsymbol{q}_2 = (1,-1)/\sqrt{2}$.

The axes of the tilted ellipse point along those eigenvectors. This explains why $S = Q\Lambda Q^{\mathrm{T}}$ is called the "principal axis theorem"—it displays the axes. Not only the axis directions (from the eigenvectors) but also the axis lengths (from the eigenvalues).

To see it all, use capital letters for the new coordinates that line up the ellipse:

$$\textbf{Lined up} \qquad \frac{x+y}{\sqrt{2}} = X \quad \text{and} \quad \frac{x-y}{\sqrt{2}} = Y \quad \text{and} \quad \mathbf{9}\boldsymbol{X}^{\mathbf{2}} + \boldsymbol{Y}^{\mathbf{2}} = \mathbf{1}.$$

The largest value of X^2 is $1/9$. The endpoint of the shorter axis has $X = 1/3$ and $Y = 0$. Notice: The *bigger* eigenvalue λ_1 gives the *shorter* axis, of half-length $1/\sqrt{\lambda_1} = 1/3$. The smaller eigenvalue $\lambda_2 = 1$ gives the greater length $1/\sqrt{\lambda_2} = 1$.

In the xy system, the axes are along the eigenvectors of S. In the XY system, the **axes are along the eigenvectors of $\boldsymbol{\Lambda}$**—the coordinate axes. All comes from $S = Q\Lambda Q^{\mathrm{T}}$.

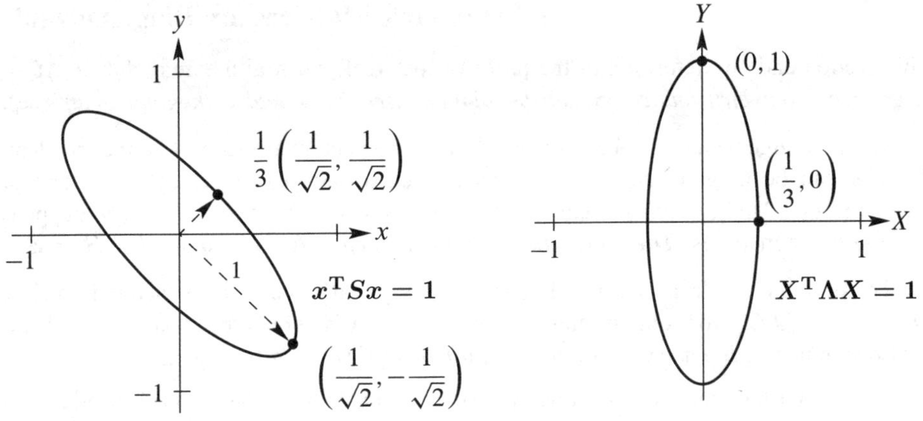

Figure 6.3: The ellipse $\boldsymbol{x}^{\mathrm{T}} S \boldsymbol{x} = 5x^2 + 8xy + 5y^2 = 1$. Lined up it is $9X^2 + Y^2 = 1$.

Optimization and Machine Learning

This book will explain **gradient descent** to minimize $f(\boldsymbol{x})$. Each step to $\boldsymbol{x}_{k+1}$ takes the **steepest direction** at the current point $\boldsymbol{x}_k$. But that steepest direction changes as we descend. This is where calculus meets linear algebra, at the minimum point $\boldsymbol{x}^*$.

Calculus	The partial derivatives of f are all zero at $\boldsymbol{x}^*$: $\dfrac{\partial f}{\partial x_i} = 0$
Linear algebra	The matrix S of second derivatives $\dfrac{\partial^2 f}{\partial x_i \, \partial x_j}$ is positive definite

If S is positive definite (or semidefinite) at all points $\boldsymbol{x} = (x_1, \ldots, x_n)$, then **the function $f(\boldsymbol{x})$ is convex**. If the eigenvalues of S stay above some positive number δ, then **$f(\boldsymbol{x})$ is strictly convex**. These are the best functions to optimize. They have only one minimum, and gradient descent will find it. Please see Chapter 9.

Machine learning produces "loss functions" with hundreds of thousands of variables. They measure the error—which we minimize. But computing all the second derivatives is barely possible. We use first derivatives to tell us a direction to move—the error drops fastest in the steepest direction. Then we take another descent step in a new direction. **This is the central computation in least squares and neural nets and deep learning.**

The key to the great success of deep learning is the special form of the learning function $\boldsymbol{F}(\boldsymbol{x}, \boldsymbol{v})$. That is described in Chapter 10. The "training data" is in the vector $\boldsymbol{v}$. The weights that decide the importance of that training data go into $\boldsymbol{x}$. If the training is well done, then $\boldsymbol{F}$ can identify new data $\boldsymbol{V}$ that it has never seen.

All Symmetric Matrices are Diagonalizable

This section ends by returning to the proof of the (real) spectral theorem $S = Q\Lambda Q^{\mathrm{T}}$. *Every real symmetric matrix $\boldsymbol{S}$ can be diagonalized by a real orthogonal matrix $\boldsymbol{Q}$.*

When no eigenvalues of A are repeated, the eigenvectors are sure to be independent. Then A can be diagonalized. But a repeated eigenvalue can produce a shortage of eigenvectors. This *sometimes* happens for nonsymmetric matrices. It *never* happens for symmetric matrices. ***There are always enough eigenvectors to diagonalize $\boldsymbol{S} = \boldsymbol{S}^{\mathrm{T}}$.***

The proof comes from **Schur's Theorem**: Every real square matrix factors into $A = QTQ^{-1} = QTQ^{\mathrm{T}}$ for some triangular matrix T. If A is symmetric, then $T = Q^{\mathrm{T}}AQ$ is also symmetric. *But a symmetric triangular matrix T is actually diagonal.*

Thus Schur has found the diagonal matrix we want. Schur's theorem is proved on the website **math.mit.edu/linearalgebra**. Here we just note that if A can be diagonalized ($A = X\Lambda X^{-1}$) then we can see that triangular matrix T. Use Gram-Schmidt from Section 4.4 to factor X into QR: R is triangular. Then

$$A = X\Lambda X^{-1} = QR\Lambda R^{-1}Q^{-1} = QTQ^{-1} \quad \text{with triangular } T = R\Lambda R^{-1}.$$

■ WORKED EXAMPLES ■

6.3 A Test these symmetric matrices S and T for positive definiteness:

$$\boldsymbol{S} = \begin{bmatrix} 2 & -1 & 0 \\ -1 & 2 & -1 \\ 0 & -1 & 2 \end{bmatrix} \quad \text{and} \quad \boldsymbol{T} = \begin{bmatrix} 2 & -1 & \boldsymbol{c} \\ -1 & 2 & -1 \\ \boldsymbol{c} & -1 & 2 \end{bmatrix}.$$

Solution The pivots of S are 2 and $\frac{3}{2}$ and $\frac{4}{3}$, all positive. Its upper left determinants are 2 and 3 and 4, all positive. The eigenvalues of S are $2-\sqrt{2}$ and 2 and $2+\sqrt{2}$, all positive. That completes three tests. Any one test is decisive!

Eigenvalues give the symmetric choice $\boldsymbol{A} = \boldsymbol{Q}\sqrt{\boldsymbol{\Lambda}}\boldsymbol{Q}^{\mathrm{T}}$. This succeeds because $\boldsymbol{A}^{\mathrm{T}}\boldsymbol{A} = \boldsymbol{Q}\boldsymbol{\Lambda}\boldsymbol{Q}^{\mathrm{T}} = \boldsymbol{S}$. This also shows that the $-1, 2, -1$ matrix S is positive definite.

Three choices A_1, A_2, A_3 give three different ways to separate S into $A^{\mathrm{T}}A$:

$$\boldsymbol{x}^{\mathrm{T}}S\boldsymbol{x} = 2x_1^2 - 2x_1x_2 + 2x_2^2 - 2x_2x_3 + 2x_3^2 \qquad \textbf{Rewrite with squares}$$

$$||A_1\boldsymbol{x}||^2 = x_1^2 + (x_2 - x_1)^2 + (x_3 - x_2)^2 + x_3^2 \qquad \boldsymbol{S} = \boldsymbol{A}_1^{\mathrm{T}}\boldsymbol{A}_1$$

$$||A_2\boldsymbol{x}||^2 = 2\left(x_1 - \tfrac{1}{2}x_2\right)^2 + \tfrac{3}{2}\left(x_2 - \tfrac{2}{3}x_3\right)^2 + \tfrac{4}{3}\,x_3^2 \qquad \boldsymbol{S} = \boldsymbol{LDL}^{\mathrm{T}} = \boldsymbol{A}_2^{\mathrm{T}}\boldsymbol{A}_2$$

$$||A_3\boldsymbol{x}||^2 = \lambda_1(\boldsymbol{q}_1^{\mathrm{T}}\boldsymbol{x})^2 + \lambda_2(\boldsymbol{q}_2^{\mathrm{T}}\boldsymbol{x})^2 + \lambda_3(\boldsymbol{q}_3^{\mathrm{T}}\boldsymbol{x})^2 \qquad \boldsymbol{S} = \boldsymbol{Q\Lambda Q}^{\mathrm{T}} = \boldsymbol{A}_3^{\mathrm{T}}\boldsymbol{A}_3$$

The determinant of T reveals when that matrix is positive definite:

Test on T $\quad$ $\det \boldsymbol{T} = 4 + 2\boldsymbol{c} - 2\boldsymbol{c}^2 = (\boldsymbol{1} + \boldsymbol{c})\,(\boldsymbol{4} - \boldsymbol{2c})$ must be positive.

At $\boldsymbol{c} = -1$ and $\boldsymbol{c} = 2$ we get $\det T = 0$. **Between $\boldsymbol{c} = -1$ and $\boldsymbol{c} = 2$ this matrix $\boldsymbol{T}$ is positive definite.** The corner entry $\boldsymbol{c} = 0$ in S was safely between -1 and 2.

Problem Set 6.3

1 Suppose $S = S^{\mathrm{T}}$. When is ASB also symmetric with the same eigenvalues as S?

(a) Transpose ASB to see that it stays symmetric when $B =$ ______.

(b) ASB is similar to S (same eigenvalues) when $B =$ ______.

Put (a) and (b) together. The symmetric matrices similar to S look like (__)S(__).

2 For S and T, find the eigenvalues and eigenvectors and the factors for $Q\Lambda Q^{\mathrm{T}}$:

$$S = \begin{bmatrix} 2 & 2 & 2 \\ 2 & 0 & 0 \\ 2 & 0 & 0 \end{bmatrix} \quad \text{and} \quad T = \begin{bmatrix} 1 & 0 & 2 \\ 0 & -1 & -2 \\ 2 & -2 & 0 \end{bmatrix}.$$

3 Find *all* orthogonal matrices that diagonalize $S = \begin{bmatrix} 9 & 12 \\ 12 & 16 \end{bmatrix}$.

4 (a) Find a symmetric matrix $\begin{bmatrix} 1 & b \\ b & 1 \end{bmatrix}$ that has a negative eigenvalue.

(b) How do you know it must have a negative pivot?

(c) How do you know it can't have two negative eigenvalues?

5 If C is symmetric prove that $A^{\mathrm{T}}CA$ is also symmetric. (Transpose it.) When A is 6 by 3, what are the shapes of C and $A^{\mathrm{T}}CA$?

6 Find an orthogonal matrix Q that diagonalizes $S = \begin{bmatrix} -2 & 6 \\ 6 & 7 \end{bmatrix}$. What is Λ?

If $A^3 = 0$ then the eigenvalues of A must be ______. Give an example that has $A \neq 0$. But if A is symmetric, diagonalize it to prove that A must be a zero matrix.

7 Write S and T in the form $\lambda_1 \boldsymbol{x}_1 \boldsymbol{x}_1^{\mathrm{T}} + \lambda_2 \boldsymbol{x}_2 \boldsymbol{x}_2^{\mathrm{T}}$ of the spectral theorem $Q\Lambda Q^T$:

$$S = \begin{bmatrix} 3 & 1 \\ 1 & 3 \end{bmatrix} \quad T = \begin{bmatrix} 9 & 12 \\ 12 & 16 \end{bmatrix} \quad (\text{keep } \|\boldsymbol{x}_1\| = \|\boldsymbol{x}_2\| = 1).$$

8 Every 2 by 2 symmetric matrix is $\lambda_1 \boldsymbol{x}_1 \boldsymbol{x}_1^{\mathrm{T}} + \lambda_2 \boldsymbol{x}_2 \boldsymbol{x}_2^{\mathrm{T}} = \lambda_1 P_1 + \lambda_2 P_2$. Explain $P_1 + P_2 = \boldsymbol{x}_1 \boldsymbol{x}_1^{\mathrm{T}} + \boldsymbol{x}_2 \boldsymbol{x}_2^{\mathrm{T}} = I$ from columns times rows of Q. Why is $P_1 P_2 = 0$?

9 What are the eigenvalues of $A = \begin{bmatrix} 0 & b \\ -b & 0 \end{bmatrix}$? Create a 4 by 4 antisymmetric matrix ($A^{\mathrm{T}} = -A$) and verify that all its eigenvalues are imaginary.

10 (Recommended) This matrix M is antisymmetric and also ______. Then all its eigenvalues are pure imaginary and they also have $|\lambda| = 1$. ($\|M\boldsymbol{x}\| = \|\boldsymbol{x}\|$ for every $\boldsymbol{x}$ so $\|\lambda \boldsymbol{x}\| = \|\boldsymbol{x}\|$ for eigenvectors.) Find all four eigenvalues from the trace of M:

$$M = \frac{1}{\sqrt{3}} \begin{bmatrix} 0 & 1 & 1 & 1 \\ -1 & 0 & -1 & 1 \\ -1 & 1 & 0 & -1 \\ -1 & -1 & 1 & 0 \end{bmatrix} \quad \text{can only have eigenvalues } i \text{ or } -i.$$

11 Show that this A (**symmetric but complex**) has only one line of eigenvectors:

$$A = \begin{bmatrix} i & 1 \\ 1 & -i \end{bmatrix} \text{ is not even diagonalizable: eigenvalues } \lambda = 0, 0.$$

$A^{\mathrm{T}} = A$ is not such a special property for complex matrices. The good property is $\overline{A}^{\mathrm{T}} = A$. Then all λ's are real and the eigenvectors are orthogonal.

12 Find the eigenvector matrices Q for S and X for B. Show that X doesn't collapse at $d = 1$, even though $\lambda = 1$ is repeated. Are those eigenvectors perpendicular?

$$S = \begin{bmatrix} 0 & d & 0 \\ d & 0 & 0 \\ 0 & 0 & 1 \end{bmatrix} \quad B = \begin{bmatrix} -d & 0 & 1 \\ 0 & 1 & 0 \\ 0 & 0 & d \end{bmatrix} \quad \text{have} \quad \lambda = 1, d, -d.$$

13 Write a 2 by 2 *complex* matrix with $\overline{S}^{\mathrm{T}} = S$ (a "Hermitian matrix"). Find λ_1 and λ_2 for your complex matrix. Check that $\overline{\boldsymbol{x}}_1^{\mathrm{T}} \boldsymbol{x}_2 = 0$ (this is complex orthogonality).

14 ***True*** (with reason) ***or false*** (with example).

(a) A matrix with n real eigenvalues and n real eigenvectors is symmetric.

(b) A matrix with n real eigenvalues and n orthonormal eigenvectors is symmetric.

(c) The inverse of an invertible symmetric matrix is symmetric.

(d) The eigenvector matrix Q of a symmetric matrix is symmetric.

(e) The main diagonal of a positive definite matrix is all positive.

15 (A paradox for instructors) If $AA^{\mathrm{T}} = A^{\mathrm{T}}A$ then A and A^{T} share the same eigenvectors (true). A and A^{T} always share the same eigenvalues. Find the flaw in this conclusion: A and A^{T} must have the same X and same Λ. Therefore A equals A^{T}.

16 Are A and B invertible, orthogonal, projections, permutations, diagonalizable? Which of these factorizations are possible: $\boldsymbol{LU}$, $\boldsymbol{QR}$, $\boldsymbol{X\Lambda X^{-1}}$, $\boldsymbol{Q\Lambda Q^{\mathrm{T}}}$?

$$A = \begin{bmatrix} 0 & 0 & 1 \\ 0 & 1 & 0 \\ 1 & 0 & 0 \end{bmatrix} \qquad B = \frac{1}{3}\begin{bmatrix} 1 & 1 & 1 \\ 1 & 1 & 1 \\ 1 & 1 & 1 \end{bmatrix}.$$

17 What number b in $A = \left[\begin{smallmatrix} 2 & b \\ 1 & 0 \end{smallmatrix}\right]$ makes $A = Q\Lambda Q^{\mathrm{T}}$ possible? What number will make it impossible to diagonalize A? What number makes A singular?

18 Find all 2 by 2 matrices that are orthogonal and also symmetric. Which two numbers can be eigenvalues of those two matrices?

19 This A is nearly symmetric. But what is the angle between the eigenvectors?

$$A = \begin{bmatrix} 1 & 10^{-15} \\ 0 & 1 + 10^{-15} \end{bmatrix} \quad \text{has eigenvectors} \quad \begin{bmatrix} 1 \\ 0 \end{bmatrix} \quad \text{and} \quad [?]$$

20 If $\lambda_{\max}$ is the largest eigenvalue of a symmetric matrix S, no diagonal entry can be larger than $\lambda_{\max}$. What is the first entry a_{11} of $S = Q\Lambda Q^{\mathrm{T}}$? Show why $a_{11} \leq \lambda_{\max}$.

21 Suppose $A^{\mathrm{T}} = -A$ (real *antisymmetric* matrix). Explain these facts about A:

(a) $\boldsymbol{x}^{\mathrm{T}} A\boldsymbol{x} = 0$ for every real vector $\boldsymbol{x}$.

(b) The eigenvalues of A are pure imaginary.

(c) The determinant of A is positive or zero (not negative).

For (a), multiply out an example of $\boldsymbol{x}^{\mathrm{T}} A\boldsymbol{x}$ and watch terms cancel. Or reverse $\boldsymbol{x}^{\mathrm{T}}(A\boldsymbol{x})$ to $-(A\boldsymbol{x})^{\mathrm{T}}\boldsymbol{x}$. For (b), $A\boldsymbol{z} = \lambda\boldsymbol{z}$ leads to $\overline{\boldsymbol{z}}^{\mathrm{T}} A\boldsymbol{z} = \lambda\overline{\boldsymbol{z}}^{\mathrm{T}}\boldsymbol{z} = \lambda\|\boldsymbol{z}\|^2$. Part (a) shows that $\overline{\boldsymbol{z}}^{\mathrm{T}} A\boldsymbol{z} = (\boldsymbol{x} - i\boldsymbol{y})^{\mathrm{T}} A(\boldsymbol{x} + i\boldsymbol{y})$ has zero real part. Then (b) helps with (c).

22 If S is symmetric and all its eigenvalues are $\lambda = 2$, how do you know that S must be $2I$? Key point: Symmetry guarantees that $S = Q\Lambda Q^{\mathrm{T}}$. What is that Λ?

23 *Which symmetric matrices S are also orthogonal*? Show why $S^2 = I$. What are the possible eigenvalues λ? Then S must be $Q\Lambda Q^{\mathrm{T}}$ for which Λ?

Problems 24–49 are about tests for positive definiteness.

24 Suppose the 2 by 2 tests $a > 0$ and $ac - b^2 > 0$ are passed by $S = \left[\begin{smallmatrix} a & b \\ c & d \end{smallmatrix}\right]$.

(i) λ_1 and λ_2 have the *same sign* because their product $\lambda_1\lambda_2$ equals ______.

(i) That sign is positive because $\lambda_1 + \lambda_2$ equals ______. So $\lambda_1 > 0$ and $\lambda_2 > 0$.

25 Which of S_1, S_2, S_3, S_4 has two positive eigenvalues? Use a test, don't compute λ's. Also find an $\boldsymbol{x}$ so that $\boldsymbol{x}^{\mathrm{T}} S_1\boldsymbol{x} < 0$, so S_1 is not positive definite.

$$S_1 = \begin{bmatrix} 5 & 6 \\ 6 & 7 \end{bmatrix} \quad S_2 = \begin{bmatrix} -1 & -2 \\ -2 & -5 \end{bmatrix} \quad S_3 = \begin{bmatrix} 1 & 10 \\ 10 & 100 \end{bmatrix} \quad S_4 = \begin{bmatrix} 1 & 10 \\ 10 & 101 \end{bmatrix}.$$

26 For which numbers b and c is positive definite? Factor S into LDL^{T}.

$$S = \begin{bmatrix} 1 & b \\ b & 9 \end{bmatrix} \qquad S = \begin{bmatrix} 2 & 4 \\ 4 & c \end{bmatrix} \qquad S = \begin{bmatrix} c & b \\ b & c \end{bmatrix}.$$

27 Write $f(x, y) = x^2 + 4xy + 3y^2$ as a *difference* of squares and find a point (x, y) where f is negative. No minimum at $(0, 0)$ even though f has positive coefficients.

28 The function $f(x, y) = 2xy$ certainly has a saddle point and not a minimum at $(0, 0)$. What symmetric matrix S produces $\begin{bmatrix} x & y \end{bmatrix} S \begin{bmatrix} x \\ y \end{bmatrix} = 2xy$? What are its eigenvalues?

29 Test to see if $A^{\mathrm{T}} A$ is positive definite in each case: A needs independent columns.

$$A = \begin{bmatrix} 1 & 2 \\ 0 & 3 \end{bmatrix} \quad \text{and} \quad A = \begin{bmatrix} 1 & 1 \\ 1 & 2 \\ 2 & 1 \end{bmatrix} \quad \text{and} \quad A = \begin{bmatrix} 1 & 1 & 2 \\ 1 & 2 & 1 \end{bmatrix}.$$

30 Which 3 by 3 symmetric matrices S and T produce these quadratics?

$$\boldsymbol{x}^{\mathrm{T}}S\boldsymbol{x} = 2(x_1^2 + x_2^2 + x_3^2 - x_1x_2 - x_2x_3). \quad \text{Why is } S \text{ positive definite?}$$

$$\boldsymbol{x}^{\mathrm{T}}T\boldsymbol{x} = 2(x_1^2 + x_2^2 + x_3^2 - x_1x_2 - x_1x_3 - x_2x_3). \quad \text{Why is } T \text{ semidefinite?}$$

31 Compute the three upper left determinants of S to establish positive definiteness. Verify that their ratios give the second and third pivots in elimination.

$$\textbf{Pivots = ratios of determinants} \qquad S = \begin{bmatrix} 2 & 2 & 0 \\ 2 & 5 & 3 \\ 0 & 3 & 8 \end{bmatrix}.$$

32 For what numbers c and d are S and T positive definite? Test their 3 determinants:

$$S = \begin{bmatrix} c & 1 & 1 \\ 1 & c & 1 \\ 1 & 1 & c \end{bmatrix} \quad \text{and} \quad T = \begin{bmatrix} 1 & 2 & 3 \\ 2 & d & 4 \\ 3 & 4 & 5 \end{bmatrix}.$$

33 Find a matrix with $a > 0$ and $c > 0$ and $a + c > 2b$ that has a negative eigenvalue.

34 *If S is positive definite then S^{-1} is positive definite.* Best proof: The eigenvalues of S^{-1} are positive because _____. Can you use another test?

35 A positive definite matrix cannot have a zero (or even worse, a negative number) on its main diagonal. Show that this matrix fails the energy test $\boldsymbol{x}^{\mathrm{T}}S\boldsymbol{x} > 0$:

$$\begin{bmatrix} x_1 & x_2 & x_3 \end{bmatrix} \begin{bmatrix} 4 & 1 & 1 \\ 1 & 0 & 2 \\ 1 & 2 & 5 \end{bmatrix} \begin{bmatrix} x_1 \\ x_2 \\ x_3 \end{bmatrix} \quad \text{is not positive when } (x_1, x_2, x_3) = (\quad,\quad,\quad).$$

36 A diagonal entry s_{jj} of a symmetric matrix cannot be smaller than all the λ's. If it were, then $S - s_{jj}I$ would have _____ eigenvalues and would be positive definite. But $S - s_{jj}I$ has a _____ on the main diagonal.

37 Give a quick reason why each of these statements is true:

(a) Every positive definite matrix is invertible.

(b) The only positive definite projection matrix is $P = I$.

(c) A diagonal matrix with positive diagonal entries is positive definite.

(d) A symmetric matrix with a positive determinant might not be positive definite!

38 For which s and t do S and T have all $\lambda > 0$ (therefore positive definite)?

$$S = \begin{bmatrix} s & -4 & -4 \\ -4 & s & -4 \\ -4 & -4 & s \end{bmatrix} \quad \text{and} \quad T = \begin{bmatrix} t & 3 & 0 \\ 3 & t & 4 \\ 0 & 4 & t \end{bmatrix}.$$

39 From $S = Q\Lambda Q^{\mathrm{T}}$ compute the positive definite symmetric square root $Q\sqrt{\Lambda}Q^{\mathrm{T}}$ of each matrix. Check that this square root gives $A^{\mathrm{T}}A = S$:

$$S = \begin{bmatrix} 5 & 4 \\ 4 & 5 \end{bmatrix} \quad \text{and} \quad S = \begin{bmatrix} 10 & 6 \\ 6 & 10 \end{bmatrix}.$$

40 Draw the tilted ellipse $x^2 + xy + y^2 = 1$ and find the half-lengths of its axes from the eigenvalues of the corresponding matrix S.

41 With positive pivots in D, the factorization $S = LDL^{\mathrm{T}}$ becomes $L\sqrt{D}\sqrt{D}L^{\mathrm{T}}$. (Square roots of the pivots give $D = \sqrt{D}\sqrt{D}$.) Then $C = \sqrt{D}L^{\mathrm{T}}$ yields the ***Cholesky factorization*** $S = C^{\mathrm{T}}C$ which is "symmetrized LU":

$$\text{From}\quad C = \begin{bmatrix} 3 & 1 \\ 0 & 2 \end{bmatrix} \quad \text{find } S. \qquad \text{From}\quad S = \begin{bmatrix} 4 & 8 \\ 8 & 25 \end{bmatrix} \quad \text{find } C = \textbf{chol}(S).$$

42 In the Cholesky factorization $S = C^{\mathrm{T}}C$, with $C = \sqrt{D}L^{\mathrm{T}}$, the square roots of the pivots are on the diagonal of C. Find C (upper triangular) for

$$S = \begin{bmatrix} 9 & 0 & 0 \\ 0 & 1 & 2 \\ 0 & 2 & 8 \end{bmatrix} \qquad \text{and} \qquad S = \begin{bmatrix} 1 & 1 & 1 \\ 1 & 2 & 2 \\ 1 & 2 & 7 \end{bmatrix}.$$

43 Without multiplying $S = \begin{bmatrix} \cos\theta & -\sin\theta \\ \sin\theta & \cos\theta \end{bmatrix} \begin{bmatrix} 2 & 0 \\ 0 & 5 \end{bmatrix} \begin{bmatrix} \cos\theta & \sin\theta \\ -\sin\theta & \cos\theta \end{bmatrix}$, find

(a) the determinant of S (b) the eigenvalues of S
(c) the eigenvectors of S (d) a reason why S is symmetric positive definite.

44 The graph of $z = x^2 + y^2$ is a bowl opening upward. *The graph* of $z = x^2 - y^2$ *is a saddle*. The graph of $z = -x^2 - y^2$ is a bowl opening downward. What is a test on a, b, c for $z = ax^2 + 2bxy + cy^2$ to have a saddle point at $(x, y) = (0, 0)$?

45 Which values of c give a bowl and which c give a saddle point for the graph of $z = 4x^2 + 12xy + cy^2$? Describe this graph at the borderline value of c.

46 When S and T are symmetric positive definite, ST might not even be symmetric. But start from $ST\boldsymbol{x} = \lambda\boldsymbol{x}$ and take dot products with $T\boldsymbol{x}$. Then prove $\lambda > 0$.

47 Suppose C is positive definite (so $\boldsymbol{y}^{\mathrm{T}}C\boldsymbol{y} > 0$ whenever $\boldsymbol{y} \neq \mathbf{0}$) and A has independent columns (so $A\boldsymbol{x} \neq \mathbf{0}$ whenever $\boldsymbol{x} \neq \mathbf{0}$). Apply the energy test to $\boldsymbol{x}^{\mathrm{T}}A^{\mathrm{T}}CA\boldsymbol{x}$ to show that ***$S = A^{\mathrm{T}}CA$ is positive definite: the crucial matrix in engineering***.

48 **Important**! Suppose S is positive definite with eigenvalues $\lambda_1 \geq \lambda_2 \geq \ldots \geq \lambda_n$.

(a) What are the eigenvalues of the matrix $\lambda_1 I - S$? Is it positive semidefinite?

(b) How does it follow that $\lambda_1\boldsymbol{x}^{\mathrm{T}}\boldsymbol{x} \geq \boldsymbol{x}^{\mathrm{T}}S\boldsymbol{x}$ for every $\boldsymbol{x}$?

(c) Draw this conclusion: **The maximum value of $\boldsymbol{x}^{\mathrm{T}}S\boldsymbol{x}/\boldsymbol{x}^{\mathrm{T}}\boldsymbol{x}$ is** _____ .

49 For which a and c is this matrix positive definite? For which a and c is it positive semidefinite (this includes definite)?

$$S = \begin{bmatrix} a & a & a \\ a & a+c & a-c \\ a & a-c & a+c \end{bmatrix}$$

All 5 tests are possible.
The energy $\boldsymbol{x}^{\mathrm{T}}S\boldsymbol{x}$ equals
$a\,(x_1 + x_2 + x_3)^2 + c\,(x_2 - x_3)^2$.

6.4 Complex Numbers and Vectors and Matrices

Everything begins with $\boldsymbol{x = i}$. By definition it solves $\boldsymbol{x^2} = \boldsymbol{-1}$. From there we discover the roots $\boldsymbol{x_1}$ and $\boldsymbol{x_2}$ of any quadratic $\boldsymbol{ax^2 + bx + c = 0}$. Then in one big step we allow **any polynomial** $\boldsymbol{c_0 x^n + c_1 x^{n-1} + \cdots + c_n = 0}$. Those coefficients $\boldsymbol{c_0}$ to $\boldsymbol{c_n}$ can be complex numbers $\boldsymbol{c = a + bi}$. *The loop is closed: All complex polynomials of degree n have n complex roots $\boldsymbol{x_1}$ to $\boldsymbol{x_n}$.* This was proved by Gauss, the greatest mathematician.

Example 1 $\quad \boldsymbol{ax^2 + bx + c = 0}$ for $\quad x = \dfrac{-b \pm \sqrt{b^2 - 4ac}}{2a} = \dfrac{-b \pm i\sqrt{4ac - b^2}}{2a}$:

Always two roots, real or complex (or one double root when $b^2 = 4ac$). For the roots of polynomials of degree $n = 3$ and $n = 4$, there are complicated formulas—not often used. A formula for the roots is *not possible* for $n \geq 5$.

Complex numbers: Addition and multiplication and division (five rules).

$$z_1 + z_2 = (a + ib) + (x + iy) = \boldsymbol{(a + x) + i(b + y)}$$

$$z_1 z_2 = (a + ib)(x + iy) = \boldsymbol{(ax - by) + i(bx + ay)} \ \text{: Notice } (ib)(iy) = -by$$

$$\overline{z} = \overline{x + iy} = \boldsymbol{x - iy} = \textbf{“Complex conjugate of } z = x + iy\textbf{”}$$

$$|z|^2 = z \ \textbf{times} \ \overline{z} = (x + iy)(x - iy) = \boldsymbol{x^2 + y^2 = r^2}$$

$$\frac{z_1}{z_2} = \frac{(a + ib)}{(x + iy)} = \frac{(a + ib)(x - iy)}{(x + iy)(x - iy)} = \frac{(ax + by) + i(bx - ay)}{x^2 + y^2}$$

You see why division is a little awkward: for $\dfrac{1}{z}$ the best we can do is $\dfrac{\overline{z}}{|z|^2}$.

Or use the polar form $\boxed{z = |z|e^{i\theta} = re^{i\theta}}$ with Euler's formula $\boxed{e^{i\theta} = \cos\theta + i\sin\theta}$

Comparing $x+iy$ with $re^{i\theta} = r\cos\theta + ir\sin\theta$ tells us that $\boldsymbol{x = r\cos\theta}$ and $\boldsymbol{y = r\sin\theta}$. The polar form is ideal for multiplying complex numbers: **Add angles and multiply $\boldsymbol{r}$'s.**

$$z_1 z_2 = |z_1|\,|z_2|e^{i(\theta_1+\theta_2)} \qquad z^n = r^n e^{in\theta} \qquad \frac{z_1}{z_2} = \frac{|z_1|}{|z_2|}e^{i(\theta_1-\theta_2)}$$

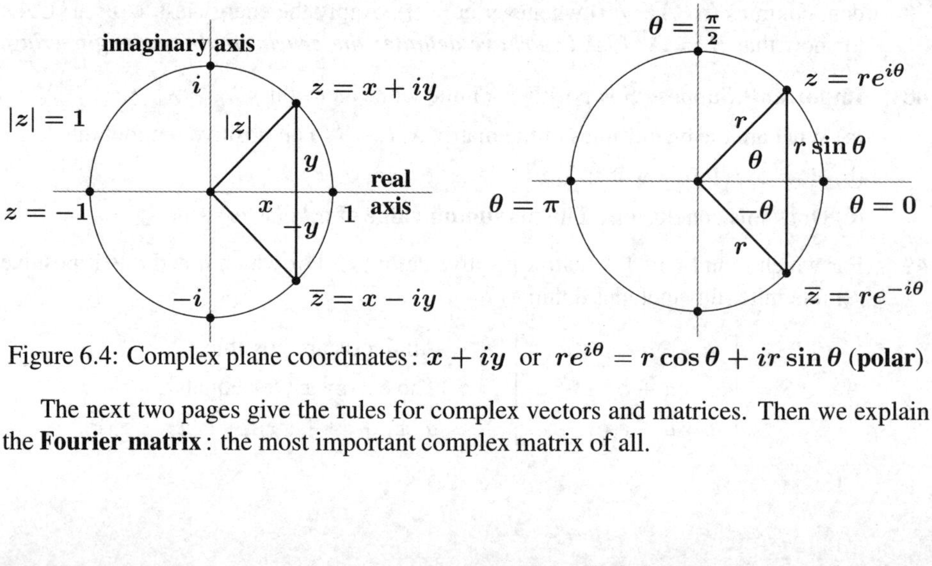

Figure 6.4: Complex plane coordinates: $\boldsymbol{x + iy}$ or $\boldsymbol{re^{i\theta}} = \boldsymbol{r\cos\theta + ir\sin\theta}$ **(polar)**

The next two pages give the rules for complex vectors and matrices. Then we explain the **Fourier matrix**: the most important complex matrix of all.

One final point about the eigenvalues of a real n by n matrix. The determinant of $A-\lambda I$ is a polynomial $P(\lambda)$ of degree n. Then $P(\lambda) = \boldsymbol{\lambda^2 - (\textbf{trace of } A)\lambda + (\textbf{det of } A)}$ for a 2 by 2 matrix. *The coefficients of that polynomial are real.* We know that real polynomials $P(\lambda)$ can produce complex roots, as in $\lambda^2+1=0$ with $\lambda = +i, -i$. Here is the point:

The complex roots of a real polynomial $P(\lambda)$ come in conjugate pairs $\lambda = a \pm bi$.

Another example is $(\lambda-1)^2+1=0$ which produces $\lambda = \mathbf{1}+\boldsymbol{i}$ and $\mathbf{1}-\boldsymbol{i}$.

The quadratic formula involves $\pm\sqrt{b^2-4ac}$. If $b^2 < 4ac$, the two square roots are a conjugate pair. In general we know:

If $P(x+iy) = a_n(x+iy)^n + \cdots + a_0 = 0$ with real coefficients a_n to a_0, then also $P(x-iy) = 0$.

For real A, the eigenvectors for λ and $\overline{\lambda}$ are $\boldsymbol{v}$ and $\overline{\boldsymbol{v}}$ because $A\boldsymbol{v} = \lambda\boldsymbol{v}$ gives $A\overline{\boldsymbol{v}} = \overline{\lambda}\overline{\boldsymbol{v}}$.

Complex Vectors : Lengths $||v||$ and Inner Products $\overline{v}^{\mathrm{T}}w$

Length $||\boldsymbol{v}||^2 = \overline{\boldsymbol{v}}^{\mathrm{T}}\boldsymbol{v} = |v_1|^2 + \cdots + |v_n|^2 \geq 0$ **Notice $\overline{v}^{\mathrm{T}}v$ and not $v^{\mathrm{T}}v$**

Dot product $= \overline{\boldsymbol{v}}^{\mathrm{T}}\boldsymbol{w} = \overline{v}_1 w_1 + \cdots + \overline{v}_n w_n$ **Notice $\overline{v}^{\mathrm{T}}w$ and not $v^{\mathrm{T}}w$**

The length of $\boldsymbol{v} = (\mathbf{1}, \boldsymbol{i})$ is $||\boldsymbol{v}|| = \sqrt{\mathbf{2}}$. The length of $\boldsymbol{w} = (\mathbf{1}+\boldsymbol{i}, \mathbf{0})$ is also $||\boldsymbol{w}|| = \sqrt{\mathbf{2}}$

Complex Matrices A and Complex Transpose $\overline{A}^{\mathrm{T}}$

The key point is to replace A^{T} or $A^{\mathrm{T}}A$ or AA^{T} by $\overline{\boldsymbol{A}}^{\mathrm{T}}$ or $\overline{\boldsymbol{A}}^{\mathrm{T}}\boldsymbol{A}$ or $\boldsymbol{A}\overline{\boldsymbol{A}}^{\mathrm{T}}$. In the same way, the test $Q^{\mathrm{T}}Q = I$ for an orthogonal matrix becomes $\overline{\boldsymbol{Q}}^{\mathrm{T}}\boldsymbol{Q} = \boldsymbol{I}$. When a complex matrix is transposed, **replace each entry $x+iy$ in the transpose by $x-iy$**. Many computer systems do this automatically.

Real **symmetric** matrices $S^{\mathrm{T}} = S$ **Orthogonal eigenvectors**	Complex **Hermitian** matrices $\overline{S}^{\mathrm{T}} = S$ **Real eigenvalues**
Real **orthogonal** matrices $\boldsymbol{Q}^{\mathrm{T}} = \boldsymbol{Q}^{-1}$ **Orthogonal eigenvectors**	Complex **unitary** matrices $\overline{\boldsymbol{Q}}^{\mathrm{T}} = \boldsymbol{Q}^{-1}$ **Eigenvalues** $|\boldsymbol{\lambda}| = \mathbf{1}$ $(\lambda = e^{i\theta})$

If A has an eigenvalue λ then $\overline{A}^{\mathrm{T}}$ has an eigenvalue $\overline{\lambda}$. Eigenvalues are real if $\overline{A}^{\mathrm{T}} = A$.

The Hermitian matrices $\overline{A}^{\mathrm{T}}A$ and $A\overline{A}^{\mathrm{T}}$ have real eigenvalues $\geq \mathbf{0}$: positive semidefinite

Eigenvectors of Hermitian matrices $S = \overline{S}^{\mathrm{T}}$ have $\overline{\boldsymbol{x}}^{\mathrm{T}}\boldsymbol{y} = 0$: **orthogonal eigenvectors**.

$\boldsymbol{S} = \overline{\boldsymbol{S}}^{\mathrm{T}} = \begin{bmatrix} \mathbf{2} & \mathbf{3}-\mathbf{3}\boldsymbol{i} \\ \mathbf{3}+\mathbf{3}\boldsymbol{i} & \mathbf{5} \end{bmatrix}$ is a typical Hermitian matrix with $\lambda = \mathbf{8}$ and $\lambda = -\mathbf{1}$ **(real $\boldsymbol{\lambda}$)**.

Its eigenvectors $\boldsymbol{x} = \begin{bmatrix} 1 \\ 1+i \end{bmatrix}$ and $\boldsymbol{y} = \begin{bmatrix} 1-i \\ -1 \end{bmatrix}$ are complex and *orthogonal*: $\overline{\boldsymbol{x}}^{\mathrm{T}}\boldsymbol{y} = \mathbf{0}$.

The Four Fundamental Subspaces of a Complex Matrix A

Suppose A is a complex matrix. Its columns span the column space $\mathbf{C}(A)$. The solutions $\boldsymbol{x}$ to $A\boldsymbol{x} = \mathbf{0}$ span the nullspace $\mathbf{N}(A)$. But the other two spaces naturally change from $\mathbf{C}(A^{\mathrm{T}})$ and $\mathbf{N}(A^{\mathrm{T}})$ to $\mathbf{C}(\overline{A}^{\mathrm{T}})$ and $\mathbf{N}(\overline{A}^{\mathrm{T}})$. Then we do again have orthogonality between the complex row space $\mathbf{C}(\overline{A}^{\mathrm{T}})$ and nullspace $\mathbf{N}(A)$. And also orthogonality between the column space $\mathbf{C}(A)$ and the left nullspace $\mathbf{N}(\overline{A}^{\mathrm{T}})$.

Complex matrices fit perfectly into linear algebra when $\overline{A}^{\mathrm{T}}$ *replaces* A^{T}.

We want the two pairs of subspaces to remain orthogonal in complex n-dimensional space $\mathbf{C}^n$. But the definition of orthogonality for complex vectors is $\overline{\boldsymbol{x}}^{\mathrm{T}}\boldsymbol{y} = \mathbf{0}$. So we must redefine two of the four fundamental subspaces, when complex numbers appear.

The row space $\mathbf{C}(A^{\mathrm{T}})$ becomes $\mathbf{C}(\overline{A}^{\mathrm{T}})$. The left nullspace $\mathbf{N}(A^{\mathrm{T}})$ becomes $\mathbf{N}(\overline{A}^{\mathrm{T}})$. The rank r and the dimensions of these subspaces stay the same.

Let me add the complex inner product in *function space*—an integral of $\overline{x}(t)y(t)\,dt$ instead of a sum of $\overline{x}_k y_k$. Please notice the most important complex basis—all the functions e^{inx} for $-\infty < n < \infty$ and $0 \le x \le 2\pi$. Those give the **Fourier basis in Hilbert space:** a complete orthogonal basis with these zero inner products between e^{inx} and e^{ikx} :

$$\int_0^{2\pi} e^{ikx}e^{-inx}\,dx = \int_0^{2\pi} e^{i(k-n)x}\,dx = \left[\frac{e^{i(k-n)x}}{i(k-n)}\right]_{x=0}^{x=2\pi} = \mathbf{0 \text{ if } k \neq n.}$$

$\overline{S}^{\mathrm{T}} = S$: Real Eigenvalues and Orthogonal Eigenvectors

Now we can complete the important steps at the beginning of Section 6.3: *Real symmetric matrices* $S = S^{\mathrm{T}}$ *have real eigenvalues.* And we can extend that statement to include **complex Hermitian matrices** $\boldsymbol{S} = \overline{\boldsymbol{S}}^{\mathrm{T}}$. Their eigenvalues are also real.

Proof: If $S\boldsymbol{x} = \lambda\boldsymbol{x}$ then $\overline{\boldsymbol{x}}^{\mathrm{T}}S\boldsymbol{x} = \lambda\overline{\boldsymbol{x}}^{\mathrm{T}}\boldsymbol{x}$. The number $\overline{\boldsymbol{x}}^{\mathrm{T}}\boldsymbol{x} = ||\boldsymbol{x}||^2$ is positive. The number $\overline{\boldsymbol{x}}^{\mathrm{T}}S\boldsymbol{x}$ is real because it equals its complex conjugate $\overline{\boldsymbol{x}}^{\mathrm{T}}\,\overline{\boldsymbol{S}}^{\mathrm{T}}\boldsymbol{x}$. Then the ratio is

$$\boldsymbol{\lambda} = \frac{\overline{\boldsymbol{x}}^{\mathrm{T}}S\boldsymbol{x}}{\overline{\boldsymbol{x}}^{\mathrm{T}}\boldsymbol{x}} = \textbf{real number} \tag{1}$$

Maximizing that "Rayleigh quotient" in (1) produces the largest eigenvalue of S. The best $\boldsymbol{x}$ will be the eigenvector for $S\boldsymbol{x} = \lambda_{\text{max}}\,\boldsymbol{x}$. Minimizing the Rayleigh quotient produces λ_{min}. The in-between eigenvalues of S are "saddle points" of the ratio in (1). First derivatives are zero and second derivatives of the quotient in (1) have mixed signs.

Finally we establish that S has orthogonal eigenvectors, so that $S = Q\Lambda Q^{-1} = Q\Lambda\overline{Q}^{\mathrm{T}}$. This requires two steps, to a triangular T and then a diagonal Λ.

1. Schur's theorem proves that every square matrix $A = QT\overline{Q}^{\mathrm{T}}$ for a **triangular** $\boldsymbol{T}$.

2. If $A = S = \overline{S}^{\mathrm{T}}$ then T is the diagonal eigenvalue matrix Λ and $\boldsymbol{S} = \boldsymbol{Q\Lambda\overline{Q}^{\mathrm{T}}}$.

1 is proved on the website **math.mit.edu/linearalgebra**. **2** is proved in the Problem Set.

Complex Eigenvalues and Eigenvectors of a Permutation P

P is an important real matrix. It has complex eigenvalues λ and eigenvectors $\boldsymbol{x}$.

$$P\boldsymbol{x} = \begin{bmatrix} 0 & \mathbf{1} & 0 & 0 \\ 0 & 0 & \mathbf{1} & 0 \\ 0 & 0 & 0 & \mathbf{1} \\ \mathbf{1} & 0 & 0 & 0 \end{bmatrix} \begin{bmatrix} x_1 \\ x_2 \\ x_3 \\ x_4 \end{bmatrix} = \begin{bmatrix} x_2 \\ x_3 \\ x_4 \\ x_1 \end{bmatrix} = \lambda \begin{bmatrix} x_1 \\ x_2 \\ x_3 \\ x_4 \end{bmatrix}. \qquad (2)$$

To find λ and $\boldsymbol{x}$, start with $x_1 = \mathbf{1}$ (first component). Then $x_2 = \lambda x_1 = \boldsymbol{\lambda}$ and $x_3 = \lambda x_2 = \boldsymbol{\lambda^2}$ and $x_4 = \lambda x_3 = \boldsymbol{\lambda^3}$. The final equation $x_1 = \lambda x_4$ becomes $\mathbf{1} = \boldsymbol{\lambda^4}$. This equation has four solutions ! They are the eigenvalues of P, all with $|\lambda| = 1$.

The determinant of $\boldsymbol{P - \lambda I}$ is $\boldsymbol{\lambda^4 - 1}$. The eigenvalues are $\boldsymbol{\lambda = 1, i, i^2 = -1, i^3 = -i}$.

Those eigenvalues add to 0 = trace of P. The imaginary roots i and $-i$ of the real polynomial $\lambda^4 - 1$ are a conjugate pair, as expected. Here is the **Fourier matrix $\boldsymbol{F}$** with columns of F = *eigenvectors of* P. The key equation is $\boldsymbol{F^{-1}PF = \Lambda}$ or $\boldsymbol{PF = F\Lambda}$.

$$\boldsymbol{PF} = \begin{bmatrix} 0 & 1 & 0 & 0 \\ 0 & 0 & 1 & 0 \\ 0 & 0 & 0 & 1 \\ 1 & 0 & 0 & 0 \end{bmatrix} \begin{bmatrix} \mathbf{1} & \mathbf{1} & \mathbf{1} & \mathbf{1} \\ \mathbf{1} & \boldsymbol{i} & \boldsymbol{i^2} & \boldsymbol{i^3} \\ \mathbf{1} & \boldsymbol{i^2} & \boldsymbol{i^4} & \boldsymbol{i^6} \\ \mathbf{1} & \boldsymbol{i^3} & \boldsymbol{i^6} & \boldsymbol{i^9} \end{bmatrix} = \begin{bmatrix} 1 & 1 & 1 & 1 \\ 1 & i & i^2 & i^3 \\ 1 & i^2 & i^4 & i^6 \\ 1 & i^3 & i^6 & i^9 \end{bmatrix} \begin{bmatrix} \mathbf{1} & & & \\ & \boldsymbol{i} & & \\ & & \boldsymbol{i^2} & \\ & & & \boldsymbol{i^3} \end{bmatrix} = \boldsymbol{F\Lambda} \quad (3)$$

When P is N by N, the same reasoning leads from $P^N = I$ to $\lambda^N = 1$. The N eigenvalues of P are again equally spaced around the circle. Now they are powers of the **complex number** $w = \boldsymbol{e^{2\pi i/N}}$ at $360/N$ degrees $= 2\pi/N$ radians.

$$\boxed{\textbf{The solutions to } \boldsymbol{\lambda^N = 1} \textbf{ are } \boldsymbol{\lambda = w, w^2, \ldots, w^{N-1}, 1} \textbf{ with } \boldsymbol{w = e^{2\pi i/N}}.} \qquad (4)$$

In the complex plane, the first eigenvalue is $e^{i\theta} = \cos\theta + i\sin\theta$ and the angle θ is $2\pi/N$. The angles for the other eigenvalues are $2\theta, 3\theta, \ldots, N\theta$. Since θ is $2\pi/N$, that last angle is $N\theta = 2\pi$ and that eigenvalue is $\lambda = e^{2\pi i}$ which is $\cos 2\pi + i \sin 2\pi = \mathbf{1}$.

$$\boldsymbol{P_8} = \begin{bmatrix} 0 & 1 & & & & & & \\ & 0 & 1 & & & & & \\ & & 0 & 1 & & & & \\ & & & 0 & 1 & & & \\ & & & & 0 & 1 & & \\ & & & & & 0 & 1 & \\ & & & & & & 0 & 1 \\ 1 & & & & & & & 0 \end{bmatrix}$$

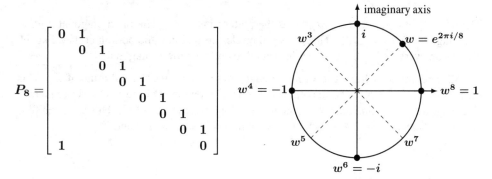

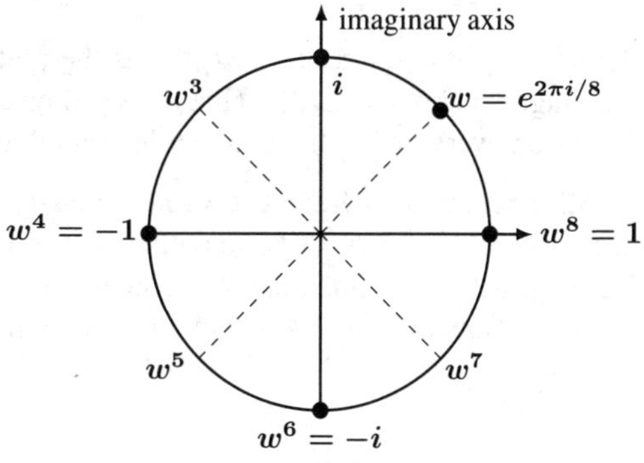

Figure 6.5: **Eigenvalues of $\boldsymbol{P_8}$** : The 8 powers of $w = \boldsymbol{e^{2\pi i/8}}$ add to zero (pair them off).

The Fourier matrix F_8 is 8 by 8. It contains the eigenvectors of the permutation P_8. If we count rows and columns from 0 to 7 (starting at 0 the way engineers do), then the eigenvalues $\lambda_k = w^k$ of the 8 by 8 permutation are $\boldsymbol{\lambda_0}$ **to** $\boldsymbol{\lambda_7}$. **The Fourier matrix** $\boldsymbol{F_N}$ (the eigenvector matrix for P_N) **has** $\boldsymbol{w^{jk}}$ **in row** $\boldsymbol{j}$**, column** $\boldsymbol{k}$.

$$\textbf{The columns of } \boldsymbol{F_N} = \textbf{eigenvectors of } \boldsymbol{P_N} \textbf{ are orthogonal: } \overline{\boldsymbol{F}}_N^{\mathrm{T}} \boldsymbol{F_N} = \boldsymbol{NI}$$

First proof: The permutation P_N is an orthogonal matrix $\Rightarrow$ *orthogonal eigenvectors.*
Second proof: The eigenvectors are complex, so orthogonality means $\overline{\boldsymbol{q}}_j^{\mathrm{T}} \boldsymbol{q}_k = 0$ if $j \neq k$.

$$(1, \overline{w}^j, \overline{w}^{2j}, \ldots, \overline{w}^{(N-1)j}) \cdot (1, w^k, w^{2k}, \ldots, w^{(N-1)k}) = 1 + a + a^2 + \cdots + a^{N-1} = \frac{\boldsymbol{a^N - 1}}{\boldsymbol{a - 1}}$$

Here $a = \overline{w}^j w^k$. If $j = k$ then $a = 1$. The dot product of every column of F with itself (the length squared) is $\boldsymbol{N}$. If $j \neq k$ then a^N is $\overline{w}^{Nj} w^{Nk} = 1$ times 1. So the sum above is $0/(a-1) = 0$. The Fourier matrix $\boldsymbol{F_N}$ has **orthogonal columns.**

All Circulant Matrices C Share the Fourier Eigenvectors

If $P\boldsymbol{x} = \lambda \boldsymbol{x}$ then $P^2 \boldsymbol{x} = \lambda P \boldsymbol{x} = \lambda^2 \boldsymbol{x}$. *Eigenvectors of P are eigenvectors of P^2.* Multiply again by P to find $P^3 \boldsymbol{x} = \lambda^2 P \boldsymbol{x} = \lambda^3 \boldsymbol{x}$. Then P^3 also has the same eigenvectors $\boldsymbol{x}$. Every power of P will have $\boldsymbol{P^n x} = \boldsymbol{\lambda^n x}$, where $\boldsymbol{x}$ is a column of the Fourier matrix F.

More than that, **every combination** $\boldsymbol{C = c_0 I + c_1 P + c_2 P^2 + c_3 P^3}$ has $C\boldsymbol{x} = (c_0 + c_1 \lambda + c_2 \lambda^2 + c_3 \lambda^3)\boldsymbol{x}$. These matrices C are called **circulants**, and you have to see how they look. Here is the eigenvector of P with $P\boldsymbol{x} = \lambda \boldsymbol{x} = i\boldsymbol{x}$. **The same $\boldsymbol{x}$ is also an eigenvector of the circulant $\boldsymbol{C}$!** The columns of F are eigenvectors of P and also of C.

$$C\boldsymbol{x} = \begin{bmatrix} c_0 & c_1 & c_2 & c_3 \\ c_3 & c_0 & c_1 & c_2 \\ c_2 & c_3 & c_0 & c_1 \\ c_1 & c_2 & c_3 & c_0 \end{bmatrix} \begin{bmatrix} 1 \\ i \\ i^2 \\ i^3 \end{bmatrix} = \underset{\textbf{eigenvalue of } \boldsymbol{C}}{(c_0 + c_1 i + c_2 i^2 + c_3 i^3)} \begin{bmatrix} 1 \\ i \\ i^2 \\ i^3 \end{bmatrix} \tag{5}$$

The name "circulant matrix" expresses the pattern inside C. The main diagonal is $\boldsymbol{c_0 I}$. The diagonal above is $\boldsymbol{c_1 P}$. That diagonal circles around into the bottom corner, like $\boldsymbol{P}$. The c's on every upper diagonal circle around to end on a lower diagonal.

All circulants $\boldsymbol{C}$ have the same eigenvectors: They are the columns of $\boldsymbol{F}$. This means that $\boldsymbol{CF} = \boldsymbol{F\Lambda}$. All circulant matrices are diagonalized by the same Fourier matrix. The diagonal eigenvalue matrix Λ contains the four eigenvalues of C. Those are the four complex numbers $c_0 + c_1\lambda + c_2\lambda^2 + c_3\lambda^3$ with $\lambda = 1$ and i and i^2 and i^3.

The Fast Fourier Transform

The discrete Fourier transform (**DFT**) of a vector $\boldsymbol{x}$ is the vector $\overline{\boldsymbol{F}}\boldsymbol{x}$. The signal is $\boldsymbol{x}$. In frequency space it is $\overline{\boldsymbol{F}}\boldsymbol{x}$. That step uses the **Fast Fourier Transform** (**FFT**). This is the fast multiplication that makes modern signal processing possible.

The key to the FFT is a factorization of $\boldsymbol{F_N}$. The middle factor has two copies of $\boldsymbol{F_{N/2}}$. That cuts the work almost in half—because the first factor only has diagonal matrices, and the third factor is just a permutation. **Here is one step in the FFT**:

$$\textbf{Three factors} \quad F_N = \begin{bmatrix} I & B \\ I & -B \end{bmatrix} \begin{bmatrix} \boldsymbol{F_{N/2}} & 0 \\ 0 & \boldsymbol{F_{N/2}} \end{bmatrix} \begin{bmatrix} \text{even } 0,2,\ldots \\ \text{odd } 1,3,\ldots \end{bmatrix} \tag{6}$$

This is simplest to see for $N=4$. Here are F_4 and two copies of F_2 and B:

$$\begin{bmatrix} 1 & 1 & 1 & 1 \\ 1 & i & i^2 & i^3 \\ 1 & i^2 & i^4 & i^6 \\ 1 & i^3 & i^6 & i^9 \end{bmatrix} = \begin{bmatrix} 1 & 0 & 1 & 0 \\ 0 & 1 & 0 & i \\ 1 & 0 & -1 & 0 \\ 0 & 1 & 0 & -i \end{bmatrix} \begin{bmatrix} 1 & 1 & & \\ 1 & -1 & & \\ & & 1 & 1 \\ & & 1 & -1 \end{bmatrix} \begin{bmatrix} 1 & 0 & 0 & 0 \\ 0 & 0 & 1 & 0 \\ 0 & 1 & 0 & 0 \\ 0 & 0 & 0 & 1 \end{bmatrix} \tag{7}$$

Suppose we apply this idea to $F_{1024}\boldsymbol{x}$. Normal matrix-vector multiplication needs $(1024)^2$ steps—more than a million. The first factor in (6) needs only N steps from B. The third factor has no multiplications. The middle factor with F's is destined to be reduced to four diagonal copies of F_{256} and then eight copies of F_{128}. This is *recursion*.

In the end there are the diagonal B's—10 matrices because $1024 = 2^{10}$. **The FFT costs about $\boldsymbol{10(1024) = N\log_2 N}$ complex multiplications instead of $\boldsymbol{N^2}$.**

A million steps have been reduced by a factor near to 100. You may need to adjust F and $\boldsymbol{x}$, to start with $N =$ power of 2. Wikipedia provides a more careful count of multiplies and adds. The FFT has many variations—from special and beautiful patterns inside F.

Multiplying Two Circulant Matrices $CD = DC$

Multiplying $\boldsymbol{C} = 2I + P + 3P^3$ times $\boldsymbol{D} = 3I + P + P^2 + 4P^3$ is like third grade multiplication of 3012 times 4113, with the important difference that $P^4 = I$:

$$\begin{array}{rrrrrrr} & & & \mathbf{3} & \mathbf{0} & \mathbf{1} & \mathbf{2} \\ & & & \mathbf{4} & \mathbf{1} & \mathbf{1} & \mathbf{3} \\ \hline & & & 9 & 0 & 3 & 6 \\ & & 3 & 0 & 1 & 2 & \\ & 3 & 0 & 1 & 2 & & \\ 12 & 0 & 4 & 8 & & & \\ \hline \mathbf{12} & \mathbf{3} & \mathbf{7} & \mathbf{18} & \mathbf{3} & \mathbf{5} & \mathbf{6} \end{array}$$

$$\boldsymbol{CD} = \mathbf{6}I + \mathbf{5}P + \mathbf{3}P^2 + \mathbf{18}P^3 + \mathbf{7}P^4 + \mathbf{3}P^5 + \mathbf{12}P^6$$

$$= (6+7)I + (5+3)P + (3+12)P^2 + 18P^3$$

$$= \mathbf{13}\boldsymbol{I} + \mathbf{8}\boldsymbol{P} + \mathbf{15}\boldsymbol{P^2} + \mathbf{18}\boldsymbol{P^3} \quad \text{since } \boldsymbol{P^4 = I}$$

Multiply $\boldsymbol{CD}$ = Convolve top rows $\boldsymbol{c} \circledast \boldsymbol{d}$

Two key points: We don't carry part of 18 into the next column. And we replace P^4, P^5, P^6 by I, P, P^2. This "multiplication of vectors" is called **cyclic convolution**:

$$\boxed{\boldsymbol{c} \circledast \boldsymbol{d} = (\mathbf{2}, \mathbf{1}, \mathbf{0}, \mathbf{3}) \circledast (\mathbf{3}, \mathbf{1}, \mathbf{1}, \mathbf{4}) = (\mathbf{13}, \mathbf{8}, \mathbf{15}, \mathbf{18}) = \boldsymbol{d} \circledast \boldsymbol{c}} \tag{8}$$

We multiply powers of P with $P^4 = I$. **The answer would be the same for $\boldsymbol{D}$ times $\boldsymbol{C}$.**

But there is no need to keep writing P's when they are understood. It is the coefficients $\mathbf{13, 8, 15, 18}$ that go into the circulant matrices $\boldsymbol{CD} = \boldsymbol{DC}$.

$$CD = \begin{bmatrix} \mathbf{2} & \mathbf{1} & \mathbf{0} & \mathbf{3} \\ 3 & 2 & 1 & 0 \\ 0 & 3 & 2 & 1 \\ 1 & 0 & 3 & 2 \end{bmatrix} \begin{bmatrix} \mathbf{3} & \mathbf{1} & \mathbf{1} & \mathbf{4} \\ 4 & 3 & 1 & 1 \\ 1 & 4 & 3 & 1 \\ 1 & 1 & 4 & 3 \end{bmatrix} = \begin{bmatrix} \mathbf{13} & \mathbf{8} & \mathbf{15} & \mathbf{18} \\ 18 & 13 & 8 & 15 \\ 15 & 18 & 13 & 8 \\ 8 & 15 & 18 & 13 \end{bmatrix} = DC. \quad (9)$$

Ordinary convolution is $c * d = (2,1,0,3) * (3,1,1,4) = (6,5,3,18,7,3,12)$. Multiply polynomials c and d of degree 3 (4 coefficients) to get a 6th degree polynomial (7 coefficients). Then $\boldsymbol{c} * \boldsymbol{d}$ changes to $\boldsymbol{c} \circledast \boldsymbol{d} = (13, 8, 15, 18)$ by using $\boldsymbol{P^4} = \boldsymbol{I}$.

The Convolution Rule in Signal Processing

Start with the formula $\boldsymbol{Fc}$ for the eigenvalues λ_0 to λ_{N-1} of any circulant matrix C. From $\boldsymbol{CF} = \boldsymbol{F\Lambda}$, the top row of C multiplies each eigenvector $\boldsymbol{q_k}$ in $\boldsymbol{F}$ to give $\boldsymbol{\lambda_k}$ times 1:

$$\begin{bmatrix} \boldsymbol{\lambda_0} \\ \boldsymbol{\lambda_1} \\ \cdot \\ \boldsymbol{\lambda_{N-1}} \end{bmatrix} = \begin{bmatrix} c_0 + c_1 + \cdots + c_{N-1} \\ c_0 + c_1 w + \cdots + c_{N-1} w^{N-1} \\ \cdot\ \cdot\ \cdot\ \cdot\ \cdot\ \cdot\ \cdot\ \cdot \\ c_0 + c_1 w^{N-1} + \cdots + c_{N-1} w^{(N-1)(N-1)} \end{bmatrix} = \boldsymbol{F} \begin{bmatrix} c_0 \\ c_1 \\ \cdot \\ c_{N-1} \end{bmatrix} = \boldsymbol{Fc} \quad (10)$$

Example for $N = 2$ with $w = e^{2\pi i/N} = -1$ in the Fourier matrix (eigenvector matrix) $\boldsymbol{F}$:

$$\boldsymbol{F} = \begin{bmatrix} 1 & 1 \\ 1 & -1 \end{bmatrix} \quad \boldsymbol{C} = \begin{bmatrix} c_0 & c_1 \\ c_1 & c_0 \end{bmatrix} \quad \begin{bmatrix} \boldsymbol{\lambda_0(C)} \\ \boldsymbol{\lambda_1(C)} \end{bmatrix} = \begin{bmatrix} c_0 + c_1 \\ c_0 - c_1 \end{bmatrix} = \begin{bmatrix} 1 & 1 \\ 1 & -1 \end{bmatrix} \begin{bmatrix} c_0 \\ c_1 \end{bmatrix} = \boldsymbol{Fc} \quad (11)$$

The other fact we need is that all the circulant matrices C and D and CD have the **same eigenvectors** $\boldsymbol{q}_0$ to $\boldsymbol{q}_{N-1}$ (the columns of F). Then the eigenvalues just multiply! $\boldsymbol{CDq_i} = \boldsymbol{C\lambda_i(D)q_i} = \boldsymbol{\lambda_i(C)\lambda_i(D)q_i}$. This is the second fact and the symbol $\boldsymbol{.*}$ is a way to express it:

$$\begin{bmatrix} \lambda_0(CD) \\ \lambda_1(CD) \\ \vdots \end{bmatrix} = \begin{bmatrix} \lambda_0(C)\lambda_0(D) \\ \lambda_1(C)\lambda_1(D) \\ \vdots \end{bmatrix} = \begin{bmatrix} \lambda_0(C) \\ \lambda_1(C) \\ \vdots \end{bmatrix} .* \begin{bmatrix} \lambda_0(D) \\ \lambda_1(D) \\ \vdots \end{bmatrix} \quad (12)$$

That Hadamard product .* multiplies vectors component by component. Now we use (10) to substitute $\boldsymbol{Fc}$ and $\boldsymbol{Fd}$ for the last two vectors in (12). The left side becomes $\boldsymbol{F}(\boldsymbol{c} \circledast \boldsymbol{d})$, when we remember from (9) that the top row of CD contains the cyclic convolution $\boldsymbol{c} \circledast \boldsymbol{d}$. Here is the beautiful result that simplifies signal processing:

Convolving vectors is multiplying transforms

$$\textbf{Convolution Rule} \quad \boldsymbol{F}(\boldsymbol{c} \circledast \boldsymbol{d}) = (\boldsymbol{Fc}) .* (\boldsymbol{Fd}) \quad (13)$$

$\boldsymbol{CD}$ for circulant matrices is $\boldsymbol{c} \circledast \boldsymbol{d}$ for top rows and $(\boldsymbol{Fc}) .* (\boldsymbol{Fd})$ for eigenvalues.

Problem Set 6.4

1 For the complex number $z = 1 - i$, find $\overline{z}$ and $r = |z|$ and $1/z$ and the angle θ.

2 Find the eigenvalues and eigenvectors of the Hermitian matrix $S = \begin{bmatrix} 1 & 1+i \\ 1-i & 2 \end{bmatrix}$.

3 If $\overline{Q}^{\mathrm{T}} Q = I$ (unitary matrix = complex orthogonal) and $Q\boldsymbol{x} = \lambda \boldsymbol{x}$, show that $|\lambda| = 1$.

4 Find the Fourier matrix F_3 with orthogonal columns = eigenvectors of $\begin{bmatrix} 0 & 1 & 0 \\ 0 & 0 & 1 \\ 1 & 0 & 0 \end{bmatrix}$.

5 (Challenge) Extend equation (7) to the 6 by 6 matrix identity that connects $\boldsymbol{F_6}$ to $\boldsymbol{F_3}$.

6 As in equation (9), multiply the 3 by 3 circulant matrices C and D with top rows $1, 1, 1$ and $1, 2, 1$. Separately find the **cyclic convolution** of those two vectors by multiplying 111 times 121 and then reducing 5 numbers to 3 numbers based on $P^3 = I$.

7 Check the Convolution Rule in equation (13) that connects those vectors $\boldsymbol{c} = (1, 1, 1)$ and $\boldsymbol{d} = (1, 2, 1)$ through the 3 by 3 Fourier matrix F_3 with $w^3 = 1$.

8 Verify Euler's great formula $e^{i\theta} = \cos\theta + i\sin\theta$ using these first terms for

$$e^{i\theta} \approx 1 + i\theta + \frac{1}{2}(i\theta)^2 + \frac{1}{6}(i\theta)^3 \qquad \cos\theta \approx 1 - \frac{1}{2}\theta^2 \qquad \sin\theta \approx \theta - \frac{1}{6}\theta^3.$$

9 Find $\cos 2\theta$ and $\sin 2\theta$ from $\cos\theta$ and $\sin\theta$ using $e^{i\theta}$ (Euler) and $(e^{i\theta})(e^{i\theta}) = e^{2i\theta}$.

10 What test decides if a circulant matrix C is invertible? Is C^{-1} also a circulant? Why?

11 Which circulant matrices are the squares of circulant matrices?

12 When would a circulant be Hermitian ($\overline{C}^{\mathrm{T}} = C$) or unitary ($\overline{C}^{\mathrm{T}} = C^{-1}$)?

13 Find the discrete Fourier transform $\boldsymbol{F x}$ of $\boldsymbol{x} = (1, 0, 1, 0)$ and $\boldsymbol{x} = (0, 1, 0, 1)$.

14 If $w = e^{2\pi i/64}$ then w^2 and $\sqrt{w}$ are among the _____ and _____ roots of 1.

15 Find the eigenvalues of the circulant second difference matrix C.

$$C = \begin{bmatrix} 2 & -1 & 0 & -1 \\ -1 & 2 & -1 & 0 \\ 0 & -1 & 2 & -1 \\ -1 & 0 & -1 & 2 \end{bmatrix} = 2I - P - P^3$$

6.5 Solving Linear Differential Equations

1 If $A\boldsymbol{x} = \lambda\boldsymbol{x}$ then $\boldsymbol{u}(t) = e^{\lambda t}\boldsymbol{x}$ will solve $\dfrac{d\boldsymbol{u}}{dt} = A\boldsymbol{u}$. Each λ and $\boldsymbol{x}$ give a solution $e^{\lambda t}\boldsymbol{x}$.

2 If $A = X\Lambda X^{-1}$ then $\boxed{\boldsymbol{u}(t) = e^{At}\boldsymbol{u}(0) = Xe^{\Lambda t}X^{-1}\boldsymbol{u}(0) = c_1 e^{\lambda_1 t}\boldsymbol{x}_1 + \cdots + c_n e^{\lambda_n t}\boldsymbol{x}_n.}$

3 Matrix exponential $\boldsymbol{e^{At}} = I + At + \cdots + (At)^n/n! + \cdots = Xe^{\Lambda t}X^{-1}$ if $A = X\Lambda X^{-1}$.

4 ***A* is stable** and $\boldsymbol{u}(t) \to \mathbf{0}$ and $e^{At} \to 0$ when all eigenvalues of A have **real part $<$ 0**.

5 **Second order eqn** / **First order system** $\boldsymbol{u'' + Bu' + Cu = 0}$ means $\begin{bmatrix} u \\ u' \end{bmatrix}' = \begin{bmatrix} 0 & 1 \\ -C & -B \end{bmatrix}\begin{bmatrix} u \\ u' \end{bmatrix}$.

Eigenvalues and eigenvectors and $A = X\Lambda X^{-1}$ are perfect for matrix powers A^k. They are also perfect for differential equations $d\boldsymbol{u}/dt = A\boldsymbol{u}$. This section is mostly linear algebra, but to read it you need one fact from calculus: *The derivative of* $e^{\lambda t}$ *is* $\lambda e^{\lambda t}$. The whole point of the section is this: **Constant coefficient differential equations can be converted into linear algebra.** Single equations in (1) and systems $\boldsymbol{du/dt = Au}$ in (2).

The ordinary equations $\dfrac{du}{dt} = u$ and $\dfrac{du}{dt} = \lambda u$ are solved by exponentials:

$$\boxed{\frac{\boldsymbol{du}}{\boldsymbol{dt}} = \boldsymbol{u} \text{ produces } \boldsymbol{u(t) = Ce^t} \qquad \frac{\boldsymbol{du}}{\boldsymbol{dt}} = \boldsymbol{\lambda u} \text{ produces } \boldsymbol{u(t) = Ce^{\lambda t}}} \tag{1}$$

At time $t = 0$ those solutions start from $u(0) = C$ because $e^0 = 1$. This "initial value" tells us C. **The solutions $\boldsymbol{u(t) = u(0)e^t}$ and $\boldsymbol{u(t) = u(0)e^{\lambda t}}$ start from $\boldsymbol{u(0)}$.**

We just solved a 1 by 1 problem. Linear algebra moves to n by n. The unknown is $\boldsymbol{u}(t)$, a vector changing with time t. It starts from $\boldsymbol{u}(0)$. The n equations contain a square matrix A. **We combine $\boldsymbol{n}$ solutions $\boldsymbol{u(t) = e^{\lambda t}x}$** from n eigenvalues $A\boldsymbol{x} = \lambda\boldsymbol{x}$:

$$\textbf{System of } \boldsymbol{n} \textbf{ equations} \quad \boxed{\frac{d\boldsymbol{u}}{dt} = A\boldsymbol{u}} \quad \text{starting from the vector } \boldsymbol{u}(0) = \begin{bmatrix} u_1(0) \\ \cdots \\ u_n(0) \end{bmatrix} \text{ at } t = 0. \tag{2}$$

These differential equations are *linear*. If $\boldsymbol{u}(t)$ and $\boldsymbol{v}(t)$ are solutions, so is $C\boldsymbol{u}(t) + D\boldsymbol{v}(t)$. We will need n constants like C and D to match the n components of $\boldsymbol{u}(0)$ at time zero. Our first job is to find n "pure exponential solutions" $\boldsymbol{u} = e^{\lambda t}\boldsymbol{x}$ by using $A\boldsymbol{x} = \lambda\boldsymbol{x}$.

Notice that A is a *constant* matrix. In other linear equations, A changes as t changes. In nonlinear equations, A changes as $\boldsymbol{u}$ changes. We don't have those difficulties, $d\boldsymbol{u}/dt = A\boldsymbol{u}$ is "linear with constant coefficients". Those and only those are the differential equations that we will convert directly to linear algebra.

Solve linear constant coefficient equations by exponentials $\boldsymbol{u} = e^{\lambda t}\boldsymbol{x}$ ***when*** $A\boldsymbol{x} = \lambda\boldsymbol{x}$.

Solution of $du/dt = Au$

Our pure exponential solution will be $e^{\lambda t}$ times a fixed vector $\boldsymbol{x}$. You may guess that λ is an eigenvalue of A, and $\boldsymbol{x}$ *is the eigenvector*. Substitute $\boldsymbol{u}(t) = e^{\lambda t}\boldsymbol{x}$ into the equation $d\boldsymbol{u}/dt = A\boldsymbol{u}$ to prove you are right. The factor $e^{\lambda t}$ will cancel to leave $\lambda\boldsymbol{x} = A\boldsymbol{x}$:

$$\textbf{Choose } \boldsymbol{u} = e^{\lambda t}\boldsymbol{x} \textbf{ when } A\boldsymbol{x} = \lambda\boldsymbol{x} \qquad \boxed{\frac{d\boldsymbol{u}}{dt} = \lambda e^{\lambda t}\boldsymbol{x}} \quad \textbf{agrees with} \quad \boxed{A\boldsymbol{u} = Ae^{\lambda t}\boldsymbol{x}} \tag{3}$$

All components of this special solution $\boldsymbol{u} = e^{\lambda t}\boldsymbol{x}$ share the same $e^{\lambda t}$. The solution grows when $\lambda > 0$. It decays when $\lambda < 0$. If λ is a complex number, its real part decides growth or decay. The imaginary part ω gives oscillation $e^{i\omega t}$ like a sine wave.

Example 1 Solve $\dfrac{d\boldsymbol{u}}{dt} = A\boldsymbol{u} = \begin{bmatrix} 0 & 1 \\ 1 & 0 \end{bmatrix}\boldsymbol{u}$ starting from $\boldsymbol{u}(0) = \begin{bmatrix} 4 \\ 2 \end{bmatrix}$.

This is a vector equation for $\boldsymbol{u}$. It contains two scalar equations for the components y and z. They are "coupled together" because the matrix A is not diagonal:

$$\frac{d\boldsymbol{u}}{dt} = A\boldsymbol{u} \qquad \frac{d}{dt}\begin{bmatrix} y \\ z \end{bmatrix} = \begin{bmatrix} 0 & 1 \\ 1 & 0 \end{bmatrix}\begin{bmatrix} y \\ z \end{bmatrix} \quad \text{means that} \quad \frac{d\boldsymbol{y}}{dt} = \boldsymbol{z} \text{ and } \frac{d\boldsymbol{z}}{dt} = \boldsymbol{y}.$$

The idea of eigenvectors is to combine those equations in a way that gets back to 1 by 1 problems. The combinations $y + z$ and $y - z$ will do it. Add and subtract equations:

$$\frac{d}{dt}(y+z) = z + y \qquad \text{and} \qquad \frac{d}{dt}(y-z) = -(y-z).$$

The combination $\boldsymbol{y} + \boldsymbol{z}$ grows like $\boldsymbol{e^t}$, because it has $\lambda = 1$. The combination $\boldsymbol{y} - \boldsymbol{z}$ decays like $\boldsymbol{e^{-t}}$, because it has $\lambda = -1$. Here is the point: We don't have to juggle the original equations $d\boldsymbol{u}/dt = A\boldsymbol{u}$, looking for these special combinations. The eigenvectors and eigenvalues of A will do it for us.

This matrix A has eigenvalues 1 and -1. The eigenvectors $\boldsymbol{x}$ are $(1, 1)$ and $(1, -1)$. The pure exponential solutions $\boldsymbol{u}_1$ and $\boldsymbol{u}_2$ take the form $e^{\lambda t}\boldsymbol{x}$ with $\lambda_1 = 1$ and $\lambda_2 = -1$:

$$\boxed{\boldsymbol{u}_1(t) = e^{\lambda_1 t}\boldsymbol{x}_1 = e^t\begin{bmatrix} 1 \\ 1 \end{bmatrix}} \quad \text{and} \quad \boxed{\boldsymbol{u}_2(t) = e^{\lambda_2 t}\boldsymbol{x}_2 = e^{-t}\begin{bmatrix} 1 \\ -1 \end{bmatrix}.} \tag{4}$$

$$\begin{array}{l}\textbf{Complete solution } \boldsymbol{u}(t) \\ \textbf{Combine } \boldsymbol{u}_1 \textbf{ and } \boldsymbol{u}_2\end{array} \qquad \boldsymbol{u}(t) = \boldsymbol{C}\boldsymbol{e^t}\begin{bmatrix} 1 \\ 1 \end{bmatrix} + \boldsymbol{D}\boldsymbol{e^{-t}}\begin{bmatrix} 1 \\ -1 \end{bmatrix} = \begin{bmatrix} \boldsymbol{C}\boldsymbol{e^t} + \boldsymbol{D}\boldsymbol{e^{-t}} \\ \boldsymbol{C}\boldsymbol{e^t} - \boldsymbol{D}\boldsymbol{e^{-t}} \end{bmatrix}. \tag{5}$$

With these two constants C and D, we can match any starting vector $\boldsymbol{u}(0) = (u_1(0), u_2(0))$. Set $t = 0$ and $e^0 = 1$. Example 1 asked for the initial value to be $\boldsymbol{u}(0) = (\boldsymbol{4}, \boldsymbol{2})$:

$$\begin{array}{c}\boldsymbol{u}(0) \textbf{ decides} \\ \boldsymbol{C} \textbf{ and } \boldsymbol{D}\end{array} \qquad C\begin{bmatrix} 1 \\ 1 \end{bmatrix} + D\begin{bmatrix} 1 \\ -1 \end{bmatrix} = \begin{bmatrix} 4 \\ 2 \end{bmatrix} \quad \text{yields} \quad \boldsymbol{C = 3} \quad \text{and} \quad \boldsymbol{D = 1}.$$

With $C = 3$ and $D = 1$ in the solution (5), the initial value problem is completely solved.

For n by n matrices we look for n eigenvectors. The numbers C and D become c_1 to c_n.

> **1.** Write $\boldsymbol{u}(0)$ as a **combination** $\boldsymbol{c_1 x_1} + \cdots + \boldsymbol{c_n x_n}$ **of the eigenvectors of** $\boldsymbol{A}$.
>
> **2.** Multiply each eigenvector $\boldsymbol{x}_i$ by **its growth factor** $\boldsymbol{e^{\lambda_i t}}$.
>
> **3.** The solution to $du/dt = Au$ is the same combination of those pure solutions $e^{\lambda t}\boldsymbol{x}$:
>
> $$\boxed{\boldsymbol{u(t) = c_1 e^{\lambda_1 t} x_1 + \cdots + c_n e^{\lambda_n t} x_n}.} \qquad (6)$$

Not included: If two λ's are equal, with only one eigenvector, another solution is needed. (It will be $te^{\lambda t}\boldsymbol{x}$.) Step 1 needs a basis of n eigenvectors to diagonalize $A = X\Lambda X^{-1}$.

Example 2 Solve $du/dt = Au$ knowing the eigenvalues $\lambda = 1, 2, 3$ of A:

Example of $du/dt = Au$
Equation for $u(t)$
Initial condition $u(0)$

$$\frac{du}{dt} = \begin{bmatrix} 1 & 1 & 1 \\ 0 & 2 & 1 \\ 0 & 0 & 3 \end{bmatrix} u \quad \text{starting from} \quad u(0) = \begin{bmatrix} 9 \\ 7 \\ 4 \end{bmatrix}.$$

The eigenvector matrix X has $\boldsymbol{x}_1 = (1, 0, 0)$ and $\boldsymbol{x}_2 = (1, 1, 0)$ and $\boldsymbol{x}_3 = (1, 1, 1)$.

> **Step 1** The vector $\boldsymbol{u}(0) = (9, 7, 4)$ is $\boldsymbol{2x_1 + 3x_2 + 4x_3}$. Then $(c_1, c_2, c_3) = (\boldsymbol{2, 3, 4})$.
>
> **Step 2** The factors $e^{\lambda t}$ give exponential solutions $\boldsymbol{u} = e^t\boldsymbol{x}_1$ and $e^{2t}\boldsymbol{x}_2$ and $e^{3t}\boldsymbol{x}_3$.
>
> **Step 3** The combination that starts from $\boldsymbol{u}(0)$ is $\boldsymbol{u}(t) = \mathbf{2}e^t\boldsymbol{x}_1 + \mathbf{3}e^{2t}\boldsymbol{x}_2 + \mathbf{4}e^{3t}\boldsymbol{x}_3$.

The coefficients 2, 3, 4 came from solving the linear equation $c_1\boldsymbol{x}_1 + c_2\boldsymbol{x}_2 + c_3\boldsymbol{x}_3 = \boldsymbol{u}(0)$:

$$\begin{bmatrix} & & \\ \boldsymbol{x}_1 & \boldsymbol{x}_2 & \boldsymbol{x}_3 \\ & & \end{bmatrix} \begin{bmatrix} c_1 \\ c_2 \\ c_3 \end{bmatrix} = \begin{bmatrix} 1 & 1 & 1 \\ 0 & 1 & 1 \\ 0 & 0 & 1 \end{bmatrix} \begin{bmatrix} 2 \\ 3 \\ 4 \end{bmatrix} = \begin{bmatrix} \mathbf{9} \\ \mathbf{7} \\ \mathbf{4} \end{bmatrix} \quad \text{which is} \quad X\boldsymbol{c} = \boldsymbol{u}(0). \qquad (7)$$

You now have the basic idea—how to solve $du/dt = Au$. The rest of this section goes further. We solve equations that contain *second* derivatives, because they arise so often in applications. We also decide whether $\boldsymbol{u}(t)$ approaches zero or blows up or just oscillates.

At the end comes the ***matrix exponential*** e^{At}. The short formula $e^{At}\boldsymbol{u}(0)$ solves the equation $du/dt = Au$ in the same way that $A^k\boldsymbol{u}_0$ solves the equation $\boldsymbol{u}_{k+1} = A\boldsymbol{u}_k$. Example 3 will show how "difference equations" help to solve differential equations.

All these steps use the λ's and the $\boldsymbol{x}$'s. This section solves the constant coefficient problems that turn into linear algebra. Those are the simplest but most important differential equations—whose solution is completely based on growth factors $e^{\lambda t}$.

Second Order Equations

The most important equation in mechanics is $\boldsymbol{my'' + by' + ky = 0}$. The first term is the mass m times the acceleration $a = y''$. Then by' is damping and ky is force.

The unknown $y(t)$ is the position of the mass m at the end of the spring. This is a second-order equation (it is *Newton's Law* $\boldsymbol{F} = \boldsymbol{ma}$). It contains the second derivative $\boldsymbol{y''} = \boldsymbol{d^2y/dt^2}$. It is still linear with constant coefficients m, b, k.

In a differential equations course, the method of solution is to substitute $y = e^{\lambda t}$. Each derivative of y brings a factor λ. We want $y = e^{\lambda t}$ to solve the spring equation:

$$\boxed{m\frac{d^2y}{dt^2} + b\frac{dy}{dt} + ky = 0 \quad \text{becomes} \quad (m\lambda^2 + b\lambda + k)\,e^{\lambda t} = 0.} \tag{8}$$

Everything depends on $m\lambda^2 + b\lambda + k = 0$. This equation (8) for λ has two roots λ_1 and λ_2. Then the equation for y has two pure solutions $y_1 = e^{\lambda_1 t}$ and $y_2 = e^{\lambda_2 t}$. Their combinations $c_1y_1 + c_2y_2$ give the complete solution. *This is not true if* $\lambda_1 = \lambda_2$.

In a linear algebra course we expect matrices and eigenvalues. Therefore we turn the scalar equation (8) (with y'') into a *vector equation for y and y'*: **First derivative only!** Suppose the mass is $m = 1$. Two equations for $\boldsymbol{u} = (y,\ y')$ give $d\boldsymbol{u}/dt = A\boldsymbol{u}$:

$$\begin{aligned} dy/dt &= y' \\ dy'/dt &= -ky - by' \end{aligned} \quad \text{converts to} \quad \frac{d}{dt}\begin{bmatrix} y \\ y' \end{bmatrix} = \begin{bmatrix} 0 & 1 \\ -k & -b \end{bmatrix}\begin{bmatrix} y \\ y' \end{bmatrix} = A\boldsymbol{u}. \tag{9}$$

The first equation $dy/dt = y'$ is trivial (but true). The second is equation (8) connecting y'' to y' and y. Together they connect $\boldsymbol{u}'$ to $\boldsymbol{u}$. Now we solve $\boldsymbol{u}' = A\boldsymbol{u}$ by eigenvalues of A:

$$A - \lambda I = \begin{bmatrix} -\lambda & 1 \\ -k & -b-\lambda \end{bmatrix} \quad \text{has determinant} \quad \lambda^2 + b\lambda + k = 0.$$

The equation for the λ's is the same as (8)! It is still $\lambda^2 + b\lambda + k = 0$, since $m = 1$.

The roots λ_1 and λ_2 are now *eigenvalues of A*. The eigenvectors and the solution are

$$\boldsymbol{x}_1 = \begin{bmatrix} 1 \\ \lambda_1 \end{bmatrix} \quad \boldsymbol{x}_2 = \begin{bmatrix} 1 \\ \lambda_2 \end{bmatrix} \quad \boldsymbol{u}(t) = c_1e^{\lambda_1 t}\begin{bmatrix} 1 \\ \lambda_1 \end{bmatrix} + c_2e^{\lambda_2 t}\begin{bmatrix} 1 \\ \lambda_2 \end{bmatrix}.$$

The first component of $\boldsymbol{u}(t)$ has $y = c_1e^{\lambda_1 t} + c_2e^{\lambda_2 t}$—the same solution as before. It can't be anything else. In the second component of $\boldsymbol{u}(t)$ you see the velocity dy/dt. The 2 by 2 matrix A in (9) is called a *companion matrix*—a companion to the equation (8).

Example 3 ***Undamped oscillation with $y'' + y = 0$ and $y = \cos t$***

This is our master equation with mass $m = 1$ and stiffness $k = 1$ and $d = 0$: no damping. Substitute $y = e^{\lambda t}$ into $y'' + y = 0$ to reach $\boldsymbol{\lambda^2 + 1 = 0}$. *The roots are $\boldsymbol{\lambda = i}$ and $\boldsymbol{\lambda = -i}$.* Then half of $e^{it} + e^{-it}$ gives the solution $y = \cos t$.

As a first-order system, the initial values $y(0) = 1$, $y'(0) = 0$ go into $\boldsymbol{u}(0) = (1, 0)$:

$$\textbf{Use } \boldsymbol{y'' = -y} \qquad \frac{d\boldsymbol{u}}{dt} = \frac{d}{dt}\begin{bmatrix} y \\ y' \end{bmatrix} = \begin{bmatrix} 0 & 1 \\ -1 & 0 \end{bmatrix}\begin{bmatrix} y \\ y' \end{bmatrix} = A\boldsymbol{u}. \tag{10}$$

The eigenvalues of A are again the same $\lambda = i$ and $\lambda = -i$ (no surprise). As $y = \cos t$ oscillates down and up, the vector $(\boldsymbol{y(t), y'(t)})\,(\boldsymbol{\cos t, -\sin t})$ goes around a circle.

A is anti-symmetric with eigenvectors $\boldsymbol{x}_1 = (1, i)$ and $\boldsymbol{x}_2 = (1, -i)$. The combination that matches $\boldsymbol{u}(0) = (1, 0)$ is $\frac{1}{2}(\boldsymbol{x}_1 + \boldsymbol{x}_2)$. Step 2 multiplies the $\boldsymbol{x}$'s by e^{it} and e^{-it}. Step 3 combines the pure oscillations e^{it} and e^{-it} to find $y = \cos t$ as expected: All good. The circular path of $\boldsymbol{u} = (\cos t, -\sin t)$ is drawn in Figure 6.6. The radius is 1 because $\cos^2 t + \sin^2 t = 1$. The "period" 2π is the time for a full circle.

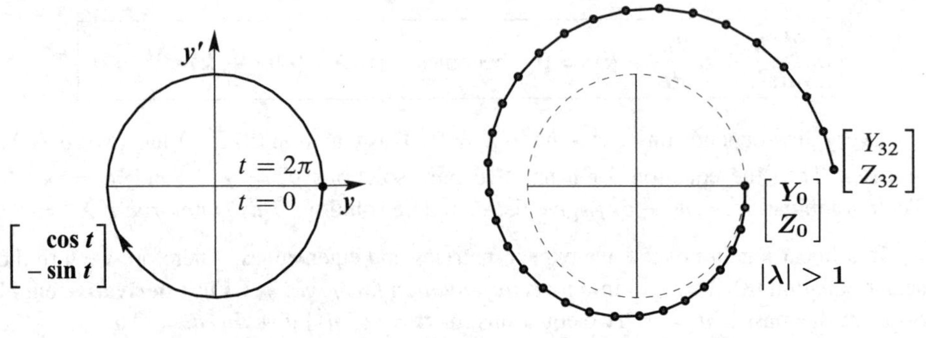

Figure 6.6: The exact solution $\boldsymbol{u} = (\cos t, -\sin t)$ stays on a circle, completed at $t = 2\pi$. Forward differences $\boldsymbol{Y_1}, \boldsymbol{Y_2}, \ldots$ in (11 **F**) spiral out in 32 steps, overshooting $y(2\pi) = 1$.

Difference Equations

To display a circle on a screen, replace $y'' = -y$ by a ***difference equation***. **Here are three choices using** $\boldsymbol{Y}(t+\Delta t) - 2\boldsymbol{Y}(t) + \boldsymbol{Y}(t - \Delta t)$. Divide by $(\Delta t)^2$ to approximate d^2y/dt^2.

F	**Forward from time $n-1$**			$-Y_{n-1}$	(11 **F**)
C	**Centered at time n**	$\dfrac{Y_{n+1} - 2Y_n + Y_{n-1}}{(\Delta t)^2}$	$=$	$-Y_n$	(11 **C**)
B	**Backward from time $n+1$**			$-Y_{n+1}$	(11 **B**)

Figure 6.6 shows the exact $y(t) = \cos t$ completing a circle at $t = 2\pi$. The three difference methods *don't* complete a perfect circle in 32 time steps of length $\Delta t = 2\pi/32$. The spirals in those pictures will be explained by eigenvalues λ for 11**F**, 11**B**, 11**C**.

Forward $|\lambda| > 1$ (spiral out) Centered $|\lambda| = 1$ (best) Backward $|\lambda| < 1$ (spiral in)

The 2-step equations (11) reduce to 1-step systems $\boldsymbol{U}_{n+1} = A\boldsymbol{U}_n$. To replace $\boldsymbol{u} = (y, y')$ the unknown is $\boldsymbol{U}_n = (Y_n, Z_n)$. **We take n time steps of size Δt starting from U_0:**

$$\begin{array}{ll} \textbf{Forward} & Y_{n+1} = Y_n + \Delta t\, Z_n \\ \textbf{(11F)} & Z_{n+1} = Z_n - \Delta t\, Y_n \end{array} \quad \text{becomes} \quad \boldsymbol{U}_{n+1} = \begin{bmatrix} 1 & \Delta t \\ -\Delta t & 1 \end{bmatrix} \begin{bmatrix} Y_n \\ Z_n \end{bmatrix} = A\boldsymbol{U}_n \quad (12)$$

Those are like $\boldsymbol{Y}' = \boldsymbol{Z}$ and $\boldsymbol{Z}' = -\boldsymbol{Y}$. They are first order equations, so we have a matrix in $\boldsymbol{U}_{n+1} = A\boldsymbol{U}_n$. Eliminating Z would recover the second order equation (11 **F**) for Y.

My question is simple. *Do the points* (Y_n, Z_n) *stay on the circle* $Y^2 + Z^2 = 1$? No, they are growing to infinity in Figure 6.6. ***We are taking powers*** A^n ***and not*** e^{At}***, so we test the magnitude*** $|\lambda| = \sqrt{1 + (\Delta t)^2}$ ***of the eigenvalues*** λ ***of A in (12)***:

$$\textbf{Eigenvalues of } A \quad \lambda = 1 \pm i\Delta t \quad \textbf{Then } |\lambda| > 1 \textbf{ and } (Y_n, Z_n) \textbf{ spirals out}$$

The backward choice in (11 **B**) will do the opposite in Figure 6.7. Notice the new matrix:

$$\begin{array}{ll}\textbf{Backward} & Y_{n+1} - Y_n = \Delta t\, Z_{n+1} \\ \textbf{Difference} & Z_{n+1} - Z_n = -\Delta t\, Y_{n+1}\end{array} \text{ is } \begin{bmatrix} 1 & -\Delta t \\ \Delta t & 1 \end{bmatrix}\begin{bmatrix} Y_{n+1} \\ Z_{n+1} \end{bmatrix} = \begin{bmatrix} Y_n \\ Z_n \end{bmatrix} = \boldsymbol{U}_n. \quad (13)$$

That matrix has eigenvalues $1 \pm i\Delta t$. But we *invert it* to reach $\boldsymbol{U}_{n+1}$ from $\boldsymbol{U}_n$. $|\lambda| < 1$ explains why *the solution spirals inward* (left figure) for backward differences.

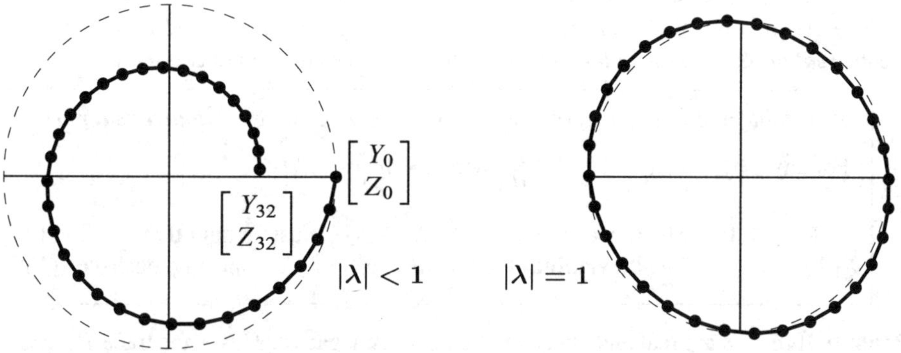

Figure 6.7: *Backward differences* (**11B**) spiral in. *Leapfrog* (**11C**) stays near the circle.

The second figure shows 32 steps with the *centered* choice (11 **C**). Now the solution stays close to the circle (Problem 28) if $\Delta t < 2$. This is the **leapfrog method**, constantly used. The second difference $\boldsymbol{Y_{n+1} - 2Y_n + Y_{n-1}}$ "leaps over" the value Y_n in (11 **C**).

This is the way a chemist follows the motion of molecules (molecular dynamics leads to giant computations). Computational science is lively because one differential equation can be replaced by many difference equations—some unstable, some stable, some neutral. Problem 26 has a fourth (very good) method that exactly completes the circle.

Real engineering and real physics deal with systems of equations (not just a single mass at one point). The unknown $\boldsymbol{y}$ is a vector. The coefficient of $\boldsymbol{y}''$ is a *mass matrix* M: n masses on the main diagonal. The coefficient of $\boldsymbol{y}$ is a *stiffness matrix* K, not a number k. The coefficient of $\boldsymbol{y}'$ is a damping matrix which might be zero.

The vector equation $\boldsymbol{My'' + Ky = f}$ is a major part of computational mechanics.

Stability of 2 by 2 Matrices

For the solution of $du/dt = Au$, there is a fundamental question. *Does the solution approach* $\boldsymbol{u} = \mathbf{0}$ *as* $t \to \infty$? *Is the problem stable*? Does energy dissipate? A solution that includes e^t is unstable. Stability depends on the eigenvalues of A.

The complete solution $\boldsymbol{u}(t)$ is built from pure solutions $e^{\lambda t}\boldsymbol{x}$. If the eigenvalue λ is real, we know exactly when $e^{\lambda t}$ will approach zero: *The number* λ *must be negative.* If the eigenvalue is a complex number $\lambda = r + is$, *the real part* r *must be negative.* When $e^{\lambda t}$ splits into $e^{rt}e^{ist}$, the factor e^{ist} has absolute value fixed at 1:

Euler's formula $\quad e^{ist} = \cos st + i \sin st \quad$ has $\quad |e^{ist}|^2 = \cos^2 st + \sin^2 st = 1.$

Then $|e^{\lambda t}| = e^{rt}$ and the real part of λ controls the growth ($r > 0$) or the decay ($r < 0$).

The question is: *Which matrices have negative eigenvalues*? More accurately, when are the ***real parts of the*** $\boldsymbol{\lambda}$***'s all negative***? 2 by 2 matrices must pass two tests.

> A is ***stable*** and $\boldsymbol{u}(t) \to \mathbf{0}$ when all eigenvalues λ of A have ***negative real parts.***
>
> For any 2 by 2 matrix $A = \begin{bmatrix} a & b \\ c & d \end{bmatrix}$ here are the two tests:
>
> $\boldsymbol{\lambda_1 + \lambda_2 < 0}$ The **trace** $T = a + d = \lambda_1 + \lambda_2$ must be **negative.** (14T)
>
> $\boldsymbol{\lambda_1 \lambda_2 > 0}$ The **determinant** $D = ad - bc = \lambda_1\lambda_2$ must be **positive.** (14D)

Reason If the λ's are real and negative, their sum is negative. This is the trace $T = a + d$. Their product is positive. This is the determinant D. The argument also goes in reverse.

If $D = \lambda_1\lambda_2$ is positive, then λ_1 and λ_2 have the same sign.

If $T = \lambda_1 + \lambda_2$ is negative, that sign will be negative.

When λ_1 and λ_2 are complex numbers, they must have the form $r + is$ and $r - is$. Otherwise T and D will not be real. The determinant D is automatically positive, since $(r + is)(r - is) = r^2 + s^2$. The trace T is $r + is + r - is = 2r$. So a negative trace T means that $r < 0$ and the matrix is stable. The two tests in (14) are correct.

The Exponential of a Matrix

We want to write the solution $\boldsymbol{u(t)}$ in a new form $\boldsymbol{e^{At}u(0)}$. Here is $\boldsymbol{e^{At}}$ for matrices!

Matrix exponential $\boldsymbol{e^{At}}$ $\quad e^{At} = I + At + \frac{1}{2}(At)^2 + \frac{1}{6}(At)^3 + \cdots \quad (15)$

Its t derivative is $\boldsymbol{Ae^{At}}$ $\quad A + A^2t + \frac{1}{2}A^3t^2 + \cdots = Ae^{At}$

Its eigenvalues are $\boldsymbol{e^{\lambda t}}$ $\quad (I + At + \frac{1}{2}(At)^2 + \cdots)\,\boldsymbol{x} = (1 + \lambda t + \frac{1}{2}(\lambda t)^2 + \cdots)\,\boldsymbol{x}$

Then $\boldsymbol{u}(t) = e^{At}\boldsymbol{u}(0)$ solves $\dfrac{d\boldsymbol{u}}{dt} = Ae^{At}\boldsymbol{u}(0) = \boldsymbol{Au(t)}$,

The number that divides $(At)^n$ is "n factorial". This is $n! = (1)(2)\cdots(n-1)(n)$. The factorials after $1, 2, 6$ are $4! = 24$ and $5! = 120$. They grow quickly. The series always converges and its derivative is always Ae^{At}. Therefore $e^{At}\boldsymbol{u}(0)$ solves the differential equation with one quick formula—*even if there is a shortage of eigenvectors.*

This chapter emphasizes how to find $\boldsymbol{u}(t) = e^{At}\boldsymbol{u}(0)$ by diagonalization. Assume A does have n independent eigenvectors, so we can substitute $A = X\Lambda X^{-1}$ into the series for e^{At}. Whenever $X\Lambda X^{-1}X\Lambda X^{-1}$ appears, cancel $X^{-1}X$ in the middle:

Use the series $$\boldsymbol{e^{At}} = I + X\Lambda X^{-1}t + \tfrac{1}{2}(X\Lambda X^{-1}t)(X\Lambda X^{-1}t) + \cdots$$

Factor out X and X^{-1} $$= \boldsymbol{X}\left[I + \Lambda t + \tfrac{1}{2}(\Lambda t)^2 + \cdots\right]\boldsymbol{X^{-1}}$$

e^{At} is diagonalized! $$\boxed{\boldsymbol{e^{At} = X\, e^{\Lambda t}\, X^{-1}}.}$$

e^{At} has the same eigenvector matrix X as A. Then Λ is a diagonal matrix and so is $e^{\Lambda t}$. The numbers $e^{\lambda_i t}$ are on the diagonal. Multiply $Xe^{\Lambda t}X^{-1}\boldsymbol{u}(0)$ to recognize $\boldsymbol{u}(t)$:

$$e^{At}\boldsymbol{u}(0) = Xe^{\Lambda t}X^{-1}\boldsymbol{u}(0) = \begin{bmatrix} & & \\ \boldsymbol{x}_1 & \cdots & \boldsymbol{x}_n \\ & & \end{bmatrix}\begin{bmatrix} e^{\lambda_1 t} & & \\ & \ddots & \\ & & e^{\lambda_n t}\end{bmatrix}\begin{bmatrix} c_1 \\ \vdots \\ c_n\end{bmatrix}. \qquad (16)$$

This solution $e^{At}\boldsymbol{u}(0)$ is the same answer that came in equation (6) from three steps:

1. $\boldsymbol{u}(0) = c_1\boldsymbol{x}_1 + \cdots + c_n\boldsymbol{x}_n = X\boldsymbol{c}$. Here we need n independent eigenvectors.
2. Multiply each $\boldsymbol{x}_i$ by its growth factor $e^{\lambda_i t}$ to follow it forward in time.
3. The best form of $\boldsymbol{u}(t) = e^{At}\boldsymbol{u}(0)$ is $\quad \boldsymbol{u}(t) = \boldsymbol{c_1 e^{\lambda_1 t}x_1 + \cdots + c_n e^{\lambda_n t}x_n}$. (17)

Example 4 Use the infinite series to find e^{At} for $A = \left[\begin{smallmatrix} 0 & 1 \\ -1 & 0\end{smallmatrix}\right]$. Notice that $A^4 = I$:

$$A = \begin{bmatrix} & 1 \\ -1 & \end{bmatrix} \quad A^2 = \begin{bmatrix} -1 & \\ & -1\end{bmatrix} \quad A^3 = \begin{bmatrix} & -1 \\ 1 & \end{bmatrix} \quad A^4 = \begin{bmatrix} 1 & \\ & 1\end{bmatrix}.$$

A^5, A^6, A^7, A^8 will be a repeat of A, A^2, A^3, A^4. The top right corner has $1, 0, -1, 0$ repeating over and over in powers of A. Then $t - \frac{1}{6}t^3$ starts the infinite series for e^{At} in that top right corner, and $1 - \frac{1}{2}t^2$ starts the top left corner:

$$e^{At} = I + At + \tfrac{1}{2}(At)^2 + \tfrac{1}{6}(At)^3 + \cdots = \begin{bmatrix} 1 - \frac{1}{2}t^2 + \cdots & t - \frac{1}{6}t^3 + \cdots \\ -t + \frac{1}{6}t^3 - \cdots & 1 - \frac{1}{2}t^2 + \cdots \end{bmatrix}.$$

That matrix $\boldsymbol{e^{At}}$ shows the infinite series for $\mathbf{cos}\,\boldsymbol{t}$ and $\mathbf{sin}\,\boldsymbol{t}$: perfect.

$$\boldsymbol{A} = \begin{bmatrix} \mathbf{0} & \mathbf{1} \\ \mathbf{-1} & \mathbf{0}\end{bmatrix} \qquad \boldsymbol{e^{At}} = \begin{bmatrix} \cos t & \sin t \\ -\sin t & \cos t\end{bmatrix}. \qquad (18)$$

This A is antisymmetric ($\boldsymbol{A}^{\mathrm{T}} = -\boldsymbol{A}$). Its exponential e^{At} is an orthogonal matrix. The eigenvalues of A are i and $-i$. Then the eigenvalues of e^{At} are e^{it} and e^{-it}.

1 ***The inverse of*** $\boldsymbol{e^{At}}$ ***is always*** $\boldsymbol{e^{-At}}$.

2 ***The eigenvalues of*** $\boldsymbol{e^{At}}$ ***are always*** $\boldsymbol{e^{\lambda t}}$.

3 ***When A is antisymmetric,*** $\boldsymbol{e^{At}}$ ***is orthogonal. Inverse = transpose =*** $\boldsymbol{e^{-At}}$.

Antisymmetric is the same as skew-symmetric: $A^{\mathrm{T}} = -A$. Now A has pure imaginary eigenvalues like i and $-i$. Then e^{At} has eigenvalues like e^{it} and e^{-it}. Their absolute value is $|\lambda| = 1$: neutral stability, pure oscillation, energy is conserved. So $||\boldsymbol{u}(t)|| = ||\boldsymbol{u}(0)||$.

If A is triangular, the eigenvector matrix X is also triangular. So are X^{-1} and e^{At}. The solution $\boldsymbol{u}(t)$ is a combination of eigenvectors. Its short form is $e^{At}\boldsymbol{u}(0)$.

Example 5 Solve $\dfrac{d\boldsymbol{u}}{dt} = A\boldsymbol{u} = \begin{bmatrix} 1 & 1 \\ 0 & 2 \end{bmatrix}\boldsymbol{u}$ starting from $\boldsymbol{u}(0) = \begin{bmatrix} 2 \\ 1 \end{bmatrix}$ at $t = 0$.

Solution The eigenvalues 1 and 2 are on the diagonal of A (since A is triangular). The eigenvectors are $(1, 0)$ and $(1, 1)$. Then e^{At} produces $\boldsymbol{u}(t)$ for every $\boldsymbol{u}(0)$:

$$\boldsymbol{u}(t) = Xe^{\Lambda t}X^{-1}\boldsymbol{u}(0) \text{ is } \begin{bmatrix} 1 & 1 \\ 0 & 1 \end{bmatrix}\begin{bmatrix} e^t & \\ & e^{2t} \end{bmatrix}\begin{bmatrix} 1 & -1 \\ 0 & 1 \end{bmatrix}\boldsymbol{u}(0) = \begin{bmatrix} e^t & e^{2t} - e^t \\ 0 & e^{2t} \end{bmatrix}\boldsymbol{u}(0).$$

That last matrix is $\boldsymbol{e^{At}}$. It is nice because A is triangular. The situation is the same as for $A\boldsymbol{x} = \boldsymbol{b}$ and inverses. We don't need A^{-1} to find $\boldsymbol{x}$, and we don't need e^{At} to solve $d\boldsymbol{u}/dt = A\boldsymbol{u}$. But as quick formulas for the answers, $A^{-1}\boldsymbol{b}$ and $e^{At}\boldsymbol{u}(0)$ are unbeatable.

Example 6 Solve $y'' + 4y' + 3y = 0$ by substituting $e^{\lambda t}$ and also by linear algebra.

Solution Substituting $y = e^{\lambda t}$ yields $(\lambda^2 + 4\lambda + 3)e^{\lambda t} = 0$. That quadratic factors into $\lambda^2+4\lambda+3 = (\lambda+1)(\lambda+3) = 0$. Therefore $\boldsymbol{\lambda_1 = -1}$ and $\boldsymbol{\lambda_2 = -3}$. The pure solutions are $y_1 = e^{-t}$ and $y_2 = e^{-3t}$. The complete solution $\boldsymbol{y} = c_1y_1 + c_2y_2$ approaches zero.

To use linear algebra we set $\boldsymbol{u} = (y, y')$. Then the vector equation is $\boldsymbol{u}' = A\boldsymbol{u}$:

$$\begin{aligned} dy/dt &= y' \\ dy'/dt &= -3y - 4y' \end{aligned} \quad \text{converts to} \quad \frac{d\boldsymbol{u}}{dt} = \begin{bmatrix} 0 & 1 \\ -3 & -4 \end{bmatrix}\boldsymbol{u}.$$

This A is a "companion matrix" and its eigenvalues are again $\boldsymbol{\lambda_1 = -1}$ and $\boldsymbol{\lambda_2 = -3}$.

$$\textbf{Same quadratic} \qquad \det(A - \lambda I) = \begin{vmatrix} -\lambda & 1 \\ -3 & -4-\lambda \end{vmatrix} = \lambda^2 + 4\lambda + 3 = 0.$$

The eigenvectors of A are $(1, \lambda_1)$ and $(1, \lambda_2)$. Either way, the decay in $y(t)$ comes from $\boldsymbol{e^{-t}}$ and $\boldsymbol{e^{-3t}}$. With constant coefficients, calculus leads to linear algebra $A\boldsymbol{x} = \lambda\boldsymbol{x}$.

Example 7 Substituting $y = e^{\lambda t}$ into $y'' - 2y' + y = 0$ gives an equation with **repeated roots**: $\lambda^2 - 2\lambda + 1 = 0$ is $(\lambda - 1)^2 = 0$ with $\boldsymbol{\lambda = 1, 1}$. A differential equations course would propose e^t and te^t as two independent solutions. Here we discover why.

Linear algebra reduces $y'' = 2y' - y$ to a vector equation for $\boldsymbol{u} = (y, y')$:

$$\frac{d}{dt}\begin{bmatrix} y \\ y' \end{bmatrix} = \begin{bmatrix} y' \\ 2y' - y \end{bmatrix} \quad \text{is} \quad \frac{d\boldsymbol{u}}{dt} = A\boldsymbol{u} = \begin{bmatrix} \mathbf{0} & \mathbf{1} \\ \mathbf{-1} & \mathbf{2} \end{bmatrix} \boldsymbol{u}. \tag{20}$$

A has a **repeated eigenvalue** $\boldsymbol{\lambda = 1, 1}$ with trace $= 2$ and $\det A = 1$. The only eigenvectors are multiples of $\boldsymbol{x} = (1, 1)$. *Diagonalization is not possible.* This matrix A has only one line of eigenvectors. So we compute e^{At} from its definition as a series:

Short series $$\boldsymbol{e^{At}} = e^{It}\, e^{(A-I)t} = \boldsymbol{e^t\,[I + (A - I)t]}. \tag{21}$$

That "infinite" series for $e^{(A-I)t}$ ended quickly because $(A - I)^2$ is the zero matrix. **You can see $\boldsymbol{te^t}$** in equation (21). The first component of $\boldsymbol{e^{At}}\, \boldsymbol{u}(0)$ is our answer $\boldsymbol{y}(t)$:

$$\begin{bmatrix} y \\ y' \end{bmatrix} = e^t \left[\boldsymbol{I} + \begin{bmatrix} -1 & 1 \\ -1 & 1 \end{bmatrix} \boldsymbol{t} \right] \begin{bmatrix} y(0) \\ y'(0) \end{bmatrix} \qquad \boldsymbol{y(t) = e^t y(0) - te^t y(0) + te^t y'(0)}.$$

Our last examples are really pushing the limits of a linear algebra course! The ordinary differential equation $d\boldsymbol{u}/dt = A\boldsymbol{u}$ changes to the **heat equation** and the **wave equation**. These are **partial differential equations** $\boldsymbol{\partial u/\partial t = \partial^2 u/\partial x^2}$ and $\boldsymbol{\partial^2 u/\partial t^2 = \partial^2 u/\partial x^2}$.

Heat equation In Figure 6.8a, the temperature at the center starts at $2\sqrt{2}$. Heat diffuses into the neighboring boxes: $d\boldsymbol{u}/dt = A\boldsymbol{u}$ (outside boxes frozen at 0°). The rate of heat flow between boxes is the temperature difference. From box 0, heat flows right and left at the rate $u_1 - u_0$ and $u_{-1} - u_0$. So the flow out is $u_1 - 2u_0 + u_{-1}$ in the second row of $A\boldsymbol{u}$.

Wave equation $d^2\boldsymbol{u}/dt^2 = A\boldsymbol{u}$ has the same eigenvectors $\boldsymbol{x}$. But now eigenvalues $\lambda < 0$ lead to **oscillations** $e^{i\omega t}\boldsymbol{x}$ and $e^{-i\omega t}\boldsymbol{x}$. The frequencies come from $\omega^2 = -\lambda$:

$$\frac{d^2}{dt^2}(e^{i\omega t}\boldsymbol{x}) = A(e^{i\omega t}\boldsymbol{x}) \qquad \text{becomes} \qquad (i\omega)^2 e^{i\omega t}\boldsymbol{x} = \lambda e^{i\omega t}\boldsymbol{x} \quad \text{and} \quad \boldsymbol{\omega^2 = -\lambda}.$$

There are two square roots of $-\lambda$, so we have $e^{i\omega t}\boldsymbol{x}$ and $e^{-i\omega t}\boldsymbol{x}$. With three eigenvectors this makes *six* solutions to $\boldsymbol{u}'' = A\boldsymbol{u}$. A combination will match the six components of $\boldsymbol{u}(0)$ and $\boldsymbol{u}'(0)$. Since $\boldsymbol{u}' = \boldsymbol{0}$ in this problem, $e^{i\omega t}\boldsymbol{x}$ and $e^{-i\omega t}\boldsymbol{x}$ produce $2\cos\omega t\,\boldsymbol{x}$.

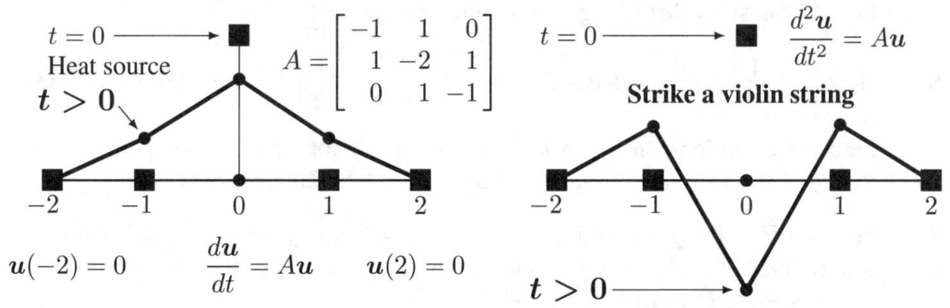

Figure 6.8: (**Heat equation**) Spike at $t=0$ diffuses away. (**Wave equation**) Waves on a string.

Problem Set 6.5

1 Find two λ's and $\boldsymbol{x}$'s so that $\boldsymbol{u} = e^{\lambda t}\boldsymbol{x}$ solves $\dfrac{d\boldsymbol{u}}{dt} = \begin{bmatrix} 4 & 3 \\ 0 & 1 \end{bmatrix} \boldsymbol{u}$.

What combination $\boldsymbol{u} = c_1 e^{\lambda_1 t}\boldsymbol{x}_1 + c_2 e^{\lambda_2 t}\boldsymbol{x}_2$ starts from $\boldsymbol{u}(0) = (5, -2)$?

2 Solve Problem 1 for $\boldsymbol{u} = (y, z)$ by back substitution, z before y:

$$\text{Solve } \frac{dz}{dt} = z \text{ from } z(0) = -2. \qquad \text{Then solve } \frac{dy}{dt} = 4y + 3z \text{ from } y(0) = 5.$$

The solution for y will be a combination of e^{4t} and e^{t}. The λ's are 4 and 1.

3 (a) If every column of A adds to zero, why is $\lambda = 0$ an eigenvalue ?

(b) With negative diagonal and positive off-diagonal adding to zero, $\boldsymbol{u}' = A\boldsymbol{u}$ will be a "continuous" Markov equation. Find the eigenvalues and eigenvectors, and the *steady state* as $t \to \infty$

$$\text{Solve } \frac{d\boldsymbol{u}}{dt} = \begin{bmatrix} -2 & 3 \\ 2 & -3 \end{bmatrix} \boldsymbol{u} \text{ with } \boldsymbol{u}(0) = \begin{bmatrix} 4 \\ 1 \end{bmatrix}. \quad \text{What is } \boldsymbol{u}(\infty)?$$

4 A door is opened between rooms that hold $v(0) = 30$ people and $w(0) = 10$ people. The movement between rooms is proportional to the difference $v - w$:

$$\frac{dv}{dt} = w - v \qquad \text{and} \qquad \frac{dw}{dt} = v - w.$$

Show that the total $v + w$ is constant (40 people). Find the matrix in $d\boldsymbol{u}/dt = A\boldsymbol{u}$ and its eigenvalues and eigenvectors. What are v and w at $t = 1$ and $t = \infty$?

5 Reverse the diffusion of people in Problem 4 to $d\boldsymbol{u}/dt = -A\boldsymbol{u}$:

$$\frac{dv}{dt} = v - w \qquad \text{and} \qquad \frac{dw}{dt} = w - v.$$

The total $v+w$ still remains constant. How are the λ's changed now that A is changed to $-A$? But show that $v(t)$ grows to infinity from $v(0) = 30$.

6 $A = \begin{bmatrix} a & 1 \\ 1 & a \end{bmatrix}$ has real eigenvalues but $B = \begin{bmatrix} b & -1 \\ 1 & b \end{bmatrix}$ has complex eigenvalues:

Find the conditions on a and b (real) so that all solutions of $d\boldsymbol{u}/dt = A\boldsymbol{u}$ and $d\boldsymbol{v}/dt = B\boldsymbol{v}$ approach zero as $t \to \infty$: $\text{Re}\,\lambda < 0$ for all eigenvalues.

7 Suppose P is the projection matrix onto the 45° line $y = x$ in $\mathbf{R}^2$. What are its eigenvalues? If $d\boldsymbol{u}/dt = -P\boldsymbol{u}$ (notice minus sign) can you find the limit of $\boldsymbol{u}(t)$ at $t = \infty$ starting from $\boldsymbol{u}(0) = (3, 1)$?

8 The rabbit population shows fast growth (from $6r$) but loss to wolves (from $-2w$). The wolf population always grows in this model ($-w^2$ would control wolves):

$$\frac{dr}{dt} = 6r - 2w \quad \text{and} \quad \frac{dw}{dt} = 2r + w.$$

Find the eigenvalues and eigenvectors. If $r(0) = w(0) = 30$ what are the populations at time t? After a long time, what is the ratio of rabbits to wolves?

9 (a) Write $(4, 0)$ as a combination $c_1\boldsymbol{x}_1 + c_2\boldsymbol{x}_2$ of these two eigenvectors of A:

$$\begin{bmatrix} 0 & 1 \\ -1 & 0 \end{bmatrix}\begin{bmatrix} 1 \\ i \end{bmatrix} = i\begin{bmatrix} 1 \\ i \end{bmatrix} \qquad \begin{bmatrix} 0 & 1 \\ -1 & 0 \end{bmatrix}\begin{bmatrix} 1 \\ -i \end{bmatrix} = -i\begin{bmatrix} 1 \\ -i \end{bmatrix}.$$

(b) The solution to $d\boldsymbol{u}/dt = A\boldsymbol{u}$ starting from $(4, 0)$ is $c_1e^{it}\boldsymbol{x}_1 + c_2e^{-it}\boldsymbol{x}_2$. Substitute $e^{it} = \cos t + i\sin t$ and $e^{-it} = \cos t - i\sin t$ to find $\boldsymbol{u}(t)$.

10 Find A to change $y'' = 5y' + 4y$ into a vector equation for $\boldsymbol{u} = (y, y')$:

$$\frac{d\boldsymbol{u}}{dt} = \begin{bmatrix} y' \\ y'' \end{bmatrix} = \begin{bmatrix} \quad & \quad \\ \quad & \quad \end{bmatrix}\begin{bmatrix} y \\ y' \end{bmatrix} = A\boldsymbol{u}.$$

What are the eigenvalues of A? Find them also from $y'' = 5y' + 4y$ with $y = e^{\lambda t}$.

11 The solution to $y'' = 0$ is a straight line $y = C + Dt$. Convert to a matrix equation:

$$\frac{d}{dt}\begin{bmatrix} y \\ y' \end{bmatrix} = \begin{bmatrix} 0 & 1 \\ 0 & 0 \end{bmatrix}\begin{bmatrix} y \\ y' \end{bmatrix} \text{ has the solution } \begin{bmatrix} y \\ y' \end{bmatrix} = e^{At}\begin{bmatrix} y(0) \\ y'(0) \end{bmatrix}.$$

This matrix A has $\lambda = 0, 0$ and it cannot be diagonalized. Find A^2 and compute $e^{At} = I + At + \frac{1}{2}A^2t^2 + \cdots$. Multiply your e^{At} times $\big(y(0), y'(0)\big)$ to check the straight line $y(t) = y(0) + y'(0)t$.

12 Substitute $y = e^{\lambda t}$ into $y'' = 6y' - 9y$ to show that $\lambda = 3$ is a repeated root. This is trouble; we need a second solution after e^{3t}. The matrix equation is

$$\frac{d}{dt}\begin{bmatrix} y \\ y' \end{bmatrix} = \begin{bmatrix} 0 & 1 \\ -9 & 6 \end{bmatrix}\begin{bmatrix} y \\ y' \end{bmatrix}.$$

Show that this matrix has $\lambda = 3, 3$ and only one line of eigenvectors. *Trouble here too.* Check that $y = te^{3t}$ is a second solution to $y'' = 6y' - 9y$.

Note In linear algebra the serious danger is a shortage of eigenvectors. Our eigenvectors $(1, \lambda_1)$ and $(1, \lambda_2)$ are the same if $\lambda_1 = \lambda_2$. Then we can't diagonalize A. In this case we don't yet have two independent solutions to $d\boldsymbol{u}/dt = A\boldsymbol{u}$.

In differential equations the danger is also a repeated λ. After $y = e^{\lambda t}$, a second solution has to be found. It turns out to be $y = \boldsymbol{te^{\lambda t}}$. This "impure" solution (with an extra $\boldsymbol{t}$) appears in the matrix exponential e^{At}. Example 7 showed how.

13 (a) Write down two familiar functions that solve the equation $d^2y/dt^2 = -9y$. Which one starts with $y(0) = 3$ and $y'(0) = 0$?

(b) This second-order equation $y'' = -9y$ produces a vector equation $\boldsymbol{u}' = A\boldsymbol{u}$:

$$\boldsymbol{u} = \begin{bmatrix} y \\ y' \end{bmatrix} \qquad \frac{d\boldsymbol{u}}{dt} = \begin{bmatrix} y' \\ y'' \end{bmatrix} = \begin{bmatrix} 0 & 1 \\ -9 & 0 \end{bmatrix} \begin{bmatrix} y \\ y' \end{bmatrix} = A\boldsymbol{u}.$$

Find $\boldsymbol{u}(t)$ by using the eigenvalues and eigenvectors of A: $\boldsymbol{u}(0) = (3, 0)$.

14 The matrix in this question is skew-symmetric ($A^{\mathrm{T}} = -A$):

$$\frac{d\boldsymbol{u}}{dt} = \begin{bmatrix} 0 & c & -b \\ -c & 0 & a \\ b & -a & 0 \end{bmatrix} \boldsymbol{u} \qquad \text{or} \qquad \begin{aligned} u_1' &= cu_2 - bu_3 \\ u_2' &= au_3 - cu_1 \\ u_3' &= bu_1 - au_2. \end{aligned}$$

(a) The derivative of $\|\boldsymbol{u}(t)\|^2 = u_1^2 + u_2^2 + u_3^2$ is $2u_1u_1' + 2u_2u_2' + 2u_3u_3'$. Substitute u_1', u_2', u_3' to get *zero*. Then $\|\boldsymbol{u}(t)\|^2$ stays equal to $\|\boldsymbol{u}(0)\|^2$.

(b) $A^{\mathrm{T}} = -A$ *makes* $Q = e^{At}$ *orthogonal*. Prove $Q^{\mathrm{T}} = e^{-At}$ from the series for Q.

15 A particular solution to $d\boldsymbol{u}/dt = A\boldsymbol{u} - \boldsymbol{b}$ is $\boldsymbol{u}_p = A^{-1}\boldsymbol{b}$, if A is invertible. The usual solutions to $d\boldsymbol{u}/dt = A\boldsymbol{u}$ give $\boldsymbol{u}_n$. Find the complete solution $\boldsymbol{u} = \boldsymbol{u}_p + \boldsymbol{u}_n$:

(a) $\displaystyle \frac{du}{dt} = u - 4$ (b) $\displaystyle \frac{d\boldsymbol{u}}{dt} = \begin{bmatrix} 1 & 0 \\ 1 & 1 \end{bmatrix} \boldsymbol{u} - \begin{bmatrix} 4 \\ 6 \end{bmatrix}.$

Questions 16–22 are about the matrix exponential e^{At}.

16 Write five terms of the infinite series for e^{At}. Take the t derivative of each term. Show that you have four terms of Ae^{At}. Conclusion: $e^{At}\boldsymbol{u}_0$ solves $\boldsymbol{u}' = A\boldsymbol{u}$.

17 The matrix $B = \left[\begin{smallmatrix} 0 & -4 \\ 0 & 0 \end{smallmatrix}\right]$ has $B^2 = 0$. Find e^{Bt} from a (short) infinite series. Check that the derivative of e^{Bt} is Be^{Bt}.

18 Starting from $\boldsymbol{u}(0)$ the solution at time T is $e^{AT}\boldsymbol{u}(0)$. Go an additional time t to reach $e^{At}\,e^{AT}\boldsymbol{u}(0)$. This solution at time $t + T$ can also be written as _____. Conclusion: e^{At} times e^{AT} equals _____.

19 If $A^2 = A$ show that the infinite series produces $e^{At} = I + (e^t - 1)A$.

20 Generally $e^Ae^B \neq e^Be^A$. They are both different from e^{A+B}. Check this for

$$A = \begin{bmatrix} 1 & 4 \\ 0 & 0 \end{bmatrix} \qquad B = \begin{bmatrix} 0 & -4 \\ 0 & 0 \end{bmatrix} \qquad A + B = \begin{bmatrix} 1 & 0 \\ 0 & 0 \end{bmatrix}.$$

21 Put $A = \left[\begin{smallmatrix} 1 & 3 \\ 0 & 0 \end{smallmatrix}\right]$ into the infinite series to find e^{At}. First compute A^2 and A^n :

$$e^{At} = \begin{bmatrix} 1 & 0 \\ 0 & 1 \end{bmatrix} + \begin{bmatrix} t & 3t \\ 0 & 0 \end{bmatrix} + \frac{1}{2}\begin{bmatrix} & \\ & \end{bmatrix} + \cdots = \begin{bmatrix} e^t & \\ 0 & \end{bmatrix}.$$

22 (Recommended) Give two reasons why the matrix exponential e^{At} is never singular:
(a) Write down its inverse. (b) Why are its eigenvalues $e^{\lambda t}$ nonzero?

23 $Y_{n+1} - 2Y_n + Y_{n-1} = -(\Delta t)^2 Y_n$ can be written as a one-step difference equation:

$$\begin{aligned} Y_{n+1} &= Y_n + \Delta t\, Z_n \\ Z_{n+1} &= Z_n - \Delta t\, Y_{n+1} \end{aligned} \qquad \begin{bmatrix} 1 & 0 \\ \Delta t & 1 \end{bmatrix}\begin{bmatrix} Y_{n+1} \\ Z_{n+1} \end{bmatrix} = \begin{bmatrix} 1 & \Delta t \\ 0 & 1 \end{bmatrix}\begin{bmatrix} Y_n \\ Z_n \end{bmatrix}$$

Invert the matrix on the left side to write this as $\boldsymbol{U}_{n+1} = A\boldsymbol{U}_n$. Show that $\det A = 1$. Choose the large time step $\Delta t = 1$ and find the eigenvalues λ_1 and $\lambda_2 = \overline{\lambda}_1$ of A:

$$\boldsymbol{A} = \begin{bmatrix} 1 & 1 \\ -1 & 0 \end{bmatrix} \textbf{ has } |\boldsymbol{\lambda_1}| = |\boldsymbol{\lambda_2}| = \mathbf{1}. \textbf{ Show that } \boldsymbol{A^6} = \boldsymbol{I} \text{ so } u_6 = u_0 \textbf{ exactly}.$$

24 That *leapfrog method* in Problem 23 is very successful for small time steps Δt. But find the eigenvalues of A for $\Delta t = \sqrt{2}$ and 2. Any time step $\Delta t > 2$ will lead to $|\lambda| > 1$, and the powers in $\boldsymbol{U}_n = A^n \boldsymbol{U}_0$ will explode.

$$A = \begin{bmatrix} 1 & \sqrt{2} \\ -\sqrt{2} & -1 \end{bmatrix} \quad \text{and} \quad A = \begin{bmatrix} 1 & 2 \\ -2 & -3 \end{bmatrix} \quad \begin{array}{l} \textbf{borderline} \\ \textbf{unstable} \end{array}$$

25 A very good idea for $y'' = -y$ is the trapezoidal method (half forward/half back). ***This may be the best way to keep*** $(\boldsymbol{Y_n}, \boldsymbol{Z_n})$ ***exactly on a circle.***

$$\textbf{Trapezoidal} \quad \begin{bmatrix} 1 & -\Delta t/2 \\ \Delta t/2 & 1 \end{bmatrix}\begin{bmatrix} Y_{n+1} \\ Z_{n+1} \end{bmatrix} = \begin{bmatrix} 1 & \Delta t/2 \\ -\Delta t/2 & 1 \end{bmatrix}\begin{bmatrix} Y_n \\ Z_n \end{bmatrix}.$$

(a) Invert the left matrix to write this equation as $\boldsymbol{U}_{n+1} = A\boldsymbol{U}_n$. *Show that A is an orthogonal matrix*: $A^{\mathrm{T}}A = I$. **These points $\boldsymbol{U}_n$ never leave the circle.** $A = (I - B)^{-1}(I + B)$ is always an orthogonal matrix if $B^{\mathrm{T}} = -B$.

(b) (Optional MATLAB) Take 32 steps from $\boldsymbol{U}_0 = (1, 0)$ to $\boldsymbol{U}_{32}$ with $\Delta t = 2\pi/32$. Is $\boldsymbol{U}_{32} = \boldsymbol{U}_0$? I think there is a small error.

26 Explain one of these three proofs that the square of e^A is e^{2A}.

1. Solving with e^A from $t = 0$ to 1 and then 1 to 2 agrees with e^{2A} from 0 to 2.
2. The squared series $(I + A + \frac{A^2}{2} + \cdots)^2$ matches $I + 2A + \frac{(2A)^2}{2} + \cdots = e^{2A}$.
3. If A can be diagonalized then $(Xe^{\Lambda}X^{-1})(Xe^{\Lambda}X^{-1}) = Xe^{2\Lambda}X^{-1}$.

Computing the Eigenvalues : The QR Algorithm

This page is about a very remarkable algorithm that finds the eigenvalues of A. The steps are easy to understand. Starting from A, we take a big step to a similar matrix $A_0 = B^{-1}AB$ (remember that similar matrices have the same eigenvalues). If A_0 were a triangular matrix, then its eigenvalues would be along its main diagonal and the problem would be solved.

This first step to A_0 can get close to triangular, but we have to allow one extra nonzero diagonal. This is Step 1 and we omit the details of B:

$$A_0 = B^{-1}AB = \begin{bmatrix} a_{11} & a_{12} & \cdot & \cdot & a_{1n} \\ \mathbf{a_{21}} & a_{22} & \cdot & \cdot & a_{2n} \\ 0 & \mathbf{a_{32}} & \cdot & \cdot & a_{3n} \\ 0 & 0 & \cdot & \cdot & \cdot \\ 0 & 0 & \cdot & a_{n\,n-1} & a_{nn} \end{bmatrix}$$

A_0 is a Hessenberg matrix : one nonzero subdiagonal. All the steps to $A_1, A_2, \ldots$ will produce Hessenberg matrices. The magic is to make that subdiagonal smaller and smaller. *Then the diagonal entries of the matrices $A_1, A_2, \ldots$ approach the eigenvalues of A.* Every step has the form $A_{k+1} = B_k^{-1}A_kB_k$, so all the A's are similar matrices—which guarantees no change in the eigenvalues.

The most beautiful and important matrices A are symmetric. In that case every step preserves the symmetry, provided B is an orthogonal matrix with $Q^{-1} = Q^{\mathrm{T}}$:

$$\text{If } A^{\mathrm{T}} = A \text{ then } A_0^{\mathrm{T}} = (Q^{-1}AQ)^{\mathrm{T}} = (Q^{\mathrm{T}}AQ)^{\mathrm{T}} = Q^{\mathrm{T}}AQ = Q^{-1}AQ = A_0. \tag{22}$$

Symmetry tells us : If there is only one nonzero subdiagonal, then there is only one nonzero superdiagonal. In other words, every step to $A_{k+1} = B_k^{-1}A_kB_k = Q_k^{-1}A_kQ_k$ will have only *three nonzero diagonals*. The computations are fast and so is the convergence.

Now we are ready to explain the step from A_k to $A_{k+1} = B_k^{-1}A_kB_k$. It begins with a Gram-Schmidt orthogonalization $A_k = Q_kR_k$. Remember from Section 4.4 that Q_k is orthogonal and R_k is upper triangular. **To find A_{k+1} we reverse those factors**:

$$\textbf{QR algorithm} \qquad \boxed{A_{k+1} = R_kQ_k = Q_k^{-1}A_kQ_k} \tag{23}$$

A_{k+1} has the same eigenvalues as A_k (*similar matrices*). By equation (22), A_{k+1} is symmetric if A_k is symmetric—because Q_k is orthogonal. The magic (that we will justify only by an example) is that the subdiagonals of $A_1, A_2, \ldots$ become smaller and smaller. The A's all have the same eigenvalues, which begin to appear quickly on the main diagonal.

$$\textbf{Example} \qquad A_0 = \begin{bmatrix} \cos\theta & \sin\theta \\ \sin\theta & 0 \end{bmatrix} = QR = \begin{bmatrix} \cos\theta & -\sin\theta \\ \sin\theta & \cos\theta \end{bmatrix} \begin{bmatrix} 1 & \cos\theta\sin\theta \\ 0 & -\sin^2\theta \end{bmatrix}$$

Reversing the factors to RQ produces A_1, with the same eigenvalues as A_0. But look how the off-diagonal entries have dropped from $\sin\theta$ to $-\sin^3\theta$:

$$A_1 = RQ = \begin{bmatrix} 1 & \cos\theta\sin\theta \\ 0 & -\sin^2\theta \end{bmatrix} \begin{bmatrix} \cos\theta & -\sin\theta \\ \sin\theta & \cos\theta \end{bmatrix} = \begin{bmatrix} \cos\theta(1+\sin^2\theta) & -\sin^3\theta \\ -\sin^3\theta & -\sin^2\theta\cos\theta \end{bmatrix}$$

Check: The trace is still $\cos\theta$ as in A_0. The determinant is still $-\sin^2\theta$, But the off-diagonal term has dropped from $\sin\theta$ in A_0 to $-\sin^3\theta$ in A_1. We are seeing cubic convergence (very rare)! The new matrix A_1 is nearly diagonal, and its last entry $-\sin^2\theta\cos\theta$ is very close to a true eigenvalue of A_0 and A_1.

It helped that the $(2,2)$ entry of A_0 was zero. If not, we can always shift A_0 by a matrix cI to achieve $(A_0)_{22}=0$, and shift back at the end of the step. The "QR algorithm with shifts" is a numerical success—very far from the original bad idea of computing the characteristic polynomial $\det(A-\lambda I)$ and finding its roots. That is a failure.

Computational linear algebra is a well-developed subject with excellent textbooks of its own. We mention two books here, and many more on the website.
Numerical Linear Algebra, SIAM, 1997, L.N. Trefethen and David Bau

And a comprehensive textbook of 756 valuable pages is
Matrix Computations, 4th Ed., Gene Golub and Charles van Loan, Johns Hopkins.

Thoughts about Differential Equations

Constant coefficient linear equations are the simplest to solve. This Section 6.5 shows you part of a differential equations course, but there is more. Here are two highlights:

1. The second order equation $mu'' + bu' + ku = 0$ has major importance in applications. The exponents λ in the solutions $u = e^{\lambda t}$ solve $m\lambda^2 + b\lambda + k = 0$.

   Underdamping $b^2 < 4mk$ **Critical damping** $b^2 = 4mk$ **Overdamping** $b^2 > 4mk$

   This decides whether λ_1 and λ_2 are real roots or repeated roots or complex roots. With complex $\lambda = a + i\omega$ the solution $u(t)$ oscillates from $e^{i\omega t}$ as it decays from e^{at}.

2. Our equations had no forcing term $f(t)$. We were finding the "nullspace solution". To $\boldsymbol{u_n(t)}$ we need to add a particular solution $\boldsymbol{u_p(t)}$ that balances the force $f(t)$. This solution can also be discovered and studied by *Laplace transform*:

   Input $f(s)$ at time s
   Growth factor $e^{A(t-s)}$
   Add up outputs at time t

$$u_{\text{particular}} = \int_0^t e^{A(t-s)} f(s)\,ds.$$

In real applications, nonlinear differential equations are solved numerically. A method with good accuracy is "Runge-Kutta". The constant solutions to $du/dt = f(u)$ are $u(t) = Y$ with $f(Y) = 0$. Then $du/dt = 0$: *no movement.* Far from Y, the computer takes over.

This basic course is the subject of my textbook (a companion to this one) on **Differential Equations and Linear Algebra**. The website is **math.mit.edu / dela**.

The individual sections of that book are described in a series of short videos, and a parallel series about numerical solutions was prepared by Cleve Moler:

ocw.mit.edu/resources/res-18-009-learn-differential-equations-up-close-with-gilbert-strang-and-cleve-moler-fall-2015/
www.mathworks.com/academia/courseware/learn-differential-equations.html

7 The Singular Value Decomposition (SVD)

7.1 Singular Values and Singular Vectors

7.2 Image Processing by Linear Algebra

7.3 Principal Component Analysis (PCA by the SVD)

This chapter develops one idea. That idea applies to every matrix, square or rectangular. It is an extension of eigenvectors. But now we need **two sets of orthonormal vectors: input vectors $\boldsymbol{v}_1$ to $\boldsymbol{v}_n$ and output vectors $\boldsymbol{u}_1$ to $\boldsymbol{u}_m$.** This is completely natural for an m by n matrix. The vectors $\boldsymbol{v}_1$ to $\boldsymbol{v}_r$ are a basis for the row space; $\boldsymbol{u}_1$ to $\boldsymbol{u}_r$ are a basis for the column space. Then we recover A from r pieces of rank one, with $r = \text{rank}(A)$ and positive singular values $\boldsymbol{\sigma_1 \geq \sigma_2 \geq \cdots \geq \sigma_r > 0}$ in the diagonal matrix $\boldsymbol{\Sigma}$.

$$\textbf{SVD} \qquad A = \boldsymbol{U\Sigma V}^{\mathrm{T}} = \sigma_1 \boldsymbol{u}_1 \boldsymbol{v}_1^{\mathrm{T}} + \sigma_2 \boldsymbol{u}_2 \boldsymbol{v}_2^{\mathrm{T}} + \cdots + \sigma_r \boldsymbol{u}_r \boldsymbol{v}_r^{\mathrm{T}}.$$

The right singular vectors $\boldsymbol{v}_i$ are eigenvectors of $A^{\mathrm{T}}A$. They give bases for the row space and nullspace of A. The left singular vectors $\boldsymbol{u}_i$ are eigenvectors of AA^{T}. They give bases for the column space and left nullspace of A. **Then $\boldsymbol{Av_i}$ equals $\boldsymbol{\sigma_i u_i}$ for $\boldsymbol{i \leq r}$.** The matrix A is diagonalized by these two orthogonal bases: $\boldsymbol{AV = U\Sigma}$.

Note that the nonzero columns of $U\Sigma$ are $\sigma_1\boldsymbol{u}_1, \ldots, \sigma_r\boldsymbol{u}_r$. Then $U\Sigma$ times V^{T} in the box above is a column times row multiplication—parallel to $S = Q\Lambda Q^{\mathrm{T}}$ for symmetric S.

Each $\boldsymbol{u}_i = \boldsymbol{v}_i$ when A is a symmetric positive definite matrix S. Those singular vectors will be the eigenvectors $\boldsymbol{q}_i$. And the singular values σ_i become the eigenvalues of S. If A is not square or not symmetric, then we need $A = U\Sigma V^{\mathrm{T}}$ instead of $S = Q\Lambda Q^{\mathrm{T}}$. Figure 7.1 will show how each step from $\boldsymbol{x}$ to $V^{\mathrm{T}}\boldsymbol{x}$ to $\Sigma V^{\mathrm{T}}\boldsymbol{x}$ to $U\Sigma V^{\mathrm{T}}\boldsymbol{x} = A\boldsymbol{x}$ acts on a circle of unit vectors $\boldsymbol{x}$. They stretch the circle into an ellipse of vectors $A\boldsymbol{x}$.

The SVD is a valuable way to understand a matrix of data. In that case AA^{T} is the **sample covariance matrix**, after centering the data and dividing by $n-1$ (Section 7.3). Its eigenvalues are σ_1^2 to σ_r^2. Its eigenvectors are the $\boldsymbol{u}$'s in the SVD. Principal Component Analysis (**PCA**) is totally based on the singular vectors of the data matrix A.

The SVD allows wonderful projects, by separating a photograph = matrix of pixels into its rank-one components. Each time you include one more piece $\sigma_i\boldsymbol{u}_i\boldsymbol{v}_i^{\mathrm{T}}$, the picture becomes clearer. Section 7.2 shows examples and a link to an excellent website. Section 7.3 describes PCA and its connection to the covariance matrix in statistics.

7.1 Singular Values and Singular Vectors

1 The Singular Value Decomposition of any matrix is $\boldsymbol{A} = \boldsymbol{U\Sigma V}^{\mathrm{T}}$ or $\boldsymbol{AV} = \boldsymbol{U\Sigma}$.

2 Singular vectors $\boldsymbol{v}_i, \boldsymbol{u}_i$ in $A\boldsymbol{v}_i = \sigma_i \boldsymbol{u}_i$ are orthonormal: $\boldsymbol{V}^{\mathrm{T}}\boldsymbol{V} = I$ and $\boldsymbol{U}^{\mathrm{T}}\boldsymbol{U} = I$.

3 The diagonal matrix $\boldsymbol{\Sigma}$ contains the singular values $\sigma_1 \geq \sigma_2 \geq \cdots \geq \sigma_r > 0$.

4 The squares σ_i^2 of those singular values are eigenvalues $\boldsymbol{\lambda}_i$ of $\boldsymbol{A}^{\mathrm{T}}\boldsymbol{A}$ and $\boldsymbol{A}\boldsymbol{A}^{\mathrm{T}}$.

The eigenvectors of symmetric matrices are orthogonal. We want to go beyond symmetric matrices to **all matrices**—without giving up orthogonality. To make this possible for every m by n matrix we need one set of n orthonormal vectors $\boldsymbol{v}_1, \ldots, \boldsymbol{v}_n$ in $\mathbf{R}^n$ and a second orthonormal set $\boldsymbol{u}_1, \ldots, \boldsymbol{u}_m$ in $\mathbf{R}^m$. **Instead of $Sx = \lambda x$ we want $Av = \sigma u$.**

This unsymmetric matrix $\boldsymbol{A}$ has orthogonal inputs $\boldsymbol{v}_1, \boldsymbol{v}_2 = (1,1)$ and $(-1,1)$:

$$\begin{bmatrix} 5 & 4 \\ 0 & 3 \end{bmatrix}\begin{bmatrix} 1 \\ 1 \end{bmatrix} = \begin{bmatrix} 9 \\ 3 \end{bmatrix} \quad \text{and} \quad \begin{bmatrix} 5 & 4 \\ 0 & 3 \end{bmatrix}\begin{bmatrix} -1 \\ 1 \end{bmatrix} = \begin{bmatrix} -1 \\ 3 \end{bmatrix}. \qquad (1)$$

The outputs $(9,3)$ and $(-1,3)$ are also orthogonal! Those $\boldsymbol{v}$'s and $\boldsymbol{u}$'s are not unit vectors but that is easily fixed. $(1,1)$ and $(-1,1)$ need to be divided by $\sqrt{2}$. $(3,1)$ and $(-1,3)$ need to be divided by $\sqrt{10}$ to give $\boldsymbol{u}_1$ and $\boldsymbol{u}_2$. **The singular values $\boldsymbol{\sigma_1}, \boldsymbol{\sigma_2}$ are $3\sqrt{5}$ and $\sqrt{5}$:**

$$\boxed{\boldsymbol{Av} = \boldsymbol{\sigma u}} \qquad \begin{bmatrix} 5 & 4 \\ 0 & 3 \end{bmatrix}\boldsymbol{v}_1 = 3\sqrt{5}\,\boldsymbol{u}_1 \quad \text{and} \quad \begin{bmatrix} 5 & 4 \\ 0 & 3 \end{bmatrix}\boldsymbol{v}_2 = \sqrt{5}\,\boldsymbol{u}_2. \qquad (2)$$

We can move from these two vector formulas to one matrix formula $\boldsymbol{AV} = \boldsymbol{U\Sigma}$. The inputs $\boldsymbol{v}_1$ and $\boldsymbol{v}_2$ go into $\boldsymbol{V}$. The outputs $\sigma_1\boldsymbol{u}_1$ and $\sigma_2\boldsymbol{u}_2$ are in $\boldsymbol{U\Sigma}$, with $\dfrac{\sqrt{5}}{\sqrt{10}} = \dfrac{1}{\sqrt{2}}$.

$$\boldsymbol{AV} = \underset{\boldsymbol{A}}{\begin{bmatrix} 5 & 4 \\ 0 & 3 \end{bmatrix}} \frac{\begin{bmatrix} 1 & -1 \\ 1 & 1 \end{bmatrix}}{\sqrt{2}} = \frac{\begin{bmatrix} 3 & -1 \\ 1 & 3 \end{bmatrix}}{\sqrt{10}} \underset{\boldsymbol{\Sigma}}{\begin{bmatrix} 3\sqrt{5} & 0 \\ 0 & \sqrt{5} \end{bmatrix}} = \boldsymbol{U\Sigma}. \qquad (3)$$

$\boldsymbol{V}$ and $\boldsymbol{U}$ are orthogonal matrices! If we multiply $\boldsymbol{AV} = \boldsymbol{U\Sigma}$ by $\boldsymbol{V}^{\mathrm{T}}$, $\boldsymbol{A}$ will be alone:

$$\boxed{\boldsymbol{AV} = \boldsymbol{U\Sigma} \quad \text{becomes} \quad \boxed{\boldsymbol{A} = \boldsymbol{U\Sigma V}^{\mathrm{T}}}. \text{ This splits into } \boldsymbol{A} = \sigma_1\boldsymbol{u}_1\boldsymbol{v}_1^{\mathrm{T}} + \sigma_2\boldsymbol{u}_2\boldsymbol{v}_2^{\mathrm{T}}} \qquad (4)$$

Those SVD equations say everything except how to find the orthogonal $\boldsymbol{v}$'s and $\boldsymbol{u}$'s.

Every matrix A is diagonalized by two *sets of singular vectors*, not one set of eigenvectors.

In this 2 by 2 example, the first piece is more important than the second piece because $\sigma_1 = 3\sqrt{5}$ is greater than $\sigma_2 = \sqrt{5}$. To recover A, add the pieces $\sigma_1\boldsymbol{u}_1\boldsymbol{v}_1^{\mathrm{T}} + \sigma_2\boldsymbol{u}_2\boldsymbol{v}_2^{\mathrm{T}}$:

$$\frac{3\sqrt{5}}{\sqrt{10}\sqrt{2}}\begin{bmatrix} 3 \\ 1 \end{bmatrix}\begin{bmatrix} 1 & 1 \end{bmatrix} + \frac{\sqrt{5}}{\sqrt{10}\sqrt{2}}\begin{bmatrix} -1 \\ 3 \end{bmatrix}\begin{bmatrix} -1 & 1 \end{bmatrix} = \frac{3}{2}\begin{bmatrix} 3 & 3 \\ 1 & 1 \end{bmatrix} + \frac{1}{2}\begin{bmatrix} 1 & -1 \\ -3 & 3 \end{bmatrix} = \begin{bmatrix} 5 & 4 \\ 0 & 3 \end{bmatrix}$$

The Geometry of the SVD

The SVD separates a matrix into $A = U\Sigma V^{\mathrm{T}}$: **(orthogonal)** × **(diagonal)** × **(orthogonal)**. In two dimensions we can draw those steps. The orthogonal matrices U and V rotate the plane. The diagonal matrix Σ stretches it along the axes. Figure 7.1 shows **rotation** times **stretching** times **rotation**. **Vectors v on the circle go to $Av = \sigma u$ on an ellipse.**

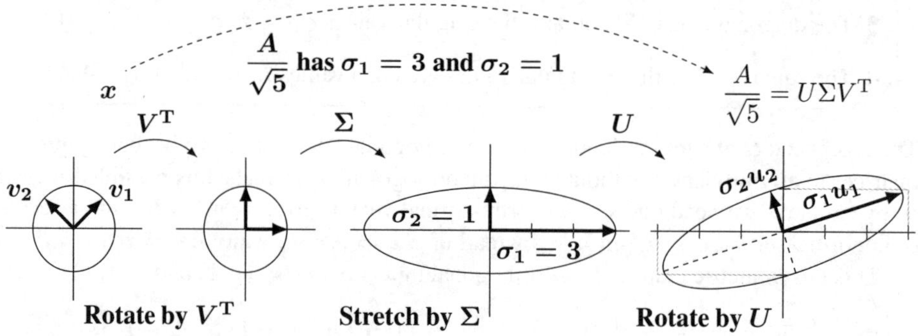

Figure 7.1: U and V are rotations and possible reflections. Σ stretches circle to ellipse.

This picture applies to a 2 by 2 invertible matrix with $\sigma_1 > 0$ and $\sigma_2 > 0$. First is a rotation of any x to $V^{\mathrm{T}}x$. Then Σ stretches that vector to $\Sigma V^{\mathrm{T}}x$. Then U rotates to $U\Sigma V^{\mathrm{T}}x$. We kept all determinants positive to avoid reflections. The four numbers a, b, c, d in the matrix connect to *two rotation angles* θ, ϕ and *two numbers* σ_1, σ_2 *in* Σ.

$$A = \begin{bmatrix} a & b \\ c & d \end{bmatrix} = \begin{bmatrix} \cos\theta & -\sin\theta \\ \sin\theta & \cos\theta \end{bmatrix} \begin{bmatrix} \sigma_1 & \\ & \sigma_2 \end{bmatrix} \begin{bmatrix} \cos\phi & \sin\phi \\ -\sin\phi & \cos\phi \end{bmatrix}. \tag{5}$$

Question 1 If the matrix is symmetric then $b = c$. Now A has 3 (not 4) parameters. *How do the 4 numbers θ, ϕ, σ_1, σ_2 reduce to 3 numbers for a symmetric matrix?*

Question 2 *If $\theta = 30°$ and $\sigma_1 = 2$ and $\sigma_2 = 1$ and $\phi = 60°$, what is A?* Check det = 2.

$$A = \begin{bmatrix} \sqrt{3}/2 & -1/2 \\ 1/2 & \sqrt{3}/2 \end{bmatrix} \begin{bmatrix} \mathbf{2} & \\ & \mathbf{1} \end{bmatrix} \begin{bmatrix} 1/2 & \sqrt{3}/2 \\ -\sqrt{3}/2 & 1/2 \end{bmatrix} = \frac{1}{4} \begin{bmatrix} \mathbf{3\sqrt{3}} & \mathbf{5} \\ \mathbf{-1} & \mathbf{3\sqrt{3}} \end{bmatrix} \tag{6}$$

Question 3 If A is **3 by 3** (9 parameters) then Σ has 3 singular values. U and V have 3 rotations each to make $3 + 3 + 3 = 9$ numbers. What are those 3 rotations for a pilot?

Answer: An airplane can "pitch" and "roll" and "yaw". (See Wolfram EulerAngles).

Question 4 **If $S = Q\Lambda Q^{\mathrm{T}}$ is symmetric positive definite, what is its SVD?**

Answer: The SVD is exactly $U\Sigma V^{\mathrm{T}} = Q\Lambda Q^{\mathrm{T}}$. The matrix $U = V = Q$ is orthogonal. And the diagonal eigenvalue matrix Λ becomes the singular value matrix Σ.

Question 5 **If $S = Q\Lambda Q^{\mathrm{T}}$ has a negative eigenvalue** ($Sx = -\alpha\, x$), what is the singular value σ and what are the vectors v and u?

Answer: The singular value will be $\sigma = +\alpha$ (positive). One singular vector (either u or v) must be $-x$ (reverse the sign). Then $Sx = -\alpha x$ is the same as $Sv = \sigma u$. The two sign changes cancel.

The Full Size Form of the SVD

The full picture includes basis vectors for the nullspaces of A and A^{T}. Those vectors $\boldsymbol{v}_{r+1}$ to $\boldsymbol{v}_n$ and $\boldsymbol{u}_{r+1}$ to $\boldsymbol{u}_m$ complete the orthogonal matrices V (n by n) and U (m by m). Then the matrix Σ is (m by n) like A. But Σ is all zero except for the r singular values $\sigma_1 \geq \sigma_2 \geq \cdots \geq \sigma_r > 0$ on its main diagonal. Those σ's multiply the vectors $\boldsymbol{u}_1$ to $\boldsymbol{u}_r$.

Full size SVD

$$\boldsymbol{AV = U\Sigma} \qquad (m \times n)\,(n \times n) = (m \times m)\,(m \times n) \qquad \boldsymbol{V^{\mathrm{T}} = V^{-1}} \;\; \boldsymbol{U^{\mathrm{T}} = U^{-1}}$$

$$A\begin{bmatrix} \boldsymbol{v}_1 \cdots \boldsymbol{v}_r & \boldsymbol{v}_{r+1} \cdots \boldsymbol{v}_n \end{bmatrix} = \begin{bmatrix} \boldsymbol{u}_1 \cdots \boldsymbol{u}_r & \boldsymbol{u}_{r+1} \cdots \boldsymbol{u}_m \end{bmatrix} \begin{bmatrix} \sigma_1 & & 0 \\ & \ddots & \\ & \sigma_r & 0 \\ 0 & 0 & 0 \end{bmatrix}$$

When we multiply on the right by $V^{\mathrm{T}} = V^{-1}$, this equation $\boldsymbol{AV} = \boldsymbol{U\Sigma}$ turns into the most famous form $\boldsymbol{A} = \boldsymbol{U\Sigma V^{\mathrm{T}}}$ of the Singular Value Decomposition. The matrices $\boldsymbol{V}$ and $\boldsymbol{U}$ are square—with bases for row space + nullspace, column space + $\mathbf{N}(A^{\mathrm{T}})$.

The Reduced Form of the SVD

That full form $AV = U\Sigma$ can have a lot of zeros in Σ when the rank of A is small and the nullspace of A is large. Those zeros contribute nothing to matrix multiplication. The heart of the SVD is in the first r $\boldsymbol{v}$'s and $\boldsymbol{u}$'s and σ's. We can change $AV = U\Sigma$ to $\boldsymbol{AV_r = U_r\Sigma_r}$ by removing the parts that are sure to produce zeros. This leaves the reduced SVD where $\boldsymbol{\Sigma_r}$ **is now square:** $\boldsymbol{(m \times n)\,(n \times r) = (m \times r)\,(r \times r)}$.

Reduced SVD

$$\boldsymbol{AV_r = U_r\Sigma_r} \qquad \boldsymbol{Av_i = \sigma_i u_i}$$

$$A\begin{bmatrix} \boldsymbol{v}_1 & \cdots & \boldsymbol{v}_r \\ & \textbf{row space} & \end{bmatrix} = \begin{bmatrix} \boldsymbol{u}_1 & \cdots & \boldsymbol{u}_r \\ & \textbf{column space} & \end{bmatrix} \begin{bmatrix} \sigma_1 & & \\ & \ddots & \\ & & \sigma_r \end{bmatrix} \tag{7}$$

We still have $V_r^{\mathrm{T}}\,V_r = I_r$ and $U_r^{\mathrm{T}}\,U_r = I_r$ from those orthogonal unit vectors $\boldsymbol{v}$'s and $\boldsymbol{u}$'s. When V_r and U_r are not square, we can't have full inverses: $V_r\,V_r^{\mathrm{T}} \neq I$ and $U_r\,U_r^{\mathrm{T}} \neq I$. **But** $\boldsymbol{A = U_r\Sigma_r V_r^{\mathrm{T}}} = \boldsymbol{u}_1\sigma_1\boldsymbol{v}_1^{\mathrm{T}} + \cdots + \boldsymbol{u_r\sigma_r v_r^{\mathrm{T}}}$ **is true**. I prefer this reduced form, because the other multiplications in the full size form $A = U\Sigma V^{\mathrm{T}}$ give only zeros.

Example 1 Our 2 by 2 matrix had $r = m = n$. The reduced form was the full form.

Example 2 $\boldsymbol{A} = \begin{bmatrix} 1 & 2 & 2 \end{bmatrix} = \begin{bmatrix} 1 \end{bmatrix}\begin{bmatrix} 3 \end{bmatrix}\begin{bmatrix} 1 & 2 & 2 \end{bmatrix}/3 = \boldsymbol{U_r\Sigma_r V_r}^{\mathrm{T}}$ has $r = 1$ and $\sigma_1 = 3$.

The rest of $U\Sigma V^{\mathrm{T}}$ contributes nothing to A, because of all the zeros in Σ. The key separation of A into $\sigma_1\boldsymbol{u}_1\boldsymbol{v}_1^{\mathrm{T}} + \cdots + \sigma_r\boldsymbol{u}_r\boldsymbol{v}_r^{\mathrm{T}}$ stops at $\sigma_1\boldsymbol{u}_1\boldsymbol{v}_1^{\mathrm{T}}$ because the rank is $\mathbf{1}$.

Proof of the SVD

The goal is $\boldsymbol{A} = \boldsymbol{U\Sigma V}^{\mathrm{T}}$. We want to identify the two sets of singular vectors, the $\boldsymbol{u}$'s and the $\boldsymbol{v}$'s. One way to find those vectors is to form the symmetric matrices $\boldsymbol{A}^{\mathrm{T}}\boldsymbol{A}$ and $\boldsymbol{A}\boldsymbol{A}^{\mathrm{T}}$:

$$\boldsymbol{A}^{\mathrm{T}}\boldsymbol{A} = (V\Sigma^{\mathrm{T}}U^{\mathrm{T}})\ (U\Sigma V^{\mathrm{T}}) = \boldsymbol{V\Sigma^{\mathrm{T}}\Sigma V^{\mathrm{T}}} \quad \text{(because } U^{\mathrm{T}}U = I\text{)} \tag{8}$$

$$\boldsymbol{A}\boldsymbol{A}^{\mathrm{T}} = (U\Sigma V^{\mathrm{T}})\ (V\Sigma^{\mathrm{T}}U^{\mathrm{T}}) = \boldsymbol{U\Sigma\Sigma^{\mathrm{T}}U^{\mathrm{T}}} \quad \text{(because } V^{\mathrm{T}}V = I\text{)} \tag{9}$$

Both (8) and (9) produced symmetric matrices. Usually $A^{\mathrm{T}}A$ and AA^{T} are different. Both right hand sides have the special form $Q\Lambda Q^{\mathrm{T}}$. **Eigenvectors of $\boldsymbol{A}^{\mathrm{T}}\boldsymbol{A}$ and $\boldsymbol{A}\boldsymbol{A}^{\mathrm{T}}$ are the singular vectors in $\boldsymbol{V}$ and $\boldsymbol{U}$.** So we know from (8) and (9) how the three pieces of $A = U\Sigma V^{\mathrm{T}}$ connect to those symmetric matrices $A^{\mathrm{T}}A$ and AA^{T}.

> **V contains orthonormal eigenvectors of $\boldsymbol{A}^{\mathrm{T}}\boldsymbol{A}$**
>
> **U contains orthonormal eigenvectors of $\boldsymbol{A}\boldsymbol{A}^{\mathrm{T}}$**
>
> **σ_1^2 to σ_r^2 are the nonzero eigenvalues of both $\boldsymbol{A}^{\mathrm{T}}\boldsymbol{A}$ and $\boldsymbol{A}\boldsymbol{A}^{\mathrm{T}}$**

We are not quite finished, for this reason. **The SVD requires that $\boldsymbol{Av_k} = \boldsymbol{\sigma_k u_k}$.** It connects each right singular vector $\boldsymbol{v}_k$ to a left singular vector $\boldsymbol{u}_k$, for $k = 1,\ldots,r$. When I choose the $\boldsymbol{v}$'s, *that choice will decide the signs of the* $\boldsymbol{u}$'s. If $S\boldsymbol{u} = \lambda\boldsymbol{u}$ then also $S(-\boldsymbol{u}) = \lambda(-\boldsymbol{u})$ and I have to know the sign to choose. More than that, there is a whole plane of eigenvectors when λ is a double eigenvalue. When I choose two $\boldsymbol{v}$'s in that plane, then $A\boldsymbol{v} = \sigma\boldsymbol{u}$ will tell me both $\boldsymbol{u}$'s.

The plan is to start with the $\boldsymbol{v}$'s. **Choose orthonormal eigenvectors $\boldsymbol{v}_1,\ldots,\boldsymbol{v}_r$ of $\boldsymbol{A}^{\mathrm{T}}\boldsymbol{A}$. Then choose $\boldsymbol{\sigma_k} = \sqrt{\boldsymbol{\lambda_k}}$.** To determine the $\boldsymbol{u}$'s we require that $\boldsymbol{Av} = \boldsymbol{\sigma u}$:

$$\textbf{v's then u's} \qquad \boldsymbol{A}^{\mathrm{T}}\boldsymbol{A}\boldsymbol{v}_k = \boldsymbol{\sigma}_k^2\boldsymbol{v}_k \ \ \textbf{and then} \ \ \boldsymbol{u}_k = \frac{\boldsymbol{A}\boldsymbol{v}_k}{\boldsymbol{\sigma_k}} \quad \textbf{for} \ \ \boldsymbol{k} = \mathbf{1},\ldots,\boldsymbol{r} \tag{10}$$

This produces the SVD! Let me check that those vectors $\boldsymbol{u}_1$ to $\boldsymbol{u}_r$ are eigenvectors of AA^{T}.

$$\boldsymbol{A}\boldsymbol{A}^{\mathrm{T}}\boldsymbol{u}_k = AA^{\mathrm{T}}\left(\frac{A\boldsymbol{v}_k}{\sigma_k}\right) = A\left(\frac{A^{\mathrm{T}}A\,\boldsymbol{v}_k}{\sigma_k}\right) = A\,\frac{\sigma_k^2\,\boldsymbol{v}_k}{\sigma_k} = \boldsymbol{\sigma}_k^2\,\boldsymbol{u}_k \tag{11}$$

The $\boldsymbol{v}$'s were chosen to be orthonormal. I must check that the $\boldsymbol{u}$'s are also orthonormal:

$$\boldsymbol{u}_j^{\mathrm{T}}\boldsymbol{u}_k = \left(\frac{A\boldsymbol{v}_j}{\sigma_j}\right)^{\mathrm{T}}\left(\frac{A\boldsymbol{v}_k}{\sigma_k}\right) = \frac{\boldsymbol{v}_j^{\mathrm{T}}(A^{\mathrm{T}}A\boldsymbol{v}_k)}{\sigma_j\,\sigma_k} = \frac{\sigma_k}{\sigma_j}\,\boldsymbol{v}_j^{\mathrm{T}}\,\boldsymbol{v}_k = \begin{cases} 1 & \text{if } j = k \\ 0 & \text{if } j \neq k \end{cases} \tag{12}$$

Notice that $(\boldsymbol{A}\boldsymbol{A}^{\mathrm{T}})\boldsymbol{A} = \boldsymbol{A}(\boldsymbol{A}^{\mathrm{T}}\boldsymbol{A})$ was the key to equation (11). The law $(AB)C = A(BC)$ is the key to a great many proofs in linear algebra. Moving those parentheses is a powerful idea. It is permitted by the *associative law*. It completes the proof.

Example 1 (completing now) **Find U and Σ and V for our original $A = \begin{bmatrix} 5 & 4 \\ 0 & 3 \end{bmatrix}$.**

With rank 2, this A has two positive singular values σ_1 and σ_2. We will see that σ_1 is larger than $\lambda_{\max} = 5$, and σ_2 is smaller than $\lambda_{\min} = 3$. Begin with $A^{\mathrm{T}}A$ and AA^{T}:

$$A^{\mathrm{T}}A = \begin{bmatrix} 25 & 20 \\ 20 & 25 \end{bmatrix} \qquad AA^{\mathrm{T}} = \begin{bmatrix} 41 & 12 \\ 12 & 9 \end{bmatrix}$$

Those have the same trace $\lambda_1 + \lambda_2 = 50$ and the same eigenvalues $\lambda_1 = \sigma_1^2 = 45$ and $\lambda_2 = \sigma_2^2 = 5$. The square roots are $\boldsymbol{\sigma_1} = \sqrt{\mathbf{45}} = \mathbf{3}\sqrt{\mathbf{5}}$ and $\boldsymbol{\sigma_2} = \sqrt{\mathbf{5}}$. Then σ_1 times σ_2 equals 15, and this is the determinant of A. The next step is to find V.

The key to V is to find the eigenvectors of $A^{\mathrm{T}}A$ (with eigenvalues 45 and 5):

$$\begin{bmatrix} 25 & 20 \\ 20 & 25 \end{bmatrix}\begin{bmatrix} \mathbf{1} \\ \mathbf{1} \end{bmatrix} = \mathbf{45}\begin{bmatrix} \mathbf{1} \\ \mathbf{1} \end{bmatrix} \qquad \begin{bmatrix} 25 & 20 \\ 20 & 25 \end{bmatrix}\begin{bmatrix} \mathbf{-1} \\ \mathbf{1} \end{bmatrix} = \mathbf{5}\begin{bmatrix} \mathbf{-1} \\ \mathbf{1} \end{bmatrix}$$

Then $\boldsymbol{v}_1$ and $\boldsymbol{v}_2$ are those orthogonal eigenvectors rescaled to length 1. Divide by $\sqrt{2}$.

$$\textbf{Right singular vectors} \quad \boldsymbol{v}_1 = \frac{1}{\sqrt{2}}\begin{bmatrix} \mathbf{1} \\ \mathbf{1} \end{bmatrix} \quad \text{and} \quad \boldsymbol{v}_2 = \frac{1}{\sqrt{2}}\begin{bmatrix} \mathbf{-1} \\ \mathbf{1} \end{bmatrix} \quad \text{(as predicted)}$$

The left singular vectors are $\boldsymbol{u}_1 = A\boldsymbol{v}_1/\sigma_1$ and $\boldsymbol{u}_2 = A\boldsymbol{v}_2/\sigma_2$. **Multiply $\boldsymbol{v}_1, \boldsymbol{v}_2$ by $\boldsymbol{A}$:**

$$\begin{aligned} A\boldsymbol{v}_1 &= \frac{3}{\sqrt{2}}\begin{bmatrix} 3 \\ 1 \end{bmatrix} = \sqrt{45}\frac{\mathbf{1}}{\sqrt{\mathbf{10}}}\begin{bmatrix} \mathbf{3} \\ \mathbf{1} \end{bmatrix} = \sigma_1 \boldsymbol{u}_1 \\ A\boldsymbol{v}_2 &= \frac{1}{\sqrt{2}}\begin{bmatrix} -1 \\ 3 \end{bmatrix} = \sqrt{5}\frac{\mathbf{1}}{\sqrt{\mathbf{10}}}\begin{bmatrix} \mathbf{-1} \\ \mathbf{3} \end{bmatrix} = \sigma_2 \boldsymbol{u}_2 \end{aligned}$$

The division by $\sqrt{10}$ makes $\boldsymbol{u}_1$ and $\boldsymbol{u}_2$ unit vectors. Then $\sigma_1 = \sqrt{45}$ and $\sigma_2 = \sqrt{5}$ as expected. The Singular Value Decomposition of A is $\boldsymbol{U}$ times $\boldsymbol{\Sigma}$ times $\boldsymbol{V}^{\mathrm{T}}$. (**Not $\boldsymbol{V}$**.)

$$\boxed{\boldsymbol{U} = \frac{1}{\sqrt{10}}\begin{bmatrix} \mathbf{3} & \mathbf{-1} \\ \mathbf{1} & \mathbf{3} \end{bmatrix} \quad \boldsymbol{\Sigma} = \begin{bmatrix} \sqrt{\mathbf{45}} & \\ & \sqrt{\mathbf{5}} \end{bmatrix} \quad \boldsymbol{V}^{\mathrm{T}} = \frac{1}{\sqrt{2}}\begin{bmatrix} \mathbf{1} & \mathbf{1} \\ \mathbf{-1} & \mathbf{1} \end{bmatrix}} \tag{13}$$

Finally we have to choose the last $n - r$ vectors $\boldsymbol{v}_{r+1}$ to $\boldsymbol{v}_n$ and the last $m - r$ vectors $\boldsymbol{u}_{r+1}$ to $\boldsymbol{u}_m$. This is easy. **These $\boldsymbol{v}$'s and $\boldsymbol{u}$'s are in the nullspaces of $\boldsymbol{A}$ and $\boldsymbol{A}^{\mathrm{T}}$.** We can choose any orthonormal bases for those nullspaces. They will automatically be orthogonal to the first $\boldsymbol{v}$'s in the row space of A and the first $\boldsymbol{u}$'s in the column space. This is because the whole spaces are orthogonal: $\mathbf{N}(A) \perp \mathbf{C}(A^{\mathrm{T}})$ and $\mathbf{N}(A^{\mathrm{T}}) \perp \mathbf{C}(A)$. *The proof of the SVD is complete* by that Fundamental Theorem of Linear Algebra.

> **To say again: Good codes do not start with $\boldsymbol{A}^{\mathrm{T}}\boldsymbol{A}$ and $\boldsymbol{A}\boldsymbol{A}^{\mathrm{T}}$.** Instead we first produce zeros in A by rotations that leave only two diagonals (and don't affect the σ's). The last page of this section describes a successful way to compute the SVD.

Now we have U and V and Σ in the full size SVD of equation (1): m $\boldsymbol{u}$'s and n $\boldsymbol{v}$'s. You may have noticed that the eigenvalues of $A^{\mathrm{T}}A$ are in $\Sigma^{\mathrm{T}}\Sigma$, and *the same numbers σ_1^2 to σ_r^2 are also eigenvalues of AA^{T} in $\Sigma\Sigma^{\mathrm{T}}$*. An amazing fact: **$BA$ always has the same nonzero eigenvalues as AB.** BA is n by n and AB is m by m.

AB and BA: Equal Nonzero Eigenvalues

If A is m by n and B is n by m, AB and BA have the same nonzero eigenvalues.

Start with $AB\boldsymbol{x} = \lambda\boldsymbol{x}$ and $\lambda \neq 0$. Multiply both sides by B, to get $BAB\boldsymbol{x} = \lambda B\boldsymbol{x}$. This says that $B\boldsymbol{x}$ is an eigenvector of BA with the same eigenvalue λ—exactly what we wanted. We needed $\lambda \neq 0$ to be sure that $B\boldsymbol{x}$ is truly a nonzero eigenvector of BA.

Notice that if B is square and invertible, then $B^{-1}(BA)B = AB$. This says that BA is *similar* to AB: *same eigenvalues*. But our first proof allows A and B to be m by n and n by m. This covers the important example of the SVD when $B = A^{\mathrm{T}}$. In that case $A^{\mathrm{T}}A$ and AA^{T} both lead to the r nonzero singular values of A.

If m is larger than n, then AB has $m - n$ extra zero eigenvalues compared to BA.

Singular Vectors for 2 by 2 Matrices

Here is a **picture proof** of the 2 by 2 SVD. We look for perpendicular vectors $\boldsymbol{v}$ and $\boldsymbol{w}$ so that $A\boldsymbol{v}$ and $A\boldsymbol{w}$ are **also** perpendicular. In our first guess $\boldsymbol{V} = (1,0)$ and $\boldsymbol{W} = (0,1)$ the angle θ from $A\boldsymbol{V}$ to $A\boldsymbol{W}$ is too small (below 90° in the first figure). Then the angle $180 - \theta$ between $A\boldsymbol{W}$ and $-A\boldsymbol{V}$ is too large (above 90° in the second figure). Therefore there must be a $\boldsymbol{v}$ between $\boldsymbol{V}$ and $\boldsymbol{W}$ that is just right. The angle from $A\boldsymbol{v}$ to $A\boldsymbol{w}$ is exactly 90° and the SVD is proved: Orthogonal inputs $\boldsymbol{v}\perp\boldsymbol{w}$, orthogonal outputs $A\boldsymbol{v}\perp A\boldsymbol{w}$.

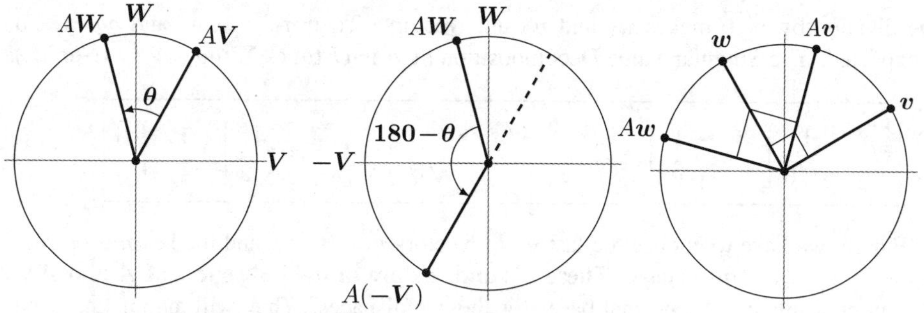

Figure 7.2: Angle θ from $A\boldsymbol{V}$ to $A\boldsymbol{W}$ is below 90°. Angle $180 - \theta$ from $A\boldsymbol{W}$ to $-A\boldsymbol{V}$ is above 90°. Somewhere in between, as $\boldsymbol{v}$ moves from $\boldsymbol{V}$ toward $\boldsymbol{W}$, the angle from $A\boldsymbol{v}$ to $A\boldsymbol{w}$ is exactly 90°. The pictures don't show vector lengths.

The next page will establish a new way to look at $\boldsymbol{v}_1$. The previous pages chose the $\boldsymbol{v}$'s as eigenvectors of $A^{\mathrm{T}}A$. Certainly that remains true. But there is a valuable way to understand these singular vectors **one at a time instead of all at once**.

The First Singular Vector v_1

$$\textbf{Maximize the ratio } \frac{||Ax||}{||x||}. \textbf{ The maximum is } \sigma_1 \textbf{ at the vector } x = v_1. \tag{14}$$

The ellipse in Figure 7.1 showed why the maximizing x is v_1. When you follow v_1 across the page, it ends at $Av_1 = \sigma_1 u_1$. The longer axis of the ellipse has length $||Av_1|| = \sigma_1$.

But we aim for an independent approach to the SVD! We are not assuming that we already know U or Σ or V. How do we recognize that the ratio $||Ax||/||x||$ is a maximum when $x = v_1$? One easy way is to square our function and work with $S = A^{\mathrm{T}}A$:

$$\textbf{Problem: Find the maximum value } \lambda \textbf{ of } \frac{||Ax||^2}{||x||^2} = \frac{x^{\mathrm{T}}A^{\mathrm{T}}Ax}{x^{\mathrm{T}}x} = \frac{x^{\mathrm{T}}Sx}{x^{\mathrm{T}}x}. \tag{15}$$

To maximize this "Rayleigh quotient", write any x as $c_1x + \cdots + c_nx_n$ with $Sx_i = \lambda_i x_i$:

$$\frac{x^{\mathrm{T}}Sx}{x^{\mathrm{T}}x} = \frac{c_1^2\lambda_1 + \cdots + c_n^2\lambda_n}{c_1^2 + \cdots + c_n^2} \leq \lambda_1 = \text{largest eigenvalue of } S. \tag{16}$$

Then the best x to maximize the ratio in (15) is an eigenvector of S!

$$Sx = \lambda x \textbf{ and the maximum of } \frac{x^{\mathrm{T}}Sx}{x^{\mathrm{T}}x} = \frac{||Ax||^2}{||x||^2} \textbf{ is } \lambda_1(S) = \sigma_1^2(A).$$

For the full SVD, we need *all* the singular vectors and singular values. To find v_2 and σ_2, we adjust the maximum problem so it looks only at vectors x orthogonal to v_1.

$$\textbf{Maximize } \frac{||Ax||}{||x||} \textbf{ under the condition } v_1^{\mathrm{T}}x = 0. \textbf{ The maximum is } \sigma_2 \textbf{ at } x = v_2.$$

"Lagrange multipliers" were invented to deal with constraints on x like $v_1^{\mathrm{T}}x = 0$. And Problem 9 gives a simple direct way to work with this condition $v_1^{\mathrm{T}}x = 0$.

Every singular vector v_{k+1} gives the maximum ratio $||Ax||/||x||$ over all vectors x that are perpendicular to the first $v_1, \ldots, v_k$. We are finding the axes of an ellipsoid and the eigenvectors of symmetric matrices $A^{\mathrm{T}}A$ or AA^{T}: all at once or separately.

Question: Why are all eigenvalues of a square matrix A less than or equal to σ_1?

Answer: Multiplying by orthogonal matrices U and V^{T} does not change vector lengths:

$$||Ax|| = ||U\Sigma V^{\mathrm{T}}x|| = ||\Sigma V^{\mathrm{T}}x|| \leq \sigma_1||V^{\mathrm{T}}x|| = \sigma_1||x|| \text{ for all } x. \tag{17}$$

An eigenvector has $||Ax|| = |\lambda|\,||x||$. Then (17) gives $|\lambda|\,||x|| \leq \sigma_1\,||x||$ and $|\lambda| \leq \sigma_1$.

Question: If $A = \boldsymbol{x}\boldsymbol{y}^{\mathrm{T}}$ has rank 1, what are $\boldsymbol{u}_1$ and $\boldsymbol{v}_1$ and σ_1? **Check that** $|\boldsymbol{\lambda}_1| \leq \boldsymbol{\sigma}_1$.

Answer: The singular vectors $\boldsymbol{u}_1 = \boldsymbol{x}/||\boldsymbol{x}||$ and $\boldsymbol{v}_1 = \boldsymbol{y}/||\boldsymbol{y}||$ have length 1. Then $\boldsymbol{\sigma}_1 = ||\boldsymbol{x}||\,||\boldsymbol{y}||$ is the only nonzero number in the singular value matrix Σ. Here is the SVD:

$$\textbf{Rank 1 matrix} \qquad A = \boldsymbol{x}\boldsymbol{y}^{\mathrm{T}} = \frac{\boldsymbol{x}}{||\boldsymbol{x}||}\left(||\boldsymbol{x}||\,||\boldsymbol{y}||\right)\frac{\boldsymbol{y}^{\mathrm{T}}}{||\boldsymbol{y}||} = \boldsymbol{u}_1\sigma_1\boldsymbol{v}_1^{\mathrm{T}}. \qquad (18)$$

Observation The only nonzero eigenvalue of $A = \boldsymbol{x}\boldsymbol{y}^{\mathrm{T}}$ is $\lambda_1 = \boldsymbol{y}^{\mathrm{T}}\boldsymbol{x}$. The eigenvector is $\boldsymbol{x}$ because $A\boldsymbol{x} = (\boldsymbol{x}\boldsymbol{y}^{\mathrm{T}})\,\boldsymbol{x} = \boldsymbol{x}\,(\boldsymbol{y}^{\mathrm{T}}\boldsymbol{x}) = \lambda_1\boldsymbol{x}$. Then the inequality $|\lambda_1| \leq \sigma_1$ becomes exactly the Schwarz inequality $|\boldsymbol{y}^{\mathrm{T}}\boldsymbol{x}| \leq ||\boldsymbol{x}||\,||\boldsymbol{y}||$.

Computing Eigenvalues and Singular Values

What is the main difference between the symmetric eigenvalue problem $S\boldsymbol{x} = \lambda\boldsymbol{x}$ and $A\boldsymbol{v} = \sigma\boldsymbol{u}$? How much can we simplify S and A before computing λ's and σ's?

Eigenvalues are the same for S and $Q^{-1}SQ = Q^{\mathrm{T}}SQ$ when Q is orthogonal.

So we have limited freedom to create zeros in $Q^{-1}SQ$ (which stays symmetric). If we try for too many zeros in $Q^{-1}S$, the final Q will destroy them. The good $Q^{-1}SQ$ will be **tridiagonal**: we can reduce S to three nonzero diagonals.
Singular values are the same for A and $Q_1^{-1}AQ_2$ *even if Q_1 is different from Q_2*.

We have more freedom to create zeros in $Q_1^{-1}AQ_2$. With the right Q's, this will be **bidiagonal** (two nonzero diagonals). We can quickly find Q and Q_1 and Q_2 so that

$$Q^{-1}SQ = \begin{bmatrix} a_1 & b_1 & & \\ b_1 & a_2 & b_2 & \\ & b_2 & \cdot & \cdot \\ & & \cdot & a_n \end{bmatrix} \leftarrow \text{for } \lambda\text{'s} \qquad \begin{matrix} Q_1^{-1}AQ_2 = \\ \text{for } \sigma\text{'s} \rightarrow \end{matrix} \begin{bmatrix} c_1 & d_1 & & \\ 0 & c_2 & d_2 & \\ & 0 & \cdot & \cdot \\ & & 0 & c_n \end{bmatrix} \qquad (19)$$

The reader will know that the singular values of A are the square roots of the eigenvalues of $S = A^{\mathrm{T}}A$. And the singular values of $Q_1^{-1}AQ_2$ are the same as the singular values of A. **Multiply (bidiagonal)^{T}(bidiagonal) to see tridiagonal.**

This offers an option that we **should not take. Don't multiply $A^{\mathrm{T}}A$** and find its eigenvalues. This is unnecessary work and the condition of the problem will be unnecessarily squared. The Golub-Kahan algorithm for the SVD works directly on A, in two steps:

1. Find Q_1 and Q_2 so that $A^* = Q_1^{-1}AQ_2$ is bidiagonal as in (19): *same σ's*
2. Adjust the shifted QR algorithm (Section 6.5, page 284) to preserve singular values of A^*

Step 1 requires $O(mn^2)$ multiplications to put an m by n matrix A into bidiagonal form. Then later steps will work only with bidiagonal matrices. Normally it then takes $O(n^2)$ multiplications to find singular values (correct to nearly machine precision). The full algorithm is described on pages 489–492 in the 4th edition of Golub-Van Loan: the bible.

When A is truly large, we turn to **random sampling**. With very high probability, *randomized linear algebra gives accurate results.* Most gamblers would say that a good outcome from careful random sampling is certain.

Problem Set 7.1

1 Find $A^{\mathrm{T}}A$ and AA^{T} and the singular vectors $\boldsymbol{v}_1, \boldsymbol{v}_2, \boldsymbol{u}_1, \boldsymbol{u}_2$ for A:

$$A = \begin{bmatrix} 0 & 1 & 0 \\ 0 & 0 & 8 \\ 0 & 0 & 0 \end{bmatrix} \quad \text{has rank } r = 2. \quad \text{The eigenvalues are } 0, 0, 0.$$

Check the equations $A\boldsymbol{v}_1 = \sigma_1 \boldsymbol{u}_1$ and $A\boldsymbol{v}_2 = \sigma_2 \boldsymbol{u}_2$ and $A = \sigma_1 \boldsymbol{u}_1 \boldsymbol{v}_1^{\mathrm{T}} + \sigma_2 \boldsymbol{u}_2 \boldsymbol{v}_2^{\mathrm{T}}$. If you remove row 3 of A (all zeros), show that σ_1 and σ_2 don't change.

2 Find the singular values and also the eigenvalues of B:

$$B = \begin{bmatrix} 0 & 1 & 0 \\ 0 & 0 & 8 \\ \frac{1}{1000} & 0 & 0 \end{bmatrix} \quad \text{has rank } r = 3 \text{ and determinant } \frac{8}{1000}.$$

Compared to A above, eigenvalues have changed much more than singular values.

3 The SVD connects $\boldsymbol{v}$'s in the row space to $\boldsymbol{u}$'s in the column space. Transpose $A = U\Sigma V^{\mathrm{T}}$ to see that $\boldsymbol{A}^{\mathrm{T}} = \boldsymbol{V}\boldsymbol{\Sigma}^{\mathrm{T}}\boldsymbol{U}^{\mathrm{T}}$ goes the opposite way, from $\boldsymbol{u}$'s to $\boldsymbol{v}$'s:

$$\boldsymbol{A}^{\mathrm{T}}\boldsymbol{u}_k = \boldsymbol{\sigma}_k \boldsymbol{v}_k \ \textbf{ for } \boldsymbol{k} = \mathbf{1}, \ldots, \boldsymbol{r} \qquad \boldsymbol{A}^{\mathrm{T}}\boldsymbol{u}_k = \mathbf{0} \ \textbf{ for } \boldsymbol{k} = \boldsymbol{r} + \mathbf{1}, \ldots, \boldsymbol{m}$$

What are the left singular vectors $\boldsymbol{u}$ and right singular vectors $\boldsymbol{v}$ for the transpose $\begin{bmatrix} 5 & 0 & ; & 4 & 3 \end{bmatrix}$ of our example matrix?

4 When $A\boldsymbol{v}_k = \sigma_k \boldsymbol{u}_k$ and $A^{\mathrm{T}}\boldsymbol{u}_k = \sigma_k \boldsymbol{v}_k$, show that S has eigenvalues σ_k and $-\sigma_k$:

$$S = \begin{bmatrix} 0 & \boldsymbol{A} \\ \boldsymbol{A}^{\mathrm{T}} & 0 \end{bmatrix} \quad \text{has eigenvectors} \quad \begin{bmatrix} \boldsymbol{u}_k \\ \boldsymbol{v}_k \end{bmatrix} \quad \text{and} \quad \begin{bmatrix} -\boldsymbol{u}_k \\ \boldsymbol{v}_k \end{bmatrix} \quad \text{and trace} = 0.$$

The eigenvectors of this symmetric matrix S tell us the singular vectors of A.

5 Find the eigenvalues and the singular values of this 2 by 2 matrix A.

$$A = \begin{bmatrix} 2 & 1 \\ 4 & 2 \end{bmatrix} \quad \text{with} \quad A^{\mathrm{T}}A = \begin{bmatrix} 20 & 10 \\ 10 & 5 \end{bmatrix} \quad \text{and} \quad AA^{\mathrm{T}} = \begin{bmatrix} 5 & 10 \\ 10 & 20 \end{bmatrix}.$$

The eigenvectors $(1, 2)$ and $(1, -2)$ of A are not orthogonal. How do you know the eigenvectors $\boldsymbol{v}_1, \boldsymbol{v}_2$ of $A^{\mathrm{T}}A$ will be orthogonal? Notice that $A^{\mathrm{T}}A$ and AA^{T} have the same eigenvalues $\lambda_1 = 25$ and $\lambda_2 = 0$.

6 Find the SVD factors U and Σ and V^{T} for $A_1 = \begin{bmatrix} 1 & 0 \\ 1 & 1 \end{bmatrix}$ and $A_2 = \begin{bmatrix} 1 & 0 & 1 & 0 \\ 1 & 1 & 1 & 1 \end{bmatrix}$.

7 The MATLAB commands $A =$ rand (20, 40) and $B =$ randn (20, 40) produce 20 by 40 random matrices. The entries of A are between 0 and 1 with uniform probability. The entries of B have a normal "bell-shaped" probability distribution. Using an svd command, find and graph their singular values σ_1 to σ_{20}. Why do they have 20 σ's?

8 Why doesn't the SVD for $A+I$ use $\Sigma+I$? If $A=\sum \sigma_i \boldsymbol{u}_i \boldsymbol{v}_i^{\mathrm{T}}$ why is $A^{+}=\sum \boldsymbol{v}_i \boldsymbol{u}_i^{\mathrm{T}}/\sigma_i$?

9 If $A=Q$ is an orthogonal matrix, why does every singular value of Q equal 1?

10 A symmetric matrix $S=A^{\mathrm{T}}A$ has eigenvalues $\lambda_1 \geq 0$ to $\lambda_n \geq 0$. Its eigenvectors $\boldsymbol{v}_1$ to $\boldsymbol{v}_n$ are orthonormal. Then any vector has the form $\boldsymbol{x}=c_1\boldsymbol{v}_1+\cdots+c_n\boldsymbol{v}_n$.

$$R(\boldsymbol{x})=\frac{\boldsymbol{x}^{\mathrm{T}}S\boldsymbol{x}}{\boldsymbol{x}^{\mathrm{T}}\boldsymbol{x}}=\frac{\lambda_1 c_1^2+\cdots+\lambda_n c_n^2}{c_1^2+\cdots+c_n^2}.$$

Which vector $\boldsymbol{x}$ gives the maximum of R? What are the numbers c_1 to c_n for that maximizing vector $\boldsymbol{x}$? Which $\boldsymbol{x}$ gives the minimum of R?

11 To find σ_2 and $\boldsymbol{v}_2$ from maximizing that ratio $R(\boldsymbol{x})$, we must rule out the first singular vector $\boldsymbol{v}_1$ by requiring $\boldsymbol{x}^{\mathrm{T}}\boldsymbol{v}_1=0$. What does this mean for c_1? Which c's give the new maximum σ_2 of $R(\boldsymbol{x})$ at the **second eigenvector** $\boldsymbol{x}=\boldsymbol{v}_2$?

12 Find $A^{\mathrm{T}}A$ and the singular vectors in $A\boldsymbol{v}_1=\sigma_1\boldsymbol{u}_1$ and $A\boldsymbol{v}_2=\sigma_2\boldsymbol{u}_2$:

$$A=\begin{bmatrix} 2 & 2 \\ -1 & 1 \end{bmatrix} \quad \text{and} \quad A=\begin{bmatrix} 3 & 3 \\ 4 & 4 \end{bmatrix}.$$

13 For this rectangular matrix find $\boldsymbol{v}_1,\boldsymbol{v}_2,\boldsymbol{v}_3$ and $\boldsymbol{u}_1,\boldsymbol{u}_2$ and σ_1,σ_2. Then write the SVD for A as $U\Sigma V^{\mathrm{T}}=(2\times 2)\,(2\times 3)\,(3\times 3)$.

$$A=\begin{bmatrix} 1 & 1 & 0 \\ 0 & 1 & 1 \end{bmatrix}$$

14 If $(A^{\mathrm{T}}A)\boldsymbol{v}=\sigma^2\boldsymbol{v}$, multiply by A. *Move the parentheses to get* $(AA^{\mathrm{T}})A\boldsymbol{v}=\sigma^2(A\boldsymbol{v})$.
If $\boldsymbol{v}$ is an eigenvector of $A^{\mathrm{T}}A$, then _____ is an eigenvector of AA^{T}.

15 Find the eigenvalues and unit eigenvectors $\boldsymbol{v}_1,\boldsymbol{v}_2$ of $A^{\mathrm{T}}A$. Then find $\boldsymbol{u}_1=A\boldsymbol{v}_1/\sigma_1$:

$$A=\begin{bmatrix} 1 & 2 \\ 3 & 6 \end{bmatrix} \text{ and } A^{\mathrm{T}}A=\begin{bmatrix} 10 & 20 \\ 20 & 40 \end{bmatrix} \text{ and } AA^{\mathrm{T}}=\begin{bmatrix} 5 & 15 \\ 15 & 45 \end{bmatrix}.$$

Verify that $\boldsymbol{u}_1$ is a unit eigenvector of AA^{T}. Complete the matrices U,Σ,V.

$$\textbf{SVD} \qquad \begin{bmatrix} 1 & 2 \\ 3 & 6 \end{bmatrix}=\begin{bmatrix} \boldsymbol{u}_1 & \boldsymbol{u}_2 \end{bmatrix}\begin{bmatrix} \sigma_1 & \\ & 0 \end{bmatrix}\begin{bmatrix} \boldsymbol{v}_1 & \boldsymbol{v}_2 \end{bmatrix}^{\mathrm{T}}.$$

16 (a) Why is the trace of $A^{\mathrm{T}}A$ equal to the sum of all a_{ij}^2?
(b) For every rank-one matrix, why is $\sigma_1^2=$ sum of all a_{ij}^2?

17 If $A=U\Sigma V^{\mathrm{T}}$ is a square invertible matrix then $A^{-1}=$ ____________.
The largest singular value of A^{-1} is therefore $1/\sigma_{\min}(A)$. The largest eigenvalue has size $1/|\lambda(A)|_{\min}$. Then equation (17) says that $\boldsymbol{\sigma_{\min}(A) \leq |\lambda(A)|_{\min}}$.

18 Suppose $A=U\Sigma V^{\mathrm{T}}$ is 2 by 2 with $\sigma_1>\sigma_2>0$. Change A by ***as small a matrix as possible*** to produce a singular matrix A_0. Hint: U and V do not change.

7.2 Image Processing by Linear Algebra

1 An image is a large matrix of grayscale values, one for each pixel and color.

2 When nearby pixels are correlated (not random) the image can be compressed.

3 Flags often give simple images. Photographs can be compressed by the SVD.

Image processing and compression are major consumers of linear algebra. Recognizing images often uses convolutional neural nets in deep learning. These topics in Chapters 7-8 represent part of the enormous intersection of computational engineering and mathematics.

This section will begin with stylized images like flags: often with *low complexity*. Then we move to photographs with many more pixels. In all cases we want efficient ways to process and transmit signals. The image is effectively a large matrix (!) that can represent light/dark and red/green/blue for every small pixel.

The SVD offers one approach to matrix approximation: *Replace* A *by* A_k. The sum A of r rank one matrices $\sigma_j \boldsymbol{u}_j \boldsymbol{v}_j^{\mathrm{T}}$ can be reduced to the sum A_k of the first k terms. This section (plus online help) will consider the effect on an image such as a flag or a photograph. Section 7.3 will explore more examples in which we need to approximate and understand a matrix of data.

Start with flags. More than 30 countries chose flags with three stripes. Those flags have a particularly simple form: easy to compress. I found a book called "Flags of the World" and the pictures range from one solid color (Libya's flag was entirely green during the Gaddafi years) to very complicated images. How would those pictures be compressed with minimum loss ?

The linear algebra answer is: *Use the* SVD. Notice that 3 stripes still produce rank 1. France has blue-white-red vertical stripes and $b\ w\ r$ in its columns. By coincidence the German flag is nearly its transpose with the colors Black-Red-Yellow:

$$\begin{bmatrix} b\,b\,w\,w\,r\,r \\ b\,b\,w\,w\,r\,r \\ b\,b\,w\,w\,r\,r \\ b\,b\,w\,w\,r\,r \\ b\,b\,w\,w\,r\,r \\ b\,b\,w\,w\,r\,r \end{bmatrix} = \begin{bmatrix} 1 \\ 1 \\ 1 \\ 1 \\ 1 \\ 1 \end{bmatrix} \begin{bmatrix} b\,b\,w\,w\,r\,r \end{bmatrix} \quad \textbf{France} \qquad \begin{bmatrix} B\,B\,B\,B\,B\,B \\ B\,B\,B\,B\,B\,B \\ R\,R\,R\,R\,R\,R \\ R\,R\,R\,R\,R\,R \\ Y\,Y\,Y\,Y\,Y\,Y \\ Y\,Y\,Y\,Y\,Y\,Y \end{bmatrix} = \begin{bmatrix} B \\ B \\ R \\ R \\ Y \\ Y \end{bmatrix} \begin{bmatrix} 1\,1\,1\,1\,1\,1 \end{bmatrix} \quad \textbf{Germany}$$

Each matrix reduces to two vectors. To transmit those images we can replace N^2 pixels by $2N$ pixels. Similarly, Italy is green-white-red and Ireland is green-white-orange. But many many countries make the problem infinitely more difficult by adding a small badge on top of those stripes. Japan has a red sun on a white background and the Maldives have an elegant white moon on a green rectangle on a white rectangle. Those curved images have infinite rank—compression is still possible and necessary, but not to rank one.

A few flags stay with finite rank but they add a cross to increase the rank. Here are two flags (Greece and Tonga) with rank 3 and rank 4.

Greece

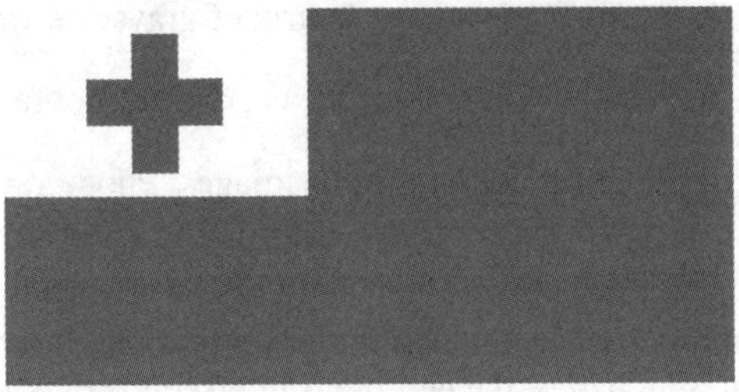

Tonga

I see four different rows in the Greek flag, but only three columns. Mistakenly, I thought the rank was 4. But now I think that row $2+$ row $3=$ row 1 and the rank of Greece is 3.

On the other hand, Tonga's flag does seem to have rank 4. The left half has four rows: all white-short red-longer red-all red. We can't produce any row from a linear combination of the other rows. The island kingdom of Tonga has the champion flag of finite rank!

Singular Values with Diagonals

Three countries have flags with only two diagonal lines: Bahamas, Czech Republic, and Kuwait. Many countries have added in stars and multiple diagonals. From my book I can't be sure whether Canada also has small curves. It is interesting to find the SVD of this matrix with lower triangular 1's—including the main diagonal—and upper triangular 0's.

Flag with a triangle
$$A=\begin{bmatrix}\mathbf{1}&0&0&0\\\mathbf{1}&\mathbf{1}&0&0\\\mathbf{1}&\mathbf{1}&\mathbf{1}&0\\\mathbf{1}&\mathbf{1}&\mathbf{1}&\mathbf{1}\end{bmatrix}\quad\text{has }A^{-1}=\begin{bmatrix}\mathbf{1}&0&0&0\\\mathbf{-1}&\mathbf{1}&0&0\\0&\mathbf{-1}&\mathbf{1}&0\\0&0&\mathbf{-1}&\mathbf{1}\end{bmatrix}.$$

A has full rank $r=N$. All eigenvalues are 1, on the main diagonal. Then A has N singular values (all positive, but not equal to 1). The SVD will produce n pieces $\sigma_i\,\boldsymbol{u}_i\,\boldsymbol{v}_i^{\mathrm{T}}$ of rank one. Perfect reproduction needs all n pieces.

In compression the small σ's can be discarded with no serious loss in image quality. We want to understand the singular values for $n=4$ and also to plot all σ's for large n. The graph on the next page will decide if A is greatly compressed by the SVD (no).

Working by hand, we begin with AA^{T} (a computer would proceed differently):

$$AA^{\mathrm{T}}=\begin{bmatrix}1&1&1&1\\1&2&2&2\\1&2&3&3\\1&2&3&4\end{bmatrix}\quad\text{and }(AA^{\mathrm{T}})^{-1}=(A^{-1})^{\mathrm{T}}A^{-1}=\begin{bmatrix}\mathbf{2}&-1&0&0\\-1&\mathbf{2}&-1&0\\0&-1&\mathbf{2}&-1\\0&0&-1&\mathbf{1}\end{bmatrix}.\quad(1)$$

That $-1,2,-1$ inverse matrix is included because all its eigenvalues have the form $2-2\cos\theta$. We know those eigenvalues! So we know the singular values of A.

$$\lambda(AA^{\mathrm{T}}) = \frac{1}{2-2\cos\theta} = \frac{1}{4\sin^2(\theta/2)} \quad \text{gives} \quad \sigma(A) = \sqrt{\lambda} = \frac{1}{2\sin(\theta/2)}. \tag{2}$$

The n different angles θ are equally spaced, which makes this example so exceptional:

$$\theta = \frac{\pi}{2n+1}, \frac{3\pi}{2n+1}, \ldots, \frac{(2n-1)\,\pi}{2n+1} \qquad \left(n = 4 \text{ includes } \theta = \frac{3\pi}{9} \text{ with } 2\sin\frac{\theta}{2} = 1\right).$$

The important point is to graph the n singular values of A. Those numbers drop off (unlike the eigenvalues of A, which are all 1). But the dropoff is not steep. So the SVD gives only moderate compression of this triangular flag. *Great compression for Hilbert.*

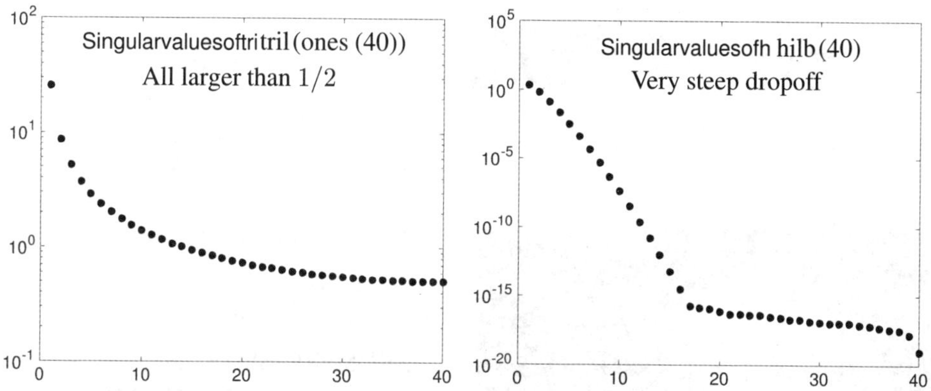

Figure 7.3: Singular values of the 40 by 40 triangle of 1's (it is not compressible). The evil Hilbert matrix $H(i,j) = (i+j-1)^{-1}$ has low effective rank: we must compress it.

The striking point about the graph is that the singular values for the triangular matrix never go below $\frac{1}{2}$. Working with Alex Townsend, we have seen this phenomenon for 0-1 matrices with the 1's in other shapes (such as circular). This has not yet been explained.

Image Compression by the SVD

Compressing photographs and images is an exceptional way to see the SVD in action. The action comes by varying the number of rank one pieces $\sigma \boldsymbol{u}\boldsymbol{v}^{\mathrm{T}}$ in the display. By keeping more terms the image improves.

We hoped to find a website that would show this improvement. By good fortune, Tim Baumann has achieved exactly what we hoped for. He has given all of us permission to use his work: **https://timbaumann.info/svd-image-compression-demo/**

The next page shows an earlier version of his website.

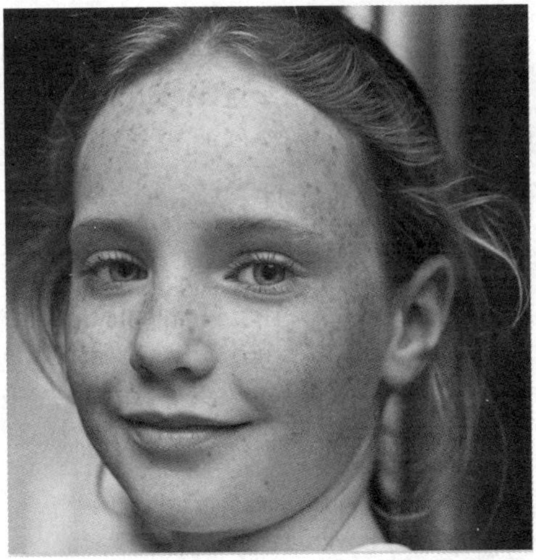

Uncompressed image. Slider at 300.

IMAGE SIZE 600×600
#PIXELS $= 360000$

UNCOMPRESSED SIZE
proportional to number of pixels

COMPRESSED SIZE
approximately proportional to
$600 \times 300 + 300 + 300 \times 600$
$= 360300$

COMPRESSION RATIO
$360000/360300 = 1.00$

Show singular values

Compressed image. Slider at 20.

IMAGE SIZE 600×600
#PIXELS $= 360000$

UNCOMPRESSED SIZE
proportional to number of pixels

COMPRESSED SIZE
approximately proportional to
$600 \times 20 + 20 + 20 \times 600$
$= 24020$

COMPRESSION RATIO
$360000/24020 = 14.99$

Show singular values

Change the number of singular values using the slider. Click on one of these images to compress it :

You can compress your own images by using the file picker or by dropping them on this page.

This is one of the five images directly available. The position of the slider determines the compression ratio. The best ratio depends on the complexity of the image—the girl and the Mondrian painting are less complex and allow higher compression than the city or the tree or the cats. When the computation of compressed size adds $600 \times 80 + 80 + 80 \times 600$, we have 80 terms $\sigma \boldsymbol{u}\boldsymbol{v}^{\mathrm{T}}$ with vectors $\boldsymbol{u}$ and $\boldsymbol{v}$ of dimension 600. The slider is set at 80.

You can compress your own images by using the "file picker" button below the six sample images provided on the site, or by dropping them directly onto the page.

One beautiful feature of Tim Baumann's site is that it operates in the browser, with instant results. This book's website **math.mit.edu/linearalgebra** can include ideas from readers. Please see that edited site for questions and comments and suggestions.

Problem Set 7.2

1 We usually think that the identity matrix I is as simple as possible. But why is I difficult to compress ? *Create the matrix for a rank* 2 *flag with a horizontal-vertical cross.* This is highly compressible.

2 These flags have rank 2. Write A and B in any way as $\boldsymbol{u}_1\boldsymbol{v}_1^{\mathrm{T}} + \boldsymbol{u}_2\boldsymbol{v}_2^{\mathrm{T}}$.

$$\boldsymbol{A}_{\textbf{Sweden}} = \boldsymbol{A}_{\textbf{Finland}} = \begin{bmatrix} 1 & 2 & 1 & 1 \\ 2 & 2 & 2 & 2 \\ 1 & 2 & 1 & 1 \end{bmatrix} \qquad \boldsymbol{B}_{\textbf{Benin}} = \begin{bmatrix} 1 & 2 & 2 \\ 1 & 3 & 3 \end{bmatrix}$$

3 Now find the trace and determinant of BB^{T} and also $B^{\mathrm{T}}B$ in Problem 2. The singular values of B are close to $\sigma_1^2 = 28 - \frac{1}{14}$ and $\sigma_2^2 = \frac{1}{14}$. Is B compressible or not?

4 Use $[U, S, V] = \mathsf{svd}\,(A)$ to find two orthogonal pieces $\sigma \boldsymbol{u}\boldsymbol{v}^{\mathrm{T}}$ of $\boldsymbol{A}_{\textbf{Sweden}}$.

5 A matrix for the Japanese flag has a circle of ones surrounded by all zeros. Suppose the center line of the circle (the diameter) has $2N$ ones. Then the circle will contain about πN^2 ones. We think of the flag as a 1-0 matrix. Its rank will be approximately CN, proportional to N. *What is that number* C ?

Hint : Remove a big square submatrix of ones, with corners at $\pm 45^\circ$ and $\pm 135^\circ$. The rows above the square and the columns to the right of the square are independent. Draw a picture and estimate the number cN of those rows. Then $C = 2c$.

6 Here is one way to start with a function $F(x, y)$ and construct a matrix A. Set $A_{ij} = \boldsymbol{F(i/N, j/N)}$. (The indices i and j go from 0 to N or from $-N$ to N.) The rank of A as N increases should reflect the simplicity or complexity of F. Find the ranks of the matrices for the functions $F_1 = \boldsymbol{xy}$ and $F_2 = \boldsymbol{x} + \boldsymbol{y}$ and $F_3 = \boldsymbol{x}^2 + \boldsymbol{y}^2$. Then find three singular values and singular vectors for F_3.

7 In Problem 6, what conditions on $F(x, y)$ will produce a symmetric matrix S ? An antisymmetric matrix A ? A singular matrix M ? A matrix of rank 2 ?

7.3 Principal Component Analysis (PCA by the SVD)

The "principal components" of A are its singular vectors, the orthogonal columns $\boldsymbol{u}_j$ and $\boldsymbol{v}_j$ of the matrices U and V. This section aims to apply the Singular Value Decomposition $A = U\Sigma V^{\mathrm{T}}$. **Principal Component Analysis (PCA)** uses the largest σ's connected to the first $\boldsymbol{u}$'s and $\boldsymbol{v}$'s to understand the information in a matrix of data.

We are given a matrix A, and we extract its most important part A_k (**largest σ's**):

$$\boldsymbol{A_k = \sigma_1 u_1 v_1^{\mathrm{T}} + \cdots + \sigma_k u_k v_k^{\mathrm{T}}} \quad \textbf{with rank } (\boldsymbol{A_k}) = \boldsymbol{k}.$$

A_k solves a matrix optimization problem—and we start there. **The closest rank $\boldsymbol{k}$ matrix to $\boldsymbol{A}$ is $\boldsymbol{A_k}$.** In statistics we are identifying the rank one pieces of A with largest variance. This puts the SVD at the center of data science.

In that world, PCA is "unsupervised" learning. Our only instructor is linear algebra—the SVD tells us to choose A_k. When the learning is supervised, we have a big set of training data. Deep Learning constructs a (nonlinear!) function F to correctly classify most of that data. Then we apply this F to new data, as you will see in Chapter 10.

Principal Component Analysis is based on matrix approximation by A_k. The proof that A_k is the best choice was begun by Schmidt (1907). He wrote about operators in function space; his ideas extend directly to matrices. Eckart and Young gave a new proof (using the Frobenius norm to measure $A - A_k$). Then Mirsky allowed any norm $||A||$ that depends only on the singular values—as in the definitions (2), (3), and (4) below.

Here is that key property of the special rank k matrix $A_k = \sigma_1 \boldsymbol{u}_1 \boldsymbol{v}_1^{\mathrm{T}} + \cdots + \sigma_k \boldsymbol{u}_k \boldsymbol{v}_k^{\mathrm{T}}$:

$$\boxed{\boldsymbol{A_k} \textbf{ is closest to } \boldsymbol{A} \qquad \textbf{If } \boldsymbol{B} \textbf{ has rank } \boldsymbol{k} \textbf{ then } ||\boldsymbol{A} - \boldsymbol{A_k}|| \le ||\boldsymbol{A} - \boldsymbol{B}||.} \tag{1}$$

Three choices for the matrix norm $||A||$ have special importance and their own names:

$$\textbf{Spectral norm} \qquad ||\boldsymbol{A}||_2 = \max \frac{||\boldsymbol{Ax}||}{||\boldsymbol{x}||} = \boldsymbol{\sigma_1} \qquad \text{(often called the } \boldsymbol{\ell^2} \textbf{ norm)} \tag{2}$$

$$\textbf{Frobenius norm} \qquad ||\boldsymbol{A}||_F = \sqrt{\boldsymbol{\sigma_1^2 + \cdots + \sigma_r^2}} \qquad \text{(7) also defines } ||\boldsymbol{A}||_F \tag{3}$$

$$\textbf{Nuclear norm} \qquad ||\boldsymbol{A}||_N = \boldsymbol{\sigma_1 + \sigma_2 + \cdots + \sigma_r} \qquad \text{(the trace norm)} \tag{4}$$

These norms have different values already for the n by n identity matrix:

$$||I||_2 = 1 \qquad ||I||_F = \sqrt{n} \qquad ||I||_N = n. \tag{5}$$

Replace I by any orthogonal matrix Q and the norms stay the same (because all $\sigma_i = 1$):

$$||Q||_2 = 1 \qquad ||Q||_F = \sqrt{n} \qquad ||Q||_N = n. \tag{6}$$

More than this, the spectral and Frobenius and nuclear norms of any matrix stay the same when A is multiplied (on either side) by an orthogonal matrix. So the norm of $A = U\Sigma V^{\mathrm{T}}$ equals the norm of Σ: $||\boldsymbol{A}|| = ||\boldsymbol{\Sigma}||$ because U and V are orthogonal matrices.

Norm of a Matrix

We need a way to measure the size of a vector or a matrix. For a vector, the most important norm is the usual length $||\boldsymbol{v}||$. For a matrix, Frobenius extended the sum of squares idea to include all the entries in A. This norm $||A||_F$ is also named Hilbert-Schmidt.

$$||\boldsymbol{v}||^2 = v_1^2 + \cdots + v_n^2 \qquad ||A||_F^2 = a_{11}^2 + \cdots + a_{1n}^2 + \cdots + a_{m1}^2 + \cdots + a_{mn}^2 \tag{7}$$

Clearly $||\boldsymbol{v}|| \geq 0$ and $||c\boldsymbol{v}|| = |c|\ ||\boldsymbol{v}||$. Similarly $||A||_F \geq 0$ and $||cA||_F = |c|\ ||A||_F$. Equally essential is the triangle inequality for $\boldsymbol{v} + \boldsymbol{w}$ and $A + B$:

Triangle inequalities $\quad ||\boldsymbol{v} + \boldsymbol{w}|| \leq ||\boldsymbol{v}|| + ||\boldsymbol{w}||$ and $||A + B||_F \leq ||A||_F + ||B||_F \quad (8)$

We use one more fact when we meet dot products $\boldsymbol{v}^{\mathrm{T}}\boldsymbol{w}$ or matrix products AB :

Schwarz inequalities $\quad |\boldsymbol{v}^{\mathrm{T}}\boldsymbol{w}| \leq ||\boldsymbol{v}||\ ||\boldsymbol{w}||$ and $||AB||_F \leq ||A||_F\ ||B||_F \quad (9)$

That Frobenius matrix inequality comes directly from the Schwarz vector inequality:

$$|(AB)_{ij}|^2 \leq ||\text{row } i \text{ of } A||^2\ ||\text{column } j \text{ of } B||^2. \text{ Add for all } i \text{ and } j \text{ to see } ||AB||_F^2.$$

This suggests that there could be a dot product of matrices. It is $A \cdot B = \text{trace}(A^{\mathrm{T}}B)$.

Note. The largest size $|\lambda|$ of the eigenvalues of A is *not an acceptable norm*! We know that a nonzero matrix could have all zero eigenvalues—but its norm $||A||$ cannot be zero. In this respect singular values are superior to eigenvalues: $\boldsymbol{\sigma_1 = ||A||_2}$.

The Eckart-Young Theorem

The theorem was in equation (1): **If $\boldsymbol{B}$ has rank $\boldsymbol{k}$ then $||\boldsymbol{A} - \boldsymbol{A_k}|| \leq ||\boldsymbol{A} - \boldsymbol{B}||$.** In all three norms $||A||$ and $||A||_F$ and $||A||_N$, we come closest to A by cutting off the SVD after k terms. The closest matrix is $A_k = \sigma_1\boldsymbol{u}_1\boldsymbol{v}_1^{\mathrm{T}} + \cdots + \sigma_k\boldsymbol{u}_k\boldsymbol{v}_k^{\mathrm{T}}$. This is the fact to use in approximating A by a low rank matrix!

We need an example and it can look extremely simple: a diagonal matrix A.

$$\text{The rank 2 matrix closest to } A = \begin{bmatrix} \mathbf{4} & 0 & 0 & 0 \\ 0 & \mathbf{3} & 0 & 0 \\ 0 & 0 & \mathbf{2} & 0 \\ 0 & 0 & 0 & \mathbf{1} \end{bmatrix} \text{ is } A_2 = \begin{bmatrix} \mathbf{4} & 0 & 0 & 0 \\ 0 & \mathbf{3} & 0 & 0 \\ 0 & 0 & \mathbf{0} & 0 \\ 0 & 0 & 0 & \mathbf{0} \end{bmatrix}$$

The difference $A - A_2$ is all zero except for the 2 and 1. Then $||A - A_2||_F = \sqrt{2^2 + 1^2}$. How could any other rank 2 matrix be closer to A than this A_2?

Please realize that this deceptively simple example includes *all matrices of the form* Q_1AQ_2. The norms and the rank are not changed by the orthogonal matrices Q_1, Q_2. So this example includes all matrices with singular values $4, 3, 2, 1$. The best approximation A_2 keeps 4 and 3. Several proofs are collected in *Linear Algebra and Learning from Data* (Wellesley-Cambridge Press). Chi-Kwong Li has simplified Mirsky's proof that A_k is closest to A for all norms like $||A||_2$ and $||A||_F$ that depend only on the σ's.

Principal Component Analysis

Now we start using the SVD. *The matrix A is full of data.* We have n samples. For each sample we measure m variables (like height and weight). The data matrix A_0 has n columns and m rows. In many applications it is a very large matrix.

The first step is to find the average (the sample mean) along each row of A_0. Subtract that mean from all m entries in the row. Now each row of the centered matrix A has *mean zero.* The columns of A are n points in $\mathbf{R}^m$. Because of centering, the sum of the n column vectors is zero. So the average column is the zero vector.

Often those n points are clustered near a line or a plane or another low-dimensional subspace of $\mathbf{R}^m$. Figure 7.3 shows a typical set of data points clustered along a line in $\mathbf{R}^2$ (after centering A_0 to shift the points left-right and up-down to have mean $(0,0)$ in A).

How will linear algebra find that closest line through $(0,0)$? **It is in the direction of the first singular vector $\boldsymbol{u}_1$ of A.** This is the key point of PCA !

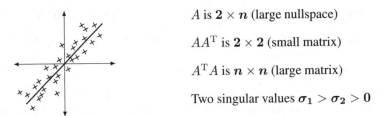

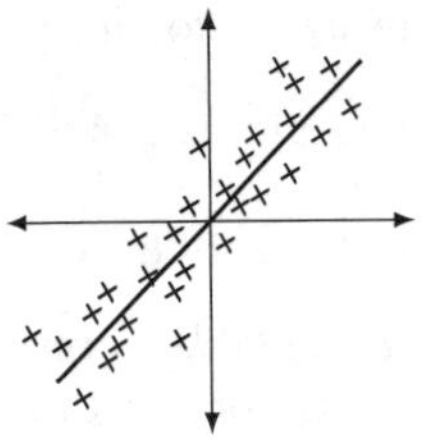

A is $\mathbf{2} \times \boldsymbol{n}$ (large nullspace)

AA^{T} is $\mathbf{2} \times \mathbf{2}$ (small matrix)

$A^{\mathrm{T}}A$ is $\boldsymbol{n} \times \boldsymbol{n}$ (large matrix)

Two singular values $\boldsymbol{\sigma_1} > \boldsymbol{\sigma_2} > \mathbf{0}$

Figure 7.4: Data points (columns of A) are often close to a line in $\mathbf{R}^2$ or a subspace in $\mathbf{R}^m$.

The Geometry Behind PCA

The best line in Figure 7.4 solves a problem in **perpendicular least squares**. This is also called *orthogonal regression.* It is different from the standard least squares fit to n data points, or the least squares solution to a linear system $A\boldsymbol{x} = \boldsymbol{b}$. That classical problem in Section 4.3 minimizes $||A\boldsymbol{x} - \boldsymbol{b}||^2$. It measures distances up and down to the best line. Our problem minimizes *perpendicular* distances. The older problem leads to a linear equation $A^{\mathrm{T}}A\widehat{\boldsymbol{x}} = A^{\mathrm{T}}\boldsymbol{b}$ for the best $\widehat{\boldsymbol{x}}$. Our problem leads to singular vectors $\boldsymbol{u}_i$ (eigenvectors of AA^{T}). Those are the two sides of linear algebra: not the same side.

The sum of squared distances from the data points to the $\boldsymbol{u}_1$ line is a minimum.

To see this, separate each column $\boldsymbol{a}_j$ of A into its components along $\boldsymbol{u}_1$ and $\boldsymbol{u}_2$:

$$\sum_1^n ||\boldsymbol{a}_j||^2 = \sum_1^n |\boldsymbol{a}_j^{\mathrm{T}}\boldsymbol{u}_1|^2 + \sum_1^n |\boldsymbol{a}_j^{\mathrm{T}}\boldsymbol{u}_2|^2. \tag{10}$$

The sum on the left is fixed by the data. The first sum on the right has terms $\boldsymbol{u}_1^{\mathrm{T}}\boldsymbol{a}_j\boldsymbol{a}_j^{\mathrm{T}}\boldsymbol{u}_1$. It adds to $\boldsymbol{u}_1^{\mathrm{T}}(AA^{\mathrm{T}})\boldsymbol{u}_1$. So when we maximize that sum in PCA by choosing the top eigenvector $\boldsymbol{u}_1$ of AA^{T}, we minimize the second sum. That second sum of squared distances from data points to the best line (or best subspace) is the smallest possible.

The Geometric Meaning of Eckart-Young

Figure 7.4 was in two dimensions and it led to the closest line. Now suppose our data matrix A_0 is 3 by n: Three measurements like age, height, weight for each of n samples. Again we center each row of the matrix, so all the rows of A add to zero. And the points move into three dimensions.

We can still look for the nearest line. It will be revealed by the first singular vector $\boldsymbol{u}_1$ of A. The best line will go through $(0,0,0)$. But if the data points fan out compared to Figure 7.4, *we really need to look for the best plane.* The meaning of "best" is still this: The sum of perpendicular distances squared to the best plane is a minimum.

That plane will be spanned by the singular vectors $\boldsymbol{u}_1$ and $\boldsymbol{u}_2$. This is the meaning of Eckart-Young. *It leads to a neat conclusion: The best plane contains the best line.*

The Statistics Behind PCA

The key numbers in probability and statistics are the **mean** and **variance**. The "mean" is an average of the data (in each row of A_0). Subtracting those means from each row of A_0 produced the centered A. The crucial quantities are the "variances" and "covariances". The variances are sums of squares of distances from the mean—along each row of A.

The variances are the diagonal entries of the matrix AA^{T}.

Suppose the columns of A correspond to a child's age on the x-axis and its height on the y-axis. (Those ages and heights are measured from the average age and height.) We are looking for the straight line that stays closest to the data points in the figure. And we have to account for the *joint age-height distribution* of the data.

The covariances are the off-diagonal entries of the matrix AA^{T}.

Those are dot products (row i of A) · (row j of A). High covariance means that increased height goes with increased age. (Negative covariance means that one variable increases when the other decreases.) Our first example has only two rows from age and height: the symmetric matrix AA^{T} is 2 by 2. As the number n of sample children increases, we divide by $n-1$ to give AA^{T} its statistically correct scale.

The sample covariance matrix is defined by $S = \dfrac{AA^{\mathrm{T}}}{n-1}$.

The factor is $n-1$ because one degree of freedom has already been used for mean $= 0$. This example with six ages and heights is already centered to make each row add to zero:

$$\textbf{Example} \qquad A = \begin{bmatrix} 3 & -4 & 7 & 1 & -4 & -3 \\ 7 & -6 & 8 & -1 & -1 & -7 \end{bmatrix}$$

For this data, the sample covariance matrix S is easily computed. It is positive definite.

$$\textbf{Variances and covariances} \qquad S = \frac{1}{6-1}AA^{\mathrm{T}} = \begin{bmatrix} 20 & 25 \\ 25 & 40 \end{bmatrix}.$$

The two orthogonal eigenvectors of S are $\boldsymbol{u}_1$ and $\boldsymbol{u}_2$. Those are the left singular vectors (often called the *principal components*) of A. The Eckart-Young theorem says that **the vector $\boldsymbol{u}_1$ points along the closest line in Figure 7.4.**

The second singular vector $\boldsymbol{u}_2$ will be perpendicular to that closest line.

Important note PCA can be described using the symmetric $S = AA^{\mathrm{T}}/(n-1)$ or the rectangular A. No doubt S is the nicer matrix. But given the data in A, computing S can be a computational mistake. For large matrices, a direct SVD of A is faster and more accurate. By going to AA^{T} we square σ_1 and σ_r and the condition number σ_1/σ_r.

In the example, S has eigenvalues near 57 and 3. Their sum is $20+40=60$, the trace of S. The first rank one piece $\sqrt{57}\boldsymbol{u}_1\boldsymbol{v}_1^{\mathrm{T}}$ is much larger than the second piece $\sqrt{3}\boldsymbol{u}_2\boldsymbol{v}_2^{\mathrm{T}}$. **The leading eigenvector $\boldsymbol{u}_1 \approx (0.6, 0.8)$ tells us that the closest line in the scatter plot has slope near $\boldsymbol{8/6}$.** The direction in the graph nearly produces a $6-8-10$ right triangle.

The Linear Algebra Behind PCA

Principal Component Analysis is a way to understand n sample points $\boldsymbol{a}_1, \ldots, \boldsymbol{a}_n$ in m-dimensional space—the data. That data plot is centered: all rows of A add to zero. The crucial connection to linear algebra is in the singular values and the left singular vectors $\boldsymbol{u}_i$ of A. Those come from the eigenvalues $\lambda_i = \sigma_i^2$ and the eigenvectors of the sample covariance matrix $S = AA^{\mathrm{T}}/(n-1)$.

The total variance in the data comes from the squared Frobenius norm of A:

$$\textbf{Total variance } \boldsymbol{T} = ||\boldsymbol{A}||_{\boldsymbol{F}}^{\boldsymbol{2}}/(\boldsymbol{n}-\boldsymbol{1}) = (||\boldsymbol{a}_1||^2 + \cdots + ||\boldsymbol{a}_n||^2)/(\boldsymbol{n}-\boldsymbol{1}). \tag{11}$$

This is the trace of S—the sum down the diagonal. Linear algebra tells us that the trace equals the **sum of the eigenvalues of the sample covariance matrix $\boldsymbol{S}$.**

The SVD is producing orthogonal singular vectors $\boldsymbol{u}_i$ that separate the data into uncorrelated pieces (with zero covariance). They come in order of decreasing variance, and the first pieces tell us what we need to know.

The trace of S connects the total variance to the sum of variances of the principal components $\boldsymbol{u}_1, \ldots, \boldsymbol{u}_r$:

$$\textbf{Total variance} \qquad \boldsymbol{T} = \boldsymbol{\sigma}_{\boldsymbol{1}}^{\boldsymbol{2}} + \cdots + \boldsymbol{\sigma}_{\boldsymbol{r}}^{\boldsymbol{2}}. \tag{12}$$

The first principal component $\boldsymbol{u}_1$ accounts for (or "*explains*") a fraction σ_1^2/T of the total variance. The next singular vector $\boldsymbol{u}_2$ of A explains the next largest fraction σ_2^2/T. Each singular vector is doing its best to capture the meaning in a matrix—and all together they succeed.

The point of the Eckart-Young Theorem is that k singular vectors (acting together) explain *more of the data than any other set of k vectors.* So we are justified in choosing $\boldsymbol{u}_1$ to $\boldsymbol{u}_k$ as a basis for the k-dimensional subspace closest to the n data points.

The "effective rank" of A and S is the number of singular values above the point where noise drowns the true signal in the data. Often this point is visible on a "**scree plot**" showing the dropoff in the singular values (or their squares σ_i^2). Look for the "elbow" in the scree plot (Figure 7.2) where the signal ends and noise takes over.

Problem Set 7.3

1 Suppose A_0 holds these 2 measurements of 5 samples:

$$A_0 = \begin{bmatrix} 5 & 4 & 3 & 2 & 1 \\ -1 & 1 & 0 & 1 & -1 \end{bmatrix}$$

Find the average of each row and subtract it to produce the centered matrix A. Compute the sample covariance matrix $S = AA^{\mathrm{T}}/(n-1)$ and find its eigenvalues λ_1 and λ_2. What line through the origin is closest to the 5 samples in columns of A?

2 Take the steps of Problem 1 for this 2 by 6 matrix A_0:

$$A_0 = \begin{bmatrix} 1 & 0 & 1 & 0 & 1 & 0 \\ 1 & 2 & 3 & 3 & 2 & 1 \end{bmatrix}$$

3 The sample variances s_1^2, s_2^2 and the sample covariance s_{12} are the entries of S. Find S after subtracting row averages from $A_0 = \begin{bmatrix} 1 & 2 & 3 \\ 5 & 2 & 2 \end{bmatrix}$. What is σ_1?

4 From the eigenvectors of $S = AA^{\mathrm{T}}$, find the line (the $\boldsymbol{u}_1$ direction through the center point) and then the plane ($\boldsymbol{u}_1$ and $\boldsymbol{u}_2$ directions) closest to these four points in three-dimensional space:

$$A = \begin{bmatrix} 1 & -1 & 0 & 0 \\ 0 & 0 & 2 & -2 \\ 1 & 1 & -1 & -1 \end{bmatrix}.$$

5 Compare ordinary least squares (Section 4.3) with PCA (perpendicular least squares). They both give a closest line $C + Dt$ to the symmetric data $\boldsymbol{b} = -1, 0, 1$ at times $t = -3, 1, 2$.

$$A = \begin{bmatrix} 1 & -3 \\ 1 & 1 \\ 1 & 2 \end{bmatrix} \qquad \boldsymbol{b} = \begin{bmatrix} -1 \\ 0 \\ 1 \end{bmatrix}$$

Least squares : $A^{\mathrm{T}} A \widehat{\boldsymbol{x}} = A^{\mathrm{T}} \boldsymbol{b}$
PCA : eigenvector of AA^{T}
(singular vector $\boldsymbol{u}_1$ of A)

6 The idea of **eigenfaces** begins with N images: same size and alignment. Subtract the average image from each of the N images. Create the sample covariance matrix $S = \Sigma A_i A_i^{\mathrm{T}}/N - 1$ and find the eigenvectors (= eigenfaces) with largest eigenvalues. They don't look like faces but their combinations come close to faces. Wikipedia gives a code for this *dimension reduction* pioneered by Turk and Pentland.

7 What are the singular values of $A - A_3$ if A has singular values $5, 4, 3, 2, 1$ and A_3 is the closest matrix of rank 3 to A?

8 If A has $\sigma_1 = 9$ and B has $\sigma_1 = 4$, what are upper and lower bounds to σ_1 for $A + B$? Why is this true?

Chapter 8

Linear Transformations

The study of linear algebra can begin with matrices, or it can begin with linear transformations. We chose matrices. Another author might say: "Matrices are only a special case of linear transformations." And going back even farther: "The real numbers are only a special case of a field." All these topics and more are proper parts of the branch of mathematics called *Algebra*.

I am not quite sure how to respond. Perhaps I would say: "Before you study languages in general (the subject of linguistics), you need to know at least one language." And working with column vectors and matrices and linear equations and inverses and vector spaces and orthogonality and linear independence and rank is so fascinating and rewarding.

To me, the variety of matrices is wonderful. It is like a community of people with different relations and different jobs. Matrices are unique, but factorization connects them to other matrices. You are allowed to have favorites. I hope you do.

P.S. Section 8.3 in this chapter discusses good bases for vector spaces of functions. The most natural extension from $\mathbf{R}^n$ to infinite dimensions is called "Hilbert space". The lengths and the inner products of functions $f(x)$ and $g(x)$ have a natural form:

$$||f||^2 = \int |f(x)|^2\,dx \quad \text{and} \quad (f,g) = \int f(x)g(x)\,dx.$$

The best bases $q_1(x), q_2(x), \ldots$ are orthogonal as always, and we will suggest three favorites. But there are many good ways to measure the length of $f(x)$ and its derivatives.

That paragraph left *Algebra* behind, and opened the way to *Functional Analysis*.

8.1 The Idea of a Linear Transformation

1 A **linear transformation** T takes vectors $\boldsymbol{v}$ to vectors $T(\boldsymbol{v})$. Linearity requires

$\boxed{T(c\boldsymbol{v}+d\boldsymbol{w})=cT(\boldsymbol{v})+dT(\boldsymbol{w})}$ Note $T(\mathbf{0})=\mathbf{0}$ so $T(\boldsymbol{v})=\boldsymbol{v}+\boldsymbol{u}_0$ is not linear.

2 The input vectors $\boldsymbol{v}$ and outputs $T(\boldsymbol{v})$ can be in $\mathbf{R}^n$ or matrix space or function space.

3 If A is m by n, $T(\boldsymbol{v})=A\boldsymbol{v}$ is linear from the input space $\mathbf{R}^n$ to the output space $\mathbf{R}^m$.

4 The derivative $T(f)=\dfrac{df}{dx}$ is linear. The integral $T^+(f)=\displaystyle\int_0^x f(t)\,dt$ is its pseudoinverse.

5 The product ST of two linear transformations is still linear : $\boxed{(ST)(\boldsymbol{v})=S(T(\boldsymbol{v})).}$

When a matrix A multiplies a vector $\boldsymbol{v}$, it "transforms" $\boldsymbol{v}$ into another vector $A\boldsymbol{v}$. ***In goes $\boldsymbol{v}$, out comes*** $T(\boldsymbol{v})=A\boldsymbol{v}$. A transformation T follows the same idea as a function. In goes a number x, out comes $f(x)$. For one vector $\boldsymbol{v}$ or one number x, we multiply by the matrix or we evaluate the function. The deeper goal is to see all vectors $\boldsymbol{v}$ at once. We are transforming the whole space $\mathbf{V}$ when we multiply every $\boldsymbol{v}$ by A.

Start again with a matrix A. It transforms $\boldsymbol{v}$ to $A\boldsymbol{v}$. It transforms $\boldsymbol{w}$ to $A\boldsymbol{w}$. Then we *know* what happens to $\boldsymbol{u}=\boldsymbol{v}+\boldsymbol{w}$. There is no doubt about $A\boldsymbol{u}$, it has to equal $A\boldsymbol{v}+A\boldsymbol{w}$. Matrix multiplication $T(\boldsymbol{v})=A\boldsymbol{v}$ gives an important ***linear transformation*** :

> A ***transformation*** T assigns an output $T(\boldsymbol{v})$ to each input vector $\boldsymbol{v}$ in $\mathbf{V}$.
>
> The transformation is ***linear*** if it meets these requirements for all $\boldsymbol{v}$ and $\boldsymbol{w}$:
>
> (a) $T(\boldsymbol{v}+\boldsymbol{w})=T(\boldsymbol{v})+T(\boldsymbol{w})$ (b) $T(c\boldsymbol{v})=cT(\boldsymbol{v})$ for all c.

If the input is $\boldsymbol{v}=\mathbf{0}$, the output must be $T(\boldsymbol{v})=\mathbf{0}$. We combine rules (a) and (b) into one :

> **Linear transformation** $T(c\boldsymbol{v}+d\boldsymbol{w})$ ***must equal*** $cT(\boldsymbol{v})+dT(\boldsymbol{w})$.

Again I can test matrix multiplication for linearity: $A(c\boldsymbol{v}+d\boldsymbol{w})=cA\boldsymbol{v}+dA\boldsymbol{w}$ is *true*.

A linear transformation is highly restricted. Suppose T adds $\boldsymbol{u}_0$ to every vector. Then $T(\boldsymbol{v})=\boldsymbol{v}+\boldsymbol{u}_0$ and $T(\boldsymbol{w})=\boldsymbol{w}+\boldsymbol{u}_0$. This isn't good, or at least *it isn't linear*. Applying T to $\boldsymbol{v}+\boldsymbol{w}$ produces $\boldsymbol{v}+\boldsymbol{w}+\boldsymbol{u}_0$. That is not the same as $T(\boldsymbol{v})+T(\boldsymbol{w})$:

Shift is not linear $\boldsymbol{v}+\boldsymbol{w}+\boldsymbol{u}_0$ is not $T(\boldsymbol{v})+T(\boldsymbol{w})=(\boldsymbol{v}+\boldsymbol{u}_0)+(\boldsymbol{w}+\boldsymbol{u}_0)$.

The exception is when $\boldsymbol{u}_0=\mathbf{0}$. The transformation reduces to $T(\boldsymbol{v})=\boldsymbol{v}$. This is the ***identity transformation*** (nothing moves, as in multiplication by the identity matrix). That is certainly linear. In this case the input space $\mathbf{V}$ is the same as the output space $\mathbf{W}$.

The linear-plus-shift transformation $T(\boldsymbol{v}) = A\boldsymbol{v} + \boldsymbol{u}_0$ is called *"affine"*. Straight lines stay straight although T is not linear. Computer graphics works with affine transformations because we must be able to shift images. This is developed on the website.

Example 1 Choose a fixed vector $\boldsymbol{a} = (1, 3, 4)$, and let $T(\boldsymbol{v})$ be the dot product $\boldsymbol{a} \cdot \boldsymbol{v}$:

The input is $\boldsymbol{v} = (v_1, v_2, v_3)$.	The output is $T(\boldsymbol{v}) = \boldsymbol{a} \cdot \boldsymbol{v} = v_1 + 3v_2 + 4v_3$.

Dot products are linear. The inputs $\boldsymbol{v}$ come from three-dimensional space, so $\mathbf{V} = \mathbf{R}^3$. The outputs are just numbers, so the output space is $\mathbf{W} = \mathbf{R}^1$. We are multiplying by the row matrix $A = [1 \;\; 3 \;\; 4]$. Then $T(\boldsymbol{v}) = A\boldsymbol{v}$.

You will get good at recognizing which transformations are linear. If the output involves squares or products or lengths, v_1^2 or $v_1 v_2$ or $\|\boldsymbol{v}\|$, then T is not linear.

Example 2 The length $T(\boldsymbol{v}) = \|\boldsymbol{v}\|$ is not linear. Requirement (a) for linearity would be $\|\boldsymbol{v} + \boldsymbol{w}\| = \|\boldsymbol{v}\| + \|\boldsymbol{w}\|$. Requirement (b) would be $\|c\boldsymbol{v}\| = c\|\boldsymbol{v}\|$. Both are false!

Not (a): The sides of a triangle satisfy an *inequality* $\|\boldsymbol{v} + \boldsymbol{w}\| \leq \|\boldsymbol{v}\| + \|\boldsymbol{w}\|$.
Not (b): The length $\|-\boldsymbol{v}\|$ is $\|\boldsymbol{v}\|$ and not $-\|\boldsymbol{v}\|$. For negative c, linearity fails.

Example 3 (Rotation) T is the transformation that *rotates every vector by* $30°$. The "*domain*" of T is the xy plane (all input vectors $\boldsymbol{v}$). The "*range*" of T is also the xy plane (all rotated vectors $T(\boldsymbol{v})$). We described T without a matrix: rotate the plane by $30°$.

Is rotation linear? *Yes it is.* We can rotate two vectors and add the results. The sum of rotations $T(\boldsymbol{v}) + T(\boldsymbol{w})$ is the same as the rotation $T(\boldsymbol{v} + \boldsymbol{w})$ of the sum. **The whole plane is turning together, in this linear transformation.**

The rule of linearity extends to combinations of three vectors or n vectors:

$$\textbf{Linearity} \quad \boldsymbol{u} = c_1\boldsymbol{v}_1 + c_2\boldsymbol{v}_2 + \cdots + c_n\boldsymbol{v}_n \quad \textbf{must transform to} \qquad T(\boldsymbol{u}) = c_1T(\boldsymbol{v}_1) + c_2T(\boldsymbol{v}_2) + \cdots + c_nT(\boldsymbol{v}_n) \tag{1}$$

The 2-vector rule starts the 3-vector proof: $T(c\boldsymbol{u} + d\boldsymbol{v} + e\boldsymbol{w}) = T(c\boldsymbol{u}) + T(d\boldsymbol{v} + e\boldsymbol{w})$. Then linearity applies to both of those parts, to give three parts: $cT(\boldsymbol{u}) + dT(\boldsymbol{v}) + eT(\boldsymbol{w})$.

The n-vector rule (1) leads to a very important fact about linear transformations:

BASIS TELLS ALL	**Suppose you know $T(\boldsymbol{v})$ for all vectors $\boldsymbol{v}_1, \ldots, \boldsymbol{v}_n$ in a basis** **Then you know $T(\boldsymbol{u})$ for every vector $\boldsymbol{u}$ in the space.**

You see the reason: Every $\boldsymbol{u}$ in the space is a combination of the basis vectors $\boldsymbol{v}_j$. Then linearity tells us that **$T(\boldsymbol{u})$ is the same combination of the outputs $T(\boldsymbol{v}_j)$.**

Lines to Lines, Triangles to Triangles

Figure 8.1 shows the line from $\boldsymbol{v}$ to $\boldsymbol{w}$ in the input space. It also shows the line from $T(\boldsymbol{v})$ to $T(\boldsymbol{w})$ in the output space. Linearity tells us: Every point on the input line goes onto the output line. And more than that: ***Equally spaced points go to equally spaced points***. The middle point $\boldsymbol{u} = \frac{1}{2}\boldsymbol{v} + \frac{1}{2}\boldsymbol{w}$ goes to the middle point $T(\boldsymbol{u}) = \frac{1}{2}T(\boldsymbol{v}) + \frac{1}{2}T(\boldsymbol{w})$.

The second figure moves up a dimension. Now we have three corners $\boldsymbol{v}_1$, $\boldsymbol{v}_2$, $\boldsymbol{v}_3$. Those inputs have three outputs $T(\boldsymbol{v}_1)$, $T(\boldsymbol{v}_2)$, $T(\boldsymbol{v}_3)$. *The input triangle goes onto the output triangle.* Equally spaced points stay equally spaced (along the edges, and then between the edges). The middle point $\boldsymbol{u} = \frac{1}{3}(\boldsymbol{v}_1 + \boldsymbol{v}_2 + \boldsymbol{v}_3)$ goes to the middle point $T(\boldsymbol{u}) = \frac{1}{3}(T(\boldsymbol{v}_1) + T(\boldsymbol{v}_2) + T(\boldsymbol{v}_3))$.

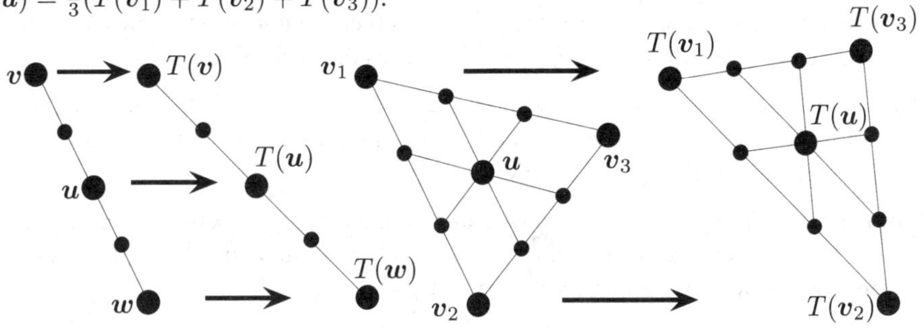

Figure 8.1: Lines to lines, equal spacing to equal spacing, $\boldsymbol{u} = \boldsymbol{0}$ to $T(\boldsymbol{u}) = \boldsymbol{0}$.

This page in the book is visual, not theoretical. We will show four houses and the matrices that produce them. The columns of H are the eleven corners of the first house. (H is 2 by 12, so **plot2d** in Problem 25 will connect the 11th corner to the first.) ***A* multiplies the columns of the house matrix to produce the corners *AH* of the other houses.**

House matrix $$\boldsymbol{H} = \begin{bmatrix} -6 & -6 & -7 & 0 & 7 & 6 & 6 & -3 & -3 & 0 & 0 & -6 \\ -7 & 2 & 1 & 8 & 1 & 2 & -7 & -7 & -2 & -2 & -7 & -7 \end{bmatrix}.$$

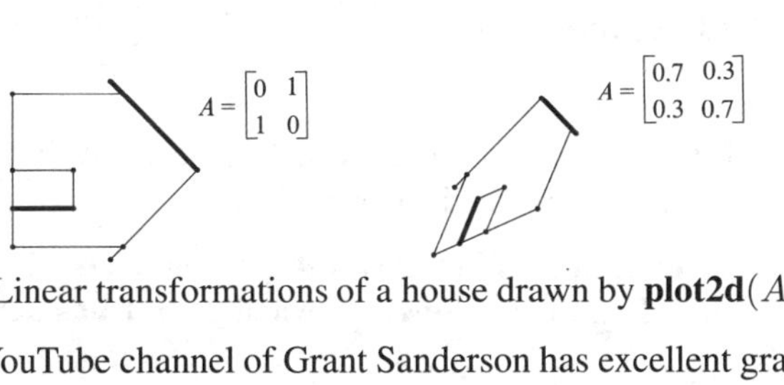

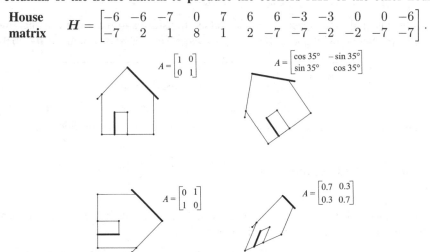

Figure 8.2: Linear transformations of a house drawn by **plot2d**$(A * H)$.

The **3Blue1Brown** YouTube channel of Grant Sanderson has excellent graphics.

Linear Transformations in Calculus

Example 4 **The transformation T takes the derivative of the input**: $T(\boldsymbol{u}) = d\boldsymbol{u}/dx$.

How do you find the derivative of $u = 6 - 4x + 3x^2$? You start with the derivatives of 1, x, and x^2. Those are the basis vectors. Their derivatives are 0, 1, and $2x$. Then you use linearity for the derivative of any combination:

$$\frac{du}{dx} = 6\,(\text{derivative of }1) - 4\,(\text{derivative of }x) + 3\,(\text{derivative of }x^2) = -4 + 6x.$$

All of calculus depends on linearity! Precalculus finds a few key derivatives, for x^n and $\sin x$ and $\cos x$ and e^x. Then linearity applies to all their combinations. The *chain rule* is special to calculus. That produces $(df/dg)\,(dg/dx)$ as the derivative of $f(g(x))$.

Nullspace of $T(u) = du/dx$. For the nullspace we solve $T(u) = 0$. The derivative is zero when u *is a constant function*. So the one-dimensional nullspace is a line in function space—all multiples of the special solution $u = 1$.

Column space of $T(u) = du/dx$. In our example the input space contains all quadratics $a + bx + cx^2$. The outputs (the column space) are all linear functions $b + 2cx$. Notice that the **Counting Theorem** is still true : $\boldsymbol{r + (n - r) = n}$.

dimension (**column space**)+dimension (**nullspace**) = **2+1 = 3** = dimension (**input space**)

What is the matrix for d/dx? I can't leave derivatives without asking for a matrix. We have a linear transformation $T = d/dx$. We know what T does to the **basis functions** :

$$\boldsymbol{v}_1, \boldsymbol{v}_2, \boldsymbol{v}_3 = \mathbf{1}, \boldsymbol{x}, \boldsymbol{x}^2 \qquad \frac{d\boldsymbol{v}_1}{dx} = \mathbf{0} \qquad \frac{d\boldsymbol{v}_2}{dx} = 1 = \boldsymbol{v}_1 \qquad \frac{d\boldsymbol{v}_3}{dx} = 2x = \mathbf{2}\boldsymbol{v}_2. \qquad (2)$$

The 3-dimensional input space **V** (= quadratics) transforms to the 2-dimensional output space **W** (= linear functions). If $\boldsymbol{v}_1, \boldsymbol{v}_2, \boldsymbol{v}_3$ were vectors, this would be the matrix :

$$\boldsymbol{A} = \begin{bmatrix} 0 & \mathbf{1} & 0 \\ 0 & 0 & \mathbf{2} \end{bmatrix} = \text{matrix form of the derivative } T = \frac{d}{dx}. \qquad (3)$$

The linear transformation du/dx is perfectly copied by the matrix multiplication $A\boldsymbol{u}$.

$$\textbf{Input } \boldsymbol{u} \quad a + bx + cx^2 \qquad \textbf{Multiplication } A\boldsymbol{u} = \begin{bmatrix} 0 & 1 & 0 \\ 0 & 0 & 2 \end{bmatrix} \begin{bmatrix} a \\ b \\ c \end{bmatrix} = \begin{bmatrix} b \\ 2c \end{bmatrix} \qquad \textbf{Output } \frac{d\boldsymbol{u}}{dx} = b + 2cx.$$

We will connect every transformation to a matrix ! The connection from T to A depended on choosing an input basis $1, x, x^2$ and an output basis $1, x$.

Next we look at integrals. They give the pseudoinverse T^+ of the derivative! I can't write T^{-1} and I can't say "*inverse*" when $f = 1$ has derivative $df/dx = 0$.

Example 5 **Integration T^+ is also linear**: $\int_0^x (D + Ex)\,dx = Dx + \frac{1}{2}Ex^2$.

The input basis is now $1, x$. The output basis is $1, x, x^2$. The matrix A^+ for T^+ is 3 by 2:

$$\begin{matrix}\textbf{Input}\\ \boldsymbol{v}=D+Ex\end{matrix} \qquad \textbf{Multiplication } A^+\boldsymbol{v} = \begin{bmatrix} 0 & 0 \\ 1 & 0 \\ 0 & \frac{1}{2} \end{bmatrix}\begin{bmatrix} D \\ E \end{bmatrix} = \begin{bmatrix} \mathbf{0} \\ \boldsymbol{D} \\ \frac{1}{2}\boldsymbol{E} \end{bmatrix} \qquad \begin{matrix}\textbf{Output} = \textbf{Integral of } \boldsymbol{v}\\ T^+(\boldsymbol{v}) = Dx + \frac{1}{2}Ex^2\end{matrix}$$

The Fundamental Theorem of Calculus says that integration is the (pseudo)inverse of differentiation. For linear algebra, the matrix A^+ is the (pseudo)inverse of the matrix A:

$$A^+A = \begin{bmatrix} 0 & 0 \\ 1 & 0 \\ 0 & \frac{1}{2} \end{bmatrix}\begin{bmatrix} 0 & 1 & 0 \\ 0 & 0 & 2 \end{bmatrix} = \begin{bmatrix} \mathbf{0} & 0 & 0 \\ 0 & \mathbf{1} & 0 \\ 0 & 0 & \mathbf{1} \end{bmatrix} \quad \text{and} \quad AA^+ = \begin{bmatrix} \mathbf{1} & 0 \\ 0 & \mathbf{1} \end{bmatrix}. \tag{4}$$

The derivative of a constant function is zero. That zero is on the diagonal of A^+A. *Calculus wouldn't be calculus without that 1-dimensional nullspace of $T = d/dx$.*

Examples of Transformations (mostly linear)

Example 6 Project every 3-dimensional vector onto the horizontal plane $z = 1$. The vector $\boldsymbol{v} = (x, y, z)$ is transformed to $T(\boldsymbol{v}) = (x, y, 1)$. This transformation is not linear. Why not? It doesn't even transform $\boldsymbol{v} = \mathbf{0}$ into $T(\boldsymbol{v}) = \mathbf{0}$.

Example 7 Suppose A is an *invertible matrix*. Certainly $T(\boldsymbol{v} + \boldsymbol{w}) = A\boldsymbol{v} + A\boldsymbol{w} = T(\boldsymbol{v}) + T(\boldsymbol{w})$. Another linear transformation is multiplication by A^{-1}. This produces the ***inverse transformation*** T^{-1}, which brings every vector $T(\boldsymbol{v})$ back to $\boldsymbol{v}$:

$$T^{-1}(T(\boldsymbol{v})) = \boldsymbol{v} \quad \text{matches the matrix multiplication} \quad A^{-1}(A\boldsymbol{v}) = \boldsymbol{v}.$$
$$\text{If } T(\boldsymbol{v}) = A\boldsymbol{v} \text{ and } S(\boldsymbol{u}) = B\boldsymbol{u}, \text{ then } T(S(\boldsymbol{u})) \text{ matches } AB\boldsymbol{u}.$$

We are reaching an unavoidable question. ***Are all linear transformations from* $\mathbf{V} = \mathbf{R}^n$ *to* $\mathbf{W} = \mathbf{R}^m$ *produced by matrices?*** When a linear T is described as a "rotation" or "projection" or "...", is there always a matrix A hiding behind T? Is $T(\boldsymbol{v})$ always $A\boldsymbol{v}$?

The answer is *yes*! This is an approach to linear algebra that doesn't start with matrices. We still end up with matrices—*after we choose an input basis and output basis.*

Note Transformations have a language of their own. For a matrix, the column space contains all outputs $A\boldsymbol{v}$. The nullspace contains all inputs for which $A\boldsymbol{v} = \mathbf{0}$. Translate those words into "*range*" and "*kernel*":

Range of T = set of *all outputs* $T(\boldsymbol{v})$. Range corresponds to **column space**.

Kernel of T = set of *all inputs for which* $T(\boldsymbol{v}) = \mathbf{0}$. Kernel corresponds to **nullspace**.

The range is in the output space $\mathbf{W}$. The kernel is in the input space $\mathbf{V}$. When T is multiplication by a matrix, $T(\boldsymbol{v}) = A\boldsymbol{v}$, range is column space and kernel is nullspace.

■ REVIEW OF THE KEY IDEAS ■

1. A transformation T takes each $\boldsymbol{v}$ in the input space to $T(\boldsymbol{v})$ in the output space.

2. T is **linear** if $T(\boldsymbol{v}+\boldsymbol{w}) = T(\boldsymbol{v}) + T(w)$ and $T(c\boldsymbol{v}) = cT(\boldsymbol{v})$: lines to lines.

3. Combinations to combinations: $T\,(c_1\boldsymbol{v}_1+\cdots+c_n\boldsymbol{v}_n) = c_1\,T(\boldsymbol{v}_1)+\cdots+c_n\,T(\boldsymbol{v}_n)$.

4. $T =$ *derivative* and $T^+ =$ *integral* are linear. So is $T(\boldsymbol{v}) = A\boldsymbol{v}$ from $\mathbf{R}^n$ to $\mathbf{R}^m$.

It is more interesting to *see* a transformation than to define it. When a 2 by 2 matrix A multiplies all vectors in $\mathbf{R}^2$, we can watch how it acts. The eleven house vectors $\boldsymbol{v}$ are transformed into eleven vectors $A\boldsymbol{v}$. Straight lines between $\boldsymbol{v}$'s become straight lines between the transformed vectors $A\boldsymbol{v}$. (The transformation from house to house is linear!) Applying A to a standard house produces a new house—possibly stretched or rotated or otherwise unlivable.

■ WORKED EXAMPLES ■

8.1 A The elimination matrix $\left[\begin{smallmatrix}1&0\\1&1\end{smallmatrix}\right]$ gives a ***shearing transformation*** from (x,y) to $\boldsymbol{T(x,y) = (x, x+y)}$. If the inputs fill a square, draw the transformed square.

Solution The points $(1,0)$ and $(2,0)$ on the x axis transform by T to $(1,1)$ and $(2,2)$ on the 45° line. Points on the y axis are *not moved*: $T(0,y)=(0,y)=$eigenvectors with $\lambda=1$.

Vertical lines slide up
This is the shearing
Squares go to parallelograms

$$A = \begin{bmatrix} 1 & 0 \\ 1 & 1 \end{bmatrix}$$

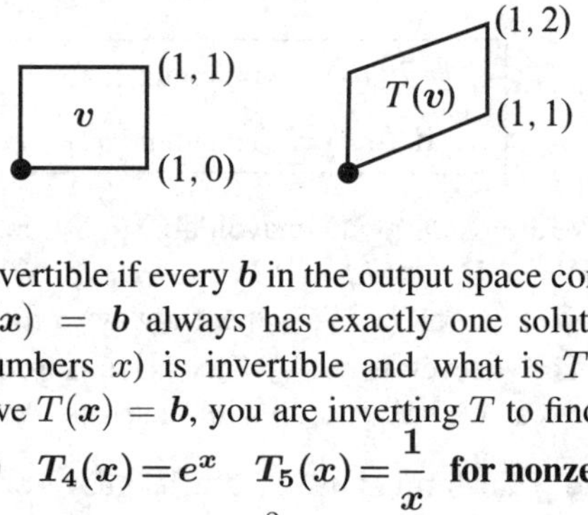

8.1 B A **nonlinear transformation** T is invertible if every $\boldsymbol{b}$ in the output space comes from exactly one $\boldsymbol{x}$ in the input space: $T(\boldsymbol{x}) = \boldsymbol{b}$ always has exactly one solution. Which of these transformations (on real numbers x) is invertible and what is T^{-1}? ***None are linear, not even*** $\boldsymbol{T_3}$. When you solve $T(\boldsymbol{x}) = \boldsymbol{b}$, you are inverting T to find $\boldsymbol{x}$.

$$\boldsymbol{T_1(x)=x^2 \quad T_2(x)=x^3 \quad T_3(x)=x+9 \quad T_4(x)=e^x \quad T_5(x)=\frac{1}{x}} \ \textbf{ for nonzero } \boldsymbol{x}\textbf{'s}$$

Solution T_1 is not invertible: $x^2 = 1$ has *two* solutions and $x^2 = -1$ has *no* solution.
T_4 is not invertible because $e^x = -1$ has no solution. (If the output space changes to *positive* b's then the inverse of $e^x = b$ is $x = \ln b$.)

Notice $T_5^2 =$ identity. But $T_3^2(x) = x + 18$. What are $T_2^2(x)$ and $T_4^2(x)$?

T_2, T_3, T_5 are invertible: $x^3 = b$ and $x + 9 = b$ and $1/x = b$ have one solution x.

$$\boldsymbol{x = T_2^{-1}(b) = b^{1/3} \qquad x = T_3^{-1}(b) = b - 9 \qquad x = T_5^{-1}(b) = 1/b}$$

Problem Set 8.1

1 A linear transformation must leave the zero vector fixed: $T(\mathbf{0}) = \mathbf{0}$. Prove this from $T(\boldsymbol{v}+\boldsymbol{w}) = T(\boldsymbol{v})+T(\boldsymbol{w})$ by choosing $\boldsymbol{w} =$ _____ (and finish the proof). Prove it also from $T(c\boldsymbol{v}) = cT(\boldsymbol{v})$ by choosing $c =$ _____.

2 Requirement (b) gives $T(c\boldsymbol{v}) = cT(\boldsymbol{v})$ and also $T(d\boldsymbol{w}) = dT(\boldsymbol{w})$. Then by addition, requirement (a) gives $T(\quad) = (\quad)$. What is $T(c\boldsymbol{v} + d\boldsymbol{w} + e\boldsymbol{u})$?

3 Which of these transformations are not linear? The input is $\boldsymbol{v} = (v_1, v_2)$:

(a) $T(\boldsymbol{v}) = (v_2, v_1)$ (b) $T(\boldsymbol{v}) = (v_1, v_1)$ (c) $T(\boldsymbol{v}) = (0, v_1)$

(d) $T(\boldsymbol{v}) = (0, 1)$ (e) $T(\boldsymbol{v}) = v_1 - v_2$ (f) $T(\boldsymbol{v}) = v_1 v_2$.

4 If S and T are linear transformations, is $T(S(\boldsymbol{v}))$ linear or quadratic?

(a) (Special case) If $S(\boldsymbol{v}) = \boldsymbol{v}$ and $T(\boldsymbol{v}) = \boldsymbol{v}$, then $T(S(\boldsymbol{v})) = \boldsymbol{v}$ or $\boldsymbol{v}^2$?

(b) (General case) $S(\boldsymbol{v}_1+\boldsymbol{v}_2) = S(\boldsymbol{v}_1)+S(\boldsymbol{v}_2)$ and $T(\boldsymbol{v}_1+\boldsymbol{v}_2) = T(\boldsymbol{v}_1)+T(\boldsymbol{v}_2)$ combine into

$$T(S(\boldsymbol{v}_1 + \boldsymbol{v}_2)) = T(\_\_\_\_) = \_\_\_\_ + \_\_\_\_.$$

5 Suppose $T(\boldsymbol{v}) = \boldsymbol{v}$ except that $T(0, v_2) = (0, 0)$. Show that this transformation satisfies $T(c\boldsymbol{v}) = cT(\boldsymbol{v})$ but does not satisfy $T(\boldsymbol{v} + \boldsymbol{w}) = T(\boldsymbol{v}) + T(\boldsymbol{w})$.

6 Which of these transformations satisfy $T(\boldsymbol{v}+\boldsymbol{w}) = T(\boldsymbol{v})+T(\boldsymbol{w})$ and which satisfy $T(c\boldsymbol{v}) = cT(\boldsymbol{v})$?

(a) $T(\boldsymbol{v}) = \boldsymbol{v}/\|\boldsymbol{v}\|$ (b) $T(\boldsymbol{v}) = v_1+v_2+v_3$ (c) $T(\boldsymbol{v}) = (v_1, 2v_2, 3v_3)$

(d) $T(\boldsymbol{v}) =$ largest component of $\boldsymbol{v}$.

7 For these transformations of $\mathbf{V} = \mathbf{R}^2$ to $\mathbf{W} = \mathbf{R}^2$, find $T(T(\boldsymbol{v}))$. Show that when $T(\boldsymbol{v})$ is linear, then also $T(T(\boldsymbol{v}))$ is linear.

(a) $T(\boldsymbol{v}) = -\boldsymbol{v}$ (b) $T(\boldsymbol{v}) = \boldsymbol{v} + (1, 1)$

(c) $T(\boldsymbol{v}) = 90^\circ$ rotation $= (-v_2, v_1)$

(d) $T(\boldsymbol{v}) =$ projection $= \frac{1}{2}\big(v_1 + v_2,\ v_1 + v_2\big)$.

8 Find the range and kernel (like the column space and nullspace) of T:

(a) $T(v_1, v_2) = (v_1 - v_2, 0)$ (b) $T(v_1, v_2, v_3) = (v_1, v_2)$

(c) $T(v_1, v_2) = (0, 0)$ (d) $T(v_1, v_2) = (v_1, v_1)$.

9 The transformation $T(v_1, v_2, v_3) = (v_2, v_3, v_1)$ is "cyclic". What is $T(T(\boldsymbol{v}))$? What is $T^3(\boldsymbol{v})$? What is $T^{100}(\boldsymbol{v})$? Apply T a hundred times to (v_1, v_2, v_3).

10 Suppose a linear T transforms $(1, 1)$ to $(2, 2)$ and $(2, 0)$ to $(0, 0)$. Find $T(\boldsymbol{v})$:

(a) $\boldsymbol{v} = (2, 2)$ (b) $\boldsymbol{v} = (3, 1)$ (c) $\boldsymbol{v} = (-1, 1)$ (d) $\boldsymbol{v} = (a, b)$.

11 A linear transformation from $\mathbf{V}$ to $\mathbf{W}$ has a linear *inverse* from $\mathbf{W}$ to $\mathbf{V}$ when the range is all of $\mathbf{W}$ and the kernel (nullspace) contains only $\boldsymbol{v} = \mathbf{0}$. Then $T(\boldsymbol{v}) = \boldsymbol{w}$ has one solution $\boldsymbol{v}$ for each $\boldsymbol{w}$ in $\mathbf{W}$. Why are these T's not invertible?

(a) $T(v_1, v_2) = (v_2, v_2)$ $\quad \mathbf{W} = \mathbf{R}^2$

(b) $T(v_1, v_2) = (v_1, v_2, v_1 + v_2)$ $\quad \mathbf{W} = \mathbf{R}^3$

(c) $T(v_1, v_2) = v_1$ $\quad \mathbf{W} = \mathbf{R}^1$

12 If $T(\boldsymbol{v}) = A\boldsymbol{v}$ and A is m by n, then T is "multiplication by A."

(a) What are the input and output spaces $\mathbf{V}$ and $\mathbf{W}$?

(b) Why is range of T = column space of A?

(c) Why is kernel of T = nullspace of A?

Problems 13-19 may be harder. The input space V contains all 2 by 2 matrices M.

13 M is any 2 by 2 matrix and $A = \begin{bmatrix} 1 & 2 \\ 3 & 4 \end{bmatrix}$. The transformation T is defined by $T(M) = AM$. What rules of matrix multiplication show that T is linear?

14 Suppose $A = \begin{bmatrix} 1 & 2 \\ 3 & 5 \end{bmatrix}$. Show that the range of T is the whole matrix space $\mathbf{V}$ and the kernel is the zero matrix:

(1) If $AM = 0$ prove that M must be the zero matrix.

(2) Find a solution to $AM = B$ for any 2 by 2 matrix B.

15 Suppose $A = \begin{bmatrix} 1 & 2 \\ 3 & 6 \end{bmatrix}$. Show that the identity matrix I is not in the range of T. Find a nonzero matrix M such that $T(M) = AM$ is zero.

16 Suppose T transposes every 2 by 2 matrix M. Try to find a matrix A which gives $AM = M^{\mathrm{T}}$. *Show that no matrix A will do it. To professors:* Is this a linear transformation that doesn't come from a matrix? The matrix should be 4 by 4!

17 The transformation T that transposes every 2 by 2 matrix is definitely linear. Which of these extra properties are true?

(a) $T^2 =$ identity transformation.

(b) The kernel of T is the zero matrix.

(c) Every 2 by 2 matrix is in the range of T.

(d) $T(M) = -M$ is impossible.

18 Suppose $T(M) = \begin{bmatrix} 1 & 0 \\ 0 & 0 \end{bmatrix} \begin{bmatrix} \mathrm{M} \end{bmatrix} \begin{bmatrix} 0 & 0 \\ 0 & 1 \end{bmatrix}$. Find a matrix with $T(M) \neq 0$. Describe all matrices with $T(M) = 0$ (the kernel) and all output matrices $T(M)$ (the range).

19 Why does every linear transformation T from $\mathbf{R}^2$ to $\mathbf{R}^2$ take squares to parallelograms? Rectangles also go to parallelograms (squashed if T is not invertible).

Questions 20–26 are about house transformations. The output is $T(H) = AH$.

20 How can you tell from the picture of T (house) that A is
(a) a diagonal matrix ? (b) a rank-one matrix ? (c) a lower triangular matrix ?

21 Draw a picture of T (house) for these matrices:

$$D = \begin{bmatrix} 2 & 0 \\ 0 & 1 \end{bmatrix} \quad \text{and} \quad A = \begin{bmatrix} .7 & .7 \\ .3 & .3 \end{bmatrix} \quad \text{and} \quad U = \begin{bmatrix} 1 & 1 \\ 0 & 1 \end{bmatrix}.$$

22 What are the conditions on $A = \begin{bmatrix} \mathrm{a} & \mathrm{b} \\ \mathrm{c} & \mathrm{d} \end{bmatrix}$ to ensure that T (house) will
(a) sit straight up ? (b) rotate the house with no change in its shape ?

23 Describe T (house) when $T(\boldsymbol{v}) = -\boldsymbol{v} + (1, 0)$. This T is "affine".

24 Change the house matrix H to add a chimney.

25 The standard house is drawn by **plot2d**(H). Circles from o and lines from –:

$$x = H(1, :)'; y = H(2, :)';$$
$$\text{axis}([-1010-1010]), \text{axis}(\text{'square'})$$
$$\text{plot}(x, y, 'o', x, y, '-');$$

Test **plot2d**(A$'*$ H) and **plot2d**(A$'*$ A $*$ H) with the matrices in Figure 8.2.

26 Without a computer sketch the houses $A * H$ for these matrices A:

$$\begin{bmatrix} 1 & 0 \\ 0 & .1 \end{bmatrix} \quad \text{and} \quad \begin{bmatrix} .5 & .5 \\ .5 & .5 \end{bmatrix} \quad \text{and} \quad \begin{bmatrix} .5 & .5 \\ -.5 & .5 \end{bmatrix} \quad \text{and} \quad \begin{bmatrix} 1 & 1 \\ 1 & 0 \end{bmatrix}.$$

27 This code creates a vector theta of 50 angles. It draws the unit circle and then it draws T (circle) = ellipse. $T(\boldsymbol{v}) = A\boldsymbol{v}$ **takes circles to ellipses.**

```
A = [2 1;1 2]    % You can change A
theta = [0:2 * pi/50:2 * pi];
circle = [cos(theta); sin(theta)];
ellipse = A * circle;
axis([–4 4 –4 4]); axis('square')
plot(circle(1,:), circle(2,:), ellipse(1,:), ellipse(2,:))
```

28 What conditions on $\det A = ad - bc$ ensure that the output house AH will
(a) be squashed onto a line?
(b) keep its endpoints in clockwise order (not reflected)?
(c) have the **same area** as the original house?

8.2 The Matrix of a Linear Transformation

1 Linearity tells us all $T(\boldsymbol{v})$ if we know $T(\boldsymbol{v}_1),\ldots,T(\boldsymbol{v}_n)$ for an **input basis** $\boldsymbol{v}_1,\ldots,\boldsymbol{v}_n$.

2 Column j in the "matrix for T" comes from applying T to the input basis vector $\boldsymbol{v}_j$.

3 Write $T(\boldsymbol{v}_j)=a_{1j}\boldsymbol{w}_1+\cdots+a_{mj}\boldsymbol{w}_m$ in the output basis of $\boldsymbol{w}$'s. **Those a_{ij} go into column j.**

4 The matrix for $T(\boldsymbol{x}) = A\boldsymbol{x}$ is A, *if* the input and output bases = columns of $I_{n\times n}$ and $I_{m\times m}$.

5 When the bases change to $\boldsymbol{v}$'s and $\boldsymbol{w}$'s, the matrix for the same T changes from A to $W^{-1}AV$.

6 **Best bases**: $\boldsymbol{V} = \boldsymbol{W} =$ **eigenvectors** and $\boldsymbol{V}, \boldsymbol{W} =$ **singular vectors** change A to Λ and Σ.

The next pages assign a matrix A to every linear transformation T. For ordinary column vectors, the input $\boldsymbol{v}$ is in $\mathbf{V} = \mathbf{R}^n$ and the output $T(\boldsymbol{v})$ is in $\mathbf{W} = \mathbf{R}^m$. The matrix A for this transformation will be m by n. Our choice of bases in $\mathbf{V}$ and $\mathbf{W}$ will decide A.

The standard basis vectors for $\mathbf{R}^n$ and $\mathbf{R}^m$ are the columns of I. That choice leads to a standard matrix. Then $T(\boldsymbol{v}) = A\boldsymbol{v}$ in the normal way. But these spaces also have other bases, so *the same transformation T is represented by other matrices.* A main theme of linear algebra is to choose the bases that give the best matrix (a diagonal matrix) for T.

All vector spaces $\mathbf{V}$ and $\mathbf{W}$ have bases. **Each choice of those bases leads to a matrix for T.** When the input basis is different from the output basis, the matrix for $T(\boldsymbol{v}) = \boldsymbol{v}$ will *not* be the identity I. It will be the "change of basis matrix". Here is the key idea:

> **Suppose we know outputs $T(\boldsymbol{v})$ for the input basis vectors $\boldsymbol{v}_1$ to $\boldsymbol{v}_n$.**
> **Columns 1 to n of the matrix will contain those outputs $T(\boldsymbol{v}_1)$ to $T(\boldsymbol{v}_n)$.**
> **A times $\boldsymbol{c}$ = matrix times vector = combination of those n columns.**
> **$A\boldsymbol{c}$ gives the correct combination $c_1T(\boldsymbol{v}_1) + \cdots + c_nT(\boldsymbol{v}_n) = T(\boldsymbol{v})$.**

Reason Every $\boldsymbol{v}$ is a unique combination $c_1\boldsymbol{v}_1 + \cdots + c_n\boldsymbol{v}_n$ of the basis vectors $\boldsymbol{v}_j$. Since T is a linear transformation (here is the moment for linearity), $T(\boldsymbol{v})$ must be **the same combination** $c_1T(\boldsymbol{v}_1) + \cdots + c_nT(\boldsymbol{v}_n)$ **of the outputs** $T(\boldsymbol{v}_j)$ **in the columns.**

Our first example gives the matrix A for the standard basis vectors in $\mathbf{R}^2$ and $\mathbf{R}^3$. The two columns of A are the outputs from $\boldsymbol{v}_1 = (1,0)$ and $\boldsymbol{v}_2 = (0,1)$.

Example 1 Suppose T transforms $\boldsymbol{v}_1 = (1,0)$ to $T(\boldsymbol{v}_1) = (2,3,4)$. Suppose the second basis vector $\boldsymbol{v}_2 = (0,1)$ goes to $T(\boldsymbol{v}_2) = (5,5,5)$. If T is linear from $\mathbf{R}^2$ to $\mathbf{R}^3$ then its "standard matrix" is 3 by 2. Those outputs $T(\boldsymbol{v}_1)$ and $T(\boldsymbol{v}_2)$ go into the columns of A:

$$A = \begin{bmatrix} 2 & 5 \\ 3 & 5 \\ 4 & 5 \end{bmatrix} \qquad c_1 = 1 \text{ and } c_2 = 1 \text{ give } T(\boldsymbol{v}_1 + \boldsymbol{v}_2) = \begin{bmatrix} 2 & 5 \\ 3 & 5 \\ 4 & 5 \end{bmatrix} \begin{bmatrix} 1 \\ 1 \end{bmatrix} = \begin{bmatrix} 7 \\ 8 \\ 9 \end{bmatrix}.$$

Change of Basis : Matrix B

Example 2 Suppose the input space $\mathbf{V} = \mathbf{R}^2$ is also the output space $\mathbf{W} = \mathbf{R}^2$. *Suppose that* $T(\boldsymbol{v}) = \boldsymbol{v}$ *is the identity transformation.* You might expect its matrix to be I, but that only happens when the input basis is the *same* as the output basis. I will choose different bases to see how the matrix is constructed.

For this special case $T(\boldsymbol{v}) = \boldsymbol{v}$, I will call the matrix B instead of A. We are just changing basis from the $\boldsymbol{v}$'s to the $\boldsymbol{w}$'s. Each $\boldsymbol{v}$ is a combination of $\boldsymbol{w}_1$ and $\boldsymbol{w}_2$.

$$\textbf{Input basis} \begin{bmatrix} \boldsymbol{v}_1 & \boldsymbol{v}_2 \end{bmatrix} = \begin{bmatrix} 3 & 6 \\ 3 & 8 \end{bmatrix} \quad \textbf{Output basis} \begin{bmatrix} \boldsymbol{w}_1 & \boldsymbol{w}_2 \end{bmatrix} = \begin{bmatrix} 3 & 0 \\ 1 & 2 \end{bmatrix} \quad \boxed{\begin{matrix} \textbf{Change} & \boldsymbol{v}_1 = \mathbf{1}\boldsymbol{w}_1 + \mathbf{1}\boldsymbol{w}_2 \\ \textbf{of basis} & \boldsymbol{v}_2 = \mathbf{2}\boldsymbol{w}_1 + \mathbf{3}\boldsymbol{w}_2 \end{matrix}}$$

Please notice! I wrote the input basis $\boldsymbol{v}_1, \boldsymbol{v}_2$ in terms of the output basis $\boldsymbol{w}_1, \boldsymbol{w}_2$. That is because of our key rule. We apply the identity transformation T to each input basis vector: $T(\boldsymbol{v}_1) = \boldsymbol{v}_1$ and $T(\boldsymbol{v}_2) = \boldsymbol{v}_2$. **Then we write those outputs $\boldsymbol{v}_1$ and $\boldsymbol{v}_2$ in the output basis $\boldsymbol{w}_1$ and $\boldsymbol{w}_2$.** Those bold numbers $\mathbf{1}, \mathbf{1}$ and $\mathbf{2}, \mathbf{3}$ tell us column 1 and column 2 of the matrix B (the change of basis matrix): $WB = V$ so $\boldsymbol{B} = \boldsymbol{W}^{-1}\boldsymbol{V}$.

$$\textbf{Matrix } B \textbf{ for change of basis} \quad \begin{bmatrix} \boldsymbol{w}_1 & \boldsymbol{w}_2 \end{bmatrix} \boxed{\begin{bmatrix} \boldsymbol{B} \end{bmatrix}} = \begin{bmatrix} \boldsymbol{v}_1 & \boldsymbol{v}_2 \end{bmatrix} \text{ is } \begin{bmatrix} 3 & 0 \\ 1 & 2 \end{bmatrix} \boxed{\begin{bmatrix} \mathbf{1} & \mathbf{2} \\ \mathbf{1} & \mathbf{3} \end{bmatrix}} = \begin{bmatrix} 3 & 6 \\ 3 & 8 \end{bmatrix}. \quad (1)$$

> When the input basis is in the columns of a matrix $\boldsymbol{V}$, and the output basis is in the columns of $\boldsymbol{W}$, the change of basis matrix for $T = I$ is $\boldsymbol{B} = \boldsymbol{W}^{-1}\boldsymbol{V}$.

The key I see a clear way to understand that rule $B = W^{-1}V$. Suppose the same vector $\boldsymbol{u}$ is written in the input basis of $\boldsymbol{v}$'s and the output basis of $\boldsymbol{w}$'s. I will do that three ways:

$$\begin{matrix} \boldsymbol{u} = c_1\boldsymbol{v}_1 + \cdots + c_n\boldsymbol{v}_n \\ \boldsymbol{u} = d_1\boldsymbol{w}_1 + \cdots + d_n\boldsymbol{w}_n \end{matrix} \quad \text{is} \quad \begin{bmatrix} \boldsymbol{v}_1 & \cdots & \boldsymbol{v}_n \end{bmatrix} \begin{bmatrix} c_1 \\ \vdots \\ c_n \end{bmatrix} = \begin{bmatrix} \boldsymbol{w}_1 & \cdots & \boldsymbol{w}_n \end{bmatrix} \begin{bmatrix} d_1 \\ \vdots \\ d_n \end{bmatrix} \quad \text{and } \boldsymbol{Vc} = \boldsymbol{Wd}.$$

The coefficients $\boldsymbol{d}$ in the new basis of $\boldsymbol{w}$'s are $\boldsymbol{d} = \boldsymbol{W}^{-1}\boldsymbol{V}\boldsymbol{c}$. Then $\boldsymbol{B}$ is $\boldsymbol{W}^{-1}\boldsymbol{V}$. (2)

This formula $\boldsymbol{B} = \boldsymbol{W}^{-1}\boldsymbol{V}$ produces one of the world's greatest mysteries: When the *standard basis* $\boldsymbol{V} = \boldsymbol{I}$ is changed to a different basis $\boldsymbol{W}$, **the change of basis matrix is not $\boldsymbol{W}$ but $\boldsymbol{B} = \boldsymbol{W}^{-1}$.** Larger basis vectors have smaller coefficients!

$$\begin{bmatrix} x \\ y \end{bmatrix} \text{ in the standard basis has coefficients } \begin{bmatrix} \boldsymbol{w}_1 & \boldsymbol{w}_2 \end{bmatrix}^{-1} \begin{bmatrix} x \\ y \end{bmatrix} \text{ in the } \boldsymbol{w}_1, \boldsymbol{w}_2 \text{ basis.}$$

Construction of the Matrix for T

Now we construct a matrix for any linear transformation. Suppose T transforms the space $\mathbf{V}$ (n-dimensional) to the space $\mathbf{W}$ (m-dimensional). We choose a basis $\boldsymbol{v}_1, \ldots, \boldsymbol{v}_n$ for $\mathbf{V}$ and we choose a basis $\boldsymbol{w}_1, \ldots, \boldsymbol{w}_m$ for $\mathbf{W}$. The matrix A will be m by n. To find the first column of A, apply T to the first basis vector $\boldsymbol{v}_1$. The output $T(\boldsymbol{v}_1)$ is in $\mathbf{W}$.

$T(\boldsymbol{v}_1)$ ***is a combination*** $a_{11}\boldsymbol{w}_1 + \cdots + a_{m1}\boldsymbol{w}_m$ ***of the output basis for*** $\mathbf{W}$.

These numbers $a_{11}, \ldots, a_{m1}$ *go into the first column of* A. Transforming $\boldsymbol{v}_1$ to $T(\boldsymbol{v}_1)$ matches multiplying $(1, 0, \ldots, 0)$ by A. It yields that first column of the matrix. When T is the derivative and the first basis vector is 1, its derivative is $T(\boldsymbol{v}_1) = \mathbf{0}$. So for the derivative matrix below, the first column of A is all zero.

Example 3 The input basis of $\boldsymbol{v}$'s is $1, x, x^2, x^3$. The output basis of $\boldsymbol{w}$'s is $1, x, x^2$. Then **T takes the derivative**: $\boldsymbol{T(v) = \dfrac{dv}{dx}}$ and A = "derivative matrix".

$$\text{If } \boldsymbol{v} = c_1 + c_2x + c_3x^2 + c_4x^3 \quad \text{then } \frac{d\boldsymbol{v}}{dx} = \mathbf{1}c_2 + \mathbf{2}c_3x + \mathbf{3}c_4x^2 \qquad A\boldsymbol{c} = \begin{bmatrix} 0 & \mathbf{1} & 0 & 0 \\ 0 & 0 & \mathbf{2} & 0 \\ 0 & 0 & 0 & \mathbf{3} \end{bmatrix} \begin{bmatrix} c_1 \\ c_2 \\ c_3 \\ c_4 \end{bmatrix} = \begin{bmatrix} c_2 \\ 2c_3 \\ 3c_4 \end{bmatrix}$$

Key rule: The jth column of A is found by applying T to the jth basis vector $\boldsymbol{v}_j$ $T(\boldsymbol{v}_j) =$ combination of output basis vectors $= a_{1j}\boldsymbol{w}_1 + \cdots + a_{mj}\boldsymbol{w}_m$. (3)

These numbers a_{ij} go into A. *The matrix is constructed to get the basis vectors right.* ***Then linearity gets all other vectors right.*** Every $\boldsymbol{v}$ is a combination $c_1\boldsymbol{v}_1 + \cdots + c_n\boldsymbol{v}_n$, and $T(\boldsymbol{v})$ is a combination of the $\boldsymbol{w}$'s. When A multiplies the vector $\boldsymbol{c} = (c_1, \ldots, c_n)$ in the $\boldsymbol{v}$ combination, $A\boldsymbol{c}$ produces the coefficients in the $T(\boldsymbol{v})$ combination. This is because matrix multiplication (combining columns) is linear like T.

Every linear transformation from V to W converts to a matrix using the bases.

Example 4 For the integral $T^+(\boldsymbol{v})$, the first basis function is again 1. Its integral is the second basis function x. So the first column of the "integral matrix" A^+ is $(0, 1, 0, 0)$.

$$\textbf{The integral of } \boldsymbol{d_1 + d_2x + d_3x^2} \quad \textbf{is } \boldsymbol{d_1x + \frac{1}{2}d_2x^2 + \frac{1}{3}d_3x^3} \qquad A^+\boldsymbol{d} = \begin{bmatrix} 0 & 0 & 0 \\ \mathbf{1} & 0 & 0 \\ 0 & \frac{1}{2} & 0 \\ 0 & 0 & \frac{1}{3} \end{bmatrix} \begin{bmatrix} d_1 \\ d_2 \\ d_3 \end{bmatrix} = \begin{bmatrix} 0 \\ \boldsymbol{d_1} \\ \frac{1}{2}\boldsymbol{d_2} \\ \frac{1}{3}\boldsymbol{d_3} \end{bmatrix}$$

If you integrate a function and then differentiate, you get back to the start. So $AA^+ = I$. But if you differentiate before integrating, the constant term is lost. So A^+A is not I.

The integral of the derivative of* 1 *is zero :

$$T^+T(1) = \text{integral of zero function } = 0.$$

This matches A^+A, whose first column is all zero. The derivative T has a kernel (the constant functions). Its matrix A has a nullspace. Main idea again: $\boldsymbol{Av}$ **copies** $\boldsymbol{T(v)}$.

The examples of the derivative and integral made three points. First, linear transformations T are everywhere—in calculus and differential equations and linear algebra. Second, spaces other than $\mathbf{R}^n$ are important—we had functions in $\mathbf{V}$ and $\mathbf{W}$. Third, **if we differentiate and then integrate, we can multiply their matrices** $\boldsymbol{A^+A}$.

Matrix Products AB Match Transformations TS

We have come to something important—the real reason for the rule to multiply matrices. *At last we discover why*! Two linear transformations T and S are represented by two matrices A and B. Now compare TS with the multiplication AB:

When we apply the transformation T to the output from S, we get TS by this rule: $(TS)(\boldsymbol{u})$ *is defined to be* $\boldsymbol{T(S(u))}$. The output $S(\boldsymbol{u})$ becomes the input to T.

When we apply the matrix A to the output from B, we multiply AB by this rule: $(AB)(\boldsymbol{x})$ is defined to be $\boldsymbol{A(Bx)}$. The output $B\boldsymbol{x}$ becomes the input to A. ***Matrix multiplication gives the correct matrix*** $\boldsymbol{AB}$ ***to represent*** $\boldsymbol{TS}$.

The transformation S is from a space $\mathbf{U}$ to $\mathbf{V}$. Its matrix B uses a basis $\boldsymbol{u}_1, \ldots, \boldsymbol{u}_p$ for $\mathbf{U}$ and a basis $\boldsymbol{v}_1, \ldots, \boldsymbol{v}_n$ for $\mathbf{V}$. That matrix is n by p. The transformation T is from $\mathbf{V}$ to $\mathbf{W}$ as before. *Its matrix A must use the same basis* $\boldsymbol{v}_1, \ldots, \boldsymbol{v}_n$ *for* $\mathbf{V}$—this is the output space for S and the input space for T. ***Then the matrix*** $\boldsymbol{AB}$ ***matches*** $\boldsymbol{TS}$.

Multiplication The linear transformation TS starts with any vector $\boldsymbol{u}$ in $\mathbf{U}$, goes to $S(\boldsymbol{u})$ in $\mathbf{V}$ and then to $T(S(\boldsymbol{u}))$ in $\mathbf{W}$. The matrix AB starts with any $\boldsymbol{x}$ in $\mathbf{R}^p$, goes to $B\boldsymbol{x}$ in $\mathbf{R}^n$ and then to $AB\boldsymbol{x}$ in $\mathbf{R}^m$. **The matrix** $\boldsymbol{AB}$ **correctly represents** $\boldsymbol{TS}$:

$$\boldsymbol{TS}: \quad \mathbf{U} \to \mathbf{V} \to \mathbf{W} \qquad \boldsymbol{AB}: \quad (m \text{ by } n)(n \text{ by } p) = (m \text{ by } p).$$

Product of transformations $\boldsymbol{TS}$ ***matches product of matrices*** $\boldsymbol{AB}$. An important case is when the spaces $\mathbf{U}, \mathbf{V}, \mathbf{W}$ are the same and their bases are the same $\Rightarrow$ **square matrices.**

Example 5 S rotates the plane by θ and T also rotates by θ. Then TS rotates by 2θ. This transformation T^2 corresponds to the rotation matrix A^2 through 2θ:

$$T = S \qquad A = B \qquad T^2 = \text{rotation by } 2\theta \qquad A^2 = \begin{bmatrix} \cos 2\theta & -\sin 2\theta \\ \sin 2\theta & \cos 2\theta \end{bmatrix}. \tag{4}$$

$$A^2 = \begin{bmatrix} \cos\theta & -\sin\theta \\ \sin\theta & \cos\theta \end{bmatrix} \begin{bmatrix} \cos\theta & -\sin\theta \\ \sin\theta & \cos\theta \end{bmatrix} = \begin{bmatrix} \cos^2\theta - \sin^2\theta & -2\sin\theta\cos\theta \\ 2\sin\theta\cos\theta & \cos^2\theta - \sin^2\theta \end{bmatrix}. \tag{5}$$

Comparing (4) with (5) produces $\cos 2\theta = \cos^2\theta - \sin^2\theta$ and $\sin 2\theta = 2\sin\theta\cos\theta$.

Example 6 S rotates by the angle θ and T rotates by $-\theta$. Then $TS = I$ leads to $AB = I$. In this case $T(S(\boldsymbol{u}))$ is $\boldsymbol{u}$. We rotate forward and back. For the matrices to match, $AB\boldsymbol{x}$ must be $\boldsymbol{x}$. *The matrices with θ and $-\theta$ are inverses.*

$$AB = \begin{bmatrix} \cos\theta & \sin\theta \\ -\sin\theta & \cos\theta \end{bmatrix} \begin{bmatrix} \cos\theta & -\sin\theta \\ \sin\theta & \cos\theta \end{bmatrix} = \begin{bmatrix} \cos^2\theta + \sin^2\theta & 0 \\ 0 & \cos^2\theta + \sin^2\theta \end{bmatrix} = I.$$

Choosing the Best Bases

Now comes the final step in this section of the book. **Choose bases that diagonalize the matrix.** With the standard basis (the columns of I) our transformation T produces some matrix A—probably not diagonal. That same T is represented by different matrices when we choose different bases. The two great choices are eigenvectors and singular vectors:

> **Eigenvectors** If T transforms $\mathbf{R}^n$ to $\mathbf{R}^n$, its matrix A is square. If there are n independent eigenvectors, *choose those as the input and output basis.* In this good basis, **the matrix for T is the diagonal eigenvalue matrix Λ.**

Example 7 **The projection matrix** T projects every $\boldsymbol{v} = (x, y)$ in $\mathbf{R}^2$ onto the line $y = -x$. Using the standard basis, $\boldsymbol{v}_1 = (1, 0)$ projects to $T(\boldsymbol{v}_1) = (\frac{1}{2}, -\frac{1}{2})$. For $\boldsymbol{v}_2 = (0, 1)$ the projection is $T(\boldsymbol{v}_2) = (-\frac{1}{2}, \frac{1}{2})$. Those are the columns of A:

Projection matrix
Standard bases
Not diagonal
$$A = \begin{bmatrix} \frac{1}{2} & -\frac{1}{2} \\ -\frac{1}{2} & \frac{1}{2} \end{bmatrix} \quad \text{has } A^{\mathrm{T}} = A \text{ and } A^2 = A.$$

Now comes the main point of eigenvectors. Make them the basis vectors ! Diagonalize !

When the basis vectors are eigenvectors, the matrix becomes diagonal.

$\boldsymbol{v}_1 = \boldsymbol{w}_1 = (1, -1)$ projects to itself : $T(\boldsymbol{v}_1) = \boldsymbol{v}_1$ and $\lambda_1 = 1$

$\boldsymbol{v}_2 = \boldsymbol{w}_2 = (1, \ \ 1)$ projects to zero : $T(\boldsymbol{v}_2) = \mathbf{0}$ and $\lambda_2 = 0$

Eigenvector bases
Diagonal matrix
$$\text{The new matrix is } \begin{bmatrix} \mathbf{1} & \mathbf{0} \\ \mathbf{0} & \mathbf{0} \end{bmatrix} = \begin{bmatrix} \boldsymbol{\lambda_1} & 0 \\ 0 & \boldsymbol{\lambda_2} \end{bmatrix} = \boldsymbol{\Lambda}. \tag{6}$$

Eigenvectors are the perfect basis vectors. They produce the eigenvalue matrix Λ.

What about other choices of *input basis* = *output basis*? Put those basis vectors into the columns of B. We saw above that the change of basis matrices (between standard basis and new basis) are $\boldsymbol{B}_{\text{in}} = \boldsymbol{B}$ and $\boldsymbol{B}_{\text{out}} = \boldsymbol{B}^{-1}$. The new matrix for T is **similar** to A:

> $\boldsymbol{A}_{\text{new}} = \boldsymbol{B}^{-1}\boldsymbol{A}\boldsymbol{B}$ **in the new basis of $\boldsymbol{b}$'s is similar to $\boldsymbol{A}$ in the standard basis :**
>
> $$A_{b\text{'s to }b\text{'s}} = B^{-1}_{\text{standard to }b\text{'s}} \quad A_{\text{standard}} \quad B_{b\text{'s to standard}} \tag{7}$$

I used the multiplication rule for the transformation ITI and the matrices $B^{-1}AB$.

Finally we allow *different spaces* $\boldsymbol{V}$ *and* $\boldsymbol{W}$, *and different bases* $\boldsymbol{v}$'s and $\boldsymbol{w}$'s. When we know T and we choose bases, we get a matrix A. Probably A is not symmetric or even square. But we can always choose $\boldsymbol{v}$'s and $\boldsymbol{w}$'s that produce a diagonal matrix. This will be the *singular value matrix* $\boldsymbol{\Sigma} = \textbf{diag}\,(\boldsymbol{\sigma_1, \ldots, \sigma_r})$ in the decomposition $A = U\Sigma V^{\mathrm{T}}$.

> **Singular vectors** The SVD says that $U^{-1}AV = \Sigma$. The right singular vectors $\boldsymbol{v}_1, \ldots, \boldsymbol{v}_n$ will be the input basis. The left singular vectors $\boldsymbol{u}_1, \ldots, \boldsymbol{u}_m$ will be the output basis. By the rule for matrix multiplication, the matrix for the same transformation in these new bases is $B_{\text{out}}^{-1} A B_{\text{in}} = U^{-1}AV = \Sigma$. Diagonal!

I can't say that Σ is "similar" to A. We are working now with two bases, input and output. But those are *orthonormal bases* and they preserve the lengths of vectors. Following a good suggestion by David Vogan, I propose that we say: **$\boldsymbol{\Sigma}$ is "isometric" to $\boldsymbol{A}$.**

Definition $\quad$ *$C = Q_1^{-1} A Q_2$ is isometric to A if Q_1 and Q_2 are orthogonal.*

Example 8 To construct the matrix A for the transformation $T = \frac{d}{dx}$, we chose the input basis $1, x, x^2, x^3$ and the output basis $1, x, x^2$. The matrix A was simple but unfortunately it wasn't diagonal. But we can take each basis *in the opposite order*.

Now the input basis is $x^3, x^2, x, 1$ and the output basis is $x^2, x, 1$. The change of basis matrices B_{in} and B_{out} are permutations. The matrix for $T(\boldsymbol{u}) = d\boldsymbol{u}/dx$ with the new bases is **the diagonal singular value matrix $\boldsymbol{B_{\text{out}}^{-1} A B_{\text{in}}}$** $= \boldsymbol{\Sigma}$ with σ's $= 3, 2, 1$:

$$\boldsymbol{B_{\text{out}}^{-1} A B_{\text{in}}} = \begin{bmatrix} & & 1 \\ & 1 & \\ 1 & & \end{bmatrix} \begin{bmatrix} 0 & \mathbf{1} & 0 & 0 \\ 0 & 0 & \mathbf{2} & 0 \\ 0 & 0 & 0 & \mathbf{3} \end{bmatrix} \begin{bmatrix} & & & 1 \\ & & 1 & \\ & 1 & & \\ 1 & & & \end{bmatrix} = \begin{bmatrix} \mathbf{3} & 0 & 0 & 0 \\ 0 & \mathbf{2} & 0 & 0 \\ 0 & 0 & \mathbf{1} & 0 \end{bmatrix}. \quad (8)$$

Well, this was a tough section. We found that x^3, x^2, x have derivatives $3x^2$, $2x$, 1.

■ REVIEW OF THE KEY IDEAS ■

1. If we know $T(\boldsymbol{v}_1), \ldots, T(\boldsymbol{v}_n)$ for a basis, linearity will determine all other $T(\boldsymbol{v})$.

2. $\left\{\begin{array}{l}\text{Linear transformation } T \\ \text{Input basis } \boldsymbol{v}_1, \ldots, \boldsymbol{v}_n \\ \text{Output basis } \boldsymbol{w}_1, \ldots, \boldsymbol{w}_m\end{array}\right\} \to$ Matrix A (m by n) represents T in these bases. Column j of A is a_{1j} to a_{mj} exactly when $T(\boldsymbol{v_j}) = a_{1j}\boldsymbol{w}_1 + \cdots + a_{mj}\boldsymbol{w}_m$

3. The change of basis matrix $B = W^{-1}V = B_{\text{out}}^{-1} B_{\text{in}}$ represents the identity $T(\boldsymbol{v}) = \boldsymbol{v}$.

4. If A and B represent T and S, and the output basis for S is the input basis for T, then the matrix AB represents the transformation $T(S(\boldsymbol{u}))$.

5. The best input-output bases are eigenvectors and/or singular vectors of A. Then
$$B^{-1}AB = \Lambda = \text{eigenvalues} \qquad B_{\text{out}}^{-1} A B_{\text{in}} = \Sigma = \text{singular values.}$$

Problem Set 8.2

Questions 1–4 extend the first derivative example to higher derivatives.

1 The transformation S takes the ***second derivative***. Keep $1, x, x^2, x^3$ as the input basis $\boldsymbol{v}_1, \boldsymbol{v}_2, \boldsymbol{v}_3, \boldsymbol{v}_4$ and also as output basis $\boldsymbol{w}_1, \boldsymbol{w}_2, \boldsymbol{w}_3, \boldsymbol{w}_4$. Write $S(\boldsymbol{v}_1), S(\boldsymbol{v}_2), S(\boldsymbol{v}_3), S(\boldsymbol{v}_4)$ in terms of the $\boldsymbol{w}$'s. Find the 4 by 4 matrix A_2 for S.

2 What functions have $S(\boldsymbol{v}) = \mathbf{0}$? They are in the kernel of the second derivative S. What vectors are in the nullspace of its matrix A_2 in Problem 1?

3 The second derivative A_2 is not the square of a rectangular first derivative matrix A_1:

$$A_1 = \begin{bmatrix} 0 & 1 & 0 & 0 \\ 0 & 0 & 2 & 0 \\ 0 & 0 & 0 & 3 \end{bmatrix} \text{ does not allow } A_1^2 = A_2.$$

Add a zero row 4 to A_1 so that output space = input space. Compare A_1^2 with A_2. Conclusion: We want output basis = ______ basis. Then $m = n$.

4 (a) The product TS of first and second derivatives produces the *third* derivative. Add zeros to make 4 by 4 matrices, then compute $A_1 A_2 = A_3$.

(b) The matrix A_2^2 corresponds to S^2 = *fourth* derivative. Why is this zero?

Questions 5–9 are about a particular transformation T and its matrix A.

5 With bases $\boldsymbol{v}_1, \boldsymbol{v}_2, \boldsymbol{v}_3$ and $\boldsymbol{w}_1, \boldsymbol{w}_2, \boldsymbol{w}_3$, suppose $T(\boldsymbol{v}_1) = \boldsymbol{w}_2$ and $T(\boldsymbol{v}_2) = T(\boldsymbol{v}_3) = \boldsymbol{w}_1 + \boldsymbol{w}_3$. T is a linear transformation. Find the matrix A and multiply by the vector $(1, 1, 1)$. What is the output from T when the input is $\boldsymbol{v}_1 + \boldsymbol{v}_2 + \boldsymbol{v}_3$?

6 Since $T(\boldsymbol{v}_2) = T(\boldsymbol{v}_3)$, the solutions to $T(\boldsymbol{v}) = \mathbf{0}$ are $\boldsymbol{v} =$ ______. What vectors are in the nullspace of A? Find all solutions to $T(\boldsymbol{v}) = \boldsymbol{w}_2$.

7 Find a vector that is not in the column space of A. Find a combination of $\boldsymbol{w}$'s that is not in the range of the transformation T.

8 You don't have enough information to determine T^2. Why is its matrix not necessarily A^2? What more information do you need?

9 Find the *rank* of A. The rank is not the dimension of the whole output space $\mathbf{W}$. It is the dimension of the ______ of T.

Questions 10–13 are about invertible linear transformations.

10 Suppose $T(\boldsymbol{v}_1) = \boldsymbol{w}_1 + \boldsymbol{w}_2 + \boldsymbol{w}_3$ and $T(\boldsymbol{v}_2) = \boldsymbol{w}_2 + \boldsymbol{w}_3$ and $T(\boldsymbol{v}_3) = \boldsymbol{w}_3$. Find the matrix A for T using these basis vectors. What input vector $\boldsymbol{v}$ gives $T(\boldsymbol{v}) = \boldsymbol{w}_1$?

11 Invert the matrix A in Problem 10. Also invert the transformation T—what are $T^{-1}(\boldsymbol{w}_1)$ and $T^{-1}(\boldsymbol{w}_2)$ and $T^{-1}(\boldsymbol{w}_3)$?

12 Which of these are true and why is the other one ridiculous?

(a) $T^{-1}T = I$ (b) $T^{-1}(T(\boldsymbol{v}_1)) = \boldsymbol{v}_1$ (c) $T^{-1}(T(\boldsymbol{w}_1)) = \boldsymbol{w}_1$.

13 Suppose the spaces **V** and **W** have the same basis $\boldsymbol{v}_1, \boldsymbol{v}_2$.

(a) Describe a transformation T (not I) that is its own inverse.

(b) Describe a transformation T (not I) that equals T^2.

(c) Why can't the same T be used for both (a) and (b)?

Questions 14–19 are about changing the basis.

14 (a) What matrix B transforms $(1,0)$ into $(2,5)$ and transforms $(0,1)$ to $(1,3)$?

(b) What matrix C transforms $(2,5)$ to $(1,0)$ and $(1,3)$ to $(0,1)$?

(c) Why does no matrix transform $(2,6)$ to $(1,0)$ and $(1,3)$ to $(0,1)$?

15 (a) What matrix M transforms $(1,0)$ and $(0,1)$ to (r,t) and (s,u)?

(b) What matrix N transforms (a,c) and (b,d) to $(1,0)$ and $(0,1)$?

(c) What condition on a, b, c, d will make part (b) impossible?

16 (a) How do M and N in Problem 15 yield the matrix that transforms (a,c) to (r,t) and (b,d) to (s,u)?

(b) What matrix transforms $(2,5)$ to $(1,1)$ and $(1,3)$ to $(0,2)$?

17 If you keep the same basis vectors but put them in a different order, the change of basis matrix B is a _____ matrix. If you keep the basis vectors in order but change their lengths, B is a _____ matrix.

18 The matrix that rotates the axis vectors $(1,0)$ and $(0,1)$ through an angle θ is Q. What are the coordinates (a,b) of the original $(1,0)$ using the new (rotated) axes? This *inverse* can be tricky. Draw a figure or solve for a and b:

$$Q = \begin{bmatrix} \cos\theta & -\sin\theta \\ \sin\theta & \cos\theta \end{bmatrix} \qquad \begin{bmatrix} 1 \\ 0 \end{bmatrix} = a\begin{bmatrix} \cos\theta \\ \sin\theta \end{bmatrix} + b\begin{bmatrix} -\sin\theta \\ \cos\theta \end{bmatrix}.$$

19 The matrix that transforms $(1,0)$ and $(0,1)$ to $(1,4)$ and $(1,5)$ is $B =$ _____. The combination $a(1,4) + b(1,5)$ that equals $(1,0)$ has $(a,b) = (\ ,\ )$. How are those new coordinates of $(1,0)$ related to B or B^{-1}?

Questions 20–23 are about the space of quadratic polynomials $y = A + Bx + Cx^2$.

20 The parabola $\boldsymbol{w}_1 = \frac{1}{2}(x^2 + x)$ equals one at $x = 1$, and zero at $x = 0$ and $x = -1$. Find the parabolas $\boldsymbol{w}_2, \boldsymbol{w}_3$, and then find $\boldsymbol{y}(x)$ by linearity.

(a) $\boldsymbol{w}_2$ equals one at $x = 0$ and zero at $x = 1$ and $x = -1$.

(b) $\boldsymbol{w}_3$ equals one at $x = -1$ and zero at $x = 0$ and $x = 1$.

(c) $\boldsymbol{y}(x)$ equals 4 at $x = 1$ and 5 at $x = 0$ and 6 at $x = -1$. Use $\boldsymbol{w}_1, \boldsymbol{w}_2, \boldsymbol{w}_3$.

21 One basis for second-degree polynomials is $\boldsymbol{v}_1 = 1$ and $\boldsymbol{v}_2 = x$ and $\boldsymbol{v}_3 = x^2$. Another basis is $\boldsymbol{w}_1, \boldsymbol{w}_2, \boldsymbol{w}_3$ from Problem 20. Find two change of basis matrices, from the $\boldsymbol{w}$'s to the $\boldsymbol{v}$'s and from the $\boldsymbol{v}$'s to the $\boldsymbol{w}$'s.

22 What are the three equations for A, B, C if the parabola $y = A + Bx + Cx^2$ equals 4 at $x = a$ and 5 at $x = b$ and 6 at $x = c$? Find the determinant of the 3 by 3 matrix. That matrix transforms values like $4, 5, 6$ to parabolas y—or is it the other way?

23 Under what condition on the numbers $m_1, m_2, \ldots, m_9$ do these three parabolas give a basis for the space of all parabolas $a + bx + cx^2$?

$$\boldsymbol{v}_1 = m_1 + m_2 x + m_3 x^2, \quad \boldsymbol{v}_2 = m_4 + m_5 x + m_6 x^2, \quad \boldsymbol{v}_3 = m_7 + m_8 x + m_9 x^2.$$

24 The Gram-Schmidt process changes a basis $\boldsymbol{a}_1, \boldsymbol{a}_2, \boldsymbol{a}_3$ to an orthonormal basis $\boldsymbol{q}_1, \boldsymbol{q}_2, \boldsymbol{q}_3$. These are columns in $A = QR$. Show that R is the change of basis matrix from the $\boldsymbol{a}$'s to the $\boldsymbol{q}$'s ($\boldsymbol{a}_2$ is what combination of $\boldsymbol{q}$'s when $A = QR$?).

25 Elimination changes the rows of A to the rows of U with $A = LU$. Row 2 of A is what combination of the rows of U? Writing $A^{\mathrm{T}} = U^{\mathrm{T}} L^{\mathrm{T}}$ to work with columns, the change of basis matrix is $B = L^{\mathrm{T}}$. We have *bases* if the matrices are _____ .

26 Suppose $\boldsymbol{v}_1, \boldsymbol{v}_2, \boldsymbol{v}_3$ are **eigenvectors** for T. This means $T(\boldsymbol{v}_i) = \lambda_i \boldsymbol{v}_i$ for $i = 1, 2, 3$. What is the matrix for T when the input and output bases are the $\boldsymbol{v}$'s?

27 Every invertible linear transformation can have I as its matrix! Choose any input basis $\boldsymbol{v}_1, \ldots, \boldsymbol{v}_n$. For output basis choose $\boldsymbol{w}_i = T(\boldsymbol{v}_i)$. Why must T be invertible?

28 Using $\boldsymbol{v}_1 = \boldsymbol{w}_1$ and $\boldsymbol{v}_2 = \boldsymbol{w}_2$ find the standard matrix for these T's:

(a) $T(\boldsymbol{v}_1) = \boldsymbol{0}$ and $T(\boldsymbol{v}_2) = 3\boldsymbol{v}_1$ (b) $T(\boldsymbol{v}_1) = \boldsymbol{v}_1$ and $T(\boldsymbol{v}_1 + \boldsymbol{v}_2) = \boldsymbol{v}_1$.

29 Suppose T reflects the xy plane across the x axis and S is reflection across the y axis. If $\boldsymbol{v} = (x, y)$ what is $S(T(\boldsymbol{v}))$? Find a simpler description of the product ST.

30 Suppose T is reflection across the 45° line, and S is reflection across the y axis. If $\boldsymbol{v} = (2, 1)$ then $T(\boldsymbol{v}) = (1, 2)$. Find $S(T(\boldsymbol{v}))$ and $T(S(\boldsymbol{v}))$. Usually $ST \neq TS$.

31 **The product of two reflections is a rotation**. Multiply these reflection matrices to find the rotation angle:

$$\begin{bmatrix} \cos 2\theta & \sin 2\theta \\ \sin 2\theta & -\cos 2\theta \end{bmatrix} \begin{bmatrix} \cos 2\alpha & \sin 2\alpha \\ \sin 2\alpha & -\cos 2\alpha \end{bmatrix}.$$

32 Suppose A is a 3 by 4 matrix of rank $r = 2$, and $T(\boldsymbol{v}) = A\boldsymbol{v}$. Choose input basis vectors $\boldsymbol{v}_1, \boldsymbol{v}_2$ from the row space of A and $\boldsymbol{v}_3, \boldsymbol{v}_4$ from the nullspace. Choose output basis vectors $\boldsymbol{w}_1 = A\boldsymbol{v}_1$, $\boldsymbol{w}_2 = A\boldsymbol{v}_2$ in the column space and $\boldsymbol{w}_3$ from the nullspace of A^{T}. What specially simple matrix represents T in these special bases?

33 The space $\mathbf{M}$ of 2 by 2 matrices has the basis $\boldsymbol{v}_1, \boldsymbol{v}_2, \boldsymbol{v}_3, \boldsymbol{v}_4$ in Worked Example **8.2 A**. Suppose T multiplies each matrix by $\left[\begin{smallmatrix} a & b \\ c & d \end{smallmatrix}\right]$. With $\boldsymbol{w}$'s equal to $\boldsymbol{v}$'s, what 4 by 4 matrix A represents this transformation T on matrix space?

34 True or False: If we know $T(\boldsymbol{v})$ for n different nonzero vectors in $\mathbf{R}^n$, then we know $T(\boldsymbol{v})$ for every vector $\boldsymbol{v}$ in $\mathbf{R}^n$.

8.3 The Search for a Good Basis

1 With a new input basis $\boldsymbol{B}_{\text{in}}$ and output basis $\boldsymbol{B}_{\text{out}}$, every matrix A becomes $\boldsymbol{B}_{\text{out}}^{-1}\boldsymbol{A}\boldsymbol{B}_{\text{in}}$.

2 $\boldsymbol{B}_{\text{in}}=\boldsymbol{B}_{\text{out}}=$**"generalized eigenvectors of A"** produces the **Jordan form $J=B^{-1}AB$.**

3 The **Fourier matrix F** $= B_{\text{in}} = B_{\text{out}}$ diagonalizes every circulant matrix (use the **FFT**).

4 Sines, cosines, e^{ikx}, Legendre and Chebyshev : those are great bases for **function space**.

This is an important section of the book. The first chapters prepared the way by explaining the idea of a **basis**. Chapter 6 introduced the eigenvectors $\boldsymbol{x}$ and Chapter 7 found singular vectors $\boldsymbol{v}$ and $\boldsymbol{u}$. Those basis vectors are two winners but many other choices are very valuable. So many computations begin with a choice of basis.

The input basis vectors will be the columns of B_{in}. The output basis vectors will be the columns of B_{out}. Always B_{in} and B_{out} are *invertible*—basis vectors are independent!

Pure algebra If A is the matrix for a transformation T in the standard basis, then

$$\boldsymbol{B}_{\text{out}}^{-1}\boldsymbol{A}\boldsymbol{B}_{\text{in}} \textbf{ is the matrix in the new bases.} \tag{1}$$

The standard basis vectors are the *columns of the identity*: $B_{\text{in}} = I_{n\times n}$ and $B_{\text{out}} = I_{m\times m}$. Now we are choosing special bases to make the matrix clearer and simpler than A. When $B_{\text{in}} = B_{\text{out}} = B$, the square matrix $B^{-1}AB$ is *similar* to A: same eigenvalues.

Applied algebra Applications are all about choosing good bases. Here are four important choices for vectors and three choices for functions. Eigenvectors and singular vectors led to Λ and Σ in Section 8.2. The Jordan form is new.

1 $\boldsymbol{B}_{\text{in}} = \boldsymbol{B}_{\text{out}} =$ **eigenvector matrix $\boldsymbol{X}$.** Then $\boldsymbol{X}^{-1}\boldsymbol{A}\boldsymbol{X}$ = **eigenvalues in Λ.**

This choice requires A to be a square matrix with n independent eigenvectors. "A must be diagonalizable." We get Λ when $B_{\text{in}} = B_{\text{out}}$ is the eigenvector matrix X.

2 $\boldsymbol{B}_{\text{in}} = \boldsymbol{V}$ and $\boldsymbol{B}_{\text{out}} = \boldsymbol{U}$: **singular vectors of $\boldsymbol{A}$.** Then $\boldsymbol{U}^{-1}\boldsymbol{A}\boldsymbol{V}$ = **diagonal Σ.**

Σ is the singular value matrix (with $\sigma_1, \ldots, \sigma_r$ on its diagonal) when B_{in} and B_{out} are the singular vector matrices V and U. Recall that those columns of B_{in} and B_{out} are orthonormal eigenvectors of $A^{\mathrm{T}}A$ and AA^{T}. Then $A = U\Sigma V^{\mathrm{T}}$ gives $\Sigma = U^{-1}AV$.

3 $\boldsymbol{B}_{\text{in}} = \boldsymbol{B}_{\text{out}} =$ **generalized eigenvectors of $\boldsymbol{A}$.** Then $\boldsymbol{B}^{-1}\boldsymbol{A}\boldsymbol{B} =$ **Jordan form $\boldsymbol{J}$.**

A is a square matrix but it may only have $\boldsymbol{s}$ **independent eigenvectors.** (If $s = n$ then B is X and J is Λ.) In all cases Jordan constructed $n - s$ additional "generalized" eigenvectors, aiming to make the Jordan form J *as diagonal as possible*:

i) There are s square blocks along the diagonal of J.

ii) Each block has one eigenvalue λ, one eigenvector, and 1's above the diagonal.

The best J has n 1×1 blocks, each containing an eigenvalue. **Then $\boldsymbol{J} = \boldsymbol{\Lambda}$** (diagonal).

Example 1 This Jordan matrix J has eigenvalues $\lambda = 2, 2, 3, 3$ (two double eigenvalues). Those eigenvalues lie along the diagonal because J is triangular. There are two independent eigenvectors for $\lambda = 2$, but there is only *one line of eigenvectors for* $\lambda = 3$. This will be true for every matrix $C = BJB^{-1}$ that is similar to J.

$$\textbf{Jordan matrix} \quad J = \begin{bmatrix} \mathbf{2} & & & \\ & \mathbf{2} & & \\ & & \begin{bmatrix} \mathbf{3} & \mathbf{1} \\ \mathbf{0} & \mathbf{3} \end{bmatrix} \end{bmatrix} \quad \begin{array}{l} \text{Two 1 by 1 blocks} \\ \text{One 2 by 2 block} \\ J \text{ has 3 eigenvectors} \\ \text{Eigenvalues } 2, 2, 3, 3 \end{array}$$

Two eigenvectors for $\lambda = 2$ are $\boldsymbol{x}_1 = (1, 0, 0, 0)$ and $\boldsymbol{x}_2 = (0, 1, 0, 0)$. One eigenvector for $\lambda = 3$ is $\boldsymbol{x}_3 = (0, 0, 1, 0)$. The "generalized eigenvector" for this Jordan matrix is the fourth standard basis vector $\boldsymbol{x}_4 = (0, 0, 0, 1)$. The eigenvectors for J (normal and generalized) are just the columns $\boldsymbol{x}_1, \boldsymbol{x}_2, \boldsymbol{x}_3, \boldsymbol{x}_4$ of the identity matrix I.

Notice $\boldsymbol{(J - 3I)x_4 = x_3}$. **The generalized eigenvector $\boldsymbol{x}_4$ connects to the true eigenvector $\boldsymbol{x}_3$.** A true $\boldsymbol{x}_4$ would have $(J - 3I)\boldsymbol{x}_4 = \mathbf{0}$, but that doesn't happen here!

Every matrix $C = BJB^{-1}$ that is similar to this J will have true eigenvectors $\boldsymbol{b}_1, \boldsymbol{b}_2, \boldsymbol{b}_3$ in the first three columns of B. The fourth column of B will be a generalized eigenvector $\boldsymbol{b}_4$ of C, tied to the true $\boldsymbol{b}_3$. Here is a quick proof that uses $B\boldsymbol{x}_3 = \boldsymbol{b}_3$ and $B\boldsymbol{x}_4 = \boldsymbol{b}_4$ to show: The fourth column $\boldsymbol{b}_4$ is tied to $\boldsymbol{b}_3$ by $\boldsymbol{(C - 3I)b_4 = b_3}$.

$$(BJB^{-1} - 3I)\,\boldsymbol{b}_4 = BJ\,\boldsymbol{x}_4 - 3B\,\boldsymbol{x}_4 = B(J - 3I)\,\boldsymbol{x}_4 = B\,\boldsymbol{x}_3 = \boldsymbol{b}_3. \tag{2}$$

The point of Jordan's theorem is that every square matrix A has a complete set of eigenvectors and generalized eigenvectors. When those go into the columns of B, the matrix $B^{-1}AB = J$ is in Jordan form. Based on Example 1, here is a description of J.

The Jordan Form

For every A, we want to choose B so that $B^{-1}AB$ is as ***nearly diagonal as possible***. When A has a full set of n eigenvectors, they go into the columns of B. Then $B = X$. The matrix $X^{-1}AX$ is diagonal, period. This is the Jordan form of A—when A can be diagonalized. In the general case, eigenvectors are missing and Λ can't be reached.

Suppose A has s independent eigenvectors. Then it is similar to a Jordan matrix with s blocks. Each block has an *eigenvalue on the diagonal with* 1's *just above it*. This block accounts for exactly one eigenvector of A. Then B contains generalized eigenvectors as well as ordinary eigenvectors.

When there are n eigenvectors, all n blocks will be 1 by 1. In that case $J = \Lambda$.

The Jordan form solves the differential equation $d\boldsymbol{u}/dt = A\boldsymbol{u}$ for **any square matrix** $\boldsymbol{A = BJB^{-1}}$. The solution $e^{At}\boldsymbol{u}(0)$ becomes $\boldsymbol{u}(t) = Be^{Jt}B^{-1}\boldsymbol{u}(0)$. J is triangular and its matrix exponential $\boldsymbol{e^{Jt}}$ involves $\boldsymbol{e^{\lambda t}}$ times powers $1, t, \ldots, t^{s-1}$. Overall the Jordan form deals optimally with *repeated eigenvalues*.

(Jordan form) If A has s independent eigenvectors, it is similar to a matrix J that has s Jordan blocks $J_1 \dots, J_s$ on its diagonal. B contains "generalized eigenvectors":

$$\textbf{Jordan form of } A \qquad B^{-1}AB = \begin{bmatrix} J_1 & & \\ & \ddots & \\ & & J_s \end{bmatrix} = J. \tag{3}$$

Each block J_i has one eigenvalue λ_i, one eigenvector, and 1's just above the diagonal:

$$\textbf{Jordan block in } J \qquad J_i = \begin{bmatrix} \lambda_i & 1 & & \\ & \cdot & \cdot & \\ & & \cdot & 1 \\ & & & \lambda_i \end{bmatrix}. \tag{4}$$

Matrices are similar if they share the same Jordan form J—not otherwise.

The Jordan form J has an off-diagonal 1 for each missing eigenvector (and the 1's are next to the eigenvalues). In every family of similar matrices, we are picking one outstanding member called J. It is nearly diagonal (or if possible completely diagonal). We can quickly solve $d\boldsymbol{u}/dt = J\boldsymbol{u}$ and take powers J^k. Every other matrix in the family has the form BJB^{-1}.

Jordan's Theorem is proved in my textbook *Linear Algebra and Its Applications*. Please refer to that book (or more advanced books) for the proof. The reasoning is rather intricate and in actual computations the Jordan form is not at all popular—its calculation is not stable. A slight change in A will separate the repeated eigenvalues and remove the off-diagonal 1's—switching Jordan to a diagonal Λ.

Proved or not, you have caught the central idea of similarity—to make A as simple as possible while preserving its essential properties. The best basis B gives $B^{-1}AB = J$.

Question Find the eigenvalues and all possible Jordan forms if $\boldsymbol{A^2}$ = **zero matrix**.

Answer The eigenvalues must all be zero, because $A\boldsymbol{x} = \lambda\boldsymbol{x}$ leads to $A^2\boldsymbol{x} = \lambda^2\boldsymbol{x} = 0\boldsymbol{x}$. The Jordan form of A has $J^2 = 0$ because $J^2 = (B^{-1}AB)(B^{-1}AB) = B^{-1}A^2B = 0$. Every block in J has $\lambda = 0$ on the diagonal. Look at J_k^2 for block sizes $1, 2, 3$:

$$\begin{bmatrix} 0 \end{bmatrix}^2 = \begin{bmatrix} 0 \end{bmatrix} \qquad \begin{bmatrix} 0 & 1 \\ 0 & 0 \end{bmatrix}^2 = \begin{bmatrix} 0 & 0 \\ 0 & 0 \end{bmatrix} \qquad \begin{bmatrix} 0 & 1 & 0 \\ 0 & 0 & 1 \\ 0 & 0 & 0 \end{bmatrix}^2 = \begin{bmatrix} 0 & 0 & \mathbf{1} \\ 0 & 0 & 0 \\ 0 & 0 & 0 \end{bmatrix}$$

Conclusion: If $J^2 = 0$ then all block sizes must be 1 or 2. J^2 is not zero for 3 by 3. **The maximum rank of $\boldsymbol{J}$ is $\boldsymbol{n/2}$**, when there are $n/2$ blocks, each of size 2 and rank 1.

Now come three great bases of applied mathematics: Fourier, Legendre, and Chebyshev. Their discrete forms are vectors in $\mathbf{R}^n$. Their continuous forms are functions. Since they are chosen once and for all, *without knowing the matrix* A, these bases $B_{\text{in}} = B_{\text{out}}$ probably don't diagonalize A. But for many important matrices A in applied mathematics, the matrices $B^{-1}AB$ are *close to diagonal.*

4 $\boldsymbol{B_{\text{in}} = B_{\text{out}} =}$ **Fourier matrix** $\boldsymbol{F}$ **Then** $\boldsymbol{Fx}$ **is a Discrete Fourier Transform of** $\boldsymbol{x}$.

Those words are telling us: The Fourier matrix with columns = eigenvectors of P in equation (6) is important. Those are good basis vectors to work with.

We ask: Which matrices are diagonalized by F? This time we are starting with the eigenvectors $(1, \lambda, \lambda^2, \lambda^3)$ and finding the matrices that have those eigenvectors:

$$\text{If } \boldsymbol{\lambda^4 = 1} \text{ then } \quad \boldsymbol{Px} = \begin{bmatrix} 0 & 1 & 0 & 0 \\ 0 & 0 & 1 & 0 \\ 0 & 0 & 0 & 1 \\ 1 & 0 & 0 & 0 \end{bmatrix} \begin{bmatrix} 1 \\ \lambda \\ \lambda^2 \\ \lambda^3 \end{bmatrix} = \lambda \begin{bmatrix} 1 \\ \lambda \\ \lambda^2 \\ \lambda^3 \end{bmatrix} = \lambda \boldsymbol{x}. \qquad (5)$$

$\boldsymbol{P}$ **is a permutation matrix**. The equation $P\boldsymbol{x} = \lambda\boldsymbol{x}$ says that $\boldsymbol{x}$ is an eigenvector and λ is an eigenvalue of P. Notice how the fourth row of this vector equation is $1 = \lambda^4$. That rule for λ makes everything work.

Does this give four different eigenvalues λ? *Yes.* The four numbers $\lambda = \boldsymbol{1, i, -1, -i}$ all satisfy $\lambda^4 = 1$. (You know $i^2 = -1$. Squaring both sides gives $i^4 = 1$.) So those four numbers are the eigenvalues of P, each with its eigenvector $\boldsymbol{x} = (1, \lambda, \lambda^2, \lambda^3)$. **The eigenvector matrix $\boldsymbol{F}$ diagonalizes the permutation matrix $\boldsymbol{P}$ above**:

$$\begin{matrix}\textbf{Eigenvalue} \\ \textbf{matrix } \boldsymbol{\Lambda} \\ \textbf{for } \boldsymbol{P}\end{matrix} \begin{bmatrix} 1 & & & \\ & i & & \\ & & -1 & \\ & & & -i \end{bmatrix} \qquad \begin{matrix}\textbf{Eigenvector} \\ \textbf{matrix is} \\ \textbf{Fourier} \\ \textbf{matrix } \boldsymbol{F}\end{matrix} \begin{bmatrix} 1 & 1 & 1 & 1 \\ 1 & i & -1 & -i \\ 1 & i^2 & 1 & (-i)^2 \\ 1 & i^3 & -1 & (-i)^3 \end{bmatrix} \qquad (6)$$

Those columns of F are orthogonal. They are eigenvectors of P and every circulant matrix $C = c_0 I + c_1 P + c_2 P^2 + c_2 P^3$. Unfortunately this Fourier matrix F is complex (it is the most important complex matrix in the world). Multiplications $F\boldsymbol{x}$ are done millions of times very quickly, by the Fast Fourier Transform.

$$\begin{matrix}\textbf{Circulant} \\ \textbf{matrix}\end{matrix} \quad \boldsymbol{C} = \begin{bmatrix} c_0 & \boldsymbol{c_1} & c_2 & c_3 \\ c_3 & c_0 & \boldsymbol{c_1} & c_2 \\ c_2 & c_3 & c_0 & \boldsymbol{c_1} \\ \boldsymbol{c_1} & c_2 & c_3 & c_0 \end{bmatrix} \quad \begin{matrix}\text{has four eigenvectors in the Fourier matrix } F \\ \text{has four eigenvalues in the vector } F\boldsymbol{c} \\ \lambda = c_0 + c_1 + c_2 + c_3 \text{ has eigenvector } (1,1,1,1)\end{matrix}$$

For more about circulant matrices please see Section 6.4.

Circulant matrices have constant diagonals. The same number c_0 goes down the main diagonal. The number c_1 is on the diagonal above, and that diagonal "wraps around" or "circles around" to the southwest corner of C. This explains the name *circulant* and it indicates that these matrices are *periodic* or *cyclic*. Even the powers of λ cycle around because $\lambda^4 = 1$ leads to $\lambda^5, \lambda^6, \lambda^7, \lambda^8 = \lambda, \lambda^2, \lambda^3, \lambda^4$.

Constancy down the diagonals is a crucial property. It corresponds to *constant coefficients* in a differential equation. This is exactly when Fourier works perfectly!

$$\text{The equation} \quad \frac{d^2u}{dt^2} = -u \quad \text{is solved by} \quad u = c_0 \cos t + c_1 \sin t.$$

$$\text{The equation} \quad \frac{d^2u}{dt^2} = tu \quad \text{cannot be solved by elementary functions.}$$

These equations are linear. The first is the oscillation equation for a simple spring. It is Newton's Law $f = ma$ with mass $m = 1$, $a = d^2u/dt^2$, and force $f = -u$. Constant coefficients produce the differential equations that you can really solve.

The equation $\boldsymbol{u'' = tu}$ has a variable coefficient t. This is Airy's equation in physics and optics (it was derived to explain a rainbow). The solutions change completely when t passes through zero, and those solutions require infinite series. *We won't go there.*

The point is that equations with constant coefficients have simple solutions like $e^{\lambda t}$. You discover λ by substituting $e^{\lambda t}$ into the differential equation. That number λ is like an eigenvalue. For $u = \cos t$ and $u = \sin t$ the number is $\lambda = i$. Euler's great formula $e^{it} = \cos t + i \sin t$ introduces complex numbers as we saw in the eigenvalues of P and C.

Bases for Function Space

For functions of x, the first basis I would think of contains the powers $\boldsymbol{1, x, x^2, x^3, \ldots}$ Unfortunately this is a terrible basis. Those functions x^n are just barely independent. x^{10} is *almost* a combination of other basis vectors $1, x, \ldots, x^9$. It is virtually impossible to compute with this "ill-conditioned" basis.

If we had vectors instead of functions, the test for a good basis would look at $B^{\mathrm{T}}B$. This matrix contains all inner products between the basis vectors (columns of B). *The basis is orthonormal when* $B^{\mathrm{T}}B = I$. That is best possible. But the basis $1, x, x^2, \ldots$ produces the evil **Hilbert matrix**: $B^{\mathrm{T}}B$ has an enormous ratio between its largest and smallest eigenvalues. A large condition number signals an unhappy choice of basis.

When the integrals go from $x = 0$ to $x = 1$, the inner product of $\boldsymbol{x^i}$ with $\boldsymbol{x^j}$ is

$$\int_0^1 x^i\, x^j\, dx = \left.\frac{x^{i+j+1}}{i+j+1}\right]_{x=0}^{x=1} = \frac{\mathbf{1}}{\boldsymbol{i+j+1}} = \text{entries of Hilbert matrix } B^{\mathrm{T}}B$$

Note Now the columns of B are functions instead of vectors. We still use $B^{\mathrm{T}}B$ to test for independence. So we need to know the dot product (inner product is a better name) of two functions—those are the numbers in $B^{\mathrm{T}}B$.

The dot product of vectors is just $\boldsymbol{x}^{\mathrm{T}}\boldsymbol{y} = x_1y_1 + \cdots + x_ny_n$. The inner product of functions will integrate instead of adding, but the idea is completely parallel:

$$\begin{aligned}
\text{Inner product } (\boldsymbol{f}, \boldsymbol{g}) &= \int f(x)g(x)\,dx \\
\text{Complex inner product } (\boldsymbol{f}, \boldsymbol{g}) &= \int \overline{f(x)}\,g(x)\,dx,\ \overline{f} = \text{ complex conjugate} \\
\text{Weighted inner product } (\boldsymbol{f}, \boldsymbol{g})_{\boldsymbol{w}} &= \int w(x)\,\overline{f(x)}\,g(x)\,dx,\ w = \text{ weight function}
\end{aligned}$$

Orthogonal Bases for Function Space

Here are the three leading even-odd bases for theoretical and numerical computations:

5. The **Fourier basis**	$\mathbf{1, \sin x, \cos x, \sin 2x, \cos 2x, \ldots}$
6. The **Legendre basis**	$\mathbf{1,\ x,\ x^2 - \frac{1}{3},\ x^3 - \frac{3}{5}x, \ldots}$
7. The **Chebyshev basis**	$\mathbf{1,\ x,\ 2x^2 - 1,\ 4x^3 - 3x, \ldots}$

The Fourier basis functions (sines and cosines) are all *periodic*. They repeat over every 2π interval because $\cos(x + 2\pi) = \cos x$ and $\sin(x + 2\pi) = \sin x$. So this basis is especially good for functions $f(x)$ that are themselves periodic: $f(x + 2\pi) = f(x)$.

This basis is also *orthogonal*. Every sine and cosine is orthogonal to every other one.

Most important, the sine-cosine basis is also *excellent for approximation*. If we have a smooth periodic function $f(x)$, then a few sines and cosines (low frequencies) are all we need. Jumps in $f(x)$ and noise in the signal are seen in higher frequencies (larger n). We hope and expect that the signal is not drowned by the noise.

The *Fourier transform* connects $f(x)$ to the coefficients a_k and b_k in its Fourier series:

Fourier series $f(x) = a_0 + b_1 \sin x + a_1 \cos x + b_2 \sin 2x + a_2 \cos 2x + \cdots$

We see that **function space is infinite-dimensional**. It takes infinitely many basis functions to capture perfectly a typical $f(x)$. But the formula for each coefficient (for example a_3) is just like the formula $\boldsymbol{b}^{\mathrm{T}}\boldsymbol{a}/\boldsymbol{a}^{\mathrm{T}}\boldsymbol{a}$ for projecting a vector $\boldsymbol{b}$ onto the line through $\boldsymbol{a}$.

Here we are projecting the function $f(x)$ onto the line in function space through $\cos 3x$:

$$\textbf{Fourier coefficient} \quad a_3 = \frac{(f(x), \cos 3x)}{(\cos 3x, \cos 3x)} = \frac{\int f(x)\cos 3x\,dx}{\int \cos 3x \cos 3x\,dx}. \tag{7}$$

Note The sine-cosine basis is not so excellent when the function $f(x)$ has a jump (like a 0–1 function). **The Fourier approximations will overshoot the jump**. This Gibbs phenomenon doesn't get better by including more sines and cosines in the approximation.

Legendre Polynomials and Chebyshev Polynomials

The Legendre polynomials are the result of applying the Gram-Schmidt idea (Section 4.4). The plan is to orthogonalize the powers $1, x, x^2, \ldots$ To start, the odd function $\boldsymbol{x}$ is already orthogonal to the even function 1 over the interval from -1 to 1. Their product $(x)(1) = x$ integrates to zero. But the inner product between x^2 and 1 is $\int x^2\,dx = 2/3$:

$$\frac{(x^2,1)}{(1,1)} = \frac{\int x^2\,dx}{\int 1\,dx} = \frac{2/3}{2} = \frac{1}{3} \qquad \text{Gram-Schmidt chooses } \boldsymbol{x^2 - \frac{1}{3}} = \textbf{Legendre}$$

Similarly the odd power x^3 has a component $3x/5$ in the direction of the odd function x:

$$\frac{(x^3,x)}{(x,x)} = \frac{\int x^4\,dx}{\int x^2\,dx} = \frac{2/5}{2/3} = \frac{3}{5} \qquad \text{Gram-Schmidt chooses } \boldsymbol{x^3 - \frac{3}{5}x} = \textbf{Legendre}$$

Continuing Gram-Schmidt for $x^4, x^5, \ldots$ produces the orthogonal Legendre functions.

Finally we turn to the Chebyshev polynomials $\boldsymbol{1, x, 2x^2 - 1, 4x^3 - 3x}$. They don't come from Gram-Schmidt. Instead they are connected to $1, \cos\theta, \cos 2\theta, \cos 3\theta$. This gives a giant computational advantage—we can use the Fast Fourier Transform. The connection of Chebyshev to Fourier appears when we set $\boldsymbol{x = \cos\theta}$:

Chebyshev	$\boldsymbol{2x^2 - 1} = 2(\cos\theta)^2 - 1 = \boldsymbol{\cos 2\theta}$
to Fourier	$\boldsymbol{4x^3 - 3x} = 4(\cos\theta)^3 - 3(\cos\theta) = \boldsymbol{\cos 3\theta}$

The nth degree Chebyshev polynomial $\boldsymbol{T_n(x)}$ converts to $\boldsymbol{\cos n\theta = T_n(\cos\theta)}$.

Note These polynomials are the basis for a big software project called **"chebfun"**. Every function $f(x)$ is replaced by a super-accurate Chebyshev approximation. Then you can integrate $f(x)$, and solve $f(x) = 0$, and find its maximum or minimum. More than that, you can solve differential equations involving $f(x)$—fast and to high accuracy.

When **chebfun** replaces $f(x)$ by a polynomial, you are ready to solve problems.

■ REVIEW OF THE KEY IDEAS ■

1. A basis is good if its matrix B is well-conditioned. Orthogonal bases are best.
2. Also good if $\Lambda = B^{-1}AB$ is diagonal. But the Jordan form J is very unstable.
3. The Fourier matrix diagonalizes constant-coefficient periodic equations: perfection.
4. The basis $1, x, x^2, \ldots$ leads to $B^{\mathrm{T}}B =$ Hillbert matrix: Terrible for computations.
5. Legendre and Chebyshev polynomials are excellent bases for function space.

Problem Set 8.3

1 In Example 1, what is the rank of $J - 3I$? What is the dimension of its nullspace? This dimension gives the number of independent eigenvectors for $\lambda = 3$.

The algebraic multiplicity is 2, because $\det(J - \lambda I)$ has the repeated factor $(\lambda - 3)^2$. The geometric multiplicity is 1, because there is only 1 independent eigenvector.

2 These matrices A_1 and A_2 are similar to J. Solve $A_1 B_1 = B_1 J$ and $A_2 B_2 = B_2 J$ to find the basis matrices B_1 and B_2 with $J = B_1^{-1} A_1 B_1$ and $J = B_2^{-1} A_2 B_2$.

$$J = \begin{bmatrix} 0 & 1 \\ 0 & 0 \end{bmatrix} \quad A_1 = \begin{bmatrix} 0 & 4 \\ 0 & 0 \end{bmatrix} \quad A_2 = \begin{bmatrix} 4 & -8 \\ 2 & -4 \end{bmatrix} \quad \begin{matrix} \text{trace } 0 + 0 \\ \text{determinant } 0 \end{matrix}$$

3 This transpose block J^{T} has the same triple eigenvalue 2 (with only one eigenvector) as J. **Find the basis change B** so that $J = B^{-1} J^{\mathrm{T}} B$ (which means $BJ = J^{\mathrm{T}} B$):

$$\begin{matrix} J^{\mathrm{T}} \text{ is similar to} \\ J = B^{-1} J^{\mathrm{T}} B \end{matrix} \qquad J = \begin{bmatrix} 2 & 1 & 0 \\ 0 & 2 & 1 \\ 0 & 0 & 2 \end{bmatrix} \qquad J^{\mathrm{T}} = \begin{bmatrix} 2 & 0 & 0 \\ 1 & 2 & 0 \\ 0 & 1 & 2 \end{bmatrix}$$

4 J and K are Jordan forms with the same zero eigenvalues and the same rank 2. But show that no invertible B solves $BK = JB$, so **K is not similar to J**:

$$J = \begin{bmatrix} 0 & 1 & 0 & 0 \\ & 0 & 0 & 0 \\ & & 0 & 1 \\ & & & 0 \end{bmatrix} \qquad K = \begin{bmatrix} 0 & 1 & 0 & 0 \\ & 0 & 1 & 0 \\ & & 0 & 0 \\ & & & 0 \end{bmatrix} \qquad \begin{matrix} \text{Different} \\ \text{block sizes!} \end{matrix}$$

5 If $A^3 = 0$ show that all $\lambda = 0$, and all Jordan blocks with $J^3 = 0$ have size $1, 2$, or 3. It follows that rank $(A) \le 2n/3$. If $A^n = 0$ why is rank $(A) < n$?

6 Show that $\boldsymbol{u}(t) = \begin{bmatrix} te^{\lambda t} \\ e^{\lambda t} \end{bmatrix}$ solves $\dfrac{d\boldsymbol{u}}{dt} = J\boldsymbol{u}$ with $J = \begin{bmatrix} \lambda & 1 \\ 0 & \lambda \end{bmatrix}$ and $\boldsymbol{u}(0) = \begin{bmatrix} 0 \\ 1 \end{bmatrix}$.

J is not diagonalizable so $te^{\lambda t}$ enters the solution to $d\boldsymbol{u}/dt = J\boldsymbol{u}$.

7 Show that the difference equation $v_{k+2} - 2\lambda v_{k+1} + \lambda^2 v_k = 0$ is solved by $v_k = \lambda^k$ and also by $v_k = k\lambda^k$. Those correspond to $e^{\lambda t}$ and $te^{\lambda t}$ in Problem 6.

8 What are the 3 solutions to $\lambda^3 = 1$? They are complex numbers $\lambda = \cos\theta + i\sin\theta = e^{i\theta}$. Then $\lambda^3 = e^{3i\theta} = 1$ when **the angle 3θ is 0 or 2π or 4π**. Write the 3 by 3 Fourier matrix F with columns $(1, \lambda, \lambda^2)$.

9 Check that any 3 by 3 circulant C has eigenvectors $(1, \lambda, \lambda^2)$ from Problem 8. If the rows of your matrix C contain c_0, c_1, c_2 then its eigenvalues are in $F\boldsymbol{c}$.

10 Using formula (7) find $a_3 \cos 3x$ in the Fourier series of $f(x) = \begin{cases} 1 \text{ for } -L \le x \le L \\ 0 \text{ for } \quad L \le |x| \le 2\pi \end{cases}$

Chapter 9

Linear Algebra in Optimization

This chapter combines calculus with linear algebra. The overall problem is **to minimize a function $F(x)$** that depends on many variables $x = (x_1, x_2, \ldots, x_n)$. Suppose we know the first derivatives of F—those are partial derivatives and they go into the gradient vector $\nabla F = (\partial F/\partial x_1, \ldots, \partial F/\partial x_n)$. Then at a "smooth" minimum point x^* of $F(x)$, calculus tells us that $\nabla F(x^*) = \mathbf{0}$. That is a fundamental problem: **To solve n equations $\nabla F(x) = \mathbf{0}$ with n unknowns x_1 to x_n.**

An important model problem has a quadratic function $Q(x) = \frac{1}{2}x^{\mathrm{T}}Sx - b^{\mathrm{T}}x$. In this case the gradient vector is $\nabla Q(x) = Sx - b$. So $\nabla Q = \mathbf{0}$ is a set of n linear equations $Sx = b$. When S is a symmetric positive definite (constant) matrix, the solution x^* will be the minimum point for $Q(x)$.

When F is not quadratic, its gradient ∇F is not linear and its second derivative matrix $S(x)$ is not constant. But if $S(x)$ is everywhere positive definite, then F is a **convex function**. Its graph is still (roughly) a bowl. Section 9.1 follows $-\nabla F$ down towards x^*. And Section 9.2 shows how to compute that gradient ∇F by "backpropagation".

There is another serious possibility to deal with in 9.3 and 9.4. There can be restrictions (constraints) on the allowed solutions x. A model problem could minimize a quadratic $Q(x)$ or a linear cost $c^{\mathrm{T}}x$ with the constraints $Ax = b$ and $x \geq \mathbf{0}$ (meaning that every $x_i \geq 0$). Now the graph of $Q(x)$ (the bowl) is chopped off by m vertical planes $x_i = 0$. The key problem is to know if $x_i > 0$ or $x_i = 0$ at the minimum.

The way to deal with a constraint like $x_1 \geq 0$ is to introduce a Lagrange multiplier λ_1. Before that, Sections 9.1-2 explain **gradient descent** and **stochastic gradient descent**. Those are the workhorses of deep learning in Chapter 10, with many variables in x.

9.1 Minimizing a Multivariable Function

Suppose $F(x_1, \ldots, x_n)$ is a function of n variables. We need basic information about the first and second derivatives of F. Those are "partial derivatives" when $n > 1$.

The important facts are in equations (1) and (2). I don't believe you need a whole course (too much about integrals) to use these facts in minimizing the function $F(x)$.

One function F
One variable x

$$F(x+\Delta x) \approx F(x) + \Delta x \frac{dF}{dx}(x) + \frac{1}{2}(\Delta x)^2 \frac{d^2F}{dx^2}(x) \qquad (1)$$

This is the beginning of a Taylor series—and we don't often go beyond that second order term. The first terms $F(x)+(\Delta x)(dF/dx)$ give a *first order* approximation to $F(x+\Delta x)$, using information at x. Then $1/2\,(\Delta x)^2 F''$ makes it a *second order* approximation.

The Δx term gives a point on the tangent line—tangent to the graph of $F(x)$. The $(\Delta x)^2$ term moves from the tangent line to the "tangent parabola". The function F will be **convex**—its slope increases and its graph bends upward, as in $y = x^2$—when the second derivative is positive: $d^2F/dx^2 > 0$. **Equation (2) lifts (1) into n dimensions.**

One function F
Variables x_1 to x_n

$$F(x+\Delta x) \approx F(x) + (\Delta x)^{\mathrm{T}}\, \nabla F + \frac{1}{2}(\Delta x)^{\mathrm{T}}\, S\, (\Delta x) \qquad (2)$$

This is the important formula! **The vector ∇F is the gradient of F**—the column vector of n partial derivatives $\partial F/\partial x_1$ to $\partial F/\partial x_n$. **S is the symmetric matrix** (Hessian matrix) **of second derivatives** $S_{ij} = \partial^2 F/\partial x_i\,\partial x_j = \partial^2 F/\partial x_j\,\partial x_i$.

The graph of $y = F(x_1, \ldots, x_n)$ is now a surface in $(n+1)$-dimensional space. The tangent line becomes a tangent plane at x. When the second derivative matrix S is positive definite, F is a *strictly convex function*: *it stays above its tangents.* A convex function F has a **minimum** at x^* if $\nabla F(x^*) = 0$: n equations for x^*.

This third formula from calculus goes directly to our vector equation $\nabla F(x) = 0$. Starting from a point x, we want to take a step Δx that brings us close to the solution.

$$\nabla F(x+\Delta x) \approx \nabla F(x) + S(x)\,\Delta x = 0 \;\; \textbf{leads to} \;\; \Delta x = -\,S^{-1}(x)\nabla F(x) \qquad (3)$$

That formula for Δx tells us one step in **Newton's Method**. We are at a point x_k and we move to $x_{k+1} = x_k + \Delta x$. The points $x_0, x_1, x_2, \ldots$ often converge quickly to the minimum point x^* where $\nabla F(x^*) = 0$. But the cost of fast convergence can be very high, to compute at each step all the second derivatives $\partial^2 F/\partial x_i \partial x_j$ in the matrix $S(x)$.

Example 1 Our model function $F(x) = \frac{1}{2}x^{\mathrm{T}}Sx - x^{\mathrm{T}}b$ has $\nabla F = Sx - b$. Newton's method using equation (3) solves $\nabla F(x^*) = Sx^* - b = 0$ in one step!

$$x_1 - x_0 = \Delta x = -S^{-1}\nabla F(x_0) = -S^{-1}(Sx_0 - b) = -x_0 + S^{-1}b$$

Cancel $-x_0$ to see the point x_1 where $\nabla F = 0$: it is $x_1 = S^{-1}b = x^*$.

Example 2 Apply Newton's method to solve $\nabla F = x^2 - 4 = 0$. The derivative is $S(x) = 2x$. Step k is $\Delta x = -(1/S(x_k))\nabla F(x_k)$:

$$x_{k+1} - x_k = (-1/2x_k)\left(x_k^2 - 4\right) \quad \text{or} \quad \boldsymbol{x_{k+1} = \frac{1}{2}\left(x_k + \frac{4}{x_k}\right)}. \tag{4}$$

The square root is $x^* = \sqrt{4} = 2 = \frac{1}{2}\left(2 + \frac{4}{2}\right)$. **The key point is fast convergence to x^*. Start from $x_0 = 4$:**

$$\boldsymbol{x_1} = \frac{1}{2}\left(x_0 + \frac{4}{x_0}\right) = \frac{1}{2}\left(4 + \frac{4}{4}\right) = \mathbf{2.5}$$

$$\boldsymbol{x_2} = \frac{1}{2}\left(x_1 + \frac{4}{x_1}\right) = \frac{1}{2}\left(2.5 + \frac{4}{2.5}\right) = \frac{1}{2}(2.5 + 1.6) = \mathbf{2.05}$$

If you take 2 more steps, you see how quickly the error (the distance from $\sqrt{4} = 2$) goes to zero. The count of zeros after the decimal point in $x_k - 2$ doubles at every step.

$$x_0 = \mathbf{4} \qquad x_1 = \mathbf{2.5} \qquad x_2 = \mathbf{2.05} \qquad x_3 = \mathbf{2.0006} \qquad x_4 = \mathbf{2.000000009}$$

The wrong decimal is twice as far out at each step. **The error $x_k - 2$ is squared:**

$$\boxed{\boldsymbol{x_{k+1}} - \mathbf{2} = \frac{1}{2}\left(x_k + \frac{4}{x_k}\right) - 2 = \frac{1}{2x_k}(\boldsymbol{x_k} - \mathbf{2})^{\mathbf{2}}} \qquad \boxed{||\boldsymbol{x}_{k+1} - \boldsymbol{x}^*|| \approx \frac{1}{4}||\boldsymbol{x}_k - \boldsymbol{x}^*||^2}$$

Squaring the error explains the speed of Newton's method—*provided $\boldsymbol{x}_k$ is close to $\boldsymbol{x}^*$.*

Summary Newton's method is fast near the true solution $\boldsymbol{x}^*$, because it uses the second derivatives of $\boldsymbol{F}(\boldsymbol{x})$. But those can be too expensive to compute—especially in high dimensions. Often neural networks are simply too large to use all the second derivatives of $\boldsymbol{F}(\boldsymbol{x})$. *Gradient descent* is the algorithm of choice for deep learning.

Gradient Descent = Steepest Descent

This section is about a fundamental problem: **Minimize a function $\boldsymbol{F(x_1, \ldots, x_n)}$.** Calculus teaches us that all the first derivatives $\partial \boldsymbol{F}/\partial x_i$ are zero at the minimum (when $\boldsymbol{F}$ is smooth). If we have $n = 20$ unknowns (a small number in deep learning), then minimizing one function $\boldsymbol{F}$ produces 20 equations $\partial \boldsymbol{F}/\partial x_i = 0$. *"Gradient descent" uses the derivatives $\partial \boldsymbol{F}/\partial x_i$ to find a direction that reduces $\boldsymbol{F}(\boldsymbol{x})$.* The steepest direction, in which $\boldsymbol{F}(\boldsymbol{x})$ decreases fastest at $\boldsymbol{x}_k$, is given by the negative gradient $-\boldsymbol{\nabla F}(\boldsymbol{x}_k)$:

$$\textbf{Steepest Direction} \qquad \boxed{\textbf{Gradient descent} \qquad \boldsymbol{x_{k+1} = x_k - s_k \nabla F(x_k)}} \tag{5}$$

(5) is a vector equation for each step $k = 1, 2, 3, \ldots$ and s_k is the *stepsize* or the *learning rate*. We hope to move toward the point $\boldsymbol{x}^*$ where the graph of $\boldsymbol{F}(\boldsymbol{x})$ hits bottom.

We are willing to assume for now that 20 first derivatives exist and can be computed. We are not willing to assume that those 20 functions also have 20 convenient derivatives $\partial/\partial x_j(\partial \boldsymbol{F}/\partial x_i)$. Those are the 210 **second derivatives** of $\boldsymbol{F}$—which go into a 20 by 20 symmetric matrix $\boldsymbol{S}$. (Symmetry reduces $n^2 = 400$ to $\frac{1}{2}(n^2+n) = 210$ derivatives.) The second derivatives would be very useful extra information, but in many problems we have to go without.

You should know that 20 first derivatives and 210 second derivatives don't multiply the computing cost by 20 and 210. The neat idea of **automatic differentiation**—rediscovered and extended as **backpropagation** in machine learning—makes the cost much smaller.

Return for a moment to equation (5). The step $-s_k \boldsymbol{\nabla f}(\boldsymbol{x_k})$ includes a minus sign (to descend) and a factor s_k (to control the the stepsize) and the downhill vector $\boldsymbol{\nabla F}$ (containing the first derivatives of $\boldsymbol{F}$ computed at the current point $\boldsymbol{x}_k$). A lot of thought and computational experience have gone into the choice of stepsize and search direction.

The Derivative of $f(x)$: $n = 1$

The derivative of $f(x)$ involves a *limit*—this is the key difference between calculus and algebra. We are comparing the values of f at two nearby points x and $x + \Delta x$, as Δx approaches zero. More accurately, we are watching the slope $\Delta f/\Delta x$ between two points on the graph of $f(x)$:

$$\textbf{Derivative of } \boldsymbol{f} \textbf{ at } \boldsymbol{x} \qquad \frac{\boldsymbol{df}}{\boldsymbol{dx}} = \textbf{limit of } \frac{\boldsymbol{\Delta f}}{\boldsymbol{\Delta x}} = \textbf{limit of } \left[\frac{\boldsymbol{f(x+\Delta x) - f(x)}}{\boldsymbol{\Delta x}}\right]. \tag{6}$$

This is a **forward** difference when $\Delta x > 0$. It is a **backward** difference when $\Delta x < 0$. When we have the same limit from both sides, that number is the slope of the graph at x.

The ramp function $\textbf{ReLU}(x) = f(x) = \textbf{max}(\boldsymbol{0}, \boldsymbol{x})$ is heavily involved in deep learning. ReLU has unequal slopes **1** to the right and **0** to the left of $x = 0$. So the derivative of ReLU does not exist at that corner point in the graph.

For the smooth function $\boldsymbol{f(x) = x^2}$, the ratio $\Delta f/\Delta x$ will safely approach the derivative df/dx from both sides. But the approach could be slow (just first order). Look again at the point $x = 0$, where the true derivative $df/dx = 2x$ is now zero:

$$\text{The ratio } \frac{\Delta f}{\Delta x} \text{ at } x = 0 \text{ is } \frac{f(\Delta x) - f(0)}{\Delta x} = \frac{(\Delta x)^2 - 0}{\Delta x} = \boldsymbol{\Delta x} \text{ Then limit} = \frac{df}{dx} = 0.$$

As explained in Section 2.5, we get a better ratio (closer to the limiting slope df/dx) by averaging the **forward difference** (where $\Delta x > 0$) with the **backward difference** (where $\Delta x < 0$). Their average is a more accurate **centered difference**.

$$\begin{matrix}\textbf{Centered} \\ \textbf{at } \boldsymbol{x}\end{matrix} \quad \frac{1}{2}\left[\frac{f(x+\Delta x)-f(x)}{\Delta x} + \frac{f(x-\Delta x)-f(x)}{-\Delta x}\right] = \frac{\boldsymbol{f(x+\Delta x) - f(x-\Delta x)}}{\boldsymbol{2\,\Delta x}} \tag{7}$$

For the example $f(x) = x^2$ this centering will produce the exact derivative $df/dx = 2x$. In the picture we are averaging plus and minus slopes to get the correct slope 0 at $x = 0$.

For all smooth functions, the centered differences reduce the error to size $(\Delta x)^2$. This is a big improvement over the error of size Δx for uncentered differences $f(x+\Delta x)-f(x)$.

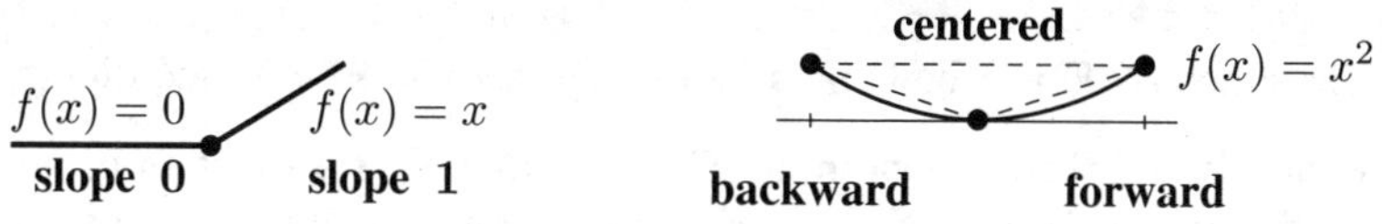

Figure 9.1: ReLU function = Ramp from deep learning. Centered slope of $f=x^2$ is exact.

Most finite difference approximations are centered for extra accuracy. But we are still dividing by a small number $2\,\Delta x$. And for a multivariable function $\boldsymbol{F}(x_1, x_2, \ldots, x_n)$ we will need ratios $\Delta \boldsymbol{F}/\Delta x_i$ in n different directions—possibly for large n. Those ratios approximate the n partial derivatives that go into the **gradient vector grad $\boldsymbol{F} = \boldsymbol{\nabla F}$**.

$$\textbf{The gradient of } \boldsymbol{F(x_1,\ldots,x_n)} \textbf{ is the column vector } \boldsymbol{\nabla F} = \left(\frac{\partial \boldsymbol{F}}{\partial \boldsymbol{x_1}}, \ldots, \frac{\partial \boldsymbol{F}}{\partial \boldsymbol{x_n}}\right).$$

Its components are the n partial derivatives of $\boldsymbol{F}$. $\boldsymbol{\nabla} F$ points in the steepest direction.

Examples 3-5 will show the value of vector notation ($\boldsymbol{\nabla F}$ is always a column vector).

Example 3 For a constant column vector $\boldsymbol{a}$, $\boldsymbol{F}(\boldsymbol{x}) = \boldsymbol{a}^{\mathrm{T}}\boldsymbol{x}$ has gradient $\boldsymbol{\nabla F} = \boldsymbol{a}$. The partial derivatives of $\boldsymbol{F} = a_1x_1 + \cdots + a_nx_n$ are the numbers $\partial \boldsymbol{F}/\partial x_k = a_k$.

Example 4 For a symmetric matrix $\boldsymbol{S}$, the gradient of $\boldsymbol{F} = \frac{1}{2}\boldsymbol{x}^{\mathrm{T}}\boldsymbol{S}\boldsymbol{x}$ is $\boldsymbol{\nabla F} = \boldsymbol{S}\boldsymbol{x}$. To see this for $n=2$, write out the function $\boldsymbol{F}(x_1, x_2)$. The matrix $\boldsymbol{S}$ is 2 by 2:

$$\boldsymbol{F} = \frac{\begin{bmatrix} x_1 & x_2 \end{bmatrix}}{2}\begin{bmatrix} a & b \\ b & c \end{bmatrix}\begin{bmatrix} x_1 \\ x_2 \end{bmatrix} = \begin{matrix} \frac{1}{2}ax_1^2 + \frac{1}{2}cx_2^2 \\ + bx_1x_2 \end{matrix} \quad \begin{bmatrix} \partial f/\partial x_1 \\ \partial f/\partial x_2 \end{bmatrix} = \begin{bmatrix} ax_1 + bx_2 \\ bx_1 + cx_2 \end{bmatrix} = \boldsymbol{S}\begin{bmatrix} \boldsymbol{x_1} \\ \boldsymbol{x_2} \end{bmatrix}.$$

Example 5 For a positive definite symmetric matrix $\boldsymbol{S}$, the minimum of a quadratic $\boldsymbol{F}(\boldsymbol{x}) = \frac{1}{2}\,\boldsymbol{x}^{\mathrm{T}}\boldsymbol{S}\boldsymbol{x} - \boldsymbol{a}^{\mathrm{T}}\boldsymbol{x}$ is the negative number $\boldsymbol{F}_{\textbf{min}} = -\frac{1}{2}\,\boldsymbol{a}^{\mathrm{T}}\boldsymbol{S}^{-1}\boldsymbol{a}$.

This is an important example! The gradient is $\boldsymbol{\nabla F} = \boldsymbol{S}\boldsymbol{x} - \boldsymbol{a}$. The minimum occurs at $\boldsymbol{x}^* = \boldsymbol{S}^{-1}\boldsymbol{a}$ where first derivatives of $\boldsymbol{F}$ are zero:

$$\boldsymbol{\nabla F} = \begin{bmatrix} \partial \boldsymbol{F}/\partial \boldsymbol{x_1} \\ \vdots \\ \partial \boldsymbol{F}/\partial \boldsymbol{x_n} \end{bmatrix} = \boldsymbol{S}\boldsymbol{x} - \boldsymbol{a} = \boldsymbol{0} \textbf{ at } \boldsymbol{x}^* = \boldsymbol{S}^{-1}\boldsymbol{a} = \textbf{arg min } \boldsymbol{F}. \tag{8}$$

As always, **that notation arg min $\boldsymbol{F}$ stands for the point $\boldsymbol{x}^*$ where the minimum of $\boldsymbol{F}(\boldsymbol{x})$ is reached.** Often we are more interested in this minimizing point $\boldsymbol{x}^*$ than in the actual minimum value $\boldsymbol{F}_{\textbf{min}} = \boldsymbol{F}(\boldsymbol{x}^*)$ at that point.

$$F_{\min} \text{ is } \frac{1}{2}(S^{-1}\boldsymbol{a})^{\mathrm{T}}S(S^{-1}\boldsymbol{a}) - \boldsymbol{a}^{\mathrm{T}}(S^{-1}\boldsymbol{a}) = \frac{1}{2}\boldsymbol{a}^{\mathrm{T}}S^{-1}\boldsymbol{a} - \boldsymbol{a}^{\mathrm{T}}S^{-1}\boldsymbol{a} = -\frac{\mathbf{1}}{\mathbf{2}}\boldsymbol{a}^{\mathbf{T}}\boldsymbol{S}^{-\mathbf{1}}\boldsymbol{a}.$$

The graph of $\boldsymbol{F}$ is a bowl passing through zero at $\boldsymbol{x} = \mathbf{0}$ and dipping to $\boldsymbol{F}_{\text{min}}$ at $\boldsymbol{x}^*$.

Example 6 The **determinant** $\boldsymbol{F}(\boldsymbol{x}) = \det \boldsymbol{X}$ is a function of all n^2 variables x_{ij}. In the formula for $\det \boldsymbol{X}$, each x_{ij} along a row is multiplied by its "cofactor" C_{ij}. This cofactor in Chapter 5 is a determinant of size $n-1$, using all the entries in $\boldsymbol{X}$ except row i and column j—and C_{ij} also includes the plus or minus sign $(-1)^{i+j}$:

$$\text{Partial derivatives } \boldsymbol{\nabla F} = \frac{\partial(\det \boldsymbol{X})}{\partial x_{ij}} = \boldsymbol{C}_{ij} \text{ in the matrix } \boldsymbol{C} \text{ of cofactors of } \boldsymbol{X}.$$

Example 7 **The logarithm of the determinant is a most remarkable function:**

$$\boldsymbol{L}(\boldsymbol{X}) = \log(\det \boldsymbol{X}) \text{ has partial derivatives } \frac{\partial \boldsymbol{L}}{\partial x_{ij}} = \frac{\boldsymbol{C}_{ij}}{\det \boldsymbol{X}} = \boldsymbol{j}, \boldsymbol{i} \textbf{ entry of } \boldsymbol{X}^{-1}.$$

The chain rule for $\boldsymbol{L} = \log \boldsymbol{F}$ is $(\partial \boldsymbol{L}/\partial \boldsymbol{F})(\partial \boldsymbol{F}/\partial x_{ij}) = (1/\boldsymbol{F})(\partial \boldsymbol{F}/\partial x_{ij}) = (1/\det \boldsymbol{X})\,\boldsymbol{C}_{ij}$. Then this ratio of cofactor C_{ij} to $\det X$ gives the j,i entry of the inverse matrix $\boldsymbol{X}^{-1}$.

It is neat that $\boldsymbol{X}^{-1}$ contains the n^2 first derivatives of $\boldsymbol{L} = \log\det \boldsymbol{X}$. The second derivatives of $\boldsymbol{L}$ are remarkable too. We have n^2 variables x_{ij} and n^2 first derivatives in $\boldsymbol{\nabla L} = (\boldsymbol{X}^{-1})^{\mathrm{T}}$. This means n^4 second derivatives! What is amazing is that the matrix of second derivatives is **negative definite** when $\boldsymbol{X} = \boldsymbol{S}$ is symmetric positive definite. So we reverse the sign of $\boldsymbol{L}$: **positive definite second derivatives $\Rightarrow$ convex function.**

$-\log(\det S)$ is a convex function of the entries of the positive definite matrix S.

The Geometry of the Gradient Vector ∇F

Start with a function $\boldsymbol{F}(x,y)$ of $n = 2$ variables. Its gradient is $\boldsymbol{\nabla F} = (\partial \boldsymbol{F}/\partial \boldsymbol{x}, \partial \boldsymbol{F}/\partial \boldsymbol{y})$. This vector changes length as we move the point x, y where the derivatives are computed:

$$\boldsymbol{\nabla F} = \left(\frac{\partial F}{\partial x}, \frac{\partial F}{\partial y}\right) \qquad \textbf{Length} = ||\boldsymbol{\nabla F}|| = \sqrt{\left(\frac{\partial F}{\partial x}\right)^2 + \left(\frac{\partial F}{\partial y}\right)^2} = \text{steepest slope of } \boldsymbol{F}$$

That length $||\boldsymbol{\nabla F}||$ tells us the steepness of the graph of $z = \boldsymbol{F}(x,y)$. The graph is normally a curved surface—like a valley in xyz space. At each point there is a slope $\partial F/\partial x$ in the x-direction and a slope $\partial F/\partial y$ in the y-direction. **The steepest slope is in the direction of $\boldsymbol{\nabla F}$ = grad $\boldsymbol{F}$. The magnitude of that steepest slope is $||\boldsymbol{\nabla F}||$.**

Example 8 The graph of a linear function $\boldsymbol{F}(x,y) = ax + by$ is the plane $z = ax + by$. The gradient is the vector $\boldsymbol{\nabla F} = \begin{bmatrix} a \\ b \end{bmatrix}$ of partial derivatives. The length of that vector is $||\boldsymbol{\nabla F}|| = \sqrt{\boldsymbol{a}^2 + \boldsymbol{b}^2}$ = **slope of the roof**. The slope is steepest in the direction of $\boldsymbol{\nabla F}$.

That steepest direction is perpendicular to the level direction. The level direction $z =$ constant has $ax + by =$ constant. It is the safe direction to walk, perpendicular to $\boldsymbol{\nabla F}$. The component of $\boldsymbol{\nabla F}$ in that flat direction is zero. Figure 9.2 shows the two perpendicular directions (level and steepest) on the plane $\boldsymbol{z} = \boldsymbol{F}(x, y) = \boldsymbol{x} + \boldsymbol{2y}$.

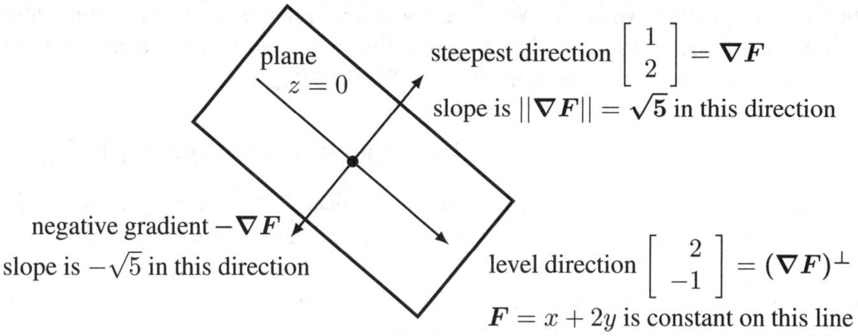

Figure 9.2: The negative gradient $-\boldsymbol{\nabla F}$ gives the direction of steepest descent.

Our basic Example 9 will be $\boldsymbol{F}(x, y) = \boldsymbol{ax^2} + \boldsymbol{by^2}$. Its gradient is $\boldsymbol{\nabla F} = \begin{bmatrix} \boldsymbol{2ax} \\ \boldsymbol{2by} \end{bmatrix}$.

$\boldsymbol{\nabla F}$ tells us the steepest direction, changing from point to point. We are on a curved surface (a bowl opening upward). The bottom of the bowl is at $x = y = 0$ where the gradient vector is zero. The slope in the steepest direction is $||\boldsymbol{\nabla F}||$. At the minimum, $\boldsymbol{\nabla F} = (2ax, 2by) = (0, 0)$ and *slope = zero*. The bowl is circular if $a = b$.

The level direction has $\boldsymbol{z} = \boldsymbol{ax^2} + \boldsymbol{by^2} =$ constant height. That plane $z =$ constant cuts through the bowl in a level curve. In this example the level curve $ax^2 + by^2 = c$ is an ellipse. But there is a serious difficulty for steepest descent:

The steepest direction changes as you go down! The gradient doesn't point to the bottom!

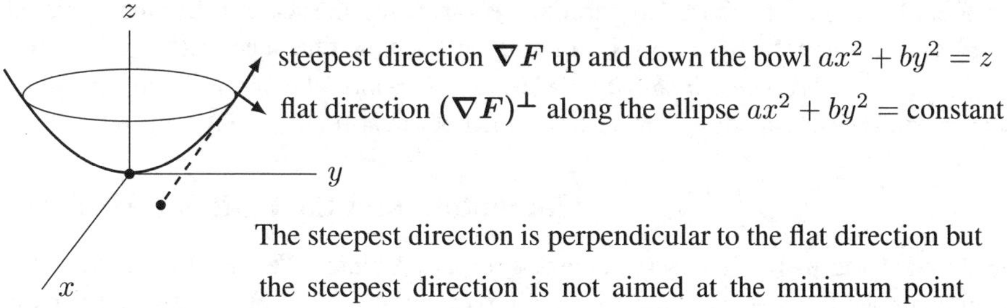

Figure 9.3: **Steepest descent moves down the bowl in the gradient direction** $\begin{bmatrix} \boldsymbol{-2ax} \\ \boldsymbol{-2by} \end{bmatrix}$.

Let me repeat. At the point x_0, y_0 the gradient direction for $\boldsymbol{F} = ax^2 + by^2$ is along $\boldsymbol{\nabla F} = (2ax_0, 2by_0)$. The steepest line is $(x, y) = (x_0, y_0) - s(2ax_0, 2by_0)$ for all s. But if $a \neq b$, the lowest point $(x, y) = (0, 0)$ **does not lie on that line** for any s.

We will not find that minimum point $(0, 0)$ in one step of "gradient descent". The steepest direction does not lead to the bottom of the bowl—except when $b = a$ and the bowl is circular.

Water changes direction as it goes down a mountain. Sooner or later, we must change direction too. In practice we keep going in the gradient direction and stop when our cost function $\boldsymbol{F}$ is not decreasing quickly. At that point Step 1 ends and we recompute the gradient $\nabla \boldsymbol{F}$. This gives a new descent direction for Step 2.

An Important Example with Zig-Zag

The example $\boldsymbol{F}(x, y) = \frac{1}{2}(\boldsymbol{x}^2 + \boldsymbol{b}\boldsymbol{y}^2)$ is extremely useful for $0 < b \leq 1$. Its gradient $\nabla \boldsymbol{F}$ has two components $\partial \boldsymbol{F}/\partial x = \boldsymbol{x}$ and $\partial \boldsymbol{F}/\partial y = \boldsymbol{b}\boldsymbol{y}$. The minimum value of $\boldsymbol{F}$ is zero. That minimum is reached at the point $(x^*, y^*) = (0, 0)$. Best of all, steepest descent with exact line search (the best s) produces a simple formula for each (x_k, y_k) in the slow progress down the bowl toward $(0, 0)$. Starting at $(x_0, y_0) = (b, 1)$ we find these points:

$$\boxed{\boldsymbol{x_k} = \boldsymbol{b}\left(\frac{\boldsymbol{b}-1}{\boldsymbol{b}+1}\right)^{\boldsymbol{k}} \quad \boldsymbol{y_k} = \left(\frac{1-\boldsymbol{b}}{1+\boldsymbol{b}}\right)^{\boldsymbol{k}} \quad \boldsymbol{F}(\boldsymbol{x_k}, \boldsymbol{y_k}) = \left(\frac{1-\boldsymbol{b}}{1+\boldsymbol{b}}\right)^{2\boldsymbol{k}} \boldsymbol{F}(\boldsymbol{x_0}, \boldsymbol{y_0})} \tag{9}$$

If $b = 1$, you see immediate success in one step. The point (x_1, y_1) is $(0, 0)$. The bowl is perfectly circular with $F = \frac{1}{2}(x^2 + y^2)$. The negative gradient direction goes exactly through $(0, 0)$. Then the first step of gradient descent finds that correct minimizing point.

The real purpose of this example is seen when b is small. The crucial ratio in equation (9) is $r = (1 - b)/(1 + b)$. For $b = \frac{1}{10}$ this ratio is $r = 9/11$. For $b = \frac{1}{100}$ the ratio is $99/101$. The ratio is approaching 1 and the progress toward $(0, 0)$ has virtually stopped when b is very small.

Figure 9.4 shows the frustrating zig-zag pattern of the steps toward $(0, 0)$. Every step is short and progress is very slow. This is a case where the stepsize s_k in $\boldsymbol{x}_{k+1} = \boldsymbol{x}_k - s_k \nabla \boldsymbol{F}(\boldsymbol{x}_k)$ was exactly chosen to minimize F (an exact line search). But the direction of $-\nabla \boldsymbol{F}$, even if steepest at (x_k, y_k), is pointing far from the final answer $(\boldsymbol{x}^*, \boldsymbol{y}^*) = (0, 0)$.

The bowl has become a narrow valley when b is small. We are uselessly crossing the valley instead of moving down the valley to the bottom at $(0, 0)$.

Momentum and the Path of a Heavy Ball

The slow zig-zag path of steepest descent is a real problem. We have to improve it. Our model example $\boldsymbol{F} = \frac{1}{2}(x^2 + by^2)$ has only two variables x, y and its second derivative matrix is diagonal—constant entries $\boldsymbol{F}_{\boldsymbol{xx}} = 1$ and $\boldsymbol{F}_{\boldsymbol{yy}} = b$. But it shows the zig-zag problem very clearly when $\boldsymbol{b} = \boldsymbol{\lambda}_{\text{min}}/\boldsymbol{\lambda}_{\text{max}}$ **is small.**

Key idea: Zig-zag would not happen for a heavy ball rolling downhill. Its momentum carries it through the narrow valley—bumping the sides but moving mostly forward. So we **add momentum with coefficient β to the gradient** (Polyak's important idea). This gives one of the most convenient and potentially useful ideas in deep learning.

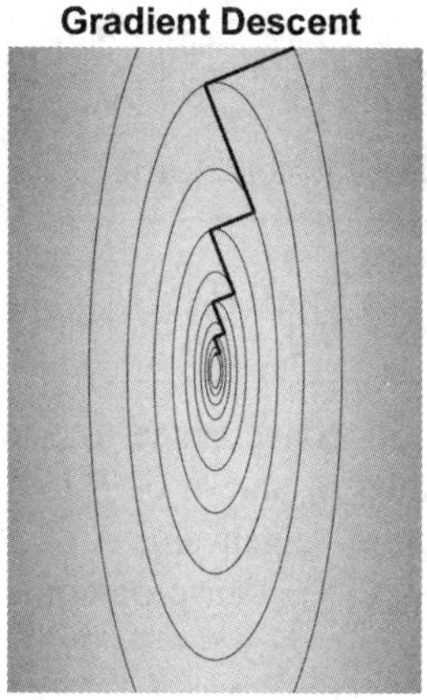

The first descent step starts out perpendicular to the level set. As it crosses through lower level sets, the function $F(x, y)$ is decreasing. **Eventually its path is tangent to a level set L.** Descent has stopped. Going further will increase F. The first step ends. The next step is perpendicular to L. So the zig-zag path took a $90°$ turn.

Note to the reader Deep learning in Chapter 10 involves a *sum of many functions* $\boldsymbol{F}(\boldsymbol{x}, \boldsymbol{v}_i)$—one for every vector $\boldsymbol{v}_i$ in the training set. The the gradient is a sum of many gradients—too many to compute at each step. The solution in Section 9.2 is to randomly (**"stochastically"**) choose only one or a few functions at every descent step.

Figure 9.4: Slow convergence on a zig-zag path to the minimum of $\boldsymbol{F} = \boldsymbol{x^2} + \boldsymbol{by^2}$.

Key idea: The direction $\boldsymbol{z}_k$ of the new step remembers the previous direction $\boldsymbol{z}_{k-1}$.

Descent with momentum $$\boxed{\boldsymbol{x}_{k+1} = \boldsymbol{x}_k - s\boldsymbol{z}_k \text{ with } \boldsymbol{z}_k = \nabla \boldsymbol{F}(\boldsymbol{x}_k) + \beta \boldsymbol{z}_{k-1}} \quad (10)$$

Now we have two coefficients to choose—the stepsize $\boldsymbol{s}$ and also $\boldsymbol{\beta}$. Most important, **the step to $\boldsymbol{x}_{k+1}$ in equation (10) involves $\boldsymbol{z}_{k-1}$.** Momentum has turned a one-step method (gradient descent) into a two-step method. To get back to one step, we have to rewrite equation (10) as **two equations** (one vector equation) for $(\boldsymbol{x}, \boldsymbol{z})$ at time $k+1$:

Descent with momentum $$\boxed{\begin{aligned} \boldsymbol{x}_{k+1} &= \boldsymbol{x}_k - s\boldsymbol{z}_k \\ \boldsymbol{z}_{k+1} - \nabla \boldsymbol{F}(\boldsymbol{x}_{k+1}) &= \beta \boldsymbol{z}_k \end{aligned}} \quad (11)$$

With those two equations, we have recovered a one-step method. This is exactly like reducing a single second order differential equation to a system of two first order equations.

The Quadratic Model

When $\boldsymbol{F}(\boldsymbol{x}) = \frac{1}{2}\boldsymbol{x}^{\mathrm{T}}S\boldsymbol{x}$ is quadratic, its gradient $\nabla \boldsymbol{F} = S\boldsymbol{x}$ is linear. This is the model problem to understand: S is symmetric positive definite and $\nabla \boldsymbol{F}(\boldsymbol{x}_{k+1})$ becomes $S\boldsymbol{x}_{k+1}$ in equation (11). Our 2 by 2 supermodel is included, when the matrix S is diagonal with entries 1 and b. For every matrix S, you will see that its largest and smallest eigenvalues determine the best choices for β and the stepsize s—so our 2 by 2 case actually contains the essence of the whole problem.

To follow the steps of accelerated descent, *track each eigenvector $\boldsymbol{q}$ of S.* Suppose $S\boldsymbol{q} = \lambda\boldsymbol{q}$ and $\boldsymbol{x}_k = c_k\,\boldsymbol{q}$ and $\boldsymbol{z}_k = d_k\,\boldsymbol{q}$ and $\nabla \boldsymbol{F}_k = S\boldsymbol{x}_k = \lambda\, c_k\,\boldsymbol{q}$. Then our equation (11) connects the numbers c_k and d_k at step k to the next c_{k+1} and d_{k+1} at step $k+1$.

Following the eigenvector q
$$\begin{aligned} c_{k+1} &= c_k - s\,d_k \\ -\lambda\,c_{k+1} + d_{k+1} &= \beta\,d_k \end{aligned} \qquad \begin{bmatrix} \mathbf{1} & 0 \\ -\boldsymbol{\lambda} & \mathbf{1} \end{bmatrix} \begin{bmatrix} c_{k+1} \\ d_{k+1} \end{bmatrix} = \begin{bmatrix} \mathbf{1} & -s \\ 0 & \boldsymbol{\beta} \end{bmatrix} \begin{bmatrix} c_k \\ d_k \end{bmatrix} \tag{12}$$

Finally we invert the first matrix ($-\lambda$ becomes $+\lambda$) to see each descent step clearly:

Descent step multiplies by R
$$\begin{bmatrix} c_{k+1} \\ d_{k+1} \end{bmatrix} = \begin{bmatrix} 1 & 0 \\ \lambda & 1 \end{bmatrix} \begin{bmatrix} 1 & -s \\ 0 & \beta \end{bmatrix} \begin{bmatrix} c_k \\ d_k \end{bmatrix} = \begin{bmatrix} \mathbf{1} & -s \\ \boldsymbol{\lambda} & \boldsymbol{\beta - \lambda s} \end{bmatrix} \begin{bmatrix} c_k \\ d_k \end{bmatrix} = \boldsymbol{R} \begin{bmatrix} c_k \\ d_k \end{bmatrix} \tag{13}$$

After k steps the starting vector is multiplied by R^k. For fast convergence to zero (which is the minimum of $\boldsymbol{F} = \frac{1}{2}\boldsymbol{x}^{\mathrm{T}} S \boldsymbol{x}$) we want both eigenvalues e_1 and e_2 of R to be as small as possible. Clearly those eigenvalues of R depend on the eigenvalue λ of S. That eigenvalue λ could be anywhere between $\lambda_{\min}(S)$ and $\lambda_{\max}(S)$. Our problem is:

$$\textbf{Choose } \boldsymbol{s} \textbf{ and } \boldsymbol{\beta} \textbf{ to minimize max} \Big[|\boldsymbol{e_1(\lambda)}|, |\boldsymbol{e_2(\lambda)}| \Big] \quad \text{for } \lambda_{\min}(S) \le \lambda \le \lambda_{\max}(S).$$

It seems a miracle that this problem has a beautiful solution. The optimal s and β are

$$s = \left(\frac{2}{\sqrt{\lambda_{\max}} + \sqrt{\lambda_{\min}}} \right)^2 \quad \text{and} \quad \beta = \left(\frac{\sqrt{\lambda_{\max}} - \sqrt{\lambda_{\min}}}{\sqrt{\lambda_{\max}} + \sqrt{\lambda_{\min}}} \right)^2. \tag{14}$$

Think of the 2 by 2 supermodel, when S has eigenvalues $\lambda_{\max} = 1$ and $\lambda_{\min} = b$:

$$s = \left(\frac{2}{1 + \sqrt{b}} \right)^2 \quad \text{and} \quad \beta = \left(\frac{1 - \sqrt{b}}{1 + \sqrt{b}} \right)^2 \tag{15}$$

These choices of stepsize and momentum give a convergence rate that looks like the rate in equation (9) for ordinary steepest descent (no momentum). But there is a crucial difference: **The number b in (9) is replaced by $\sqrt{b}$ in (15).**

$$\textbf{Ordinary descent factor} \quad \left(\frac{\mathbf{1} - \boldsymbol{b}}{\mathbf{1} + \boldsymbol{b}} \right)^{\mathbf{2}} \qquad \textbf{Accelerated descent factor} \quad \left(\frac{\mathbf{1} - \sqrt{\boldsymbol{b}}}{\mathbf{1} + \sqrt{\boldsymbol{b}}} \right)^{\mathbf{2}} \tag{16}$$

So similar but so different. The real test comes when b is very small. Then the ordinary descent factor is essentially $1 - 4b$, very close to 1. The accelerated descent factor is essentially $1 - 4\sqrt{b}$, *much further below* 1.

To emphasize the improvement that momentum brings, suppose $b = 1/100$. Then $\sqrt{b} = 1/10$ (ten times larger than b). The convergence factors in equation (16) are

$$\textbf{Steepest descent} \quad \left(\frac{.99}{1.01} \right)^2 \approx \mathbf{.96} \qquad \textbf{Accelerated descent} \quad \left(\frac{.9}{1.1} \right)^2 \approx \mathbf{.67}$$

Ten steps of ordinary descent multiply the starting error by 0.67. This is matched by a single momentum step. Ten steps with the momentum term multiply the error by 0.018.

Notice that the **condition number** $\lambda_{\max}/\lambda_{\min} = 1/b$ of S controls everything.

Note 1 Nesterov also created an improvement on gradient descent—different from momentum.

Note 2 "Adaptive" methods use *all earlier choices* of the step direction to modify the current direction $\boldsymbol{\nabla L_k}$ for the step to $\boldsymbol{x}_{k+1}$. The formula for the current direction becomes $D_k = \delta D_{k-1} + (1-\delta)\boldsymbol{\nabla L}_k$. Class projects confirmed the speedup that comes from using the earlier directions which went into D_{k-1}.

The batch size B in stochastic gradient descent can increase at later steps.

Problem Set 9.1

1 For which functions $\boldsymbol{F}(x)$ are equations (1) and (2) exactly correct for all x ?

2 Write down Newton's method for the equation $x^2+1=0$. It can't converge (there is no real solution). If $x_n = \cos\theta/\sin\theta$ show that $x_{n+1} = \cos 2\theta/sin 2\theta$. This is an example of "chaos" in College Math J. 22 (1991) 3-12.

3 The determinant of a 1 by 1 matrix is just $\det X = x_{11}$. Find the first and second derivatives of $F(X) = -\log(\det X) = -\log x_{11}$ for $x_{11} > 0$. Sketch the graph of $F = -\log x$ to see that this function F is convex.

4 A symmetric 2 by 2 matrix has $\det(A) = ac - b^2$. Show from the matrix of second derivatives that $F = -\log\left(ac - b^2\right)$ is a convex function of a, b, c.

5 What is the gradient descent equation $\boldsymbol{x}_{k+1} = \boldsymbol{x}_k - s_k \nabla F(\boldsymbol{x}_k)$ for the least squares problem of minimizing $F(\boldsymbol{x}) = \frac{1}{2}||A\boldsymbol{x} - \mathbf{b}||^2$?

6 Find the gradient of F and the gradient descent equations for $F(x,y) = \frac{1}{2}\left(x^2 + \frac{1}{4}y^2\right)$. Starting from $(x_0, y_0) = \left(\frac{1}{4}, 1\right)$ confirm or improve equation (9) for (x_1, y_1).

7 Add momentum to Problem 6 and take one "accelerated" step : equations (10)–(11).

8 Show that this non-quadratic example has its minimum at $x=0$ and $y=+\infty$:

$$\boldsymbol{F}(x,y) = \frac{\mathbf{1}}{\mathbf{2}}\boldsymbol{x}^{\mathbf{2}} + \boldsymbol{e}^{-\boldsymbol{y}} \qquad \nabla \boldsymbol{F} = \begin{bmatrix} \boldsymbol{x} \\ -\boldsymbol{e}^{-\boldsymbol{y}} \end{bmatrix} \quad \boldsymbol{S} = \begin{bmatrix} 1 & 0 \\ 0 & e^{-y} \end{bmatrix}$$

Take one gradient descent step from $(x_0, y_0) = (1, 1)$.

9.2 Backpropagation and Stochastic Gradient Descent

This short section explains the key computational step in deep learning. The goal of that step is to optimize the set of **weights** $\boldsymbol{x}$ in the learning function $\boldsymbol{F}(\boldsymbol{x}, \boldsymbol{v})$. We are given the "training data"—a large set of input vectors $\boldsymbol{v}_1, \ldots, \boldsymbol{v}_N$. For each input vector $\boldsymbol{v}_i$ we know the correct output vector $\boldsymbol{w}_i$. Our goal is to choose the weights $\boldsymbol{x}$ so that the error vectors $\boldsymbol{e}_i = \boldsymbol{F}(\boldsymbol{x}, \boldsymbol{v}_i) - \boldsymbol{w}_i$ are as small as possible. We are fitting the learning function $\boldsymbol{F}$ to the N points of training data, by choosing weights $\boldsymbol{x}$ that **minimize the total loss $\boldsymbol{L}(\boldsymbol{x})$**:

$$\text{Minimize} \quad \boldsymbol{L}(\boldsymbol{x}) = \frac{1}{N}\sum_{i=1}^{N} \boldsymbol{\ell}(\boldsymbol{e}_i) = \frac{1}{N}\sum_{i=1}^{N} \boldsymbol{\ell}(\boldsymbol{F}(\boldsymbol{x}, \boldsymbol{v}_i) - \boldsymbol{w}_i). \tag{1}$$

The best weights $\boldsymbol{x}$ will depend on the form we choose for the learning function $\boldsymbol{F}(\boldsymbol{x}, \boldsymbol{v})$. The weights also depend on the loss function $\boldsymbol{\ell}$ that we choose to measure the errors $\boldsymbol{e}_i$. If we choose the "square loss" $\boldsymbol{\ell}(\boldsymbol{e}) = ||\boldsymbol{e}||^2$, our problem is to find a least squares fit of the training data $(\boldsymbol{v}_1, \boldsymbol{w}_1), \ldots, (\boldsymbol{v}_N, \boldsymbol{w}_N)$.

Here is a big picture of the problem. We are creating a function $\boldsymbol{F}$ that approximates or even "interpolates" the training data: $\boldsymbol{F}(\boldsymbol{v}_i) \approx \boldsymbol{w}_i$ for $i = 1, \ldots, N$. Then when we have **new test data $\boldsymbol{V}$**, we will estimate the unknown output $\boldsymbol{W}$ by our function $\boldsymbol{F}(\boldsymbol{V})$. Suppose we are trying to read a handwritten number $\boldsymbol{W}$. We have N pictures $\boldsymbol{v}_1, \ldots, \boldsymbol{v}_N$ of known numbers $\boldsymbol{w}_1, \ldots, \boldsymbol{w}_N$. Based on that training data, how do we weight the pixels in the new picture $\boldsymbol{V}$ to learn the number $\boldsymbol{W}$ in that picture?

Interpolation is an old problem. Deep learning is a new approach—much more successful than past approaches. One of the worst approaches is to fit the data by a very high degree polynomial F. The result is extremely unstable. A small change in the input $\boldsymbol{V}$ produces a large change in the output $\boldsymbol{W} = \boldsymbol{F}(\boldsymbol{V})$. We have chosen the wrong form (a polynomial) for the learning function $\boldsymbol{F}$. Chapter 10 will describe a much better choice for the form of $\boldsymbol{F}$. This section is about **"fitting the function $\boldsymbol{F}$ to the training data"**. We are choosing the weights $\boldsymbol{x}$ to achieve $\boldsymbol{F}(\boldsymbol{x}, \boldsymbol{v}_i) \approx \boldsymbol{w}_i$ for each data point.

Three big steps are involved in constructing and using the learning function $\boldsymbol{F}$:

1 Choose the form of $\boldsymbol{F}(\boldsymbol{x}, \boldsymbol{v})$. It includes a nonlinear function, often "ReLU".

2 Compute the weights $\boldsymbol{x}$ in $\boldsymbol{F}$ that best fit the training data $\boldsymbol{F}(\boldsymbol{x}, \boldsymbol{v}_i) \approx \boldsymbol{w}_i$.

3 Using those weights, evaluate $\boldsymbol{W} = \boldsymbol{F}(\boldsymbol{x}, \boldsymbol{V})$ for new test data $\boldsymbol{V}$.

Step **1** is the goal of Chapter 10 in this book. We will create a "neural net" to describe the form of $\boldsymbol{F}$. The key point is that $\boldsymbol{F}$ is produced by **a chain of simpler functions $\boldsymbol{F}_1$ to $\boldsymbol{F}_L$**:

$$\boldsymbol{F}(\boldsymbol{x}, \boldsymbol{v}) = \boldsymbol{F}_L(\boldsymbol{F}_{L-1}(\ldots(\boldsymbol{F}_2(\boldsymbol{F}_1(\boldsymbol{v}))))) \tag{2}$$

Each function $\boldsymbol{F}_k$ in that chain has its own set of weights $\boldsymbol{x}_k$. You will see in Chapter 10 that the individual functions $\boldsymbol{F}_k$ have a simple piecewise linear form. It is the *chain of functions* $\boldsymbol{F}_1$ to $\boldsymbol{F}_L$ that produces a powerful learning function $\boldsymbol{F}$.

Section 10.1 will count the (large) number of linear pieces of $\boldsymbol{F}$ created by ReLU.

This section describes the **backpropagation algorithm** that efficiently finds the derivatives of $\boldsymbol{F}$ with respect to the weight matrices $\boldsymbol{x}_1$ to $\boldsymbol{x}_L$ in the functions $\boldsymbol{F}_1$ to $\boldsymbol{F}_L$. By knowing those derivatives $\partial \boldsymbol{F}_k/\partial \boldsymbol{x}_k$, we can use **gradient descent** (or **stochastic** gradient descent) to optimize the weights. So once the form of $\boldsymbol{F}$ is chosen, deep learning is all about computing the weights $\boldsymbol{x}$ that minimize a sum of losses $\boldsymbol{\ell}(\boldsymbol{x}, \boldsymbol{v}_i)$:

$$\textbf{Loss function} \qquad \boldsymbol{\ell}(\boldsymbol{x}, \boldsymbol{v}_i) = \boldsymbol{F}(\boldsymbol{x}, \boldsymbol{v}_i) - \boldsymbol{w}_i = \textbf{approximate} - \textbf{true}$$

A personal note. I am greatly in debt to two friends who explained the computation of the gradients $\boldsymbol{\nabla F}$ and $\boldsymbol{\nabla \ell}$ (the derivatives of the loss with respect to the weights). One is Alexander Craig, who was a student in my Math 18.065 class at MIT. His project for the class was backpropagation, and his report was wonderful. *Alex shared in the writing of this section* 9.2. My other source of support was Andreas Griewank, one of the principal creators of "*automatic differentiation*". This algorithm finds the derivatives of $\boldsymbol{F}$ and $\boldsymbol{\ell}$ with respect to each of the weights with amazing speed. Applied recursively to the chain of $\boldsymbol{L}$ functions $\boldsymbol{F} = \boldsymbol{F}_L(\boldsymbol{F}_{L-1}(\ldots \boldsymbol{F}_1))$ this is backpropagation.

The reader will know that based on this computational engine, deep learning has been an overwhelming success.

The Multivariable Chain Rule

We begin with the chain rule for a composite function $\boldsymbol{h}(\boldsymbol{x}) = \boldsymbol{f}(\boldsymbol{g}(\boldsymbol{x}))$ of several input variables $x_1, \ldots, x_m$. The outputs from $\boldsymbol{g}(\boldsymbol{x})$ have n components $g_1, \ldots, g_n$. Then the outputs from $\boldsymbol{h}(\boldsymbol{x}) = \boldsymbol{f}(\boldsymbol{g}(\boldsymbol{x}))$ have p components $h_1, \ldots, h_p$. The Jacobian matrices $\boldsymbol{\partial f/\partial g}$ and $\boldsymbol{\partial g/\partial x}$ contain the partial derivatives of f and g separately:

$$\frac{\boldsymbol{\partial f}}{\boldsymbol{\partial g}} = \begin{bmatrix} \dfrac{\partial f_1}{\partial g_1} & \cdots & \dfrac{\partial f_1}{\partial g_n} \\ \vdots & \ddots & \vdots \\ \dfrac{\partial f_p}{\partial g_1} & \cdots & \dfrac{\partial f_p}{\partial g_n} \end{bmatrix} \qquad \frac{\boldsymbol{\partial g}}{\boldsymbol{\partial x}} = \begin{bmatrix} \dfrac{\partial g_1}{\partial x_1} & \cdots & \dfrac{\partial g_1}{\partial x_m} \\ \vdots & \ddots & \vdots \\ \dfrac{\partial g_n}{\partial x_1} & \cdots & \dfrac{\partial g_n}{\partial x_m} \end{bmatrix} \tag{3}$$

Each h_i depends on the g's and each g_j depends on the x's. Therefore h_i depends on $x_1, \ldots, x_m$. The chain rule aims to find all the derivatives $\partial h_i/\partial x_k$. The scalar rule is a dot product (row i of $\boldsymbol{\partial f/\partial g}$) $\cdot$ (column k of $\boldsymbol{\partial g/\partial x}$). The multivariable chain rule for $\boldsymbol{\partial h/\partial x}$ is exactly the **matrix product** $(\boldsymbol{\partial f/\partial g})\,(\boldsymbol{\partial g/\partial x})$:

$$\frac{\boldsymbol{\partial h_i}}{\boldsymbol{\partial x_k}} = \frac{\partial h_i}{\partial g_1}\frac{\partial g_1}{\partial x_k} + \cdots + \frac{\partial h_i}{\partial g_n}\frac{\partial g_n}{\partial x_k} = \left(\textbf{row } \boldsymbol{i} \textbf{ of } \frac{\boldsymbol{\partial f}}{\boldsymbol{\partial g}}\right) \cdot \left(\textbf{column } \boldsymbol{j} \textbf{ of } \frac{\boldsymbol{\partial g}}{\boldsymbol{\partial x}}\right) \tag{4}$$

The learning function $\boldsymbol{F}$ is a chain of functions $\boldsymbol{F}_1$ to $\boldsymbol{F}_L$. Then the matrix chain rule for the x-derivatives of $\boldsymbol{F_L}(\boldsymbol{F_{L-1}}(\ldots \boldsymbol{F_1}(\boldsymbol{v})))$ becomes a product of L matrices:

$$\boxed{\textbf{Matrix Chain Rule} \qquad \frac{\boldsymbol{\partial F}}{\boldsymbol{\partial x}} = \frac{\boldsymbol{\partial F_L}}{\boldsymbol{\partial F_{L-1}}}\,\frac{\boldsymbol{\partial F_{L-1}}}{\boldsymbol{\partial F_{L-2}}} \cdots \frac{\boldsymbol{\partial F_1}}{\boldsymbol{\partial x}}} \tag{5}$$

Stochastic Gradient Descent

Before we compute derivatives by backpropagation, there is an important decision to make. The loss function $\boldsymbol{L}(\boldsymbol{x})$ that we minimize is a sum of individual errors $\boldsymbol{\ell}(\boldsymbol{F}(\boldsymbol{x}, \boldsymbol{v}_i) - \boldsymbol{w}_i)$ for every sample $\boldsymbol{v}_i$ in the training set. The derivative of $\boldsymbol{F}(\boldsymbol{x}, \boldsymbol{v}_i)$ for one sample is hard enough! Computing derivatives for *every sample* with respect to *every weight* at *every step* of steepest descent is an enormous task. We have to reduce this computation.

Stochastic gradient descent randomly selects one sample input or a small batch of B samples, to improve the weights from $\boldsymbol{x}_k$ to $\boldsymbol{x}_{k+1}$. This works surprisingly well. Often the first descent steps show big improvements in the weights $\boldsymbol{x}$. Close to the optimal weights, the iterations can become erratic and we may not see fast convergence—but high precision is not needed or expected. Often we can stop the descent iterations early. SGD reduces the cost of each gradient step (from weights $\boldsymbol{x}_k$ to $\boldsymbol{x}_{k+1}$) to an acceptable level.

SGD for Ordinary Least Squares

Least squares provides an excellent (and optional) test of this "stochastic" idea. We are minimizing $||A\boldsymbol{x} - \boldsymbol{b}||^2$. At each step we randomly choose one sample row $\boldsymbol{a}_i$ of A. We adjust the current approximation $\boldsymbol{x}_k$ so the column $\boldsymbol{x}_{k+1}$ solves equation i of $A\boldsymbol{x} = \boldsymbol{b}$:

$$\boldsymbol{x}_{k+1} = \boldsymbol{x}_k + \frac{\boldsymbol{b}_i - \boldsymbol{a}_i\boldsymbol{x}_k}{||\boldsymbol{a}_i||^2}\,\boldsymbol{a}_i^{\mathrm{T}} \quad \text{solves} \quad \boldsymbol{a}_i\boldsymbol{x}_{k+1} = \boldsymbol{b}_i. \tag{6}$$

Kaczmarz introduced this method, thinking to cycle over and over from the first row of A to the last row. Strohmer and Vershynin changed to randomly chosen rows $\boldsymbol{a}_i$ (**stochastic descent**). They proved a form of exponential convergence to the solution of the least squares equation $A^{\mathrm{T}}A\widehat{\boldsymbol{x}} = A^{\mathrm{T}}\boldsymbol{b}$. A key point in the convergence shows up in Figure 9.5: *Fast start toward the best* $\boldsymbol{x}^*$ *and oscillating finish.* So stop early.

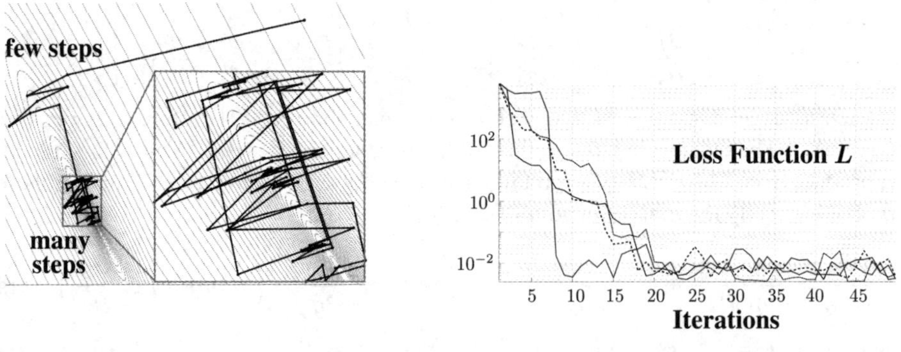

Figure 9.5: The left figure shows a trajectory of stochastic gradient descent with two unknowns. The early iterations succeed but later iterations oscillate (as shown in the inset). On the right, the quadratic cost function decreases quickly at first and then fluctuates instead of converging. The four paths in that graph start from the same $\boldsymbol{x}_0$ with random choices of i as in equation (6).

The Key Ideas of Backpropagation

For the present, we keep the idea simple and the notation simple. Our learning function $\boldsymbol{F}$ depends on the weights $\boldsymbol{x}$ (the matrices A_1 to A_L and the vectors $\boldsymbol{b}_1$ to $\boldsymbol{b}_L$ at layers 1 to L). The special simplifying feature is that each weight enters only one of the functions $\boldsymbol{F}_1$ to $\boldsymbol{F}_L$ in the chain. We will write $\boldsymbol{x}_j$ for the matrix A_j and vector b_j that enter layer j: $\boldsymbol{F}_j = \text{ReLU}(A_j\boldsymbol{F}_{j-1} + b_j)$. ReLU is a famously simple function—see Figure 10.1. The chain rule for the derivative of the loss $\boldsymbol{\ell}$ with respect to the weights $\boldsymbol{x}_j$ in $\boldsymbol{F}_j$ is

$$\frac{\partial \ell}{\partial x_j} = \frac{\partial \ell}{\partial F_L} \cdot \frac{\partial F_L}{\partial F_{L-1}} \cdot \frac{\partial F_{L-1}}{\partial F_{L-2}} \cdots \frac{\partial F_{j+1}}{\partial F_j} \cdot \frac{\partial F_j}{\partial x_j} \tag{7}$$

The key point is that this involves only the weights x_j to x_L—not earlier weights. So the efficient computation starts with $\partial\ell/\partial F_L \cdot \partial F_L/\partial F_{L-1}$ (involving A_L and b_L). Then the next step introduces only the weights A_{L-1} and b_{L-1} in $\boldsymbol{x}_{L-1}$. We have every reason to choose the backward mode in this differentiation:

Computing $\partial F/\partial x_j$ involves only the last weights from x_j to x_L

Remember that the weights $\boldsymbol{x}_j$ include the matrix A_j and the vector b_j . The derivative of F_j with respect to A_j will need three components (this makes it a tensor). We will compute that derivative a few pages onward. Here we take the final step from the gradient of F to the gradient of the loss $\boldsymbol{\ell}$. This step displays the second big reward for computing successive derivatives *backward from the output.*

Second key point: The loss $\boldsymbol{\ell}(\boldsymbol{x}, \boldsymbol{v})$ is a scalar function of the output vector $\boldsymbol{F}(\boldsymbol{x}, \boldsymbol{v})$. In the simplest case of a "square loss", that function is the squared length of the error:

$$\textbf{Loss:} \quad \boldsymbol{\ell}(\boldsymbol{x}) = (\boldsymbol{w} - \boldsymbol{F}(\boldsymbol{x},\boldsymbol{v}))^{\mathrm{T}}(\boldsymbol{w} - \boldsymbol{F}(\boldsymbol{x},\boldsymbol{v})) \quad \text{and} \quad \frac{\partial \boldsymbol{\ell}}{\partial \boldsymbol{F}} = 2(\boldsymbol{w} - \boldsymbol{F}(\boldsymbol{x},\boldsymbol{v})) \tag{8}$$

Here $\boldsymbol{w}$ is the true (desired) output from the original training sample $\boldsymbol{v}$, and $\boldsymbol{F}(\boldsymbol{x}, \boldsymbol{v})$ is the computed output at layer L using the weights. The key point for this and other choices of $\boldsymbol{\ell}$ is that $\partial\boldsymbol{\ell}/\partial F$ is a vector. Then the backward chain of matrices $M_k = \partial F_k/\partial F_{k-1}$ multiplies that vector. At every step this multiplication becomes **a vector times a matrix!**

$$\begin{array}{c}\text{(vector)(matrix)}\\ \textbf{not}\text{ (matrix)(matrix)}\end{array} \qquad \frac{\partial \ell}{\partial x_j} = \left(\left(\left(\frac{\partial \ell}{\partial F} M_L\right) M_{L-1}\right) \ldots M_{j+1}\right) \tag{9}$$

Suppose for example that each matrix M is n by n. Then each vector-matrix step in the chain needs $\boldsymbol{n^2}$ multiplications. If we choose the opposite order, with matrix-matrix steps, those would have $\boldsymbol{n^3}$ multiplications. The backward order (L first) is very much faster.

Summary

Reverse-mode differentiation is more efficient than forward-mode for two key reasons:

1. **Most operations are vector-matrix multiplications (or vector-tensor).**

2. **Work will be reused for the remaining derivative computations.**

Reverse mode differentiation achieves these efficiencies by rearranging the parentheses in the chain rule from left to right—from the loss function at the output back to layer j:

$$\textbf{Back-Prop} \qquad \frac{\partial \ell}{\partial b_j} = \left(\left(\left(\frac{\partial \ell}{\partial F_L} \cdot \frac{\partial F_L}{\partial F_{L-1}}\right) \cdot \frac{\partial F_{L-1}}{\partial F_{L-2}}\right) \cdots \frac{\partial F_{j+1}}{\partial F_j}\right) \cdot \frac{\partial F_j}{\partial b_j} \tag{10}$$

Since $\partial \ell / \partial F_L$ is a vector, our intermediate results are mostly vectors. So we are performing vector-matrix multiplications. When we compute $\partial \ell / \partial A_j$ we cannot avoid a single vector-tensor contraction at the end—because A is a matrix. Section 10.2 computes that matrix $\partial F_j / \partial A_j$.

Summary In addition to this speed-up for every operation, reverse-mode also requires fewer total operations because it allows us to reuse work in a remarkably elegant way. The key insight here is that the chain rule expressions for all derivatives at layers $j \leq k$ share the same prefix $\partial \ell / \partial F_k$. Since reverse-mode groups the terms from the left, **it can reuse $\boldsymbol{\partial \ell / \partial F_k}$ in all computations for layers $\boldsymbol{j \leq k}$.** Backpropagation uses this optimization at every layer, resulting in the following algorithm:

To initialize, perform a forward pass through the chain using input $\boldsymbol{v}$, and store all intermediate results F_j . Also, compute $\partial \ell / \partial F_L$. Then,

for $j = L \to 1$ **do**

Compute and store

$$\frac{\partial \ell}{\partial b_j} = \frac{\partial \ell}{\partial F_j} \cdot \frac{\partial F_j}{\partial b_j} \quad \text{and} \quad \frac{\partial \ell}{\partial A_j} = \frac{\partial \ell}{\partial F_j} \cdot \frac{\partial F_j}{\partial A_j}$$

if $j \neq 1$ **then**

Compute $\partial \ell / \partial F_{j-1}$ to be used at the next iteration:

$$\frac{\partial \ell}{\partial F_{j-1}} = \frac{\partial \ell}{\partial F_j} \cdot \frac{\partial F_j}{\partial F_{j-1}}$$

end if

end for

Thus backpropagation computes all the required derivatives in a single backward pass through the network. It requires only $2L$ vector-matrix multiplies and L vector-tensor contractions. To compute the same values, forward-mode differentiation would require $O(L^2)$ matrix-tensor contractions and matrix-matrix multiplies. Backpropagation is more efficient by multiple orders of magnitude.

Backpropagation in Keras and TensorFlow

Keras is one of the most popular machine learning libraries available today. It provides an simple interface to neural networks in Python, and it executes backpropagation internally using the TensorFlow library.

TensorFlow's implementation of backpropagation (see arXiv 1610.01178) is a significantly more complex and powerful generalization of the algorithm we described above. TensorFlow is designed to differentiate arbitrarily complex chains of functions. To do this, it constructs a computational graph representing the function to be differentiated. Then it inserts into that graph the operations representing the derivatives, usually working backwards from output to input. At that point, TensorFlow can perform the derivative computation for arbitrary input to the network. In the case where the network has the structure we describe above, TensorFlow's computational graph with derivatives would produce the same operations as our pseudocode.

The essay **https://www.simplilearn.com/keras-vs-tensorflow-vs-pytorch-article** may be useful in coding backbpropagation.

Backpropagation on the Website

Coding for deep learning is discussed on the website **https://math.mit.edu/linearalgebra**. It has its own section on that site—we hope it allows you to create a learning function $\boldsymbol{F}(\boldsymbol{x}, \boldsymbol{v})$ and test it on examples. Python and Julia and MATLAB are the website languages.

Derivatives of $Av + b$

We now compute the derivatives of each layer F_j with respect to the weights A_j and $\boldsymbol{b}_j$ introduced at that layer. The layer equation begins with the vector $\boldsymbol{w}_j = A_j \boldsymbol{v}_{j-1} + \boldsymbol{b}_j$ and then computes $\boldsymbol{v}_j = \text{ReLU}(\boldsymbol{w}_j)$. We will take the easiest step first, and **we drop the index j for now**. Our function is $\boldsymbol{w}(A, \boldsymbol{b}) = A\boldsymbol{v} + \boldsymbol{b}$.

First of all, what is the derivative of the ith output $\boldsymbol{w}_i = (A\boldsymbol{v} + \boldsymbol{b})_i$ with respect to the jth input $\boldsymbol{b}_j$? An unusual question with an easy answer:

$$\text{If } i = j, \text{ then } \frac{\partial w_i}{\partial b_i} = 1 \qquad \text{If } i \neq j \text{ then } \frac{\partial w_i}{\partial b_j} = 0. \tag{11}$$

This $1 - 0$ alternative appears so often in mathematics that it has its own notation δ_{ij} :

$$\delta_{ij} = \left\{ \begin{array}{ll} 1 & \text{if } i = j \\ 0 & \text{if } i \neq j \end{array} \right\} = \text{ the entries in the identity matrix } I$$

Next, what is the derivative of the ith output w_i with respect to the (j, k) input A_{jk} ? Suddenly we have three indices i, j, k. This indicates that our derivative is a **tensor** (more exactly, a 3-way tensor). Tensors are like matrices but with more indices—instead of a rectangle of numbers in a matrix, we now have a 3-dimensional **"box of numbers"**.

Note We remark here that tensors are important, theoretically and also computationally. (They enter **multilinear algebra**.) But they are not trivial extensions of the idea of matrices. Even the rank of a tensor is not easy to define, and the SVD factorization (for example) fails. At least rank one seems straightforward: The tensor entries T_{ijk} have the product form $a_i b_j c_k$ for three vectors $\boldsymbol{a}, \boldsymbol{b}, \boldsymbol{c}$—like $A_i b_j$ for a rank one matrix.

Here is the tensor derivative formula when $\boldsymbol{w} = A\boldsymbol{v} + \boldsymbol{b}$:

$$\boldsymbol{T_{ijk}} = \frac{\boldsymbol{\partial w_i}}{\boldsymbol{\partial A_{jk}}} = \boldsymbol{v_k \delta_{ij}} \text{ as in } T_{111} = v_1 \text{ and } T_{122} = 0. \tag{12}$$

This says: Row j of A affects component i of $\boldsymbol{w} = A\boldsymbol{v} + \boldsymbol{b}$ only if $j = i$. And when we do have $j = i$, the number A_{jk} in that row is multiplying v_k. So the derivative of the output w_i with respect to the matrix entry A_{jk} is just v_k when $i = j$.

Example There are six b's and a's in $\begin{bmatrix} \boldsymbol{w_1} \\ w_2 \end{bmatrix} = \begin{bmatrix} \boldsymbol{b_1} \\ b_2 \end{bmatrix} + \begin{bmatrix} \boldsymbol{a_{11}v_1 + a_{12}v_2} \\ a_{21}v_1 + a_{22}v_2 \end{bmatrix}.$

Derivatives of $\boldsymbol{w_1}$ $\dfrac{\partial w_1}{\partial b_1} = \mathbf{1}, \dfrac{\partial w_1}{\partial b_2} = \mathbf{0}, \dfrac{\partial w_1}{\partial a_{11}} = \boldsymbol{v_1}, \dfrac{\partial w_1}{\partial a_{12}} = \boldsymbol{v_2}, \dfrac{\partial w_1}{\partial a_{21}} = \dfrac{\partial w_1}{\partial a_{22}} = \mathbf{0}.$

Derivatives for Hidden Layers

Now suppose there is one hidden layer, so $L = 2$. The output is $\boldsymbol{w} = \boldsymbol{v}_L = \boldsymbol{v}_2$, the hidden layer contains $\boldsymbol{v}_1$, and the input is $\boldsymbol{v}_0 = \boldsymbol{v}$.

$$\boldsymbol{v}_1 = \text{ReLU}\,(\boldsymbol{b}_1 + A_1\boldsymbol{v}_0) \quad \text{and} \quad \boldsymbol{w} = \boldsymbol{b}_2 + A_2\boldsymbol{v}_1 = \boldsymbol{b}_2 + A_2\text{ReLU}\,(\boldsymbol{b}_1 + A_1\boldsymbol{v}_0).$$

Equations (11)–(12) give the derivatives of $\boldsymbol{w}$ with respect to the last weights $\boldsymbol{b}_2$ and A_2. The function ReLU is absent at the output and $\boldsymbol{v}$ is $\boldsymbol{v}_1$. But the derivatives of $\boldsymbol{w}$ with respect to $\boldsymbol{b}_1$ and A_1 do involve the nonlinear function ReLU acting on $\boldsymbol{b}_1 + A_1\boldsymbol{v}_0$.

So the derivatives in $\partial\boldsymbol{w}/\partial A_1$ need the chain rule $\partial f/\partial x = (\partial f/\partial g)(\partial g/\partial x)$:

$$\textbf{Chain rule} \quad \frac{\partial \boldsymbol{w}}{\partial A_1} = \frac{\partial[A_2\text{ReLU}\,(\boldsymbol{b}_1 + A_1\boldsymbol{v}_0)]}{\partial A_1} = A_2\,\text{ReLU}\,'(\boldsymbol{b}_1 + A_1\boldsymbol{v}_0)\,\frac{\partial(\boldsymbol{b}_1 + A_1\boldsymbol{v}_0)}{\partial A_1}. \tag{13}$$

That chain rule has three factors. Starting from $\boldsymbol{v}_0$ at layer $L-2 = 0$, the weights $\boldsymbol{b}_1$ and A_1 bring us toward the layer $L - 1 = 1$. The derivatives of that step are exactly like (11) and (12). But the output of that partial step is not $\boldsymbol{v}_{L-1}$. To find that hidden layer we first have to apply ReLU. So the chain rule includes its derivative **ReLU** $'$. Then the final step (to $\boldsymbol{w}$) multiplies by the last weight matrix A_2.

The Problem Set extends these formulas to L layers. They could be useful. But with pooling and batch normalization, automatic differentiation seems to defeat hard coding.

Very important Notice how formulas like (13) go **backwards from $\boldsymbol{w}$ to $\boldsymbol{v}$**. Automatic backpropagation will do this too. "**Reverse mode**" starts with the output.

Details of the Derivatives $\partial w/\partial A_1$

We feel some responsibility to look more closely at equation (13). Its nonlinear part **ReLU** $'$ comes from the derivative of the nonlinear activation function. The usual choice is the ramp function ReLU $(x) = (x)_+ =$ limiting case of an S-shaped sigmoid function.

$$\frac{d(\text{ReLU})}{dx} = \begin{cases} 0 & x < 0 \\ 1 & x > 0 \end{cases} \qquad \textbf{ReLU}\,'(\boldsymbol{b}_1 + A_1\boldsymbol{v}_0) = \begin{cases} 0 & \text{when } (\boldsymbol{b}_1 + A_1\boldsymbol{v}_0)_i < 0 \\ 1 & \text{when } (\boldsymbol{b}_1 + A_1\boldsymbol{v}_0)_i > 0 \end{cases}$$

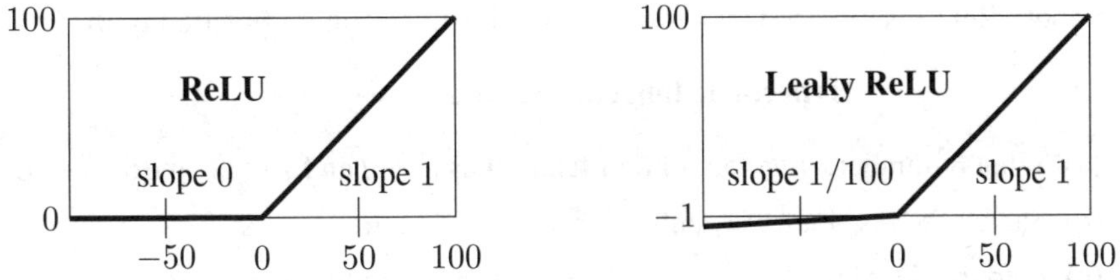

Figure 9.6: The graphs of ReLU and Leaky ReLU (two options for nonlinear activation).

Returning to formula (13), write A and $\boldsymbol{b}$ for the matrix A_{L-1} and the vector $\boldsymbol{b}_{L-1}$ that produce the last hidden layer. Then ReLU and A_L and $\boldsymbol{b}_L$ produce the final output $\boldsymbol{w} = \boldsymbol{v}_L$. Our interest is in $\partial w/\partial A$, the dependence of $\boldsymbol{w}$ on the next to last matrix of weights.

$$\boxed{\boldsymbol{w} = A_L(\text{ReLU}(A\boldsymbol{v}+\boldsymbol{b})) + \boldsymbol{b}_L \quad \text{and} \quad \frac{\partial \boldsymbol{w}}{\partial A} = A_L \text{ReLU}'(A\boldsymbol{v}+\boldsymbol{b})\frac{\partial(A\boldsymbol{v}+\boldsymbol{b})}{\partial A}} \tag{14}$$

We think of ReLU as a diagonal matrix of ReLU functions acting component by component on $A\boldsymbol{v}+\boldsymbol{b}$. Then $J = \text{ReLU}'(A\boldsymbol{v}+\boldsymbol{b})$ is a diagonal matrix with 1's for positive components and 0's for negative components. (Don't ask about the derivative of ReLU at 0):

$$\boldsymbol{w} = A_L \,\text{ReLU}\,(A\boldsymbol{v}+\boldsymbol{b}) \quad \text{and} \quad \frac{\partial \boldsymbol{w}}{\partial A} = A_L\, J\, \frac{\partial(A\boldsymbol{v}+\boldsymbol{b})}{\partial A} \tag{15}$$

We know every component (v_k or zero) of the third factor from the derivatives in (11)–(12).

When the sigmoid function ReLU_a replaces the ReLU function, the diagonal matrix $J = \text{ReLU}_a'(A\boldsymbol{v}+\boldsymbol{b})$ no longer contains 1's and 0's. Now we evaluate the derivative dR_a/dx at each component of $A\boldsymbol{v}+\boldsymbol{b}$.

In practice, backpropagation finds the derivatives with respect to all A's and $\boldsymbol{b}$'s. It creates those derivatives automatically and very effectively.

Problem Set 9.2

1 For the equations $x+2y = 3$ and $2x+3y = 5$, test the Kaczmarz iteration in equation (6) starting with $(x_0, y_0) = (0, 0)$. Compare the cyclic order (equation $1, 2, 1, 2, \ldots$) with random order.

2 Write down the first backpropagation step to minimize $\ell(x) = (1 - F(x))^2$ and $F(x) = F_2(F_1(x))$ with $F_2 = (F_1(x)) = \cos(\sin x)$ and $x_0 = 0$. This 1-variable question can't show the virtues of backpropagation.

3 Suppose ReLU is replaced by the smooth nonlinear activation function **tanh**:

$$\textbf{Hyperbolic tangent} \quad \textbf{tanh}(x) = \frac{e^x - e^{-x}}{e^x + e^{-x}}$$

(a) What 3 numbers give the limit of tanh(x) as $x \to 0$ and $x \to \infty$ and $x \to -\infty$?

(b) Sketch the graph of $y = \tanh(x)$ from $x = -\infty$ to $x = +\infty$.

(c) Find the derivative dy/dx to show that tanh is steadily increasing.

4 Before ReLU, tanh was a popular nonlinear "activation function" in deep learning. It is an option in **playground.tensorflow.org**. Because it is a smooth function, its derivatives can be computed for all inputs (unlike ReLU). The input b could be a scalar or a vector:

Find the gradient with respect to $\boldsymbol{b}$ of the vector with components $\tanh(A\boldsymbol{x} + \boldsymbol{b})$.

5 $F = F_2(x, F_1(y))$ is a function of x and y. Find its partial derivatives $\partial F/\partial x$ and $\partial F/\partial y$. Which of these derivatives would you compute first?

9.3 Constraints, Lagrange Multipliers, Minimum Norms

We start with the line $\mathbf{3x + 4y = 1}$ in the xy plane. That is the constraint—our solution must be a point that lies on that line. *The problem is to find the point* (x, y) *with minimum norm.* The solution (x^*, y^*) will depend on the choice of norm. Here are the three most popular ways to measure the size $||\boldsymbol{v}||$ of a vector $\boldsymbol{v} = (x, y)$.

ℓ^1 norm $\;||\boldsymbol{v}||_1 = |x| + |y|$	All norms must obey $||c\boldsymbol{v}|| = |c|\,||\boldsymbol{v}||$ and the
ℓ^2 norm $\;||\boldsymbol{v}||_2 = \sqrt{x^2 + y^2}$	**triangle inequality** $||\boldsymbol{v} + \boldsymbol{w}|| \le ||\boldsymbol{v}|| + ||\boldsymbol{w}||$.
ℓ^∞ norm $\;||\boldsymbol{v}||_\infty =$ max of $|x|$ and $|y|$	**ℓ^p norms** $\;||\boldsymbol{v}||_p = (|x|^p + |y|^p)^{1/p}$ for $p \ge 1$

For the vector $\boldsymbol{v} = (1, 1)$, those three norms are $||\boldsymbol{v}|| = 2, \sqrt{2}$, and 1. A good way to visualize each norm is to draw the regions where $||\boldsymbol{v}||_1 \le 1$ and $||\boldsymbol{v}||_2 \le 1$ and $||\boldsymbol{v}||_\infty \le 1$. The picture below shows a **diamond** and a **circle** and a **square**. For the ℓ^p norms, the shapes of the regions $||\boldsymbol{v}||_p \le 1$ would gradually approach the square as $p \to \infty$.

To see why $p < 1$ fails to give a norm, we include the picture for $p = \frac{1}{2}$. In that case $\boldsymbol{v} = (1, 0)$ and $\boldsymbol{w} = (0, 1)$ have norm 1, but $\boldsymbol{v} + \boldsymbol{w}$ has norm $(1 + 1)^2 = 4$. This violates the triangle inequality $||\boldsymbol{v} + \boldsymbol{w}|| \le ||\boldsymbol{v}|| + ||\boldsymbol{w}||$ and shows that $p = \frac{1}{2}$ does *not* give a legal norm.

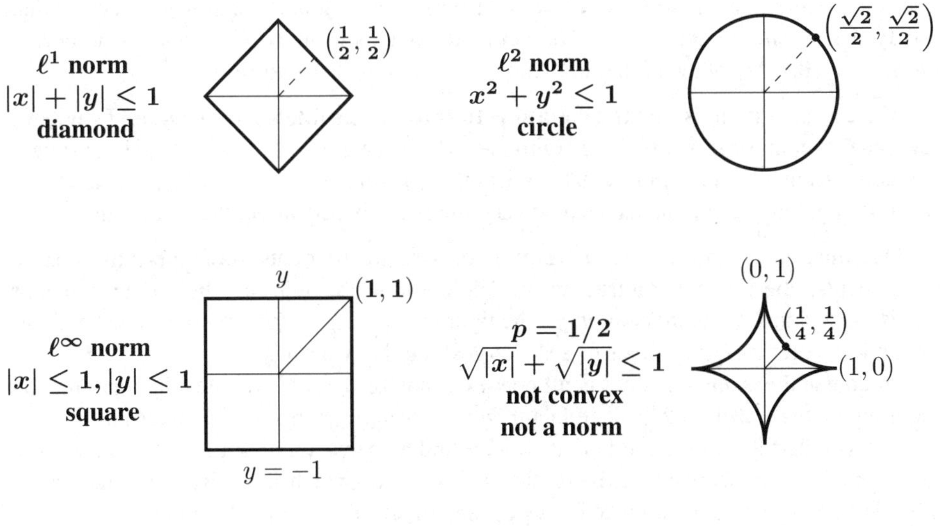

Figure 9.7: The important vector norms $||\boldsymbol{v}||_1, ||\boldsymbol{v}||_2, ||\boldsymbol{v}||_\infty$ and a failure for $p < 1$.

If we move up to $\boldsymbol{v} = (x, y, z)$ in three dimensions, the circle will become a ball $x^2 + y^2 + z^2 \le 1$. The square $||\boldsymbol{v}||_\infty \le 1$ will become a cube $-1 \le x, y, z \le 1$. The diamond $|x| + |y| + |z| \le 1$ from the $\boldsymbol{\ell_1}$ norm will now have six corners. A sharp corner of that diamond will still be first to touch a line or a plane in 3D, as the diamond grows.

The next figure shows the winning vectors $\boldsymbol{v}^*$ in the three important norms.

Now we are ready to minimize $||\boldsymbol{v}||_1$ and $||\boldsymbol{v}||_2$ and $||\boldsymbol{v}||_\infty$ with the linear constraint $\mathbf{3}\boldsymbol{x} + \mathbf{4}\boldsymbol{y} = \mathbf{1}$. A picture shows the line and the three minimum points, depending on the choice of norm. We are scaling down the pictures of $||\boldsymbol{v}|| = 1$ **until they barely touch the constraint line**. The touching points $\boldsymbol{v}^*$ are the solutions for the $\boldsymbol{\ell}^1$ and $\boldsymbol{\ell}^2$ and $\boldsymbol{\ell}^\infty$ norms.

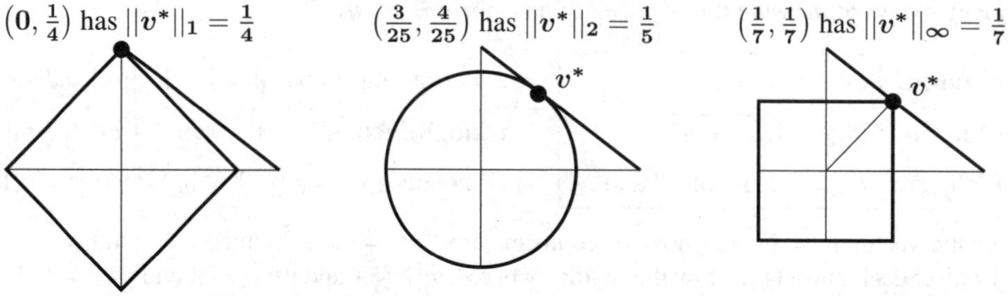

Figure 9.8: The solutions $\boldsymbol{v}^*$ to the $\boldsymbol{\ell}^1$ and $\boldsymbol{\ell}^2$ and $\boldsymbol{\ell}^\infty$ minimizations. The first is **sparse**.

The first figure displays a highly important property of the minimizing solution to the $\boldsymbol{\ell}^1$ problem: **That solution $\boldsymbol{v}^* = (0, \frac{1}{4})$ has only one nonzero component**. The vector $\boldsymbol{v}^*$ is **sparse**. This is because **a diamond touches a line at a sharp point**. The line (or a plane like $3x + 4y + 5z = 1$ in three dimensions) contains the vectors that satisfy the constraints $A\boldsymbol{v} = \boldsymbol{b}$. The diamond expands to meet that plane at the corner $(x, y, z) = (0, 0, \frac{1}{5})$ of the diamond! And $\boldsymbol{v}^* = (0, 0, \frac{1}{5})$ has two zeros.

The essential point is that **the solutions to those $\boldsymbol{\ell}_1$ problems are sparse.** They have few nonzero components, and those components have meaning. By contrast the least squares solution (using $\boldsymbol{\ell}^2$) has many small and non-interesting components in n dimensions. By squaring, those components become very small and hardly affect the $\boldsymbol{\ell}^2$ distance.

One final observation: **The "$\boldsymbol{\ell}^0$ norm"** of a vector $\boldsymbol{v}$ counts the number of nonzero components. But this is not a true norm. The points with $||\boldsymbol{v}||_0 = 1$ lie on the x axis or y axis—one nonzero component only. The figure for $p = \frac{1}{2}$ on the previous page becomes even more extreme—just a cross or a skeleton along the two axes.

Of course this skeleton is not at all convex. The "zero norm" violates the fundamental requirement that $||\mathbf{2}\boldsymbol{v}|| = 2\,||\boldsymbol{v}||$. In fact $||\mathbf{2}\boldsymbol{v}||_0 = ||\boldsymbol{v}||_0 =$ number of nonzeros in $\boldsymbol{v}$.

The wonderful observation is that we can find a sparse solution to $A\boldsymbol{v} = \boldsymbol{b}$ *by using the $\boldsymbol{\ell}^1$ norm*. We have "convexified" that $\boldsymbol{\ell}^0$ skeleton (which lies only along the axes). We filled out the skeleton to make it convex, and the result was the $\boldsymbol{\ell}^1$ diamond.

Summary These vector problems show the importance of the choice of norm. For the matrix problems of Chapter 10, we will again have an important decision—this time for a **matrix norm**. Convexity will again be a crucial property—tied to the triangle inequality $||A + B|| \leq ||A|| + ||B||$ for matrix norms.

Lagrange Multipliers = Derivatives of the Cost

Now we will redo the ℓ^2 problem using Lagrange multipliers, because that idea extends to all "quadratic programs" and even to "convex programs". You will see how Lagrange multipliers deal with constraints. We want to bring out the meaning of those multipliers $\lambda_1, \ldots, \lambda_m$. After introducing them and using them, it is a big mistake to discard them.

Our first example is in two dimensions. The function F is quadratic. The set K is linear.

$$\textbf{Minimize } F(\boldsymbol{x}) = x_1^2 + x_2^2 \textbf{ on the line } K: a_1x_1 + a_2x_2 = b$$

On the line K, we are looking for the point that is nearest to $(0,0)$. The cost $F(x)$ is distance squared. In Figure 9.7, the constraint line was **tangent to the circle** at the winning point $\boldsymbol{x}^* = (x_1^*, x_2^*)$. We discover this from simple calculus, *after we bring the constraint equation* $a_1x_1 + a_2x_2 = b$ *into the function* $F = x_1^2 + x_2^2$.

This was Lagrange's beautiful idea.

Subtract from $F(\boldsymbol{x})$ an unknown multiplier $\boldsymbol{\lambda}$ times $a_1x_1 + a_2x_2 - b$

$$\textbf{Lagrangian } L(\boldsymbol{x}, \boldsymbol{\lambda}) = F(x) - \boldsymbol{\lambda}(a_1x_1 + a_2x_2 - b)$$
$$= x_1^2 + x_2^2 - \lambda(a_1x_1 + a_2x_2 - b) \qquad (1)$$

Set the derivatives $\partial L/\partial x_1$ and $\partial L/\partial x_2$ and $\partial L/\partial \lambda$ to zero.

Solve those three equations for x_1, x_2, λ.

$$\partial L/\partial x_1 = 2x_1 - \lambda a_1 = 0 \qquad (2a)$$

$$\partial L/\partial x_2 = 2x_2 - \lambda a_2 = 0 \qquad (2b)$$

$$\partial L/\partial \lambda = -(a_1x_1 + a_2x_2 - b) = 0 \quad \text{(the constraint !)} \qquad (2c)$$

The first equations give $x_1 = \frac{1}{2}\lambda a_1$ and $x_2 = \frac{1}{2}\lambda a_2$. Substitute into $a_1x_1 + a_2x_2 = b$:

$$\frac{1}{2}\lambda a_1^2 + \frac{1}{2}\lambda a_2^2 = b \text{ and } \lambda = \frac{2b}{a_1^2 + a_2^2}. \qquad (3)$$

Substituting λ into (2a) and (2b) reveals the closest point (x_1^*, x_2^*) and the minimum cost $(x_1^*)^2 + (x_2^*)^2$:

$$x_1^* = \frac{1}{2}\lambda a_1 = \frac{a_1 b}{a_1^2 + a_2^2} \qquad x_2^* = \frac{1}{2}\lambda a_2 = \frac{a_2 b}{a_1^2 + a_2^2} \qquad (x_1^*)^2 + (x_2^*)^2 = \frac{b^2}{a_1^2 + a_2^2}$$

The derivative of the minimum cost with respect to the constraint level b is the Lagrange multiplier $\boldsymbol{\lambda}$!

$$\frac{d}{db}\left(\frac{b^2}{a_1^2 + a_2^2}\right) = \frac{2b}{a_1^2 + a_2^2} = \lambda. \qquad (4)$$

Quadratic Programming with Positive Definite S

We now move that example from the plane $\mathbf{R}^2$ to the space $\mathbf{R}^n$. Instead of one constraint on $\boldsymbol{x}$ we have m **constraints** $A^{\mathrm{T}}\boldsymbol{x} = \boldsymbol{b}$. **The matrix** A^{T} **will be** m **by** n. There will be m Lagrange multipliers $\boldsymbol{\lambda}_1, \ldots, \boldsymbol{\lambda}_m$: one for each constraint. The cost function $F(\boldsymbol{x}) = \frac{1}{2}\boldsymbol{x}^{\mathrm{T}}S\boldsymbol{x}$ allows any symmetric positive definite matrix S. Up to now we had $S = I$.

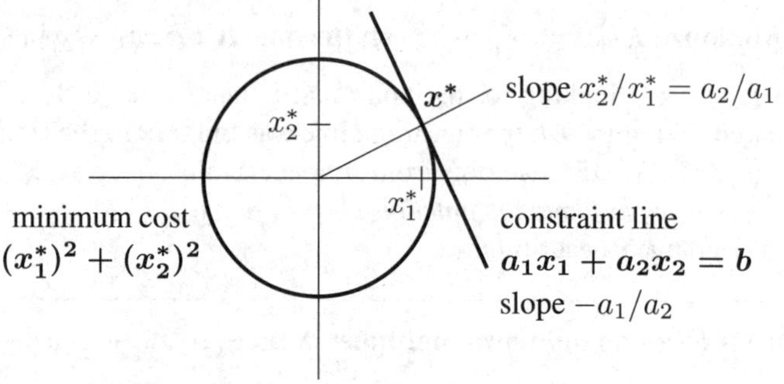

Figure 9.9: The constraint line is tangent to the minimum cost circle at the solution $\boldsymbol{x}^*$. In three dimensions the circle and line become a sphere and plane.

$$\textbf{Problem: Minimize } F = \tfrac{1}{2}\boldsymbol{x}^{\mathrm{T}}S\boldsymbol{x} \textbf{ subject to } A^{\mathrm{T}}\boldsymbol{x} = \boldsymbol{b}. \tag{5}$$

With m constraints there will be m Lagrange multipliers $\boldsymbol{\lambda} = (\lambda_1, \ldots, \lambda_m)$. They build the constraints $A^{\mathrm{T}}\boldsymbol{x} = \boldsymbol{b}$ into the Lagrangian $L(\boldsymbol{x}, \boldsymbol{\lambda}) = \frac{1}{2}\boldsymbol{x}^{\mathrm{T}}S\boldsymbol{x} - \boldsymbol{\lambda}^{\mathrm{T}}(A^{\mathrm{T}}\boldsymbol{x} - \boldsymbol{b})$. The $n+m$ derivatives of L give $n+m$ equations for a vector $\boldsymbol{x}$ in $\mathbf{R}^n$ and $\boldsymbol{\lambda}$ in $\mathbf{R}^m$:

$$\begin{array}{ll} x\text{-derivatives of } L: & S\boldsymbol{x} - A\boldsymbol{\lambda} = \mathbf{0} \\ \lambda\text{-derivatives of } L: & A^{\mathrm{T}}\boldsymbol{x} \quad\;\; = \boldsymbol{b} \end{array} \tag{6}$$

The first equations give $\boldsymbol{x} = S^{-1}A\boldsymbol{\lambda}$. Then the second equations give $A^{\mathrm{T}}S^{-1}A\boldsymbol{\lambda} = \boldsymbol{b}$. This determines the optimal multipliers $\boldsymbol{\lambda}^*$ and then the optimal $\boldsymbol{x}^* = S^{-1}A\boldsymbol{\lambda}^*$:

$$\textbf{Solution } \boldsymbol{\lambda}^*, \boldsymbol{x}^* \qquad \boldsymbol{\lambda}^* = (A^{\mathrm{T}}S^{-1}A)^{-1}\boldsymbol{b} \qquad \boldsymbol{x}^* = S^{-1}A(A^{\mathrm{T}}S^{-1}A)^{-1}\boldsymbol{b}. \tag{7}$$

Minimum cost $F^* = \dfrac{1}{2}(\boldsymbol{x}^*)^{\mathrm{T}}S\boldsymbol{x}^* = \dfrac{1}{2}\boldsymbol{b}^{\mathrm{T}}(A^{\mathrm{T}}S^{-1}A)^{-1}A^{\mathrm{T}}S^{-1}SS^{-1}A(A^{\mathrm{T}}S^{-1}A)^{-1}\boldsymbol{b}$.
This simplifies a lot !

$$\begin{array}{ll} \textbf{Minimum cost } F^* = \dfrac{1}{2}\boldsymbol{b}^{\mathrm{T}}(A^{\mathrm{T}}S^{-1}A)^{-1}\boldsymbol{b} \\ \textbf{Gradient of cost } \dfrac{\partial F^*}{\partial \boldsymbol{b}} = (A^{\mathrm{T}}S^{-1}A)^{-1}\boldsymbol{b} = \boldsymbol{\lambda}^* \end{array} \tag{8}$$

This is truly a model problem. Please note that $F(\boldsymbol{x}) = \frac{1}{2}\boldsymbol{x}^{\mathrm{T}}S\boldsymbol{x} - \boldsymbol{c}^{\mathrm{T}}\boldsymbol{x}$ can include a linear term, and that some authors reverse λ to $-\lambda$. When the constraint changes to m inequalities $A^{\mathrm{T}}\boldsymbol{x} \leq \boldsymbol{b}$, the multipliers become $\lambda_i \geq 0$ and the problem becomes distinctly harder! A zero component $\lambda_i^* = 0$ in the solution signals that the ith constraint $\boldsymbol{a}_i^{\mathrm{T}}\boldsymbol{x} \leq b_i$ is *not active*. The minimum of F would be the same without that ith constraint.

Here is the "saddle point matrix" or "Karush-Kuhn-Tucker matrix" in (6):

$$\textbf{KKT matrix} \qquad K\begin{bmatrix} \boldsymbol{x} \\ -\boldsymbol{\lambda} \end{bmatrix} = \begin{bmatrix} S & A \\ A^{\mathrm{T}} & 0 \end{bmatrix}\begin{bmatrix} \boldsymbol{x} \\ -\boldsymbol{\lambda} \end{bmatrix} = \begin{bmatrix} \boldsymbol{0} \\ \boldsymbol{b} \end{bmatrix}. \tag{9}$$

That symmetric matrix K is not positive definite or negative definite. Suppose you multiply the first block row $[S \quad A]$ by $A^{\mathrm{T}}S^{-1}$ to get $[A^{\mathrm{T}} \quad A^{\mathrm{T}}S^{-1}A]$. Subtract from the second block row to see a zero block:

$$\begin{bmatrix} S & A \\ 0 & -A^{\mathrm{T}}S^{-1}A \end{bmatrix}\begin{bmatrix} \boldsymbol{x} \\ -\boldsymbol{\lambda} \end{bmatrix} = \begin{bmatrix} \boldsymbol{0} \\ \boldsymbol{b} \end{bmatrix}. \tag{10}$$

This is just elimination on the 2 by 2 block matrix K. That new block in the 2, 2 position is called the **Schur complement** (named after the greatest linear algebraist of all time).

We reached the same equation $A^{\mathrm{T}}S^{-1}A\boldsymbol{\lambda} = \boldsymbol{b}$ as before. Elimination is simply an organized way to solve linear equations. The first n pivots were positive because S is **positive** definite. Now there will be m negative pivots because $-A^{\mathrm{T}}S^{-1}A$ is **negative** definite. This is the unmistakable sign of a **saddle point in the Lagrangian** $\boldsymbol{L(x,\lambda)}$.

That function $L = \frac{1}{2}\boldsymbol{x}^{\mathrm{T}}S\boldsymbol{x} - \boldsymbol{\lambda}^{\mathrm{T}}(A^{\mathrm{T}}\boldsymbol{x} - \boldsymbol{b})$ is **convex in $\boldsymbol{x}$ and concave in $\boldsymbol{\lambda}$!**

Minimax = Maximin

There is more to learn from this problem. The $\boldsymbol{x}$-derivative and the $\boldsymbol{\lambda}$-derivative of L were set to zero in equation (6). We solved those two equations for $\boldsymbol{x}^*$ and $\boldsymbol{\lambda}^*$. That pair $(\boldsymbol{x}^*, \boldsymbol{\lambda}^*)$ is a saddle point of $\boldsymbol{L} = \frac{1}{2}\boldsymbol{x}^{\mathrm{T}}S\boldsymbol{x} - \boldsymbol{\lambda}^{\mathrm{T}}(A^{\mathrm{T}}\boldsymbol{x} - \boldsymbol{b})$. By solving (6) we found the minimum cost and its derivative in (8).

Suppose we separate this into two problems: a minimum and a maximum problem. First minimize $L(\boldsymbol{x}, \boldsymbol{\lambda})$ for each fixed λ. The minimizing $\boldsymbol{x}^*$ depends on $\boldsymbol{\lambda}$. Then find the $\boldsymbol{\lambda}^*$ that maximizes $L(\boldsymbol{x}^*(\boldsymbol{\lambda}), \boldsymbol{\lambda})$.

Minimize $\boldsymbol{L}$ at $\boldsymbol{x}^* = S^{-1}A\,\boldsymbol{\lambda}$ At that point $\boldsymbol{x}^*$, $\min L = -\dfrac{1}{2}\boldsymbol{\lambda}^{\mathrm{T}}A^{\mathrm{T}}S^{-1}A\,\boldsymbol{\lambda} + \boldsymbol{\lambda}^{\mathrm{T}}\boldsymbol{b}$

Maximize that minimum $\boldsymbol{\lambda}^* = (A^{\mathrm{T}}S^{-1}A)^{-1}\boldsymbol{b}$ gives $L = \dfrac{1}{2}\boldsymbol{b}^{\mathrm{T}}(A^{\mathrm{T}}S^{-1}A)^{-1}\boldsymbol{b}$

$$\boxed{\max_{\boldsymbol{\lambda}} \ \min_{\boldsymbol{x}} \ \boldsymbol{L} = \frac{1}{2}\boldsymbol{b}^{\mathrm{T}}(A^{\mathrm{T}}S^{-1}A)^{-1}\,\boldsymbol{b}}$$

This *maximin* was $\boldsymbol{x}$ first and $\boldsymbol{\lambda}$ second. The reverse order is *minimax*: $\boldsymbol{\lambda}$ first, $\boldsymbol{x}$ second.

The maximum over $\boldsymbol{\lambda}$ of $L(\boldsymbol{x}, \boldsymbol{\lambda}) = \dfrac{1}{2}\boldsymbol{x}^{\mathrm{T}}S\boldsymbol{x} - \boldsymbol{\lambda}^{\mathrm{T}}(A^{\mathrm{T}}\boldsymbol{x} - \boldsymbol{b})$ is $\begin{cases} +\infty \text{ if } A^{\mathrm{T}}\boldsymbol{x} \neq \boldsymbol{b} \\ \frac{1}{2}\boldsymbol{x}^{\mathrm{T}}S\boldsymbol{x} \text{ if } A^{\mathrm{T}}\boldsymbol{x} = \boldsymbol{b} \end{cases}$

The minimum over $\boldsymbol{x}$ of that maximum over $\boldsymbol{\lambda}$ is our same answer $\frac{1}{2}\boldsymbol{b}^{\mathrm{T}}(A^{\mathrm{T}}S^{-1}A)^{-1}\boldsymbol{b}$.

$$\boxed{\min_{\boldsymbol{x}} \; \max_{\boldsymbol{\lambda}} \; L = \frac{1}{2}\boldsymbol{b}^{\mathrm{T}}(A^{\mathrm{T}}S^{-1}A)^{-1}\boldsymbol{b}}$$

At the saddle point $(\boldsymbol{x}^*, \boldsymbol{\lambda}^*)$ **we have** $\dfrac{\partial L}{\partial \boldsymbol{x}} = \dfrac{\partial L}{\partial \boldsymbol{\lambda}} = 0$ **and** $\max\limits_{\boldsymbol{\lambda}} \min\limits_{\boldsymbol{x}} L = \min\limits_{\boldsymbol{x}} \max\limits_{\boldsymbol{\lambda}} L.$

Quadratic Programming : General Case

Now we have reached a dominant problem in applied and computational optimization. The quantity to minimize is still a **quadratic** starting with $\frac{1}{2}\boldsymbol{x}^{\mathrm{T}}S\boldsymbol{x}$. The constraints to satisfy are still **linear**. The difference is that S may be only **semidefinite** and some or all of the constraints may be **inequalities**.

In this case the Lagrange multipliers also obey linear inequalities ! Normally, either the constraint on $\boldsymbol{x}$ or the constraint on λ turns out to be tight (meaning that equality holds). The difficulty is : We don't know which of those inequalities is the tight one. And even just 10 constraint pairs would lead to 2^{10} configurations of $(\lambda = 0)$ or $(\lambda > 0)$. We cannot try all of those 1024 possibilities. We won't have to, if we depend properly on linear algebra.

Our reference for this topic is the excellent textbook "*Numerical Optimization*" by Wright and Nocedal. Here is the basic problem of quadratic programming :

$$\boxed{\begin{array}{llll} \text{Minimize} & q(x) = \dfrac{1}{2}\boldsymbol{x}^{\mathrm{T}}S\boldsymbol{x} + \boldsymbol{x}^{\mathrm{T}}\boldsymbol{c} & & \\ \text{subject to} & \boldsymbol{a}_i^{\mathrm{T}}\boldsymbol{x} = \boldsymbol{b}_i & i = 1 \text{ to } m & \text{(equations)} \\ & \boldsymbol{a}_j^{\mathrm{T}}\boldsymbol{x} \geq \boldsymbol{c}_j & j = 1 \text{ to } p & \text{(inequalities)} \end{array}}$$

The problem is convex if $S = S^{\mathrm{T}}$ is positive semidefinite, as we assume. Nonconvex problems could have several local minima.

Start with only equality constraints $A\boldsymbol{x} = \boldsymbol{b}$ and no inequalities. Introduce Lagrange multipliers λ_1 to λ_m as before, for those m constraints on $\boldsymbol{x}$:

The "Lagrangian" is $\boldsymbol{L} = \dfrac{1}{2}\boldsymbol{x}^{\mathrm{T}}S\boldsymbol{x} + \boldsymbol{x}^{\mathrm{T}}\boldsymbol{c} - \boldsymbol{\lambda}^{\mathrm{T}}(A\boldsymbol{x} - \boldsymbol{b})$.

Its $\boldsymbol{x}$ derivatives give $S\boldsymbol{x}^* + \boldsymbol{c} - A^{\mathrm{T}}\boldsymbol{\lambda}^* = \boldsymbol{0}$.

Its $\boldsymbol{\lambda}$ derivatives give $A\boldsymbol{x}^* = \boldsymbol{b}$ (reproducing the constraints)

As before, those equations yield the **KKT matrix**. This is the central matrix of constrained optimization : size $\boldsymbol{n+m}$:

$$K\begin{bmatrix} \boldsymbol{x}^* \\ -\boldsymbol{\lambda}^* \end{bmatrix} = \begin{bmatrix} S & A^{\mathrm{T}} \\ A & 0 \end{bmatrix}\begin{bmatrix} \boldsymbol{x}^* \\ -\boldsymbol{\lambda}^* \end{bmatrix} = \begin{bmatrix} -\boldsymbol{c} \\ \boldsymbol{b} \end{bmatrix} \tag{11}$$

Notice that many authors define the problem so that $-A^{\mathrm{T}}$ replaces A^{T} in the matrix equation. Wright and Nocedal show how a simple restatement of the problem leads to the symmetric (but not positive definite or even semidefinite !) KKT matrix in (11).

Crucial question : **When is the KKT matrix invertible ?** We can certainly assume that A has full rank m—the constraints are independent. Then we see an easy case that assumes all independent columns in K :

If S is positive definite, then K is invertible.

But now S might be only semidefinite ! In that case S needs help from A, to make those first n columns of K independent. Here is the key condition that allows A to give full support to S—making K invertible even if S is not :

If the columns of N are a basis for the nullspace of A, then the reduced matrix $N^{\mathrm{T}}SN$ must be positive definite for K to be invertible.

Note that N has $n - m$ columns, so $N^{\mathrm{T}}SN$ has size $n - m$. The assistance from A is producing independence of the first n columns of K, when S by itself has dependent columns. Here is the proof.

Suppose the KKT matrix is *not* invertible : $\begin{bmatrix} S & A^{\mathrm{T}} \\ A & 0 \end{bmatrix}\begin{bmatrix} \boldsymbol{w} \\ \boldsymbol{v} \end{bmatrix} = \begin{bmatrix} \mathbf{0} \\ \mathbf{0} \end{bmatrix}.$

Then $A\boldsymbol{w} = \mathbf{0}$. So $\boldsymbol{w} = N\boldsymbol{y}$ for some $\boldsymbol{y}$:

$$0 = \begin{bmatrix} \boldsymbol{w} \\ \boldsymbol{v} \end{bmatrix}^{\mathrm{T}} \begin{bmatrix} S & A^{\mathrm{T}} \\ A & 0 \end{bmatrix} \begin{bmatrix} \boldsymbol{w} \\ \boldsymbol{v} \end{bmatrix} = \boldsymbol{w}^{\mathrm{T}} S \boldsymbol{w} = \boldsymbol{y}^{\mathrm{T}} N^{\mathrm{T}} S N \boldsymbol{y}.$$

This proves : If K is not invertible then $N^{\mathrm{T}}SN$ is not positive definite. Some vector $\boldsymbol{w} \neq \mathbf{0}$ is in the nullspace of both S and A.

In the good case, when the KKT matrix is invertible, how should we solve equation (11) for the optimal $\boldsymbol{x}$ and $\boldsymbol{\lambda}$? This is a numerical question. The easy choice is elimination, with row exchanges to avoid small pivots. But row exchanges will ruin the symmetry of K. (Symmetry normally makes elimination go twice as fast, because we only need half of the matrix.) Wright and Nocedal develop stable algorithms that find $\boldsymbol{x}$ and $\boldsymbol{\lambda}$ quickly.

Inequality Constraints

Finally we come to problems with p inequalities $\boldsymbol{a}_j^{\mathrm{T}}\boldsymbol{x} \geq \boldsymbol{c}_j$ in addition to the m equality constraints $\boldsymbol{a}_i^{\mathrm{T}}\boldsymbol{x} = \boldsymbol{b}_i$. Here at the start is the crucial question : **Which of those p inequalities are "active" at the optimal solution x^* of the problem ?** If we knew that "active set" in advance, we could turn those inequalities into equations—and just ignore the "inactive" inequalities. But we don't know active from inactive until we solve the problem.

Here is a rough description of a successful *active set method.*

As we solve the problem, we have a "working set" of equality constraints at each step. If our current solution $\boldsymbol{x}_k$ does not minimize the quadratic for this working set problem, we temporarily ignore all constraints outside the working set and solve that subproblem. Now move as far as possible toward that solution, until running into a "blocking constraint". Add that constraint to the new working set and solve again.

Wright and Nocedal turn that rough idea into a very useful active set algorithm. The final working set is the correct active set of constraints that become equalities at the optimal $\boldsymbol{x}^*$. When we describe the simplex method for linear programming, you will see a similar idea. **Each step moves in a cost-reducing direction until it hits a new constraint.**

Dual Problems in Science and Engineering

Minimizing a quadratic $\frac{1}{2}\boldsymbol{x}^{\mathrm{T}}S\boldsymbol{x}$ with a linear constraint $A^{\mathrm{T}}\boldsymbol{x} = \boldsymbol{b}$ is not just an abstract exercise. **It is the central problem in physical applied mathematics**—when a linear differential equation is made discrete. Here are two major examples in engineering.

1 **Network equations for electrical circuits**

Unknowns: Voltages at nodes, currents along edges

Equations: Kirchhoff's Laws (balance of currents at each node)

Constraint: Ohm's Law (current proportional to voltage drop)

Matrices: $A^{\mathrm{T}}S^{-1}A$ is the **conductance matrix**.

2 **Finite element method for structures**

Unknowns: Displacements at nodes, stresses in the structure

Equations: Balance of forces at each node

Constraint: Stress-strain relations (Hooke's Law for a spring)

Matrices: $A^{\mathrm{T}}S^{-1}A$ is the **stiffness matrix**.

The full list would extend to every field of engineering. For stable problems, the stiffness matrix and the conductance matrix are symmetric positive definite. Normally the constraints are equations and not inequalities. Then mathematics offers three approaches to the modeling of the physical problem:

(i) Linear equations with the **stiffness matrix** or **conductance matrix** $\boldsymbol{A^{\mathrm{T}}S^{-1}A}$

(ii) Minimization with currents or stresses as the unknowns $\boldsymbol{x}$

(iii) Maximization with voltages or displacements as the unknowns $\boldsymbol{\lambda}$

In the end, the linear equations (i) are the popular choice. We reduce equation (9) to equation (10). Those network equations account for Kirchhoff and Ohm together. The structure equations account for force balance and elastic properties of the material. All electrical and mechanical laws are built into the final system with $A^{\mathrm{T}}S^{-1}A$.

For problems of fluid flow, that system of equations is often in its saddle point form. The unknowns $\boldsymbol{x}$ and $\boldsymbol{\lambda}$ are velocities and pressures. The numerical analysis is well described in *Finite Elements and Fast Iterative Solvers* by Elman, Silvester, and Wathen.

For network equations and finite element equations leading to conductance matrices and stiffness matrices $\boldsymbol{A^{\mathrm{T}}CA}$, one reference is my textbook on *Computational Science and Engineering*. The video lectures for **Math 18.085** are on OpenCourseWare **ocw.mit.edu**.

In statistics and least squares (linear regression), the matrix $A^{\mathrm{T}}\Sigma^{-1}A$ includes $\Sigma =$ covariance matrix. We divide by variances σ^2 to whiten the noise: $\Sigma = I$.

For nonlinear problems, the energy is no longer a quadratic $\frac{1}{2}\boldsymbol{x}^{\mathrm{T}}S\boldsymbol{x}$. *Geometric nonlinearities* appear in the matrix A. *Material nonlinearities* (usually simpler) appear in the matrix S. Large displacements and large stresses are a typical source of nonlinearity.

Problem Set 9.3

1 For $\boldsymbol{v} = (x, y)$ on the constraint line $3x + 4y = 1$, find the point $\boldsymbol{v}_p^* = (x_p^*, y_p^*)$ that minimizes $||\boldsymbol{v}||^p = |x|^p + |y|^p$. As p increases from $p = 1$ to $p = \infty$, that point $\boldsymbol{v}_p^*$ should move down the line from $\boldsymbol{v}_1^* = \left(0, \frac{1}{4}\right)$ to $\boldsymbol{v}_\infty^* = \left(\frac{1}{7}, \frac{1}{7}\right)$.

2 "The triangle inequality $||\boldsymbol{v} + \boldsymbol{w}|| \le ||\boldsymbol{v}|| + ||\boldsymbol{w}||$ is satisfied when **the unit ball** $||\boldsymbol{v}|| \le \mathbf{1}$ **has a convex shape"—and not otherwise**. This means that between any two points $\boldsymbol{v}_1, \boldsymbol{v}_2$ in the ball where $||\boldsymbol{v}|| \le \mathbf{1}$, the straight line must stay in the ball. Connect this convexity requirement to the triangle inequality.

3 Minimize $F(\boldsymbol{x}) = \frac{1}{2}\boldsymbol{x}^{\mathrm{T}} S \boldsymbol{x} = \frac{1}{2}x_1^2 + 2x_2^2$ subject to $A^{\mathrm{T}}\boldsymbol{x} = x_1 + 3x_2 = b$.

(a) What is the Lagrangian $L(\boldsymbol{x}, \lambda)$ for this problem ?

(b) Find and solve for $\boldsymbol{x}^*$ and $\boldsymbol{\lambda}^*$ the 3 equations "derivative of $L =$ zero".

(c) Draw Figure 9.7 for this problem with constraint line tangent to cost circle.

(d) Verify that the derivative of the minimum cost is $\boldsymbol{\partial F^* / \partial b} = \boldsymbol{\lambda}^*$.

4 Minimize $F(\boldsymbol{x}) = \frac{1}{2}\left(x_1^2 + 4x_2^2\right)$ subject to $2x_1 + x_2 = 5$. Find and solve the three equations $\partial L/\partial x_1 = 0$ and $\partial L/\partial x_2 = 0$ and $\partial L/\partial \lambda = 0$. Draw the constraint line $2x_1 + x_2 = 5$ tangent to the ellipse $\frac{1}{2}\left(x_1^2 + 4x_2^2\right) = F_{\text{min}}$ at the minimum (x_1^*, x_2^*).

5 The saddle point matrix K in equation (9) reduces to U by elimination. How many positive pivots for K ? How many positive eigenvalues for K ?

$$\boldsymbol{K} = \begin{bmatrix} S & A \\ A^{\mathrm{T}} & 0 \end{bmatrix} = \begin{bmatrix} 1 & 0 & 1 \\ 0 & 4 & 3 \\ 1 & 3 & 0 \end{bmatrix} \quad \rightarrow \quad \boldsymbol{U} = \begin{bmatrix} S & A \\ 0 & -A^{\mathrm{T}} S^{-1} A \end{bmatrix}$$

6 For any invertible symmetric matrix S, *the number of positive pivots equals the number of positive eigenvalues*. The pivots appear in $\boldsymbol{S} = \boldsymbol{LDL}^{\mathrm{T}}$ (triangular L) The eigenvalues appear in $\boldsymbol{S} = \boldsymbol{Q\Lambda Q}^{\mathrm{T}}$ (orthogonal Q). A nice proof sends L and D to Q and Λ. The eigenvalues don't cross zero: **The same signs in** $\boldsymbol{D}$ **and** $\boldsymbol{\Lambda}$.

Prove this **"Law of Inertia"** for any 2 by 2 invertible symmetric matrix S: $\boldsymbol{S}$ **has 0 or 1 or 2 positive eigenvalues when it has 0 or 1 or 2 positive pivots.**

1. Take the determinant of $LDL^{\mathrm{T}} = Q\Lambda Q^{\mathrm{T}}$ to show that $\det D$ and $\det \Lambda$ have the same sign. If the determinant is negative then S has __ positive eigenvalue in Λ and __ positive pivot in D.
2. If the determinant is positive, S could be positive definite or negative definite. Show that both pivots of S are positive when both eigenvalues are positive.

7 Find the minimum value of $F(\boldsymbol{x}) = \frac{1}{2}\left(x_1^2 + x_2^2 + x_3^2\right)$ with one constraint $x_1 + x_2 + x_3 = 3$ and then with an additional constraint $x_1 + 2x_2 + 3x_3 = 12$. The first minimum value should be less than the second minimum value: Why ? The two problems have a __ and __ tangent to a sphere in $\mathbf{R}^3$.

8 Describe the points $\boldsymbol{x} = \begin{bmatrix} x_1 & x_2 \end{bmatrix}^{\mathrm{T}}$ in the square $|x_1| \le 1, |x_2| \le 1$ by a system $A\boldsymbol{x} \le \boldsymbol{b}$ of linear inequalities. The problem is to find A and $\boldsymbol{b}$.

9.4 Linear Programming, Game Theory, and Duality

This section is about piecewise linear optimization problems—not quadratics. The standard LP problem has linear cost $\boldsymbol{c}^{\mathrm{T}}\boldsymbol{x}$ (to be minimized). It has linear constraints $A\boldsymbol{x} = \boldsymbol{b}$. And it requires $\boldsymbol{x} \geq \mathbf{0}$ (every component has $x_k \geq 0$). It is those inequalities that make life interesting. My favorite is the max flow / min cut problem coming below.

An inequality constraint $\boldsymbol{x_k} \geq \mathbf{0}$ has two states—active and inactive. If the minimizing solution ends up with $\boldsymbol{x}_k^* > \mathbf{0}$, then that requirement was inactive—it didn't change anything. Its Lagrange multiplier will have $\boldsymbol{\lambda}_k^* = \mathbf{0}$. The minimum cost is not affected by that constraint on x_k. But if the constraint $\boldsymbol{x_k} \geq \mathbf{0}$ actively forces the best $\boldsymbol{x}^*$ to have $x_k^* = 0$, then the multiplier will have $\lambda_k^* > 0$. So the optimality condition is $\boldsymbol{x}_k^* \boldsymbol{\lambda}_k^* = \mathbf{0}$ for each k.

Our approach here will be to see the "duality" between a minimum problem and a maximum problem— *two linear programs that are solved at the same time*. Dantzig invented the simplex algorithm to find the optimal solution.

One more point about linear programming. It solves all **2-person zero sum games.** Profit to one player is loss to the other player. The game begins with a payoff matrix A, and the rules are simple. At every step, player R chooses a row of A and player C chooses a column. **Usually not the same choices every turn**—a mixed strategy tends to be better. Then R pays C the number in that row and column of A.

The best strategies $\boldsymbol{x}$ and $\boldsymbol{y}$ (min for R, max for C) give a saddle point of $\boldsymbol{x}^{\mathrm{T}} A \boldsymbol{y}$. Three person games (like Jeopardy !) are another world entirely, much more difficult.

Linear Programming

Linear programming starts with a cost vector $\boldsymbol{c} = (c_1, \ldots, c_n)$. The problem is to minimize the cost $F(\boldsymbol{x}) = c_1 x_1 + \cdots + c_n x_n = \boldsymbol{c}^{\mathrm{T}}\boldsymbol{x}$. The constraints are m linear equations $A\boldsymbol{x} = \boldsymbol{b}$ and n inequalities $x_1 \geq 0, \ldots, x_n \geq 0$. We just write $\boldsymbol{x} \geq 0$ to include all n components:

Linear Program $\boxed{\textbf{Minimize } \boldsymbol{c}^{\mathrm{T}}\boldsymbol{x} \textbf{ subject to } A\boldsymbol{x} = \boldsymbol{b} \textbf{ and } \boldsymbol{x} \geq \mathbf{0}}$ (1)

If A is 1 by 3, $A\boldsymbol{x} = \boldsymbol{b}$ gives a plane like $x_1 + x_2 + 2x_3 = 4$ in 3-dimensional space. That plane will be chopped off by the constraints $x_1 \geq 0, x_2 \geq 0, x_3 \geq 0$. This leaves a triangle on a plane, with corners at $(x_1, x_2, x_3) = (4,0,0)$ and $(0,4,0)$ and $(0,0,2)$. Our problem is to find the point $\boldsymbol{x}^*$ in this triangle K that minimizes the cost $\boldsymbol{c}^{\mathrm{T}}\boldsymbol{x}$.

Because the cost is linear, **its minimum is reached at a corner of K**. Linear programming has to find that minimum cost corner $\boldsymbol{x}^*$. Computing all corners is exponentially impractical when m and n are large. So Dantzig's **simplex method** finds one starting corner that satisfies $A\boldsymbol{x} = \boldsymbol{b}$ and $\boldsymbol{x} \geq \mathbf{0}$. Then it moves along an edge of the triangle K to another (lower cost) corner. The cost $\boldsymbol{c}^{\mathrm{T}}\boldsymbol{x}$ drops at every step.

It is a linear algebra problem to find the steepest edge and the next corner (where that edge ends). The simplex method repeats this step many times, from corner to corner.

New starting corner, new steepest edge, new lower cost corner in the simplex method. Each step reduces the cost $\boldsymbol{c}^{\mathrm{T}}\boldsymbol{x}$. The method stops when the corner has minimum cost.

Our first interest is to identify the **dual problem**—a maximum problem for $\boldsymbol{y}$ in $\mathbf{R}^m$. It is standard to use $\boldsymbol{y}$ instead of $\boldsymbol{\lambda}$ for the dual unknowns—the Lagrange multipliers.

$$\textbf{Dual Problem} \quad \boxed{\textbf{Maximize } \boldsymbol{y}^{\mathrm{T}}\boldsymbol{b} \textbf{ subject to } A^{\mathrm{T}}\boldsymbol{y} \le \boldsymbol{c}.} \tag{2}$$

This is another linear program for the simplex method to solve. It has the same inputs $A, \boldsymbol{b}, \boldsymbol{c}$ as before. When the matrix A is m by n, the matrix A^{T} is n by m. So $A^{\mathrm{T}}\boldsymbol{y} \le \boldsymbol{c}$ has n constraints. A beautiful fact: $\boldsymbol{y}^{\mathrm{T}}\boldsymbol{b}$ in the maximum problem is never larger than $\boldsymbol{c}^{\mathrm{T}}\boldsymbol{x}$ in the minimum problem. **Maximum of $\boldsymbol{y}^{\mathrm{T}}\boldsymbol{b}$ in (2) $\le$ minimum of $\boldsymbol{c}^{\mathrm{T}}\boldsymbol{x}$ in (1)**

$$\textbf{Weak duality} \quad \boldsymbol{y}^{\mathrm{T}}\boldsymbol{b} = \boldsymbol{y}^{\mathrm{T}}(A\boldsymbol{x}) = (A^{\mathrm{T}}\boldsymbol{y})^{\mathrm{T}}\boldsymbol{x} \le \boldsymbol{c}^{\mathrm{T}}\boldsymbol{x} \tag{3}$$

Maximizing pushes $\boldsymbol{y}^{\mathrm{T}}\boldsymbol{b}$ upward. Minimizing pushes $\boldsymbol{c}^{\mathrm{T}}\boldsymbol{x}$ downward. The great duality theorem (**the minimax theorem**) says that they meet at the best $\boldsymbol{x}^*$ and the best $\boldsymbol{y}^*$.

$$\boxed{\textbf{Duality} \quad \textbf{The maximum of } \boldsymbol{y}^{\mathrm{T}}\boldsymbol{b} \textbf{ in (2) equals the minimum of } \boldsymbol{c}^{\mathrm{T}}\boldsymbol{x} \textbf{ in (1).}}$$

The simplex method will solve both problems at once. For many years that method had no competition. Now this has changed. Newer algorithms go directly *through* the set of allowed $\boldsymbol{x}$'s instead of traveling around its edges (from corner to corner). *Interior point methods* are competitive because they can use calculus to achieve steepest descent.

The situation right now is that either method could win—along the edges or inside.

Key points about linear programming and the simplex method

1. The set of vectors with $A\boldsymbol{x} = \boldsymbol{b}$ is chopped off by the planes $x_i \ge 0$.

2. The minimum cost $\boldsymbol{c}^{\mathrm{T}}\boldsymbol{x}$ occurs at a corner of that **feasible set K**.

3. The simplex method travels from corner to corner on the boundary of K.

4. It stops at the best corner, where all outgoing edges increase the cost $\boldsymbol{c}^{\mathrm{T}}\boldsymbol{x}$.

An Example is Online

This book's website **math.mit.edu/linearalgebra** develops an example: A Ph.D and a friend and a computer charge only \$5, \$3, \$8 for each hour of homework help, to solve 4 hard problems. They work x_1, x_2, x_3 hours and they solve $1, 1, 2$ problems per hour.

Minimize cost $\boldsymbol{c}^{\mathrm{T}}\boldsymbol{x} = 5x_1+3x_2+8x_3$ with $\boldsymbol{x} \ge \mathbf{0}$ and $A\boldsymbol{x} = x_1+x_2+2x_3 = 4$ problems.

The plane $A\boldsymbol{x} = 4$ cuts the x_1, x_2, x_3 axes at the corners $(4,0,0)$ and $(0,4,0)$ and $(0,0,2)$: all by Ph.D or all by friend or all by computer. One of those corners is the optimal solution!

In this small example you can find the cost of all three corners. In a realistic problem, the simplex method moves to a lower cost corner as one x_i drops to zero and another x_i increases from zero. Duality holds at the eventual winning corner.

Max Flow-Min Cut

Here is a special linear program. The matrix A will be the incidence matrix of a graph. The equation $A^{\mathrm{T}}\boldsymbol{y} = \mathbf{0}$ means that **flow in equals flow out** at every node. Each edge of the graph has a **capacity constraint** M_j—which the flow y_j along that edge cannot exceed.

The maximum problem sends the greatest possible flow from the source node s to the sink node t. This flow is returned from t to s on a special edge with unlimited capacity—drawn on the next page. The constraints on $\boldsymbol{y}$ are Kirchhoff's Current Law $A^{\mathrm{T}}\boldsymbol{y} = 0$ and the capacity bounds $|y_j| \leq M_j$ on each edge of the graph. The beauty of this example is that you can solve it by common sense (for a small graph). In the process, you discover and solve the dual minimum problem, which finds the **"mincut" of smallest capacity**.

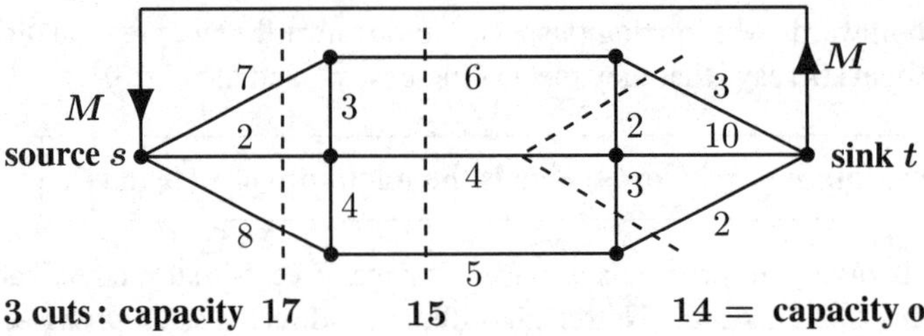

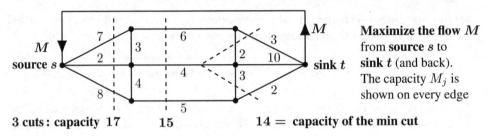

Maximize the flow M from **source** s to **sink** t (and back). The capacity M_j is shown on every edge

Figure 9.10: The max flow M is bounded by the capacity of any cut (dotted line). *By duality, the* ***capacity of the minimum cut*** *equals the* ***maximum flow*** : Here $M = \mathbf{14}$.

Begin by sending flow out of the source. The three edges going out from s have capacity $7 + 2 + 8 = \mathbf{17}$. Is there a tighter bound than $M \leq 17$?

Yes, a cut through the three middle edges only has capacity $6 + 4 + 5 = \mathbf{15}$. Therefore 17 cannot reach the sink. Is there a tighter bound than $M \leq 15$?

Yes, a cut through five later edges only has capacity $\mathbf{3 + 2 + 4 + 3 + 2} = \mathbf{14}$. The total flow M cannot exceed 14. Is that flow of 14 achievable and is this the tightest cut ?

Yes, 14 is the min cut (*it is an ℓ^1 problem* !). By duality **14 is also the max flow**.

Wikipedia shows a list of faster and faster algorithms to solve this important problem. It has many applications. If the capacities M_j are integers, the optimal flows y_j are integers. Normally integer programs are extra difficult, but not here.

A special max flow problem has all capacities $M_j = 1$ or 0. The graph is **bipartite** (all edges go from a node in part 1 down to a node in part 2). We are matching people in part 1 to jobs in part 2 (at most one person per job and one job per person). Then the maximum matching is M = max flow in the graph = max number of matchings.

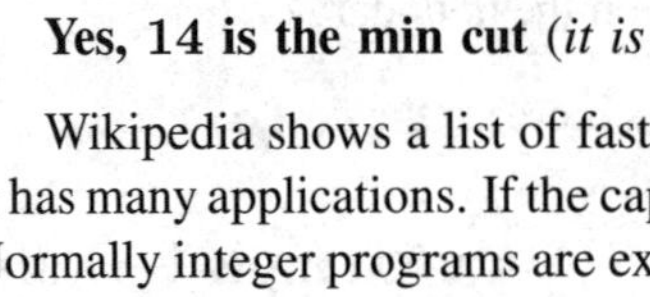

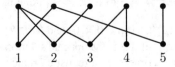

This bipartite graph allows a perfect matching: $\boldsymbol{M} = \mathbf{5}$. Remove the edge from 2 down to 1. Now only $M = 4$ assignments are possible, because 2 and 5 will only be qualified for job 5.

For bipartite graphs, max flow = min cut is König's theorem and Hall's marriage theorem.

Two Person Games

Games with three or more players are very difficult to solve. Groups of players can combine against the others, and those alliances are unstable. New teams will often form. It took John Nash to make good progress, leading to his Nobel Prize (in economics!). But two-person zero-sum games were completely solved by von Neumann. We will see their close connection to linear programming and duality. **Here are the rules.**

> The players are R and C. There is a **payoff matrix** $\boldsymbol{A}$. At every turn, player R chooses a *row* of A and player C chooses a *column*. The number in that row and column of A is the payoff *from* R *to* C. Then R and C take another turn.

$\boldsymbol{R}$ **wants to minimize** the payoff and $\boldsymbol{C}$ **wants to maximize**. Here is a very small payoff matrix. It has two rows for R to choose and three columns for C:

Payoff matrix

	y_1	y_2	y_3
x_1	**1**	**0**	**2**
x_2	**3**	**−1**	**4**

C likes those large numbers in column 3. R sees that the smallest number in that column is 2 (in row 1). Both players have no reason to move from this simple strategy of column 3 for C and row 1 for R. The payoff from R to C is 2 on every turn.

2 is smallest in its column and largest in its row

This is a **saddle point**. C cannot expect to win more than 2. R cannot expect to lose less than 2. The optimal strategies $\boldsymbol{x}^*$ and $\boldsymbol{y}^*$ are clear: row 1 for R and column 3 for C.

But a change in column 3 will require new thinking by both players.

New payoff matrix

	y_1	y_2	y_3
x_1	**1**	**0**	**4**
x_2	**3**	**−1**	**2**

R likes those small and favorable numbers in column 2. But C will never choose that column. Column 3 looks best (biggest) for C, and R should counter by choosing row 2 (to avoid paying 4). But then column 1 becomes better than column 3 for C, because winning 3 in column 1 is better than winning 2.

You are seeing that C still wants column 3 but must go sometimes to column 1. Also R must have a **mixed strategy**: choose rows 1 and 2 with probabilities x_1 and x_2. The choice at each turn must be unpredictable, or the other player will take advantage. So the decision for R is **two probabilities** $\boldsymbol{x_1 \geq 0}$ **and** $\boldsymbol{x_2 \geq 0}$ that add to $x_1 + x_2 = 1$.

row 1	1	0	4
row 2	3	−1	2
$\boldsymbol{x_1}$**(row 1)** $+$ $\boldsymbol{x_2}$**(row 2)**	$\boldsymbol{x_1 + 3x_2}$	$\boldsymbol{-x_2}$	$\boldsymbol{4x_1 + 2x_2}$

R will choose fractions x_1 and x_2 to make the worst (*largest*) payoff as small as possible. Remembering $x_2 = 1 - x_1$, this will happen when the two largest payoffs are equal:

$$x_1 + 3x_2 = 4x_1 + 2x_2 \;\; \textbf{means} \;\; \boldsymbol{x_1 + 3(1 - x_1) = 4x_1 + 2(1 - x_1).}$$

That equation gives $x_1^* = \frac{1}{4}$ **and** $x_2^* = \frac{3}{4}$. **The new mixed row is 2.5, −.75, 2.5.**

Similarly C will choose columns $1, 2, 3$ with probabilities y_1, y_2, y_3. Again they add to 1. That mixed strategy combines the three original columns into a new column for C.

column 1	column 2	column 3	**mix 1, 2, 3**
1	0	4	$y_1 + 4y_3$
3	−1	2	$3y_1 - y_2 + 2y_3$

C will choose the fractions $y_1 + y_2 + y_3 = 1$ to make the worst (*smallest*) payoff as large as possible. That happens when $y_2 = 0$ and $y_3 = 1 - y_1$. The two mixed payoffs are equal :

$$y_1 + 4(1 - y_1) = 3y_1 + 2(1 - y_1) \;\; \text{gives} \;\; -3y_1 + 4 = y_1 + 2 \;\; \text{and} \;\; y_1^* = y_3^* = \tfrac{1}{2}.$$

The new mixed column has 2.5 in both components. These optimal strategies identify 2.5 as the value of the game. With the mixed strategy $x_1^* = \frac{1}{4}$ and $x_2^* = \frac{3}{4}$, Player R can guarantee to pay **no more than 2.5**. Player C can guarantee to receive **no less than 2.5**. We have found the saddle point (best mixed strategies), with payoff = 2.5 both ways.

Von Neumann's **minimax theorem** for games gives a solution for every payoff matrix. It is equivalent to the duality theorem *min* $\boldsymbol{c}^{\mathrm{T}}\boldsymbol{x}$ = *max* $\boldsymbol{y}^{\mathrm{T}}\boldsymbol{b}$ for linear programming.

Semidefinite Programming (SDP)

The cost to minimize is still $\boldsymbol{c}^{\mathrm{T}}\boldsymbol{x}$: linear cost. But now the constraints on $\boldsymbol{x}$ involve symmetric matrices S. We are given S_0 to S_n and $S(\boldsymbol{x}) = S_0 + x_1 S_1 + \cdots + x_n S_n$ is required to be *positive semidefinite (or definite)*. Fortunately this is a convex set of $\boldsymbol{x}$'s—the average of two semidefinite matrices is semidefinite. (Just average the two energies $\boldsymbol{v}^{\mathrm{T}} S \boldsymbol{v} \geq 0$.)

Now the set of allowed $\boldsymbol{x}$'s could have curved sides instead of flat sides :

$$S_0 + x_1 S_1 + x_2 S_2 = \begin{bmatrix} x_1 & 1 \\ 1 & x_2 \end{bmatrix} \text{ is positive semidefinite when } \boldsymbol{x_1 \geq 0} \text{ and } \boldsymbol{x_1 x_2 \geq 1}.$$

Minimizing the maximum eigenvalue of $S(\boldsymbol{x})$ is also included with an extra variable t :

$$\text{Minimize } t \text{ so that } tI - S(\boldsymbol{x}) \text{ is positive semidefinite.}$$

For those and most semidefinite problems, interior-point methods are the best. Essentially we are solving a least squares problem at each iteration—usually 5 to 50 iterations.

As in linear programming, there is a **dual problem** (a maximization). The value of this dual is never above the value $\boldsymbol{c}^{\mathrm{T}}\boldsymbol{x}$ of the original. When we maximize in the dual and minimize $\boldsymbol{c}^{\mathrm{T}}\boldsymbol{x}$ in the primal, we hope to make those answers equal. But this might not happen for semidefinite programs with matrix inequalities.

SDP gives a solution method for matrix problems that previously looked too difficult.

Problem Set 9.4

1 Is the constraint $\boldsymbol{x} \geq \mathbf{0}$ needed in equation (3) for weak duality? Is the inequality $A^{\mathrm{T}}\boldsymbol{y} \leq \boldsymbol{c}$ already enough to prove that $(A^{\mathrm{T}}\boldsymbol{y})^{\mathrm{T}}\boldsymbol{x} \leq \boldsymbol{c}^{\mathrm{T}}\boldsymbol{x}$? I don't think so.

2 Suppose the constraints are $x_1 + 2x_2 + 2x_3 = 4$ and $x_1 \geq 0, x_2 \geq 0, x_3 \geq 0$. Find the three corners of this triangle in $\mathbf{R}^3$. Which corner minimizes the cost $\boldsymbol{c}^{\mathrm{T}}\boldsymbol{x} = 5x_1 + 3x_2 + 8x_3$?

3 What maximum problem for y is the dual to Problem 2? One constraint in the primal problem means one unknown y in the dual problem. Solve this dual problem.

4 Suppose the constraints are $\boldsymbol{x} \geq \mathbf{0}$ and $x_1 + 2x_3 + x_4 = 4$ and $x_2 + x_3 - x_4 = 2$. Two equality constraints on four unknowns, so a corner like $\boldsymbol{x} = (0, 6, 0, 4)$ has $4 - 2 = 2$ zeros. Find another corner with $\boldsymbol{x} = (x_1, x_2, 0, 0)$ and show that it costs more than the first corner.

5 Find the optimal (minimizing) strategy for R to choose rows. Find the optimal (maximizing) strategy for C to choose columns. What is the payoff from R to C in each game, at this optimal minimax point $\boldsymbol{x}^*, \boldsymbol{y}^*$?

$$\textbf{Payoff matrices} \quad \begin{bmatrix} 1 & 2 \\ 4 & 8 \end{bmatrix} \quad \begin{bmatrix} 1 & 4 \\ 8 & 2 \end{bmatrix}$$

6 If $A^{\mathrm{T}} = -A$ (antisymmetric payoff matrix), why is this a fair game for R and C with minimax payoff equal to zero?

7 Suppose the payoff matrix is a diagonal matrix Σ with entries $\sigma_1 > \sigma_2 > \ldots > \sigma_n$. What strategies are optimal for player R and player C?

8 Convert $||(x_1, x_2, x_3)||_1 \leq 2$ in the $\boldsymbol{\ell^1}$ **norm** to eight linear inequalities $A\boldsymbol{x} \leq \boldsymbol{b}$. The constraint $||\boldsymbol{x}|| \leq 2$ in the $\boldsymbol{\ell^\infty}$ **norm** also produces eight linear inequalities.

9 In the $\boldsymbol{\ell^2}$ **norm,** $||\boldsymbol{x}|| \leq 2$ is a quadratic inequality $x_1^2 + x_2^2 + x_3^2 \leq 4$. But in semidefinite programming (SDP) this becomes one matrix inequality $\boldsymbol{X}\boldsymbol{X}^{\mathrm{T}} \leq 4I$.

Why is this constraint $\boldsymbol{X}\boldsymbol{X}^{\mathrm{T}} \leq 4I$ *equivalent to* $\boldsymbol{x}^{\mathrm{T}}\boldsymbol{x} \leq 4$?

10 Suppose the constraints on $x_1, \ldots, x_n$ are $A\boldsymbol{x} \leq \boldsymbol{b}$ and $\boldsymbol{x} \geq \mathbf{0}$ (every $x_i \geq 0$). If two vectors $\boldsymbol{x}$ and $\boldsymbol{X}$ satisfy these constraints, show that their midpoint $\boldsymbol{M} = \frac{1}{2}\boldsymbol{x} + \frac{1}{2}\boldsymbol{X}$ satisfies the constraints. The constraint set K is *convex*.

Note Duality offers an important option: Solve the primal *or the dual.* That applies to optimization in machine learning.

Chapter 10

Learning from Data

The training set for deep learning is a collection of inputs $\boldsymbol{v}$ and the corresponding outputs $\boldsymbol{w}$. Each input is a vector with p components v_1 to v_p (the "features" of that input). Each output is a vector with components w_1 to w_q (the "classifiers" of that output). The form of the learning function $\boldsymbol{F}(\boldsymbol{x}, \boldsymbol{v})$ is chosen in advance to include "weights" $\boldsymbol{x}$ so that $\boldsymbol{F}(\boldsymbol{x}, \boldsymbol{v})$ is very close to (or equal to) $\boldsymbol{w}$.

With good training and good weights, the function $\boldsymbol{F}$ can be applied to unseen **test data** $\boldsymbol{V}$ from a statistically similar population. Success comes if $\boldsymbol{F}(\boldsymbol{x}, \boldsymbol{V})$ is close to the correct output $\boldsymbol{W}$ from that test data—which the learning function $\boldsymbol{F}$ has not seen.

Section 10.1 describes $\boldsymbol{F}(\boldsymbol{x}, \boldsymbol{v})$ as a chain of L functions $F_L(F_{L-1}(\ldots(F_2(F_1))))$. Each function F_k has its own weights $\boldsymbol{x}_k$. Those weights consist of a matrix A_k and a vector $\boldsymbol{b}_k$. The input to F_k is the output $\boldsymbol{v}_{k-1}$ from F_{k-1}. The linear part of F_k uses the weights to form $A_k\boldsymbol{v}_{k-1} + \boldsymbol{b}_k$. To each component of this vector we apply a nonlinear *"activation function"*. A simple and successful choice is

$$\text{ReLU}\,(y) = \max(y, 0) = \begin{cases} y & \text{if } \; y \geq 0 \\ 0 & \text{if } \; y \leq 0 \end{cases}$$

By applying ReLU to each component of $A_k\boldsymbol{v}_{k-1} + \boldsymbol{b}_k$, we have $\boldsymbol{v}_k = F_k(\boldsymbol{v}_{k-1})$. This vector $\boldsymbol{v}_k$ lies on layer k, and it is the input to the next function F_{k+1} in the chain (with its own weights A_{k+1} and $\boldsymbol{b}_{k+1}$).

Notice that ReLU is a piecewise linear function (two pieces). Then each component of $\boldsymbol{v}_k$ is also a piecewise linear function. Section 10.1 will attempt to count the pieces in the first function $\boldsymbol{v}_1 = F_1(\boldsymbol{v}_0)$ and in the chain of functions $\boldsymbol{v}_L = F(\boldsymbol{v}_0)$. Section 10.2 describes (among other things) a wonderful website for experiments.

Page 385 includes a link to Python and Julia codes on math.mit.edu/linearalgebra.

Generalization : Why is Deep Learning So Effective on New Data ?

A central question for deep learning is **generalization**. This refers to the behavior of a neural network on test data that it has not seen. If we construct a function $F(\boldsymbol{x}, \boldsymbol{v})$ that successfully classifies the known training data $\boldsymbol{v}$, will F continue to give correct results when $\boldsymbol{v}$ is outside the training set ?

The answer must lie in the stochastic gradient descent algorithm (Chapter 9) that chooses weights. Those weights $\boldsymbol{x}$ minimize a loss function $\boldsymbol{L}(\boldsymbol{x}, \boldsymbol{v})$ over the training data. The question is : **Why do the computed weights do so well on the test data ?**

Often we have more free parameters in $\boldsymbol{x}$ than data in $\boldsymbol{v}$. In that case we can expect many sets of weights (many vectors $\boldsymbol{x}$) to be equally accurate on the training set. Those weights could be good or bad. They could generalize well or poorly. Our algorithm chooses a particular $\boldsymbol{x}$ and applies those weights to new test data $\boldsymbol{V}$.

An unusual experiment produced unexpectedly positive results. The components of each input vector $\boldsymbol{v}$ were randomly shuffled. So the individual features represented by $\boldsymbol{v}$ suddenly had no meaning. Nevertheless the deep neural net learned those randomized samples. This random labeling of the training samples (the experiment has become famous) is described in arXiv : 1611.03530.

Patterns in Data

This part of the book is a great adventure—hopefully for the reader, certainly for the author, and it involves the whole science of thought and intelligence. You could call it Machine Learning (ML) or Artificial Intelligence (AI). Human intelligence created it (but we don't fully understand what we have done). Out of some combination of ideas and failures, attempting at first to imitate the neurons in the brain, a successful approach has emerged to finding **patterns in data**.

What is important to understand about deep learning is that those data-fitting computations, of almost unprecedented size, are often heavily underdetermined. There are a great many points in the training data, but there can be far more weights to be computed in a deep network. The art of deep learning is to find, among many possible solutions, one that **will generalize to new data.**

It is a remarkable observation that learning on deep neural nets with many weights leads to a successful tradeoff: F is accurate on the training set *and* the unseen test set. We depend on Chapter 9 for the key algorithm of **stochastic gradient descent** to compute good weights $\boldsymbol{x}$ in the learning function $\boldsymbol{F}(\boldsymbol{x}, \boldsymbol{v})$.

10.1 Piecewise Linear Learning Functions

Suppose one of the digits $0, 1, \ldots, 9$ is drawn in a square. How does a person recognize which digit it is? That neuroscience question is not answered here. How can a computer recognize which digit it is? This is a machine learning question. Probably both answers begin with the same idea: *Learn from examples.*

So we start with M different images (the training set). An image will be a set of p small pixels—or a vector $\boldsymbol{v} = (v_1, \ldots, v_p)$. The component v_i tells us the "grayscale" of the ith pixel in the image: how dark or light it is (or Red-Green-Blue for a color image). So we have M images each with p features: M vectors $\boldsymbol{v}$ in p-dimensional space. For every $\boldsymbol{v}$ in that training set we know the digit it represents.

In a way, we know a function. We have M inputs in $\mathbf{R}^p$ each with an output from 0 to 9. But we don't have a "rule". We are helpless with a new input. Machine learning proposes to create a rule that succeeds on (most of) the training images. But "succeed" means much more than that: The rule should give the correct digit for a much wider set of test images, taken from the same population. This essential requirement is called *generalization*.

What form shall the rule take? Here we meet the fundamental question. Our first answer might be: $F(\boldsymbol{v})$ could be a linear function from $\mathbf{R}^p$ to $\mathbf{R}^{10}$ (a 10 by p matrix). The M inputs $\boldsymbol{v}$ would be vectors with p features each, from the samples in the training set. The 10 outputs would be probabilities of the numbers 0 to 9.

The difficulty is: **Linearity is far too limited**. Artistically, two zeros could make an 8. Images don't add. In recognizing faces instead of numbers, we will need a lot of pixels—and the input-output rule from $\boldsymbol{F}$ is nowhere near linear.

Artificial intelligence languished for a generation, waiting for new ideas. There is no claim that the absolutely best class of functions has now been found. That class needs to allow a great many parameters (called weights). And it must remain feasible to compute all those weights (in a reasonable time) from knowledge of the training set.

The choice that has succeeded beyond expectation—and has turned shallow learning into deep learning—is *Continuous Piecewise Linear (CPL) functions.* **Linear** for simplicity, **continuous** to model an unknown but reasonable rule, and **piecewise** to achieve the nonlinearity that is an absolute requirement for real images and data.

This leaves the crucial question of computability. What parameters will quickly describe a large family of CPL functions? In approximating differential equations, finite elements start with a triangular mesh. But specifying many individual nodes in $\mathbf{R}^p$ is expensive. Much better if those nodes are the *intersections* of a smaller number of lines (or hyperplanes). Please know that a regular grid is too simple.

Here is a first construction of a piecewise linear function of the data vector $\boldsymbol{v}$. Choose a matrix A_1 and vector $\boldsymbol{b}_1$. Then set to zero (this is the nonlinear ReLU step) all negative components of $A_1\boldsymbol{v} + \boldsymbol{b}_1$. Then multiply by a matrix A_2 to produce 10 outputs in $\boldsymbol{w} = F(\boldsymbol{v}) = A_2(\text{ReLU}\,(A_1\boldsymbol{v} + \boldsymbol{b}))$. That vector ReLU $(A_1\boldsymbol{v} + \boldsymbol{b})$ forms a "hidden layer" between the input $\boldsymbol{v}$ and the output $\boldsymbol{w}$ in Figure 10.1.

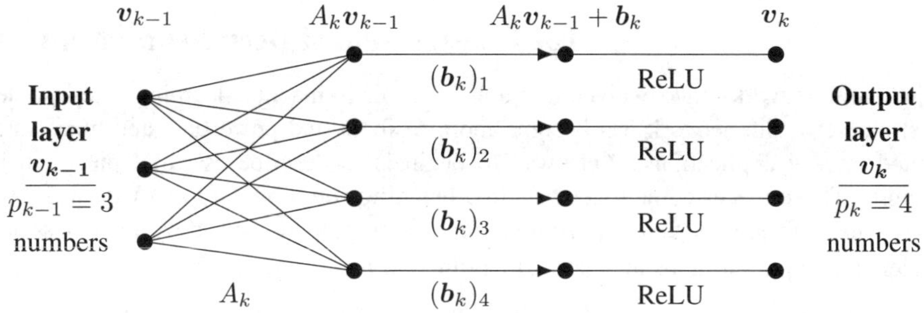

Figure 10.1: **Step k to layer** v_k has three parts: $v_k = F_k(v_{k-1}) = \text{ReLU}(A_k v_{k-1} + b_k)$
1 Multiply $A_k v_{k-1}$ (p_k rows) **2** Add vector b_k **3** Apply ReLU to each component

The ReLU function produces two linear pieces: either output $=$ input (if input ≥ 0) or output $=$ zero (if input ≤ 0). Originally, a smooth curve like $1/(1+e^{-x})$ was expected to help in optimizing the weights A_1, b_1, A_2. That assumption proved to be wrong.

The graph of each component of ReLU $(A_1 v + b_1)$ has two halfplanes (one is flat, from the zeros where $A_1 v + b_1$ is negative). If A_1 is p_1 by p_0, the input space is sliced by p_1 hyperplanes into r pieces. We can count those pieces ! This measures the "expressivity" of the function F_1. The count in this section is a sum of binomial coefficients $p!/n!(p-n)!$

$$r(p_1, p_0) = \binom{p_1}{0} + \binom{p_1}{1} + \cdots + \binom{p_1}{p_0} \quad \text{linear pieces of } F_1(v_0)$$

This number gives an impression of the graph of F_1—a mosaic of r flat pieces. But our function is not yet sufficiently expressive, and one more idea is needed: **add more layers.**

Here is the indispensable ingredient in the learning function F. The best way to create complex functions from simple functions is by **composition**. Each F_k uses A_k and b_k followed by the nonlinear ReLU : $F_k(v) = \text{ReLU}(A_k v + b_k)$. **The composition of L functions is the chain** $F(v) = F_L(F_{L-1}(\ldots F_2(F_1(v))))$. We have $L-1$ hidden layers before the final output layer. The network becomes "deeper" as L increases.

The great optimization problem of deep learning is to compute weights A_k and b_k that will make the outputs $F(v)$ nearly correct—close to the digit $w(v)$ that the image v represents. This problem of minimizing some measure of $F(v) - w$ is solved by following a gradient downhill. The gradient of this complicated function is computed by *backpropagation*—the workhorse of deep learning that executes the chain rule.

A historic competition in 2012 was to identify the 1.2 million images collected in ImageNet. The breakthrough neural network in AlexNet had 60 million weights. Its accuracy (after 5 days of stochastic gradient descent) cut in half the next best error rate. Deep learning had arrived.

Our goal here was to identify continuous piecewise linear functions as powerful approximators. That family is also convenient—closed under addition $F + G$ and maximization **max**(F, G) and composition $F(G)$. The magic is that the learning function gives accurate results on images v that F has never seen.

The Construction of Deep Neural Networks

Deep neural networks have evolved into a major force in machine learning. Step by step, the structure of the network has become more resilient and powerful—and more easily adapted to new applications. One way to begin is to describe essential pieces in the structure. Those pieces come together into a **learning function** $F(x, v)$ *with weights* x *that capture information from the training data* v—to prepare for use with new test data.

Here are important steps in creating that function F :

1	Key operation	**Composition** $F(x, v) = F_3(F_2(F_1(v)))$
2	Key rule	**Chain rule for x-derivatives of F**
3	Key algorithm	**Stochastic gradient descent to find the best weights x**
4	Key subroutine	**Backpropagation to execute the chain rule**
5	Key nonlinearity	**ReLU**(y) = **max**$(y, 0)$ = **ramp function**

Our first step is to describe the pieces $F_1, F_2, F_3, \ldots$ for one layer of neurons at a time. The weights x that connect the layers v are optimized in creating F. The vector $v = v_0$ comes from the training set, and the function F_k produces the vector v_k at layer k. The whole success is to build the power of F from those pieces F_k in equation (1).

F_k is a Piecewise Linear Function of v_{k-1}

The input to F_k is a vector v_{k-1} of length p_{k-1}. The output is a vector v_k of length p_k, ready for input to F_{k+1}. This function F_k has two parts, first linear and then nonlinear:

1. The linear part of F_k yields $A_k v_{k-1} + b_k$ (that bias vector b_k makes this "affine")

2. A fixed nonlinear function like ReLU is applied to *each component* of $A_k v_{k-1} + b_k$

$$\boxed{v_k = F_k(v_{k-1}) = \text{ReLU}\,(A_k v_{k-1} + b_k)} \tag{1}$$

The training data for each sample is in a feature vector v_0. The matrix A_k has shape p_k by p_{k-1}. The column vector b_k has p_k components. **These A_k and b_k are weights** constructed by the optimization algorithm. Frequently *stochastic gradient descent* computes optimal weights $x = (A_1, b_1, \ldots, A_L, b_L)$ in the central computation of deep learning. It relies on backpropagation to find the x-derivatives of F, to solve $\nabla F = 0$.

The activation function $\text{ReLU}(y) = \max(y, 0)$ gives flexibility and adaptability. Linear steps alone were of limited power and ultimately they were unsuccessful.

ReLU is applied to every "neuron" in every internal layer. There are p_k neurons in layer k, containing the p_k outputs from $A_k v_{k-1} + b_k$. Notice that ReLU itself is continuous and piecewise linear, as its graph shows. (The graph is just a ramp with slopes 0 and 1. Its derivative is the usual step function.) When we choose ReLU, the composite function $F = F_L \cdots (F_2(F_1(v)))$ has an important and attractive property:

The learning function F is continuous and piecewise linear in v.

One Internal Layer ($L = 2$)

Suppose we have measured $p_0 = 3$ features of one sample in the training set. Those features are the 3 components of the input vector $\boldsymbol{v} = \boldsymbol{v}_0$. Then the first function F_1 in the chain multiplies $\boldsymbol{v}_0$ by a matrix A_1 and adds an offset vector $\boldsymbol{b}_1$ (bias vector). If A_1 is 4 by 3 and the vector $\boldsymbol{b}_1$ is 4 by 1, we have $p_1 = 4$ components of $A_1\boldsymbol{v}_0 + \boldsymbol{b}_1$.

That step found 4 combinations of the 3 original features in $\boldsymbol{v} = \boldsymbol{v}_0$. The 12 weights in Figure 10.1 were optimized over many feature vectors $\boldsymbol{v}_0$ in the training set, to choose a 4 by 3 matrix (and a 4 by 1 bias vector) that would find 4 insightful combinations.

The final step to reach $\boldsymbol{v}_1$ is to apply the nonlinear "activation function" to each of the 4 components of $A_1\boldsymbol{v}_0 + \boldsymbol{b}_1$. Historically, the graph of that nonlinear function was often given by a smooth "*S-curve*". Particular choices then and now are in Figure 10.2.

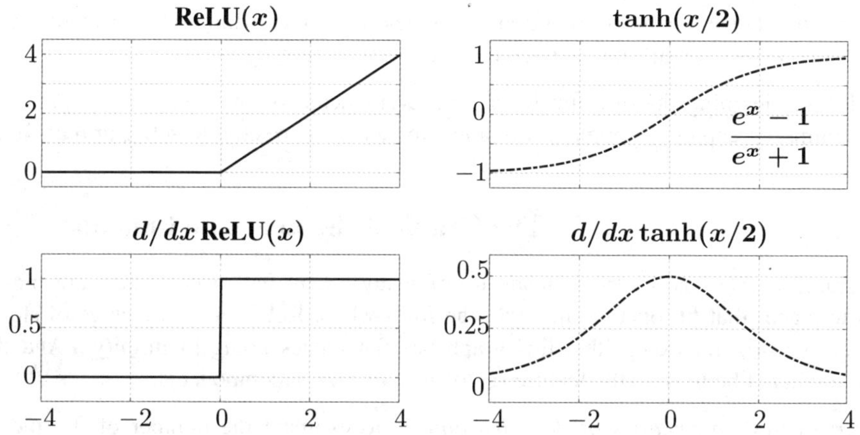

Figure 10.2: The **R**ectified **L**inear **U**nit and a sigmoid option for nonlinearity.

Previously it was thought that a sudden change of slope would be dangerous and possibly unstable. But large scale numerical experiments indicated otherwise ! A better result was achieved by the **ramp function** ReLU$(y) = \max(y, 0)$. We will work with ReLU :

$$\textbf{Substitute } A_1\boldsymbol{v}_0 + \boldsymbol{b}_1 \textbf{ into ReLU to find } \boldsymbol{v}_1 \qquad (\boldsymbol{v}_1)_k = \max((A_1\boldsymbol{v}_0 + \boldsymbol{b}_1)_k, 0). \tag{2}$$

Now we have the components of $\boldsymbol{v}_1$ at the four "neurons" in layer 1. The input layer held the three components of this particular sample of training data. We may have thousands or millions of samples. The optimization algorithm found A_1 and $\boldsymbol{b}_1$, usually by stochastic gradient descent using backpropagation to compute gradients of the overall loss.

Suppose our neural net is shallow instead of deep. It only has this first layer of 4 neurons. Then the final step will multiply the 4-component vector $\boldsymbol{v}_1$ by a 1 by 4 matrix A_2 (a row vector). It can add a single number $\boldsymbol{b}_2$ to reach the value $\boldsymbol{v}_2 = A_2\boldsymbol{v}_1 + \boldsymbol{b}_2$. The nonlinear function ReLU is not applied to the output.

$$\begin{array}{l}\text{Overall we compute } \boldsymbol{v}_2 = F(\boldsymbol{x}, \boldsymbol{v}_0) \text{ for each feature vector } \boldsymbol{v}_0 \text{ in the training set.} \\ \text{The steps are } \boldsymbol{v}_2 = A_2\boldsymbol{v}_1 + \boldsymbol{b}_2 = A_2\left(\text{ReLU}\left(A_1\boldsymbol{v}_0 + \boldsymbol{b}_1\right)\right) + \boldsymbol{b}_2 = F(\boldsymbol{x}, \boldsymbol{v}_0).\end{array} \tag{3}$$

The goal in optimizing $\boldsymbol{x} = (A_1, \boldsymbol{b}_1, A_2, \boldsymbol{b}_2)$ is that the output values $\boldsymbol{v}_\ell = \boldsymbol{v}_2$ at the last layer $\ell = 2$ should correctly capture the important features of the training data $\boldsymbol{v}_0$.

For a **classification problem** each sample $\boldsymbol{v}_0$ of the training data is assigned **1 or** $\mathbf{-1}$. We want the output $\boldsymbol{v}_2$ to have that correct sign (most of the time). For a **regression problem** we use the numerical value (**not just the sign**) of $\boldsymbol{v}_2$. We do not choose enough weights A_k and $\boldsymbol{b}_k$ to get every sample correct. And we do not necessarily want to! **Overfitting the training data** could give erratic results when $\boldsymbol{F}$ is applied to new and unknown test data. Depending on our choice of loss function $\boldsymbol{L}(\boldsymbol{x}, \boldsymbol{v}_2)$ to minimize, this can be least squares or entropy minimization: "square loss" or "cross-entropy loss".

Our hope is that **the function $\boldsymbol{F}$ has learned the data**. This is machine learning. We want a balance where the function $\boldsymbol{F}$ has learned what is important in recognizing *dog versus cat*—or identifying an oncoming car versus a turning car.

Machine learning doesn't aim to capture every detail of the numbers $0, 1, 2 \ldots, 9$. It just aims to capture enough information to decide correctly **which number it is**.

The Graph of the Learning Function $\boldsymbol{F}(\boldsymbol{v})$

The graph of $F(\boldsymbol{v})$ is a surface made up of many, many flat pieces—they are planes or hyperplanes that fit together along all the folds where ReLU gives a change of slope. This is like origami except that this graph has flat pieces going to infinity. And the graph might not be in $\mathbf{R}^3$—the feature vector $\boldsymbol{v} = \boldsymbol{v}_0$ has p_0 components.

Part of the *mathematics of deep learning* is to estimate the number of flat pieces and to visualize how they fit into one piecewise linear surface. That estimate comes after an example of a neural net with one internal layer. Each feature vector $\boldsymbol{v}_0$ contains p_0 measurements like height, weight, age of a sample in the training set.

In the example, F had three inputs in $\boldsymbol{v}_0$ and one output $\boldsymbol{v}_2$. Its graph will be a piecewise flat surface in 4-dimensional space. The height of the graph is $\boldsymbol{v}_2 = F(\boldsymbol{v}_0)$, over the point $\boldsymbol{v}_0$ in 3-dimensional space. Limitations of space in the book (and severe limitations of imagination in the author) prevent us from drawing that graph in $\mathbf{R}^4$. Nevertheless we can try to count the flat pieces, based on 3 inputs and 4 neurons and 1 output.

Note 1 With only $m = 2$ inputs (2 features for each training sample) the graph of F is a surface in 3D. We can attempt to describe it.

Note 2 You actually see points on the graph of F when you run examples on **playground.tensorflow.org. It is pictured in Section 10.2.**

That website offers four options for the training set of points $\boldsymbol{v}_0$. You choose the number of layers and neurons. Please choose the ReLU activation function! Then the program counts epochs as gradient descent optimizes the weights. (An *epoch* sees all samples on average once.) If you have allowed enough layers and neurons to correctly classify the blue and orange training samples, you will see a polygon separating them. **That polygon shows where $\boldsymbol{F} = \mathbf{0}$.** It is the cross-section of the graph of $\boldsymbol{F}(\boldsymbol{v})$ at height zero.

We will discuss experiments on this **playground.tensorflow.org** site in Section 10.2.

Important Note : Fully Connected versus Convolutional

We don't want to mislead the reader. Those "fully connected" nets are often not the most effective. If the weights around one pixel in an image can be repeated around all pixels (why not ?), then one row of A is all we need. The row can assign zero weights to faraway pixels. Local **convolutional neural nets** (**CNN's**) are also described in Section 10.2.

You will see that the count grows exponentially with the number of neurons and layers. That is a useful insight into the power of deep learning. We badly need insight because the size and depth of the neural network make it difficult to visualize in full detail.

Counting Flat Pieces in the Graph : One Internal Layer

It is easy to count entries in the weight matrices A_k and the bias vectors $\boldsymbol{b}_k$. Those numbers determine the function $\boldsymbol{F}$. But it is far more interesting to count the number of flat pieces in the graph of $\boldsymbol{F}$. This number measures the **expressivity** of the neural network. $F(\boldsymbol{x}, \boldsymbol{v})$ is a more complicated function than we fully understand (at least so far). The system is deciding and acting on its own, without explicit approval of its "thinking". For driverless cars we will see the consequences fairly soon.

Suppose $\boldsymbol{v}_0$ has p_0 components and $A_1\boldsymbol{v}_0 + \boldsymbol{b}_1$ has p_1 components. We have p_1 functions of $\boldsymbol{v}_0$. Each of those linear functions is zero along a hyperplane (dimension $p_0 - 1$). When we apply ReLU to that linear function it becomes piecewise linear, with a fold along that hyperplane. On one side of the fold its graph is sloping, on the other side the function changes from negative to zero.

Then the next matrix A_2 combines those p_1 piecewise linear functions of v_0, so we now have folds along p_1 *different hyperplanes*. This describes each piecewise linear component of the next layer $A_2(\text{ReLU}(A_1\boldsymbol{v}_0 + \boldsymbol{b}_1))$ in the typical case.

You could think of p_1 straight folds in the plane (the folds are actually along p_1 hyperplanes in p_0-dimensional space). The first fold separates the plane in two pieces. The next fold from ReLU will leave us with four pieces. The third fold is more difficult to visualize, but Figure 10.3 will show that there are *seven* (*not eight*) *pieces*.

In combinatorial theory, we have a **hyperplane arrangement**—and a theorem of Tom Zaslavsky counts the pieces. The proof is presented in Richard Stanley's great textbook on *Enumerative Combinatorics* (2001). But that theorem is more complicated than we need, because it allows the fold lines to meet in all possible ways. Our task is simpler because we assume that the fold lines are in "general position". For this case we now apply the neat counting argument given by Raghu, Poole, Kleinberg, Gangul, and Dickstein : *On the Expressive Power of Deep Neural Networks*, arXiv : 1606.05336.

Theorem Suppose the graph of $\boldsymbol{F}(\boldsymbol{v})$ has folds along p_1 hyperplanes. Those come from p_1 linear equations $\boldsymbol{a}_i^{\mathrm{T}}\boldsymbol{v} + b_i = 0$, in other words from ReLU at p_1 neurons. If $\boldsymbol{v}$ has p_0 components, then the p_1 folds produce $\boldsymbol{r}$ linear regions in $\boldsymbol{F}(\boldsymbol{v})$:

$$\boldsymbol{r}(\boldsymbol{p_1}, \boldsymbol{p_0}) = \binom{\boldsymbol{p_1}}{\boldsymbol{0}} + \binom{\boldsymbol{p_1}}{\boldsymbol{1}} + \cdots + \binom{\boldsymbol{p_1}}{\boldsymbol{p_0}} \tag{4}$$

These binomial coefficients are

$$\binom{\boldsymbol{p_1}}{\boldsymbol{i}} = \frac{p_1\,!}{i\,!(p_1 - i)\,!} \quad \text{with } 0\,! = 1 \text{ and } \binom{\boldsymbol{p_1}}{\boldsymbol{0}} = 1 \text{ and } \binom{\boldsymbol{p_1}}{\boldsymbol{i}} = 0 \text{ for } i > p_1.$$

Example The function $F(x, y, z) = \text{ReLU}\,(x) + \text{ReLU}\,(y) + \text{ReLU}\,(z)$ has 3 folds along the 3 planes $x = 0, y = 0, z = 0$. Those planes divide $\mathbf{R}^3$ into $r(3, 3) = 8$ pieces where $F = x + y + z$ and $x + z$ and x and 0 (and 4 more). Adding $\text{ReLU}\,(x + y + z - 1)$ gives a fourth fold and $r(4, 3) = 15$ pieces of $\mathbf{R}^3$. Not 16 because the new fold plane $x + y + z = 1$ does not meet the 8th original piece where $x < 0, y < 0, z < 0$.

George Polya's famous YouTube video *Let Us Teach Guessing* cut a cake by 5 planes. He helps the class to find $\boldsymbol{r}(5, 3) = 26$ pieces. Our cakes can be high-dimensional.

One hyperplane in $\mathbf{R}^m$ produces $\binom{1}{0} + \binom{1}{1} = 2$ regions. And 2 hyperplanes will produce $\boldsymbol{r}(2, p_0) = 1 + 2 + 1 = 4$ regions provided $p_0 > 1$. When $p_0 = 1$ we have two folds in a line, which only separates the line into $\boldsymbol{r}(2, 1) = 3$ pieces.

The count $\boldsymbol{r}$ of linear pieces will follow from the recursive formula

Proof of formula (4)

$$\boldsymbol{r}(p_1, p_0) = \boldsymbol{r}(p_1 - 1, p_0) + \boldsymbol{r}(p_1 - 1, p_0 - 1). \tag{5}$$

To understand that recursion, start with $p_1 - 1$ hyperplanes and $\boldsymbol{r}(p_1 - 1, p_0)$ regions. *Add one more hyperplane* H (dimension $p_0 - 1$). The established $p_1 - 1$ hyperplanes cut H into $\boldsymbol{r}(p_1 - 1, p_0 - 1)$ regions. Each of those pieces of H divides one existing region into two, adding $\boldsymbol{r}(p_1 - 1, p_0 - 1)$ regions to the original $\boldsymbol{r}(p_1 - 1, p_0)$; see Figure 10.3. So the recursion is correct, and we now apply equation (5) to compute $\boldsymbol{r}(p_1, p_0)$.

The count starts at $\boldsymbol{r}(1, 0) = \boldsymbol{r}(0, 1) = \mathbf{1}$. Then (4) is proved by induction on $p_1 + p_0$:

$$\begin{aligned}
\boldsymbol{r}(p_1 - 1, p_0) + \boldsymbol{r}(p_1 - 1, p_0 - 1) &= \sum_0^{p_0} \binom{p_1 - 1}{i} + \sum_0^{p_0 - 1} \binom{p_1 - 1}{i} \\
&= \binom{p_1 - 1}{0} + \sum_0^{p_0 - 1} \left[\binom{p_1 - 1}{i} + \binom{p_1 - 1}{i + 1}\right] \\
&= \binom{p_1}{0} + \sum_0^{p_0 - 1} \binom{p_1}{i + 1} = \sum_0^{p_0} \binom{\boldsymbol{p_1}}{\boldsymbol{i}}.
\end{aligned} \tag{6}$$

The two terms in brackets (second line) became one term because of a useful identity:

$$\binom{p_1 - 1}{i} + \binom{p_1 - 1}{i + 1} = \binom{p_1}{i + 1} \quad \text{and the induction is complete.} \tag{7}$$

Mike Giles suggested Figure 10.3 to show the effect of the last hyperplane H.

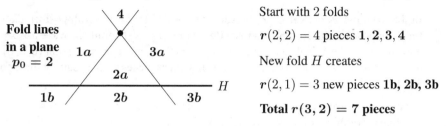

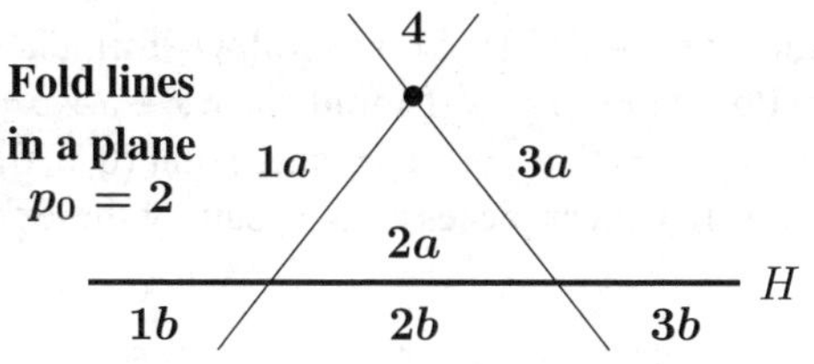

Start with 2 folds

$r(2,2) = 4$ pieces $\mathbf{1, 2, 3, 4}$

New fold H creates

$r(2,1) = 3$ new pieces **1b, 2b, 3b**

Total $r(3,2) = 7$ pieces

Figure 10.3: The $r(2,1) = 3$ pieces of H create 3 new regions. Then the count becomes $r(3,2) = 4 + 3 = 7$ flat regions in the continuous piecewise linear surface $v_2 = F(v_0)$. A fourth fold will cross all 3 existing folds and create 4 new regions, so $r\,(4,2) = \mathbf{11}$.

Flat Pieces of $F(v)$ with More Hidden Layers

Counting the linear pieces of $F(\boldsymbol{v})$ is much harder with 2 internal layers in the network. Again $\boldsymbol{v}_0$ and $\boldsymbol{v}_1$ have p_0 and p_1 components. Now $A_2\boldsymbol{v}_1 + \boldsymbol{b}_2$ will have p_2 components before ReLU. Each one is like the function F for one layer, described above. Then application of ReLU will create new folds in its graph. Those folds are along the lines where a component of $A_2\boldsymbol{v}_1 + \boldsymbol{b}_2$ is zero.

Remember that each component of $A_2\boldsymbol{v}_1 + \boldsymbol{b}_2$ is piecewise linear, not linear. So it crosses zero (if it does) along a piecewise linear surface, not a hyperplane. The straight lines in Figure 10.3 for the folds in $\boldsymbol{v}_1$ will only be *piecewise straight* for the folds in $\boldsymbol{v}_2$. So the count becomes variable, depending on the details of $\boldsymbol{v}_0, A_1, \boldsymbol{b}_1, A_2$, and $\boldsymbol{b}_2$.

Still we can estimate the number of linear pieces. We have p_2 piecewise straight lines (or piecewise hyperplanes) from p_2 ReLU's at the second hidden layer. If those lines were actually straight, we would have a total of $p_1 + p_2$ folds in each component of $\boldsymbol{v}_3 = \boldsymbol{F}(\boldsymbol{v}_0)$. *Then the formula* (4) *to count the pieces would have* $p_1 + p_2$ *in place of* p_1. This estimate is confirmed by Hanin and Rolnick (*arXiv*: 1906.00904). So the count of neurons, not layers and depth, decides the number of linear pieces in $F(\boldsymbol{x}, \boldsymbol{v})$.

Boris Hanin has shown that the **depth–width ratio L/W** is critical for neural nets.

The "chain" or "composition" of F_k's would simply represent matrix multiplication if all our functions were linear: $F_k(\boldsymbol{v}) = A_k\boldsymbol{v}$. Then $F_3(\boldsymbol{v}_0) = A_3A_2A_1\boldsymbol{v}_0$: just one matrix. For nonlinear F_k the meaning is the same: Compute $\boldsymbol{v}_1 = F_1(\boldsymbol{v}_0)$, then $\boldsymbol{v}_2 = F_2(\boldsymbol{v}_1)$, and finally $\boldsymbol{v}_3 = F_3(\boldsymbol{v}_2)$. This operation of **composition $F_3(F_2(F_1(\boldsymbol{v}_0)))$** is far more powerful in creating functions than addition or multiplication.

Dropout

Dropout is the removal of randomly selected neurons in the network. Those are components of the input layer $\boldsymbol{v}_0$ or of hidden layers $\boldsymbol{v}_n$ before the output layer $\boldsymbol{v}_L$. All weights in the A's and $\boldsymbol{b}$'s connected to those dropped neurons disappear from the net. Typically hidden layer neurons might be given probability $p = 0.5$ of surviving, and input components might have $p \geq 0.8$. *The main objective of random dropout is to avoid overfitting.* It is an inexpensive averaging method compared to combining predictions from many networks.

Problem Set 10.1

1 In the example F = ReLU (x) + ReLU (y) + ReLU (z) that follows formula (4) for r, suppose the 4th fold comes from ReLU $(x+y+z)$. Its fold plane $x+y+z=0$ now meets the 3 original fold planes $x=0, y=0, z=0$ at a single point $(0,0,0)$—*an exceptional case*. Describe the 12 (not 15) linear pieces of G = sum of these four ReLU's.

2 Suppose we have $m = 2$ inputs and N neurons on a hidden layer, so $F(x, y)$ is a linear combination of N ReLU's. Write out the formula for $r(N, 2)$ to show that the count of linear pieces of F has leading term $\frac{1}{2}N^2$.

3 Suppose we have $N = 18$ lines in a plane. If 9 are vertical and 9 are horizontal, how many pieces of the plane? Compare with $r(18, 2)$ when the lines are in general position and no three lines meet.

4 What weight matrix A_1 and bias vector $\boldsymbol{b}_1$ will produce ReLU $(x + 2y - 4)$ and ReLU $(3x - y + 1)$ and ReLU $(2x + 5y - 6)$ as the $N = 3$ components of the first hidden layer? (The input layer has 2 components x and y.) If the output w is the sum of those three ReLU's, how many pieces in $w(x, y)$?

5 Folding a line four times gives $r(4, 1) = 5$ pieces. Folding a plane four times gives $r(4, 2) = 11$ pieces. According to formula (4), how many flat subsets come from folding $\mathbf{R}^3$ four times? The flat subsets of $\mathbf{R}^3$ meet at 2D planes (like a door frame).

6 The binomial theorem finds the coefficients $\binom{N}{k}$ in $(\boldsymbol{a}+\boldsymbol{b})^N = \sum_{0}^{N} \binom{N}{k} a^k b^{N-k}$.

For $a=b=1$ what does this reveal about those coefficients and $r(N, m)$ for $m \geq N$?

7 In Figure 10.3, one more fold will produce 11 flat pieces in the graph of $z = F(x, y)$. Check that formula (4) gives $r(4, 2) = 11$. How many pieces after five folds?

8 Explain with words or show with graphs why each of these statements about Continuous Piecewise Linear functions (CPL functions) is true:

> ***M*** The maximum $M(x, y)$ of two CPL functions $F_1(x, y)$ and $F_2(x, y)$ is CPL.
> ***S*** The sum $S(x, y)$ of two CPL functions $F_1(x, y)$ and $F_2(x, y)$ is CPL.
> ***C*** If the one-variable functions $y = F_1(x)$ and $z = F_2(y)$ are CPL, so is the composition $C(x) = z = F_2(F_1(x))$.

9 How many weights and biases are in a network with $p_0 = 4$ inputs in each feature vector $\boldsymbol{v}_0$ and $N = 6$ neurons on each of the 3 hidden layers? How many activation functions (ReLU) are in this network, before the final output?

10 What is the smallest number of pieces that 20 fold lines can produce in a plane?

11 How many pieces are produced from 10 vertical and 10 horizontal folds?

12 What is the maximum number of pieces from 20 fold lines in a plane?

10.2 Creating and Experimenting

playground.tensorflow.org

This section begins with a website that beautifully illustrates deep learning. Layers can be added or removed; so can nodes (neurons) on each layer. Four examples are open for experiment—three are relatively easy and the fourth is not. This educational website came from Daniel Smilkov at Google, and it is highly recommended for simplicity and clarity.

Here is the picture that appears when you go to **playground.tensorflow.org**. We have added names in bold letters for the decisions to be made by the user, including

1 The number of layers and nodes (neurons) on each layer—not really hidden

2 The "learning rate" to control the stepsize in gradient descent (try .003)

3 The nonlinear activation (ReLU is recommended) applied at each node

4 The batch size (1 or more) and regularization in stochastic gradient descent

5 A choice of four sets of training data (three easy and one with difficult spirals)

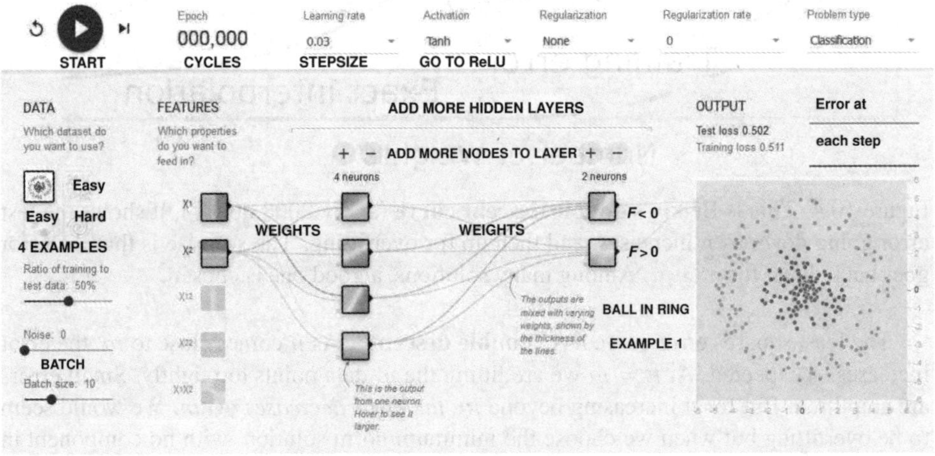

When you choose an example and press 'Start', the system begins to cycle through the blue and orange data points—aiming to create weights in a learning function $\boldsymbol{F}$ that classifies those points correctly. Heavier curves between nodes indicate greater weights.

Each cycle is an *epoch*. You could start with the first two features x_1 and x_2 (the position of every data point). The output on the right shows the loss function approaching a minimum (hopefully near zero). And you see how $\boldsymbol{F}$ is separating the blue and orange points more or less correctly. The separator is a white line where $\boldsymbol{F} = 0$.

The Problem Set suggests additional experiments. Dataset 4 needs more neurons!

Generalization and Double Descent

Here is a big question about deep learning: **Why does it succeed so well**? And why does it sometimes fail (as in the playground website)? Success is measured by computing the matrix weights x so that the function $F(x, v)$ is accurate for the training data v—and then applying F to new test data V. Often but not always the results are good. The output $F(x, V)$ is accurate, with weights that were chosen for the v's. Without help, the system has generalized to the new data.

This success needs explanation, which has begin to come. For linear problems, the classical test is **"stability"** of the approximation. Small changes of input bring only small changes of output. But our F is emphatically nonlinear as we add new data.

Statistics has issued a strong warning against **overfitting** the data. In a typical example of least squares, the results become less stable as the number n of fitting parameters approaches the number m of fitted data. But our F is a nonlinear function of the data. Experiments are needed to point the way.

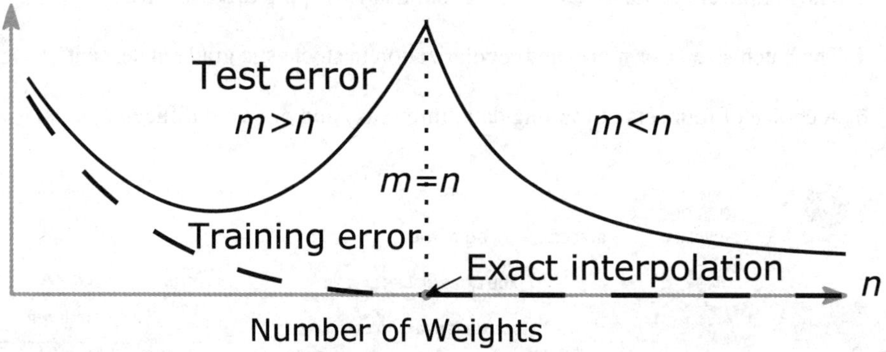

Figure 10.4: This is Belkin's **double descent curve** (arXiv 2003.00307). It shows the test error going down as n increases, and then up for overfitting. The surprise is that the error goes back down for $n > m$. Among many solutions, a good one is chosen.

The least squares error curve has **"double descent"**. As n comes close to m, the error increases as expected. At $n = m$ we are fitting the m data points too tightly. Small errors are amplified. But for n increasing beyond m, *the error decreases again*. We would seem to be overfitting but when we choose the minimum norm solution, with no component in the nullspace of A, the accuracy improves and the error curve begins another descent.

Belkin has led the analysis (still evolving) of this phenomenon.

Convolutional Neural Nets : Sharing the Weights

Up to now, the weight matrices A_k between layers of the neural net have been "dense". All entries in A_k were independent parameters. Our object was to fit the data as closely as possible, using every entry in A_k as an available parameter.

But for some input data (like images), this process is too expensive. An image has too many pixels and their interdependence is far from random. Nearby pixels are highly correlated. Distant pixels are weakly correlated. By restricting the weight matrices A_k, the training is faster and much more efficient. Convolutional nets were an important idea.

The restriction to **local convolution matrices** is successful in many problems. A convolution applies the *same weights* around each point in the image. It is *shift invariant*. And the convolution is *local* if the weight applied to all faraway pixels is zero. In signal processing, these are band matrices with constant diagonals : known as "filters" or "Toeplitz matrices". In image processing (2-dimensional, with column vectors A, B, C), the local convolution might have $3^2 = 9$ free parameters in A, B, C to be repeated around every point.

$$\begin{array}{c}\textbf{Filter in 1D}\\ =\\ \textbf{Toeplitz matrix}\end{array} \begin{bmatrix} a & b & c & & & \\ & a & b & c & & \\ & & a & b & c & \\ & & & a & b & c \end{bmatrix} \qquad \begin{array}{c}\textbf{Filter in 2D}\\ =\\ \textbf{Block Toeplitz matrix}\end{array} \begin{bmatrix} A & B & C & & & \\ & A & B & C & & \\ & & A & B & C & \\ & & & A & B & C \end{bmatrix}$$

Figure 10.5: **2D** convolutions replace a row of 3 weights by a square ABC of 9 weights.

Max-pooling

Multiply the previous layer $\boldsymbol{v}$ by A, as before. Then from each even-odd pair of outputs like components 2 and 3 of $A\boldsymbol{v}+\boldsymbol{b}$, *keep the maximum*. Please notice right away : Max-pooling is simple and fast, **but taking the maximum is not a linear operation**. It is a fast route to dimension reduction, and it may also reduce the danger of overfitting.

For an image (a 2-dimensional signal) we might use max-pooling over 2 by 2 squares of pixels. Each dimension is reduced by 2, The image dimension is reduced by 4. This speeds up the training, when the number of neurons on a hidden layer is divided by 4.

Normally a max-pooling step is given its own separate place in the overall architecture of the neural net. A part of that architecture might look like this :

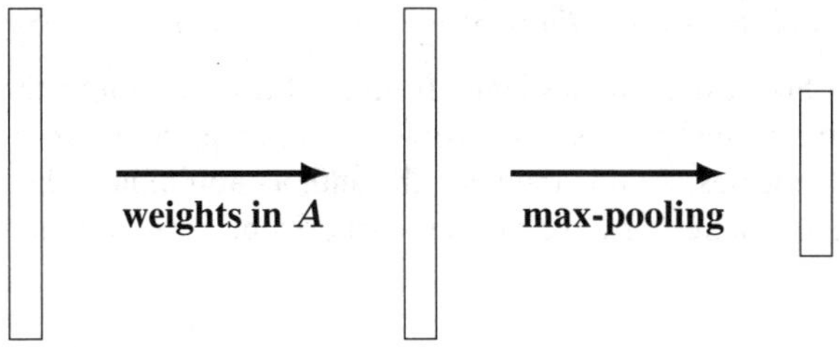

$\boldsymbol{v}_n$ in layer n $\quad \boldsymbol{v}_{n+1} = \textbf{ReLU}\,(A\boldsymbol{v}_n + \boldsymbol{b}_n) \quad \boldsymbol{v}_{n+2} = \textbf{max2}\,(\boldsymbol{v}_{n+1})$

Four Big Successes of Deep Learning

One day there will be a proper history of this subject—the ups and downs, successes and failures, and the ideas that made successes out of the failures. Here I report on two great achievements of the DeepMind team in London, plus a success with language translation and a new code that can specially affect students : **by solving all your Problem Sets** !

1 **AlphaGo Zero's** neural network was trained using TensorFlow. At the start it knew nothing about Go beyond the rules. It only saw the stones on the board. The key was reinforcement learning, playing 4.9 million games against itself. By evaluating the strength of each position, it developed the skills required to beat top humans (The earlier AlphaGo took months of training to achieve the same level.)

2 **AlphaFold** was named the Scientific Breakthrough of the Year in 2021. It predicts the 3D structure of millions of proteins. Knowledge of that structure had been a serious bottleneck in drug discovery, and AlphaFold predicts the fine details of a protein with phenomenal accuracy. This work (by many authors at DeepMind) ought to be a candidate for a Nobel Prize.

   Demis Hassabis founded DeepMind in 2010 and remains the CEO. He spoke in 2022 about the big problem of understanding intelligence, and the specific problem of protein structure. I think you would enjoy his lecture on **cbmm.mit.edu/video/using-ai-accelerate-scientific-discovery**.

3 **GoogleTranslate** Language translation is a perfect example of the Golden Rule of Machine Learning : *Don't code every grammatical rule and exception*. Let the computer discover the rules. If you are teaching your friend a game, just start to play.

   In a way, that is the whole principle of machine learning. Start with a training set, not a learning function. It is examples that produce the function. For language translation just as for chess and Go, the system will learn what it needs to know.

4 **A Neural Network Solves, Explains, and Generates University Math Problems** This network can understand and solve typical problems in math homeworks—over a wide range of courses. *What will become of Problem Sets* ? The code is automatically synthesized using OpenAI's Codex transformer in a highly original way.

   Iddo Drori led a large team to a code that far surpasses its competitors—the problem was recognized and he found a solution. The paper is found at arXiv : 2112.15594. It appeared in PNAS : *Proceedings of the National Academy of Sciences* (2022).

   By chance, the first test examples came from the Fifth Edition of the textbook you are reading. Linear algebra was first ! Now the examples come from a whole range of college math courses. At this moment, the authors don't know the eventual effect on teaching mathematics. I have no idea—but Problem Sets won't be the same !

Problem Set 10.2

Problems 1–2 use the blue ball, orange ring example on playground.tensorflow.org. When learning succeeds, a white polygon separates blue from orange in the figure.

1 Does learning succeed for $N = 4$ neurons ? What is the count $r(N, 2)$ of flat pieces in $F(\boldsymbol{v})$? The white polygon shows where flat pieces in the graph of $F(\boldsymbol{v})$ change sign as they go through the base plane $z = 0$. How many sides in the polygon ?

2 Reduce to $N = 3$ neurons in one layer. Does F still separate blue from orange ? How many flat pieces $r(3, 2)$ in $F(\boldsymbol{v})$? Reduce further to $N = 2$.

3 Example 2 has blue and orange in two quadrants each. With one layer, do $N = 3$ neurons and even $N = 2$ neurons classify that training data correctly ? How many flat pieces are needed for success ? Describe the unusual graph of $F(\boldsymbol{v})$ when $N = 2$.

4 Example 4 with blue and orange spirals is much more difficult ! With one hidden layer, can the network learn this training data ? Describe the results as N increases. How many neurons bring complete separation of the spirals with two hidden layers ?

I found that $4 + 4 + 2$ and $4 + 4 + 4$ neurons give very unstable iterations for that spiral graph. There were spikes in the training loss until the algorithm stopped trying. playground.tensorflow.org was a gift to the world from Daniel Smilkov.

Python and Julia Codes: Recognizing Handwritten Digits

The codes on this book's website are a gift from Ruochen Li, who took the *Learning from Data* course 18.065 in 2022. I am very grateful ! Image recognition is an excellent example based on measuring the grayscale of each pixel in the training set taken from MNIST (a typical size is 60,000 images). The problem is made more difficult by irregular handwriting and uncentered placement.

You can experiment with different sizes of the training set. Try activation functions like tanh in addition to ReLU. The batch size and the number of cycles (epochs) and the learning rate (stepsize) determine the training accuracy and speed. An appropriate loss function to minimize is categorical_crossentropy, when there are 10 categories 0 to 9 for the output number. And "softmax" is the final step when converting 10 numbers w_0 to w_9 into 10 probabilities p_0 to p_9 that add to 1 :

$$\textbf{Softmax} \qquad \textbf{Probabilities} \qquad \boldsymbol{p_j = \frac{1}{S}\, e^{w_j}} \quad \text{where} \quad \boldsymbol{S = \sum_{k=0}^{9} e^{w_k}}$$

I hope your experiments are successful. Please see **math.mit.edu/linearalgebra** and **https://www.simplilearn.com/keras-vs-tensorflow-vs-pytorch-article**.

10.3 Mean, Variance, and Covariance

1 The **probabilities** $\boldsymbol{p_1}$ to $\boldsymbol{p_n}$ of outcomes $\boldsymbol{x_1}$ to $\boldsymbol{x_n}$ are positive numbers adding to 1.

2 The **mean** m is the *average value* or expected value of the outcome: $\boldsymbol{m = \sum p_i x_i}$.

3 The **variance** $\boldsymbol{\sigma^2}$ is the average *squared distance* from the mean $\boldsymbol{\sum p_i (x_i - m)^2}$.

4 The **covariance** of random x_i, y_j with means m_x, m_y is $\sum \sum \boldsymbol{p_{ij}} (x_i - m_x)(y_j - m_y)$.

The mean is simple and we will start there. Right away we have two different situations. We may have the results (**sample values**) from a completed trial, or we may have the expected results (**expected values**) from future trials. Examples will show the difference:

Sample values Five random freshmen have ages $\mathbf{18, 17, 18, 19, 17}$

Sample mean $\frac{1}{5}(18 + 17 + 18 + 19 + 17) = \mathbf{17.8}$ based on the data

Probabilities The ages in a freshmen class are 17 (**20%**), 18 (**50%**), or 19 (**30%**)

Expected age E $[x]$ **of a random freshman** $= (0.2)\,17 + (0.5)\,18 + (0.3)\,19 = \mathbf{18.1}$

Both numbers 17.8 and 18.1 are correct averages. The sample mean starts with N samples x_1 to x_N from a completed trial. Their mean is the *average* of the N observed samples:

$$\textbf{Sample mean} \qquad m = \mu = \frac{1}{N}(x_1 + x_2 + \cdots + x_N) \tag{1}$$

The **expected value of** $\boldsymbol{x}$ starts with the probabilities $p_1, \ldots, p_n$ of the ages $x_1, \ldots, x_n$:

$$\textbf{Expected value} \quad m = \mathrm{E}[x] = p_1 x_1 + p_2 x_2 + \cdots + p_n x_n \tag{2}$$

This is $\boldsymbol{p \cdot x}$. The number $\mathbf{E}[x]$ **tells us what to expect,** $\boldsymbol{m = \mu}$ **tells us what we got.**

A fair coin has probability $p_0 = \frac{1}{2}$ of tails and $p_1 = \frac{1}{2}$ of heads. Then $\mathbf{E}[x] = \frac{1}{2}\,(0) + \frac{1}{2}(1)$. The fraction of heads in N coin flips is the sample mean. The **"Law of Large Numbers"** says that with probability 1, the sample mean will converge to its expected value $\mathbf{E}[x] = \frac{1}{2}$ as the sample size N increases. This does *not* mean that if we have seen more tails than heads, the next sample is likely to be heads. The odds remain 50-50.

The first 1000 flips do affect the sample mean. *But* 1000 *flips will not affect its limit*— because you are dividing by $N \to \infty$.

Note Probability and statistics are essential for modern applied mathematics. With multiple experiments, the mean $\boldsymbol{m}$ is a vector. The variances/covariances go into a matrix. Then linear algebra diagonalizes that covariance matrix and we can understand it.

Variance (around the mean)

The **variance** $\boldsymbol{\sigma^2}$ measures expected distance (squared) from the expected mean $\mathrm{E}[x]$. The **sample variance** $\boldsymbol{S^2}$ measures actual distance (squared) from the actual sample mean. The square root is the **standard deviation** $\boldsymbol{\sigma}$ **or** $\boldsymbol{S}$. After an exam, I email the results μ and S to the class. I don't know the expected m and σ^2 because I don't know the probabilities p_0 to p_{100} for each score. (After 60 years, I still have no idea what to expect.)

The distance is always *from the mean*—sample or expected. We are looking for the size of the "spread" around the mean value $x = m$. Start with N samples.

$$\textbf{Sample variance} \qquad \boldsymbol{S^2 = \frac{1}{N-1}\left[(x_1 - m)^2 + \cdots + (x_N - m)^2\right]} \tag{3}$$

The sample ages $x = 18, 17, 18, 19, 17$ have mean $m = 17.8$. That sample has variance 0.7 :

$$S^2 = \frac{1}{4}\left[(.2)^2 + (-.8)^2 + (.2)^2 + (1.2)^2 + (-.8)^2\right] = \frac{1}{4}(2.8) = \mathbf{0.7}$$

The minus signs disappear when we compute squares. Please notice ! Statisticians divide by $N - 1 = 4$ (and not $N = 5$) so that S^2 is an unbiased estimate of σ^2. One degree of freedom is already accounted for in the sample mean.

An important identity comes from splitting each $(x - m)^2$ into $x^2 - 2mx + m^2$:

$$\begin{aligned}
\text{sum of } (x_i - m)^2 &= (\text{sum of } x_i^2) - 2m(\text{sum of } x_i) + (\text{sum of } m^2) \\
&= (\text{sum of } x_i^2) - 2m(Nm) + Nm^2 \\
\textbf{sum of } \boldsymbol{(x_i - m)^2} &= \boldsymbol{(\textbf{sum of } x_i^2) - Nm^2}.
\end{aligned} \tag{4}$$

This is an equivalent way to find $(x_1 - m)^2 + \cdots + (x_N - m^2)$ by adding $x_1^2 + \cdots + x_N^2$.

Now start with probabilities p_i (never negative !) instead of samples. We find expected values instead of sample values. **The variance** $\boldsymbol{\sigma^2}$ **is the crucial number in statistics.**

$$\textbf{Variance} \qquad \boldsymbol{\sigma^2 = \mathrm{E}\left[(x - m)^2\right] = p_1(x_1 - m)^2 + \cdots + p_n(x_n - m)^2}. \tag{5}$$

We are squaring the distance from the expected value $m = \mathrm{E}[x]$. We don't have samples, only expectations. We know probabilities but we don't know the experimental outcomes. Equation (3) for the sample variance S^2 extends directly to equation (6) for the variance σ^2 :

$$\text{Sum of } p_i(x_i - m)^2 = (\text{Sum of } p_i x_i^2) - (\text{Sum of } p_i x_i)^2 \quad \text{or} \quad \boxed{\boldsymbol{\sigma^2 = \mathrm{E}[x^2] - (\mathrm{E}[x])^2}} \tag{6}$$

Example 1 Coin flipping has outputs $x = 0$ and 1 with probabilities $p_0 = p_1 = \frac{1}{2}$.

Mean $\boldsymbol{m}$ $= \frac{1}{2}(0) + \frac{1}{2}(1) = \boldsymbol{\frac{1}{2}}$ = average outcome = $\mathrm{E}[x]$

Variance $\boldsymbol{\sigma^2}$ $= \frac{1}{2}\left(0 - \frac{1}{2}\right)^2 + \frac{1}{2}\left(1 - \frac{1}{2}\right)^2 = \frac{1}{8} + \frac{1}{8} = \boldsymbol{\frac{1}{4}}$ = average of (distance from m)^2

For the average distance (not squared) from m, what do we expect ? $\mathbf{E}\boldsymbol{[x - m]}$ **is zero !**

Example 2 Find the variance σ^2 of the ages of college freshmen.

Solution The probabilities of ages $x_i = 17, 18, 19$ were $p_i = 0.2$ and 0.5 and 0.3. The expected value was $\boldsymbol{m = \sum p_i x_i = 18.1}$. The variance uses those same probabilities:

$$\begin{aligned}\boldsymbol{\sigma^2} &= (0.2)(\mathbf{17 - 18.1})^2 + (0.5)(\mathbf{18 - 18.1})^2 + (0.3)(\mathbf{19 - 18.1})^2 \\ &= (0.2)(\mathbf{1.21}) + (0.5)(\mathbf{0.01}) + (0.3)(\mathbf{0.81}) = \mathbf{0.49}. \ \textit{Then } \boldsymbol{\sigma} = \mathbf{0.7}.\end{aligned}$$

This measures the spread of 17, 18, 19 around $\mathrm{E}[x]$, weighted by probabilities 0.2, 0.5, 0.3.

Continuous Probability Distributions

Up to now we have allowed for ages 17, 18, 19: 3 outcomes. If we measure age in days instead of years, there will be too many possible ages. Better to allow *every number between 17 and 20*—a continuum of possible ages. Then probabilities p_1, p_2, p_3 for ages x_1, x_2, x_3 change to a **probability distribution $p(x)$ for a range of ages** $17 \le x \le 20$.

The best way to explain probability distributions is to give you two examples. They will be the **uniform distribution** and the **normal distribution**. The first (uniform) is easy. The normal distribution is all-important.

Uniform distribution Suppose ages are uniformly distributed between 17.0 and 20.0. All those ages are "equally likely". Of course any one exact age has no chance at all. There is zero probability that you will hit the exact number $x = 17.1$ or $x = 17 + \sqrt{2}$. But you can provide **the chance $F(x)$ that a random freshman has age less than x**:

The chance of age less than $x = 17$ is $F(17) = \mathbf{0}$	$x \le 17$ won't happen
The chance of age less than $x = 20$ is $F(20) = \mathbf{1}$	$x \le 20$ will surely happen
The chance of age less than x is $\boldsymbol{F(x) = \frac{1}{3}(x - 17)}$	**17 to 20 : $\boldsymbol{F}$ goes from 0 to 1**

From 17 to 20, the **cumulative distribution $F(x)$** increases linearly because p is constant.

You could say that $p(x)\,dx$ is the probability of a sample falling in between x and $x + dx$. This is "infinitesimally true": $p(x)\,dx$ is $F(x + dx) - F(x)$. Here is calculus:

$p(x)$ is the **derivative of $F(x)$** and $F(x)$ is the **integral of $p(x)$**.

$$\textbf{Probability of } \boldsymbol{a \le x \le b = \int_a^b p(x)\,dx = F(b) - F(a)} \qquad (7)$$

$F(b)$ is the probability of $x \le b$. Subtract $F(a)$ to keep $x \ge a$. That leaves $\mathrm{Prob}\{a \le x \le b\}$.

Mean and Variance of $p(x)$

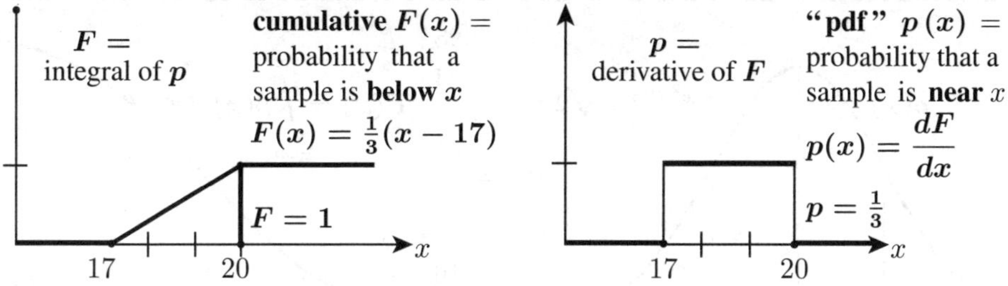

Figure 10.6: $F(x)$ is the **cumulative distribution** and its derivative $p(x) = dF/dx$ is the **probability density function (pdf)**. The area up to x under the graph of $p(x)$ is $F(x)$.

What are the mean m and variance σ^2 for a probability distribution? Previously we added $p_i x_i$ to get the mean (expected value). With a continuous distribution we **integrate $xp(x)$**:

Mean	$\boldsymbol{m = \mathrm{E}[x] = \int x\, p(x)\, dx}$
Variance	$\boldsymbol{\sigma^2 = \mathrm{E}\left[(x-m)^2\right] = \int p(x)\,(x-m)^2\, dx}$

When ages are uniform between 17 and 20, the mean is $m = 18.5$ with variance $\sigma^2 = \dfrac{3}{4}$.

$$\sigma^2 = \int_{17}^{20} \frac{1}{3}(x-18.5)^2\, dx = \int_0^3 \frac{1}{3}(x-1.5)^2\, dx = \frac{1}{9}(x-1.5)^3\Big]_{x=0}^{x=3} = \frac{2}{9}(1.5)^3 = \frac{3}{4}.$$

That is a typical example, and here is the complete picture for a uniform $p(x)$, 0 to a.

Uniform for $0 \le x \le a$ **Density** $\boldsymbol{p(x) = \dfrac{1}{a}}$	**Cumulative** $\boldsymbol{F(x) = \dfrac{x}{a}}$
Mean $\boldsymbol{m = \int x\; p(x)\, dx = \dfrac{a}{2}}$	**Variance** $\boldsymbol{\sigma^2} = \displaystyle\int_0^a \frac{1}{a}\left(x - \frac{a}{2}\right)^2 dx = \boldsymbol{\frac{a^2}{12}}$

For one random number between 0 and 1 $\left(\text{mean } \frac{1}{2}\right)$ the variance is $\sigma^2 = \frac{1}{12}$.

Normal Distribution : Bell-shaped Curve

The normal distribution is also called the "Gaussian" distribution. It is the most important of all probability density functions $p(x)$. The reason for its overwhelming importance comes from repeating an experiment and averaging the outcomes. Each experiment has its own distribution (like heads and tails). *The average approaches a normal distribution.*

Central Limit Theorem (informal) The average of N samples of "any" probability distribution approaches a normal distribution as $N \to \infty$.

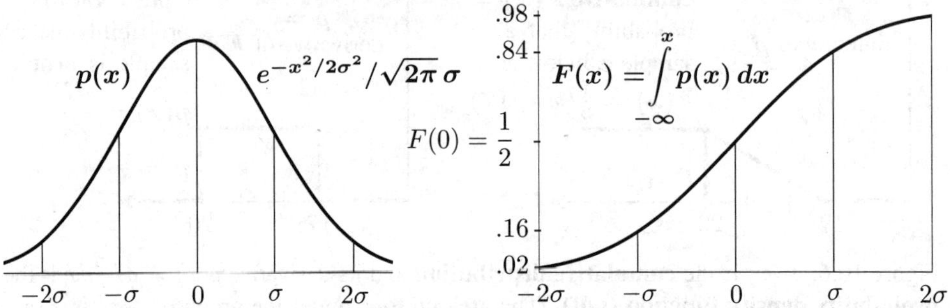

Figure 10.7: The standard normal distribution $p(x)$ has mean $m = 0$ and $\sigma = 1$.

The "standard normal distribution" $p(x)$ is symmetric around $x = 0$, so its mean value is $\boldsymbol{m = 0}$. It is chosen to have a standard variance $\boldsymbol{\sigma^2 = 1}$. It is called $\mathbf{N}(0, 1)$.

The graph of $\boldsymbol{p(x)} = \dfrac{\mathbf{1}}{\sqrt{\mathbf{2}\pi}} e^{-x^2/2}$ is the **bell-shaped curve** with variance $\sigma^2 = 1$.

The integral to compute σ^2 uses the idea in Problem 11 to reach 1. Figure 10.7 shows a graph of $p(x)$ for $\mathbf{N}(0, \sigma^2)$ and also its cumulative distribution $F(x) =$ integral of $p(x)$. From $\boldsymbol{F(x)} =$ **area under** $\boldsymbol{p(x)}$ you see a very important approximation for opinion polling:

The probability that a random sample falls between $-\sigma$ and σ is $\boldsymbol{F(\sigma) - F(-\sigma) \approx \frac{2}{3}}$.

Similarly, the probability that a random x lies between -2σ and 2σ ("*less than two standard deviations from the mean*") is $F(2\sigma) - F(-2\sigma) \approx 0.95$. If you have an experimental result further than 2σ from the mean, it is fairly sure to be not accidental. The normal distribution with any mean m and standard deviation σ comes by shifting and stretching $e^{-x^2/2}/\sqrt{2\pi}$. **Shift $\boldsymbol{x}$ to $\boldsymbol{x - m}$. Stretch $\boldsymbol{x - m}$ to $\boldsymbol{(x - m)/\sigma}$.**

Gaussian density $\boldsymbol{p(x)}$ **Normal distribution** $\mathbf{N}(\boldsymbol{m, \sigma^2})$	$\boldsymbol{p(x) = \dfrac{1}{\sigma\sqrt{2\pi}} e^{-(x-m)^2/2\sigma^2}}$	(8)

The integral of $p(x)$ is $F(x)$—the probability that a random sample will fall below x. There is no simple formula to integrate $e^{-x^2/2}$, so $F(x)$ is computed very carefully.

Mean and Variance: N Coin Flips and $N \to \infty$

Example 3 **Suppose x is 1 or -1 with equal probabilities $p_1 = p_{-1} = \frac{1}{2}$.**

The mean value is $\boldsymbol{m} = \frac{1}{2}(1) + \frac{1}{2}(-1) = \mathbf{0}$. The variance is $\boldsymbol{\sigma^2} = \frac{1}{2}(1)^2 + \frac{1}{2}(-1)^2 = \mathbf{1}$.

The key question is the *average* $A_N = (x_1 + \cdots + x_N)/N$. The independent x_i are ± 1 and we are dividing their sum by N. The expected mean of A_N is still zero. The law of large numbers says that this sample average $A_N =$ **(# heads − # tails)/N** approaches zero with probability 1. How fast does A_N approach zero? **What is its variance σ_N^2?**

$$\text{By linearity} \quad \boldsymbol{\sigma_N^2} = \frac{\sigma^2}{N^2} + \frac{\sigma^2}{N^2} + \cdots + \frac{\sigma^2}{N^2} = N\frac{\sigma^2}{N^2} = \frac{\mathbf{1}}{\boldsymbol{N}} \quad \text{since } \sigma^2 = 1.$$

> Here are the results from three numerical tests: random 0 or 1 averaged over N trials.
>
> [**48** 1's from $N =$ **100**] [**5035** 1's from $N =$ **10000**] [**19967** 1's from $N =$ **40000**].
>
> **The standardized $\boldsymbol{X} = (\boldsymbol{x} - \boldsymbol{m})/\boldsymbol{\sigma} = \left(A_N - \frac{1}{2}\right) \times 2\sqrt{N}$** was [**−.40**] [**.70**] [**−.33**].

The Central Limit Theorem says that the average of many coin flips will approach a normal distribution. Let us begin to see how that happens: **binomial approaches normal.**

The "binomial" probabilities $p_0, \ldots, p_N$ count the number of heads in N coin flips.

For each (fair) flip, the probability of heads is $\frac{1}{2}$. For $N = 3$ flips, the probability of heads all three times is $\left(\frac{1}{2}\right)^3 = \frac{1}{8}$. The probability of heads twice and tails once is $\frac{3}{8}$, from three sequences HHT and HTH and THH. These numbers $\frac{\mathbf{1}}{\mathbf{8}}$ and $\frac{\mathbf{3}}{\mathbf{8}}$ are pieces of $\left(\frac{1}{2} + \frac{1}{2}\right)^3 = \frac{\mathbf{1}}{\mathbf{8}} + \frac{\mathbf{3}}{\mathbf{8}} + \frac{\mathbf{3}}{\mathbf{8}} + \frac{\mathbf{1}}{\mathbf{8}} = 1$. *The average number of heads in* 3 *flips is* 1.5.

$$\textbf{Mean } m = (3 \text{ heads})\frac{\mathbf{1}}{\mathbf{8}} + (2 \text{ heads})\frac{\mathbf{3}}{\mathbf{8}} + (1 \text{ head})\frac{\mathbf{3}}{\mathbf{8}} + \mathbf{0} = \frac{3}{8} + \frac{6}{8} + \frac{3}{8} = \mathbf{1.5 \text{ heads}}$$

With N flips, Example 3 (or common sense) gives a mean of $\boldsymbol{m} = \boldsymbol{\Sigma x_i p_i} = \frac{\boldsymbol{N}}{\mathbf{2}}$ heads.

The variance σ^2 is based on the *squared distance* from this mean $N/2$. With $N = 3$ the variance is $\sigma^2 = \frac{\mathbf{3}}{\mathbf{4}}$ (*which is* $N/4$). To find σ^2 we add $(x_i - m)^2 p_i$ with $m = 1.5$:

$$\boldsymbol{\sigma^2} = (3 - 1.5)^2 \frac{1}{8} + (2 - 1.5)^2 \frac{3}{8} + (1 - 1.5)^2 \frac{3}{8} + (0 - 1.5)^2 \frac{1}{8} = \frac{9 + 3 + 3 + 9}{32} = \frac{\mathbf{3}}{\mathbf{4}}.$$

For any N, the variance for a binomial distribution is $\boldsymbol{\sigma_N^2} = \boldsymbol{N/4}$. Then $\sigma_N = \sqrt{N}/2$.

Figure 10.8 will show how the probabilities of 0, 1, 2, 3, 4 heads in $N = 4$ flips come close to a bell-shaped Gaussian. We are seeing the Central Limit Theorem in action. That Gaussian is centered at the mean value $m = N/2 = 2$.

To reach the standard Gaussian (mean 0 and variance 1) we shift and rescale. If x is the number of heads in N flips—the average of N zero-one outcomes—then x is shifted by its mean $m = N/2$ and rescaled by $\sigma = \sqrt{N}/2$ to produce the standard X:

Shifted and scaled $\quad X = \dfrac{x-m}{\sigma} = \dfrac{x - \frac{1}{2}N}{\sqrt{N}/2} \quad (N = 4 \text{ has } X = x - 2)$

Subtracting m is "centering" or "detrending". The mean of X is zero.

Dividing by σ is "normalizing" or "standardizing". The variance of X is 1.

It is fun to see the Central Limit Theorem giving the right answer at the center point $X = 0$. At that point, the factor $e^{-X^2/2}$ equals 1. We know that the variance for N coin flips is $\sigma^2 = N/4$. **The center of the bell-shaped curve has height $1/\sqrt{2\pi}\,\sigma = \sqrt{2/\pi N}$.**

What is the height at the center of the coin-flip distribution (binomial distribution)? For $N = 4$, the probabilities p_0 to p_4 for $0, 1, 2, 3, 4$ heads come from $\left(\frac{1}{2} + \frac{1}{2}\right)^4$.

$$\textbf{Center probability} = \frac{\mathbf{6}}{\mathbf{16}} \qquad \left(\frac{1}{2}+\frac{1}{2}\right)^4 = \frac{1}{16} + \frac{4}{16} + \frac{\mathbf{6}}{\mathbf{16}} + \frac{4}{16} + \frac{1}{16} = 1. \tag{9}$$

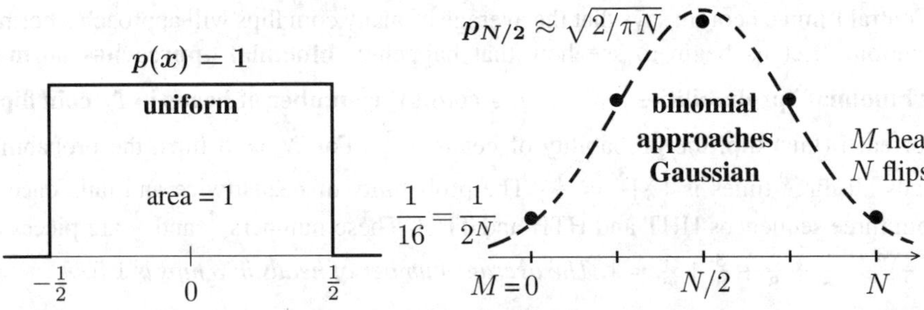

Figure 10.8: The probabilities $p = (1, 4, 6, 4, 1)/16$ for the number of heads in 4 flips. These p_i approach a Gaussian distribution with variance $\sigma^2 = N/4$ centered at $m = N/2$. For X, the Central Limit Theorem gives convergence to the normal distribution $\mathbf{N}(0, 1)$.

The **binomial** $\left(\frac{1}{2} + \frac{1}{2}\right)^N$ tells us the probabilities for $0, 1, \ldots, N$ heads

The **center term** is the probability of $\frac{N}{2}$ heads, $\frac{N}{2}$ tails $\quad \dfrac{\mathbf{1}}{\mathbf{2}^N} \dfrac{\mathbf{N!}}{\mathbf{(N/2)!\,(N/2)!}}$

For $N = 4$, those factorials produce $4!/2!\,2! = 24/4 = 6$. For large N, *Stirling's formula* $\sqrt{2\pi N}(N/e)^N$ *is a close approximation to* $N!$. Use this formula for N and twice for $N/2$:

$$\begin{array}{l}\textbf{Limit of binomial}\\ \textbf{Center probability}\end{array} \qquad p_{N/2} \approx \frac{1}{2^N}\frac{\sqrt{2\pi N}(N/e)^N}{\pi N (N/2e)^N} = \frac{\sqrt{2}}{\sqrt{\pi N}} = \frac{1}{\sqrt{2\pi}\,\sigma}. \tag{10}$$

The last step used the variance $\sigma^2 = N/4$ for N coin-flips. The result $1/\sqrt{2\pi}\sigma$ matches the center value ($\sqrt{2/N\pi}$ above) for the Gaussian. The Central Limit Theorem is true:

The centered binomial distribution approaches the normal distribution $\mathbf{N}(0, \sigma^2)$.

Covariance Matrices and Joint Probabilities

Linear algebra enters when we run M different experiments at once. We might measure age and height ($M = 2$ measurements of N children). Each experiment has its own mean value. So we have a vector $\boldsymbol{m} = (\boldsymbol{m_a}, \boldsymbol{m_h})$ containing two mean values. Those could be *sample means* of age and height. Or m_1 and m_2 could be *expected values* of age and height based on known probabilities.

A matrix becomes involved when we look at variances in age and in height. Each experiment will have a sample variance S_i^2 and an expected $\sigma_i^2 = \mathrm{E}\left[(x_i - m_i)^2\right]$ based on the squared distance from its mean. Those variances σ_1^2 and σ_2^2 will go on the main diagonal of the "variance-covariance matrix". So far we have made no connection between the two parallel experiments. They measure different random variables, but age and height are not independent!

If we measure age and height for children, the results will be strongly correlated. Older children are generally taller. Suppose the means m_a and m_h are known. Then σ_a^2 and σ_h^2 are the separate variances in age and height. **The new number is the covariance $\boldsymbol{\sigma_{ah}}$, which measures the connection of each possible age to each possible height.**

$$\textbf{Covariance} \quad \sigma_{ah} = \mathbf{E}\left[(\textbf{age} - \textbf{mean age})\ (\textbf{height} - \textbf{mean height})\right]. \tag{11}$$

This definition needs a close look. To compute σ_{ah}, it is not enough to know the probability of each age and the probability of each height. We have to know the **joint probability $\boldsymbol{p_{ah}}$ of each pair (age and height)**. This is because age is related to height.

$\boldsymbol{p_{ah}}$ = probability that a random child has age $= \boldsymbol{a}$ **and** height $= \boldsymbol{h}$: both at once

$\boldsymbol{p_{ij}}$ = **probability that experiment 1 produces $\boldsymbol{x_i}$ and experiment 2 produces $\boldsymbol{y_j}$**

Suppose experiment 1 (age) has mean m_1. Experiment 2 (height) has its own mean m_2. The covariance between experiments 1 and 2 looks at **all pairs** of ages $\boldsymbol{x_i}$ and heights $\boldsymbol{y_j}$. To match equation (11), we multiply by the joint probability $\boldsymbol{p_{ij}}$ of that age-height pair.

Expected value of
$(\boldsymbol{x - m_1})(\boldsymbol{y - m_2})$

$$\textbf{Covariance} \quad \sigma_{12} = \sum_{\textbf{all } i, j}\sum p_{ij}(x_i - m_1)(y_j - m_2) \tag{12}$$

To capture this idea of "joint probability p_{ij}" we begin with two small examples.

Example 4 **Flip two coins separately**. With 1 for heads and 0 for tails, the results can be $(1,1)$ or $(1,0)$ or $(0,1)$ or $(0,0)$. Those four outcomes all have probability $\left(\frac{1}{2}\right)^2 = \frac{1}{4}$. For independent experiments we multiply probabilities: *The covariance will be zero.*

$$p_{ij} = \textbf{Probability of } (i,j) = (\textbf{Probability of } i) \textbf{ times } (\textbf{Probability of } j).$$

Example 5 *Glue the coins together*, facing the same way. The only possibilities are $(1,1)$ and $(0,0)$. Those have probabilities $\frac{1}{2}$ and $\frac{1}{2}$. The probabilities p_{10} and p_{01} are zero. $(1,0)$ and $(0,1)$ won't happen because the coins stick together: both heads or both tails.

Joint probability matrices $\boldsymbol{p_{ij}}$ **for Examples 4 and 5**
$$\boldsymbol{P_4} = \begin{bmatrix} \frac{1}{4} & \frac{1}{4} \\ \frac{1}{4} & \frac{1}{4} \end{bmatrix} \quad \text{and} \quad \boldsymbol{P_5} = \begin{bmatrix} \frac{1}{2} & 0 \\ 0 & \frac{1}{2} \end{bmatrix}.$$

Let me stay longer with P, to show it in good matrix notation. The matrix shows the probability p_{ij} of each pair (x_i, y_j)—starting with $(x_1, y_1) =$ (heads, heads) and $(x_1, y_2) =$ (heads, tails). Notice the row sums p_1, p_2 and column sums P_1, P_2 and the total sum $= 1$.

$$\textbf{Probability matrix} \quad P = \begin{bmatrix} p_{11} & p_{12} \\ p_{21} & p_{22} \end{bmatrix} \quad \begin{matrix} p_{11} + p_{12} = \boldsymbol{p_1} \\ p_{21} + p_{22} = \boldsymbol{p_2} \end{matrix} \begin{pmatrix} \text{first} \\ \text{coin} \end{pmatrix}$$

$$\text{(second coin) column sums} \quad \boldsymbol{P_1} \quad \boldsymbol{P_2} \qquad \text{4 entries add to 1}$$

Those sums p_1, p_2 and P_1, P_2 are the **marginals** of the joint probability matrix P:

$p_1 = p_{11} + p_{12} =$ chance of heads from **coin 1** (coin 2 can be heads or tails)

$P_1 = p_{11} + p_{21} =$ chance of heads from **coin 2** (coin 1 can be heads or tails)

Example 4 showed *independent* random variables. Every probability p_{ij} equals p_i times p_j $\left(\frac{1}{2}\text{ times }\frac{1}{2}\text{ gave } p_{ij} = \frac{1}{4}\text{ in that example}\right)$. In this case **the covariance σ_{12} will be zero.** Heads or tails from the first coin gave no information about the second coin.

Zero covariance $\boldsymbol{\sigma_{12}}$ for independent trials
$$\boldsymbol{V} = \begin{bmatrix} \sigma_1^2 & 0 \\ 0 & \sigma_2^2 \end{bmatrix} = \textbf{diagonal covariance matrix } \boldsymbol{V}.$$

Independent experiments have $\sigma_{12} = 0$ because every p_{ij} equals $(p_i)(p_j)$ in equation (12).

$$\sigma_{12} = \sum_i \sum_j (p_i)(p_j)(x_i - m_1)(y_j - m_2) = \left[\sum_i (p_i)(x_i - m_1)\right]\left[\sum_j (p_j)(y_j - m_2)\right] = [0][0].$$

Example 6 The glued coins show perfect correlation. Heads on one means heads on the other. **The covariance $\boldsymbol{\sigma_{12}}$ moves from 0 to $\boldsymbol{\sigma_1}$ times $\boldsymbol{\sigma_2}$.** This is the largest possible value of σ_{12}. Here it is $\left(\frac{1}{2}\right)\left(\frac{1}{2}\right) = \sigma_{12} = \left(\frac{1}{4}\right)$, as a separate computation confirms:

$$\textbf{Means} = \frac{1}{2} \qquad \boldsymbol{\sigma_{12}} = \frac{1}{2}\left(1 - \frac{1}{2}\right)\left(1 - \frac{1}{2}\right) + \mathbf{0} + \mathbf{0} + \frac{1}{2}\left(0 - \frac{1}{2}\right)\left(0 - \frac{1}{2}\right) = \frac{1}{4}$$

Heads or tails from coin 1 gives complete information about heads or tails from coin 2:

Glued coins give largest possible covariances
Singular covariance matrix: determinant = 0
$$V_{\text{glue}} = \begin{bmatrix} \sigma_1^2 & \sigma_1\sigma_2 \\ \sigma_1\sigma_2 & \sigma_2^2 \end{bmatrix}$$

Always $\sigma_1^2\sigma_2^2 \geq (\sigma_{12})^2$. Thus σ_{12} is *between* $-\sigma_1\sigma_2$ *and* $\sigma_1\sigma_2$. The matrix V is **positive definite** (or in this singular case of glued coins, V is **positive semidefinite**). Those are important facts about all M by M covariance matrices V for M experiments.

Note that the **sample covariance matrix S** from N trials is certainly semidefinite. Every sample X = (age, height) contributes to the **sample mean $\overline{X}$** $= (m_a, m_h)$. Each rank-one term $(X_i - \overline{X})(X_i - \overline{X})^{\mathrm{T}}$ is positive semidefinite and we just add to reach the sample matrix S. No probabilities in S, use the actual outcomes:

$$\boxed{\overline{\boldsymbol{X}} = \frac{X_1 + \cdots + X_N}{N} \qquad \boldsymbol{S} = \frac{(X_1 - \overline{X})(X_1 - \overline{X})^{\mathrm{T}} + \cdots + (X_N - \overline{X})(X_N - \overline{X})^{\mathrm{T}}}{N-1}} \tag{13}$$

The Covariance Matrix V is Positive Semidefinite

Come back to the *expected* covariance σ_{12} between two experiments 1 and 2 (two coins):

$$\boxed{\begin{aligned} \sigma_{12} &= \text{expected value of } [(\textit{output}\,1 - \textit{mean}\,1) \text{ times } (\textit{output}\,2 - \textit{mean}\,2)] \\ \boldsymbol{\sigma_{12}} &= \sum\sum \boldsymbol{p_{ij}\,(x_i - m_1)\,(y_j - m_2)}. \text{ The sum includes all pairs } i, j. \end{aligned}} \tag{14}$$

$p_{ij} \geq 0$ is the probability of seeing output x_i in experiment 1 **and** y_j in experiment 2. Some pair of outputs must appear. Therefore the n^2 joint probabilities p_{ij} add to 1.

$$\textbf{Total probability (all pairs) is 1} \qquad \sum_{\text{all } i,j}\sum p_{ij} = 1. \tag{15}$$

Here is another fact we need. *Fix on one particular output* x_i in experiment 1. Allow *all outputs* y_j in experiment 2. Add the probabilities of $(x_i, y_1), (x_i, y_2), \ldots, (x_i, y_n)$:

$$\textbf{Row sum } \boldsymbol{p_i} \textbf{ of } \boldsymbol{P} \qquad \sum_{j=1}^{n} \boldsymbol{p_{ij}} = \textbf{probability } \boldsymbol{p_i} \textbf{ of } \boldsymbol{x_i} \textbf{ in experiment 1.} \tag{16}$$

Some y_j must happen in experiment 2 ! Whether the two coins are completely separate or glued, we get the same answer $\frac{1}{2}$ for the probability $p_H = p_{HH} + p_{HT}$ that coin 1 is heads:

$$\text{(separate) } P_{HH} + P_{HT} = \frac{1}{4} + \frac{1}{4} = \frac{\mathbf{1}}{\mathbf{2}} \qquad \text{(glued) } P_{HH} + P_{HT} = \frac{1}{2} + 0 = \frac{\mathbf{1}}{\mathbf{2}}.$$

That basic reasoning allows us to write one matrix formula that includes the covariance σ_{12} along with the separate variances σ_1^2 and σ_2^2 for experiment 1 and experiment 2. We get the whole covariance matrix V by adding the matrices V_{ij} for each pair (i, j):

$$\boxed{\begin{array}{l}\textbf{Covariance matrix} \\ \boldsymbol{V} = \textbf{sum of all } \boldsymbol{V_{ij}}\end{array} \quad V = \sum_{\text{all } i,j}\sum p_{ij} \begin{bmatrix} (x_i - m_1)^2 & (x_i - m_1)(y_j - m_2) \\ (x_i - m_1)(y_j - m_2) & (y_j - m_2)^2 \end{bmatrix}} \tag{17}$$

Off the diagonal, this is equation (12) for the covariance σ_{12}. On the diagonal, we are getting the ordinary variances σ_1^2 and σ_2^2. I will show in detail how we get $V_{11} = \sigma_1^2$ by using equation (16). Allowing all j just leaves the probability p_i of x_i in experiment 1:

$$\boldsymbol{V_{11}} = \sum_{\text{all } i,j}\sum p_{ij}(x_i - m_1)^2 = \sum_{\text{all } i} (\text{probability of } x_i)\,(x_i - m_1)^2 = \boldsymbol{\sigma_1^2}. \qquad (18)$$

Please look at that twice. It is the key to producing the whole covariance matrix by one formula (17). The beauty of that formula is that it combines 2 by 2 matrices V_{ij}. And the matrix V_{ij} in (17) for each pair of outcomes i, j is **positive semidefinite**:

V_{ij} has diagonal entries $p_{ij}(x_i - m_1)^2 \geq 0$ and $p_{ij}(y_j - m_2)^2 \geq 0$ and $\det(V_{ij}) = 0$.

That matrix V_{ij} has rank 1. Equation (17) multiplies p_{ij} *times column* $\boldsymbol{U}$ *times row* $\boldsymbol{U}^{\mathrm{T}}$:

$$\begin{bmatrix} (x_i - m_1)^2 & (x_i - m_1)(y_j - m_2) \\ (x_i - m_1)(y_j - m_2) & (y_j - m_2)^2 \end{bmatrix} = \begin{bmatrix} x_i - m_1 \\ y_j - m_2 \end{bmatrix} \begin{bmatrix} x_i - m_1 & y_j - m_2 \end{bmatrix} \qquad (19)$$

Every matrix $\boldsymbol{p_{ij}UU^{\mathrm{T}}}$ *is positive semidefinite.* So the whole matrix V (the sum of those rank 1 matrices) is **at least semidefinite**—and usually V is positive definite:

The covariance matrix V is positive definite unless the experiments are dependent.

Now we move from two variables x and y to M variables like age-height-weight. Each child has an age-height-weight vector $\boldsymbol{X}$ with $M = 3$ components. The covariance matrix $\boldsymbol{V}$ is now M by M. The matrix $\boldsymbol{V}$ is created from the output vectors $\boldsymbol{X}$ and their average $\overline{\boldsymbol{X}} = \mathbf{E}[\boldsymbol{X}]$:

Covariance matrix	$\boldsymbol{V} = \mathbf{E}\left[(\boldsymbol{X} - \overline{\boldsymbol{X}})\,(\boldsymbol{X} - \overline{\boldsymbol{X}})^{\mathrm{T}}\right]$	$\boldsymbol{V_{ij} = p_{ij}\,(X_i - m_i)\,(X_j - m_j)}$	(20)

Remember that $\boldsymbol{XX^{\mathrm{T}}}$ and $\overline{\boldsymbol{X}}\,\overline{\boldsymbol{X}}^{\mathrm{T}}$ = (column)(row) are M by M matrices.

For $M = 1$ (one variable) you see that $\overline{X}$ is the mean m and V is the variance σ^2. For $M = 2$ (two coins) you see that $\overline{X}$ is (m_1, m_2) and V matches equation (17). The expectation always adds up outputs times their probabilities. For age-height-weight the output could be $X = $ (5 years, 31 inches, 48 pounds) and its probability is $\boldsymbol{p_{5,31,48}}$.

Now comes a new idea. *Take any linear combination* $\boldsymbol{c}^{\mathrm{T}}\boldsymbol{X} = c_1X_1 + \cdots + c_MX_M$. With $\boldsymbol{c} = (6, 2, 5)$ this would be $\boldsymbol{c}^{\mathrm{T}}\boldsymbol{X} = 6 \times \text{age} + 2 \times \text{height} + 5 \times \text{weight}$. By linearity we know that **mean of $\boldsymbol{c}^{\mathrm{T}}\boldsymbol{X}$** = expected value $\mathrm{E}[\boldsymbol{c}^{\mathrm{T}}\boldsymbol{X}] = \boldsymbol{c}^{\mathrm{T}}\mathrm{E}[\boldsymbol{X}] = \boldsymbol{c}^{\mathrm{T}}\overline{\boldsymbol{X}}$:

$$\mathbf{E}[\boldsymbol{c}^{\mathrm{T}}\boldsymbol{X}] = \boldsymbol{c}^{\mathrm{T}}\mathbf{E}[\boldsymbol{X}] = 6\,(\text{expected age}) + 2\,(\text{expected height}) + 5\,(\text{expected weight}).$$

More than the mean of $\boldsymbol{c}^{\mathrm{T}}\boldsymbol{X}$, we also know its *variance* $\boldsymbol{\sigma^2 = c^{\mathrm{T}} V c}$:

$$\text{Variance of } \boldsymbol{c}^{\mathrm{T}}\boldsymbol{X} = \boldsymbol{c}^{\mathrm{T}}\mathbf{E}\left[\left(\boldsymbol{X}-\overline{\boldsymbol{X}}\right)\left(\boldsymbol{X}-\overline{\boldsymbol{X}}\right)^{\mathrm{T}}\right]\boldsymbol{c} = \boldsymbol{c}^{\mathrm{T}}\boldsymbol{V}\boldsymbol{c} \tag{21}$$

Now the key point: *The variance of* $\mathbf{c}^{\mathrm{T}}\boldsymbol{X}$ *can never be negative.* So $\mathbf{c^{T}}\boldsymbol{V}\mathbf{c} \geq \mathbf{0}$.
New proof: *The covariance matrix* V *is positive semidefinite by the energy test* $\mathbf{c^{T}}\boldsymbol{V}\mathbf{c} \geq \mathbf{0}$.

Covariance matrices V open up the link between probability and linear algebra: $\boldsymbol{V}$ **equals** $\boldsymbol{Q\Lambda Q^{\mathrm{T}}}$ with eigenvalues $\lambda_i \geq 0$ and orthonormal eigenvectors $\boldsymbol{q}_1$ to $\boldsymbol{q}_M$.

> **Diagonalizing the covariance matrix V means finding M *independent* experiments as combinations of the original M experiments.**

The Covariance Matrix for $Z = AX$

Here is a good way to see σ_z^2 when $z = x + y$. Think of (x, y) as a column vector $\boldsymbol{X}$. Think of the 1 by 2 matrix $\boldsymbol{A} = \begin{bmatrix} 1 & 1 \end{bmatrix}$ multiplying that column vector $\boldsymbol{X} = (x, y)$. Then $\boldsymbol{AX}$ is the sum $z = x + y$. The variance $\boldsymbol{\sigma_z^2}$ goes into matrix notation as

$$\boldsymbol{\sigma_z^2} = \begin{bmatrix} 1 & 1 \end{bmatrix}\begin{bmatrix} \sigma_x^2 & \sigma_{xy} \\ \sigma_{xy} & \sigma_y^2 \end{bmatrix}\begin{bmatrix} 1 \\ 1 \end{bmatrix} \quad \text{which is } \boldsymbol{\sigma_z^2 = AVA^{\mathrm{T}}}. \tag{22}$$

Now for the main point. The vector $\boldsymbol{X}$ could have M components coming from M experiments (instead of only 2). Those experiments will have an M by M covariance matrix $\boldsymbol{V_X}$. The matrix A could be K by M. Then $A\boldsymbol{X}$ is a vector with K combinations of the M outputs (instead of one combination $x + y$ of two outputs).

That vector $\boldsymbol{Z} = A\boldsymbol{X}$ of length K has a K by K covariance matrix $V_{\boldsymbol{Z}}$. Then the great rule for covariance matrices—of which equation (22) was only a 1 by 2 example—is this beautiful formula: The covariance matrix of $\boldsymbol{AX}$ is $\boldsymbol{A}$ **(covariance matrix of** $\boldsymbol{X}$**)** $\boldsymbol{A}^{\mathrm{T}}$:

$$\textbf{The covariance matrix of } \boldsymbol{Z = AX} \qquad \boxed{V_{\boldsymbol{Z}} = \boldsymbol{A}V_{\boldsymbol{X}}\boldsymbol{A}^{\mathrm{T}}} \tag{23}$$

To me, this neat formula shows the beauty of matrix multiplication. I won't prove this formula, just admire it. It is constantly used in applications.

The Correlation ρ

Correlation $\boldsymbol{\rho_{xy}}$ is closely related to covariance $\boldsymbol{\sigma_{xy}}$. They both measure dependence or independence. Start by rescaling or "standardizing" the random variables x and y **The new** $\boldsymbol{X = x/\sigma_x}$ **and** $\boldsymbol{Y = y/\sigma_y}$ **have variance** $\boldsymbol{\sigma_X^2 = \sigma_Y^2 = 1}$. This is just like dividing a vector $\boldsymbol{v}$ by its length to produce a unit vector $\boldsymbol{v}/||\boldsymbol{v}||$ of length 1.

The correlation of x and y is the covariance of X and Y. If the original covariance of x and y was σ_{xy}, then rescaling to X and Y gives correlation $\boldsymbol{\rho_{xy} = \sigma_{xy}/\sigma_x\sigma_y}$.

$$\boxed{\textbf{Correlation } \boldsymbol{\rho_{xy}} = \frac{\boldsymbol{\sigma_{xy}}}{\boldsymbol{\sigma_x\sigma_y}} = \textbf{covariance of } \frac{\boldsymbol{x}}{\boldsymbol{\sigma_x}} \textbf{ and } \frac{\boldsymbol{y}}{\boldsymbol{\sigma_y}}} \quad \textbf{Always } \boldsymbol{-1 \leq \rho_{xy} \leq 1}$$

Problem Set 10.3

1 If all 24 samples from a population produce the same age $x = 20$, what are the sample mean μ and the sample variance S^2? What if $x = 20$ or 21, 12 times each?

2 Add 7 to every output x. What happens to the mean and the variance? What are the new sample mean, the new expected mean, and the new variance?

3 We know: $\frac{1}{3}$ of all integers are divisible by 3 and $\frac{1}{7}$ of integers are divisible by 7. What fraction of integers will be divisible by 3 or 7 or both?

4 Suppose you sample from the numbers 1 to 1000 with equal probabilities $1/1000$. What are the probabilities p_0 to p_9 that the last digit of your sample is $0, \ldots, 9$? What is the expected mean m of that last digit? What is its variance σ^2?

5 Sample again from 1 to 1000 but look at the last digit of the sample *squared*. That square could end with $x = 0, 1, 4, 5, 6$, or 9. What are the probabilities $p_0, p_1, p_4, p_5, p_6, p_9$? What are the (expected) mean m and variance σ^2 of that number x?

6 (a little tricky) Sample again from 1 to 1000 with equal probabilities and let x be the *first* digit ($x = 1$ if the number is 15). What are the probabilities p_1 to p_9 (adding to 1) of $x = 1, \ldots, 9$? What are the mean and variance of x?

7 Suppose you have $N = 4$ samples $157, 312, 696, 602$ in Problem 5. What are the first digits x_1 to x_4 of the squares? What is the sample mean μ? What is the sample variance S^2? Remember to divide by $N - 1 = 3$ and not $N = 4$.

8 Equation (4) gave a second equivalent form for S^2 (the variance using samples):

$$\boldsymbol{S^2} = \frac{1}{N-1} \text{ sum of } (x_i - m)^2 = \frac{1}{N-1}\left[(\text{sum of } x_i^2) - Nm^2\right].$$

Verify the matching identity for the expected variance σ^2 (using $m = \Sigma\, p_i\, x_i$):

$$\boldsymbol{\sigma^2 = \textbf{sum of } p_i\, (x_i - m)^2 = (\textbf{sum of } p_i\, x_i^2) - m^2.}$$

9 Computer experiment: Find the average $A_{1000000}$ of a million random 0-1 samples! What is your value of the standardized variable $X = \left(A_N - \frac{1}{2}\right) \times 2\sqrt{N}$?

10 For any function $f(x)$ the expected value is $\mathrm{E}[f] = \sum p_i\, f(x_i)$ or $\int p(x)\, f(x)\, dx$ (discrete or continuous probability). The function can be x or $(x - m)^2$ or x^2.

If the mean is $\mathrm{E}[x] = m$ and the variance is $\mathrm{E}[(x - m)^2] = \sigma^2$, **what is $\mathbf{E}[x^2]$**?

11 Show that the standard normal distribution $p(x)$ has total probability $\int \boldsymbol{p(x)\, dx = 1}$ as required. A famous trick multiplies $\int p(x)\, dx$ by $\int p(y)\, dy$ and computes the integral over all x and all y ($-\infty$ to ∞). The trick is to replace $dx\, dy$ in that double integral by $r\, dr\, d\theta$ (polar coordinates with $x^2 + y^2 = r^2$). Explain each step:

$$2\pi \int_{-\infty}^{\infty} p(x)\, dx \int_{-\infty}^{\infty} p(y)\, dy = \iint_{-\infty}^{\infty} e^{-(x^2+y^2)/2} dx\, dy = \int_{\theta=0}^{2\pi} \int_{r=0}^{\infty} \boldsymbol{e^{-r^2/2}\, r\, dr\, d\theta} = \boldsymbol{2\pi}.$$

Thoughts about the success of Deep Learning

A mathematician naturally asks, "What question has deep learning answered ? Is it a unique and complete answer, or will there be new ideas that replace or extend it ? What is the key to its success ?" I don't have fully satisfying answers, but the questions are important.

The framework seems clear. We are given training data $\boldsymbol{v}$, and we know what object $\boldsymbol{w}$ it describes. We want to create a function $\boldsymbol{F}$ so that $\boldsymbol{F}(\boldsymbol{v})$ is close to or equal to $\boldsymbol{w}$. (Close means an **approximation** $\boldsymbol{F}(\boldsymbol{v}) \approx \boldsymbol{w}$. Equal means an **interpolation** $\boldsymbol{F}(\boldsymbol{v}) = \boldsymbol{w}$.) The essential requirement is that $\boldsymbol{F}$ generalizes successfully to new test data $\boldsymbol{V}$. The assumption is that $\boldsymbol{v}$ and $\boldsymbol{V}$ come from populations that are statistically similar or even identical. Success is achieved if $\boldsymbol{F}(\boldsymbol{V})$ is near the actual result $\boldsymbol{W}$ from the test data $\boldsymbol{V}$.

In brief, we look for a function that (nearly) interpolates the training data and successfully extrapolates to the test data.

The hard question is: **What shall be the form of $\boldsymbol{F}$?** Successful generalization is asking for "stability"—if $\boldsymbol{V}$ is near to $\boldsymbol{v}$, then $\boldsymbol{F}(\boldsymbol{V})$ is near to $\boldsymbol{F}(\boldsymbol{v})$. We can also hope for simplicity—constructing $\boldsymbol{F}(\boldsymbol{v})$ and evaluating $\boldsymbol{F}(\boldsymbol{V})$ must not take forever.

I think it is correct to say that classical answers failed. If $\boldsymbol{F}$ is a polynomial—or any linear combination of preselected functions—then good results in a high-dimensional space have been very elusive. At the other extreme is the "finite element method"—which uses low degree piecewise polynomials and more or less preselected interpolation points. That idea works well in structural engineering. It seems to be less successful for the tremendous variety of applications of deep learning—which often has no visible structure at all.

The key to success (unexpected success) could lie in **composition** $\boldsymbol{F_L(\ldots(F_2(F_1(v))))}$ **of simple functions**. That process quickly brings an amazing complexity. The functions in this book were continuous and piecewise linear (provided ReLU is chosen).

Other authors have explored the use of rational functions (known to provide good approximation to a wide class of functions but dangerous near poles). And the crucial understanding of *generalization* continues to move forward.

It is truly remarkable that the composition of such simple functions has revolutionized our power of approximation and our boldness in new applications.

A1 The Ranks of AB and $A+B$

From Chapters 1 to 3, we know that rank of A = rank of A^{T}. This page establishes more key facts about ranks: **When we multiply matrices, the rank cannot increase.** You will see this by looking at column spaces and row spaces. **3** shows one special situation $B = A^{\mathrm{T}}$ when the rank stays the same. Then you know the rank of $A^{\mathrm{T}}A$. Statement 4 becomes important when data science factors A into $U\Sigma V^{\mathrm{T}}$ or CR.
Here are five key facts in one place. The most important fact is **rank of $\boldsymbol{A}$ = rank of $\boldsymbol{A}^{\mathrm{T}}$.**

1 **Rank of $AB \le$ rank of A Rank of $AB \le$ rank of B**
2 **Rank of $A+B \le$ (rank of A) + (rank of B)**
3 **Rank of $A^{\mathrm{T}}A$ = rank of AA^{T} = rank of A = rank of A^{T}**
4 **If A is m by r and B is r by n**—both with rank r—**then AB also has rank r**

Statement 1 involves the column space of AB and the row space of AB:

$\mathbf{C}(AB)$ is contained in $\mathbf{C}(A)$ so the dimension of $\mathbf{C}(AB)$ cannot be larger

Every column of AB is a combination of the columns of A (*matrix multiplication*)
Every row of AB is a combination of the rows of B (*matrix multiplication*)

Remember from Section 1.4 that **row rank = column rank**. We can use rows or columns. *The rank cannot grow when we multiply* AB. Statement **1** in the box is frequently used.

Statement 2 Each column of $A+B$ is the sum of (column of A) + (column of B).

rank $(A+B) \le$ rank (A) + rank (B) is **true**. Bases for $\mathbf{C}(A)$ and $\mathbf{C}(B)$ span $\mathbf{C}(A+B)$.

rank $(A+B)$ = rank (A) + rank (B) is **not** always true. It is certainly false if $A=B=I$.

Statement 3 A and $A^{\mathrm{T}}A$ both have n columns. **They also have the same nullspace.** (See Problem 4.19.) So $n-r$ is the same for both, and *the rank r is the same for both.* Then rank$(A^{\mathrm{T}}) \ge$ rank$(A^{\mathrm{T}}A)$ = rank(A). Exchange A and A^{T} to show **equal ranks.**

Statement 4 We are told that A and B have rank r. By Statement **3**, $A^{\mathrm{T}}A$ and BB^{T} have rank r. Those are r by r matrices so they are invertible. So is their product $A^{\mathrm{T}}ABB^{\mathrm{T}}$. Then

$$r = \textbf{rank of } (\boldsymbol{A}^{\mathrm{T}}\boldsymbol{A}\boldsymbol{B}\boldsymbol{B}^{\mathrm{T}}) \le \textbf{rank of } (\boldsymbol{AB}) \le \textbf{rank of } \boldsymbol{A} = r. \text{ So } AB \text{ has rank } r.$$

Note This does not mean that every product of rank r matrices will have rank r. Statement **4** assumes that **$\boldsymbol{A}$ has exactly $\boldsymbol{r}$ columns** and **$\boldsymbol{B}$ has $\boldsymbol{r}$ rows.** BA can easily fail.

$$A = \begin{bmatrix} 1 \\ 1 \\ 1 \end{bmatrix} \qquad B = \begin{bmatrix} 1 & 2 & -3 \end{bmatrix} \qquad \textbf{\textit{AB} has rank 1} \qquad \textbf{But \textit{BA} is zero!}$$

A2 Matrix Factorizations

1. $A = CR =$ (basis for column space of A) (basis for row space of A)

   Requirements: C is m by r and R is r by n. Columns of A go into C if they are not combinations of earlier columns of A. R contains the nonzero rows of the reduced row echelon form $R_0 = \textbf{rref}(A)$. Those rows begin with an r by r identity matrix so $A = CR = \begin{bmatrix} C & CF \end{bmatrix} P = \begin{bmatrix} \text{indep columns} & \text{dep columns} \end{bmatrix}$ permute columns.

2. $A = CW^{-1}B \begin{pmatrix} C = \text{first } r \\ \text{independent columns} \end{pmatrix} \begin{pmatrix} W = \text{first } r \text{ by } r \\ \text{invertible submatrix} \end{pmatrix} \begin{pmatrix} B = \text{first } r \\ \text{independent rows} \end{pmatrix}$

   Requirements: C and B contain r columns and r rows of A. Those columns meet those rows in the square invertible matrix W (Section 3.2). Then $W^{-1}B = R$.

3. $A = LU = \begin{pmatrix} \text{lower triangular } L \\ \text{1's on the diagonal} \end{pmatrix} \begin{pmatrix} \text{upper triangular } U \\ \text{pivots on the diagonal} \end{pmatrix}$

   Requirements: No row exchanges as Gaussian elimination reduces square A to U.

4. $A = LDU = \begin{pmatrix} \text{lower triangular } L \\ \text{1's on the diagonal} \end{pmatrix} \begin{pmatrix} \text{pivot matrix} \\ D \text{ is diagonal} \end{pmatrix} \begin{pmatrix} \text{upper triangular } U \\ \text{1's on the diagonal} \end{pmatrix}$

   Requirements: No row exchanges. The pivots in D are divided out from rows of U to leave 1's on the diagonal of U. If A is symmetric then U is L^{T} and $A = LDL^{\mathrm{T}}$.

5. $PA = LU$ (permutation matrix P to avoid zeros in the pivot positions).

   Requirements: A is invertible. Then P, L, U are invertible. P does all of the row exchanges on A in advance, to allow normal LU. Alternative: $A = L_1 P_1 U_1$.

6. $S = C^{\mathrm{T}}C =$ (lower triangular) (upper triangular) with $\sqrt{D}$ on both diagonals

   Requirements: S is symmetric and positive definite (all n pivots in D are positive). This *Cholesky factorization* $C = \mathsf{chol}(S)$ has $C^{\mathrm{T}} = L\sqrt{D}$, so $S = C^{\mathrm{T}}C = LDL^{\mathrm{T}}$.

7. $A = QR =$ (orthonormal columns in Q) (upper triangular matrix R).

   Requirements: A has independent columns. Those are *orthogonalized* in Q by the Gram-Schmidt or Householder process. If A is square then $Q^{-1} = Q^{\mathrm{T}}$.

8. $A = X\Lambda X^{-1} =$ (eigenvectors in X) (eigenvalues in Λ) (eigenvectors of A^{T} in X^{-1}).

   Requirements: A must have n linearly independent eigenvectors in X : $AX = X\Lambda$.

9. $S = Q\Lambda Q^{\mathrm{T}} =$ (orthogonal matrix Q) (real eigenvalue matrix Λ) (Q^{T} is Q^{-1}).

   Requirements: S is *real and symmetric*: $S^{\mathrm{T}} = S$. This is the Spectral Theorem.

10. $\boldsymbol{A = BJB^{-1}}$ = (generalized eigenvectors of $\boldsymbol{A}$ in $\boldsymbol{B}$) (Jordan blocks in $\boldsymbol{J}$) ($\boldsymbol{B^{-1}}$).

Requirements: A is square. This **Jordan form** $\boldsymbol{J}$ has a block for each linearly independent eigenvector of A. Every block has only one eigenvalue: *Appendix* **A5**.

11. $\boldsymbol{A = U\Sigma V^{\mathrm{T}}} = \begin{pmatrix} \text{orthogonal} \\ \boldsymbol{U} \text{ is } m \times m \end{pmatrix} \begin{pmatrix} m \times n \text{ singular value matrix} \\ \boldsymbol{\sigma_1, \ldots, \sigma_r} \text{ on its diagonal} \end{pmatrix} \begin{pmatrix} \text{orthogonal} \\ \boldsymbol{V} \text{ is } n \times n \end{pmatrix}.$

Every $\boldsymbol{A}$ is included. This ***Singular Value Decomposition*** (SVD) has eigenvectors of AA^{T} in $\boldsymbol{U}$ and eigenvectors of $A^{\mathrm{T}}A$ in $\boldsymbol{V}$; $\boldsymbol{\sigma_i} = \sqrt{\boldsymbol{\lambda_i}(A^{\mathrm{T}}A)} = \sqrt{\boldsymbol{\lambda_i}(AA^{\mathrm{T}})}$.

Those singular values are $\sigma_1 \geq \sigma_2 \geq \cdots \geq \sigma_r > 0$. By column-row multiplication

$$\boldsymbol{A = U\Sigma V^{\mathrm{T}}} = \sigma_1 \boldsymbol{u}_1 \boldsymbol{v}_1^{\mathrm{T}} + \cdots + \sigma_r \boldsymbol{u}_r \boldsymbol{v}_r^{\mathrm{T}} = \text{ sum of } r \text{ matrices of rank 1.}$$

If S is symmetric positive definite then $U = V = Q$ and $\Sigma = \Lambda$ and $S = Q\Lambda Q^{\mathrm{T}}$.

12. $\boldsymbol{A^{+} = V\Sigma^{+}U^{\mathrm{T}}} = \begin{pmatrix} \text{orthogonal} \\ \boldsymbol{V} \text{ is } n \times n \end{pmatrix} \begin{pmatrix} n \times m \text{ pseudoinverse of } \boldsymbol{\Sigma} \\ 1/\sigma_1, \ldots, 1/\sigma_r \text{ on diagonal} \end{pmatrix} \begin{pmatrix} \text{orthogonal} \\ \boldsymbol{U} \text{ is } m \times m \end{pmatrix}.$

Requirements: None. The **pseudoinverse** $\boldsymbol{A^{+}}$ has $A^{+}A$ = projection onto row space of A and AA^{+} = projection onto column space. $A^{+} = A^{-1}$ if A is invertible. The shortest least-squares solution to $A\boldsymbol{x} = \boldsymbol{b}$ is $\boldsymbol{x}^{+} = A^{+}\boldsymbol{b}$. This solves $A^{\mathrm{T}}A\boldsymbol{x}^{+} = A^{\mathrm{T}}\boldsymbol{b}$.

13. $\boldsymbol{A = QS}$ = (orthogonal matrix $\boldsymbol{Q}$) (symmetric positive definite matrix $\boldsymbol{S}$).

Requirements: A is invertible. This **polar decomposition** has $S^2 = A^{\mathrm{T}}A$. The factor S is semidefinite if A is singular. The reverse polar decomposition $A = KQ$ has $K^2 = AA^{\mathrm{T}}$. Both have $Q = UV^{\mathrm{T}}$ from the SVD.

14. $\boldsymbol{A = U\Lambda U^{-1}}$ = (unitary $\boldsymbol{U}$) (eigenvalue matrix $\boldsymbol{\Lambda}$) ($\boldsymbol{U^{-1}}$ which is $\overline{\boldsymbol{U}}^{\mathrm{T}}$).

Requirements: **$\boldsymbol{A}$ is normal**: $\overline{\boldsymbol{A}}^{\mathrm{T}}\boldsymbol{A} = \boldsymbol{A}\overline{\boldsymbol{A}}^{\mathrm{T}}$. Its **orthonormal** (possibly complex) **eigenvectors** are columns of U. Complex Λ unless $\boldsymbol{A} = \overline{\boldsymbol{A}}^{\mathrm{T}}$ = Hermitian matrix.

15. $\boldsymbol{A = QTQ^{-1}}$ = (unitary $\boldsymbol{Q}$) (triangular $\boldsymbol{T}$ with λ's on diagonal) ($\boldsymbol{Q^{-1}} = \overline{\boldsymbol{Q}}^{\mathrm{T}}$).

Requirements: *Schur triangularization* of any square A. There is a matrix Q with orthonormal columns that makes $Q^{-1}AQ$ triangular: Section 6.4.

16. $\boldsymbol{F_n} = \begin{bmatrix} I & D \\ I & -D \end{bmatrix} \begin{bmatrix} \boldsymbol{F_{n/2}} & \\ & \boldsymbol{F_{n/2}} \end{bmatrix} \begin{bmatrix} \text{even-odd} \\ \text{permutation} \end{bmatrix} =$ **one FFT step in its matrix form: $\boldsymbol{n}$ to $\boldsymbol{n/2 = m}$**

Requirements: $\boldsymbol{F_n}$ = **Fourier matrix** with entries w^{jk} where $w^n = 1$: $\boldsymbol{F_n \overline{F}_n = nI}$. D has $1, w, \ldots, w^{n/2-1}$ on its diagonal. For $n = 2^{\ell}$ the *Fast Fourier Transform* will compute $\boldsymbol{F_n x}$ with only $\frac{1}{2}n\ell = \frac{1}{2}n\log_2 n$ multiplications from ℓ stages of D's.

The FFT combines the $\frac{1}{2}$-size transforms $\boldsymbol{y' = F_m c'}$ and $\boldsymbol{y'' = F_m c''}$ into $\boldsymbol{y} = \boldsymbol{F_n c}$.

$$y_j = y_j' + (w_n)^j y_j'' \quad \text{and} \quad y_{j+m} = y_j' - (w_n)^j y_j'' \quad \text{for} \quad j = 0, \ldots, m-1$$

A3 Counting Parameters in the Basic Factorizations

$$\boldsymbol{A = LU} \quad \boldsymbol{A = QR} \quad \boldsymbol{S = Q\Lambda Q^{\mathrm{T}}} \quad \boldsymbol{A = X\Lambda X^{-1}} \quad \boldsymbol{A = QS} \quad \boldsymbol{A = U\Sigma V^{\mathrm{T}}}$$

This is a review of key ideas in linear algebra. The ideas are expressed by those factorizations and our plan is simple: *Count the parameters in each matrix.* We hope to see that in each equation like $A = LU$, the two sides have the same number of parameters.

For $A = LU$, both sides have n^2 free parameters from $\frac{1}{2}n(n-1) + \frac{1}{2}n(n+1) = n^2$.

$\boldsymbol{L}$: **Triangular** $n \times n$ matrix with 1's on the diagonal	$\frac{1}{2}\boldsymbol{n(n-1)}$	**in** $\boldsymbol{L}$
$\boldsymbol{U}$: **Triangular** $n \times n$ matrix with free diagonal	$\frac{1}{2}\boldsymbol{n(n+1)}$	**in** $\boldsymbol{U}$
$\boldsymbol{Q}$: **Orthogonal** $n \times n$ matrix	$\frac{1}{2}\boldsymbol{n(n-1)}$	**in** $\boldsymbol{Q}$
$\boldsymbol{S}$: **Symmetric** $n \times n$ matrix	$\frac{1}{2}\boldsymbol{n(n+1)}$	**in** $\boldsymbol{S}$
$\boldsymbol{\Lambda}$: **Diagonal** $n \times n$ matrix	$\boldsymbol{n}$	**in** $\boldsymbol{\Lambda}$
$\boldsymbol{X}$: $n \times n$ matrix of independent **eigenvectors**	$\boldsymbol{n^2 - n}$	**in** $\boldsymbol{X}$

Comments are needed for Q. Its first column $\boldsymbol{q}_1$ is a unit vector. The requirement $||\boldsymbol{q}_1|| = 1$ has used up one of the n parameters in $\boldsymbol{q}_1$. Then $\boldsymbol{q}_2$ has $n - 2$ parameters—it is a unit vector and it is orthogonal to $\boldsymbol{q_1}$. The sum $(n-1) + (n-2) + \cdots + 1$ equals $\frac{1}{2}\boldsymbol{n(n-1)}$ free parameters in Q: n columns averaging $(n-1)/2$ free parameters.

The eigenvector matrix X has only $n^2 - n$ parameters, not n^2. If $\boldsymbol{x}$ is an eigenvector then so is $c\boldsymbol{x}$ for any $c \neq 0$. We could require the largest component of every $\boldsymbol{x}$ to be 1. This leaves $n-1$ parameters for each eigenvector (and no free parameters for X^{-1}). The count for the two sides now agrees for the *first five* factorizations in line 1 above.

For the SVD, use the reduced form $\boldsymbol{A_{m\times n} = U_{m\times r}\Sigma_{r\times r}V^{\mathrm{T}}_{r\times n}}$ (known zeros are not free parameters!) Suppose that $m \leq n$ and A is a full rank matrix with $\boldsymbol{r = m}$. The parameter count for A is $\boldsymbol{mn}$. *The total count for U, Σ, V is also $\boldsymbol{mn}$.* The reasoning for orthonormal columns in U and V is the same as it was above for the columns of Q.

$$\boldsymbol{U} \text{ has } \frac{\mathbf{1}}{\mathbf{2}}\boldsymbol{m(m-1)} \quad \boldsymbol{\Sigma} \text{ has } \boldsymbol{m} \quad \boldsymbol{V} \text{ has } (n-1)+\cdots+(n-m) = \boldsymbol{mn} - \frac{\mathbf{1}}{\mathbf{2}}\boldsymbol{m(m+1)}$$

Finally, suppose that A is an m *by* n *matrix of rank* r. **How many free parameters in a rank $\boldsymbol{r}$ matrix?** We can count again for $U_{m\times r}\Sigma_{r\times r}V^{\mathrm{T}}_{r\times n}$:

$$\boldsymbol{U} \text{ has } (m-1)+\cdots+(m-r) = \boldsymbol{mr} - \frac{\mathbf{1}}{\mathbf{2}}\boldsymbol{r(r+1)} \quad \boldsymbol{V} \text{ has } \boldsymbol{nr} - \frac{\mathbf{1}}{\mathbf{2}}\boldsymbol{r(r+1)} \quad \boldsymbol{\Sigma} \text{ has } \boldsymbol{r}$$

The total parameter count for a matrix A of rank r is $\boldsymbol{mr + nr - r^2 = (m + n - r)\,r}$.

We reach the same total for $A = CR$ in Section 1.4. The r columns of C were taken directly from A. The row matrix R includes an r by r identity matrix (fixed not free!). Then the count for $A = CR$ agrees with the previous count for $A = U\Sigma V^{\mathrm{T}}$, when the rank is r:

C has $\boldsymbol{mr}$ parameters R has $\boldsymbol{nr - r^2}$ parameters Total $\boldsymbol{mr + nr - r^2}$.

A4 Codes and Algorithms for Numerical Linear Algebra

LAPACK	is the first choice for dense linear algebra codes.
ScaLAPACK	achieves high performance for very large problems.
COIN/OR	provides high quality codes for the optimization problems of operations research.

Here are sources for specific algorithms.

Direct solution of linear systems	
Basic matrix-vector operations	BLAS
Elimination with row exchanges	LAPACK
Sparse direct solvers (UMFPACK)	SuiteSparse, SuperLU
QR by Gram-Schmidt and Householder	LAPACK
Eigenvalues and singular values	
Shifted QR method for eigenvalues	LAPACK
Golub-Kahan method for the SVD	LAPACK
Iterative solutions	
Preconditioned conjugate gradients for $S\boldsymbol{x} = \boldsymbol{b}$	Trilinos
Preconditioned GMRES for $A\boldsymbol{x} = \boldsymbol{b}$	Trilinos
Krylov-Arnoldi for $A\boldsymbol{x} = \lambda x$	ARPACK, Trilinos, SLEPc
Extreme eigenvalues of S	see also BLOPEX
Optimization	
Linear programming	CLP in COIN/OR
Semidefinite programming	CSDP in COIN/OR
Interior point methods	IPOPT in COIN/OR
Convex Optimization	CVX, CVXR
Randomized linear algebra	
Randomized factorizations via pivoted QR	users.ices.utexas.edu/
$A = CMR$ columns/mixing/rows	~pgm/main_codes.html
Fast Fourier Transform	FFTW.org
Repositories of high quality codes	GAMS and Netlib.org
ACM Transactions on Mathematical Software	TOMS
Deep learning software	
Deep learning in Julia	Fluxml.ai/Flux.jl/stable
Deep learning in MATLAB	Mathworks.com/learn/tutorials/deep–learning–onramp.html
Deep learning in Python and JavaScript	Tensorflow.org, Tensorflow.js
Deep learning in R	Keras, KerasR

A5 The Jordan Form of a Square Matrix

Some square matrices A with repeated eigenvalues don't have n independent eigenvectors. Therefore they can't be diagonalized by $X^{-1}AX = \Lambda$: *X is not invertible.* Jordan established a *nearly diagonal form* $J = BAB^{-1}$ for every matrix A. The "Jordan form" J has k Jordan blocks $J_1, \ldots J_k$ when A has k independent eigenvectors. J has the same eigenvalues as A.

$$J = BAB^{-1} = \begin{bmatrix} J_1 & & & \\ & \cdot & & \\ & & \cdot & \\ & & & J_k \end{bmatrix} \text{ has Jordan blocks } J_i = \begin{bmatrix} \lambda_i & 1 & 0 & 0 \\ & \lambda_i & 1 & 0 \\ & & \cdot & 1 \\ & & & \lambda_i \end{bmatrix}$$

If A can be diagonalized, then $k = n$ and $B = X$. So J is the eigenvalue matrix Λ (*all n blocks J_1 to J_n are* 1 *by* 1). If A can't be diagonalized, then one or more blocks will be larger. Each block J_i of size n_i has only *one true eigenvector* $\boldsymbol{x}_i = (1, 0, \ldots, 0)$ and one eigenvalue λ_i. The matrix B contains eigenvectors of A along with "generalized eigenvectors" of A. Here is an example rather than a proof.

Example $$J = \left[\begin{array}{cc|c} 3 & 1 & 0 \\ 0 & 3 & 0 \\ \hline 0 & 0 & 3 \end{array}\right] \text{ has } \lambda = 3,3,3 \text{ with only two genuine eigenvectors } \begin{bmatrix} 1 \\ 0 \\ 0 \end{bmatrix} \text{ and } \begin{bmatrix} 0 \\ 0 \\ 1 \end{bmatrix} \text{ and } A = B^{-1}JB.$$

This Jordan form makes $A^N = B^{-1}J^NB$ and $e^{tA} = B^{-1}e^{tJ}B$ as simple as possible. For powers of J, we just compute the powers of each Jordan block:

$$\begin{bmatrix} 3 & 1 \\ 0 & 3 \end{bmatrix}^N = \begin{bmatrix} 3^N & N3^{N-1} \\ 0 & 3^N \end{bmatrix} \qquad \exp\left(\begin{bmatrix} 3 & 1 \\ 0 & 3 \end{bmatrix} t\right) = \begin{bmatrix} e^{3t} & te^{3t} \\ 0 & e^{3t} \end{bmatrix}$$

That exponential formula is telling us the missing solution to the differential equation $d\boldsymbol{U}/dt = J\boldsymbol{U}$ (and also $d\boldsymbol{u}/dt = A\boldsymbol{u}$). The usual solution has e^{3t}. We can't just use that twice, when $\lambda = 3$ is repeated. **The missing solution is $\boldsymbol{te^{3t}}$.** And a triple eigenvalue $\lambda = 3$ with only one eigenvector (and one Jordan block) would also involve $\boldsymbol{t^2e^{3t}}$.

The Cayley-Hamilton Theorem $p(A)$ = Zero Matrix

"Every matrix A satisfies its own characteristic equation $p(A) = 0$." The determinant of $A - \lambda I$ is a polynomial $p(\lambda)$. The n solutions to $p(\lambda) = 0$ are the eigenvalues of A. Our example above has $p(\lambda) = (\lambda - 3)^3$ with a triple eigenvalue $\lambda = 3, 3, 3$. Then Cayley-Hamilton says that $p(A) = (A - 3I)^3$ has to be the zero matrix. Jordan makes this easy, because $(A - 3I)^3 = B(J - 3I)^3B^{-1}$ and $(J - 3I)^3$ is certainly the zero matrix:

Example $$(J - 3I)^3 = \begin{bmatrix} 0 & 1 & 0 \\ 0 & 0 & 0 \\ 0 & 0 & 0 \end{bmatrix}^3 = \begin{bmatrix} 0 & 0 & 0 \\ 0 & 0 & 0 \\ 0 & 0 & 0 \end{bmatrix} \quad \textbf{Then } (A - 3I)^3 = 0.$$

A6 Tensors

In linear algebra, **a tensor is a multidimensional array**. (To Einstein, a tensor was a function that followed certain transformation rules.) An ordinary matrix is a 2-way tensor. *A 3-way tensor T is a stack of matrices.* Its elements T_{ijk} have three indices: row number i and column number j and "tube number" k.

An example is a color image. It has 3 slices corresponding to red-green-blue. The slices T_1, T_2, T_3 show the density of one of those primary colors RGB ($k = 1$ to 3), at each pixel (i, j) in the image.

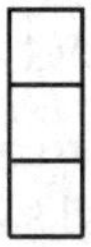

vector

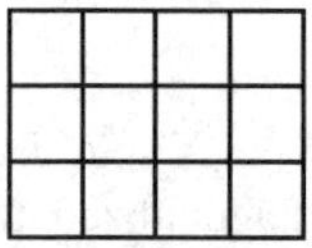

matrix (2-way array)

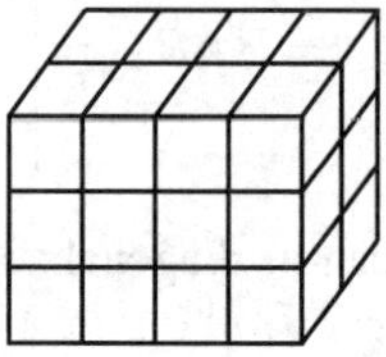

tensor (3-way array)

Another example is a joint probability tensor. Now p_{ijk} is the probability that a random individual has (for example) age i and height j and weight k. The sum of all those numbers $\boldsymbol{p_{ijk}}$ will be 1. For $i = 9$, the sum of all p_{9jk} would be the fraction of individuals that have age 9—the sum over one slice of the tensor.

A fundamental problem—with tensors as with matrices—is to decompose the tensor T into simpler pieces. For a matrix A that was accomplished by the SVD. The pieces that add to A are matrices (we should now say 2-tensors), with the special property that each piece is a rank-one matrix $\boldsymbol{u}\sigma\boldsymbol{v}^{\mathrm{T}}$. Linear algebra allowed us to require that the $\boldsymbol{u}$'s from different pieces were orthogonal, and the $\boldsymbol{v}$'s were also orthogonal and that there were only r pieces ($r \leq m$ and $r \leq n$).

Sad to say, this SVD format is not possible for a 3-way tensor. We can still ask for R rank-one pieces that approximately add to T:

$$\textbf{\textit{CP} Decomposition} \qquad T \approx \boldsymbol{a}_1 \circ \boldsymbol{b}_1 \circ \boldsymbol{c}_1 + \cdots + \boldsymbol{a}_R \circ \boldsymbol{b}_R \circ \boldsymbol{c}_R. \tag{1}$$

Orthogonality of the $\boldsymbol{a}$'s and of the $\boldsymbol{b}$'s and of the $\boldsymbol{c}$'s is generally impossible. The number of pieces is not set by T (its "rank" is not well defined). But an approximate decomposition of this kind is still useful in computations with tensors. One option is to solve alternately for the $\boldsymbol{a}_i$ (with fixed $\boldsymbol{b}_i$ and $\boldsymbol{c}_i$) and then for the $\boldsymbol{b}_i$ (fixed $\boldsymbol{a}_i$ and $\boldsymbol{c}_i$) and then for the $\boldsymbol{c}_i$ (fixed $\boldsymbol{a}_i$ and $\boldsymbol{b}_i$). Those subproblems can be reduced to least squares. Other approximate decompositions of T are possible.

The theory of tensor decompositions (multilinear algebra) is driven by applications. We must be able to compute with T. So the algorithms for multiway tensors are steadily improving, even without the orthogonality properties of an SVD.

A7 The Condition Number of a Matrix Problem

The condition number measures the ratio of (*change in solution*) to (*change in data*). The most common problem is to solve n linear equations $A\boldsymbol{x} = \boldsymbol{b}$ in n unknowns $\boldsymbol{x}$. In this case the data is $\boldsymbol{b}$ and the solution is $\boldsymbol{x} = A^{-1}\boldsymbol{b}$. The matrix A is fixed. The change in the data is $\Delta\boldsymbol{b}$ and the change in the solution is $\boldsymbol{\Delta x} = A^{-1}\boldsymbol{\Delta b}$.

We have to decide the meaning of "change". Do we compute *the absolute change* $||\Delta\boldsymbol{b}||$ *or the relative change* $||\Delta\boldsymbol{b}|| \,/\, ||\boldsymbol{b}||$? That decision for the data $\boldsymbol{b}$ will bring a similar decision for the condition number of A.

$$\begin{matrix}\textbf{Absolute}\\\textbf{condition}\end{matrix} = \max_{\boldsymbol{b},\,\Delta\boldsymbol{b}} \frac{||\Delta\boldsymbol{x}||}{||\Delta\boldsymbol{b}||} = \max \frac{||A^{-1}\Delta\boldsymbol{b}||}{||\Delta\boldsymbol{b}||} \qquad \begin{matrix}\textbf{Relative}\\\textbf{condition}\end{matrix} = \max_{\boldsymbol{b},\,\Delta\boldsymbol{b}} \frac{||\Delta\boldsymbol{x}|| \,/\, ||\boldsymbol{x}||}{||\Delta\boldsymbol{b}|| \,/\, ||\boldsymbol{b}||} \tag{1}$$

The absolute choice looks good but it has a problem. If we divide the matrix A by 10, we are multiplying A^{-1} by 10. The absolute condition goes up by 10. But solving $A\boldsymbol{x} = \boldsymbol{b}$ did not become 10 times harder. **The relative condition number is the right choice.**

$$\boxed{\text{cond}(A) = \max_{\boldsymbol{b},\,\Delta\boldsymbol{b}} \frac{||A^{-1}\Delta\boldsymbol{b}||}{||\Delta\boldsymbol{b}||}\,\frac{||A\boldsymbol{x}||}{||\boldsymbol{x}||} = ||\boldsymbol{A}^{-1}||\,||\boldsymbol{A}|| = \frac{\boldsymbol{\sigma}_{\max}}{\boldsymbol{\sigma}_{\min}}} \tag{2}$$

If A is the simple diagonal matrix Σ with entries $\sigma_1 \geq \cdots \geq \sigma_r = \sigma_{\min}$, then its norm is $\sigma_{\max} = \sigma_1$. The norm of A^{-1} is $1/\sigma_{\min}$. The orthogonal matrices U and V in the SVD leave the norms unchanged. So the ratio $\boldsymbol{\sigma}_{\max} \,/\, \boldsymbol{\sigma}_{\min}$ is cond(A). We are using the usual measure of length $||\boldsymbol{x}||^2 = x_1^2 + \cdots + x_n^2$.

Notice that $\sigma_{\min}$ (*not* $\lambda_{\min}$) measures the distance from A to the nearest singular matrix. At first we might expect to see $A - \lambda_{\min}I$, bringing the smallest eigenvalue to zero. **Wrong.** The nearest singular matrix to $A = U\Sigma V^{\mathrm{T}}$ is $U(\Sigma - \sigma_{\min}I)V^{\mathrm{T}}$ because the orthogonal matrices U and V^{T} don't affect the norm. Bring the smallest **singular value** to zero.

Each eigenvalue of A also has a condition number. Suppose λ is a simple root (not a repeated root) of the equation $\det(A - \lambda I) = 0$. Then $A\boldsymbol{x} = \lambda\boldsymbol{x}$ and $A^{\mathrm{T}}\boldsymbol{y} = \lambda\boldsymbol{y}$ for unit eigenvectors $||\boldsymbol{x}|| = ||\boldsymbol{y}|| = 1$. **The condition number of $\boldsymbol{\lambda}$ is $\mathbf{1}/|\boldsymbol{y}^{\mathrm{T}}\boldsymbol{x}|$.** In other words it is $1/|\cos\theta|$, where θ is the angle between the right eigenvector $\boldsymbol{x}$ and the left eigenvector $\boldsymbol{y}$. (The name left comes from the equation $\boldsymbol{y}^{\mathrm{T}}A = \lambda\boldsymbol{y}^{\mathrm{T}}$, with $\boldsymbol{y}^{\mathrm{T}}$ on the left side of A.)

Notice that a symmetric matrix A will have $\boldsymbol{y} = \boldsymbol{x}$ with $\cos\theta = 1$. The eigenvalue problem is perfectly conditioned for symmetric matrices just as $Q\boldsymbol{x} = \boldsymbol{b}$ is perfectly conditioned for orthogonal matrices with $||Q||\,||Q^{-1}|| = \mathbf{1} =$ condition number of Q.

The formula $1/|\boldsymbol{y}^{\mathrm{T}}\boldsymbol{x}|$ comes from the change $\boldsymbol{\Delta\lambda} \approx (\boldsymbol{y}^{\mathrm{T}}\,\boldsymbol{\Delta A}\,\boldsymbol{x})/\boldsymbol{y}^{\mathrm{T}}\boldsymbol{x}$ in the eigenvalue created by a small change $\boldsymbol{\Delta A}$ in the matrix.

A8 Markov Matrices and Perron-Frobenius

This appendix is about positive matrices (all $a_{ij} > 0$) and nonnegative matrices (all $a_{ij} \geq 0$). Markov matrices M are important examples, when *every column of M adds to* 1. Positive numbers adding to 1 makes you think of probabilities.

A useful fact about any Markov matrix M: *The largest eigenvalue is always* $\lambda = 1$. We know that every column of $M - I$ adds to zero. So the rows of $M - I$ add to a row of zeros, and $M - I$ is not invertible: $\boldsymbol{\lambda = 1}$ **is an eigenvalue of** $\boldsymbol{M}$. Here are two examples:

$$A = \begin{bmatrix} \mathbf{0.8} & \mathbf{0.3} \\ \mathbf{0.2} & \mathbf{0.7} \end{bmatrix} \text{ has eigenvalues 1 and } \frac{1}{2} \qquad B = \begin{bmatrix} \mathbf{0} & \mathbf{1} \\ \mathbf{1} & \mathbf{0} \end{bmatrix} \text{ has eigenvalues 1 and } -1$$

That matrix A is typical of Markov. The eigenvectors are $\boldsymbol{x}_1 = (0.6, 0.4)$ and $\boldsymbol{x}_2 = (1, -1)$:

$$\begin{bmatrix} 0.8 & 0.3 \\ 0.2 & 0.7 \end{bmatrix}\begin{bmatrix} 0.6 \\ 0.4 \end{bmatrix} = \mathbf{1}\begin{bmatrix} 0.6 \\ 0.4 \end{bmatrix} \text{ is a steady state} \qquad \begin{matrix} \boldsymbol{A > 0 \quad \lambda_{\max} > 0} \\ \textbf{Eigenvector} > \mathbf{0} \end{matrix}$$

$$\begin{bmatrix} 0.8 & 0.3 \\ 0.2 & 0.7 \end{bmatrix}\begin{bmatrix} 1 \\ -1 \end{bmatrix} = \frac{\mathbf{1}}{\mathbf{2}}\begin{bmatrix} 1 \\ -1 \end{bmatrix} \text{ is a "transient" that disappears}$$

Our favorite example is based on rental cars in Chicago and Denver. We start with 100 cars in Chicago and no cars in Denver: $\boldsymbol{y}_0 = (100, 0)$. Every month we multiply the current vector $\boldsymbol{y}_n$ by A to find $\boldsymbol{y}_{n+1}$: the number in Chicago and Denver after $n + 1$ months:

$$\boldsymbol{y}_0 = \begin{bmatrix} 100 \\ 0 \end{bmatrix} \quad \boldsymbol{y}_1 = \begin{bmatrix} 80 \\ 20 \end{bmatrix} \quad \boldsymbol{y}_2 = \begin{bmatrix} 70 \\ 30 \end{bmatrix} \quad \boldsymbol{y}_3 = \begin{bmatrix} 65 \\ 35 \end{bmatrix} \quad \cdots \quad \boldsymbol{y}_\infty = \begin{bmatrix} \mathbf{60} \\ \mathbf{40} \end{bmatrix}.$$

That steady state $(\mathbf{60}, \mathbf{40})$ is an eigenvector of A for $\lambda = 1$. If we had started with $\boldsymbol{y}_0 = (60, 40)$ then we would have stayed there forever. Starting at $(100, 0)$ we needed to get 40 cars to Denver. You see that number 40 at time zero reduced to 20 at time 1, 10 at time 2, and 5 at time 3. That is the effect of the other eigenvalue $\lambda = \frac{1}{2}$, dividing its eigenvector by 2 at every step:

$$\boldsymbol{y}_0 = \begin{bmatrix} 60 \\ 40 \end{bmatrix} + \begin{bmatrix} 40 \\ -40 \end{bmatrix} \quad \boldsymbol{y}_1 = \begin{bmatrix} 60 \\ 40 \end{bmatrix} + \begin{bmatrix} 20 \\ -20 \end{bmatrix} \quad \cdots \quad \boldsymbol{y}_n = \begin{bmatrix} 60 \\ 40 \end{bmatrix} + \left(\frac{1}{2}\right)^n \begin{bmatrix} 40 \\ -40 \end{bmatrix}.$$

This is $\boldsymbol{y}_n = A^n \boldsymbol{y}_0$ coming from the single step equation $\boldsymbol{y}_{n+1} = A\boldsymbol{y}_n$. In matrix notation, A^n approaches rank one!

$$A^n = (X\Lambda X^{-1})^n = X\Lambda^n X^{-1} = \begin{bmatrix} 0.6 & 1 \\ 0.4 & -1 \end{bmatrix}\begin{bmatrix} 1 & \\ & (\frac{1}{2})^n \end{bmatrix}\begin{bmatrix} 1 & 1 \\ 0.4 & -0.6 \end{bmatrix} \to A^\infty = \begin{bmatrix} 0.6 & 0.4 \\ 0.4 & 0.4 \end{bmatrix}$$

You have now seen a typical Markov matrix with $\lambda_{\max} = 1$. Its eigenvector $(0.6, 0.4)$ is the survivor as time goes forward. All small eigenvalues have $\lambda^n \to 0$. But our second Markov example has a second eigenvalue $\lambda = -1$. *Now we don't approach a steady state*:

$$\boldsymbol{B} = \begin{bmatrix} \mathbf{0} & \mathbf{1} \\ \mathbf{1} & \mathbf{0} \end{bmatrix} \text{ has eigenvalue } \lambda_1 = \mathbf{1} \text{ with } \boldsymbol{x}_1 = \begin{bmatrix} 1 \\ 1 \end{bmatrix} \text{ and } \lambda_2 = \mathbf{-1} \text{ with } \boldsymbol{x}_2 = \begin{bmatrix} 1 \\ -1 \end{bmatrix}.$$

The zeros in B allow that second eigenvalue $\boldsymbol{\lambda_2 = -1}$ to have the same size as $\boldsymbol{\lambda_1 = 1}$. All cars switch cities every month (Chicago to Denver and Denver to Chicago). If we start at $\boldsymbol{y}_0 = (60, 40)$ then the next month reverses to $\boldsymbol{y}_1 = B\boldsymbol{y}_0 = (40, 60)$:

$$\boldsymbol{y}_0 = \begin{bmatrix} 60 \\ 40 \end{bmatrix} = \begin{bmatrix} 50 \\ 50 \end{bmatrix} + \begin{bmatrix} 10 \\ -10 \end{bmatrix} \quad \boldsymbol{y}_1 = \begin{bmatrix} 40 \\ 60 \end{bmatrix} = \begin{bmatrix} 50 \\ 50 \end{bmatrix} - \begin{bmatrix} 10 \\ -10 \end{bmatrix} \quad \boldsymbol{y}_2 = \boldsymbol{y}_0 \quad \boldsymbol{y}_3 = \boldsymbol{y}_1 \cdots$$

No steady state because $\lambda_2 = -1$ also has size $|\lambda_2| = 1$. This will not happen when the Markov matrix A has all $a_{ij} > 0$. $|\lambda_1| = |\lambda_2|$ *might happen* when B has some $B_{ij} = 0$.

Perron found the proof in the first case $A_{ij} > 0$. Then Frobenius allowed $B_{ij} = 0$. In this short appendix we stay with Perron. Every positive matrix A is allowed.

Theorem (Perron) All numbers in $A\boldsymbol{x} = \lambda_{\max}\,\boldsymbol{x}$ are strictly positive.

Proof Start with $A > 0$. The key idea is to look at all numbers t such that $A\boldsymbol{x} \geq t\boldsymbol{x}$ for some nonnegative vector $\boldsymbol{x}$ (other than $\boldsymbol{x} = \mathbf{0}$). We are allowing inequality in $A\boldsymbol{x} \geq t\boldsymbol{x}$ in order to have many small positive candidates t. For the **largest value $t_{\max}$** (which is attained), we will show that ***equality holds***: $A\boldsymbol{x} = t_{\max}\,\boldsymbol{x}$. Then $t_{\max}$ is our eigenvalue $\lambda_{\max}$ and $\boldsymbol{x}$ is the positive eigenvector—which we now prove.

If $A\boldsymbol{x} \geq t_{\max}\boldsymbol{x}$ is not an equality, multiply both sides by A. Because $A > 0$, that produces a strict inequality $A^2\boldsymbol{x} > t_{\max}A\boldsymbol{x}$. Therefore the positive vector $\boldsymbol{y} = A\boldsymbol{x}$ satisfies $A\boldsymbol{y} > t_{\max}\,\boldsymbol{y}$. This means that $t_{\max}$ could be increased. This contradiction forces the equality $A\boldsymbol{x} = t_{\max}\boldsymbol{x}$, and *we have an eigenvalue*. Its eigenvector $\boldsymbol{x}$ is positive because on the left side of that equality, $A\boldsymbol{x}$ is sure to be positive.

To see that no eigenvalue can be larger than $t_{\max}$, suppose $A\boldsymbol{z} = \lambda\boldsymbol{z}$. Since λ and $\boldsymbol{z}$ may involve negative or complex numbers, we take absolute values: $|\lambda||\boldsymbol{z}| = |A\boldsymbol{z}| \leq A|\boldsymbol{z}|$ by the "triangle inequality." This $|\boldsymbol{z}|$ is a nonnegative vector, so this $|\lambda|$ is one of the possible candidates t. Therefore $|\lambda|$ cannot exceed $t_{\max}$—which must be $\lambda_{\max}$.

Many Markov examples start with a zero in the matrix (Frobenius) but then A^2 or some higher power A^m is strictly positive (Perron). So we can apply Perron. These "primitive matrices" also have one steady state eigenvector from the eigenvalue $\lambda = 1$.

The big example is the **Google matrix *G*** that is the basis for the PageRank algorithm. (A major company survives entirely on linear algebra.) The matrix starts with $A_{ij} = 1$ when page j has a link to page i. Then divide each column j of A by the number of outgoing links: now column sums $= 1$. Finally $G_{ij} = \alpha A_{ij} + (1 - \alpha)/N$ is Google's choice so that every $G_{ij} > 0$ (Perron). See Wikipedia and the book by Amy Langville and Carl Meyer (which is quickly found using Google).

Reference

Amy Langville and Carl Meyer, *Google's Page Rank and Beyond: The Science of Search Engine Rankings*, Princeton University Press (2011).

A9 Elimination and Factorization

If a matrix A has rank r, then its row echelon form (from elimination) contains the identity matrix in its first r independent columns. How do we *interpret the matrix* F that appears in the remaining columns of that echelon form? F multiplies those first r independent columns of A to give its $n-r$ dependent columns. Then F reveals bases for the row space and the nullspace of the original matrix A. And F is the key to the column-row factorization $\boldsymbol{A}=\boldsymbol{CR}$.

1. Elimination must be just about the oldest algorithm in linear algebra. By systematically producing zeros in a matrix, it simplifies the solution of m equations $A\boldsymbol{x}=\boldsymbol{b}$. We take as example this 3 by 4 matrix A, with row $1+$ row $2=$ row 3. Then its rank is $r=2$, and we execute the familiar elimination steps to find its *reduced row echelon form* Z:

$$\boldsymbol{A}=\begin{bmatrix}1&2&11&17\\3&7&37&57\\4&9&48&74\end{bmatrix}\rightarrow\boldsymbol{Z}=\begin{bmatrix}1&0&3&5\\0&1&4&6\\0&0&0&0\end{bmatrix}.$$

At this point, we pause the algorithm to ask a question: **How is $\boldsymbol{Z}$ related to $\boldsymbol{A}$?** One answer comes from the fundamental subspaces associated with $\boldsymbol{A}$:

1) The two nonzero rows of $\boldsymbol{Z}$ (call them $\boldsymbol{R}$) are a basis for the row space of $\boldsymbol{A}$.

2) The first two columns of $\boldsymbol{A}$ (call them $\boldsymbol{C}$) are a basis for the column space of $\boldsymbol{A}$.

3) The nullspace of $\boldsymbol{Z}$ equals the nullspace of $\boldsymbol{A}$ (orthogonal to the same row space).

Those were our reasons for elimination in the first place. "Simplify the matrix A without losing the information it contains." By applying the same steps to the right hand side of $A\boldsymbol{x}=\boldsymbol{b}$, we reach $Z\boldsymbol{x}=\boldsymbol{d}$—with the same solutions $\boldsymbol{x}$ and the simpler matrix Z.

The object of this short note is to express the result of elimination in a different way. This factorization cannot be new, but it deserves new emphasis.

Elimination factors $\boldsymbol{A}$ into $\boldsymbol{C}$ times $\boldsymbol{R}=(\boldsymbol{m}\times\boldsymbol{r})$ times $(\boldsymbol{r}\times\boldsymbol{n})$

$$\boldsymbol{A}=\begin{bmatrix}1&2&11&17\\3&7&37&57\\4&9&48&74\end{bmatrix}=\begin{bmatrix}1&2\\3&7\\4&9\end{bmatrix}\begin{bmatrix}1&0&3&5\\0&1&4&6\end{bmatrix}=\boldsymbol{CR}$$

C has full column rank $r=2$ and R has full row rank $r=2$. When we establish that $A=CR$ is true for every matrix A, this factorization brings with it a proof of the first great theorem in linear algebra: **Column rank equals row rank.**

2. Here is a description of C and R that is independent of the algorithm (row operations) that computes them.

Suppose the first r independent columns of A go into C. Then the other $n-r$ columns of A must be combinations CF of those independent columns in C. That key matrix F is part of the row factor $R = \begin{bmatrix} I & F \end{bmatrix} P$, with r independent rows. Then right away we have $A = CR$:

$$\boldsymbol{A} = \boldsymbol{CR} = \begin{bmatrix} C & CF \end{bmatrix} \boldsymbol{P} = \begin{bmatrix} \textbf{Indep cols} & \textbf{Dep cols} \end{bmatrix} \textbf{Permute cols} \qquad (1)$$

If the r independent columns come first in A, that permutation matrix will be $P = I$. Otherwise we need P to permute the n columns of C and CF into correct position in A. Here is an example of $A = C\begin{bmatrix} I & F \end{bmatrix} P$ in which P exchanges columns 2 and 3:

$$\boldsymbol{A} = \begin{bmatrix} 1 & 2 & 3 & 4 \\ 1 & 2 & 4 & 5 \end{bmatrix} = \begin{bmatrix} 1 & 3 \\ 1 & 4 \end{bmatrix} \begin{bmatrix} 1 & 2 & 0 & 1 \\ 0 & 0 & 1 & 1 \end{bmatrix} = \begin{bmatrix} 1 & 3 \\ 1 & 4 \end{bmatrix} \begin{bmatrix} 1 & 0 & 2 & 1 \\ 0 & 1 & 0 & 1 \end{bmatrix} P = \boldsymbol{CR}. \qquad (2)$$

The essential information in **rref**(A) is the list of r independent columns of A, and the matrix F (r by $n-r$) that combines those independent columns to give the $n-r$ dependent columns CF in A. This uniquely defines **rref**(A).

3. The factorization $A = CR$ is confirmed. But how do we determine the first r independent columns in A (going into C) and the dependencies of the remaining $n-r$ columns CF? This is the moment for **row operations** on A. Three operations are allowed, all invertible, to put A into its reduced row echelon form $Z =$ **rref**(A):

(a) Subtract a multiple of one row from another row (below or above)

(b) Exchange two rows

(c) Divide a row by its first nonzero entry

All linear algebra teachers and a positive fraction of students know those steps and their outcome **rref**(A). It contains an r by r identity matrix I (only zeros can precede those 1's). The position of I reveals the first r independent columns of A. And equation (1) above reveals the meaning of F! It tells us how the $n-r$ *dependent* columns CF of A come from the independent columns in C. The remaining $m-r$ dependent rows of A must become zero rows in Z:

$$\textbf{Elimination reduces } \boldsymbol{A} \textbf{ to } \boldsymbol{Z} = \textbf{rref}(A) = \begin{bmatrix} \boldsymbol{I} & \boldsymbol{F} \\ \mathbf{0} & \mathbf{0} \end{bmatrix} \boldsymbol{P} \qquad (3)$$

This is Gauss-Jordan elimination: operating on rows above the pivot row as well as below. But the matrix that reduces A to this echelon form is less important than the **factorization** $\boldsymbol{A} = \boldsymbol{CR}$ that it uncovers in equation (1).

4. Before we apply $A = CR$ to solving $A\boldsymbol{x} = \mathbf{0}$, we want to give a column by column (left to right) construction of **rref**(A) from A. After elimination on k columns, that part of the matrix is in its own **rref** form. We are ready for elimination on the current column $k+1$. This new column has an upper part $\boldsymbol{u}$ and a lower part $\boldsymbol{\ell}$:

$$\textbf{First } \boldsymbol{k+1} \textbf{ columns} \qquad \begin{bmatrix} I_k & F_k \\ 0 & 0 \end{bmatrix} P_k \text{ followed by } \begin{bmatrix} \boldsymbol{u} \\ \boldsymbol{\ell} \end{bmatrix}. \tag{4}$$

The big question is : **Does this new column $\boldsymbol{k+1}$ join with $\boldsymbol{I_k}$ or $\boldsymbol{F_k}$?**

If $\boldsymbol{\ell}$ is all zeros, the new column is **dependent** on the first k columns. Then $\boldsymbol{u}$ joins with F_k to produce F_{k+1} in the next step to column $k+2$.

If $\boldsymbol{\ell}$ is not all zero, the new column is **independent** of the first k columns. Pick any nonzero in $\boldsymbol{\ell}$ (preferably the largest) as the pivot. Move that row of A to the top of $\boldsymbol{\ell}$. Then use that pivot row to zero out (by standard elimination) all the rest of column $k+1$. (That step is expected to change the columns after $k+1$.) Column $k+1$ joins with I_k to produce I_{k+1}. We adjust P_k and we are ready for column $k+2$.

At the end of elimination, we have a most desirable list of column numbers. They tell us the **first $\boldsymbol{r}$ independent columns of $\boldsymbol{A}$**. Those are the columns of C. They led to the identity matrix $I_{r \text{ by } r}$ in the row factor R of $A = CR$.

5. What is achieved by reducing A to **rref**(A) ? The row space is not changed ! Then its orthogonal complement (**the nullspace of $\boldsymbol{A}$**) is not changed. Each column of CF tells us how a dependent column of A is a combinaton of the independent columns in C. Effectively, **the columns of $\boldsymbol{F}$ are telling us $\boldsymbol{n-r}$ solutions to $\boldsymbol{A\boldsymbol{x} = 0}$.**

This is easiest to see by the example in Section 1 :

$$\begin{matrix} x_1 + 2x_2 + 11x_3 + 17x_4 = 0 \\ 3x_1 + 7x_2 + 37x_3 + 57x_4 = 0 \end{matrix} \quad \text{reduces to} \quad \begin{matrix} x_1 + 3x_3 + 5x_4 = 0 \\ x_2 + 4x_3 + 6x_4 = 0 \end{matrix}$$

The solution with $x_3 = 1$ and $x_4 = 0$ is $\boldsymbol{x} = \begin{bmatrix} -3 & -4 & 1 & 0 \end{bmatrix}^{\mathrm{T}}$. Notice 3 and 4 from F. The second solution with $x_3 = 0$ and $x_4 = 1$ is $\boldsymbol{x} = \begin{bmatrix} -5 & -6 & 0 & 1 \end{bmatrix}^{\mathrm{T}}$. Those solutions are the two columns of $\boldsymbol{X}$ in $\boldsymbol{AX} = \mathbf{0}$. (This example has $P = I$.) **The $\boldsymbol{n-r}$ columns of $\boldsymbol{X}$ are a natural basis for the nullspace of $\boldsymbol{A}$**:

$$\boldsymbol{A} = \boldsymbol{C}\begin{bmatrix} \boldsymbol{I} & \boldsymbol{F} \end{bmatrix}\boldsymbol{P} \text{ multiplies } \boldsymbol{X} = \boldsymbol{P}^{\mathrm{T}} \begin{bmatrix} -\boldsymbol{F} \\ \boldsymbol{I_{n-r}} \end{bmatrix} \text{ to give}$$

$$\boldsymbol{AX} = -\boldsymbol{CF} + \boldsymbol{CF} = \mathbf{0}. \tag{5}$$

Each column of X solves $A\boldsymbol{x} = \mathbf{0}$ (note that $PP^{\mathrm{T}} = I$). Those $n-r$ solutions in X tell us what we know : Each dependent column of A is a combination of the independent columns in C. In the example, column 3 $=$ 3 (column 1) $+$ 4 (column 2).

Gauss-Jordan elimination leading to $A = CR$ is less efficient than the Gauss process that directly solves $A\boldsymbol{x} = \boldsymbol{b}$. The latter stops at a triangular system $U\boldsymbol{x} = \boldsymbol{c}$: back substitution produces $\boldsymbol{x}$. Gauss-Jordan has the extra cost of eliminating upwards. If we only want to solve n by n equations, stopping at a triangular factorization is faster.

6. Block elimination The row operations that reduce A to its echelon form produce one zero at a time. A key part of that echelon form is an r by r identity matrix. If we think on a larger scale—instead of one row at a time—**that output I tells us that some r by r matrix has been inverted**. Following that lead brings a "matrix understanding" of elimination.

Suppose that the matrix $\boldsymbol{W}$ in the first r rows and columns of A is invertible. Then *elimination takes all its instructions from* W ! One entry at a time—or all at once by "block elimination"—**W will change to I**. In other words, the first r rows of A will yield I and F. **This identifies F as $W^{-1}H$**. And the last $m - r$ rows will become zero rows.

$$\textbf{Block elimination} \quad A = \begin{bmatrix} W & H \\ J & K \end{bmatrix} \text{ reduces to } \begin{bmatrix} I & F \\ 0 & 0 \end{bmatrix} = \textbf{rref}(A). \tag{6}$$

That just expresses the facts of linear algebra: If A begins with r independent rows and its rank is r, then the remaining $m - r$ rows are combinations of those first r rows : $\begin{bmatrix} J & K \end{bmatrix} = JW^{-1} \begin{bmatrix} W & H \end{bmatrix}$. Those $m - r$ rows become zero rows in elimination.

In general that first r by r block might not be invertible. But elimination will find W. We can move W to the upper left corner by row and column permutations $\boldsymbol{P_r}$ and $\boldsymbol{P_c}$.

Then the full expression of block elimination to reduced row echelon form is

$$\boldsymbol{P_r A P_c} = \begin{bmatrix} W & H \\ J & K \end{bmatrix} \rightarrow \begin{bmatrix} I & W^{-1}H \\ 0 & 0 \end{bmatrix} \tag{7}$$

7. This raises an interesting point. Since A has rank r, we know that it has r independent rows and r independent columns. Suppose those rows are in a submatrix B and those columns are in a submatrix C. Is it always true that the r by r "intersection" W of those rows with those columns will be **invertible**? (Everyone agrees that somewhere in A there is an r by r invertible submatrix. The question is whether $B \cap C$ can be counted on to provide such a submatrix.) Is W automatically full rank?

The answer is *yes*. **The intersection of r independent rows of A with r independent columns does produce a matrix W of rank r. W is invertible.**

Proof: Every column of A is a combination of the r columns of C.
Every column of the r by n matrix B is the same combination of the r columns of W.
Since B has rank r, its column space is all of $\mathbf{R}^r$.
Then the column space of W is also $\mathbf{R}^r$ and **the square submatrix W has rank r**.

A10 Computer Graphics

Computer graphics deals with images. The images are moved around. Their scale is changed. Three dimensions are projected onto two dimensions. All the main operations are done by matrices—but the shape of these matrices is surprising.

The transformations of three-dimensional space are done with 4 *by* 4 *matrices.* You would expect 3 by 3. The reason for the change is that one of the four key operations cannot be done with a 3 by 3 matrix multiplication. Here are the four operations:

Translation	**(shift the origin to another point $P_0 = (x_0, y_0, z_0)$)**
Rescaling	**(by c in all directions or by different factors c_1, c_2, c_3)**
Rotation	**(around an axis through the origin or an axis through P_0)**
Projection	**(onto a plane through the origin or a plane through P_0).**

Translation is the easiest—just add (x_0, y_0, z_0) to every point. But this is not linear! No 3 by 3 matrix can move the origin. So we change the coordinates of the origin to $(0, 0, 0, 1)$. This is why the matrices are 4 by 4. The "*homogeneous coordinates*" of the point (x, y, z) are $(x, y, z, 1)$ and we now show how they work.

1. Translation Shift the whole three-dimensional space along the vector $\boldsymbol{v}_0$. The origin moves to (x_0, y_0, z_0). This vector $\boldsymbol{v}_0$ is added to every point $\boldsymbol{v}$ in $\mathbf{R}^3$. Using homogeneous coordinates, the 4 by 4 matrix T shifts the whole space by $\boldsymbol{v}_0$:

$$\textbf{\textit{Translation matrix}} \qquad T = \begin{bmatrix} 1 & 0 & 0 & 0 \\ 0 & 1 & 0 & 0 \\ 0 & 0 & 1 & 0 \\ x_0 & y_0 & z_0 & 1 \end{bmatrix}.$$

Important: *Computer graphics works with row vectors.* We have row times matrix instead of matrix times column. You can quickly check that $[0\ 0\ 0\ 1]\,T = [x_0\ y_0\ z_0\ 1]$.

To move the points $(0, 0, 0)$ and (x, y, z) by $\boldsymbol{v}_0$, change to homogeneous coordinates $(0, 0, 0, 1)$ and $(x, y, z, 1)$. Then multiply by T. A row vector times T gives a row vector. ***Every*** $\boldsymbol{v}$ ***moves to*** $\boldsymbol{v} + \boldsymbol{v}_0$: $[x\ y\ z\ 1]\,T = [x + x_0\ \ y + y_0\ \ z + z_0\ \ 1]$.

The output tells where any $\boldsymbol{v}$ will move. (It goes to $\boldsymbol{v} + \boldsymbol{v}_0$.) Translation is now achieved by a matrix, which was impossible in $\mathbf{R}^3$.

2. Scaling To make a picture fit a page, we change its width and height. A copier will rescale a figure by 90%. In linear algebra, we multiply by .9 times the identity matrix. That matrix is normally 2 by 2 for a plane and 3 by 3 for a solid. In computer graphics, with homogeneous coordinates, the matrix is *one size larger:*

$$\textbf{\textit{Rescale the plane:}} \quad S = \begin{bmatrix} .9 & & \\ & .9 & \\ & & 1 \end{bmatrix} \qquad \textbf{\textit{Rescale a solid:}} \quad S = \begin{bmatrix} c & 0 & 0 & 0 \\ 0 & c & 0 & 0 \\ 0 & 0 & c & 0 \\ 0 & 0 & 0 & 1 \end{bmatrix}.$$

Important: S is not cI. We keep the "1" in the lower corner. Then $[x, y, 1]$ times S is the correct answer in homogeneous coordinates. The origin stays in its normal position because $[0\ 0\ 1]S = [0\ 0\ 1]$.

If we change that 1 to c, the result is strange. ***The point*** (cx, cy, cz, c) ***is the same as*** $(x, y, z, 1)$. The special property of homogeneous coordinates is that *multiplying by cI does not move the point.* The origin in $\mathbf{R}^3$ has homogeneous coordinates $(0, 0, 0, 1)$ and $(0, 0, 0, c)$ for every nonzero c. This is the idea behind the word "homogeneous."

Scaling can be different in different directions. To fit a full-page picture onto a half-page, scale the y direction by $\frac{1}{2}$. To create a margin, scale the x direction by $\frac{3}{4}$. The graphics matrix is diagonal but not 2 by 2. It is 3 by 3 to rescale a plane and 4 by 4 to rescale a space:

$$\textbf{\textit{Scaling matrices}} \quad S = \begin{bmatrix} \frac{3}{4} & & \\ & \frac{1}{2} & \\ & & 1 \end{bmatrix} \quad \text{and} \quad S = \begin{bmatrix} c_1 & & & \\ & c_2 & & \\ & & c_3 & \\ & & & 1 \end{bmatrix}.$$

That last matrix S rescales the x, y, z directions by positive numbers c_1, c_2, c_3. The extra column in all these matrices leaves the extra 1 at the end of every vector.

Summary The scaling matrix S is the same size as the translation matrix T. They can be multiplied. To translate and then rescale, multiply $\boldsymbol{v}TS$. To rescale and then translate, multiply $\boldsymbol{v}ST$. Are those different? *Yes.*

The point (x, y, z) in $\mathbf{R}^3$ has homogeneous coordinates $(x, y, z, 1)$ in $\mathbf{P}^3$. This "projective space" is not the same as $\mathbf{R}^4$. It is still three-dimensional. To achieve such a thing, (cx, cy, cz, c) is the same point as $(x, y, z, 1)$. Those points of projective space $\mathbf{P}^3$ are really lines through the origin in $\mathbf{R}^4$.

Computer graphics uses ***affine*** transformations, *linear plus shift.* An affine transformation T is executed on $\mathbf{P}^3$ by a 4 by 4 matrix with a special fourth column:

$$A = \begin{bmatrix} a_{11} & a_{12} & a_{13} & 0 \\ a_{21} & a_{22} & a_{23} & 0 \\ a_{31} & a_{32} & a_{33} & 0 \\ a_{41} & a_{42} & a_{43} & 1 \end{bmatrix} = \begin{bmatrix} T(1,0,0) & 0 \\ T(0,1,0) & 0 \\ T(0,0,1) & 0 \\ T(0,0,0) & 1 \end{bmatrix}.$$

The usual 3 by 3 matrix tells us three outputs, this tells four. The usual outputs come from the inputs $(1, 0, 0)$ and $(0, 1, 0)$ and $(0, 0, 1)$. When the transformation is linear, three outputs reveal everything. When the transformation is affine, the matrix also contains the output from $(0, 0, 0)$. Then we know the shift.

3. Rotation A rotation in $\mathbf{R}^2$ or $\mathbf{R}^3$ is achieved by an orthogonal matrix Q. The determinant is $+1$. (With determinant -1 we get an extra reflection through a mirror.) Include the extra column when you use homogeneous coordinates!

$$\textbf{\textit{Plane rotation}} \quad Q = \begin{bmatrix} \cos\theta & -\sin\theta \\ \sin\theta & \cos\theta \end{bmatrix} \quad \text{becomes} \quad \boxed{R = \begin{bmatrix} \cos\theta & -\sin\theta & 0 \\ \sin\theta & \cos\theta & 0 \\ 0 & 0 & 1 \end{bmatrix}}.$$

This matrix rotates the plane around the origin. ***How would we rotate around a different point*** $(4,5)$? The answer brings out the beauty of homogeneous coordinates. ***Translate*** $(4,5)$ ***to*** $(0,0)$, ***then rotate by*** θ, ***then translate*** $(0,0)$ ***back to*** $(4,5)$:

$$\boldsymbol{v}T_-RT_+ = \begin{bmatrix} x & y & 1 \end{bmatrix} \begin{bmatrix} 1 & 0 & 0 \\ 0 & 1 & 0 \\ -4 & -5 & 1 \end{bmatrix} \begin{bmatrix} \cos\theta & -\sin\theta & 0 \\ \sin\theta & \cos\theta & 0 \\ 0 & 0 & 1 \end{bmatrix} \begin{bmatrix} 1 & 0 & 0 \\ 0 & 1 & 0 \\ 4 & 5 & 1 \end{bmatrix}.$$

I won't multiply. The point is to apply the matrices one at a time: $\boldsymbol{v}$ translates to $\boldsymbol{v}T_-$, then rotates to $\boldsymbol{v}T_-R$, and translates back to $\boldsymbol{v}T_-RT_+$. Because each point $\begin{bmatrix} x & y & 1 \end{bmatrix}$ is a row vector, T_- acts first. The center of rotation $(4,5)$—otherwise known as $(4,5,1)$—moves first to $(0,0,1)$. Rotation doesn't change it. Then T_+ moves it back to $(4,5,1)$. All as it should be. The point $(4,6,1)$ moves to $(0,1,1)$, then turns by θ and moves back.

In three dimensions, every rotation Q turns around an axis. The axis doesn't move—it is a line of eigenvectors with $\lambda = 1$. Suppose the axis is in the z direction. The 1 in Q is to leave the z axis alone, the extra 1 in R is to leave the origin alone:

$$Q = \begin{bmatrix} \cos\theta & -\sin\theta & 0 \\ \sin\theta & \cos\theta & 0 \\ 0 & 0 & 1 \end{bmatrix} \quad \text{and} \quad R = \begin{bmatrix} & & & 0 \\ & Q & & 0 \\ & & & 0 \\ 0 & 0 & 0 & 1 \end{bmatrix}.$$

Now suppose the rotation is around the unit vector $\boldsymbol{a} = (a_1, a_2, a_3)$. With this axis $\boldsymbol{a}$, the rotation matrix Q which fits into R has three parts:

$$Q = (\cos\theta)I + (1-\cos\theta)\begin{bmatrix} a_1^2 & a_1a_2 & a_1a_3 \\ a_1a_2 & a_2^2 & a_2a_3 \\ a_1a_3 & a_2a_3 & a_3^2 \end{bmatrix} - \sin\theta \begin{bmatrix} 0 & a_3 & -a_2 \\ -a_3 & 0 & a_1 \\ a_2 & -a_1 & 0 \end{bmatrix}. \quad (1)$$

The axis doesn't move because $\boldsymbol{a}Q = \boldsymbol{a}$. When $\boldsymbol{a} = (0,0,1)$ is in the z direction, this Q becomes the previous Q—for rotation around the z axis.

The linear transformation Q always goes in the upper left block of R. Below it we see zeros, because rotation leaves the origin in place. When those are not zeros, the transformation is affine and the origin moves.

4. Projection In a linear algebra course, most planes go through the origin. In real life, most don't. A plane through the origin is a vector space. The other planes are affine spaces, sometimes called "flats." An affine space is what comes from translating a vector space.

We want to project three-dimensional vectors onto planes. Start with a plane through the origin, whose unit normal vector is $\boldsymbol{n}$. (We will keep $\boldsymbol{n}$ as a column vector.) The vectors in the plane satisfy $\boldsymbol{n}^{\mathrm{T}}\boldsymbol{v} = 0$. ***The usual projection onto the plane is the matrix*** $\boldsymbol{I - nn^{\mathrm{T}}}$. To project a vector, multiply by this matrix. The vector $\boldsymbol{n}$ is projected to zero, and the in-plane vectors $\boldsymbol{v}$ are projected onto themselves:

$$(I - \boldsymbol{n}\boldsymbol{n}^{\mathrm{T}})\boldsymbol{n} = \boldsymbol{n} - \boldsymbol{n}(\boldsymbol{n}^{\mathrm{T}}\boldsymbol{n}) = \boldsymbol{0} \quad \text{and} \quad (I - \boldsymbol{n}\boldsymbol{n}^{\mathrm{T}})\boldsymbol{v} = \boldsymbol{v} - \boldsymbol{n}(\boldsymbol{n}^{\mathrm{T}}\boldsymbol{v}) = \boldsymbol{v}.$$

In homogeneous coordinates the projection matrix becomes 4 by 4 (but the origin doesn't move):

$$\textbf{\textit{Projection onto the plane}} \quad \boldsymbol{n}^{\mathrm{T}}\boldsymbol{v} = 0 \qquad P = \begin{bmatrix} & & & 0 \\ & I - \boldsymbol{n}\boldsymbol{n}^{\mathrm{T}} & & 0 \\ & & & 0 \\ 0 & 0 & 0 & 1 \end{bmatrix}.$$

Now project onto a plane $\boldsymbol{n}^{\mathrm{T}}(\boldsymbol{v} - \boldsymbol{v}_0) = 0$ that does *not* go through the origin. One point on the plane is $\boldsymbol{v}_0$. This is an affine space (or a ***flat***). It is like the solutions to $A\boldsymbol{v} = \boldsymbol{b}$ when the right side is not zero. One particular solution $\boldsymbol{v}_0$ is added to the nullspace—to produce a flat.

The projection onto the flat has three steps. Translate $\boldsymbol{v}_0$ to the origin by T_-. Project along the $\boldsymbol{n}$ direction, and translate back along the row vector $\boldsymbol{v}_0$:

$$\textbf{\textit{Projection onto a flat}} \qquad T_-PT_+ = \begin{bmatrix} I & 0 \\ -\boldsymbol{v}_0 & 1 \end{bmatrix} \begin{bmatrix} I - \boldsymbol{n}\boldsymbol{n}^{\mathrm{T}} & 0 \\ 0 & 1 \end{bmatrix} \begin{bmatrix} I & 0 \\ \boldsymbol{v}_0 & 1 \end{bmatrix}.$$

I can't help noticing that T_- and T_+ are inverse matrices: translate and translate back. They are like the elementary matrices of Chapter 2.

The exercises will include reflection matrices, also known as *mirror matrices*. These are the fifth type needed in computer graphics. A reflection moves each point twice as far as a projection—***the reflection goes through the plane and out the other side***. So change the projection $I - \boldsymbol{n}\boldsymbol{n}^{\mathrm{T}}$ to $I - 2\boldsymbol{n}\boldsymbol{n}^{\mathrm{T}}$ for a mirror matrix.

The matrix P gave a "*parallel*" projection. All points move parallel to $\boldsymbol{n}$, until they reach the plane. The other choice in computer graphics is a "*perspective*" projection. This is more popular because it includes foreshortening. With perspective, an object looks larger as it moves closer. Instead of staying parallel to $\boldsymbol{n}$ (and parallel to each other), the lines of projection come *toward the eye*—the center of projection. This is how we perceive depth in a two-dimensional photograph.

The basic problem of computer graphics starts with a scene and a viewing position. Ideally, the image on the screen is what the viewer would see. The simplest image assigns just one bit to every small picture element—called a ***pixel***. It is light or dark. This gives a black and white picture with no shading. You would not approve. In practice, we assign shading levels between 0 and 2^8 for three colors like red, green, and blue. That means $8 \times 3 = 24$ bits for each pixel. Multiply by the number of pixels, and a lot of memory is needed!

Physically, a *raster frame buffer* directs the electron beam. It scans like a television set. The quality is controlled by the number of pixels and the number of bits per pixel. In this area, the standard text is *Computer Graphics: Principles and Practice* by Hughes, Van Dam, McGuire, Skylar, Foley, Feiner, and Akeley (3rd edition, Addison-Wesley, 2014). Notes by Ronald Goldman and by Tony DeRose were excellent references.

■ REVIEW OF THE KEY IDEAS ■

1. Computer graphics needs shift operations $T(\boldsymbol{v}) = \boldsymbol{v} + \boldsymbol{v}_0$ as well as linear operations $T(\boldsymbol{v}) = A\boldsymbol{v}$.

2. A shift in $\mathbf{R}^n$ can be executed by a matrix of order $n+1$, using homogeneous coordinates.

3. The extra component 1 in $[x\ y\ z\ 1]$ is preserved when all matrices have the numbers $0, 0, 0, 1$ as last column.

Problem Set A10

1 A typical point in $\mathbf{R}^3$ is $x\boldsymbol{i} + y\boldsymbol{j} + z\boldsymbol{k}$. The coordinate vectors $\boldsymbol{i}$, $\boldsymbol{j}$, and $\boldsymbol{k}$ are $(1,0,0)$, $(0,1,0)$, $(0,0,1)$. The coordinates of the point are (x, y, z).

This point in computer graphics is $x\boldsymbol{i} + y\boldsymbol{j} + z\boldsymbol{k} +$ **origin**. Its homogeneous coordinates are (, , ,). Other coordinates for the same point are (, , ,).

2 A linear transformation T is determined when we know $T(\boldsymbol{i}), T(\boldsymbol{j}), T(\boldsymbol{k})$. For an affine transformation we also need $T($______$)$. The input point $(x, y, z, 1)$ is transformed to $xT(\boldsymbol{i}) + yT(\boldsymbol{j}) + zT(\boldsymbol{k}) +$ ______.

3 Multiply the 4 by 4 matrix T for translation along $(1,4,3)$ and the matrix T_1 for translation along $(0,2,5)$. The product TT_1 is translation along ______.

4 Write down the 4 by 4 matrix S that scales by a constant c. Multiply ST and also TS, where T is translation by $(1,4,3)$. To blow up the picture around the center point $(1,4,3)$, would you use $\boldsymbol{v}ST$ or $\boldsymbol{v}TS$?

5 What scaling matrix S (in homogeneous coordinates, so 3 by 3) would produce a 1 by 1 square page from a standard 8.5 by 11 page?

6 What 4 by 4 matrix would move a corner of a cube to the origin and then multiply all lengths by 2? The corner of the cube is originally at $(1,1,2)$.

7 When the three matrices in equation 1 multiply the unit vector $\boldsymbol{a}$, show that they give $(\cos\theta)\boldsymbol{a}$ and $(1-\cos\theta)\boldsymbol{a}$ and $\mathbf{0}$. Addition gives $\boldsymbol{a}Q = \boldsymbol{a}$ and the rotation axis is not moved.

8 If $\boldsymbol{b}$ is perpendicular to $\boldsymbol{a}$, multiply by the three matrices in 1 to get $(\cos\theta)\boldsymbol{b}$ and $\mathbf{0}$ and a vector perpendicular to $\boldsymbol{b}$. So $Q\boldsymbol{b}$ makes an angle θ with $\boldsymbol{b}$. ***This is rotation***.

9 What is the 3 by 3 projection matrix $I - \boldsymbol{n}\boldsymbol{n}^{\mathrm{T}}$ onto the plane $\frac{2}{3}x + \frac{2}{3}y + \frac{1}{3}z = 0$? In homogeneous coordinates add $0, 0, 0, 1$ as an extra row and column in P.

Index of Equations

$A = CR$, 22, 93, 97, 98, 142, 401
$A = CR = \begin{bmatrix} C & CF \end{bmatrix} P$, 142, 410
$A = CW^{-1}B$, 108, 401
$A = C^{\mathrm{T}}C$, 261
$A = LDU$, 63, 401
$A = LU$, 53, 59, 83, 401
$A = QR$, 143, 176, 182, 185, 197, 401
$A = QS$, 197, 402
$A = QTQ^{-1}$: Schur, 256
$A = U\Sigma V^{\mathrm{T}}$, 286, 287
$A = X\Lambda X^{-1}$, 232, 233, 277, 401
$A = \lambda_1 \boldsymbol{x}_1 \boldsymbol{y}_1^{\mathrm{T}} + \cdots + \lambda_n \boldsymbol{x}_n \boldsymbol{y}_n^{\mathrm{T}}$, 245
$A = \begin{bmatrix} W & H \\ J & K \end{bmatrix} \rightarrow \begin{bmatrix} I & F \\ 0 & 0 \end{bmatrix}$, 410
$A = uv^{\mathrm{T}}$, 230
$(AB)C = A(BC)$, 27, 29, 290
$(AB)^{\mathrm{T}} = B^{\mathrm{T}}A^{\mathrm{T}}$, 67
$(A^{-1})^{\mathrm{T}} = (A^{\mathrm{T}})^{-1}$, 67
$(\det AB) = (\det A)(\det B)$, 205
$A(B + C) = AB + AC$, 29
$AB - BA = I$, 244
$AB = BA$: Same Λ, 244
$AB = I$ and $BA = I$, 147
$AC^{\mathrm{T}} = (\det A)I$, 204
$AV = U\Sigma$, 286, 287
$AV_r = U_r\Sigma_r$, 289
$AX = X\Lambda$, 233
$A \approx CMR$, 108
$A\backslash \boldsymbol{b} = A^{-1}\boldsymbol{b}$, 105
$A\boldsymbol{v} = \sigma \boldsymbol{u}$, 287
$A^{+} = (A^{\mathrm{T}}A)^{-1}A^{\mathrm{T}}$, 190
$A^{+} = A^{\mathrm{T}}(AA^{\mathrm{T}})^{-1}$, 190
$A^{+} = R^{+}C^{+}$, 193
$A^{+} = R^{\mathrm{T}}(C^{\mathrm{T}}AR^{\mathrm{T}})^{-1}C^{\mathrm{T}}$, 185, 195
$A^{+} = V\Sigma^{+}U^{\mathrm{T}}$, 195, 402
$A^{\mathrm{T}}(\boldsymbol{b} - A\widehat{\boldsymbol{x}}) = \mathbf{0}$, 156
$A^{\mathrm{T}}A = S$, 256
$A^{\mathrm{T}}A\widehat{\boldsymbol{x}} = A^{\mathrm{T}}\boldsymbol{b}$, 155, 163
$A^{\mathrm{T}} = -A$, 282
$A^2 = 0$, 329
$A^k = X\Lambda^k X^{-1}$, 233
$A^k \rightarrow 0$, 235
$A^{-1} = C^{\mathrm{T}} / \det A$, 208
$A_{\text{new}} = B^{-1}AB$, 322
$A_{ij} = F(i/N, j/N)$, 301
$||A - A_k|| \leq ||A - B||$, 302, 303
$||AB||_F \leq ||A||_F\, ||B||_F$, 303
$||A\boldsymbol{x} - \boldsymbol{b}||^2 = ||A\boldsymbol{x} - \boldsymbol{p}||^2 + ||\boldsymbol{e}||^2$, 164
$||A\widehat{\boldsymbol{x}} - \boldsymbol{b}||^2 =$ error, 164
$||A|| = ||\Sigma||$, 302
$||A||_F = \sqrt{\sigma_1^2 + \cdots + \sigma_r^2}$, 302
$B = W^{-1}V$, 319
$BA = I$, 103, 147
$BA \neq AB$, 28
$B_{\text{out}}^{-1}AB_{\text{in}}$, 327
$C = \sqrt{D}L^{\mathrm{T}}$, 261
$E = \boldsymbol{x}^{\mathrm{T}}S\boldsymbol{x}$, 252
$EA = R_0 = \text{rref}(A)$, 142
$EA = U$, 49, 53
$E^{-1} = L$, 53
$F(c \circledast d) = (Fc).*(Fd)$, 268
$F_{\min} = \frac{1}{2}\boldsymbol{b}^{\mathrm{T}}(A^{\mathrm{T}}S^{-1}A)^{-1}\boldsymbol{b}$, 358
$F_{k+2} = F_{k+1} + F_k$, 236
$H = WF$, 95
$J = B^{-1}AB$, 327
$\text{KKT} = \begin{bmatrix} \boldsymbol{S} & \boldsymbol{A} \\ \boldsymbol{A}^{\mathrm{T}} & \mathbf{0} \end{bmatrix}$, 359
$K = \text{toeplitz}[2, -1, 0, 0]$, 78
$KU/h^2 = F$, 77
$L = \frac{1}{2}\boldsymbol{x}^{\mathrm{T}}S\boldsymbol{x} - \boldsymbol{\lambda}^{\mathrm{T}}(A^{\mathrm{T}}\boldsymbol{x} - \boldsymbol{b})$, 358
$My'' + Ky = f$, 275
$P = A(A^{\mathrm{T}}A)^{-1}A^{\mathrm{T}}$, 151, 155, 185
$P = QQ^{\mathrm{T}}$, 185
$P = \boldsymbol{a}\boldsymbol{a}^{\mathrm{T}}/\boldsymbol{a}^{\mathrm{T}}\boldsymbol{a}$, 154
$PA = LU$, 45, 65, 401
$P^2 = P = P^{\mathrm{T}}$, 151, 154, 157, 159
$P^{-1} = P^{\mathrm{T}}$, 64
$P_{\text{column}} = AA^{+}$, 191
$P_{\text{row}} = A^{+}A$, 191
$||Q\boldsymbol{x}|| = ||\boldsymbol{x}||$, 178
$Q = I - 2\boldsymbol{u}\boldsymbol{u}^{\mathrm{T}}$, 178
$Q^{+} = Q^{\mathrm{T}}$, 185

$Q^TQ = I$, 176, 177
$(Qx)^T(Qy) = x^Ty$, 178
$R(x) = x^TSx/x^Tx$, 296
$R = \begin{bmatrix} I & F \end{bmatrix}$, 94
$R = \begin{bmatrix} I & F \end{bmatrix} P$, 93, 94
$R\widehat{x} = Q^Tb$, 182
$R_0 = \text{rref}(A)$, 93
$R_0 = \text{rref}(A) = \begin{bmatrix} I & F \\ 0 & 0 \end{bmatrix} P$, 94
$R_0x = d$, 104
$S = A^TA$, 249, 250, 256
$S = A^TCA$, 261
$S = C^TC$, 401
$S = Q\Lambda Q^T$, 246, 254, 256, 401
$S = Q\Lambda Q^{-1}$, 246
$S = S^T$, 69
$S = \lambda_1 x_1 x_1^T + \lambda_2 x_2 x_2^T$, 257
$(ST)(v) = S(T(v))$, 309
$T(M) = AM$, 316
$T(u) = c_1T(v_1) + \cdots + c_nT(v_n)$, 310
$T(u) = du/dx$, 312
$T(cv) = cT(v)$, 315
$T(cv + dw) = cT(v) + dT(w)$, 309
$T(f) = df/dx$, 309
$T^+(f) = \int_0^x f(t)\,dt$, 309
$T^+T(1) = 0$, 321
$T_{ijk} = \dfrac{\partial w_i}{\partial A_{jk}} = v_k\delta_{ij}$, 352
$U = K\backslash F$, 78
$Ux = c$, 41
$X^{-1}AX = \Lambda$, 232
$Y_{n+1} - 2Y_n + Y_{n-1}/(\Delta t)^2$, 274
$[X, E] = \textbf{eig}(A)$, 226
$\Delta x = -S^{-1}(x)\nabla F(x)$, 336
$\Delta y = y(x + h) - y(x)$, 74
$A^{-1} = C^T / \det A$, 202
$F(x, v) = F_L(\ldots(F_2(F_1(v))))$, 346
$F(x, v_i) \approx w_i$, 346
$F(x, y) = \frac{1}{2}(x^2 + by^2)$, 342
$F_{\text{min}} = -a^TS^{-1}a/2$, 339
$L(X) = \log(\det X)$, 340
$L(x) = \frac{1}{N}\sum_{i=1}^N \ell(e_i)$, 346
$\ell(x, v_i) = F(x, v_i) - w_i$, 347
$\nabla F(x^*) = 0$, 336
$\nabla F = (\partial F/\partial x_1, \ldots, \partial F/\partial x_n)$, 335
$\nabla F = Sx - b$, 336
$\nabla Q(x) = Sx - b$, 335
$\overline{z} = \overline{x + iy} = x - iy$, 262
$|z|^2 = z$ times $\overline{z} = r^2$, 262
$e^{i\theta} = \cos\theta + i\sin\theta$, 262
$e = b - p$, 151
$p = \dfrac{a^Tb}{a^Ta}a$, 153
$x_{k+1} = x_k + \Delta x$, 336
$x = (-b \pm \sqrt{b^2 - 4ac})/2a$, 262
$x = A\backslash b$, 45
$x = x_p + x_n$, 105, 107
$x = x_{\text{part}} +$ any x_{null}, 104
$x = x_{\text{row}} + x_{\text{null}}$, 145
$x = \frac{1}{N}\sum_{i=1}^N \ell(F(x, v_i) - w_i)$, 346
$x^TSx = \lambda x^Tx$, 248
$x^TSx > 0$, 248, 253, 260
$x_{\text{complete}} = x_p + c_1s_1 + c_2s_2$, 110
$y^Tb = y^T(Ax) = (A^Ty)^Tx \le c^Tx$, 365
$y^TA = \lambda y^T$, 245
$z = |z|e^{i\theta} = re^{i\theta}$, 262
$z^n = r^ne^{in\theta}$, 262
$z_1z_2 = |z_1|\,|z_2|e^{i(\theta_1+\theta_2)}$, 262
$u(t) = Ce^{\lambda t}$, 270
$u(t) = Ce^{\lambda t}x$, 271
$u(t) = u(0)e^{\lambda t}$, 270
$u(t) = e^{At}u(0)$, 277
$u(t) = e^{At}u(0) = Xe^{\Lambda t}X^{-1}u(0)$, 270
$u = e^{\lambda t}x$ when $Ax = \lambda x$, 270, 271
$u_k = A^ku_0$, 238
$u_k = X\Lambda^kX^{-1}u_0$, 232, 238
$u_{k+1} = Au_k$, 238
$\cos n\theta = T_n(\cos\theta)$, 333
$\cos\theta \approx 1 - \theta^2/2$, 269
$\det A =$ sum of all D_P, 208
$\det A = \lambda_1\lambda_2\cdots\lambda_n$, 244
$\det A = \pm$ (product of the pivots), 205
$\det AB = (\det A)(\det B)$, 198, 206
$\det A^T = \det A$, 204, 205
$\det E \le ||e_1||\ldots||e_n||$, 214
$\det = \lambda_1\,\lambda_2\cdots\lambda_n$, 228

$\frac{\partial F}{\partial x} = \frac{\partial F_L}{\partial F_{L-1}} \frac{\partial F_{L-1}}{\partial F_{L-2}} \cdots \frac{\partial F_1}{\partial x}$, 347
$\text{Re}\lambda < 0$, 270
$\text{cond}(A) = ||A^{-1}||\,||A|| = \sigma_{\max}/\sigma_{\min}$, 407
$\mathbf{N}(A^{\mathrm{T}}A) = \mathbf{N}(A)$, 151, 161
$\lambda(AB) = \lambda(BA)$, 292
$\lambda = (1 \pm \sqrt{5})/2$, 237
$\lambda = \overline{\boldsymbol{x}}^{\mathrm{T}} S\boldsymbol{x}/\overline{\boldsymbol{x}}^{\mathrm{T}}\boldsymbol{x}$, 264
$\lambda = w, w^2, \ldots, w^{N-1}, 1$, 265
$\lambda^N = 1$, 265
$\begin{bmatrix} A & I \end{bmatrix} \to \begin{bmatrix} I & A^{-1} \end{bmatrix}$, 57
$\begin{bmatrix} A & \boldsymbol{b} \end{bmatrix} \to \begin{bmatrix} R_0 & \boldsymbol{d} \end{bmatrix}$, 104
$\nabla F(x + \Delta x) \approx \nabla F(x) + S(x)\,\Delta x = 0$, 336
$\overline{Q}^{\mathrm{T}} Q = I$, 263, 269
$\overline{S}^{\mathrm{T}} = S$, 258, 264
$\overline{\boldsymbol{x}}^{\mathrm{T}} S\boldsymbol{x} = \lambda\overline{\boldsymbol{x}}^{\mathrm{T}}\boldsymbol{x}$, 247
$\partial F^*/\partial b = (A^{\mathrm{T}}S^{-1}A)^{-1}\boldsymbol{b} = \lambda^*$, 358
$\sin\theta \approx \theta - \theta^3/6$, 269
$|\boldsymbol{v}^{\mathrm{T}}\boldsymbol{w}| \le ||\boldsymbol{v}||\,||\boldsymbol{w}||$, 303
$|\lambda(Q)| = 1$, 263
$||c\boldsymbol{v}|| = |c|\,||\boldsymbol{v}||$, 355
$||f||^2 = \int |f(x)|^2\,dx$, 308
$||\boldsymbol{x}_{k+1} - \boldsymbol{x}^*|| \approx 1/4\,||\boldsymbol{x}_k - \boldsymbol{x}^*||^2$, 337
$||\boldsymbol{v} + \boldsymbol{w}|| \le ||\boldsymbol{v}|| + ||\boldsymbol{w}||$, 363
$||\boldsymbol{v} + \boldsymbol{w}|| \le ||\boldsymbol{v}|| + ||\boldsymbol{w}||$, 303, 355
$||\boldsymbol{v}||^2 = \overline{\boldsymbol{v}}^{\mathrm{T}}\boldsymbol{v}$, 263
$a_i^{\mathrm{T}}x = b_i$ and $a_j^{\mathrm{T}}x \ge c_j$, 360
$b = \lambda_{\min}/\lambda_{\max}$, 342
$c\boldsymbol{v} + d\boldsymbol{w}$, 2
$c_1 e^{\lambda_1 t}\boldsymbol{x}_1 + \cdots + c_n e^{\lambda_n t}\boldsymbol{x}_n$, 272, 277
$-d^2u/dx^2 = f(x)$, 77
$d\boldsymbol{u}/dt = \lambda e^{\lambda t}x = A\boldsymbol{u}$, 271
$d\boldsymbol{u}/dt = \lambda\boldsymbol{u}$, 270
$d^2\boldsymbol{F}/dx^2 > 0$, 336
$d^2\boldsymbol{u}/dt^2 = A\boldsymbol{u}$, 279
$e^{At} = I + At + \frac{1}{2}(At)^2 + \cdots$, 276
e^t and te^t, 278
$e^{At} = X\,e^{\Lambda t}\,X^{-1}$, 277
$e^{At}\boldsymbol{x} = e^{\lambda t}\boldsymbol{x}$, 276
$e^A e^B \ne e^B e^A$, 282
$(e^A)^2 = e^{2A}$, 283
$(e^{At})^{-1} = e^{-At}$, 278
$(f, g) = \int f(x)g(x)dx$, 308, 332
$h^2\,d^2y/2\,dx^2$, 74
$m\,\frac{d^2y}{dt^2} + b\,\frac{dy}{dt} + ky = 0$, 273
$m\lambda^2 + b\lambda + k = 0$, 273
$mu'' + bu' + ku = 0$, 285
$n > m$, 98
$p(A) =$ zero matrix, 244
$p = A\widehat{\boldsymbol{x}}$, 155
$p = A\widehat{\boldsymbol{x}} = A(A^{\mathrm{T}}A)^{-1}A^{\mathrm{T}}\boldsymbol{b}$, 155
$q(x) = 1/2x^{\mathrm{T}}Sx + x^{\mathrm{T}}c$, 360
$w = e^{2\pi i/N}$, 265
$x_1 = \det B_1/\det A$, 207
$x_{k+1} = x_k - s_k\nabla F(x_k)$, 337
$y'' = -y$, 273
$y(t) = c_1y_1 + c_2y_2$, 273
$1/\boldsymbol{z} = \overline{\boldsymbol{z}}/|\boldsymbol{z}|^2$, 262
$5x^2 + 8xy + 5y^2 = 1$, 254
$9X^2 + Y^2 = 1$, 254
$\text{GM} \le \text{AM}$, 239
$F(x + \Delta x) \approx F(x) + (\Delta x)^{\mathrm{T}}\nabla F + \frac{1}{2}(\Delta x)^{\mathrm{T}}\boldsymbol{S}(\Delta x)$, 336
$\dim(\mathbf{V}) + \dim(\mathbf{W}) = \dim(\mathbf{V}\cap\mathbf{W}) + \dim(\mathbf{V}+\mathbf{W})$, 128
Circulants $CD = DC$, 268
Dot product $\overline{\boldsymbol{v}}^{\mathrm{T}}\boldsymbol{w}$, 263
Hermitian $\overline{S}^{\mathrm{T}} = S$, 263
Markov $\lambda_{\max} = 1$, 408
Max = min, 365, 367
Max flow = Min cut, 366
Pivot $= D_k/D_{k-1}$, 249
Rank $(AB) \le$ rank(A), 137, 400
Rank $(AB) \le$ rank(B), 103, 137, 400
Rank of $A^{\mathrm{T}}A$ = rank of A, 400
Row rank = Column rank, 32, 400
Schur $A = QTQ^{-1}$, 402
Stability of ODE $\text{Re}\lambda < 0$, 275
Steepest slope $= ||\nabla F||$, 340
Trace XY = trace YX, 244
Transpose of $A = d/dt$, 69
Unitary matrix $\overline{Q}^{\mathrm{T}} = Q^{-1}$, 263

Index of Notations

Index

For computer codes please see the Index of Notations and the website

Six Great Theorems of Linear Algebra

Dimension Theorem All bases for a vector space have the same number of vectors.

Counting Theorem Dimension of column space + dimension of nullspace = number of columns.

Rank Theorem Dimension of column space = dimension of row space. This is the rank.

Fundamental Theorem The row space and nullspace of A are orthogonal complements in $\mathbf{R}^n$.

SVD There are orthonormal bases ($\boldsymbol{v}$'s and $\boldsymbol{u}$'s for the row and column spaces) so that $A\boldsymbol{v}_i = \sigma_i \boldsymbol{u}_i$.

Spectral Theorem If $S^{\mathrm{T}} = S$ there are orthonormal $\boldsymbol{q}$'s so that $S\boldsymbol{q}_i = \lambda_i \boldsymbol{q}_i$ and $S = Q\Lambda Q^{\mathrm{T}}$.

Linear Algebra Websites

math.mit.edu/linearalgebra This book's website has problem solutions and extra help

math.mit.edu/learningfromdata Linear Algebra and Learning from Data (2019)

ocw.mit.edu MIT's OpenCourseWare includes video lectures in Math 18.06, 18.065, 18.085

math.mit.edu/weborder.php To order all books published by Wellesley-Cambridge Press